The Neurophysiological Bases of Auditory Perception

Enrique A. Lopez-Poveda · Alan R. Palmer
Ray Meddis
Editors

The Neurophysiological Bases of Auditory Perception

MW

Editors
Enrique A. Lopez-Poveda
Universidad de Salamanca
Institute of Neurociencias de Castilla
y León
C/Pintor Fernando Gallego 1
37007 Salamanca
Spain
ealopezpoveda@usal.es

Alan R. Palmer
MRC Institute of Hearing Research
University Park
Nottingham
United Kingdom
alan@ihr.mrc.ac.uk

Ray Meddis
University of Essex
Wivenhoe Park
Colchester, Essex
United Kingdom
rmeddis@essex.ac.uk

ISBN 978-1-4419-5685-9 e-ISBN 978-1-4419-5686-6
DOI 10.1007/ 978-1-4419-5686-6
Springer New York Dordrecht Heidelberg London

Library of Congress Control Number: 2009943543

Printed on acid-free paper

Springer is part of Springer Science+Business Media (www.springer.com)

Preface

This volume contains the papers presented at the 15th International Symposium on Hearing (ISH), which was held at the Hotel Regio, Santa Marta de Tormes, Salamanca, Spain, between 1st and 5th June 2009.

Since its inception in 1969, this Symposium has been a forum of excellence for debating the neurophysiological basis of auditory perception, with computational models as tools to test and unify physiological and perceptual theories. Every paper in this symposium includes two of the following: auditory physiology, psychophysics or modeling. The topics range from cochlear physiology to auditory attention and learning. While the symposium is always hosted by European countries, participants come from all over the world and are among the leaders in their fields. The result is an outstanding symposium, which has been described by some as a "world summit of auditory research."

The current volume has a bottom-up structure from "simpler" physiological to more "complex" perceptual phenomena and follows the order of presentations at the meeting. Parts I to III are dedicated to information processing in the peripheral auditory system and its implications for auditory masking, spectral processing, and coding. Part IV focuses on the physiological bases of pitch and timbre perception. Part V is dedicated to binaural hearing. Parts VI and VII cover recent advances in understanding speech processing and perception and auditory scene analysis. Part VIII focuses on the neurophysiological bases of novelty detection, attention, and learning. Finally, Part IX describes novel results and ideas on hearing impairment. Some chapters have appended a written discussion by symposium participants; a form of online review that significantly enhances the quality of the content. In summary, the volume describes state-of-the-art knowledge on the most current topics of auditory science and will hopefully act as a valuable resource to stimulate further research.

It is not possible to organize a meeting of this size and importance without a considerable amount of help. We would like to express our most sincere thanks to the organizing team: Almudena Eustaquio-Martín, Jorge Martín Méndez, Patricia Pérez González, Peter T. Johannesen, and Christian Sánchez Belloso, whose expertise and willing help were essential to the smooth running of the meeting and preparation of this volume. Many thanks also to the staff of the Fundación General de la Universidad de Salamanca for their skillful and unconditional support with the administrative aspects of the organization. We are very grateful for the generosity of our sponsors: the Institute of Neuroscience of Castilla y León, the University of Salamanca, GAES S.A, Oticon España S.A, and, very specially, MED-EL.

Finally, we would like to thank all authors and participants for the high quality of their scientific contributions and for their cheerful conviviality during the meeting. The first editor wishes to thank all participants for generously and enthusiastically allowing this extraordinary symposium to take part in Spain for the first time ever.

Salamanca, Spain — Enrique A. Lopez-Poveda
Nottingham, UK — Alan R. Palmer
Colchester, UK — Ray Meddis

Sponsored by

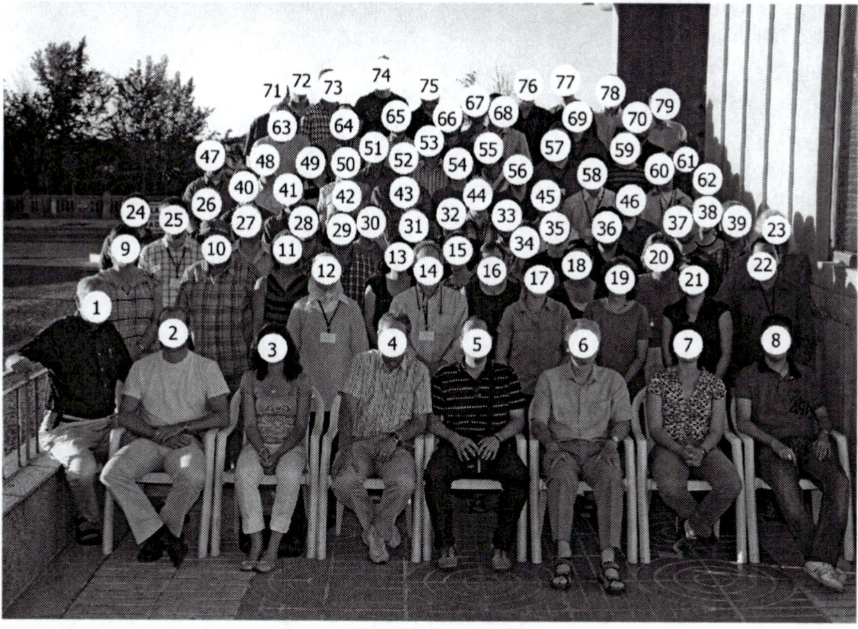

Participants of the 15th International Symposium on Hearing

Key to photograph

Agus, Trevor	48	Martín Méndez, Jorge	8
Alves-Pinto, Ana	13	Mc Laughlin, Myles	24
Balaguer-Ballester, Emili	55	McAlpine, David	53
Bernstein, Joshua	57	McKinney, Martin	47
Bruce, Ian	43	Meddis, Ray	6
Carlyon, Robert	37	Nelson, Paul	72
Carr, Catherine	19	Oberfeld-Twistel, Daniel	35
Colburn, Steve	14	Oxenham, Andrew	38
Culling, John	62	Palmer, Alan	4
Dau, Torsten	74	Panda R. Manasa	22
Delgutte, Bertrand	44	Patterson, Roy	45
Denham, Susan	20	Pérez González, Patricia	7
Elhilali, Mounya	18	Plack, Chris	31
Eustaquio-Martin, Almudena	3	Pressnitzer, Daniel	51
Ewert, Stephan	54	Recio-Spinoso, Alberto	70
Ferry, Robert	25	Roberts, Brian	27
Ghitza, Oded	63	Rupp, André	56
Goupell, Matthew	28	Shamma, Shihab	66
Hancock, Kenneth	71	Simon, Jonathan	46
Heinz, Michael	33	Spitzer, Philipp	69
Henning, G. Bruce	79	Strahl, Stefan	68
Ibrahim, Rasha	12	Strelcyk, Olaf	34
Irino, Toshio	41	Strickland, Elizabeth	16
Jennings, Skyler	64	Sumner, Christian	59
Johannesen, Peter	2	Tan M. Christine	21
Johnsrude, Ingrid	15	Tollin, Daniel	73
Joris, Philip	26	Trahiotis, Constantine	29
Kidani, Shunsuke	58	Unoki, Masashi	36
Klinge, Astrid	67	Uppenkamp, Stefan	39
Klump, Georg M.	75	van de Par, Steven	40
Kohlrausch, Armin	23	van der Heijden, Marcel	30
Laback, Bernhard	50	Viemeister, Neal	32
Lecluyse, Wendy	17	Wang, Xiaoqin	65
LeGoff, Nicolas	78	Watkins, Anthony	10
Li, Feipeng	49	Wiegrebe, Lutz	52
Lopez-Poveda, Enrique	5	Witton, Caroline	9
Lüddemann, Helge	60	Wojtczak, Magdalena	11
Lütkenhöner, Bernd	76	Yasin, Ifat	61
Majdak, Piotr	42	Young, Eric	1
Marquardt, Torsten	77		

Not in photograph

Antunes, Flora
Brand, Thomas
Devore, Sasha
King, Andrew
Lopéz-García, Mª Dolores
Neher, Tobias
Nodal, Fernando
Russell, Ian
Sánchez Belloso, Christian
Schnupp, Jan
Xia, Jing

Contents

Part IX Hearing Impairment

About the Editors

Enrique A. Lopez-Poveda, Ph.D. is the director of the Auditory Computation and Psychoacoustics Unit of the Neuroscience Institute of Castilla y León (University of Salamanca, Spain). His research focuses on the understanding and modeling of human cochlear nonlinear signal processing and the role of the peripheral auditory system in normal and impaired auditory perception. He has authored over 45 scientific papers and book chapters and is the co-editor of the book Computational Models of the Auditory System (*Springer Handbook of Auditory Research*). He has been principal investigator, participant, and consultant on numerous research projects. He is a fellow of the Acoustical Society of America and a member of the Association for Research in Otololaryngology.

Alan R. Palmer, Ph.D. is Deputy Director of the MRC Institute of Hearing Research and holds a Special Professorship in neuroscience at the University of Nottingham, UK. He received his first degree in Biological Sciences from the University of Birmingham, UK and his PhD in Communication and Neuroscience from the University of Keele, UK. After postdoctoral research at Keele, he established his own laboratory at the National Institute for Medical Research in London. This was followed by a Royal Society University Research Fellowship at the University of Sussex before taking a program leader position at the Medical Research Council Institute for Hearing Research in 1986. He heads a research team that uses neurophysiological, computational, and neuroanatomical techniques to study the way the brain processes sound.

Ray Meddis, Ph.D. is the director of the Hearing Research Laboratory at the University of Essex, England. His research has concentrated on the development of computer models of the physiology of the auditory periphery and how these can be incorporated into models of psychophysical phenomena such as pitch and auditory scene analysis. He has published extensively in this area. He is the co-editor of the book *Computational Models of the Auditory System* (*Springer Handbook of Auditory Research*). His current research concerns the application of computer models to an understanding of hearing impairment. He is a fellow of the Acoustical Society of America and a member of the Association of Research in Otolaryngololgy.

Contributors

Trevor R. Agus
CNRS & Université Paris Descartes & École normale supérieure, Paris, France

Jont B. Allen
ECE Department, University of Illinois at Urbana-Champaign, Urbana, IL, USA

Ana Alves-Pinto
MRC Institute of Hearing Research, University Park, Nottingham, UK

Flora Antunes
Auditory Neurophysiology Unit, Institute for Neuroscience of Castilla y Leon, Spain

Yoshie Aoki
Faculty of Systems Engineering, Wakayama University, Wakayama, Japan

Go Ashida
Department of Biology, University of Maryland, College Park, MD, USA

Victoria M. Bajo
Department of Physiology, Anatomy and Genetics,University of Oxford, Oxford, UK

Emili J. Balaguer-Ballester
Computational Neuroscience group, Central Institute for Mental Health (ZI), Ruprecht-Karls, Germany

Sylvie Baudoux
MRC Institute of Hearing Research, Nottingham, UK

Marion Beauvais
CNRS & Université Paris Descartes & École normale supérieure, Paris, France

Joshua Bernstein
Army Audiology and Speech Center, Walter Reed Army Medical Center, Washington, DC, USA

Leslie R. Bernstein
Departments of Neuroscience and Surgery (Otolaryngology), University of Connecticut Health Center, Farmington, CT, USA

Rainer Beutelmann
Medical Physics, University of Oldenburg, Oldenburg, Germany

Jennifer Bizley
Department of Physiology, Anatomy and Genetics, University of Oxford, Oxford, UK

Jonathan D. Boley
Weldon School of Biomedical Engineering, Purdue University, West Lafayette, IN, USA

Thomas Brand
Medical Physics, University of Oldenburg, Oldenburg, Germany

Jeroen Breebaart
Philips Research Laboratories, Eindhoven, The Netherlands

Guy J. Brown
Department of Computer Science, University of Sheffield, Sheffield, UK

Ian C. Bruce
Department of Electrical & Computer Engineering, McMaster University, West Hamilton, ON, Canada

Andrew Brughera
Department of Biomedical Engineering, Boston University, Boston, MA, USA

Andrew Byrne
Department of Psychology, University of Minnesota, Minneapolis, MN, USA

Robert P. Carlyon
MRC Cognition & Brain Sciences Unit, Cambridge, UK

Catherine Carr
Department of Biology, University of Maryland, College Park, MD, USA

Orlando Castellano
Departamento de Biología Celular y Patología, Instituto de Neurociencias de Castilla y León, Universidad de Salamanca, Salamanca, Spain

Nicholas R. Clark
MRC Institute of Hearing Research, University Park, Nottingham, UK

Martin Coath
Centre for Theoretical and Computational Neuroscience, University of Plymouth, Devon, UK

Harry Steven Colburn
Department of Biomedical Engineering, Boston University, Boston, MA, USA

Ellen Covey
Department of Psychology, University of Washington, Seattle, WA, USA

John Francis Culling
School of Psychology, Cardiff University, Cardiff, UK

Rhodri Cusack
MRC Cognition & Brain Sciences Unit, Cambridge, UK

Christopher J. Darwin
Department of Psychology, University of Sussex, Brighton, UK

Torsten Dau
Department of Electrical Engineering, Centre for Applied Hearing Research, Denmark

Stephen David
Institute for Systems Research, Electrical and Computer Engineering Department, University of Maryland, College Park, MD, USA

Matthew H. Davis
MRC Cognition & Brain Sciences Unit, Cambridge, UK

José Anchieta de Castro e Horta Júnior
Departamento de Anatomia, Instituto de Biociências, UNESP, Brazil

Bertrand Delgutte
Eaton-Peabody Laboratory, Massachusetts Eye & Ear Infirmary, Boston, MA, USA

Susan L. Denham
Centre for Theoretical and Computational Neuroscience, University of Plymouth, Plymouth, Devon, UK

Sasha Devore
Eaton-Peabody Laboratory, Massachusetts Eye & Ear Infirmary, Boston, MA, USA

Mathias Dietz
Medizinische Physik, Universität Oldenburg, Oldenburg, Germany

Darren Edwards
MRC Institute of Hearing Research, University Park, Nottingham, UK

Mounya Elhilali
Department of Electrical and Computer Engineering, Johns Hopkins University, Baltimore, MD, USA

Stephan D. Ewert
Medizinische Physik, Universität Oldenburg, Oldenburg, Germany

Robert Ferry
University of Essex, Colchester, UK

Suzanne Fitzpatrick
Department of Psychology, Lancaster University, Lancaster, UK

Jonathan Fritz
Institute for Systems Research, Electrical and Computer Engineering Department, University of Maryland, College Park, MD, USA

Etienne Gaudrain
Centre for the Neural Basis of Hearing, University of Cambridge, Cambridge, UK

Oded Ghitza
Sensimetrics Corporation and Boston University, Malden, MA, USA

Hedwig E. Gockel
MRC Cognition & Brain Sciences Unit, Cambridge, UK

Ricardo Gómez-Nieto
Departamento de Biología Celular y Patología, Instituto de Neurociencias de Castilla y León, Universidad de Salamanca, Salamanca, Spain

Matthew Goupell
Acoustics Research Institute, Austrian Academy of Sciences, Vienna, Austria

Steven Greenberg
Silicon Speech, Santa Venetia, CA, USA

Benedikt Grothe
Biocenter, University of Munich, Germany

Kinga Gyimesi
Department of General Psychology, Institute for Psychology, Budapest, Hungary

Kenneth Hancock
Eaton-Peabody Laboratory, Massachusetts Eye & Ear Infirmary, Boston, MA, USA

Nicol S. Harper
UCL Ear Institute, London, UK

Phillipp Hehrmann
UCL Gatsby Computational Neuroscience Centre, London, UK

Antje Heinrich
MRC Cognition & Brain Sciences Unit, Cambridge, UK

Michael G. Heinz
Department of Speech, Language, and Hearing Sciences, Purdue University, West Lafayette, IN, USA

G.B. Henning
The Institute of Ophthalmology, London, UK

Javier M. Herrero-Turrión
Instituto de Neurociencias de Castilla y León, Universidad de Salamanca, Salamanca, Spain

Arjan Hillebrand
Department of Clinical Neurophysiology, VU University Medical Center, Amsterdam, The Netherlands

Volker Hohmann
Medizinische Physik, Universität Oldenburg, Oldenburg, Germany

Stephen D. Holmes
Psychology, School of Life and Health Sciences, Aston University, Birmingham, UK

Rasha A. Ibrahim
Department of Electrical & Computer Engineering, McMaster University, West Hamilton, ON, Canada

Toshio Irino
Faculty of Systems Engineering, Wakayama University, Wakayama, Japan

Naoya Itatani
Animal Physiology and Behaviour Group, Institute for Biology and Environmental Sciences, Oldenburg, Germany

Skyler Jennings
Department of Speech Language and Hearing Sciences, Purdue University, West Lafayette, IN, USA

Peter Johannesen
Instituto de Neurociencias de Castilla y León, Universidad de Salamanca, Salamanca, Spain

Ingrid Johnsrude
Department of Psychology, Queen's University, Kingston, ON, Canada

Philip X. Joris
Laboratory of Auditory Neurophysiology, Medical School, Leuven, Belgium

Tim Jürgens
Medical Physics, University of Oldenburg, Oldenburg, Germany

Katharina Kaiser
Biocenter, University of Munich, Munich, Germany

Sushrut Kale
Weldon School of Biomedical Engineering, Purdue University, West Lafayette, IN, USA

Hideki Kawahara
Faculty of Systems Engineering, Wakayama University, Wakayama, Japan

Shunsuke Kidani
School of Information Science, Nomi Ishikawa, Japan

Andrew J. King
Department of Physiology, Anatomy and Genetics, University of Oxford, Oxford, UK

Martin Klein-Henning
Medizinische Physik, Universität Oldenburg, Oldenburg, Germany

Astrid Klinge
Animal Physiology and Behaviour Group, Institute for Biology and Environmental Sciences, Oldenburg, Germany

Georg M. Klump
Animal Physiology and Behaviour Group, Institute for Biology and Environmental Sciences, Oldenburg, Germany

Armin Kohlrausch
Philips Research Laboratories, Eindhoven, The Netherlands

Kanthaiah Koka
Department of Physiology and Biophysics, University of Colorado Health Sciences Center, Aurora, CO, USA

Birger Kollmeier
University of Oldenburg, Medical Physics, Germany

Katrin Krumbholz
MRC Institute of Hearing Research, University Park, Nottingham, UK

Bernhard Laback
Acoustics Research Institute, Austrian Academy of Sciences, Vienna, Austria

Nicolas Le Goff
Technical University, Eindhoven, The Netherlands

Wendy Lecluyse
University of Essex, Colchester, UK

Christian Leibold
Biocenter, University of Munich, Munich, Germany

Feipeng Li
ECE Department, University of Illinois at Urbana-Champaign, Urbana, IL, USA

Mª Dolores Estilita López García
Departamento de Biología Celular y Patología, Instituto de Neurociencias de Castilla y León, Universidad de Salamanca, Salamanca, Spain

Enrique A. Lopez-Poveda
Instituto de Neurociencias de Castilla y León, Universidad de Salamanca, Salamanca, Spain

Helge Lüddemann
Medizinische Physik, Institut für Physik, Universität Oldenburg, Oldenburg, Germany

Andrei N. Lukashkin
School of Life Sciences, University of Sussex, Falmer, Brighton, UK

Victoria Lukashkina
School of Life Sciences, University of Sussex, Falmer, Brighton, UK

Bernd Lütkenhöner
Section of Experimental Audiology, Münster University Hospital, Münster, Germany

Ling Ma
Department of Bioengineering, University of Maryland, College Park, MD, USA

Katrina MacLeod
Department of Biology, University of Maryland, College Park, MD, USA

Julia K. Maier
UCL Ear Institute, London, UK

Piotr Majdak
Acoustics Research Institute, Austrian Academy of Sciences, Vienna, Austria

Simon J. Makin
Department of Psychology, The University of Reading, Reading, UK

Manuel S. Malmierca
Auditory Neurophysiology Unit, Institute for Neuroscience of Castilla y Leon, University of Salamanca, Spain

Torsten Marquardt
UCL Ear Institute, London, UK

Myles Mc Laughlin
Laboratory of Auditory Neurophysiology, K.U. Leuven, Belgium

David McAlpine
UCL Ear Institute, London, UK

Ray Meddis
University of Essex, Colchester, UK

Ralf M. Meyer
Medical Physics, University of Oldenburg, Oldenburg, Germany

Christophe Micheyl
Department of Psychology, University of Minnesota, Minneapolis, MN, USA

Jessica J.M. Monaghan
Centre for the Neural Basis of Hearing, University of Cambridge, Cambridge, UK

Paul Nelson
Department of Biomedical Engineering, Johns Hopkins University, Baltimore, MD, USA

Fernando R. Nodal
Department of Physiology, Anatomy and Genetics, University of Oxford, Oxford, UK

Daniel Oberfeld-Twistel
Department of Psychology, Johannes Gutenberg-Universität, Germany

Andrew Oxenham
Department of Psychology, University of Minnesota, Minneapolis, MN, USA

Alan Palmer
MRC Institute of Hearing Research, Nottingham, UK

Manasa R. Panda
University of Essex, Colchester, UK

Roy D. Patterson
Centre for the Neural Basis of Hearing, University of Cambridge, Cambridge, UK

Christian S. Pedersen
UCL Ear Institute, London, UK

Christopher J. Plack
Division of Human Communication and Deafness, University of Manchester, Manchester, UK

Daniel Pressnitzer
CNRS & Université Paris Descartes & École normale supérieure, Paris, France

Friedemann Pulvermuller
MRC Cognition & Brain Sciences Unit, Cambridge, UK

Andrew Raimond
Department of Psychology, The University of Reading, Reading, UK

Alberto Recio-Spinoso
Leiden University Medical Center, Leiden, The Netherlands

Guy P. Richardson
School of Life Sciences, University of Sussex, Brighton, UK

Helmut Riedel
Medizinische Physik, Institut für Physik, Universität Oldenburg, Oldenburg, Germany

Brian Roberts
Psychology, School of Life and Health Sciences, Aston University, Birmingham, UK

María E. Rubio
Department of Physiology and Neurobiology, University of Connecticut, Farmington, CT, USA

Andre Rupp
Sektion Biomagnetismus, Abteilung Neurologie, Universität Heidelberg, Heidelberg, Germany

Ian Russell
School of Life Sciences, University of Sussex, Brighton, UK

Maneesh Sahani
UCL Gatsby Computational Neuroscience Centre, London, UK

Jan Schnupp
Department of Physiology, Anatomy and Genetics, University of Oxford, Oxford, UK

Andrew Schwartz
Eaton-Peabody Laboratory, Massachusetts Eye & Ear Infirmary, Boston, MA, USA

Shihab A. Shamma
Institute for Systems Research, Electrical and Computer Engineering Department, University of Maryland, College Park, MD, USA

Barbara Shinn-Cunningham
Department of Biomedical Engineering, Boston University, Boston, MA, USA

Yury Shtyrov
MRC Cognition & Brain Sciences Unit, Cambridge, UK

Jonathan Z. Simon
Department of Electrical & Computer Engineering, University of Maryland, College Park, MD, USA

Donal G. Sinex
Department of Psychology and Biology, Utah State University, Logan, UT, USA

Ida Siveke
Biocenter, University of Munich, Munich, Germany

Gábor Stefanics
Department of General Psychology, Institute for Psychology, Budapest, Hungary

Mark Stellmack
Department of Psychology, University of Minnesota, Minneapolis, MN, USA

Olaf Strelcyk
Centre for Applied Hearing Research, Department of Electrical Engineering, Lyngby, Denmark

Elizabeth Strickland
Department of Speech Language and Hearing Sciences, Purdue University, West Lafayette, IN, USA

Christian Sumner
MRC Institute of Hearing Research, Nottingham, UK

Jayaganesh Swaminathan
Department of Speech, Language, and Hearing Sciences, Purdue University, West Lafayette, IN, USA

Christine M. Tan
University of Essex, Colchester, UK

Sarah K. Thompson
MRC Cognition & Brain Sciences Unit, Cambridge, UK

Simon J. Thorpe
Centre de Recherche Cerveau et Cognition (CERCO), CNRS, Toulouse, France

Daniel Tollin
Department of Physiology and Biophysics, University of Colorado Health Sciences Center, Aurora, CO, USA

Constantine Trahiotis
Departments of Neuroscience and Surgery, University of Connecticut Health Center, Farmington, CT, USA

Masashi Unoki
Advanced Institute of Science and Technology, Nomi Ishikawa, Japan

Stefan Uppenkamp
Medizinische Physik, Universität Oldenburg, Oldenburg, Germany

Steven van de Par
Philips Research Laboratories, Eindhoven, The Netherlands

Marcel van der Heijden
Laboratory of Auditory Neurophysiology, Medical School, Leuven, Belgium

Neal Viemeister
Department of Psychology, University of Minnesota, Minneapolis, MN, USA

Kerry Walker
Department of Physiology, Anatomy and Genetics, University of Oxford, Oxford, UK

Thomas C. Walters
Centre for the Neural Basis of Hearing, University of Cambridge, Cambridge, UK

Xiaoqin Wang
Laboratory of Auditory Neurophysiology, Department of Biomedical Engineering, Baltimore, MD, USA

Anthony J. Watkins
Department of Psychology, The University of Reading, Reading, UK

Lutz Wiegrebe
Biocenter, University of Munich, Munich, Germany

Hagen Wierstorf
Medizinische Physik, Universität Oldenburg, Oldenburg, Germany

István Winkler
Department of General Psychology, Institute for Psychology, Budapest, Hungary

Daniel Winkowski
Institute for Systems Research, Electrical and Computer Engineering Department, University of Maryland, College Park, MD, USA

Caroline Witton
Wellcome Trust Laboratory for MEG Studies, Aston University, Birmingham, UK

Magdalena Wojtczak
Department of Psychology, University of Minnesota, Minneapolis, MN, USA

Jing Xia
Department of Cognitive and Neural Systems, Boston University, Boston, MA, USA

Juanjuan Xiang
Starkey Laboratories Inc., Eden, Prairie, MN, USA

Pingbo Yin
Institute for Systems Research, Electrical and Computer Engineering Department, University of Maryland, College Park, MD, USA

Eric Young
Department of Biomedical Engineering, Johns Hopkins University, Baltimore, MD, USA

Part I
Peripheral/Cochlear Processing

Chapter 1
Otoacoustic Emissions Theories Can Be Tested with Behavioral Methods

Enrique A. Lopez-Poveda and Peter T. Johannesen

Abstract When two pure tones (or primaries) of slightly different frequencies (f_1 and f_2; $f_2 > f_1$) are presented to the ear, new frequency components not present in the stimulus may be recorded in the ear canal. These new components are termed distortion product otoacoustic emissions (DPOAEs) and are generated by nonlinear interaction of the primaries within the cochlea. It has been conjectured that the level of the $2f_1 - f_2$ DPOAE component is maximal when the primaries produce approximately equal excitation at the f_2 cochlear region because this is where and when the overlap between the traveling waves evoked by the two primaries is maximal. This region, however, almost certainly shifts as the level of the primaries increases following the well-known level-dependent shift of the cochlear traveling-wave peak. Furthermore, mutual suppression between the primaries may also affect the combination of primary levels that maximizes the DPOAE levels. This report summarizes our attempts to test these conjectures using psychophysical masking methods that are commonly applied to infer human cochlear responses. Test frequencies of 0.5, 1 and 4 kHz and a fixed frequency ratio of $f_2/f_1 = 1.2$ were considered. Results supported that maximal-level DPOAEs occur when the primaries produce comparable excitation at the cochlear site with CF ~ f_2. They also suggest that the site of maximum interaction hardly shifts with increasing primary level and that mutual suppression between the primaries does not affect significantly the optimal DPOAE primary level rule.

Keywords Human • Otoacoustic emission generation • Cochlear nonlinearity • Masking • Temporal masking curve • Growth of masking • Suppression • Level-dependent shifts

---

E.A. Lopez-Poveda (✉)
Instituto de Neurociencias de Castilla y León, Universidad de Salamanca,
Calle Pintor Fernando Gallego 1, 37007 Salamanca, Spain
e-mail: ealopezpoveda@usal.es

E.A. Lopez-Poveda et al. (eds.), *The Neurophysiological Bases of Auditory Perception*,
DOI 10.1007/978-1-4419-5686-6_1,

1.1 Introduction

It has been conjectured that the level of the $2f_1-f_2$ component of the distortion product otoacoustic emission (DPOAE) is maximal when the two primary tones produce equal responses at the f_2 cochlear site (Kummer et al. 2000). This chapter reports on an attempt to test this conjecture and various refinements in *human* using psychoacoustical methods.

For listeners with normal hearing, the amplitude of the $2f_1-f_2$ DPOAE component depends on the frequencies and levels of the primary tones (Johnson et al. 2006). We will call *optimal* rule to the combination of primary levels (L_1, L_2) that produces the highest levels of the $2f_1-f_2$ DPOAE component for any given pair of primary frequencies (f_1, f_2). Up to now, optimal rules have been obtained empirically; that is, by searching the (L_1, L_2) space for the combinations that maximize the level of the $2f_1-f_2$ DPOAE (Johnson et al. 2006; Kummer et al. 1998, 2000; Neely et al. 2005). Several studies have concluded that for $L_2 \leq 65$ dB SPL, the optimal rule conforms to a linear relationship of the form $L_1 = aL_2 + b$, with L_1 and L_2 being the levels of the f_1 and f_2 primary tones, respectively (Johnson et al. 2006). This relationship is commonly known as the "scissors" rule. The actual values of a and b, however, are still a matter of controversy. Also controversial is whether the optimal rule differs across f_2 frequencies (Johnson et al. 2006; Kummer et al. 2000). The controversy is due to uncertainties on the cochlear generation mechanisms and conditions that maximize DPOAEs. Kummer et al. (2000) conjectured that the level of the $2f_1-f_2$ DPOAE is highest when the two primary tones produce *equal* excitation at the f_2 cochlear site. This seems theoretically reasonable because these are the conditions of maximum interaction between the cochlear traveling waves evoked by the two primaries (Fig. 1.1).

We have recently tested the conjecture of Kummer et al. (2000) in *human* using psychoacoustical methods (Lopez-Poveda and Johannesen, submitted). Our approach was based on comparing *empirical* optimal rules with *behavioral* rules inferred from temporal masking curves (TMCs). A TMC is a plot of the levels of a

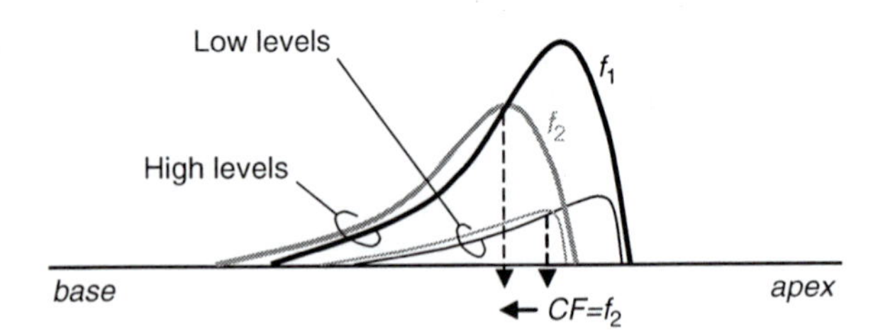

Fig. 1.1 Simplified excitation patterns evoked by the f_1 (*black*) and f_2 (*gray*) DPOAE primary tones at low (*thinner lines*) and high (*thicker lines*) levels. At low levels, the site of maximum interaction between both traveling waves (denoted by *vertical dashed lines*) occurs at the cochlear site with a CF ~ f_2. At high levels, however, the place of maximum interaction shifts toward the cochlear base. The direction of the shift may be different for different cochlear regions (Carney et al. 1999)

pure-tone forward masker required to just mask a fixed pure-tone probe as a function of the time gap between the masker and the probe. The probe level is *fixed* just above the absolute threshold for the probe. The masker level at the masking threshold increases with increasing time gap, but the rates of increase are typically different for different masker frequencies (Nelson et al. 2001). The current interpretation is that the masker level for any gap reflects the residual effect of the masker at a cochlear site tuned approximately at the probe frequency after it has decayed in time until the probe is presented (Nelson et al. 2001; Lopez-Poveda et al. 2003). There is evidence that the form and rate of recovery from forward masking is independent of masker frequency, at least for levels below around 80 dB SPL (Lopez-Poveda and Alves-Pinto 2008; Wojtczak and Oxenham 2009). Further, the residual effect at the time of the probe must be the same for different masker frequencies at the masking threshold because the probe level is fixed. Therefore, it seems reasonable to *assume* that for any given masker-probe time gap, two different masker tones at their masking thresholds produce identical degrees of excitation at a cochlear site with a characteristic frequency (CF) approximately equal to the probe frequency. This method is now widely used to infer human cochlear responses.

Based on the above ideas, we measured two TMCs, both for a probe frequency equal to the DPOAE test frequency (f_2), and for masker frequencies equal to the DPOAE primary tones (f_1, f_2). We then plotted the resulting levels for the f_1 masker against those for the f_2 masker, paired according to masker-probe time gap. This plot should illustrate the $L_1 - L_2$ combination for which f_1 and f_2 produce comparable degrees of excitation at the f_2 cochlear site. We demonstrated a reasonable match between individual TMC and optimal rules for f_2 frequencies of 1 and 4 kHz, for a primary frequency ratio of $f_2/f_1 = 1.2$, and for L_2 below around 65 dB SPL. This supported the conjecture of Kummer et al. (2000), at least for these conditions.

The correspondence between the TMC and optimal rules tended to be less, however, for L_2 above around 65 dB SPL. Furthermore, the degree of correspondence between the two rules was variable across subjects. The reasons for these (subtle) discrepancies are uncertain. We hypothesize that the conjecture of Kummer et al. (2000) may be valid for low levels only (below around 65 dB SPL) and/or that the TMC method is not the most adequate psychoacoustical method to test it. Indeed, the probe level is fixed in the TMC method in an attempt to minimize off-frequency listening associated to (1) the broadening of cochlear excitation by the probe and (2) the likely shifts of the traveling-wave peak (Robles and Ruggero 2001). The cochlear site of maximum interaction between the two DPOAE primary tones, however, almost certainly shifts with increasing the level of the primaries (Fig. 1.1). If the generation theory of maximal-level DPOAEs was correct then optimal level rules, unlike TMC rules, would be affected by this shift (Lopez-Poveda and Johannesen, submitted). If this were the case, then optimal rules should match more closely with behavioral rules inferred from growth-of-forward-maskability (GOFM) curves for probe and masker frequencies equal to f_2 and f_1, respectively (Oxenham and Plack 1997). GOFM curves are graphical representations of the levels of a pure-tone *forward* masker required to just mask a pure-tone probe as a function of the probe level. They are assumed to reflect the level required for the masker to produce approximately the same excitation as the probe at the cochlear site most sensitive to the probe. Because the probe level varies, the cochlear place in question almost certainly *shifts* with increasing probe level, just as it would happen hypothetically for the place of maximum interaction between the traveling waves evoked by the DPOAE primaries (Fig. 1.1).

On the other hand, optimal rules are based on actual DPOAE measurements, which involve the simultaneous presentations of the two primaries, and thus they implicitly encompass possible nonlinear interactions (most notably suppression) of the primaries within the cochlea (Kummer et al. 2000). The effects of these interactions are disregarded, however, by TMC- or GOFM-based behavioral rules because neither method requires the simultaneous presentation of the relevant tones. If cochlear shifts *and* suppression affected simultaneously optimal rules, then optimal rules should match more closely with behavioral rules inferred from growth-of-*simultaneous*-maskability (GOSM) curves for probe and masker frequencies equal to f_2 and f_1, respectively, where the probe and the masker are presented simultaneously.

We report here on a first attempt to test these refinements of the conjecture of Kummer et al. (2000) by comparing optimal level rules with behavioral rules inferred from TMCs (as in Lopez-Poveda and Johannesen, submitted), GOFM, and GOSM curves. Test frequencies of 0.5, 1, and 4 kHz were considered, and the primary frequency ratio (f_2/f_1) was always fixed at 1.2. For convenience, three of the participants in the study of Lopez-Poveda and Johannesen (submitted) were invited back to collect the complementary data necessary for the present study; specifically, to measure GOFM and GOSM functions at the three test frequencies and optimal rules at 0.5 kHz.

1.2 Methods

1.2.1 Listeners

Three of the normal-hearing participants in the related study of Lopez-Poveda and Johannesen (submitted) (S1, S2 and S8) took part in the present experiments. There was no particular reason to choose this subset of listeners other than they were available to run the complementary conditions. Their ages were between 24 and 40, and their thresholds for the different probes and maskers considered in the study are shown in Table 1.1. S1 was author PTJ.

Table 1.1 Listeners' absolute thresholds (in dB SPL) for the various pure tones used as probes and maskers in the present study

Frequency (kHz)	0.5	1	4
Tone duration (ms)	300/110/10	300/110/10	300/110/10
S1	13/15/39	6/8/30	13/13/33
S2	14/21/43	10/15/30	21/21/40
S8	11/18/34	5/7/38	11/10/30

1.2.2 Optimal Rules

The magnitude (in dB SPL) of the $2f_1-f_2$ DPOAE was measured for f_2 frequencies of 0.5, 1 and 4 kHz and for a fixed primary frequency ratio of $f_2/f_1=1.2$. Individual optimal L_1-L_2 rules were obtained by searching the (L_1, L_2) parameter space to find the (L_1, L_2) combinations that produced the highest DPOAE response levels. L_2 was varied in 5-dB steps within the range 35–75 dB SPL. For each fixed L_2, L_1 was varied in 3-dB steps and the individual optimal value (the level that produced the strongest DPOAE) was noted. In an attempt to reduce the potential variability of the DPOAE levels due to the DPOAE fine structure, optimal L_1 levels were obtained for three f_2 frequencies close to the frequency of interest (e.g., f_2=3,960, 4,000, 4,040 Hz), and their mean was taken as the "true" optimal L_1 level. The rationale for this averaging is described elsewhere (Johannesen and Lopez-Poveda 2008). All other methods and procedures, including the strict criterion for rejection of system's artifact, were described in Johannesen and Lopez-Poveda (2008).

1.2.3 Behavioral Rules

1.2.3.1 Common Procedures

TMCs, GOFM, and GOSM curves were measured for probe frequencies (f_P) equal to the DPOAE test frequencies (f_2). Masker frequencies were equal to the frequencies of the DPOAE primaries f_1 and/or f_2 (see below). All masker levels at the probe masking threshold were measured using a two-interval, two-alternative, forced-choice adaptive procedure with feedback. The resulting masker levels corresponded to the values at which the probe could be detected on 70.9% of the occasions (Levitt 1971). For the TMCs, the initial masker level step was 6 dB for the first three reversals, and then it was 2 dB for the last 12 reversals. Threshold was calculated as the mean level at the last 12 reversals. For the GOFM and GOSM curves, the step sizes were 6 and 2 dB for the first two and the last ten reversals, respectively, and threshold was calculated as the mean level at the last 10 reversals. At least three threshold estimates were measured per condition, and the mean was regarded as the 'true' threshold. The procedure and equipment are described in detail elsewhere (Johannesen and Lopez-Poveda 2008).

1.2.3.2 TMC Stimuli

For each probe frequency (f_P), two TMCs were measured for masker frequencies of f_P and $f_P/1.2$. The durations of the masker and the probe were 110 and 10 ms, respectively, including 5-ms cosine-squared onset and offset ramps. The probe had no steady state portion. The masker-probe time gaps ranged from 5 to 100 ms in

5-ms steps with an additional gap of 2 ms. The level of the probe was fixed at 9 dB above the individual absolute threshold for the probe (shown in Table 1.1). Individual $L_1 - L_2$ rules were obtained by plotting the levels for the f_1 masker against those for the f_2 masker, paired according to the masker-probe time gaps. These rules will be referred to as TMC rules.

1.2.3.3 GOFM Stimuli

A GOFM curve was measured for each of probe frequency and for a masker frequency of $f_P/1.2$. The durations of the masker and the probe were 110 and 10 ms, respectively, including 5-ms cosine-squared onset and offset ramps. The probe had no steady state portion and was presented immediately after the masker offset. Probe levels ranged from 20 to 90 dB SPL in 5-dB steps. To allow GOFM curves reveal the effects of increasing probe level, particularly the contribution of the likely shifts of the traveling-wave peak, no high-pass noise was used, unlike would be otherwise commonly the case (Oxenham and Plack 1997). Further, it was deemed unnecessary to measure on-frequency GOFM curves (i.e., curves for a masker frequency equal to the probe frequency) on the assumption that their slope would be close to unity (Oxenham and Plack 1997; but see Nelson et al. 2001). Therefore, the function illustrating masker level at masking threshold against probe level was regarded directly as the individual GOFM $L_1 - L_2$ level rule.

1.2.3.4 GOSM Stimuli

GOSM curves were measured for the same probe and masker frequencies as considered for the GOFM curves. Masker and probe were presented simultaneously. They both had durations of 110 ms, including 5-ms cosine-squared onset and offset ramps. Probe levels ranged from 20 to 90 dB SPL in 5-dB steps. The initial masker level was fixed at 14 dB SPL. The function illustrating masker level at masking threshold against probe level was regarded as the individual GOSM level rule.

1.3 Results and Discussion

Figure 1.2 illustrates the individual TMCs. These are a subset of the TMCs reported in the related study of Lopez-Poveda and Johannesen (submitted) and are reproduced here for completeness. For any given probe frequency and gap, the difference in masker level is assumed to reflect the level difference required for the two maskers to produce an identical cochlear response at the cochlear site with a CF approximately equal to the probe frequency (Nelson et al. 2001).

Figure 1.3 illustrates individual $L_1 - L_2$ level rules obtained with the three behavioral methods as well as optimal rules obtained empirically. TMC rules and optimal rules at 1 and 4 kHz are reproduced from the related study of Lopez-Poveda and Johannesen (submitted). Figure 1.4 illustrates straight-line fits to the mean data across subjects (not shown) of Fig. 1.3 (the corresponding equations are given in Table 1.2). These fits very clearly illustrate the main trends observed in the individual results.

A first striking result was that optimal (circles) and TMC (open triangles) rules *nearly overlapped* individually (Fig. 1.3) as well as on average (Fig. 1.4), for the three test frequencies. Lopez-Poveda and Johannesen (submitted) had already reported the overlap at 1 and 4 kHz, but Fig. 1.3 provides the first evidence of an overlap also at 0.5 kHz. Thus, the present results extend to 0.5 kHz, the support that TMC rules provide to the conjecture of Kummer et al. (2000).

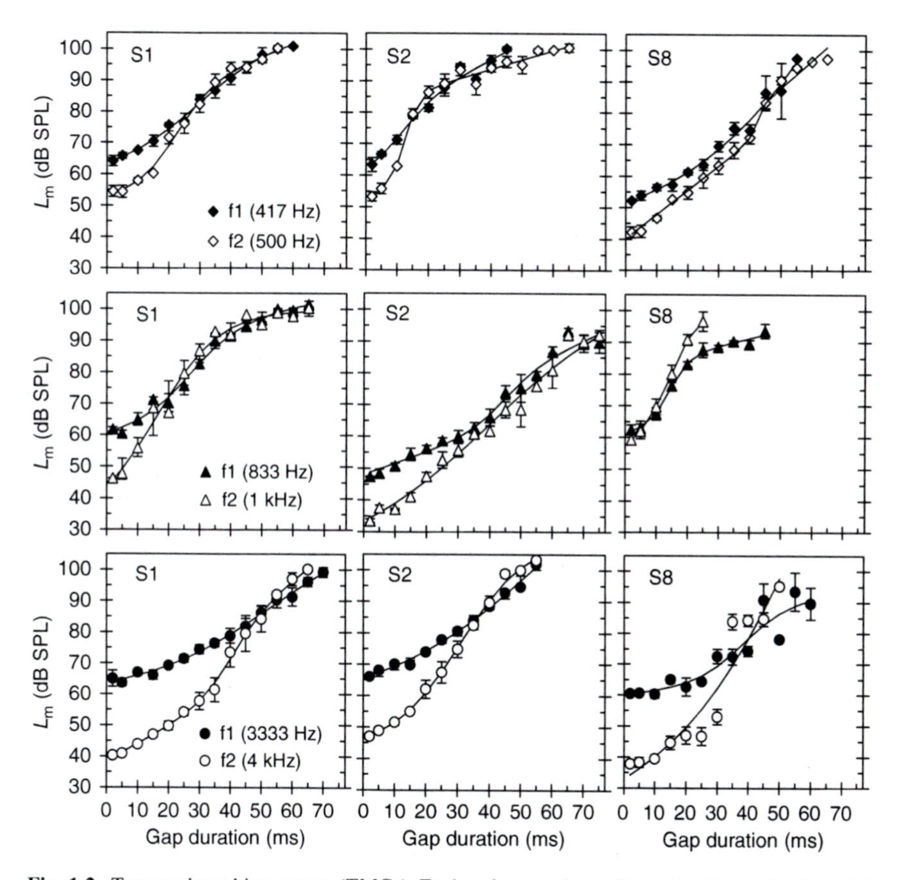

Fig. 1.2 Temporal masking curves (TMCs). Each *column* and *row* illustrates the results for a different listener (S1, S2, and S8) and probe frequency ($f_P = f_2$), respectively. *Filled* and *open symbols* illustrate TMCs for the f_1 (=f2/1.2) and f_2 maskers, respectively. *Error bars* illustrate one standard deviation. *Lines* illustrate ad hoc function fits to the data. L_m stands for masker level. Data are from Lopez-Poveda and Johannesen (submitted)

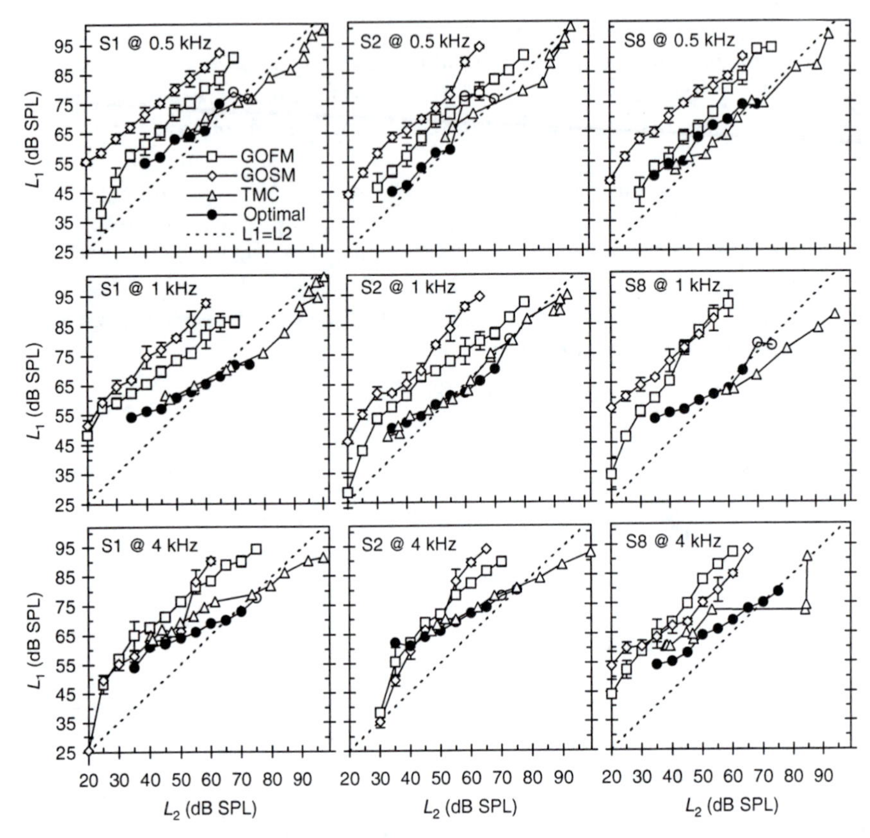

Fig. 1.3 A comparison of individual optimal $L_1 - L_2$ rules with behavioral rules inferred from TMCs, GOFM curves, and GOSM curves. Each *row* and *column* illustrates results for a different test frequency and subject, respectively. *Error bars* illustrate one standard deviation and are only shown for the GOFM and GOSM functions. L_1 and L_2 denote different things for different curves: for GOSM and GOFM functions, they denote masker and probe levels, respectively; for TMC-based functions they denote the levels of the f_1 and f_2 maskers, respectively, for corresponding gaps (Fig. 1.1); for the optimal rules, they denote the levels of the f_1 and f_2 primaries. *Open circles* denote possible suboptimal L_1 values; that is, the actual optimal values would have been almost certainly higher than the values denoted by *open circles*, but they were beyond the highest L_1 value allowed by the DPOAE system. *Dotted lines* illustrate $L_1 = L_2$ relationships

Table 1.2 Equations ($L_1 = aL_2 + b$) of linear regression least-squares fits to the mean data (not shown) of Fig. 1.3 for the various test frequencies and methods considered in the study

Frequency (kHz)	0.5	1	4
Optimal	$L_1 = 0.84L_2 + 18$	$L_1 = 0.60L_2 + 29$	$L_1 = 0.55L_2 + 36$
TMC	$L_1 = 0.76L_2 + 21$	$L_1 = 0.74L_2 + 20$	$L_1 = 0.46L_2 + 44$
GOFM	$L_1 = 1.01L_2 + 18$	$L_1 = 1.01L_2 + 23$	$L_1 = 1.08L_2 + 22$
GOSM	$L_1 = 0.89L_2 + 33$	$L_1 = 0.94L_2 + 33$	$L_1 = 1.24L_2 + 12$

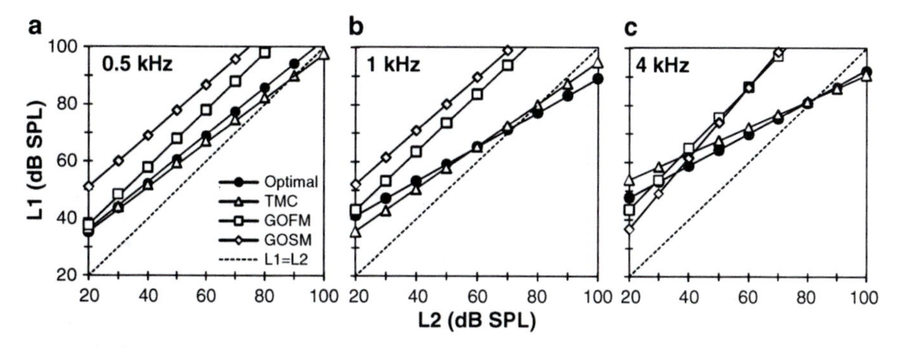

Fig. 1.4 *Straight lines* fitted by the method of least squares to the mean data of Fig. 1.3 over the level range available in the data. Each *panel* is for a different test frequency. *Symbols* are as in Fig. 1.3.

The match between optimal and TMC rules should not be taken as evidence that the level of the $2f_1-f_2$ DPOAE as measured in the ear canal originate solely or even mostly at the f_2 cochlear *site*. A similar match would have been obtained even if the DPOAE originated at multiple cochlear locations so long as the DPOAEs coming from those other cochlear sites were proportional to the DPOAE generated at the f_2 cochlear site. In fact, there is evidence that ear-canal DPOAEs are the vector sum of DPOAE contributions from several cochlear regions (Shaffer et al. 2003; Shera and Guinan 2007).

Both TMC and optimal rules tended to approach an equal level relationship ($L_1=L_2$) as the test frequency decreased from 4 to 0.5 kHz (Fig. 1.4 and Table 1.2). Assuming that both TMC and optimal rules reflect L_1-L_2 combinations that produce equal primary excitations at the f_2 cochlear site, this implies that the cochlear input/output functions (thus cochlear gain and compression) for the two primary tones are more similar for apical than for basal cochlear sites. This result is consistent with the conclusions of earlier behavioral (Lopez-Poveda et al. 2003) and physiological studies (Johnson et al. 2006). The present results also support to the notion that optimal L_1-L_2 rules should differ across test frequencies, as has been previously suggested based on empirical measures (Johnson et al. 2006; Neely et al. 2005).

A second striking result was that GOFM curves differed significantly from optimal rules and, contrary to the hypotheses (see Sect. 1.1), the difference was more pronounced at high L_2 levels, particularly at 1 and 4 kHz (Figs. 1.3 and 1.4). Increasing probe level, as in the GOFM method, almost certainly has two effects: excite a broader cochlear region *and* shift the peak of the traveling-wave to a cochlear region with a CF different from f_P. Both effects could have facilitated probe detection by off-frequency listening: the first of them by combining information over more than one cochlear (frequency) channel; the second, by listening through a (single) cochlear channel with CF$\neq f_P$, where the traveling-wave peak occurs. An ideal experimental procedure would have allowed canceling the first of the two contributions to off-frequency listening to reveal the true effect of traveling-wave peak shifts upon the GOFM curve. Unfortunately, it was not possible to design such procedure. The use of a high-pass noise to minimize off-frequency listening, as is customary when the GOFM method is applied to infer basilar membrane input/output functions (Oxenham and Plack 1997; Nelson et al. 2001), would have simultaneously obscured both contributions. Indeed, there is evidence that the TMC and the GOFM procedures produce generally similar cochlear input/output curves when a high-pass noise is used to minimize off-frequency listening (Rosengard et al. 2005; Nelson et al. 2001).

An alternative procedure would have been to measure on-frequency GOFM and plot the levels for the off-frequency (f_p/1.2) maskers against those for the on-frequency (f_p) maskers, paired according to probe levels. This might have helped calibrating for any nonlinear effects of forward masking per se (Oxenham and Plack 1997). We, however, deemed this unnecessary because the GOFM curves for on-frequency maskers typically have slopes of 1 dB/dB (Oxenham and Plack 1997; Rosengard et al. 2005) (Note, however, that there is evidence for higher slopes in the absence of high-pass noise, as was the case in the present study; Nelson et al. 2001). Furthermore, we deemed this procedure inadequate because both contributions to off-frequency listening would have facilitated probe detection similarly for on- and off-frequency GOFMS and hence the resulting level rule would have canceled out peak-shift effects.

It is possible, therefore, that the present GOFM curves were actually steeper (particularly at high probe levels) that they would have been without the potentially facilitatory effect of off-frequency listening due to a broader probe excitation pattern. Given the trends in the individual curves (Fig. 1.3), however, it seems unlikely that GOFM and optimal curves would have overlapped at high L_2, even considering that some of the reported L_1 values (depicted by open circles in Fig. 1.3) were probably suboptimal (actual optimal values could not be measured because the procedure called for levels beyond the maximum output of the DPOAE system). Most importantly, considering that GOFM functions, unlike TMC rules, were likely influenced by traveling-wave peak shifts at high levels *and* that optimal rules were significantly closer to TMC than to GOFM rules (Figs. 1.3 and 1.4), it seems unlikely that the cochlear site of DPOAE generation shifted with increasing level at any of the frequencies tested.

An incidental, but nevertheless interesting, result was that the divergence between GOFM and TMC rules at high levels was gradually more pronounced with increasing test frequency (Fig. 1.4). Since GOFM curves, unlike TMCs, are likely affected by off-frequency listening (see the preceding paragraphs), this suggests that off-frequency listening is more pronounced for high than for low probe frequencies.

GOSM curves also differed significantly from optimal rules at the three test frequencies in two aspects: they were overall steeper and, except at 4 kHz, their actual L_1 values were significantly higher (Figs. 1.3 and 1.4). At the same time, GOSM and GOFM curves had similar slopes across test frequencies and levels (Fig. 1.4 and Table 1.2). Two-tone suppression could have affected GOSM curves but not GOFM curves. Since both set of curves turned out to have very similar mean slopes

(Fig. 1.4), it seems unlikely that suppression actually affected probe detection in the GOSM paradigm. Furthermore, if the masker had suppressed the excitation to the probe, as is commonly thought (Oxenham and Plack 1997), then masker levels should have been lower for the GOSM than for the GOFM maskers and this was not the case. In fact, the opposite was generally true (Figs. 1.3 and 1.4). Finally, since optimal and TMC rules overlapped at nearly all levels, and the latter were almost certainly unaffected by suppression, it may be concluded that mutual suppression between the primaries does not affect optimal level rules significantly. This is consistent with previous observations (Kummer et al. 2000; Shera and Guinan 2007).

1.4 Conclusions

The main conclusions are:

1. The present results support the notion that the highest levels $2f_1-f_2$ DPOAE component occur when the two primary tones produce comparable degrees of excitation at the cochlear region with CF ~ f_2.
2. Assuming that the highest DPOAE levels occur at the region of maximum interaction between the traveling waves evoked by the two primaries, the present results suggest that this region does not change significantly with increasing primary level.
3. The present results further suggest that mutual suppression between the two DPOAE primaries does not affect significantly optimal L_1-L_2 rules.

The present conclusions apply to test frequencies of 0.5, 1, and 4 kHz and a fixed primary frequency ratio of $f_2/f_1=1.2$. A more comprehensive study should confirm the present conclusions to other test frequencies and frequency ratios.

Acknowledgments Work supported by The Oticon Foundation, the Spanish Ministry of Science and Technology (BFU2006-07536), and Junta de Castilla y León (GR221).

References

Carney LH, McDuffy MJ, Shekhter I (1999) Frequency glides in the impulse response of auditory-nerve fibers. J Acoust Soc Am 105:2384–2391

Johannesen PT, Lopez-Poveda EA (2008) Cochlear nonlinearity in normal-hearing subjects as inferred psychophysically and from distortion-product otoacoustic emissions. J Acoust Soc Am 124:2149–2163

Johnson TA, Neely ST, Garner CA, Gorga MP (2006) Influence of primary-level and primary-frequency ratios on human distortion product otoacoustic emissions. J Acoust Soc Am 119:418–428

Kummer P, Janssen T, Arnold W (1998) The level and growth behavior of the 2 f1–f2 distortion product otoacoustic emission and its relationship to auditory sensitivity in normal hearing and cochlear hearing loss. J Acoust Soc Am 103:3431–3444

Kummer P, Janssen T, Hulin P, Arnold W (2000) Optimal L(1)-L(2) primary tone level separation remains independent of test frequency in humans. Hear Res 146:47–56

Levitt H (1971) Transformed up-down methods in psychoacoustics. J Acoust Soc Am 49:467–477

Lopez-Poveda EA, Alves-Pinto A (2008) A variant temporal-masking-curve method for inferring peripheral auditory compression. J Acoust Soc Am 123:1544–1554

Lopez-Poveda EA, Johannesen PT (2009) Otoacoustic emission theories and behavioral estimates of human basilar membrane motion are mutually consistent. J Assoc Res Otolaryngol 10:511–523

Lopez-Poveda EA, Plack CJ, Meddis R (2003) Cochlear nonlinearity between 500 and 8000 Hz in listeners with normal hearing. J Acoust Soc Am 113:951–960

Neely ST, Johnson TA, Gorga MP (2005) Distortion-product otoacoustic emission measured with continuously varying stimulus level. J Acoust Soc Am 117:1248–1259

Nelson DA, Schroder AC, Wojtczak M (2001) A new procedure for measuring peripheral compression in normal-hearing and hearing-impaired listeners. J Acoust Soc Am 110:2045–2064

Oxenham AJ, Plack CJ (1997) A behavioral measure of basilar-membrane nonlinearity in listeners with normal and impaired hearing. J Acoust Soc Am 101:3666–3675

Robles L, Ruggero MA (2001) Mechanics of the mammalian cochlea. Physiol Rev 81:1305–1352

Rosengard PS, Oxenham AJ, Braida LD (2005) Comparing different estimates of cochlear compression in listeners with normal and impaired hearing. J Acoust Soc Am 117:3208–3241

Shaffer LA, Withnell RH, Dhar S, Lilly DJ, Goodman SS, Harmon KM (2003) Sources and mechanisms of DPOAE generation: Implications for the prediction of auditory sensitivity. Ear Hear 24:367–379

Shera CA, Guinan JJ Jr (2007) Cochlear traveling-wave amplification, suppression, and beamforming probed using noninvasive calibration of intracochlear distortion sources. J Acoust Soc Am 121:1003–1016

Wojtczak M, Oxenham AJ (2009) Pitfalls in behavioral estimates of basilar-membrane compression in humans. J Acoust Soc Am 125:270–281

Chapter 2
Basilar Membrane Responses to Simultaneous Presentations of White Noise and a Single Tone

Alberto Recio-Spinoso and Enrique A. Lopez-Poveda

Abstract One of the fundamental problems in auditory physiology and psychophysics is the detection of tones in noise. In spite of its importance, little is known about its mechanical basis, namely the response of the basilar membrane (BM) to the simultaneous presentation of broadband noise and a pure tone. BM responses to simultaneous presentations of white noise and a single tone were recorded at the base of the chinchilla cochlea using a displacement-sensitive heterodyne laser interferometer. The characteristic frequency (CF) of this region was in the 8–10 kHz range. Fourier analysis showed that the response of the BM to a single tone whose frequency is near or at CF decreases in proportion to the level of the white noise stimulus. Mechanical suppression was larger for softer single tones than for louder ones. Suppression also depended on the tone frequency, being less pronounced as the frequency of the stimulus was set away from CF. In fact, suppression of tones whose frequency was more than 0.5 octaves away from CF – in the linear region of the cochlea – was nonexistent. The results of these experiments were modeled using a nonlinear model of the human cochlea. Suppression in this model was similar to the one observed in our experiments.

Keywords Cochlear mechanics • Basilar membrane • Noise stimulation

2.1 Introduction

One of the basic problems in hearing science is the detection of tones in a noisy environment. Masking effects by noise have been widely studied in auditory psychophysics research ["Noise has no rival in psychoacoustical experiments as a masking agent..." (Green 1960)] and, to a lesser extent, in auditory physiology

A. Recio-Spinoso (✉)
Leiden University Medical Center, P.O. Box 9600, 2300 RC, Leiden, The Netherlands
e-mail: reciospinoso@ymail.com

E.A. Lopez-Poveda et al. (eds.), *The Neurophysiological Bases of Auditory Perception*,
DOI 10.1007/978-1-4419-5686-6_2,

(e.g., Rhode et al. 1978; Costalupes 1985; Young and Barta 1986). As yet, we do not know the response of the basilar membrane in the cochlea to tone plus noise stimuli, i.e., the mechanical bases of the effects of noise on the detection of tone signals.

Responses of auditory nerve fibers (ANFs) to the simultaneous presentation of Gaussian noise and weak single tones at the fiber's characteristic frequency (CF) are dominated by the noise signal (Rhode et al. 1978) until the level of the tone increases beyond a certain level. This led Rhode to conclude that noise acted as a masker of weak CF tones. On the other hand, single tones with frequencies well below the CF can *decrease* the rate evoked by the noise stimulus. Synchronization coefficients-vs.-tone level functions, obtained from the response to CF tones of low-frequency neurons, were paralleled and shifted toward higher levels as the level of the noise masker increased. This shift resembles the shift to the right of rate-level curves of ANFs reported by Young and Barta (1986). The shift of the rate curves toward higher levels has the effect of increasing the threshold of the ANF. The size of the threshold increase is dependent on the noise level.

Ruggero et al. (1992) have shown that the phenomenon of two-tone rate suppression in the auditory nerve is a consequence primarily of nonlinear phenomena in the basilar membrane. In this work, we report a reduction in the response of the basilar membrane to a single tone – evaluated at the tone's frequency – because of the simultaneous presentation of white noise. Mechanical suppression by white noise shares many of the characteristics of two-tone suppression.

Suppression effects reported here can be accounted for by nonlinear models of the cochlea such as the one used in this report (Lopez-Poveda and Meddis 2001) or, to some extent, by a linear filter followed by an instantaneous nonlinearity. Other aspects of the response, such as the response statistics, are more difficult to account for by these types of models.

2.2 Methods

2.2.1 Physiological Recordings and Data Analysis

Three chinchillas (average weight=500 g) were anesthetized using pentobarbital sodium (70 mg/kg, i.p.) Supplementary smaller doses were given as needed to maintain a deeply areflexive state. Tracheotomies were performed, but artificial ventilation was used only when necessary. Core body temperature, measured using a rectal probe, was maintained at 38°C using a thermostatically controlled heating pad. The left pinna was resected, the bulla was widely opened, and the stapedius muscle was usually detached from its anchoring. A silver-wire electrode was placed on the round window to record compound action potentials evoked by tone bursts. Gold-coated polystyrene microspheres (25 mm diameter) were introduced into the cochlea in the area called the hook region, after removing the round window membrane, or in a region approximately 2–3 mm away from the hook, after thinning the bone with a microchisel and then removing bone fragments with a metal pick. BM recordings – using a displacement-sensitive heterodyne laser interferometer (Cooper and Rhode 1992) – were made after the hole was covered with a piece of coverslip glass. The output of the laser system was sampled using a 16-bit A/D system (Analogic FAST-16) at a rate of 250 kHz.

Both noise and single tones were generated using an IBM-type computer in conjunction with a Tucker-Davis Technologies system. The acoustic system was a reverse-driven condenser microphone cartridge (Bruel and Kjær model 1434 with squared-root precompensation). Stimuli were delivered to the tympanic membrane through a closed-field sound system. In the experiments reported here, the noise and sinusoidal stimuli were delivered via separate speakers. In the following text, noise levels are expressed in decibels of attenuation (dBA) with 0 dBA being approximately equal to 100 dB SPL.

Noise stimuli consisted of a frozen sample with a Gaussian distribution that was generated using MATLAB. In two experiments, the frozen sample presented to the animal was the same. In one experiment (m34), however, the polarity of the noise sample was alternated before being presented.

BM responses were analyzed using MATLAB. The analysis usually consisted in performing Fourier transformation to extract amplitude and phase information.

2.2.2 *Computer Modeling*

The *human* basilar membrane model of Lopez-Poveda and Meddis (2001) was used as published. It incorporates an outer-middle ear filter stage, followed by a dual-resonance nonlinear (DRNL) filter (Meddis et al. 2001). Its parameters were set for a CF of 8 kHz (as specified in Table I of their study). Stimulus conditions were similar to those used for the physiological recordings. The duration of the single tones and the noise was long (2 s including 10-ms onset/offset ramps) to minimize spurious results due to noise randomness. The model response was analyzed over the last half of the response waveform to minimize transient effects. The sample frequency was set to 64 kHz.

2.3 Results

Mechanical suppression is defined as the amplitude reduction in the response evoked by a tone signal (the *probe*) due to the presence of a noise stimulus (the *masker*). The abovementioned reduction is evaluated at the frequency of the probe signal after Fourier analysis of the BM mechanical response waveform. Mechanical suppression was observed in two animals exhibiting nonlinear cochlear behavior, as evidenced from their BM responses to single tones, and did not occur in a damaged (i.e., linear) preparation.

Figure 2.1a displays the response amplitudes evoked by a single tone with frequency equal to CF (=8.5 kHz) as a function of probe level. Dashed lines also represent BM responses measured at CF but during simultaneous presentation of the probe and the noise. There is a shift to the right in the input/output functions which is proportional to the noise level, provided that the level of the noise is beyond certain value. Data shown in Fig. 2.1a indicate that BM motion evoked by a single tone at CF is reduced during the simultaneous presentation of white noise. Probe and noise levels determine the amount of suppression.

Results shown in Fig. 2.1b are from a different preparation but, basically, show the same effects as in Fig. 2.1a. (The polarity of the noise sample, however, was kept constant, which explains the "background activity" around 1 nm.) In other words, suppression effects increase in proportion to the noise level and also are more obvious for probes with levels below 60–70 dB SPL.

Figure 2.2 illustrates the response of the computer model to a single tone alone (continuous lines with square symbols) and in the presence of white noise (dotted lines) evaluated at the frequency of the tone (8.5 kHz). The computer model reproduces the main characteristics of the experimental responses, at least qualitatively. The noise suppressor reduces the range of levels over which compression occurs for tones at CF (Fig. 2.2). The reduction is stronger for probe levels below around 60–70 dB SPL and increases with increasing noise level (Fig. 2.2), just as observed experimentally. The model also reproduces the "background activity" observed experimentally for low tone levels in the presence of the noise (cf. Fig. 2.1b).

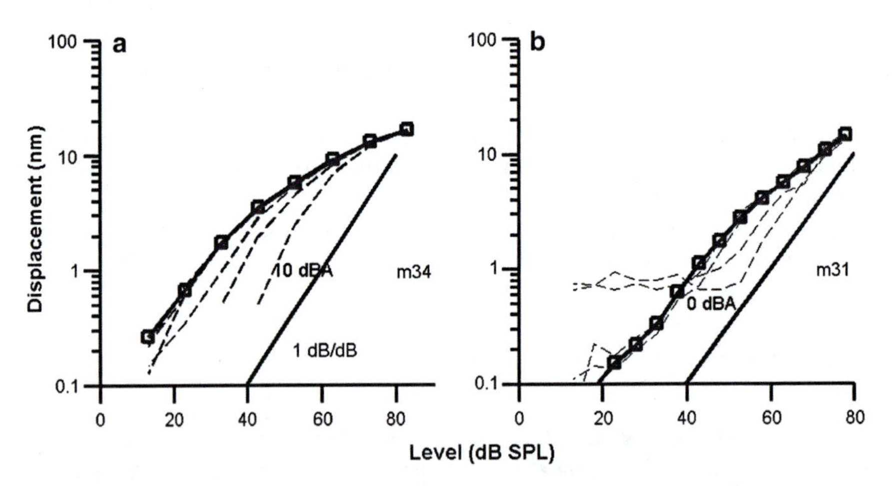

Fig. 2.1 Basilar membrane (BM) responses evoked by simultaneous presentations of white noise and a single tone [*dashed lines* in (**a**) and (**b**)]. The tone frequency matched the CF of the BM recording site [8.5 kHz in (**a**) and 9.75 kHz in (**b**)]. *Continuous lines with open squares* in (**a**) and (**b**) represent BM responses to single CF tone alone. In the results shown in (**a**) and (**b**), the most intense noise level (0 or 10 dBA) produces the largest suppression in the amplitude evoked by the single tone

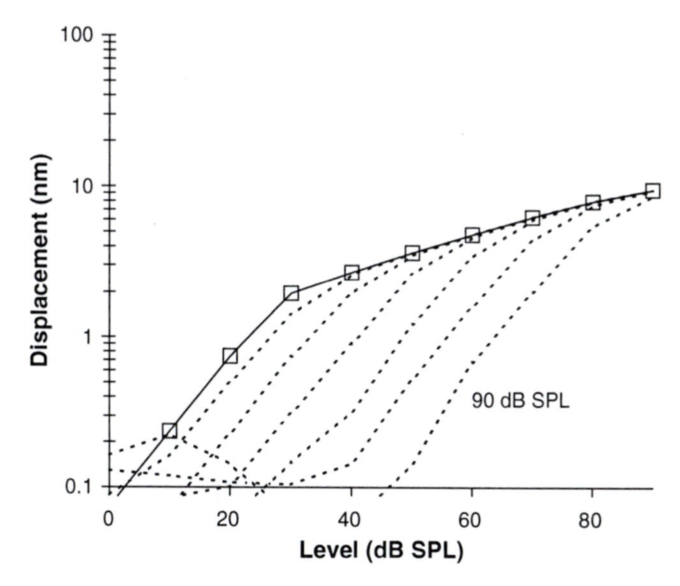

Fig. 2.2 Input/output curves obtained from the model to pure tones in the absence (*squares continuous line*) and in the presence of Gaussian random noise (*dotted lines*). Each *dotted line* represents the input/output curve for a different noise level, decreasing from 90 to 40 dB SPL in 10-dB steps

To highlight the dependence of the BM response to the probe signal on the noise masker, some data points shown in Fig. 2.1a are plotted again in Fig. 2.3a as a function of noise level. Figure 2.3a shows that BM response to the probe signal remained relatively constant until the masker level reached a certain value or "threshold." Suppression thresholds – defined as the masker level necessary to produce 10% reduction in the probe response (Cooper 1996) – varied with the intensity of the probe signal. That is, for a fixed masker level, suppression thresholds decreased with decreasing probe level.

The simultaneous presence of the noise masker also produced small increases in phase *lags* in probe responses, as shown in Fig. 2.3b. Each data point in any of the three curves displayed in Fig. 2.3b represents the phase – evaluated at CF – in response to the probe plus masker signals relative to the phase measured in response to the probe alone. Such results indicate that the phase did not change for noise levels below 40 dBA. For noise levels above 40 dBA, phase lag values increase in proportion to the noise level. In conclusion, reductions in the amplitude of the BM response to a probe signal – because of the presence of a simultaneous masker – go together with increases in phase lags.

The effects of the masker level on the responses evoked by probes with frequencies other than CF are shown in Fig. 2.4. Figure 2.4a illustrates response area curves (i.e., iso-level curves displaying displacement as a function of probe frequency) in the presence of a noise masker whose level is 70 dBA. Figure 4b displays responses

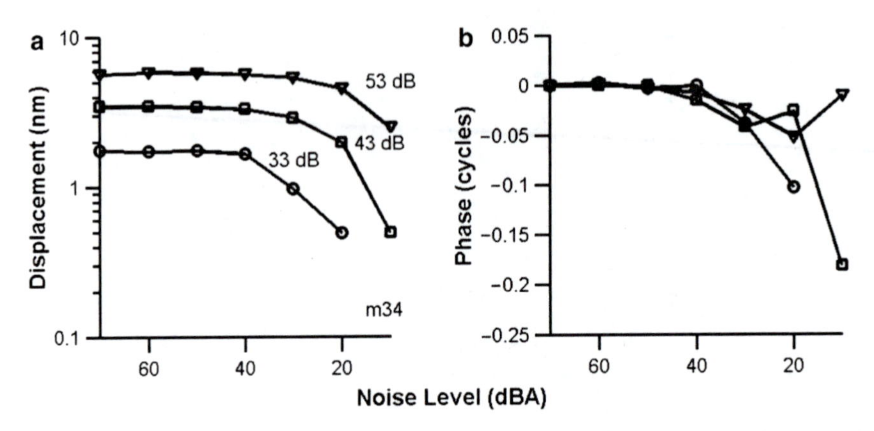

Fig. 2.3 Dependence of suppression on noise level. (**a**) BM response as a function of noise level. Tone level indicated for each curve. *Asterisks* indicate a 10% reduction in the tone signal response. (**b**) *Symbols* represent the corresponding phases relative to the response phase measured when the noise level =70 dBA. *Negative values* indicate phase lags

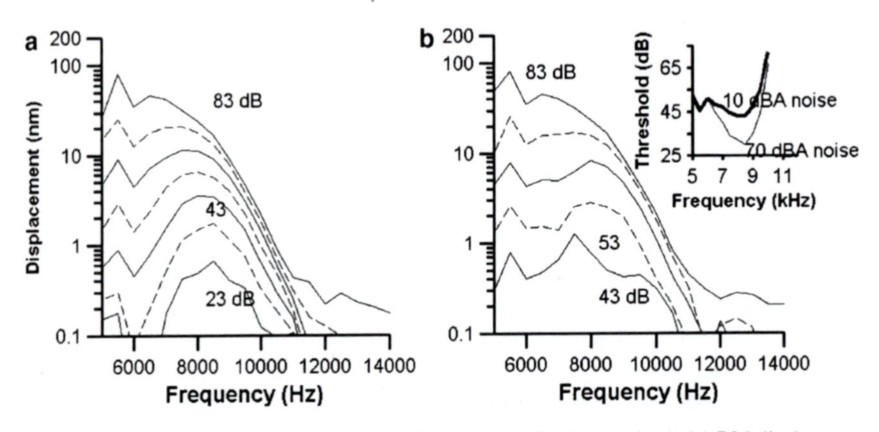

Fig. 2.4 Response area curves measured in the presence of noise maskers. (**a**) BM displacement measured at the frequency of the probe signal (abscissa) in the presence of a 70 dBA masker. Probe levels are 10 dB apart from each other. (**b**) Same as in (**a**) but displacements measured in the presence of a 10 dBA masker. *Inset* in (**b**) represents tuning curves obtained in the presence of 70-dBA (*thin lines*) and a 0 dBA (*thick lines*) maskers

area curves measured in the presence of a 0 dBA masker. The strongest effects occurred for tones with frequencies around CF and levels below 70 dB SPL. Suppression effects were also minimal for probe frequencies ½ octave below or above CF. The inset in Fig. 2.4b displays two iso-response curves, or tuning curves, obtained from the response area data shown in Fig. 2.4a, b. A given point in each of the tuning curves indicates the level of the probe signal, or threshold, required to achieve a given velocity criterion (100 μm/s; Ruggero et al. 1992) in the presence of a masker. A velocity criterion was used to compute the tuning curves to facilitate comparison with similar curves shown by Ruggero et al. (1992). Thresholds for

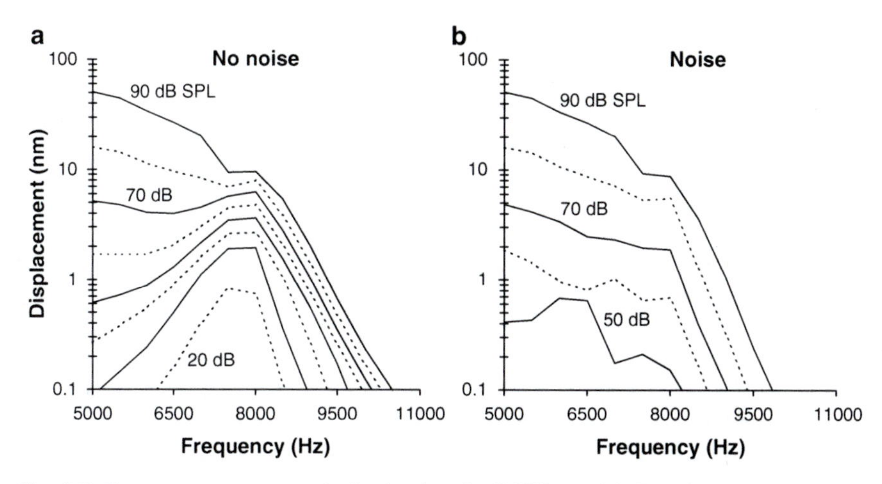

Fig. 2.5 Response area curves obtained using the DNRL model. Data in (**a**) shows results obtained with single tones only and (**b**) was obtained during the simultaneous presentation of a 90-dB noise masker

probe signals with frequencies around CF are larger when the maker level is a 10 dBA than at 70 dBA. The tip of the tuning curves shifts to a lower frequency as the masker level increases. There is also a widening of the tuning curves.

Similar effects to those shown in Fig. 2.4 were also obtained using our computer model (Fig. 2.5). Our computer model reflects, to a certain extent, most of the amount of masking dependence on the frequency and amplitude of the probe signal.

2.4 Discussion

Although direct evidence of mechanical suppression by a type of broadband stimuli (Schroeder-phase complexes) has been previously presented (Recio and Rhode 2000), this chapter shows the first direct evidence of mechanical suppression of a single tone by white noise stimuli, which is a fundamental phenomenon in sensory physiology and psychophysics.

Suppression of a single tone by white noise shares many of the properties of two-tone suppression (Ruggero et al. 1992):

- For a constant masker level, mechanical suppression magnitude decreases with increasing probe intensity.
- Magnitude of suppression increases with increasing noise level.
- Suppression reaches a maximum for probe frequencies near CF.
- Suppression yields small phases lags in responses to CF probe tones.
- Mechanical suppression depends on the physiological state of the cochlea, being nonexistent for linear preparations.

An out-of-the-box *human* version of the computer model was used for convenience, instead of a version specifically tuned to reproduce the present *chinchilla* BM data. This, however, does not undermine the validity of our conclusions. The suppression characteristics of the model are intrinsic to the architecture of its component DRNL filter and are qualitatively consistent with the experimental observations.

The reduction in the amplitude of the BM response at the probe tone observed when a noise signal is applied simultaneously with the probe could be due to a change in cochlear gain. It has recently been shown (Recio-Spinoso et al. 2009) that BM responses to noise yield gain functions that are related to the noise level (i.e., BM responses to noise become more sensitive as the level of the noise decreases.). However, the present computer simulations show that the effects may be accounted for with a DRNL filter (Figs. 2.2 and 2.5) whose gain is fixed (i.e., does not depend on the output level) and thus unaltered by the suppressor. Therefore, the gain reduction suggested by the experimental data (Fig. 2.1) may be only apparent.

It is important to distinguish between suppression as a reduction of the spectral component at the probe frequency and suppression as a reduction of the peak waveform amplitude (c.f., Fig. 6 of Cooper 1996). *Any* frequency independent compressive nonlinearity reduces the amplitude of the spectral component at the probe frequency both for noise and tone suppressors. The DRNL filter incorporates a broken-stick nonlinear gain (Meddis et al. 2001), hence it is not surprising that our model accounts for the present experimental results. The peculiarity of our model is that it accounts for the dependence of the amount of suppression on the probe frequency relative to the CF (Fig. 2.5). Although not shown here, the model also accounts for the reduction of *peak* BM velocity observed experimentally for some high-side suppressors (e.g., Fig. 6D of Cooper 1996). This is because the broken-stick nonlinearity of the DRNL filter is followed by a narrowly-tuned filter centered at the CF. This filter lets the suppressed on-CF probe through unaltered while it attenuates the high-side suppressor. Therefore, its net effect is to reduce the *peak* output amplitude.

The statistical distribution of BM responses to Gaussian white noise is also Gaussian and its envelope is Rayleigh distributed (Recio-Spinoso et al. 2009). The distribution of the BM response to noise of the DRNL filter, however, is not Gaussian (results not shown). This could be a consequence of the type of nonlinearity used in the model (i.e., instantaneous). The study and characterization of a finite-time nonlinearity in the cochlea will be the subject of future work.

Acknowledgments Experimental work supported by NIH grant RO1 DC01910 to W.S. Rhode. Modeling work supported by Spanish Ministry of Science and Technology (BFU2006-07536 and CIT390000-2005-4) and The Oticon Foundation to E.A. Lopez-Poveda.

References

Cooper NP (1996) Two-tone suppression in cochlear mechanics. J Acoust Soc Am 99: 3087–3098

Cooper NP, Rhode WS (1992) Basilar membrane mechanics in the hook region of cat and guinea pig cochlea: sharp tuning and nonlinearity in the absence of baseline shifts. Hear Res 63:163–190

Costalupes JA (1985) Representation of tones in noise in the response of auditory nerve fibers in cats. I. Comparison with detection thresholds. J Neurosci 5:3261–3269

Green DM (1960) Auditory detection of a noise signal. J Acoust Soc Am 32:121–131

Lopez-Poveda EA, Meddis R (2001) A human nonlinear cochlear filterbank. J Acoust Soc Am 110:3107–3118

Meddis R, O'Mard LP, Lopez-Poveda EA (2001) A computational algorithm for computing nonlinear auditory frequency selectivity. J Acoust Soc Am 109:2852–2861

Recio A, Rhode WS (2000) Basilar membrane response to broadband stimuli. J Acoust Soc Am 108:2281–2298

Recio-Spinoso A, Narayan SS, Ruggero MA (2009) Basilar-membrane responses to noise at a basal site of the chinchilla cochlea: quasi-linear filtering. J Assoc Res Otolaryngol 10(4):471–484

Rhode WS, Geisler CD, Kennedy DT (1978) Auditory nerve fiber responses to wide-band noise and tone combinations. J Neurophysiol 41:692–704

Ruggero MA, Robles L, Rich NC (1992) Two-tone suppression in the basilar membrane of the cochlea: Mechanical basis of auditory-nerve rate suppression. J Neurophysiol 68:1087–1099

Young ED, Barta PE (1986) Rate responses of auditory nerve fibers to tones in noise near masked threshold. J Acoust Soc Am 79:426–442

Chapter 3
The Influence of the Helicotrema on Low-Frequency Hearing

Torsten Marquardt and Christian Sejer Pedersen

Abstract Below a certain stimulus frequency, the travelling wave reaches the apical end of the cochlea and differential pressure across the basilar membrane is shunted by the helicotrema. The effect on the forward-middle-ear-transfer function (fMETF) could be measured on both ears of five human subjects by a noninvasive technique, based on the suppression of otoacoustic emissions. All fMETFs show a pronounced resonance feature with a centre frequency that is similar between left and right ears, but differs individually between 40 and 65 Hz. Below this resonance, the shunting causes a 6-dB/octave-increase in the slope of the generally rising fMETF (20–250 Hz). The subject's individual fMETF was then compared with their behaviourally obtained equal-loudness-contours (ELCs) using a 100-Hz reference tone at 20 dB SL. Surprisingly there, the resonance is only reflected in two of the five subjects. Nevertheless, the transition frequency of the slope appears to correlate between individual fMETF and ELC.

Keywords Low-frequency cochlear acoustic • Middle ear transfer function • Equal-loudness contours

3.1 Introduction

Below a certain stimulus frequency, the travelling wave reaches the apical end of the cochlea, and differential pressure across the basilar membrane (BM) is shunted by the helicotrema. This does not only prevent BM displacement to static pressure changes, but can also alter the hearing sensitivity to low-frequency sounds. Dallos (1970) showed that the forward-middle-ear-transfer function (fMETF) at low frequencies has species-dependent characteristics. In cat and chinchilla, the

T. Marquardt (✉)
UCL Ear Institute, 332 Gray's Inn Rd., London, WC1X8EE, UK
e-mail: t.marquardt@ucl.ac.uk

E.A. Lopez-Poveda et al. (eds.), *The Neurophysiological Bases of Auditory Perception*,
DOI 10.1007/978-1-4419-5686-6_3, © Springer Science+Business Media, LLC 2010

oscillatory perilymph flow through cochlear ducts and helicotrema at frequencies below approximately 100 Hz is inertia dominated, and indeed, the slope of the fMETF increases by 6 dB/octave below this frequency. In contrast, cochleae of guinea pig and kangaroo rat exhibit more turns, more tapering, and a smaller helicotrema, so that the perilymph movement here is dominated by viscous friction. Therefore, the fMETF of these two species has the same slope at frequencies below and within the existence region of the resistive travelling wave. These two frequency regions can be easily distinguished by a resonance feature between them, which is prominent in all four species. The resonance can be modelled by a lumped-element circuit that is thought to simulate the interaction between the inertia of the oscillatory fluid movement through the helicotrema and the apical compliance of the BM (Marquardt and Hensel 2008).

Recently, a noninvasive technique, based on the suppression of otoacoustic emissions, enabled the measurement of the fMETF shape below 400 Hz, and allowed to compare the cochlear impedance features between guinea pig and human cochleae (Marquardt et al. 2007). The results confirm Dallos' guinea pig data and support his anatomy-based conjecture that the human cochlea is inertia dominated at the lowest frequencies. It was found that the resonance in humans is approximately an octave lower than in the four species studied by Dallos, implying that the shunting by the helicotrema becomes effective only below approximately 40 Hz.

In our literature review, we were unable to find a convincing behavioural homologue to the resonance, neither in low-frequency hearing threshold, nor equal-loudness-contours (ELCs; for a review see Møller and Pedersen 2004). This negative finding is, however, not entirely conclusive. Often, the data had insufficient frequency resolution to be able to reveal the resonance. Furthermore, the frequency of the resonance might vary individually, and consequently the feature could have been "averaged out" in the published means over many subjects. We decided to obtain ELC and fMETF from a small number of human subjects, and compare their features on an individual basis.

3.2 Methods

3.2.1 Forward-Middle-Ear-Transfer Function

The noninvasive measurement of the shape of the fMETF has been described in detail by Marquardt et al. (2007), although the hardware has slightly changed since. The principle will be only briefly summarized here. Inverted fMETFs were obtained experimentally by adjusting the level of a tonal low-frequency stimulus so as to evoke a constant BM displacement amplitude, independent of its frequency. Constant BM displacement was monitored by simultaneously measuring distortion product ototacustic emissions (DPOAEs), which were suppressed periodically with the frequency of the BM displacement. The method is based on the assumption that a constant DPOAE suppression depth indicates a constant BM displacement (independent of the suppressor frequency). Because the BM displacement is monitored at a location that is far basal from the characteristic place of the suppressor tone, the BM displacement caused by this tone is stiffness-controlled, and therefore proportional to the pressure difference across the BM. Consequently, the fMETF measured here is defined as the ratio between this pressure difference and the pressure in the ear canal (The illustrated raw data are, like the behavioural data, obtained as an iso-output function, and represent therefore the inverse of the fMETF. Nevertheless, they are in the text referred to as fMETF). The chosen DPOAE suppression depth, in these iso-suppression experiments, differed between subject and ears (for individual DPOAE primary parameter, and suppression depth see Table 3.1). It has been shown previously that the shape of the fMETF, which is of interest here, is independent of suppression depth and DPOAE primary parameters (Marquardt et al. 2007).

DPOAE were measured with an Etymotics ER-10C probe. The high-pass cut-off frequency of its microphone amplifier was increased to 1 kHz in order to avoid overloading the A/D converter of the multi-channel sound card (MOTU UltraLite) with the comparatively intense suppressor tone. This tone was produced by a DT48 earphone (Beyerdynamic) that was directly driven by the headphone amplifier of the soundcard. The earphone output was delivered to the probe's ear plug via a narrow silicone tube (300 mm in length, ~0.5 mm i.d.), constituting an acoustic low-pass filter that prevents accidental sound delivery above 100 phon (given the maximum voltage of the headphone amplifier). Stimulus waveforms and DPOAE signal analysis was computed by custom-made software implemented MatLab running under Windows XP. After displaying the $2f_1-f_2$ suppression pattern of a 20-s long recording, the suppressor tone level for the next 20-s recording could be

Table 3.1 Individual DPOAE primary parameters and chosen DPOAE suppression depth

Subject ear	f_1 (Hz)	f_2 (Hz)	l_1	l_2	$2f_1-f_2$	Iso-suppress.
A, left	2,095	2,515	62	50	16	11
A, right	1,610	1,915	65	50	14	4
B, left	2,060	2,515	65	50	3	13
B, right	1,580	1,915	62	50	2	10
C, left	1,845	2,215	62	50	1	11
C, right	1,830	2,215	65	50	6	11
D, left	2,060	2,515	65	50	3	5
D, right	1,815	2,215	62	50	7	7
E, left	2,080	2,515	62	50	6	11
E, right	2,080	2,515	62	50	8	13

Out of the 18 combinations tested in each ear, the primary parameters given here resulted in the largest unsuppressed $2f_1-f_2$ DPOAE level (given in column "$2f_1-f_2$" in dB SPL; parameters l_1, l_2, also given in dB SPL). These parameters were subsequently used during the measurement of this ear's fMETF. The measurement involved the low-frequency suppression the $2f_1-f_2$ DPOAE by a constant, but arbitrarily chosen amount (given in column "iso-suppress.")

adjusted by the experimenter. This was repeated until the desired DPOAE suppression was achieved (typically 3 or 4 repetitions). Iso-suppression levels for the following suppressor tone frequencies have been obtained in ascending order: 20, 30, 35, 40, 45, 50, 55, 60, 65, 70, 75, 80, 90, 100, 125, and 250 Hz.

3.2.2 Hearing Thresholds and Equal-Loudness-Contours

A special low-frequency test facility (Santillan et al. 2007) was used for the threshold and ELC measurements. The signal to each of the 40 loudspeakers mounted in two walls is individually filtered (digitally precomputed) in a manner that avoids standing waves within the test room, so that a large homogeneous sound field in its centre is created (frequency range 2–350 Hz). The facility is equipped with a ventilation system that gives sufficient airflow for continuous occupation of the room, while still maintaining a background noise level lower than 10 dB below the normal hearing threshold (ISO 389-7 2005) for each 1/3-octave frequency band. A Pentium 4 3.2 GHz PC runs the psychophysical protocols and controls the output of two RME 9652 Hammerfall soundcards connected via optical cables to five eight-channel Swissonic 24-bit 48 kHz D/A converters. The D/A converters are connected to eight six-channel Rotel RB-976 MK II power amplifiers (modified for lower noise and frequency range) that drive the forty 13-in. Seas 33F-WKA woofers.

The pure-tone low-frequency hearing thresholds were measured using a slightly modified version of the standard ascending method (ISO 8253-1 1989). The modification consists of having level steps of −7.5 dB rather than −10 dB after each ascend, a modification that was proposed by Lydolf et al. (1999) to give interlaced presentation levels and thus a higher resolution of the psychometric function. Each frequency was measured once except for 100 Hz, where a second measurement was tested for repeatability. Their mean value was then used to determine the 20 dB SL of the 100-Hz reference tone used in the subsequent ELC measurement. A complete ELC measurement, having the same frequency resolution than the fMETF, was obtained within a day, but consisted usually of two separate sessions. The results of four of such measurements, spread over several days, were averaged. A two-alternative forced-choice maximum-likelihood procedure was applied as described by Møller and Andresen (1984). The tone durations for both threshold and equal-loudness determinations were 1 s plus linear fade in/out ramps of 250 ms each. Pilot measurements on subject A were more extensive, with details given in the result section.

3.2.3 Subjects

Ten subjects (8 male, 2 female, aged 22–40) were recruited and initially tested for high levels of the $2f_1-f_2$ DPAOE. In search for optimum stimulus conditions, three

f_2-tones were tested (1,915, 2,215, and 2,515 Hz; l_2 = 50 dB SPL) in combination with various parameter settings for the other primary tone (f_1 and l_1; 18 combinations in total). When the $2f_1 - f_2$ component exceeded 0 dB SPL in any of these, the fMETF of this ear was obtained immediately, using the best combination found, without replacement of the DPOAE probe. As known from previous experiments, a minimum $2f_1 - f_2$ level of 0 dB SPL is required for a sufficient signal-to-noise ratio to reliably apply the fMETFs analysis. Seven of the ten subjects fulfilled this criterion. For two of these seven subjects, the fMETF could not be obtained because their DPOAE was not sufficiently suppressible by even a 90-phon low-frequency suppressor, the loudest tones approved by the UCL Ethics commission for this study. Where the fMETF could be measured in one ear, they could usually also be obtained in the other ear, so that altogether ten fMETFs have been obtained (five subjects, both ears). ELC from these five subjects have then been obtained within three month of their fMETF measurements.

3.3 Results

Detailed low-frequency hearing threshold and ELC measurements were obtained in one subject (first author), before other subjects have been recruited for a subset of those measurements. The results of these pilot measurements are shown in Fig. 3.1. His hearing threshold has been measured with high resolution and six repeats were averaged (three of increasing frequencies, interleaved with three of decreasing frequencies). In accordance with the isophon curves (ISO 389-7 2005, thin dashed lines) the averaged hearing threshold curve (bold dotted line) is much steeper than his fMETFs, obtained with the DPOAE iso-suppression technique at higher intensities (left ear: bold solid line, right ear: bold dashed line). Both fMETFs show a pronounced resonance feature with a local minimum at approximately 55 Hz. (That is, a higher intensity is required here to achieve the iso-BM-displacement amplitude. Remember that in analogy to the behavioural data, the fMETF is measured as iso-output functions therefore plotted inversely here.) This resonance separates the fMETF into regions of different slopes. Below the resonance the fMETF slope is steeper by approximately 6 dB/ octave. Note that only the shape of the fMETF can be assessed with the DPOAE iso-suppression technique. Their vertical position cannot be compared across ears because it is largely influenced by the DPOAE primary parameter and the chosen suppression depth. Also the modelled fMETF (Marquardt and Hensel 2008), has an arbitrary vertical position in the figure (bold dash-dot line).

Surprisingly, subject A's hearing threshold is only little influenced by the fMETF resonance. It shows, however, a slight consistent dip at 80 Hz, which is not seen in his fMETFs. Next we obtained an ELC using a 100-Hz reference tone at 20 dB SL. Here, the fMETF resonance, and also the 80-Hz dip showed much better. We should note here that this subject has occasionally a low-frequency tinnitus in his right ear that was audible before, during and after all measurements in the quiet

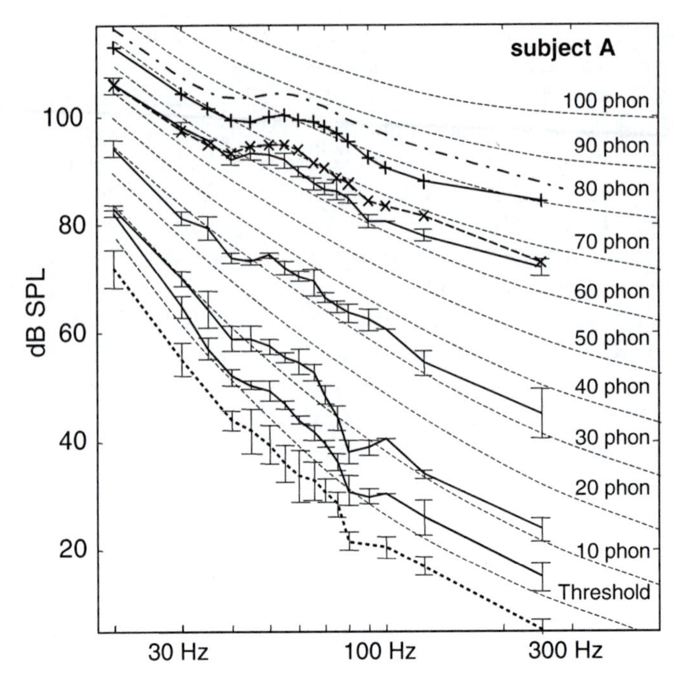

Fig. 3.1 Prior to the assessment the other subjects, Subject A has undergone a more extensive set of behavioural measurements, including high-resolution hearing thresholds (*bold dotted line*), and ELCs with 100-Hz reference tone of 10, 20, 40 and 60 dB SL (*thin solid lines*; error bars: SEM). They have then been compared with the shape of his left and right inverse fMETF (*bold solid line* and *bold dashed line*, respectively), and the inverse fMETF calculated using a lumped-element model (*bold dash-dot line*), simulating the cochlear acoustics at low frequencies (Marquardt and Hensel 2008). The *thin dashed lines* represent standardized isophons (ISO 389-7 2005)

test chamber (including the fMETF measurements). He describes its percept as fluctuating, like narrow-band noise, centred at approximately 80 Hz.

Higher intensity ELCs, one with 40 dB SL and another with 60 dB SL reference, showed less of the resonance and no 80-Hz dip. Measurements above these intensities were impossible due to the audible distortions of the sound system that would have influenced the subjective loudness judgements. With the expectation to find both features most pronounced just above hearing threshold, a further ELC was measured with a reference at 10 dB SL. This curve's shape, however, was just somewhere in between those of the threshold curve and ELC, 20 dB above.

Because of the excessive time these behavioural measurements take, we decided to obtain for the other subjects hearing threshold at fewer frequencies, and to focus on only one ELC obtained with a reference at 20 dB SL since here the fMETF resonance was most pronounced in our pilot study. Their results are shown in Fig. 3.2. It can be said that the fMETFs of all subjects measured so

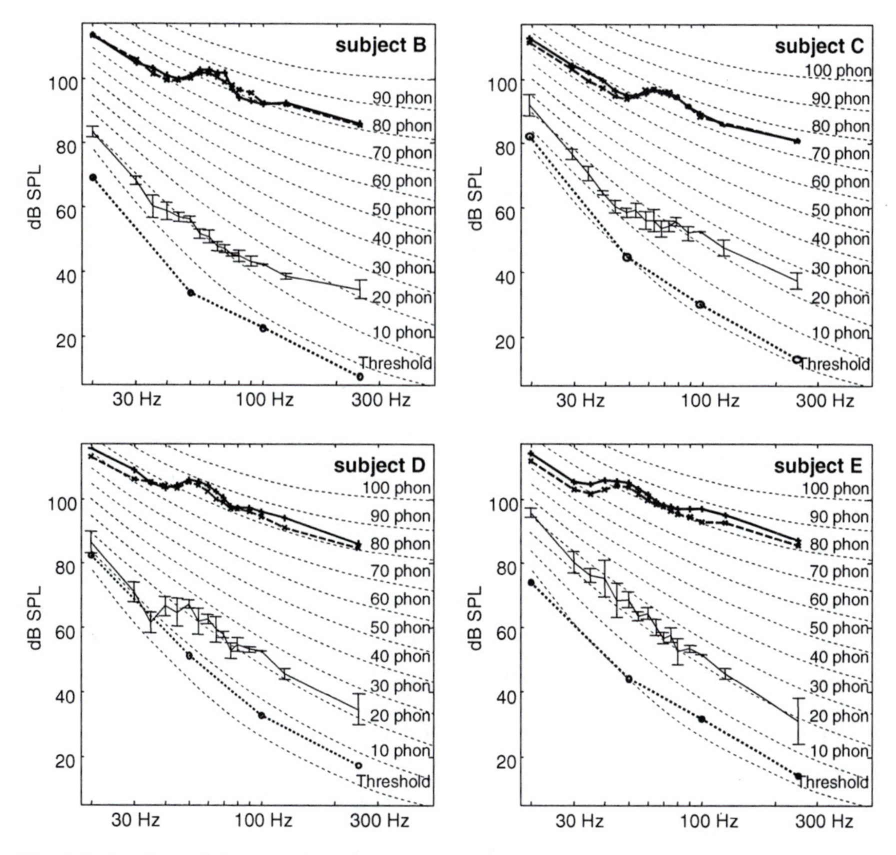

Fig. 3.2 A subset of the behavioural measurements of Fig. 3.1 was obtained from further four subjects and compared with their individual left and right inverse fMETF (for details, see caption of Fig.3.1)

far (five in this study, plus two different subjects from a previous study; see Marquardt et al. 2007) are very similar in shape, including the resonance feature that separates the regions of different slopes. In comparison to its inter-subject variability, the frequency of the resonance is almost identical in an individual's left and right ear.

Whereas most of the behaviourally obtained ELC also shows a distinct transition in slope, the ELCs of only two subjects (A and D) show a behavioural homologue to their fMETF resonance. Also subject C's ELC exhibits a pronounced step between the two slope regions where his fMETF shows the resonance. In addition to the resonance, such large offset is also apparent in subject D's ELC. The subject B (second author) and E have rather smooth ELCs with little offset. Apart from subject A, none of the other subjects had low-frequency tinnitus, or had otherwise ever experienced a particular sensitivity to low-frequency noise.

3.4 Discussion

3.4.1 Comparison Between fMETF and ELC

The fMETF, as defined here, describes the gain between the pressure in the ear canal and the differential pressure across the BM as a function of frequency. Since the pressure across the BM drives the BM, which itself leads to the depolarisation of the sensory cells, one would expect that any irregularities seen in the fMETF, especially such a narrow oscillation as observed here, would be also reflected in the neural output of the cochlea and show consequently an effect on auditory perception. We are therefore surprised that the resonance, visible in all fMETFs, is not consistently seen in the behavioural ELC data. Given the similarity of the fMETFs across all subjects, the diversity amongst the ELCs is rather astonishing, and at this point, we cannot offer a final explanation for this. Before we attempt to address some potential reasons, we would first like to emphasize one fMETF feature that is also reflected in most of the behaviourally obtained ELCs: Dallos observed already in 1970 that in some species, the slope of the fMETF below the resonance is approximately 6 dB/octave steeper than above it. He suggests that this phenomenon reflects the inertia dominance of the oscillatory perilymph flow through the helicotrema in cochleae with large cross-sectional apexes. This conclusion is consistent with the previously observed change in fMETF phase from 180 to 90° where the slope transition occurs (Dallos 1970; Marquardt et al. 2007). The latter can be clearly identified in the ELC of most of our subjects, at a frequency that apparently correlates between individual fMETF and ELC (see e.g. subject C, who has the highest transition frequency in both).

As expected from the systematic change in overall slope of the standardized isophons with stimulus level, the ELCs obtained for low sound intensities are steeper than the fMETF measured at much higher intensities. Such intensity difference is, however, unlikely to underlay differences in the shape of fMETF and ELC, as subject A's ELC, measured at 60 dB SL, shows. It is almost identical in slope, shape and resonance to his right fMETF, which has been obtained using lower biasing tone levels (chosen DPOAE suppression of only 4 dB). A further difference in the methods is that the ELCs are obtained binaurally, but fMETFs are measured monaurally. However, the resonance is very similar between an individual's left and right ear fMETF, so that the binaural behavioural assessment cannot have obliterated this feature. The fact that a similar resonance feature has been observed in both, the ELC and fMETFs of some subjects, weakens generally the possibility that differences in stimulation, like the different coupling of the sound source to the ear (occluded/ open ear canal) might explain the disagreement between both shapes seen in the other subjects.

The difference in the cochlear location, where the measurements are taken, bears in our opinion the most likely explanation. During the fMETF experiments, the pressure difference across the BM is monitored within of second cochlear turn, where the applied DPOAE primary tones have their characteristic places. The behavioural assessments are most likely based on neural activity stemming from the most apical part of the BM, where its low compliance leads to the largest displacement in response to the low-frequency tones. The minimalist model by Marquardt and Hensel (2008) is inadequate to study, where along the BM the simulated resonance has its largest effect. By extending the model by a transmission line, we intend to investigate whether, and under which condition, the resonance affects also the most apical BM, and hope so to be able to simulate the behavioural differences seen between our subjects. Animal recordings from auditory nerve fibres (or more central neurons), in response to low-frequency stimulation, might later confirm, or dismiss propositions derived from these modelling studies by analysing their discharge magnitude and discharge phase in the low-frequency tails of their frequency-tuning curves as a function of their best frequency.

3.4.2 Agreement with Isophons (ISO 389-7 2005)

In general, the individual ELC, measured here with high frequency resolution, show large deviations from the standardized isophon curves. Based on previously published population averages that have been often obtained with lower frequency resolution, the standard shows featureless isophons of smooth curvature (thin dashed lines in both figures). Although a pronounced oscillation in the ELC can be regarded as an exception in our sample, the ELCs of all subjects show an offset, between their lower and higher frequency part. In other words, at frequencies below the slope transition, the ELC follows a lower isophon than at frequencies above it. The offset differs between individuals in magnitude and frequency, but is typically between five and ten phon, and happens at frequencies between 40 and 80 Hz. This offset is better seen in the fMETFs, and is also in agreement with the lumped-element model by Marquardt and Hensel (2008), which simulates the effect of the helicotrema on cochlear acoustics (Fig. 3.1, bold dash-dot line). Although these results need to be confirmed by a larger normal population first, they might impact on future revision of the standard ISO 389-7 (ELC), and possibly ISO 226 (2003; auditory threshold). Care must be taken not to obliterate the sensitivity steps in the ELC of the individual ears by averaging across the population because its frequency-location differs between individuals. One should keep in mind that the subjects in this study might not represent such normal population because they have been selected with regard to high DPAOE levels, and DPAOEs that can be modulated by low-frequency tones. For future objective measurements, steady-state ABR measurements might become a possible alternative to the DPOAE iso-suppression technique applied here.

3.4.3 A Possible Cochlear Origin of Low-Frequency Hypersensitivity or Tinnitus

Low-frequency sounds are almost impossible to shield against, are often very intrusive, and can cause extreme distress to some people who are sensitive to it. Environmental agencies are frequently confronted by individuals complaining about low-frequency environmental noise. The frequency of the annoying noise described by the complainant is commonly at odds with spectral peaks measured by the environmental officer in the complainant's acoustic environment. Often, other people, living in the same location, are undisturbed by, or even unable to hear the humming noise. This leads to situations where the complainant feels misunderstood or not taken seriously.

The complainants often report effects on their life quality that are similar to those described by sufferers of tinnitus. In a study by Pedersen et al. (2008), one third of their cohort, consisting of low-frequency noise complainers, turned out to have low-frequency tinnitus. In this context, it is worth to report about the low-frequency tinnitus of subject A in more detail. He reported to perceive this narrow-band-like sound at a frequency that turned out to be just above his resonance. There, also his behavioural results (below 20 phon) indicate an increased sensitivity. Subject A reported a strong lateralisation of the 80-Hz tone towards the tinnitus ear, whereas the 100-Hz reference tone was always perceived towards the other ear (also when, as a control, the subject was turned 180°, thus facing the rear speakers). All other frequencies tested were perceived by him far more centrally. This suggests that his tinnitus is linked with a true hypersensitivity at 80 Hz which appears to be at the cost of sensitivity at 100 Hz. Monaural ELC measurements using earphones, probing the right and ear separately, will hopefully give further insight here. Despite the fact that the loudness judgement of subject D was also influenced by the resonance in her fMETF, she did not report low-frequency tinnitus, or any past experience of hypersensitivity to low-frequency noise. However, note that her ELC lacks the dip located above her resonance.

The frequency-relationship between subject A's sensitivity peak (80 Hz) and his resonance might of course be entirely coincident. Further tests of individuals with known low-frequency hypersensitivity and/or tinnitus are needed to confirm such connection. The observed resonance and possible effects, associated with it (but not revealed here, e.g., local anti-resonances), could potentially boost hearing sensitivity within a narrow frequency regions. The complainant's cochlea might so create its own spectral peaks, to which he or she has developed a hypersensitivity, not just acoustically, but possibly also neuro-physiologically, or psychologically. Such kind of cochlear filtering could explain the frequency mismatch between their perceived noise and the spectral peaks measured in their acoustic environment. We hope that our initial experiments will evolve eventually into some form of diagnostic test that can objectively reveal possible frequency regions to which the individual low-frequency noise sufferer is particularly sensitive to. This might help them understand their condition, and possibly aid them in controlling it.

3.5 Summary

Previous animal studies have concluded that the geometry of the cochlea, including that of the helicotrema, influences the low-frequency cochlear impedance, and consequently the fMETF. A recently developed non-invasive technique, based on

DPAOE suppression, can assess the shape of the fMETF of the human ear at low frequencies. In this study, we investigated to what extend the features seen in the fMETF shape influence auditory perception, i.e. are reflected in behavioural measurements like low-frequency hearing threshold, and ELC. The fMETFs of all five subjects measured show a pronounced resonance at a frequency that is similar in left and right ear, but varies among them between 40 and 65 Hz. This feature is, however, only reflected clearly in the subjective loudness judgment of two subjects. At this point, we have little explanation why this resonance affects hearing in some subjects, but leaves others unaffected. It is possible that we are tapping on a mechanism that underlies, or triggers a frequency-specific hypersensitivity, that can cause severe suffering from low-frequency noise for some individuals, while others can barely hear it.

References

Dallos P (1970) Low-frequency auditory characteristics: species dependence. J Acoust Soc Am 48:489–499

ISO 226 (2003) Normal equal-loudness level contours. International Organization for Standardization, Genéve

ISO 389-7 (2005) Reference zero for the calibration of audiometric equipment – part 7: reference threshold of hearing under free-field and diffuse-field listening conditions. International Organization for Standardization, Genéve

ISO 8253-1 (1989) Acoustics – audiometric test methods – part 1: basic pure tone air and bone conduction threshold audiometry. International Organization for Standardization, Genève

Lydolf M et al (1999) Binaural free-field threshold of hearing for pure tones between 20 Hz and 16 kHz. Ph.D. thesis, Aalborg University

Marquardt T, Hensel J (2008) A lumped-element model of the apical cochlea at low frequencies. In: Cooper NP, Kemp DT (eds) Concepts and challenges in the biophysics of hearing. World Scientific, Singapore

Marquardt T, Hensel J, Mrowinski D, Scholz G (2007) Low-frequency characteristics of human and guinea pig cochleae. J Acoust Soc Am 121:3628–3638

Møller H, Andresen J (1984) Loudness of pure tones at low and infrasonic frequencies. J Low Freq Noise Vib 3(2):78–87

Møller H, Pedersen CS (2004) Hearing at low and infrasonic frequencies. Noise Health 6(23):37–57

Pedersen CS, Møller H, Persson-Waye K (2008) A detailed study of low-frequency noise complaints. J Low Freq Noise Vib 27:1–33

Santillan AO, Pedersen CS, Lydolf M (2007) Experimental implementation of a low-frequency global sound equalization method based on free field propagation. Appl Acoust 68:1063–1085

Chapter 4
Mechanisms of Masking by Schroeder-Phase Complexes

Magdalena Wojtczak and Andrew J. Oxenham

Abstract Effects of phase curvature in forward masking by Schroeder-phase harmonic tone complexes have been attributed to the interaction between the phase characteristics of the masker and the basilar-membrane (BM) filter, and to BM compression. This study shows that some of these effects may rely on additional mechanisms that have not been previously considered. First, the effect of masker phase curvature was observed for a masker with components well below the signal frequency that was presumably processed without compression by the BM filter. Second, the magnitude of the effect of masker phase curvature depended on masker duration, for durations between 30 and 200 ms. A model implementing the time course and level dependence of the medial olivocochlear reflex (MOCR) produced predictions consistent with the data from masking by on- and off-frequency Schroeder-phase maskers. The results suggest that MOCR effects may play a greater role in forward masking than previously thought.

Keywords Schroeder-phase maskers • Compression • Efferents

4.1 Introduction

Masking by Schroeder-phase complexes has been used to probe basilar-membrane (BM) compression and phase response in humans. These harmonic complexes can produce different amounts of masking, despite having identical power spectra, depending on the value of their constant phase curvature. The components of a Schroeder-phase complex have phases described by

$$\Phi_n = \frac{C\pi n(n-1)}{N} \tag{4.1}$$

M. Wojtczak (✉)
Department of Psychology, University of Minnesota, Minneapolis, MN, USA
e-mail: wojtc001@umn.edu

E.A. Lopez-Poveda et al. (eds.), *The Neurophysiological Bases of Auditory Perception*,
DOI 10.1007/978-1-4419-5686-6_4, © Springer Science+Business Media, LLC 2010

where n is the component number, N is the number of components in the masker, and C is a constant (Lentz and Leek 2001). For a given number of components and a given fundamental frequency, the phase curvature of the complex depends on the value of C. Two factors are known to contribute to the effect of the masker phase curvature on the amount of masking: (1) the interaction between the phase curvature of the BM filter tuned to the signal frequency and that of the Schroeder-phase masker, and (2) compression of the BM waveform in response to the masker at the place corresponding to the signal frequency (Carlyon and Datta 1997a; Oxenham and Dau 2001a). When the phase curvature of the masker is equal in magnitude but opposite in sign to the phase curvature of the BM filter, the BM phase response cancels relative phase shifts between the masker components. This results in the internal masker waveform with all components starting with the same phase. Such a waveform is peakier than waveforms for the maskers with other constant phase curvatures (Kohlrausch and Sander 1995; Carlyon and Datta 1997b; Summers et al. 2003). In simultaneous masking, a peakier waveform provides an opportunity to detect the signal during the valleys of the fluctuating masker. However, listening in the valleys of the waveform cannot be used to improve signal detection in forward masking, where phase effects have also been observed (Carlyon and Datta 1997a). Models of cochlear processing predict that for stimuli with the same rms amplitude, compression leads to differences between the power of internal waveforms with fluctuating versus flatter envelopes (Carlyon and Datta 1997a; Oxenham and Dau 2001a). These differences result in different amounts of BM excitation for maskers with different phase curvatures, and thus different masked thresholds.

Data presented in this chapter suggest that the phase interactions and compression alone may not be sufficient to account for all aspects of masking by Schroeder-phase complexes. The data suggest an involvement of a mechanism with a relatively slow time course that operates at the high frequencies used in this study. A simple model implementing the time course and level dependence of the medial olivocochlear reflex (MOCR) effects in humans (Backus and Guinan 2006) generates predictions that are in good agreement with the data.

4.2 Experiment: Effects of Masker Duration in Masking by On- and Off-Frequency Schroeder-Phase Complexes

4.3 Methods

Forward masking of a 10-ms 6-kHz signal was measured as a function of the phase curvature of a harmonic masker. The signal was gated with 5-ms raised-cosine ramps. The masker was a Schroeder-phase complex with components between 4,800 and 7,200 Hz (on-frequency masker) in one condition, and with components between 1,600 and 4,000 Hz (off-frequency masker), in another condition. In both conditions, the fundamental frequency was 100 Hz. The masker-signal offset–onset delay was 0 ms. Different masker phase curvatures were obtained by using values of parameter C in equation (4.1) from the range between -1 and 1 in steps of 0.25. Three levels, 45, 65, and 85 dB SPL were used for the on-frequency masker. The off-frequency masker was presented at one level, 85 dB SPL. The experiment was run for two masker durations, 200 and 30 ms.

Thresholds were measured using a three-interval forced-choice procedure coupled with an adaptive 2-down 1-up technique. The masker was presented in all three intervals and the signal was presented in one randomly chosen interval. Feedback indicating the correct response was provided following the listener's response. The signal level was initially varied in 8-dB steps. The step size was halved after every other reversal until it reached 2 dB. Eight reversals were completed with a 2-dB step and the signal levels at these reversals were averaged to compute threshold. Four to six single-run thresholds were averaged to compute the final threshold estimate.

The stimuli were generated digitally on a PC with a 48-kHz sampling rate. Before each trial, the masker with the appropriate duration was cut out from a waveform that was longer than the desired duration by one cycle. For each trial, the starting point was chosen randomly within the first cycle of that waveform. The stimuli were played out via a 24-bit LynxStudio Lynx22 sound card and were presented to the left earphone of a Sennheiser HD580 headset.

Six listeners with normal hearing participated in the study. Their hearing thresholds were below 15 dB HL at audiometric frequencies between 250 and 8,000 Hz.

4.4 Results and Discussion

Figure 4.1 shows the mean data for the 200-ms (left panel) and 30-ms (right panel) on-frequency maskers. In each panel, the symbols represent the data and the dashed lines represent predictions by a model discussed below. For the 200-ms masker, the effect of the phase curvature was significant for all masker levels, as determined by a repeated-measures ANOVA ($p<0.01$). The minimum threshold corresponded to a C value between 0 and 0.25, which agrees with the value of 0.13 calculated from the normalized phase curvature of a filter tuned to 6 kHz, estimated by Oxenham and Dau (2001b). The position of the minimum threshold did not depend on level, consistent with earlier reports suggesting a lack of level effects on the phase curvature of BM filters (Shera 2001). For the 30-ms masker (right panel), the effect of the phase curvature was reduced when compared with that for the longer masker. The effect became insignificant for the 85-dB SPL 30-ms masker. This result is inconsistent with time-invariant BM nonlinearity. If the size of the effect were determined by the amount of compression of the BM response to the masker, one would have to conclude that the response for the 200-ms masker was more compressive than that for the 30-ms masker. Such a conclusion has no support in direct mechanical measurements of BM responses, which show that compression is nearly instantaneous and does not build up over time (Ruggero et al. 1997). The effect of masker duration is consistent with an involvement of some additional mechanism with a relatively slow build-up time.

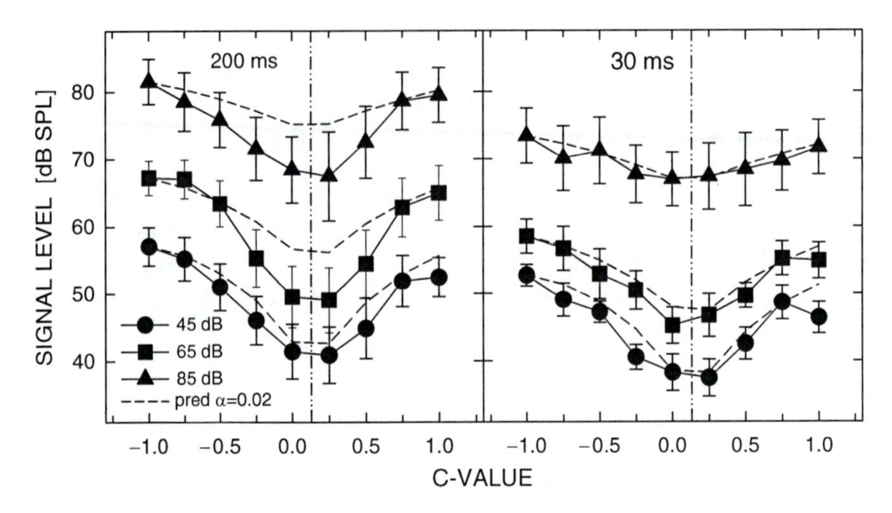

Fig. 4.1 Masked thresholds for the on-frequency Schroeder-phase masker, plotted as a function of parameter C in equation (4.1). Data for the 200- and 30-ms maskers are shown in the *left* and *right panel*, respectively. *Dashed lines* represent model predictions. The *vertical lines* denote the position of the expected minimum threshold based on the phase curvature of the 6-kHz filter in humans

Data for the off-frequency masker, shown in Fig. 4.2, also cannot be explained in terms of compression acting upon the internal masker waveform.

For the 200-ms masker (left panel), the data showed a significant effect of the masker phase curvature (ANOVA, $p<0.01$). This result was surprising because all masker components were more than half-an-octave below the signal frequency. Since the signal frequency was high (6 kHz), it is unlikely that the response to the off-frequency masker was compressed at the BM place with the characteristic frequency (CF) equal to the signal frequency. The minimum threshold was found for $C=0$, consistent with an earlier study which showed that the phase curvature of a BM filter is zero for stimuli more than half-an-octave below the CF (Oxenham and Ewert 2005). For the 30-ms masker (right panel), the effect of the masker phase curvature diminished and was not statistically significant ($p=0.06$). Thus, the results for the off-frequency masker are inconsistent with previous explanations of masker phase effects for two reasons. First, no effects of the masker phase curvature should be observed in forward masking for a masker that is processed linearly in the filter tuned to the signal frequency. Second, masker duration should not influence the effect of the masker phase curvature, or the lack thereof. Thus, the data for the off-frequency masker also suggest the involvement of an additional, relatively slow, mechanism in masking by Schroeder-phase complexes. In the next section, model predictions are generated by assuming that the MOCR activation produces the two unexpected effects observed in the data.

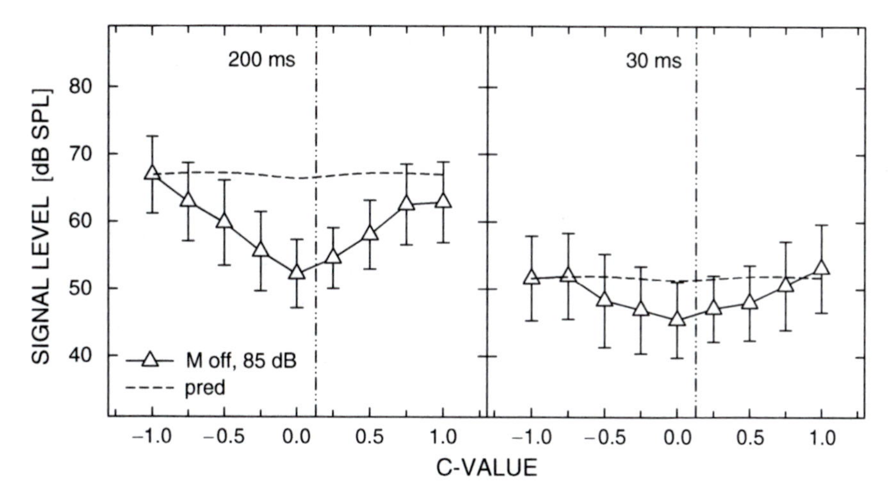

Fig. 4.2 As Fig. 4.1, but for the off-frequency masker

4.5 Model Predictions

4.5.1 Model Description

A model combining the magnitude response of a level-dependent gammachirp filter tuned to 6 kHz (Irino and Patterson 1997) with the phase response derived from the phase curvature estimated from human masking data by Oxenham and Dau (2001b) was used to predict the data in Figs. 4.1 and 4.2. Since the masked thresholds were measured for a 0-ms delay, it was assumed that masked threshold is directly proportional to the power of the internal response to the masker. For each masker, the output of the gammachirp filter was half-wave rectified, low-pass filtered at 0.5 kHz to extract the envelope, and squared. For the on-frequency masker, the squared envelope was then subjected to a 0.2 compression, whereas an exponent of 1 (linear processing) was assumed for the off-frequency masker. The response to the signal was also subjected to the compression of 0.2, and the signal-to-masker ratio necessary for threshold was chosen to make the predicted and measured thresholds match for $C=-1$.

4.5.2 Model Predictions

Predictions by the model are shown by dashed lines in Figs. 4.1 and 4.2. For the on-frequency masker (Fig. 4.1), the predictions fall very close to the data for the 30-ms masker (right panel), but the effect of the masker phase curvature is underestimated for the 200-ms masker. The overall shape of the masking functions and the *C* value corresponding to the minimum threshold are well predicted by the model, but the data for both masker durations cannot be accounted for using the same compression exponent. The model also fails to predict data for the off-frequency masker (Fig. 4.2). While the measured threshold depends on the masker phase curvature, the model predicts no such dependence. Even if a residual compression of the masker were assumed (Lopez-Poveda et al. 2003), the same compression exponent could not predict the data for both masker durations of the off-frequency maker (left and right panels). Thus, the interaction between the phase curvatures of the Schroeder-phase maskers and that of the BM filter and compression cannot account for two effects in our data: (1) the effect of masker duration, and (2) the effect of the phase curvature of the off-frequency masker.

4.5.3 Simulating the Effect of the MOCR

The most parsimonious approach would be to attempt to account for the two unexpected results in terms of the same mechanism. Since the data for the 30-ms masker were accurately predicted by the model simulating interactions of the masker and BM filter phase curvatures and compression, the hypothesized mechanism should play a negligible role in masking by short maskers, and should become important for longer durations. This implies a mechanism with a relatively long time constant. To account for the effects observed for the off-frequency masker, the mechanism should exhibit tuning that is at least as broad as BM tuning. Finally, the mechanism should operate at high frequencies because its effect was observed for a 6-kHz signal. These characteristics resemble some known characteristics of the effect of MOCR activation (for a review see Guinan 2006). The approach presented below was based on the, so far speculative, assumption that the MOCR contributes to the effects of the phase curvature of Schroeder-phase maskers when the masker duration is sufficiently long. It was assumed that the effect of the MOCR elicited by the masker is proportional to the amount of masker excitation on the BM. Maskers yielding BM waveforms with flatter envelopes produce a greater excitation, and thus a stronger effect of the MOCR. It was further assumed that the effect of the MOCR is to reduce cochlear gain of the response to the signal. To overcome the gain reduction, a higher level of the signal was required for these maskers to reach masked threshold. Since the off-frequency masker was presumably processed linearly, it was assumed that the effect of the phase curvature observed for that masker was entirely due to efferent activation. This assumption is not unreasonable given a recent study by Lilaonitkul and Guinan (2009), which showed that an MOCR elicitor can produce strong effects at frequencies half-an-octave to one octave above the frequency of the elicitor. The masker phase effects observed for the on-frequency maskers were assumed to reflect the combined effect of BM compression and the effect of the MOCR. Under these assumptions, it was possible to make predictions regarding the effect of the phase curvature for the 200-ms on-frequency masker, by considering the size of the effect for the 30-ms on-frequency masker, for which the MOCR played a negligible role, and for the 200-ms off-frequency masker, for which the masker phase effect was entirely attributed to the MOCR. First, the size of the effect was quantified by fitting a sine function separately to each set of data in Figs. 4.1 and 4.2 and computing a difference between the highest and lowest thresholds, ΔL_S, on the fitted function. Next, the difference in gain of the response to the signal due to the MOCR elicited by the most and least effective 200-ms off-frequency maskers was estimated by multiplying ΔL_S by the 0.2 compression of the response to the signal. The gain reduction due to the MOCR elicited by the 30-ms off-frequency masker was obtained from the time course of the MOCR effect estimated by Backus and Guinan (2006). Backus and Guinan found that the fast effect of the MOCR activation in humans builds up with a time constant of about 70 ms. For simplicity, it was assumed that the off-frequency masker elicits the same MOCR effect as the on-frequency masker of the same level. Since only one level of the off-frequency masker was tested, predictions for different levels of the on-frequency masker were made assuming a 2% decrease in the MOCR effect per 1 dB decrease in elicitor level, as shown by Backus and Guinan (2006). Figure 4.3 shows differences between the highest and lowest threshold obtained from sine-function fits to the data for the three levels of the on-frequency masker, for the 30-ms masker (white bars), and the 200-ms masker (hatched bars). The differences predicted from the combined effect of compression and efferent activation computed as described above are shown by the black bars.

The measured effect of the masker phase curvature and the effect predicted using the time course and level dependence of the MOCR effect are in very good agreement, despite the obvious oversimplifications of the model. Although the agreement between the predicted and measured differences in threshold is promising, further,

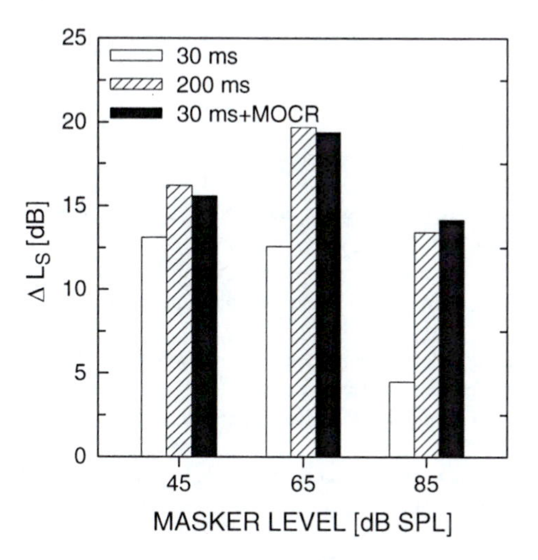

Fig. 4.3 Differences between the highest and lowest masked thresholds for the 30-ms (*white bars*) and 200-ms (*hatched bars*) on-frequency maskers, for three masker levels. The *black bars* represent the difference predicted using the level dependence and the time course of the MOCR effect

more direct measurements of the effect of the phase curvature of Schroeder-phase complexes on the MOCR activation are needed to make strong conclusions about the involvement of efferents.

4.6 Conclusions

The effect of the masker phase curvature in forward masking by Schroeder-phase complexes depends on masker duration, for durations between 30 and 200 ms. The dependence on masker duration suggests an involvement of a mechanism with a long time constant that has not been considered in previous interpretations and analyses of Schroeder-phase masking data. In addition, the effect of the phase curvature is observed in the absence of compression, possibly reflecting the involvement of the same mechanism. A model that incorporates filtering with a realistic phase curvature and compression could not account for either of the findings from this study. Simulations obtained by implementing some known characteristics of the MOCR produce predictions that are consistent with the data. The data and the modeling results indicate that further research is needed to fully understand all the mechanisms involved in masking by Schroeder-phase complexes. Until all the mechanisms have been identified, it may not be prudent to use results from masking by these complexes to infer the amount of BM compression.

Acknowledgments This work was supported by grant R01 DC 03909 from the National Institutes of Health. We thank Amy Olund for her help with data collection. These data and the model predictions are also reported in Wojtczak and Oxenham (2009).

References

Backus BC, Guinan JJ Jr (2006) Time-course of the human medial olivocochlear reflex. J Acoust Soc Am 119:2889–2904

Carlyon RP, Datta AJ (1997a) Excitation produced by Schroeder-phase complexes: evidence for fast-acting compression in the auditory system. J Acoust Soc Am 101:3636–3647

Carlyon RP, Datta AJ (1997b) Masking period patterns of Schroeder-phase complexes: effects of level, number of components, and phase of flanking components. J Acoust Soc Am 101:3648–3657

Guinan JJ Jr (2006) Olivocochlear efferents: anatomy, physiology, function, and the measurement of efferent effects in humans. Ear Hear 27:589–607

Irino T, Patterson RD (1997) A time-domain, level-dependent auditory filter: the gammachirp. J Acoust Soc Am 101:412–419

Kohlrausch A, Sander A (1995) Phase effects in masking related to dispersion in the inner ear. II. Masking period patterns of short targets. J Acoust Soc Am 97:1817–1829

Lentz JJ, Leek MR (2001) Psychophysical estimates of cochlear phase response: masking by harmonic complexes. J Assoc Res Otolaryngol 2:408–422

Lilaonitkul W, Guinan JJ Jr (2009) Reflex control of the human inner ear: a half-octave offset in medial efferent feedback that is consistent with an efferent role in the control of masking. J Neurophysiol 101(3):1394–1406
Lopez-Poveda EA, Plack CJ, Meddis R (2003) Cochlear nonlinearity between 500 and 8000 Hz in listeners with normal hearing. J Acoust Soc Am 113:951–960
Oxenham AJ, Dau T (2001a) Reconciling frequency selectivity and phase effects in masking. J Acoust Soc Am 110:1525–1538
Oxenham AJ, Dau T (2001b) Towards a measure of auditory-filter phase response. J Acoust Soc Am 110:1525–1538
Oxenham AJ, Ewert S (2005) Estimates of auditory filter phase response at and below characteristic frequency. J Acoust Soc Am 117:1713–1716
Ruggero MA, Rich NC, Recio A, Narayan SS, Robles L (1997) Basilar-membrane responses to tones at the base of the chinchilla cochlea. J Acoust Soc Am 101:2151–2163
Shera C (2001) Intensity invariance of fine structure in basilar-membrane click response: implications for cochlear mechanics. J Acoust Soc Am 110:332–348
Smith BK, Sieben UK, Kohlrausch A, Schroeder MR (1986) Phase effects in masking related to dispersion in the inner ear. J Acoust Soc Am 80:1631–1637
Summers V, de Boer E, Nuttal AL (2003) Basilar-membrane responses to multicomponent (Schroeder-phase) signals: understanding intensity effects. J Acoust Soc Am 114:294–306
Wojtczak M, Oxenham AJ (2009) On- and off-frequency masking by Schroeder-phase complexes. J Assoc Res Otolaryngol 10:595-607

Chapter 5
The Frequency Selectivity of Gain Reduction Masking: Analysis Using Two Equally-Effective Maskers

Skyler G. Jennings and Elizabeth A. Strickland

Abstract The "temporal effect" occurs when masked threshold is shifted as a result of the signal being preceded by sound (i.e., a "precursor") instead of silence. Several authors have suggested that the temporal effect may be mediated by the medial olivocochlear reflex (MOCR), which reduces the gain of the cochlear amplifier. We recently measured an analogous temporal effect in forward masking (Jennings et al., J Acoust Soc Am 125:2172–2181, 2009). This study estimated the basilar membrane input–output (I/O) function using psychophysical methods. When an on-frequency precursor was present, the gain of the I/O function decreased, consistent with the MOCR hypothesis. Here, we present data on the same forward masking temporal effect, but with specific interest on the tuning of the precursor's effect. In experiment 1, off-frequency GOM was measured to estimate the I/O function. In experiment 2, psychophysical tuning curves (PTCs) were measured to estimate the tuning of the precursor for two signal levels. Finally, in experiment 3 we combined the masker levels from experiment 1 and the precursor levels from experiment 2 to measure the shift in signal threshold. In the discussion, we model these data in terms of gain reduction and additivity of masking. The results and modeling suggest that the precursor reduced the gain of the I/O function and that the resulting PTCs reflect the tuning of the gain reduction mechanism.

Keywords Olivocochlear efferents • Temporal masking • Models of forward masking

S.G. Jennings (✉)
Department of Speech Language and Hearing Sciences, Heavilon Hall, Purdue University, 500 Oval Dr. West Lafayette, IN 47907, USA
e-mail: sgjennin@purdue.edu

E.A. Lopez-Poveda et al. (eds.), *The Neurophysiological Bases of Auditory Perception*,
DOI 10.1007/978-1-4419-5686-6_5,

5.1 Introduction

Like other sensory systems, the auditory system changes in response to stimulation. One mechanism producing this change is the medial olivocochlear reflex (MOCR). This reflex has been studied anatomically and physiologically, and has been shown to adjust the gain of the active process in the cochlea in response to sound. In a series of studies in our laboratory, we have explored psychoacoustic effects that appear to be consistent with MOCR involvement. These studies started with a phenomenon called "the temporal effect" in simultaneous masking. The temporal effect refers to the fact that a short signal presented at the onset of a masker may be detected at a lower signal-to-masker ratio if the signal is preceded by acoustic stimulation. This preceding acoustic stimulation has been shown to be consistent with (1) a decrease in the gain of the cochlear input–output (I/O) function and (2) a decrease in frequency selectivity (Strickland 2001). Both of these results suggest that the temporal effect is consistent with a reduction in the gain of the active process.

Recently, we have moved to a forward masking technique that – similar to the temporal effect technique – takes advantage of the "sluggishness" of the MOCR by introducing preceding acoustic stimulation. In this technique, the short signal is presented immediately after a short masker. Although this masker may elicit the MOCR, it is hypothesized that the gain adjustment occurs after the offset of the signal due to MOCR sluggishness. [For example, Backus and Guinan (2006) reported a time delay of approximately 20 ms (post stimulus onset) for the MOCR to have any effect.] This short masker is preceded by a precursor that has two roles depending on its frequency. The on-frequency precursor (i.e., same frequency as the signal) has the role of eliciting MOCR gain reduction during the presentation of the signal, while the off-frequency precursor (i.e., well below the signal frequency) acts as a control against attention-related effects. When compared to the control condition, data with an on-frequency precursor were consistent with a decrease in cochlear gain as measured by psychophysical tuning curves (PTCs) and off-frequency growth-of-masking (GOM) functions (Krull and Strickland 2008; Jennings et al. 2009). In addition, fitting these data with a model that assumed gain was reduced by the precursor produced better fits than a model assuming additivity of masking.

The purpose of the present study was twofold. One goal was to measure frequency selectivity for the precursor by measuring a PTC for the precursor alone. We expected that this would reveal frequency selectivity similar to that of excitatory masking. This would be consistent with data showing that the frequency selectivity of efferent fibers (Liberman 1988) and of the MOCR (Warren and Liberman 1989) is roughly similar to that of afferent fibers.

The second goal was to more thoroughly compare the gain reduction and the additivity of masking models. For clarity and ease in discussing these models, we adopt the terminology used by Plack et al. (2006) when referring to the precursor and masker. This terminology labels the two equally effective, nonoverlapping forward maskers as masker 1 (M1) and masker 2 (M2) with M1 (i.e., the precursor) occurring prior to M2 (i.e., the masker). This "combined masker" technique involves comparing signal thresholds in the combined masker case ($S_{\mathrm{M1+M2}}$) with thresholds in the presence of each

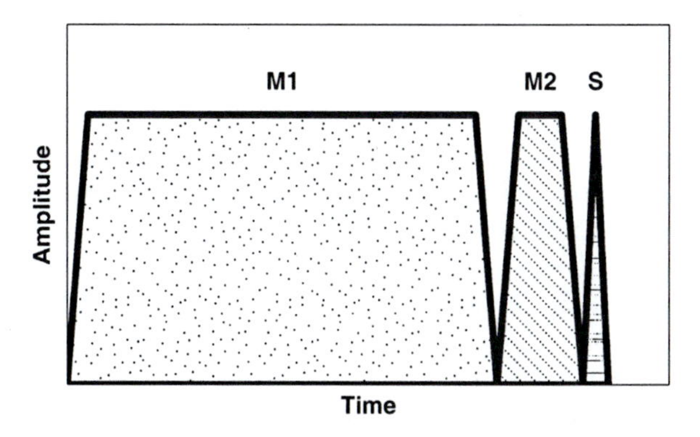

Fig. 5.1 Schematic of the temporal characteristics of the stimuli. Masker 1 (M1), masker 2 (M2), and the signal (S) are displayed

masker individually (S_{M1}, S_{M2}). Figure 5.1 provides a schematic representation of the temporal characteristics of M1, M2 and the signal. Additivity of masking assumes the internal representation of the maskers add at the output of the basilar membrane (BM). Hence, at low signal levels (where the BM response is linear with a slope of 1) this addition leads to a 3 dB increase in signal threshold, while at high signal levels, the increase is appreciably greater due to BM compression (Penner and Shiffrin 1980; Plack and O'Hanlon 2003; Plack et al. 2006; Plack et al. 2007). In contrast, the gain reduction model assumes M1 reduces the gain of the cochlear I/O function. As a result, the signal operates on an "adapted" version of the original I/O function. The S_{M1+M2} thresholds are then detected at a constant signal-to-M2 ratio along this function.

Since the effectiveness of each model (additivity and gain reduction) relies on BM compression, we estimated the cochlear I/O function via the off-frequency GOM technique (Oxenham and Plack 1997) using M2. To evaluate the influence of compression on model predictions, we chose two signal thresholds from the GOM function – one above the compression breakpoint and the other below it. A PTC was built for each of these signal levels using M1. The S_{M1+M2} thresholds were then predicted based on the individual assumptions of each model and the estimated I/O function.

5.2 Method

5.2.1 Subjects

Five subjects (ages: 26, 23, 22, 50, 21; gender: 4 female, 1 male) participated in the experiments. They had normal auditory function based on measures of acoustic immittance, distortion-product otoacoustic emissions and puretone audiometry (quiet thresholds were ≤15 dB HL from 250 to 8,000 Hz for all subjects). Subjects were paid for their participation and were previously inexperienced with psychoacoustic tasks (except S4 who is an author). A training period, lasting 4–6 h, preceded data collection.

5.2.2 *Stimuli*

All stimuli were sinusoids whose temporal characteristics are schematized in Fig. 5.1. The 4 kHz signal was 6 ms in total duration with 3 ms onset/offset ramps. M1 had a total duration of 100 ms and occurred prior to M2 and the signal. M2 had a total duration of 20 ms and occurred between the M1 and the signal. Onset and offset ramps were 5 ms for both maskers. The frequency of M2 was always 2 kHz, while the frequency of M1 took on values between 2 kHz and 4.25 kHz depending on the experimental condition. High-pass noise was present in all conditions to prevent off-frequency listening. This noise started 50 ms before the onset of M1 and finished 50 ms after the offset of the signal and its spectrum level was set at 50 dB below the signal level. The lower cutoff frequency for the high-pass noise was 4.8 kHz.

Stimuli were generated digitally (sampling frequency=25 kHz), output to four separate D/A channels (TDT DA3-4), low pass filtered at 10 kHz (TDT FT5 and FT6-2), mixed (TDT SM3), and sent to an ER-2 insert earphone via a headphone buffer (TDT HB6). The frequency response of these earphones is flat between 250 and 8,000 Hz.

5.2.3 *Procedures*

Thresholds were measured in a double-walled sound-attenuating booth using a three-interval forced-choice adaptive procedure (2-down, 1-up stepping rule) that converged on the 70.7% correct point on the psychometric function (Levitt 1971). Observation intervals were marked by visual stimuli, and the subject's task was to indicate the interval containing the signal after which feedback indicated a correct or incorrect response. Following the subject's response, the level was adjusted by 5 dB until after the second reversal where this value was reduced to 2 dB. Each threshold estimate was computed from a 50-trial block by averaging an even number of reversals after the first two. A final threshold estimate was obtained by averaging the thresholds over at least two blocks. Blocks with a standard deviation greater than 5 dB were excluded. Thresholds for the 6 ms signal in quiet for S1–S5 were: 27.90, 28.28, 23.92, 27.63, and 21.5 dB SPL.

5.2.4 *Experiments*

5.2.4.1 Experiment 1: Off-Frequency GOM

We estimated the cochlear I/O function by measuring GOM with an off-frequency masker. This method is similar to the technique originally described by Oxenham and Plack 1997, except that we did not measure on-frequency GOM, which they used to control for non-linearities in forward masking. Since on-frequency forward masking is generally linear (with a slope of 1) for short masker–signal intervals, the off-frequency masking data alone serve as a reasonable estimate of the I/O function.

We measured the levels of M2 needed to mask a series of signal levels ranging from 25 to 60 dB SPL. M1 was not present in this experiment; however, a control stimulus was presented in the temporal location of M1. This control stimulus was a 0.8-kHz sinusoid presented at 60 dB SPL and was intended to make this experiment temporally similar to experiments 2 and 3. Also, we assumed that this stimulus did not elicit the MOCR at the signal place. A similar technique was employed by Jennings et al. (2009).

5.2.4.2 Experiment 2: PTCs

We measured PTCs for two signal levels. We selected these levels based on the results of experiment 1. One level was below the compression breakpoint ("below *BP*") of the GOM function, while the other was at or above it ("above *BP*"). For all listeners, these levels were 35 and 40 dB SPL, respectively. Although these levels are only 5 dB apart, the additivity and gain reduction models predict the 40 dB SPL signal should produce a relatively higher S_{M1+M2} threshold (as measured in Expt. 3) due to BM compression. Only M1 was present during this experiment. A silent interval occupied the temporal location of M2. We constructed the PTCs by measuring M1 thresholds for several masker frequencies ranging from 2 kHz to 4.25 kHz.

5.2.4.3 Experiment 3: Combined Maskers

Experiment 3 measured S_{M1+M2} threshold for several M1+M2 combinations. We combined the maskers by matching the M1 threshold levels in experiment 1 corresponding to the 35 dB SPL signal, with M2 threshold values from experiment 2 corresponding to the PTC for the 35 dB SPL signal for three M2 frequencies (3, 3.5, and 4 kHz). Thresholds were combined similarly for the 40 dB SPL signal. We measured S_{M1+M2} for each M1, M2 pair, resulting in a total of six data points per subject (2 signal levels × 3 M1 frequencies). One subject (S3) reported difficulty with this experiment due to the high-pass noise. This difficulty was verified by unreliable estimates of S_{M1+M2} thresholds. For this subject, the high-pass noise was removed.

5.2.4.4 Experiment 4: Control Experiment

The levels used for M1 and M2 in experiment 3 were assumed to mask a 35 or 40 dB SPL signal depending on the condition. In order to validate this assumption, we measured signal thresholds in the presence of each M1 or M2 level individually (i.e., the masker level was constant, while the signal level varied).

5.3 Results

5.3.1 GOM Data (masking data for M2)

The first column in Fig. 5.2 depicts the off-frequency GOM data, with each row representing a subject (S1–S5). M2 level is plotted as a function of signal level. We fit a model to each subject's data (solid lines) to estimate the cochlear I/O function. The model was defined by

$$
\begin{aligned}
L_{\text{out}} &= 10\times\log_{10}(10^{(L_{\text{in}}/10)} - 10^{(\alpha/10)}) + G \quad L_{\text{in}} \le BP \\
L_{\text{out}} &= G + 10\times c\times\log_{10}(10^{(L_{\text{in}}/10)} - 10^{(\alpha/10)}) \\
&\quad -10\times\log_{10}(10^{(BP/10)} - 10^{(\alpha/10)})(c+1) \quad L_{\text{in}} > BP,
\end{aligned}
$$

where L_{out} is the predicted M2 level, L_{in} is the input signal level and G, c, *BP,* and α are model parameters representing gain, compression, breakpoint, and internal noise, respectively. Including the internal noise parameter (α) assumed that the signal is, in part, simultaneously masked by a constant low-level physiological noise, which adds (internally) with the masking effect of M2. Qualitatively, the internal noise parameter produced the steep roll off observed at lower signal levels. The mean and standard deviations of the model parameters across listeners were: G [42.80 dB (6.34)], c [0.36 dB/dB (0.20)], *BP* [39.34 dB SPL (1.0)], and α [29.06 dB SPL (4.32)]. The average rms error across subjects for the GOM fits was 0.46 dB.

5.3.2 PTC Data (Masking Data for M1)

The second column of Fig. 5.2 displays the PTC data, where M1 level is plotted as a function of M1 frequency. Open and filled circles represent the "below *BP*" and "above *BP*" signal conditions. We successfully fit roex filter shapes to each PTC except for S1 in the above *BP* condition. The estimated filter sharpness (Q_{10}) from the roex fits and was on average 7.01(1.93) and 6.81(1.26) for the below *BP* and above *BP* conditions, respectively. This similarity in Q_{10} estimates is unsurprising given the signal levels were only 5 dB different across PTCs. The rms error for the roex fits was on average 1.10 dB.

5.3.3 Combined Masker Data and Control Experiment

The rightmost column of Fig. 5.2 displays the combined masker data, where signal level is plotted against M1 frequency. Open diamonds represent $S_{\text{M1+M2}}$ thresholds

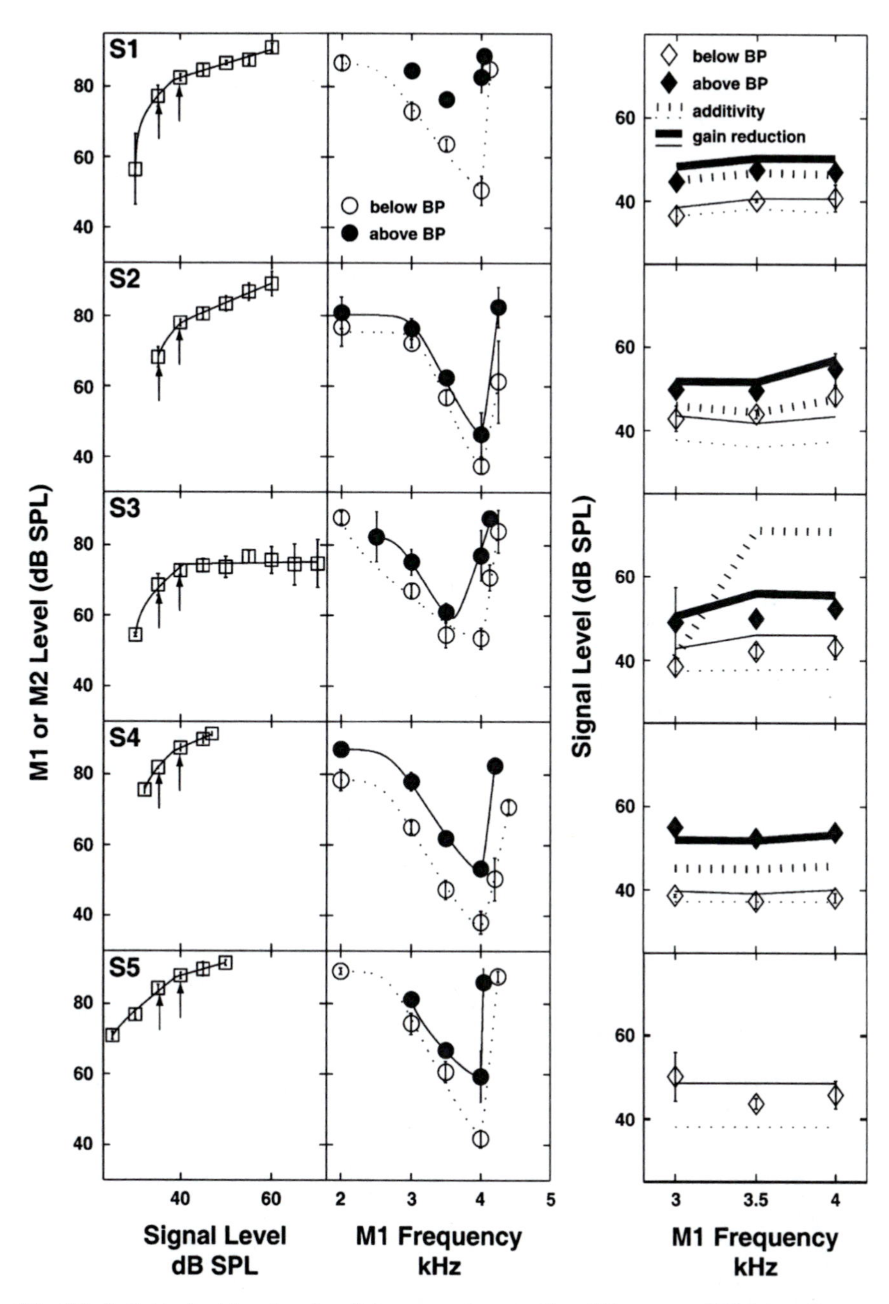

Fig. 5.2 Individual subject data for all three experiments. The off-frequency GOM data (Expt. 1) are presented in the left-most column. The arrows represent the two signal levels used in Expt. 2 (i.e., below *BP* [35 dB SPL] and above *BP* [40 dB SPL]). The PTC data (Expt. 2) are presented in the center column with the open and filled circles representing data from these two signal levels. The combined masker data (Expt. 3) are presented in the right-most column with open and filled diamonds representing M1 and M2 combinations corresponding to these two signal levels. Error bars represent one standard deviation about the mean. Lines are model fits to the data (see text)

for the 35 dB SPL (below *BP*) signal. Similarly, the filled circles represent S_{M1+M2} thresholds for the 40 dB SPL (above *BP*) signal. S5 was unavailable for testing in this condition. The signal threshold shift resulting from these two equally-effective maskers can be observed as the vertical distance between the data and the threshold for each masker individually (i.e., 35 dB and 40 dB SPL for the open and filled symbols, respectively). Dashed and solid lines are predictions for the additivity and gain reduction models which are discussed below in detail. In the control experiment, we measured threshold for the signal in the presence of each individual masker used in experiment 3. In theory, these thresholds should equal the signal levels used in experiment 2 (i.e., 35 and 40 dB SPL). Although individual differences existed, the grand average across subjects was 34.57 dB SPL (1.44) and 40.37 dB SPL (2.46) for the below and above *BP* conditions.

5.4 Modeling

5.4.1 Additivity Model

We designed the additivity model (Fig. 5.3a) based on the results of the control experiment and the estimated cochlear I/O function. The two inputs to the model were the signal thresholds measured in the presence of M1 and M2 individually (S_{M1}, S_{M2}). Considering the experimental design, S_{M1} and S_{M2} were measured using two different methods, and therefore the threshold estimates may vary slightly.

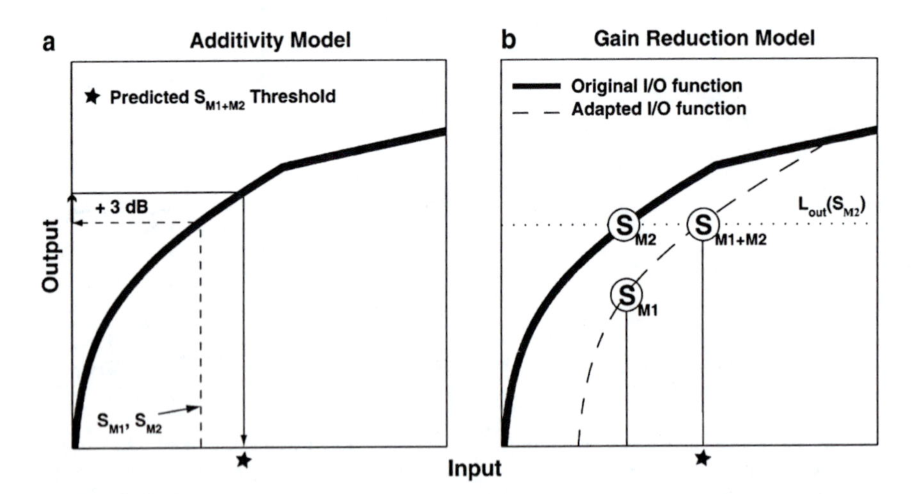

Fig. 5.3 Schematics of the (**a**) additivity and (**b**) gain reduction models. Signal thresholds in the presence of masker 1 (S_{M1}), masker 2 (S_{M2}), and both maskers combined (S_{M1+M2}) are represented for each model

The first method fixed the signal level at 35 or 40 dB SPL and varied the masker (Expts. 1 and 2); therefore S_{M1} and S_{M2} was either 35 or 40 dB SPL using this method. The second method fixed the masker level (i.e., those measured in Expts. 1 and 2) and varied the signal level to estimate threshold (Expt. 4). Although these thresholds were near there expected values of 35 and 40 dB SPL on average, individual differences existed. To optimize the model predictions, we allowed the model to pick the best of four possible S_{M1}, S_{M2} combinations (i.e., the combination producing the lowest sum of squares error). The schematic of the additivity model (Fig. 5.3a) depicts the case where S_{M1} and S_{M2} were equal. The two inputs were then passed through the estimated cochlear I/O function, converted to intensity units, added, converted back to dB, and then passed through the inverse I/O function to obtain the predicted S_{M1+M2} threshold.

5.4.2 Gain Reduction Model

The gain reduction model made the following assumptions: (1) M1 masked the signal by decreasing the gain of the I/O function until the signal's output was equal to the output at quiet threshold (i.e., masking by M1 was not excitatory), and (2) M2 masked the signal at a constant signal-to-masker ratio by means of excitatory masking. Under these assumptions, S_{M1+M2} was a function of an adapted version of the original unadapted I/O function the gain, breakpoint and, internal noise of which were defined by

$$G_{\text{adapt}}(S_{M1}) = G - (S_{M1} - \theta)$$

$$BP_{\text{adapt}} = \frac{G + BP - c \times BP - G_{\text{adapt}}}{1 - c}$$

$$\alpha_{\text{adapt}} = \alpha + G - G_{\text{adapt}}$$

where θ is the subject signal threshold in quiet. As evident in the equation above, G_{adapt} assumed that the difference between G and G_{adapt} (i.e., the change in gain) was equal to the difference between S_{M1} and θ. The compression slope of the adapted I/O function was the same as the original estimate (i.e., $c_{\text{adapt}} = c$). Maintaining the same compression slope is consistent with Plack et al. (2004) who showed that hearing impaired listeners, despite having less gain, had similar psychophysical estimates of compression relative to normal hearing listeners. A schematic of the original and adapted I/O functions are presented in Fig. 5.3b by the solid and dashed lines, respectively.

The predicted S_{M1+M2} threshold was then calculated by passing $L_{\text{out}}(S_{M2})$ through the inverse of the adapted I/O function, where $L_{\text{out}}(S_{M2})$ is the output of the original I/O function evaluated at S_{M2}. Similar to the additivity model, we allowed this model to pick the best of four possible S_{M1}, S_{M2} combinations. The schematic of the gain reduction model (Fig. 5.3b) depicts the case where S_{M1} and S_{M2} were equal.

5.4.3 *Modeling Results*

The rms error value for the additivity and gain reduction models was 7.90 and 2.86 dB, respectively. For many of the predictions, the models are similarly accurate (e.g., S1-all data, and S3 & S4-below *BP* data). This is interesting considering the assumptions are quite different between models. For example, the additivity model assumed the S_{M1}, S_{M2}, and S_{M1+M2} thresholds were a result of a single unadapted I/O function, while the gain reduction model assumed these thresholds depended on two interrelated I/O functions. For the prediction where the models were appreciably different, the additivity model tended to under-predict the data (e.g., S2-all data, S3-below *BP* data, S4-above *BP* data, S5-below *BP* data). This finding is consistent with previous studies, which assumed additivity of masking and predicted S_{M1+M2} thresholds based on the off-frequency GOM function (Krull and Strickland 2008; Jennings et al. 2009).

5.4.3.1 Additivity Model

Two predictions (S3 at 3.5 and 4 kHz) account for a moderate amount of the rms error for the additivity model. Among the entire data set (all subjects included), these two predictions were the only extrapolations (i.e., all other predictions were interpolated from the original I/O function). These extrapolations were made because the additivity model predicted a signal output above the outputs available from the estimated I/O function. In order to obtain the predictions for these data, we assumed the I/O function slope returned to unity at the highest level measured for this subject (70 dB SPL). When we excluded these predictions, the rms error became 5.63 dB.

Our additivity predictions are relatively poorer than studies, which predict the I/O function parameters (e.g., compression) based on the assumption of additivity of masking (Plack and O'Hanlon 2003, Plack et al. 2006). Although there are several explanations for this difference, the likely cause relates to the constraints imposed by the estimated I/O function. These constraints restrict the model from having any true "free" parameters. The only liberty afforded to the model was the option of picking the best combination of S_{M1} and S_{M2} values from a limited set of measured thresholds. Better fitting additivity models in previous studies had more free parameters and may account for the difference in rms error between these different modeling approaches.

5.4.3.2 Gain Reduction Model

The average decrease in gain (i.e., $G–G_{adapt}$) was 9.15 and 13.06 dB for the below and above *BP* conditions, respectively. These values are consistent with previous studies, which measured a decrease in gain using the off-frequency GOM technique and an on-frequency precursor (Krull and Strickland 2008; Jennings et al. 2009). Similar to the additivity model, this model was highly constrained by the estimated I/O function resulting in no true "free" parameters.

5.5 Discussion

Based on the frequency specificity and temporal properties of the MOCR, we hypothesized that M1 masked the signal via a reduction in gain. If this hypothesis were true, it suggests that the PTCs measured in experiment 2 represent the tuning of this gain reduction mechanism. We tested this hypothesis with the gain reduction model. Despite this model being very constrained, it predicted the data reasonably well. Given the Q_{10} values and shape of the PTCs, the tuning of the gain reduction mechanism appears to be similar to the tuning of excitatory masking as originally hypothesized. For example, Jennings et al. (2009) measured tuning for excitatory masking and reported an average Q_{10} value of 6.92, which is very close to the values reported in this study (i.e., 7.01 and 6.81).

The modeling results suggest that the gain reduction model is a viable alternative for interpreting the masking effect of two consecutive forward maskers. This model has also been effective at predicted simultaneous masking data related to the temporal effect (Strickland 2001, 2004, 2008). The ability of the model to predict simultaneous and forward masking data is a strong point, along with the fact that it is consistent with a known physiological mechanism (i.e., the MOCR), which could mediate the gain reduction.

The results of this study have implications beyond providing an alternative framework for modeling the masking effect of two consecutive forward maskers. In a broader context, these psychoacoustic results are consistent with the idea that the auditory periphery is not time invariant. Specifically, they suggest that cochlear gain is decreased over the course of acoustic stimulation – an effect perhaps mediated by the MOCR. Although reducing cochlear gain appears to increase forward masking, it produces the opposite effect in simultaneous masking as illustrated by the temporal effect discussed in the introduction. In simultaneous masking, a reduction in gain produces a release from masking, which may extend to stimuli beyond sinusoids. For example, this release from masking could occur when listening to speech in a noisy background.

Acknowledgments This research was funded by NIH/NIDCD grants R01-DC008327 & T32-DC00030. We thank Michael G. Heinz for providing helpful comments on an earlier version of the manuscript.

References

Backus BC, Guinan JJ (2006) Time-course of the human medial olivocochlear reflex. J Acoust Soc Am 119:2889–2904

Jennings SG, Strickland EA, Heinz MG (2009) Precursor effects on behavioral estimates of frequency selectivity and gain. J Acoust Soc Am 125:2172–2181

Krull V, Strickland EA (2008) The effect of a precursor on growth of forward masking. J Acoust Soc Am 123:4352–4357

Levitt H (1971) Transformed up-down methods in psychoacoustics. J Acoust Soc Am 49:467–477

Liberman MC (1988) Response properties of cochlear efferent neurons: monaural vs. binaural stimulation and the effects of noise. J Neurophysiol 60:1779–1798

Oxenham AJ, Plack CJ (1997) A behavioral measure of basilar-membrane non-linearity in listeners with normal and impaired hearing. J Acoust Soc Am 101:3666–3675

Plack CJ, O'Hanlon CG (2003) Forward masking additivity and auditory compression at low and high frequencies. J Assoc Res Otolaryngol 4:405–415

Plack CJ, Drga V, Lopez-Poveda EA (2004) Inferred basilar-membrane response functions for listeners with mild to moderate sensorineural hearing loss. J Acoust Soc Am 115:1684–1695

Plack CJ, Oxenham AJ, Drga V (2006) Masking by inaudible sounds and the linearity of temporal summation. J Neurosci 26:8767–8773

Plack CJ, Carcagno S, Oxenham AJ (2007) A further test of the linearity of temporal summation in forward masking. J Acoust Soc Am 122:1880–1883

Penner MJ, Shiffrin RM (1980) Nonlinearities in the coding of intensity within the context of a temporal summation model. J Acoust Soc Am 67:617–627

Strickland EA (2001) The relationship between frequency selectivity and overshoot. J Acoust Soc Am 109:2062–2073

Strickland EA (2004) The temporal effect with notched-noise maskers: analysis in terms of input–output functions. J Acoust Soc Am 115:2234–2245

Strickland EA (2008) The relationship between precursor level and the temporal effect. J Acoust Soc Am 123:946–954

Warren EH, Liberman MC (1989) Effects of contralateral sound on auditory-nerve responses. II. Dependence on stimulus variables. Hear Res 37:105–122

Chapter 6
Investigating Cortical Descending Control of the Peripheral Auditory System

Darren Edwards and Alan Palmer

Abstract Corticofugal projections may modulate auditory signal processing at all levels of the ascending auditory system. The final link in this pathway, extending descending control to the cochlea, is the olivocochlear bundle, which originates in the olivary complex and consists of lateral and medial (MOC) systems. Classically, activation of the MOC system leads to an increase in cochlear microphonic and a decrease in cochlear action potential amplitudes.

Here, we investigated the effect of reversible cortical inactivation on the guinea pig cochlea. During cortical inactivation, CM and CAP amplitudes were decreased to varying extents. Reactivation of the cortex led to a recovery to control levels. These results indicate that even in the anaesthetized animal, the cortex is exercising control over the function of the cochlea.

Keywords Descending control • Cochlear Potentials

6.1 Introduction

The ascending auditory system consists of converging and diverging pathways that transform the auditory information as it is transferred from the cochlea to the cortex. Alongside this ascending system is a profuse descending system that directly reaches all the way down to the cochlear nucleus and indirectly to the cochlea. The functions of these descending connections are not well understood, but they presumably serve to modulate the ascending activity and have been implicated in altering subcortical functions such as sharpness of tuning and neural plasticity.

A. Palmer (✉)
MRC Institute of Hearing Research, Science Road, University Park,
Nottingham, NG7 2RD, UK
e-mail: alan@ihr.mrc.ac.uk

E.A. Lopez-Poveda et al. (eds.), *The Neurophysiological Bases of Auditory Perception*,
DOI 10.1007/978-1-4419-5686-6_6, © Springer Science+Business Media, LLC 2010

Projections originating in the cortex are found terminating in all nuclei of the auditory system. The two largest corticofugal pathways terminate in the medial geniculate body and the inferior colliculus (IC) (Winer and Lee 2007). Further, descending pathways project between sub-cortical auditory nuclei, for example, descending pathways from the IC project to the superior olivary complex (SOC) and cochlear nucleus (Winer 2005). The olivocochlear bundle (OCB) provides the final link in the descending auditory system. It originates in the SOC and projects to the cochlea.

The olivocochlear projection is divided into medial and lateral systems. The lateral olivocochlear (LOC) neurons have unmyelinated axons and terminate on the afferent dendrites contacting the inner hair cells (IHCs), while medial olivocochlear (MOC) neurones have myelinated axons and project directly to the outer hair cells (OHCs) (Guinan 2006). Most experimental work on the function and physiology of the OCB has involved the MOC system. Activation of the MOC system inhibits cochlear neural output, demonstrated by a decrease in the amplitude of the cochlear action potential (CAP) (Galambos 1956), an increase in amplitude of the cochlear microphonic (CM) (Fex 1959), and a decrease in distortion product otoacoustic emissions (Mountain 1980). These effects are caused by a conductance increase in OHCs and a reduction in cochlear amplification of the basilar membrane's response to sound (Guinan 2006).

Electrical stimulation of the IC leads to effects that mimic the classical CAP suppression and CM enhancement (that results from directly stimulating the MOC) and have therefore been attributed to the activation of the MOC system (Mulders and Robertson 2000). Effects consisting of either suppression or enhancement of the CAP associated with no change in the CM are hypothesized to occur as a result of activation of the LOC system and changes in the IHCs (Groff and Liberman 2003). However, this interpretation is questioned by Mulders and Robertson (2006) who abolished all enhancement and suppression effects in the cochlea through the application of the MOC blocker gentamicin.

Xiao and Suga (2002) found a decrease in CM amplitude in the bat cochlea when stimulating the contralateral auditory cortex, while stimulation of the ipsilateral resulted in an increase in CM amplitude. Interestingly, when the cortex was deactivated either contralaterally or ipsilaterally (using muscimol) the CM amplitude was reduced.

An obvious role for the descending connections is to direct auditory attention and a number of studies have looked at the effect of attention on cochlear potentials. In cats, a decrease in CAP amplitude to click stimuli (most pronounced at low sound levels) occurs during periods of visual attention, with no effect on the CM (Oatman 1971, 1976). More recently, this decrease in CAP amplitude has been found associated with an increase in CM amplitude using a similar visual attention protocol in chinchillas (Delano et al. 2007).

In light of these findings,, it seems highly probable that the cortex is directly influencing the cochlea. Here, we demonstrate that the cortex, even in the anaesthetized guinea pig is modulating cochlear function. When the cortex is deactivated, there are changes in the amplitude of both CM and CAP that appear to depend on the side that is deactivated.

6.2 Methods

6.2.1 Anaesthesia and Surgical Preparation

Guinea pigs were initially anaesthetized with urethane (1.1 g/kg in 20% solution, intraperitoneal) and supplemented as necessary by Hypnorm (fentanyl citrate, 0.315 mg/ml; fluanisone, 10 mg/ml, intramuscular) to maintain areflexia. A single dose of atropine sulphate (0.06 mg/kg, subcutaneous) was given to reduce bronchial secretions. All animals were tracheotomized and respired with oxygen to maintain normal end-tidal CO_2 levels. Core temperature was maintained at 38°C by a heating blanket and rectal probe. Animals were placed in a stereotaxic frame, with hollow ear bars. A craniotomy was performed over the right auditory cortex and a cryoloop (Lomber et al. 1999) positioned to cover the full extent of the primary auditory cortex. A microelectrode was inserted through the loop into the deep layers to record cortical activity. The surface of the brain was covered in 1.5% agar to stabilize the recordings and to prevent desiccation. The auditory bulla was opened and an insulated silver wire ball electrode was placed on the bony shelf close to the round window to record CAP and CM. The bulla was resealed, including a long thin plastic tube that allowed closed-field sound presentation equalizing the pressure across the eardrum.

6.2.2 Stimulus Presentation and Neuronal Recordings

Auditory stimuli were generated using in-house software and delivered through sealed acoustic systems. A single stimulus set was used in each experiment, played once during each warm and cool condition. The set consisted of ten repetitions with a single repetition comprising 50 ms tones presented in a pseudo-random order ranging in half octave steps between 0.5 and up to 16 or 32 kHz in level between 10 and 95 dB SPL. All stimuli were delivered to the ear ipsilateral to the round window recording.

Neurophysiological recording was undertaken within a sound attenuated chamber on a floating table. The gross cochlea output was recorded with an insulated silver wire ball electrode placed near the round window either contralaterally, ipsilaterally or bilaterally to the exposed cortex. Output was amplified by 60 dB then filtered between 300 and 33,000 Hz for the CM, and between 300 and 3,000 Hz for the CAP. Alternate stimuli were inverted to allow cancellation of the CM when measuring the CAP. Raw waveforms for each channel were recorded and stored.

6.2.3 Cortical Cooling

A cryoloop (Lomber et al. 1999) was used to reversibly deactivate the cortex. A 4 mm diameter hypodermic loop was placed over the right primary auditory cortex and the cortical surface temperature reduced to 1–2°C by pumping cold (−70°C) ethanol through the loop. The cortical surface temperature was recorded by a thermister. At this surface temperature, all layers of the cortex and associated descending projections are inactivated. Cortical reactivation was achieved by simply switching off the cryoloop pump and allowing the normal blood flow to rewarm the cortex, cortical responses returned after 20–30 min.

6.3 Results

During cortical inactivation there is a reduction in the amplitude of the CAP and CM at suprathreshold sound levels. Although clearly present, these amplitude reductions vary across animals in terms of size, timing, the frequencies at which it occurs and the side at which the measurements are made (contralateral or ipsilateral to the inactivation).

6.3.1 Effect of Cortical Inactivation on the Contralateral Cochlea

Of the 13 animals tested, eight demonstrated a reduction in CM during inactivation of the cortex contralateral to the round window electrode. The reduction in CM amplitude was largest at suprathreshold sound levels (above 70 dB SPL) where it varied in different animals between 2 and 15 dB. The CM recovers toward the control level once the cortex is reactivated. In all animals the recovery of the CM amplitude after the first inactivation was incomplete. However, whatever nonrecoverable deterioration was caused by this first inactivation cycle seemed quite stable and recovery after subsequent inactivations were complete to the new stable level. These features are clearly visible in the data shown in Fig. 6.1a, b, which show changes in the cochlear microphonic potentials in a single animal where recording was contralateral to the inactive cortex. The largest CM amplitude functions in this figure are the initial measurements. Subsequently, the recovery was to a slightly lower level as shown by the two overlapping continuous functions. In this animal, the decline in CM amplitude (Fig. 6.1a, b) during cortical cooling occurred maximally at 2 and 2.8 kHz with a decrease of approximately 5 dB. Over all experiments a reduction/recovery was seen at all frequencies tested between 0.5 and 11.3 kHz, although the effects were largest between 0.5 and 4 kHz.

Of the eight animals in which cortical cooling reduced the CM, six showed no effect on CAP (Fig. 6.1c, d). Two animals demonstrated a decrease in amplitude of up to 7 dB at frequencies from 0.5 to 8 kHz. This decrease often occurred at sound levels lower than those bringing about effects on CM (between 30 and 60 dB SPL). A single animal CAP was reduced at all frequencies up to 16 kHz by up to 5 dB, but showed no effect on CM.

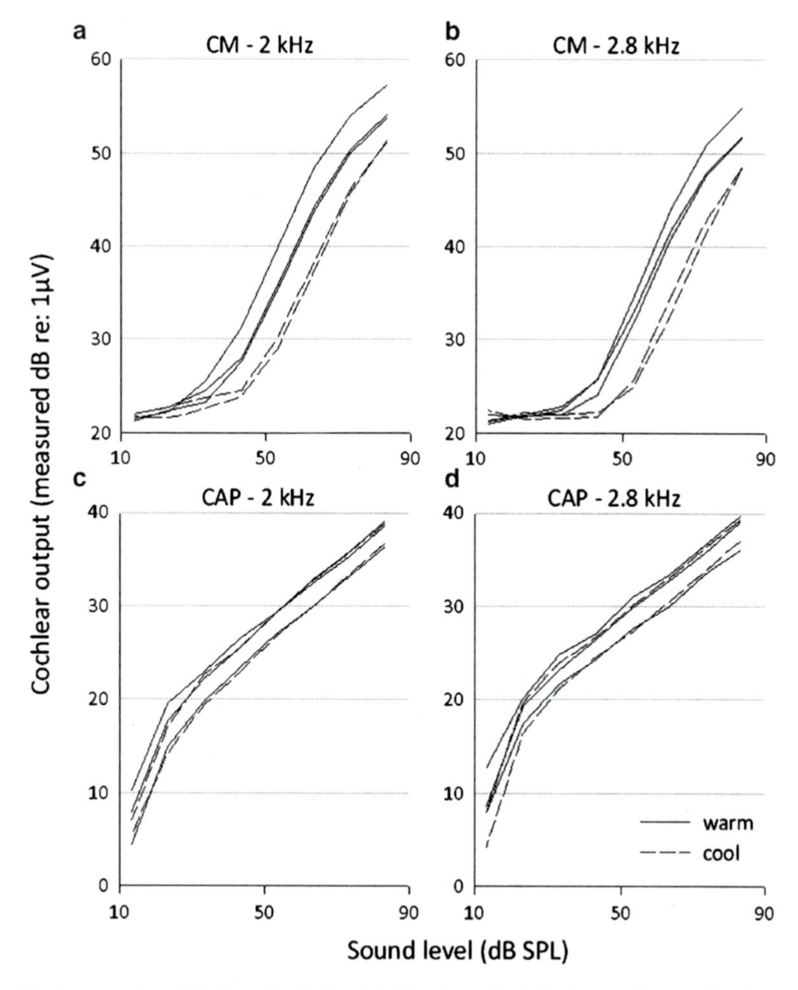

Fig. 6.1 Input–output functions for CM and CAP at 2 and 2.8 kHz in experiment 838. CM amplitude is reduced by up to 5 dB during the cooled conditions (**a**, **b**) whereas there is no difference in CAP amplitude between the two conditions (**c**, **d**). The recovery after the first cycle of cooling is not complete, although it recovers to the new level after the second cycle. A progressive decrease in CAP amplitude occurs as the experiment advances

The reduction/recovery effects are shown clearly when CM and CAP amplitudes for a single frequency and levels are plotted as a function of time. Figure 6.2 demonstrates in three animals at a number of frequencies, a decrease in CM followed by an increase occurring at various cycles during the experiment. Figure 6.2a at 7 kHz shows an initial decrease in CM amplitude of 10 dB followed by a 6 dB recovery toward the control level, in the second cycle a decrease of 6 dB is followed by a recovery of 6 dB. This example demonstrates the effect in both cycles of cooling. A number of experiments have shown a decline and subsequent recovery in the first

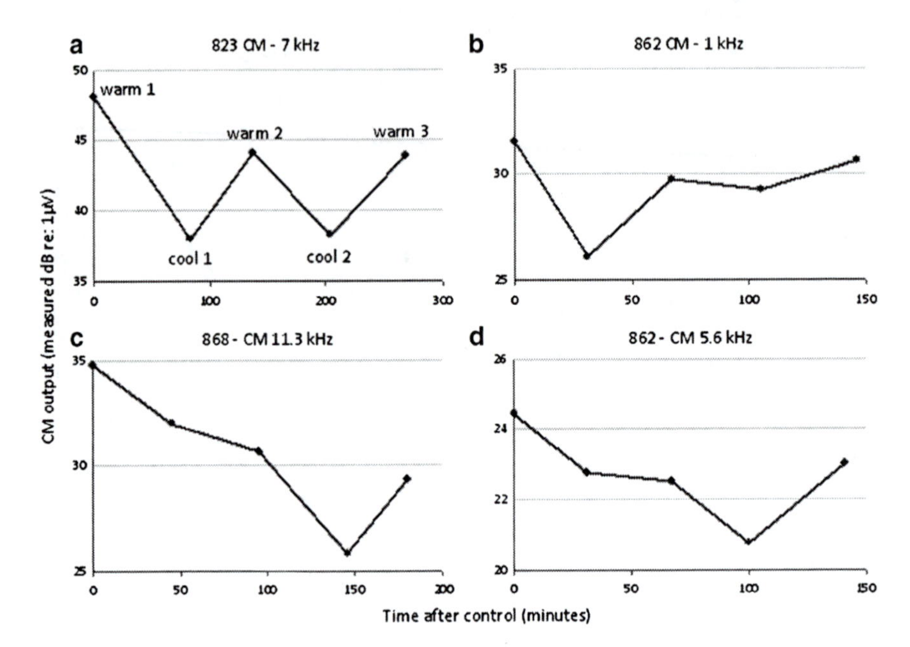

Fig. 6.2 CM time-amplitude curves at 80 dB sound level. (**a**) Experiment 823 at 7 kHz: a reduction in CM output during cool 1 and cool 2 which recovers upon re-warming. (**b**) Experiment 862 at 1 kHz: reduction/recovery in the first cooling cycle. (**c**) Experiment 868 at 11.3 kHz and (**d**) Experiment 862 at 5.6 kHz: reduction/recovery in the second cooling cycle

cycle of cooling but no effect in the second cycle (Fig. 6.2b), while others have displayed the effect only on the second cycle of cooling (Fig. 6.2c, d).

6.3.2 *Effect of Cortical Inactivation on the Ipsilateral Cochlea*

The effect of inactivation of the cortex ipsilateral to the round window electrode was tested in four animals. In all four, the cortical cooling resulted in a reduction in both CM and CAP amplitude. In three animals the size of the decrease in CAP varied from 5 to 25 dB and occurred at all frequencies tested between 0.5 and 16 kHz. The largest effects on the CAP amplitude were found at medium sound levels (30–60 dB SPL). In some experiments the largest effect was at low frequencies, in others it was at high frequencies (at 8 kHz and above). In some experiments the largest effect was in the second cycle of inactivation, while in others it was in the third cycle. Figure 6.3a, b demonstrate the reduction/recovery in CAP amplitude in one animal at 0.5 kHz. The reduction in CAP over the three cycles of cooling was always greater than 10 dB and even on the last cycle the 15 dB reduction in amplitude fully recovers. The decrease in CM that accompanied these large CAP changes was mostly quite small, but varied between 2 and 14 dB. These CM reductions were found at all

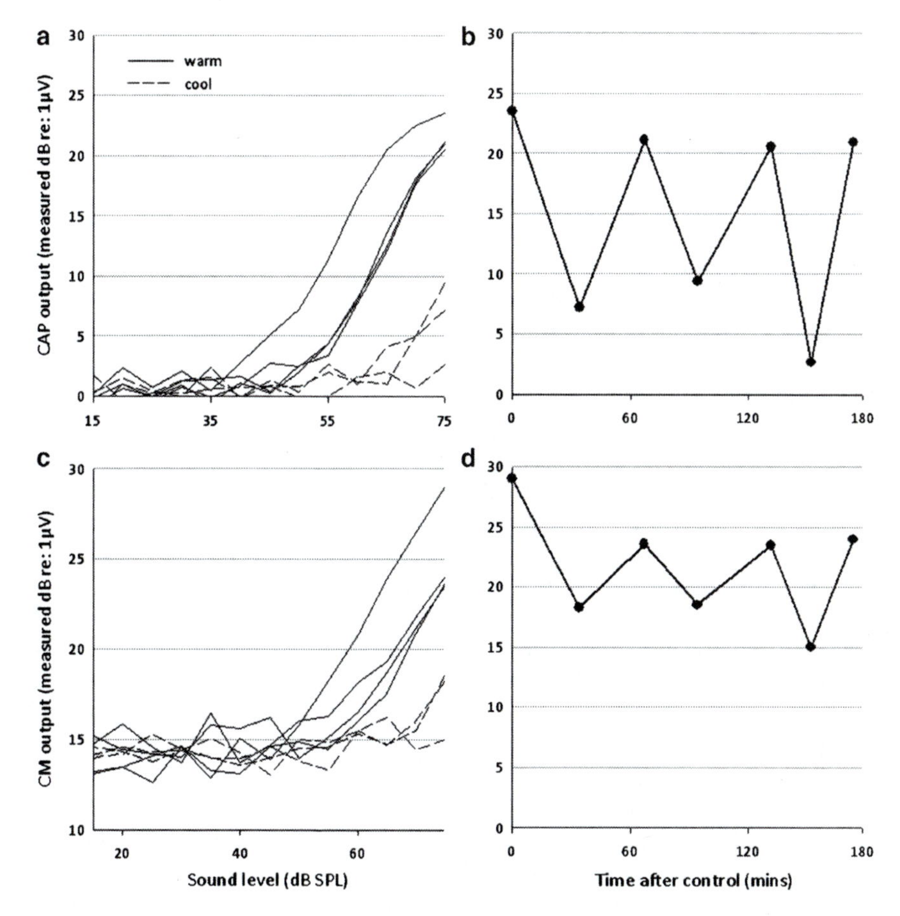

Fig. 6.3 Input–output function and time-amplitude curve at 80 dB for CAP (**a**, **b**) and CM (**c**, **d**) at 0.5 kHz in experiment 905

frequencies between 0.5 and 16 kHz and varied in size during different cycles of cooling (Fig. 6.3c, d). Largest effects were found at high sound levels (above 70 dB SPL).

Of the four animals tested with ipsilateral cooling CM reductions that were larger than the CAP reductions were found in only one. In this animal, the CM was reduced by up to 7 dB occurred at 0.5, 0.7 and 1 kHz in both cycles of cooling and only in the second cycle at frequencies above 4 kHz. CAP showed a similar size reduction at all frequencies between 0.5 and 16 kHz, but only in the second cycle of cooling.

Although it may vary in its appearance (size and timing), this reduction/recovery effect has been identified at least once at all the frequencies tested throughout the study for CM and CAP up to 16 kHz.

Overall, cooling the cortex ipsilateral to the round window electrode seemed to produce more profound and more reliable reductions in the CAP than cooling the contralateral cortex. The effect of such cooling on the CM appeared to be about the same whichever cortex was cooled.

6.4 Discussion

The results indicate that, even in the anaesthetized animal, the auditory cortex is exercising a level of direct descending control over the cochlea. Cochlear potentials are reduced in amplitude by cortical inactivation: the inactivation appears to have a greater influence on CM than on CAP contralaterally and on CAP than on CM ipsilaterally.

In a previous study in the bat on the influence of the cortex on the cochlea both increases and decreases in CM from ipsilateral and contralateral electrical stimulation, respectively were found. However, when the cortex was inactivated with muscimol a reduction in CM was seen ipsilaterally and contralaterally (Xiao and Suga 2002), a result consistent with the present data. The Xiao and Suga (2002) study used both cortical activation and inactivation, which did not, under some conditions, produce opposite changes in the CM amplitude. The descending auditory system from cortex to cochlea is complex. Our data do not replicate the classical effects of stimulating the MOC. This is perhaps unsurprising given the anatomical complexity, the data of Xiao and Suga, and the fact that the majority of electrical stimulation experiments have focused on nuclei below the cortex, and therefore may take a different path to the cochlea.

Inferior colliculus stimulation experiments (Groff and Liberman 2003; Mulders and Robertson 2000) found suppression of the CAP and enhancement of the CM, which is also found with direct MOC stimulation (Galambos 1956; Fex 1959) and has therefore been attributed to indirect activation of the MOC system. In contrast, the present study demonstrates parallel changes in the cochlear potentials (i.e., both CAP and CM are suppressed or unaffected). However, nonclassical effects where CAP suppression or enhancement was associated with no change in CM have also been observed (Groff and Liberman 2003) and attributed to activation of the LOC system. This raises the question as to whether the effects of cortical inactivation shown here occur via the medial or lateral olivocochlear systems or both.

The MOC system terminates on the OHCs directly causing increased conductance and therefore the increased current passing through them (Guinan 1996). Because of this, it is the most likely candidate for bringing about the alterations in CM found in this study although it would require a reduction in OHC conductance to reduce the CM. Since the increased conductance is considered to be due to a shunting action, which reduces the OHC motility, we might perhaps expect suppressed CM to represent the inverse. The reduction in the CAP would argue that things are not so simple. Given the global nature of the cortical inactivation and the diversity of the descending pathways, it seems likely that it is not solely affecting the MOC system. The lateral system terminates on the dendrites of the afferent fibers leaving the IHC, and therefore has the potential to alter the firing rate. As the CAP represents the

synchronous firing of these axons, any alteration in the firing could affect the CAP amplitude. Although, as described earlier, the medial system has the potential to affect both CAP and CM at the same time, it suppresses the CAP and increases the CM. The present study found a parallel decrease in CM and CAP, so it is conceivable that both the medial and lateral systems are contributing to this result.

The differences between contralateral and ipsilateral recording poses interesting questions as to the laterality of the effect. Efferents from the cortex to the SOC are largely ipsilateral (Coomes and Schofield 2004); at the OCB, the majority of MOC neurons project contralaterally while more LOC neurons project ipsilaterally. The fact that there appears to be a greater effect on CAP when recording ipsilaterally along with the ipsilateral LOC projection could suggest that these effects are brought about through the LOC system.

The global nature of the inactivation brought about by our relatively large cryoloop clearly demonstrates effects on the cochlea. However, such global inactivation or indeed activation are unlikely to provide any insight into the normal functioning of this system. A more sophisticated approach will eventually be needed to fully understand this cortical control system.

6.5 Comment by Stefan Strahl

Did you observe any changes to the compound action potential waveform? In the context that the second negative potential (N2) can disappear in lession studies (e.g., Meurice et al. 1991) it might be interesting if any effects (e.g., to the latencies) of N1, P1 and N2 are observable in your experiments.

6.6 Reply Alan R. Palmer

We have normalized and compared the CAP waveforms measured before and during cooling. There is a change in the shape of the CAP waveform as the N1 component is reduced by a larger proportion than the N2 component and there is an increase in latency of about 0.5 ms. There was no gross change in the shape such as the loss of the N2 peak that occurs when the auditory nerve is sectioned as it leaves the internal auditory canal (Meurice et al. 1991).

References

Coomes DL, Schofield BR (2004) Projections from the auditory cortex to the superior olivary complex in guinea pigs. Eur J Neurosci 19:2188–2200

Delano PH, Elgueda D, Hamame CM, Robles L (2007) Selective attention to visual stimuli reduces cochlear sensitivity in chinchillas. J Neurosci 27:4146–4153

Fex J (1959) Augmentation of cochlear microphonic by stimulation of efferent fibres to the cochlea; preliminary report. Acta Otolaryngol 50:540–541
Galambos R (1956) Suppression of auditory nerve activity by stimulation of efferent fibers to cochlea. J Neurophysiol 19:424–437
Groff JA, Liberman MC (2003) Modulation of cochlear afferent response by the lateral olivocochlear system: activation via electrical stimulation of the inferior colliculus. J Neurophysiol 90:3178–3200
Guinan JJ (1996) Physiology of the oivocochlear efferents. In: Dallos P, Popper AN, Fay RR (eds) The cochlea. Springer-Verlag, New York, pp 435–502
Guinan JJ Jr (2006) Olivocochlear efferents: anatomy, physiology, function, and the measurement of efferent effects in humans. Ear Hear 27:589–607
Lomber SG, Payne BR, Horel JA (1999) The cryoloop: an adaptable reversible cooling deactivation method for behavioral or electrophysiological assessment of neural function. J Neurosci Methods 86:179–194
Meurice JC, Paquereau J, Marillaud A (1991) Same location of the source of P1 of BAEPs and N1 of CAP in guinea pig. Hear Res 53:209–219
Mountain DC (1980) Changes in endolymphatic potential and crossed olivocochlear bundle stimulation alter cochlear mechanics. Science 210:71–72
Mulders WH, Robertson D (2000) Effects on cochlear responses of activation of descending pathways from the inferior colliculus. Hear Res 149:11–23
Mulders WH, Robertson D (2006) Gentamicin abolishes all cochlear effects of electrical stimulation of the inferior colliculus. Exp Brain Res 174:35–44
Oatman LC (1971) Role of visual attention on auditory evoked potentials in unanesthetized cats. Exp Neurol 32:341–356
Oatman LC (1976) Effects of visual attention on the intensity of auditory evoked potentials. Exp Neurol 51:41–53
Winer JA (2005) Decoding the auditory corticofugal systems. Hear Res 207:1–9
Winer JA, Lee CC (2007) The distributed auditory cortex. Hear Res 229:3–13
Xiao Z, Suga N (2002) Modulation of cochlear hair cells by the auditory cortex in the mustached bat. Nat Neurosci 5:57–63

Chapter 7
Exploiting Transgenic Mice to Explore the Role of the Tectorial Membrane in Cochlear Sensory Processing

Guy P. Richardson, Victoria Lukashkina, Andrei N. Lukashkin, and Ian J. Russell

Abstract Recent observations have changed our understanding of tectorial membrane function. Transgenic mice have shown that the tectorial membrane is a structure that can influence the sensitivity and tuning properties of the cochlea in several ways. It ensures that the gain and timing of cochlear feedback are optimal; that the hair bundles of the inner hair cells are driven efficiently by the outer hair cells, and it may influence the extent to which different elements are coupled along the length of the cochlea.

Keywords Cochlea • Deafness genes • Hearing loss • TECTA • Tectorial membrane • Cochlear amplification

7.1 Introduction

The organ of Corti is the complex, cellular filling that is sandwiched between two complex, extracellular matrices, the basilar membrane (BM) and tectorial membrane (TM) (Richardson et al. 2008). These extracellular matrices fundamentally shape the responses of cochlea to sound. The gradation in thickness, length, and packing density of the collagen fibrils of the BM provides the stiffness gradient that determines the gradient in frequency tuning along the length of the cochlea (von Békésy 1960). The outer hair cells (OHCs) are coupled directly to the TM through the tallest rows of stereocilia in the hair bundles (Lim 1986). Shear displacement between the TM and the reticular lamina, caused by acoustic stimulation, leads to hair bundle displacement. Hair bundle displacements toward the tallest row of stereocilia increase the opening probability of the transducer channels located near the tips of the stereocilia, and hence the flow of current into the hair cells (Hudspeth 1989).

I.J. Russell (✉)
School of Life Sciences, University of Sussex,
Falmer, Brighton, BN1 9QG, UK
e-mail: i.j.russell@sussex.ac.uk

E.A. Lopez-Poveda et al. (eds.), *The Neurophysiological Bases of Auditory Perception*,
DOI 10.1007/978-1-4419-5686-6_7, © Springer Science+Business Media, LLC 2010

Through interaction principally with the basilar membrane, reticular laminar, and TM, OHCs feedback energy into the cochlea to amplify the vibrations of the cochlear partition and overcome the effects of viscous damping on sensitivity and frequency tuning (reviewed by Dallos 2008, Hudspeth 2008). The hair bundles of inner hair cells (IHCs) are not attached to the TM, and they sense the highly tuned consequences of the miniscule, mechanical interactions between the reticular laminar and the TM to sound (Nowotny and Gummer 2006; Fridberger et al. 2006). IHCs relay their responses to this interaction via the auditory nerve to the brain (Russell 2008).

The TM is relatively inaccessible to experimental investigation. For this reason, there have been a very few in vivo measurements of its physiological properties (Zwislocki et al. 1988) and the major experimental approach has been to make measurements from isolated cochlea. A more recent approach is to deduce the role of TM in cochlear processing through specific deletions and mutations of the three glycoproteins, α-tectorin (Tecta), β-tectorin (Tectb), and otogelin that account for approximately 50% of the protein present in the tectorial membrane (Richardson et al. 2008). It is with the outcomes of this latter approach that this chapter is concerned.

7.2 Three Tectorin Mutants

Experimental data obtained in vivo from mice with mutations in Tecta and Tectb have furthered our understanding of how the TM operates, confirming many roles suggested from in-vitro studies and revealing additional functions.

7.2.1 Tecta Mice

The TM of *Tecta*$^{\Delta ENT/\Delta ENT}$ mice, which lacks α-Tectorin, also lacks all noncollagenous matrix and is completely detached from both the surface of the organ of Corti and the spiral limbus, although the architecture of the organ of Corti is otherwise normal (Legan et al. 2000). Measurements from *Tecta*$^{\Delta ENT/\Delta ENT}$ mice reveal that the presence of the TM is important for exciting the OHCs, although it would appear that OHCs of the *Tecta*$^{\Delta ENT/\Delta ENT}$ mice become excited in response to high-level, high-frequency tones when, at very high levels, the relationship between BM displacement and SPL is compressive (Fig. 7.1a, b). The BMs of wild type and *Tecta*$^{\Delta ENT/\Delta ENT}$ mice are tuned, but those of the *Tecta*$^{\Delta ENT/\Delta ENT}$ mice are 35 dB less sensitive (Fig. 7.1c, d). A second peak of sensitivity that is normally observed at approximately 0.5 octaves below the best frequency in BM tuning curves is not detectable in these mice, confirming that it is dependent on the presence of the TM. It is suggested that this second cochlear resonance, attributed to the TM, provides an inertial mass against which OHCs can exert forces (Gummer et al. 1996; Legan et al. 2000; Lukashkin and Russell 2003; Lukashkin et al. 2007).

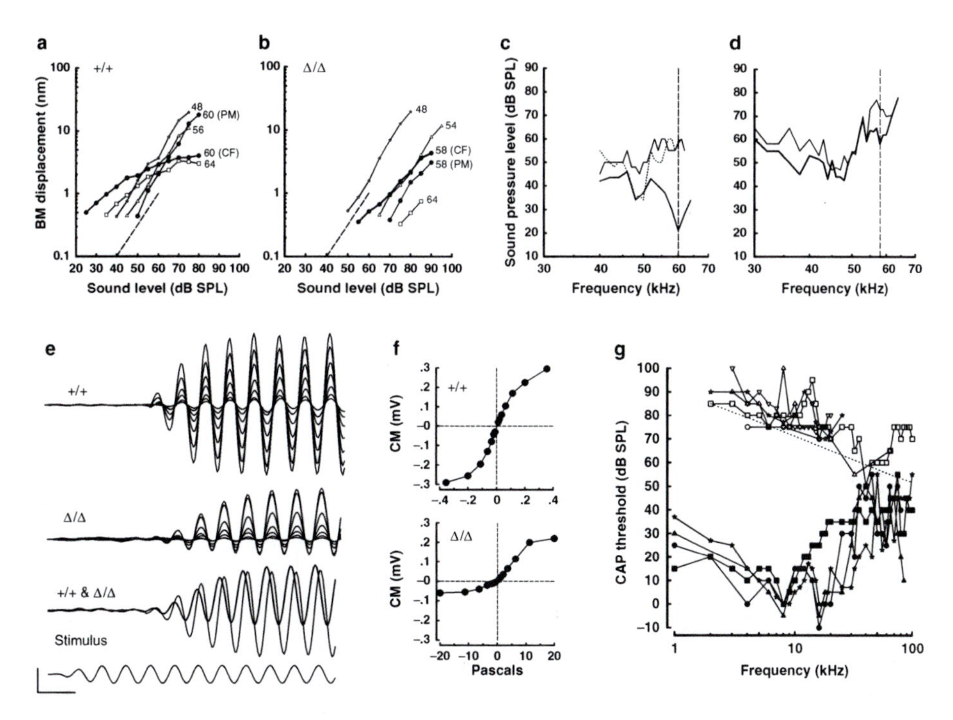

Fig. 7.1 Mechanical and electrical responses of the cochlea. BM responses from *Tecta*$^{+/+}$ mice (**a**, **c**; $^{+/+}$; CF=60 kHz) and *Tecta*$^{\Delta ENT/\Delta ENT}$ mice (**b**, **d** [$^{\Delta/\Delta}$]; CF=58 kHz). (**a**, **b**) BM displacement as a function of sound levels for tones with frequencies at and close to the CF. *Thick lines* show responses at the CF. The measurement frequency in kHz is indicated adjacent to the level functions, and PM indicates post-mortem measurements. The *dashed line* indicates a slope of 1 dB/dB. (**c**, **d**) BM frequency threshold curves for a 0.4 nm displacement criterion. *Thick lines* are from the living preparation. *Dashed lines* are at decreased sensitivity. *Thin lines* show post-mortem responses. The *asterisk* indicates the second resonance. *Vertical dashed* and *dotted lines* indicate the CF. The *asterisk* indicates the second threshold minimum. (**e**) CM measured in response to an 18 kHz tone at high temporal resolution. ($^{+/+}$), *Tecta*$^{+/+}$ mouse at levels from 60 to 90 dB SPL in 5 dB SPL steps; ($^{\Delta/\Delta}$), *Tecta*$^{\Delta ENT/\Delta ENT}$ mouse at levels from 95 to 125 dB SPL in 5 dB SPL steps; ($^{+/+}$ & $^{\Delta/\Delta}$, *thin line*), *Tecta*$^{+/+}$ mouse at 95 dB SPL; ($^{+/+}$ & $^{\Delta/\Delta}$, *thick line*), *Tecta*$^{\Delta ENT/\Delta ENT}$ at 115 dB SPL; (Stimulus), control voltage to microphone driver. *Vertical line*, 0.1 mV for three *upper sets of traces*; *horizontal line*, 0.1 ms. (**f**) Peak CM as function of RMS sound pressure at the tympanic membrane. ($^{+/+}$), *Tecta*$^{+/+}$ mouse; ($^{\Delta/\Delta}$), *Tecta*$^{\Delta ENT/\Delta ENT}$ mouse. (**g**) CAP threshold as a function of stimulus frequency recorded from *Tecta*$^{+/+}$ (*closed symbols*) and *Tecta*$^{\Delta ENT/\Delta ENT}$ mice (*open symbols*). Modified with permission from Legan et al. (2000)

Measurements of the cochlear microphonic potentials (Legan et al. 2000) and distortion product otoacoustic emissions (DPOAEs) (Lukashkin et al. 2004) reveal that the OHCs in *Tecta*$^{\Delta ENT/\Delta ENT}$ mice can be driven at high sound pressure levels in the absence of a tectorial membrane and that they are fluid coupled, responding to velocity rather than to displacement (Fig. 7.1e, f). Accordingly, the neural threshold audiograms reveal a frequency-dependent hearing loss in these mice (Fig. 7.1g).

There is an approximately 60–80 dB hearing loss in the 20 kHz region, but this loss reduces as the stimulus frequency increases, presumably as the thickness of the fluid boundary layer over the organ of Corti decreases and allows the hair bundles to be stimulated by fluid flow. Cochlear microphonic measurements also reveal that the mechano-electrical transducer of the free-standing hair bundles of the OHCs in the *Tecta*$^{\Delta ENT/\Delta ENT}$ mice are not biassed toward the most sensitive region of their operating range, as they are in wild-type mice (Fig. 7.1f). As a consequence, the ability of the OHCs to operate as efficient amplifiers of BM motion is compromised (see also Lukashkin et al. 2004). The cochlear responses of the *Tecta*$^{\Delta ENT/\Delta ENT}$ mouse therefore demonstrate that the TM is, in addition to acting as a second resonator, also crucial for the sensitive operation of the OHCs.

7.2.2 Y1870C Missense Mutation in TECTA

Mice heterozygous for the Tecta Y1870C mutation, a missense mutation that is found in TECTA and causes a stable, moderate-to-severe (60–80 dB) hearing loss in an Austrian family, have a TM with a severely reduced limbal attachment zone, an unusual hump-backed shape, an enlarged subtectorial space, a loss of striated-sheet matrix in the sulcal region, and a detachment of Kimura's membrane (Legan et al. 2005). The tips of the tallest stereocilia of the OHC bundles are, however, firmly imbedded in the TM. Despite the rather abnormal morphology of the TM, most attributes of OHC function and BM motion appear normal in the *Tecta*$^{Y187oC/+}$ mice. The BM is sharply tuned with negligible loss in sensitivity, the second resonance is present in the BM response (compare Fig. 7.2a, b), and the phase and symmetry of the cochlear microphonic potentials are normal (Fig. 7.2c, d). Neural threshold audiograms, however, reveal a 60 dB loss in sensitivity in the auditory nerve (Fig. 7.2d, e) indicating that the TM plays a critical role in ensuring that the motion of the OHC hair bundles drives the hair bundles of the IHCs. The enlarged subtectorial space seen in the *Tecta*$^{Y187oC/+}$ mice may compromise the fluid coupling between IHC and OHC bundles that has been proposed to result from the counter-phase motion of the TM and the reticular lamina (Nowotny and Gummer 2006).

Masker tuning curves (Dallos and Cheatham 1976) recorded from *Tecta*$^{Y187oC/+}$ mice, differ from those of *Tecta*$^{+/+}$ mice (Fig. 7.2f). The most notable feature is the presence of a sharp notch of insensitivity at the characteristic frequency. This loss in neural sensitivity (60–70 dB) at the characteristic frequency is similar in magnitude to the hearing impairment reported for the affected members of the Austrian DFNA8/12 family with the Y1870C a-tectorin mutation (Verhoeven et al. 1998). Two other major features of the masker tuning curves in *Tecta*$^{Y187oC/+}$ mice are the absence of a notch of insensitivity on the low-frequency shoulder and the appearance of a lobe of sensitivity on the high-frequency side. If BM responses were passively transmitted to the IHCs, a loss in coupling of the IHC bundles to the TM caused by an enlargement of the subtectorial space would simply desensitize, and possibly broaden, the neural frequency tuning curves. We suggest, the IHC hair

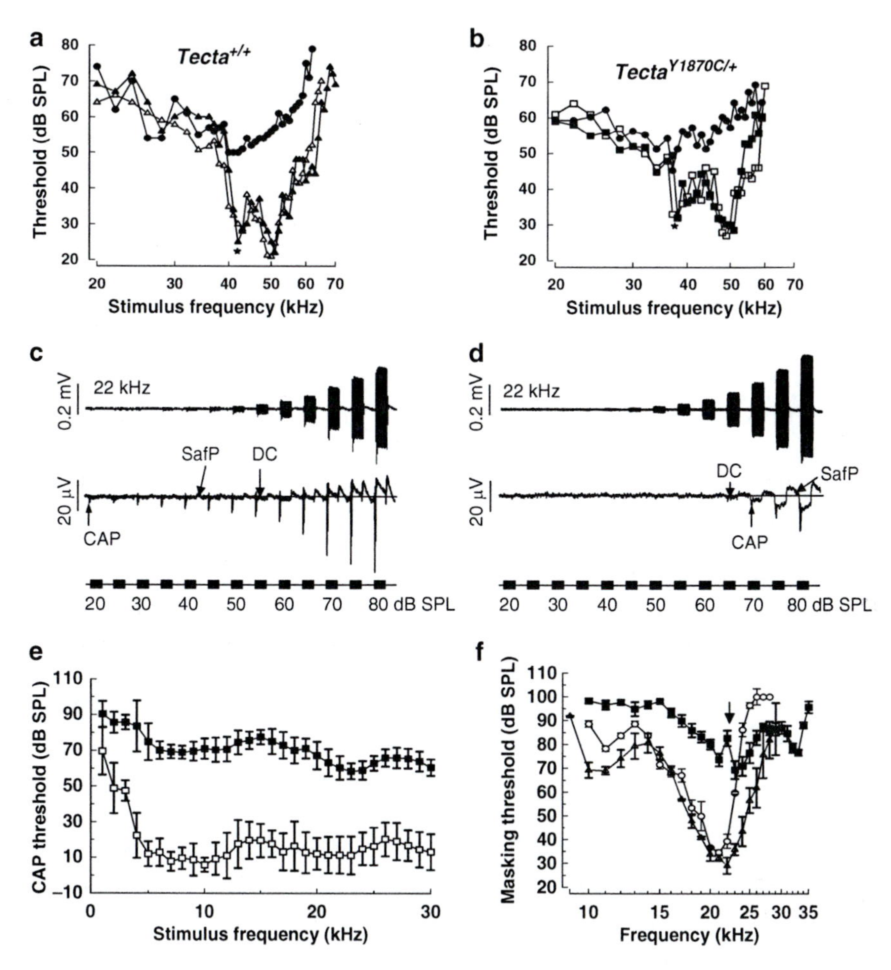

Fig. 7.2 Basliar membrane, cochlear microphonics, CAPs and neural masking tuning curves in *Tecta*^{Y187oC} mice. (**a, b**) BM displacement tuning curves for two *Tecta*$^{+/+}$ mice (**c**, *open and filled triangles*; $Q_{10dB} = 10.0$ and 10.2, respectively) and two *Tecta*$^{Y187oC/+}$ mice (**d**, *open and filled squares*; $Q_{10dB} = 10.9$ and 8.5, respectively). *Solid circles* in **b** and **d**: post-mortem measurements. Threshold criterion is 0.2 nm; *stars* indicate low-frequency resonance. (**c, d**) Cochlear microphonics (CM) and the tonic components (DC) of the cochlear voltage response recorded from the round window to 22 kHz tones, at the levels indicated, from *Tecta*$^{+/+}$ (**c**) and *Tecta*$^{Y187oC/+}$ (**d**) mice. *Upper traces*: unfiltered CM response, bandwidth 0–125 kHz. *Middle traces*: CM response filtered at 0–3 kHz. *Lower traces*: timing of tone bursts. *DC* tonic components of the cochlear voltage response, *CAP* compound action potential, *SafP* slow after-potential. (**e**) CAP thresholds as a function of stimulus frequency for *Tecta*$^{+/+}$ (*open squares*) and *Tecta*$^{Y187oC/+}$ (*filled squares*) mice (mean 7 s.d. from ten *Tecta*$^{+/+}$ and ten *Tecta*$^{Y187oC/+}$ mice). (**f**) Simultaneous masking tuning curves of CAP recorded from the round window. *Open triangles*: masking tuning curve from *Tecta*$^{+/+}$ mice (frequency and level of probe: 22 kHz, 30 dB SPL), Q_{10dB} (bandwidth measured 10 dB from tip/probe frequency) ¼ 5.6. *solid squares*: masking tuning curve from *Tecta*$^{Y187oC/+}$ mice, (frequency and level of probe: 22 kHz, 70 dB SPL), Q_{10dB} ¼ 3.9. *Open circles*: masking tuning curve from *Tecta*$^{+/+}$ mice, (frequency and level of probe: 22 kHz, 70 dB SPL), Q_{10dB} ¼ 7.2. *Arrow* indicates notch at probe frequency in *Tecta*$^{Y187oC/+}$ mice. *Open symbols* represent mean 7 s.d. for measurements from ten mice; *solid squares* represent mean 7 s.d. from four mice (Modified with permission from Legan et al. 2005)

bundles in *Tecta*$^{Y1870C/+}$ mice are displaced through radial motion of the reticular laminar, when the radial velocity is sufficiently large enough for the tips of the hair bundles to penetrate above the unstirred fluid level in contact with reticular laminar surface (legan et al. 2005).

The appearance of a sharp notch rather than a peak at the characteristic frequency in the neural masking tuning curves of *Tecta*$^{Y1870C/+}$ mice may indicate, therefore, that IHC bundle movements, and hence displacement of the reticular lamina is specifically reduced at the characteristic frequency. BM vibrations, therefore, are not faithfully relayed to the IHCs through vibration of the reticular lamina over the entire frequency range. If this is so, then radial vibrations of the TM must, in wild-type mice, normally dominate the input to IHCs at their characteristic frequency. These observations suggest that the vibrations of the BM are normally transmitted to the IHCs by means of the TM rather than through the reticular lamina at the characteristic frequency. Insensitive notches at frequencies below the characteristic frequency are a feature of neural tuning curves from fibres that innervate the high-frequency region of the cochlea (Liberman and Kiang 1978), and high-frequency plateaus have been described in the BM tuning curves from a variety of mammals (Robles and Ruggero 2001). The absence of a notch on the low-frequency side of the *Tecta*$^{Y1870C/+}$ neural tuning curve and the appearance of a lobe on the high-frequency side can both be attributed to a failure in phase cancellation. Enlargement of the subtectorial space in the IHC region of the *Tecta*$^{Y1870C/+}$ mouse will abrogate interactions between the TM and the reticular lamina and prevent such phase cancellation.

7.2.3 Beta Tectorin Mice: Sharpened Cochlear Tuning in a Mouse with a Genetically Modified Tectorial Membrane

Mice homozygous for a functional null deletion in the Tectb gene have a TM that remains attached to the spiral limbus and the surface of the organ of Corti (Russell et al. 2007). The TMs of the *Tectb*$^{-/-}$ mice lack any sign of the organized striated-sheet matrix that is characteristic of the TMs of wild-type mice. Instead the collagen fibrils are imbedded in a matrix formed by apparently randomly dispersed, rather irregular looking filaments that are probably formed by Tecta. There is a marked loss in cochlear sensitivity for frequencies below 20 kHz in the *Tectb*$^{-/-}$ mouse (Fig. 7.3a), presumably due to the distinct enlargement and swelling of the TM that is seen in the apical end of the cochlea in these mutants. There is no significant change in the dimensions of the TM in the basal, high-frequency end of the cochlea of *Tectb*$^{-/-}$ mice and, according to measurements of BM displacements; the threshold is only about 10 dB higher. Remarkably, however, the high-frequency BM frequency tuning curves are actually sharper than those of wild-type littermates (Fig. 7.3b). Furthermore, the neural masking tuning curves in this cochlear region are also much sharper with the Q10 dB being almost three times greater than that in wild types (Fig. 7.3c).

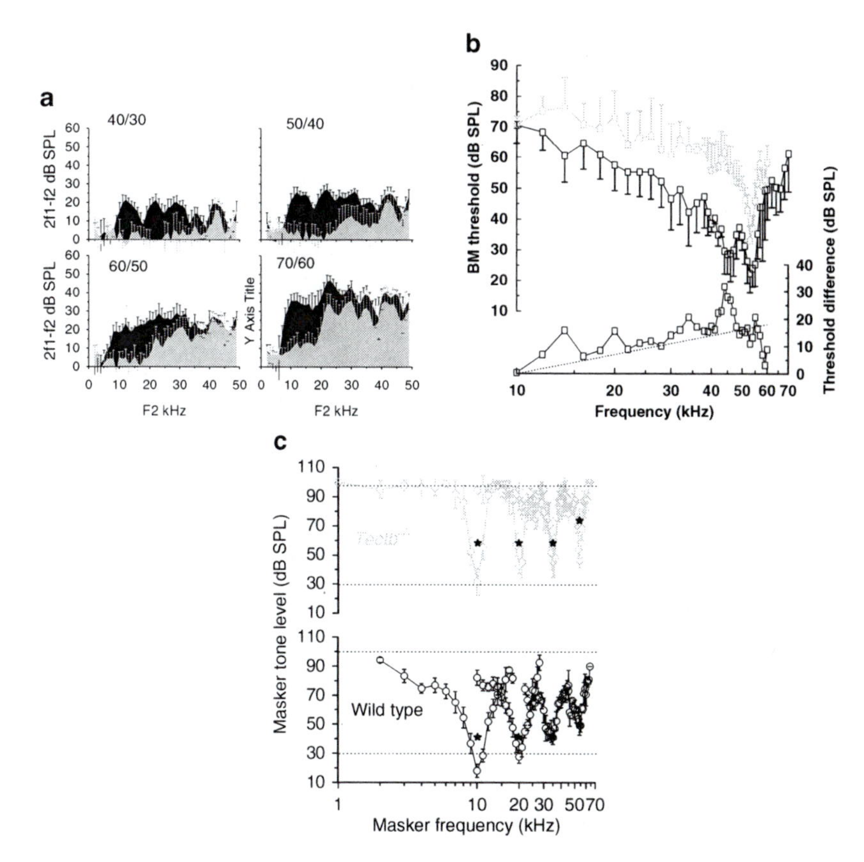

Fig. 7.3 Acoustical, mechanical, and neural recordings from the cochleae of wild type (*black*) and *Tectb*$^{-/-}$ (*grey*) mice. (**a**) Mean ± s.d. (n = 10) of the magnitude of 2f1 − f2 DPOAEs as functions of f2 frequency. Levels of f2/f1 tones in dB SPL are shown in *upper left* of each panel. (**b**) Iso-response displacement frequency tuning curves (mean ± s.d., n = 8 for each genotype) response criterion (0.2 nm, 53 kHz location). *Arrow*, low-frequency resonance. *Bottom curve*: difference between wild type and *Tectb*$^{-/-}$ tuning curves. (**c**) Simultaneous masking tuning curves (n = 30, mean ± s.d.) of compound action potentials in response to 10, 20, 35, and 54 kHz probe tones. The levels and frequencies of the probe tone are indicated by *stars*. *Dotted lines* indicate SPLs of 30 and 100 dB. (Modified with permission from Russell et al. 2007)

The finding that both neural and BM frequency tuning are enhanced by the loss of Tectb has led to the suggestion that tuning in the active cochlea depends on the degree of longitudinal elastic coupling within the TM, and that such coupling is reduced in the absence of the organized striated-sheet matrix within which the collagen fibrils are normally imbedded. Two mechanisms could be responsible for sharpening of the tuning. Reduced elastic coupling may lead to a decrease in the TM space constant, and hence a decrease in the number of OHCs acting in synchrony along the cochlea to boost mechanical responses locally. Sharpening of the tuning should also be observed if reduced TM elasticity leads to a decrease in the wavelength of the TM travelling wave that has been described in vitro (Ghaffari et al. 2007). A shorter wavelength would mean that effective interaction between the basilar and TMs occurs over a short stretch of the cochlea. The responses of *Tectb*$^{-/-}$ mutants reveal the counter demands of cochlear tuning and sensitivity, and the need to compromise between these requirements in the mammalian cochlea.

7.3 Conclusions

Recent observations have changed our understanding of TM function. Transgenic mice have shown that the TM is a structure that can influence the sensitivity and tuning properties of the cochlea in several ways. It ensures that the gain and timing of cochlear feedback are optimal, and that the hair bundles of the IHCs are driven efficiently by the OHCs. The TM provides the principal drive to the IHCs at the characteristic frequency, interacting with the reticular lamina to shape the frequency responses of IHCs, and hence determining the sharpness and sensitivity of the neural tuning curves.

The TM may influence the extent to which different elements are coupled along the length of the cochlea.

Acknowledgements We thank James Hartley for technical assistance. This research was supported by grants from the MRC and Wellcome Trust.

References

Dallos P (2008) Cochlear amplification, outer hair cells and prestin. Curr Opin Neurobiol 18:370–376

Dallos P, Cheatham MA (1976) Compound action potential (AP) tuning curves. J Acoust Soc Am 59:591–597

Fridberger A, Tomo I, Ulfendahl M, Boutet de Monvel J (2006) Imaging hair cell transduction at the speed of sound: dynamic behavior of mammalian stereocilia. Proc Natl Acad Sci U S A 103:1918–1923

Ghaffari R, Aranyosi AJ, Freeman DM (2007) Longitudinally propagating travelling waves of the mammalian tectorial membrane. Proc Natl Acad Sci U S A 104:16510–16515

Gummer AW, Hemmert W, Zenner HP (1996) Resonant tectorial membrane motion in the inner ear: its crucial role in frequency tuning. Proc Natl Acad Sci U S A 93:8727–8732

Hudspeth AJ (1989) How the ear's works work. Nature 341:397–404

Hudspeth AJ (2008) Making an effort to listen: mechanical amplification in the ear. Neuron 59:530–545

Legan PK, Lukashkina VA, Lukashkin AN, Goodyear RJ, Richardson GP (2000) A targeted deletion in alphatectorin reveals that the tectorial membrane is required for the gain and timing of cochlear feedback. Neuron 28:273–285

Legan PK, Lukashkina VA, Goodyear RJ, Lukashkin AN, Verhoeven K, Van Camp G, Russell IJ, Richardson GP (2005) A deafness mutation isolates a second role for the tectorial membrane in hearing. Nat Neurosci 8(8):1035–1042

Liberman M, Kiang NY-S (1978) Acoustic trauma in cats. Acta Otolaryngol Suppl 358:1–63
Lim D (1986) Functional structure of the organ of Corti: a review. Hear Res 22:117–146
Lukashkin AN, Russell IJ (2003) A second, low frequency mode of vibration in the intact mammalian cochlea. J Acoust Soc Am 113:1544–1550
Lukashkin AN, Lukashkina VA, Legan PK, Richardson GP, Russell IJ (2004) Role of the tectorial membrane revealed by otoacoustic emissions recorded from wild-type and transgenic Tecta(deltaENT/deltaENT) mice. J Neurophysiol 91:163–171
Lukashkin AN, Smith JK, Russell IJ (2007) Properties of distortion product otoacoustic emissions and neural suppression tuning curves attributable to the tectorial membrane resonance. J Acoust Soc Am 121:337–343
Nowotny M, Gummer AW (2006) Nanomechanics of the subtectorial space caused by electromechanics of cochlear outer hair cells. Proc Natl Acad Sci U S A 103:2120–2125
Richardson GP, Lukashkin AN, Russell IJ (2008) The tectorial membrane: one slice of a complex cochlear sandwich. Curr Opin Otolaryngol Head Neck Surg 16:458–464
Robles L, Ruggero MA (2001) Mechanics of the mammalian cochlea. Physiol Rev 81:1305–1352
Russell IJ (2008) Cochlear receptor potentials. In: Dallos P, Oertel D (ed) Audition. The senses: a comprehensive reference, vol 3. Academic, San Diego
Russell IJ, Legan PK, Lukashkina VA, Lukashkin AN, Goodyear RJ, Richardson GP (2007) Sharpened cochlear tuning in a mouse with a genetically modified tectorial membrane. Nat Neurosci 10:215–223
Verhoeven K, Van Laer L, Kirschhofer K, Legan PK, Hughes DC, Schatteman I, Verstreken M, Van Hauwe P, Coucke P, Chen A, Smith RJ, Somers T, Offeciers FE, Van de Heyning P, Richardson GP, Wachtler F, Kimberling WJ, Willems PJ, Govaerts PJ, Van Camp G (1998) Mutations in human a-tectorin cause autosomal dominant nonsyndromic hearing impairment. Nat Genet 19:60–62
von Békésy G (1960) Experiments in hearing. McGraw-Hill, New York
Zwislocki JJ, Chamberlain SC, Slepecky NB (1988) Tectorial membrane. I: static mechanical properties in vivo. Hear Res 33:207–222

Liberman M, Kiang NY-S (1978) Acoustic trauma in cats. Acta Otolaryngol Suppl 358:1–63
Lim D (1986) Functional structure of the organ of Corti: a review. Hear Res 22:117–146
Lukashkin AN, Russell IJ (2003) A second, low frequency mode of vibration in the intact mammalian cochlea. J Acoust Soc Am 113:1544–1550
Lukashkin AN, Lukashkina VA, Legan PK, Richardson GP, Russell IJ (2004) Role of the tectorial membrane revealed by otoacoustic emissions recorded from wild-type and transgenic Tecta(deltaENT/deltaENT) mice. J Neurophysiol 91:163–171
Lukashkin AN, Smith JK, Russell IJ (2007) Properties of distortion product otoacoustic emissions and neural suppression tuning curves attributable to the tectorial membrane resonance. J Acoust Soc Am 121:337–343
Nowotny M, Gummer AW (2006) Nanomechanics of the subtectorial space caused by electromechanics of cochlear outer hair cells. Proc Natl Acad Sci U S A 103:2120–2125
Richardson GP, Lukashkin AN, Russell IJ (2008) The tectorial membrane: one slice of a complex cochlear sandwich. Curr Opin Otolaryngol Head Neck Surg 16:458–464
Robles L, Ruggero MA (2001) Mechanics of the mammalian cochlea. Physiol Rev 81:1305–1352
Russell IJ (2008) Cochlear receptor potentials. In: Dallos P, Oertel D (ed) Audition. The senses: a comprehensive reference, vol 3. Academic, San Diego
Russell IJ, Legan PK, Lukashkina VA, Lukashkin AN, Goodyear RJ, Richardson GP (2007) Sharpened cochlear tuning in a mouse with a genetically modified tectorial membrane. Nat Neurosci 10:215–223
Verhoeven K, Van Laer L, Kirschhofer K, Legan PK, Hughes DC, Schatteman I, Verstreken M, Van Hauwe P, Coucke P, Chen A, Smith RJ, Somers T, Offeciers FE, Van de Heyning P, Richardson GP, Wachtler F, Kimberling WJ, Willems PJ, Govaerts PJ, Van Camp G (1998) Mutations in human a-tectorin cause autosomal dominant nonsyndromic hearing impairment. Nat Genet 19:60–62
von Békésy G (1960) Experiments in hearing. McGraw-Hill, New York
Zwislocki JJ, Chamberlain SC, Slepecky NB (1988) Tectorial membrane. I: static mechanical properties in vivo. Hear Res 33:207–222

Chapter 8
Auditory Prepulse Inhibition of Neuronal Activity in the Rat Cochlear Root Nucleus

Ricardo Gómez-Nieto, J.A.C. Horta-Júnior, Orlando Castellano, Donal G. Sinex, and Dolores E. López

Abstract Cochlear root neurons (CRNs) provide short-latency acoustic inputs to the caudal pontine reticular nucleus that elicit an acoustic startle reflex (ASR). Auditory prepulse inhibition (PPI) of the acoustic startle response is the reduction in ASR magnitude that is observed when a strong acoustic startling stimulus (pulse) is shortly preceded by a weak sound (prepulse). It has been suggested that a short descending auditory pathway conveys auditory prepulses to the level of the CRNs to mediate the inhibition of ASR. However, no electrophysiological data is available to confirm such inhibition. We here investigated the effects of auditory prepulses on the neuronal activity of CRNs using extracellular recordings in vivo from single CRNs. Our results show that CRN responses are strongly inhibited by auditory prepulses. As occurs in the behavioral paradigm, the inhibition of the CRN responses depended on parameters of the auditory prepulse, such as intensity and interstimulus interval, showing their strongest inhibition at high intensity level and short interstimulus intervals. Furthermore, we tested the auditory PPI on the activity of different neuron types in the ventral cochlear nucleus. Of all neuron types tested, only CRNs exhibited a strong attenuation of activity. Our results corroborate our previous hypothesis that CRNs might be involved in the neural circuit of the inhibition of the ASR and suggest that several neuronal pathways participate in that circuit.

Keywords Sound processing • Cochlear root neurons • Extracellular recordings • Ventral nucleus of the trapezoid body

---

D.E. López (✉)
Departamento de Biología Celular y Patología,
Instituto de Neurociencias de Castilla y León,
Universidad de Salamanca, Salamanca 37007, Spain
e-mail: lopezde@usal.es

E.A. Lopez-Poveda et al. (eds.), *The Neurophysiological Bases of Auditory Perception*,
DOI 10.1007/978-1-4419-5686-6_8,

8.1 Introduction

The acoustic startle reflex (ASR) is a defensive behavior that is elicited by sudden intense acoustic stimuli and is composed of a twitch of facial, neck, and limb muscles (Landis and Hunt 1939; Hoffman and Ison 1980). In rodents, the ASR is mediated by a neuronal pathway in which the cochlear root neurons (CRNs) provide short-latency acoustic inputs to the caudal pontine reticular nucleus (PnC), which in turn projects to brainstem and spinal motor neurons (Lee et al. 1996; López et al. 1999; Nodal and López 2003; Horta-Júnior et al. 2008). Auditory prepulse inhibition (PPI) of the acoustic startle response is the reduction in ASR magnitude that is observed when a strong acoustic startling stimulus (pulse) is shortly preceded by a weak sound (prepulse). Most authors agree that PPI is mediated by a long multimodal circuit at the level of the PnC (reviewed in Fendt et al. 2001; Li et al. 2009). Nevertheless, this circuit does not explain all of the properties of PPI such as the effectiveness of interstimulus intervals as short as 20 ms (Hoffman and Ison 1980). Building upon morphological findings, our group suggested the existence of another short neuronal pathway to mediate PPI at the level of the CRNs (Gómez-Nieto et al. 2008a). To date, no electrophysiological data is available to support the inhibition of CRNs by auditory prepulses. In this study, we investigated the effects of auditory prepulses on the neuronal activity of CRNs using extracellular recordings in vivo. Our electrophysiological results cohered with our previously morphological studies (Gómez-Nieto et al. 2008a, b) and support the idea that CRNs might participate in the PPI of the ASR.

8.2 Materials and Methods

8.2.1 Animals, Surgery, and Stereotaxic Approach

Four adult female Sprague-Dawley rats (Charles River Laboratories) weighing 290–320 g were used in this study. The animal protocols were approved by the Utah State University Institutional Animal Care and Use Committee (IACUC). All experiments conformed to local and international guidelines on the ethical use of animals, and all efforts were made to minimize the number of animals used. For the surgical procedures, the animals were deeply anesthetized with a mixture of ketamine (70 mg/kg body weight) and xylazine (4.0 mg/kg body weight) and maintained in this state by supplementary doses, as required, throughout the duration of the experiment. The rats were mounted in a stereotaxic instrument (#900, Kopf, Tujunga, CA, USA) using hollow ear bars. The recording/marking microelectrode was aimed to pass through the rat's cochlear root nucleus, which lies within the extranuclear portion of the cochlear nerve root, located ventral to the ventral cochlear nucleus (VCN) (Merchán et al. 1988; Osen et al. 1991). To target the corresponding location in the brainstem, we followed identical coordinates and surgical procedures to those used in our previous studies (López et al. 1999; Sinex et al. 2001; Nodal and López 2003; Gómez-Nieto et al. 2008a).

8.2.2 *Stimulation, Data Collection, and Analyses*

The stimulation was carried out in a systematic manner using devices and methods that were nearly identical to those used by Sinex et al. (2001). The frequency response of the acoustic system was measured in a coupler with a calibrated microphone (#4136, B&K, GA, USA). This standard calibration curve was used to set the levels of tones in all experiments. Furthermore, we measured auditory evoked potentials prior to and after each experiment to monitor the status of the animal and the acoustic system during the experiment. To record the neuronal activity, we used carbon fiber electrodes (#E1011, Carbostar-1, Kation Scientific, MN, USA) with impedance at 1 kHz of 0.4–0.8 MΩ and tip length of 15 mm. The electrode was fixed to the stereotaxic frame by a Kopf carrier, positioned as described by López et al. (1999), then advanced from outside the sound booth (IAC) by a microdrive (Trent-Wells). Then, we followed the procedure described by Sinex et al. (2001) to search for and identify single CRNs units. Up to seven electrode passes, each varying slightly from the initial pass in its point of entry into the cerebellum, were made in each animal.

Stimulus presentation and data collection were controlled by a PC-compatible computer and custom software written for MATLAB (MathWorks, Natick MA, USA). Stimulus waveforms were digitally synthesized online. Tone bursts were used as search stimuli, and data were recorded only from single units. Frequency tuning curves were measured with an automated procedure that estimated threshold at many closely spaced frequencies. Frequency response maps were obtained by presenting, in random order, a set of frequency/intensity pairings specifically chosen to define the complete details of a unit's frequency selectivity. The response map consisted of a set of spike counts obtained at 6 sound pressure levels (SPLs) and 63 frequencies. Characteristic frequency and threshold were measured from the automated tuning curve. Following the collection of tuning curve or response map data, separate, more detailed measurements of the responses to tones were made to access the modifications by prepulse presentation at different conditions (duration, intensity and interstimulus intervals). In the probes designed to test the auditory PPI (Fig. 8.1a), two consecutive pure tones of the characteristic frequency (CF) for the units investigated were delivered to an earphone (#1310B, Radio Shack Super Tweeter, Korea) coupled to a hollow ear bar in the stereotaxic apparatus. The responses to tones were used to calculate first-spike latencies and peristimulus time (PST) histograms. Each PST histograms and dot rasters displayed the responses to 50–100 repetitions of the tones. The prepulse and pulse parameters (intensity, duration, and interstimulus intervals) used in the probes were showed in each PST histogram. The binwidth for the PST histograms was of 1.4 ms and was reduced to 0.3 ms in the expanded views.

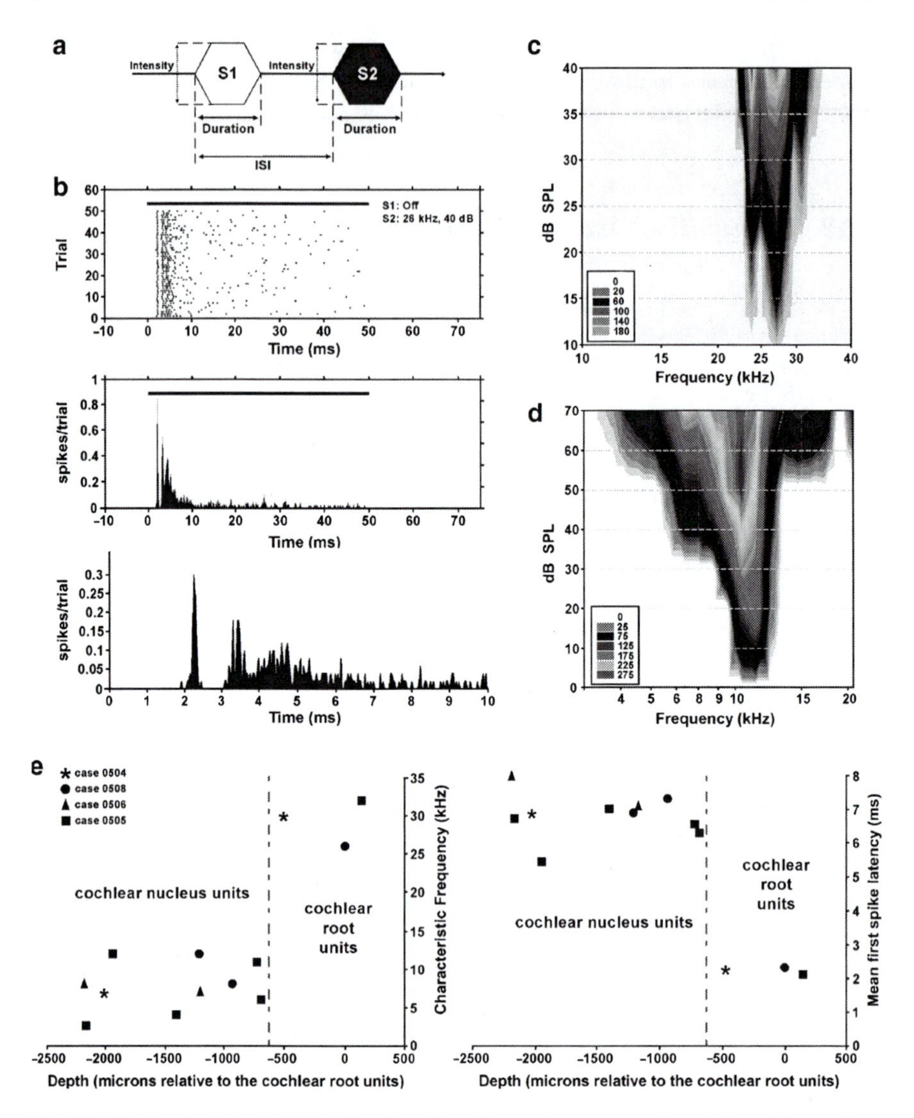

Fig. 8.1 Electrophysiological identification of cochlear root neurons and ventral cochlear nucleus neurons. (**a**) Probe designed to test the auditory prepulse stimulation. S1: prepulse tone, S2: pulse tone, ISI: interstimulus interval. (**b**) Dots raster (*top*) and PST histograms (*middle* and *bottom*) of spikes elicited by a 26 kHz, 40 dB SPL tone. Notice the typical discharge pattern "primary-like with notch" of a cochlear root neuron. The expanded view (*bottom*) of the first 10 ms after tone onset of the response shows a first spike latency of 2.3 ms approximately. No correction for the acoustic delay of approximately 0.5 ms was made. (**c, d**) Frequency response map of a cochlear root unit (**c**) and a ventral cochlear nucleus unit (**d**) analyzed in the study. *Gray-scale plots* indicate the magnitude of response to tones at many different frequencies and SPLs. (**e**) Recording locations of the units analyzed in the study. Characteristic frequency (*left*) and latency (*right*) are related to the depth in microns where cochlear root units were found. The *symbols* indicate the units obtained from different cases. Locations dorsal to the target are indicated by *negative numbers*. The *vertical dashed line* represents the separation of the two groups of units. Cochlear root neurons exhibit high CF (~30 kHz) and short first spike latency (~2 ms) compared to cochlear nucleus units

8.3 Results

8.3.1 *Electrophysiological Identification of Cochlear Root Neurons*

A total of 13 auditory units were recorded in four different animals as the electrode was advanced through the cochlear nucleus and into the cochlear nerve root. We found three units that fulfilled the electrophysiological criteria established by Sinex et al. (2001), for the identification of CRNs. The remaining units were encountered in the VCN and were fitted within the category of the major classes of VCN units (primary-like, chopper, pauser, and on-sustained neurons). The discharge pattern of CRNs was characterized by a highly reliable first-spike discharge followed by a brief refractory period of (~1 ms). Thus, the PST histogram shape was clearly identified as primary-like with notch (Fig. 8.1b). CRN responses did not show habituation to the stimuli presentation, even after as many as 100 repetitions of a stimulus in relatively rapid succession. CRNs exhibited short first-spikes latencies of approximately 2.2 ms (uncorrected for acoustic delay of about 0.5 ms; Fig. 8.1b, e), high characteristic frequencies of approximately 30 kHz, and sharp tuning curves with threshold of 10 dB SPL (Fig. 8.1c, e). Those particular features distinguished CRNs' response from the response of VCN units that were recorded at a more dorsal location (Fig. 8.1e). The consistently high characteristic frequencies of CRNs (Fig. 8.1c, e) deviated markedly from those of the overlying portions of the VCN (Fig. 8.1d, e). First-spike latencies of CRN were approximately 4 ms shorter than those of VCN units (Fig. 8.1e). CRNs were always deeper (≥500 μm) than, and spatially separate from, VCN units recorded in the same animals (Fig. 8.1e). Such unique properties point to the CRNs as the source of short-latency acoustic inputs to the reticular formation that leads to an acoustic startle response (Lee et al. 1996; López et al. 1999; Sinex et al. 2001).

8.3.2 *Auditory Prepulse Inhibition of Cochlear Root Neurons Response*

To test whether or not CRN responses are reduced by auditory prepulse stimulation, we delivered weak auditory stimuli (S1, prepulse) presented before strong auditory stimuli (S2, pulse) that evoked a full CRN response. The time between the prepulse and the main pulse (interstimulus intervals, ISIs) was also considered in the analysis (Fig. 8.1a). Our results showed that CRNs' responses were strongly inhibited by prepulse stimulation (Fig. 8.2). This inhibition was more evident in the first spike discharge, and hence led a reduction in the overall response. Even though the prepulse was just 15 dB SPL over the threshold, the CRNs' response evoked by pulses of 45 dB SPL was drastically reduced by half (Fig. 8.2a). In this probe, the fact that the unit barely generated spikes to the onset of the prepulse and the inhibition observed with an interstimulus interval of 150 ms provides evidence of the strong influence that the prepulse had on the CRN's response.

In an attempt to reveal the effects of prepulse intensity, prepulse duration, and interstimulus intervals on the CRN responses, we developed three probes. In probe-1, we tested the effect of the prepulse intensity by increasing intensities of the prepulse up to the pulse intensity (Fig. 8.2b). A prepulse duration of 50 ms and an interstimulus interval of 100 ms were kept fixed during this probe. Our results showed that the reduction of spikes to the onset of the pulse was more dramatic as prepulse intensity increased (Fig. 8.2b). The strongest inhibition of the CRNs' response (90%) occurred with a prepulse intensity equal to the pulse intensity (45 dB SPL). In probe-2, we tested the effect of the prepulse duration by increasing durations of the prepulse up to the pulse duration (Fig. 8.2c). A prepulse intensity of 45 dB SPL and an interstimulus interval of 150 ms were kept fixed during this probe. PST histograms showed that the reduction of spikes to the onset of the pulse was more significant as prepulse duration increased (Fig. 8.2c). The strongest inhibition of the CRNs' response (80%) was observed with a prepulse duration equal to the pulse duration (50 ms). In probe-3, we tested the effect of the interstimulus interval by decreasing the time between the prepulse and the pulse (Fig. 8.2d). In this probe, we maintained a prepulse intensity of 25 dB SPL and a duration of 50 ms. PST histograms showed that a consistent reduction of spikes to the onset of the pulse as the interstimulus interval decreased (Fig. 8.2d). The CRN's response to the pulse was totally blocked when the prepulse was delivered just before the pulse (interstimulus interval of 50 ms).

Taken together, the auditory PPI of the CRN responses is strongly dependent on the prepulse intensity, the prepulse duration, and the interstimulus intervals.

8.3.3 Does Auditory Prepulse Inhibition Occur in Neurons Types of the Ventral Cochlear Nucleus?

We have presented data indicating that CRN responses are strongly reduced by auditory prepulses. To determine whether or not auditory PPI also occurs in neurons of the VCN, we used the same stimulation probes used previously for the CRNs (Fig. 8.2e). A total of ten single units in the VCN were tested for the auditory prepulse stimulation. None of them showed the reduction of neuronal activity observed in the CRNs. A representative example to highlight the differences between CRNs and units in the VCN is presented in Fig. 8.2e, f. In this case, a CRN unit and a VCN chopper unit were encountered along the same electrode track and received the auditory prepulse stimulation using the same parameter settings. The parameters' settings were adjusted to those which the CRNs exhibited the strongest inhibition (see detailed information in Fig. 8.2e, f). Our results showed that the number of spikes per trial and PST histogram-shape in the VCN unit remained constant as we delivered the prepulse or the prepulse intensity increased (Fig. 8.2e). On the contrary, the CRN unit showed the characteristic inhibition of neuronal activity (Fig. 8.2f). Furthermore, we tested auditory prepulse stimulation at many ISIs and prepulse intensities (Fig. 8.3). As expected, CRNs showed inhibition of spikes as ISIs decreased and prepulse

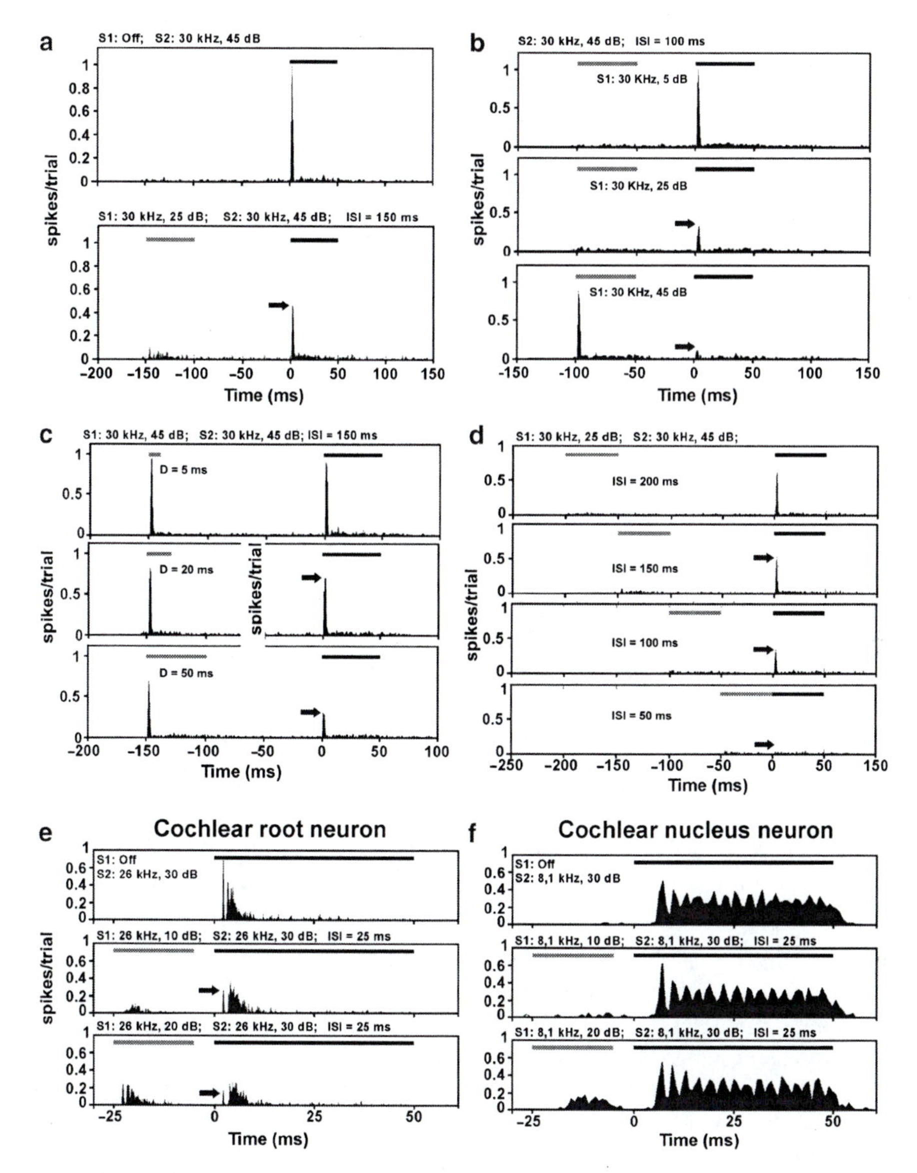

Fig. 8.2 Auditory prepulse inhibition of neuronal activity in the rat cochlear root nucleus. The prepulse and pulse parameters (intensity, duration, and interstimulus intervals) used in the probes were showed in the upper part of each PST histogram. (**a**) PST histograms showing the effects of prepulse tones (S1, *gray line*) on the CRN response to pulse tones (S2, *black line*). Notice the reduction of the spikes per trial (*arrow*) after prepulse stimulation. (**b**) PST histograms for three prepulse stimulation probes show the effects of the prepulse intensity level on the discharge of a cochlear root neuron. *Arrows* show the reduction of the spikes per trial with the increase of prepulse intensity. (**c**) PST histograms for three prepulse stimulation probes show the effects of the prepulse duration on the discharge of a cochlear root neuron. *Arrows* indicate the reduction of the spikes per trial as increase the prepulse duration increased. (**d**) PST histograms for four prepulse stimulation probes show the effects of the interstimulus interval (ISI) on the discharge of a cochlear root neuron. *Arrows* indicate the reduction of the spikes per trial as ISIs decreased. (**e**) PST histograms for a CRN unit and a chopper unit from the ventral cochlear nucleus (**f**). The prepulse stimulation probes are the same for both neuron types (except that each unit was stimulated by a tone at its own CF). The CRN reduced its number of spikes during prepulse stimulation, but the chopper unit did not

intensity increased, showing their strongest inhibition at high intensity level and short interstimulus intervals (Fig. 8.3a). Nevertheless, inhibition was not observed in VCN units such as chopper, primary-like, pauser, and long-latency neurons (Fig. 8.3b–e), suggesting that the auditory PPI occurs only in CRNs.

8.4 Discussion

We showed that the neuronal activity of CRNs is strongly inhibited by auditory prepulses. As occurs in the behavioral paradigm (Fendt et al. 2001), the inhibition of CRNs depended on the parameters of the auditory prepulse such as intensity, duration, and interstimulus interval. Furthermore, our study revealed that CRNs behave in a distinct manner when compared to VCN units. Auditory prepulses induced inhibition of CRNs' responses in an intensity- and ISI-dependent fashion, whereas the neuronal activity of VCN units remained essentially constant without significant dependence of auditory prepulses. Therefore, our results indicate that a specialized mechanism of neuronal inhibition was carried out in the cochlear root nucleus. PPI of CRNs' neuronal activity is consistent with the idea that CRNs are part of the neuronal circuit that mediates the PPI of the ASR. The effect of interstimulus interval suggests that other neuronal pathways participate in that circuit.

8.4.1 Auditory Prepulse Inhibition as a Specialized Mechanism of Neuronal Inhibition in the Cochlear Root Nucleus

The dramatic drop in the discharge pattern of CRNs was observed after stimuli of two consecutive pure tones, a prepulse preceding the testing pulse, separated by an interstimulus interval (ISI). For that reason, it might be inferred that the reduction of the discharge pattern to the pulse is due to forward masking or two-tone suppression of the primary neuron input, especially at the shortest interstimulus intervals. However, several lines of evidence argue for other, more important sources of inhibition. First, any effect that directly reflects the responses of primary neurons should also be observable in the responses of other second-order neurons, but no meaningful response reductions were observed in CN neurons in this study. Also, inhibition in response to the pulse was obtained after stimulation with prepulses of weak intensities and short durations (less than the pulse parameters) that would not be expected to elicit strong suppression of primary neurons. This suggests that the sound processing of weak prepulses was carried out by a mechanism that enhances the effect of the prepulses on the CRNs' response; descending inhibition could provide that kind of mechanism. Finally, inhibition of CRN responses was observed at interstimulus intervals that were long enough to avoid forward masking. Several studies relating forward masking in auditory nerve fibers and VCN units have stated

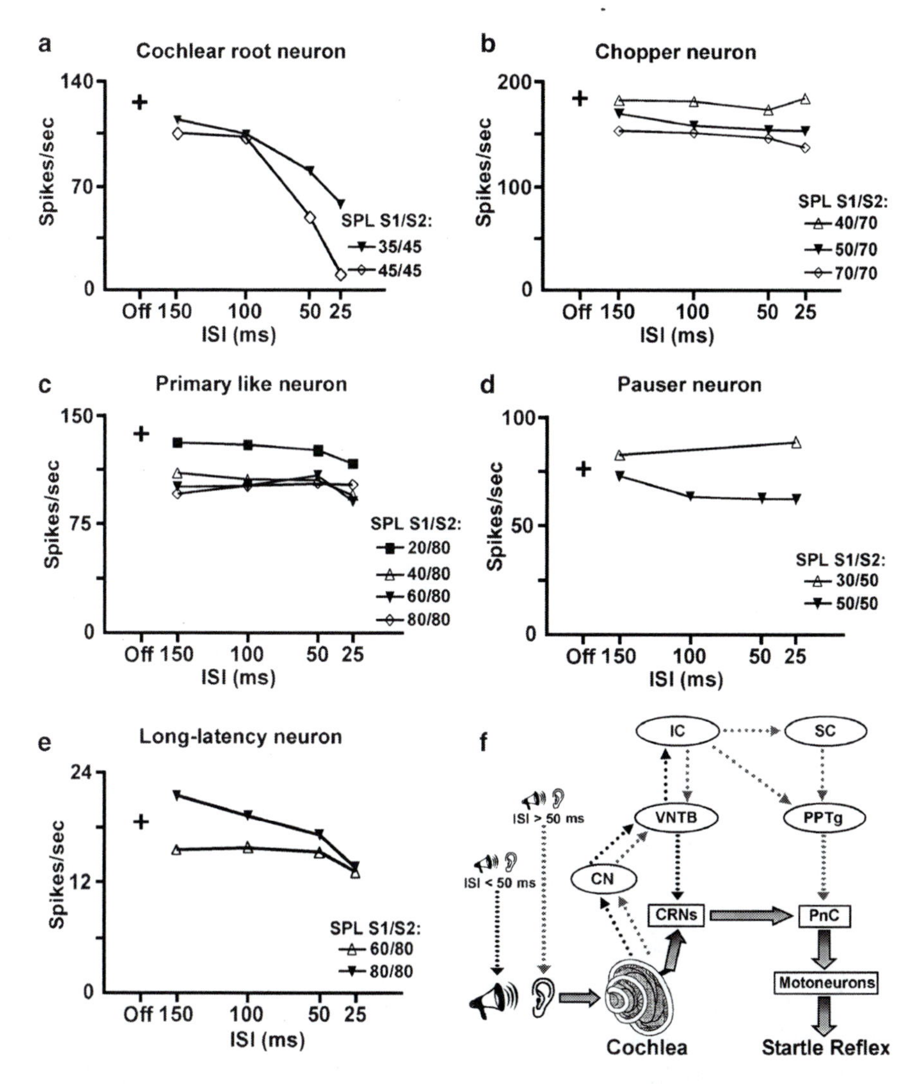

Fig. 8.3 Auditory prepulse inhibition tested in different cochlear neuron types. (**a**) Plot shows number of spikes vs. interstimulus intervals (ISIs) for one CRN unit (**a**), and four CN units: a chopper unit (**b**), a primary-like unit (**c**), a pauser unit (**d**), and a long latency unit (**e**) after different stimulation parameters indicated in the legend at lower left corner of each plot. The CRN reduced its response at shorter interstimulus intervals, and the reductions were large. In contrast, the responses of all other ventral cochlear units were independent of prepulse intensity and interstimulus interval. (**f**) Hypothetical mediating circuit of the auditory prepulse inhibition based on ISIs. The VNTB-to-CRN pathway mediates fast acoustic inputs to reduce CRN responses at short ISIs (<50 ms, *black dashed arrows*), while the IC, SC to PPTg pathway mediates a slower pathway for auditory prepulse inhibition at long ISIs (>50 ms, *gray dashed arrows*). *CN* cochlear nucleus; *CRNs* cochlear root neurons; *IC* inferior colliculus; *PnC* caudal pontine reticular formation; *PPTg* pedunculopontine tegmental nucleus; *SC* superior colliculus; *VNTB* ventral nucleus of the trapezoid body

that there is no significant inhibition at ISIs of 100 ms (Harris and Dallos 1979; Shore 1995). In our experiments, we showed auditory PPI of CRNs' response with a range of ISIs up to 200 ms, indicating that a mechanism of inhibition was operating to produce the observed effects. Even if forward masking contributed to the inhibition of the CRN responses at short ISIs, these response decrements are likely to have a more complex basis. In our study, we also revealed that CRNs differed significantly from all other VCN units after auditory prepulse stimulation. The reduction of CRNs' response depended on prepulse parameters, whereas VCN units showed little or no prepulse-dependence. Together, our results suggest that a specialized inhibitory pathway reaches the cochlear root nucleus.

8.4.2 *Proposed Mediating Circuit for the Auditory Prepulse Inhibition of the ASR Based on Interstimulus Intervals*

Building upon the findings of the present study, and considering the important role of the CRNs in the initiation and full expression of the ASR (Lee et al. 1996; López et al. 1999; Nodal and López 2003), we conclude that the inhibition of CRNs' response by auditory prepulses might be involved in behavioral PPI. The auditory prepulses influenced the inhibition of the CRNs' responses in a similar manner than those in the inhibition of the ASR (our results, Fendt et al. 2001). In both the physiological response of the CRNs and the PPI behavioral paradigm, the strongest inhibition occurred as the intensity and duration level increased up to the pulse or startle threshold. As noted earlier, the auditory PPI of CRNs' response is closely related to a specialized mechanism of neuronal inhibition. The numerous and diverse afferent inputs of CRNs might be involved in such an inhibitory mechanism (Merchán et al. 1988; Osen et al. 1991; López et al. 1993; Gómez-Nieto et al. 2008a, b). Since acetylcholine participated in the mediation of the PPI (Koch et al. 1993; Swerdlow and Geyer 1993; Fendt and Koch 1999), the cholinergic pathway from the ventral nucleus of the trapezoid body to the CRNs might be particularly relevant (Gómez-Nieto et al. 2008a). Supporting this, CRNs contain the neurotransmitter receptor types that are necessary to inhibit the ASR, including the M2 and M4 cholinergic receptor types that are involved in the PPI behavioral paradigm (Bosch and Schmid 2006; Gómez-Nieto et al. 2008b).

It could be argued that the prepulse activates the medial olivocochlear neurons (MOC), which exert an inhibitory effect on the cochlea, reducing cochlear gain and the response of primary neurons (Summers et al. 2003). That would decrease the auditory input to CRNs and produce a decrease in the CRNs' firing (Pilz et al. 1987). However, again, if that is a factor, then the same phenomenon should also produce a decrease in the responses of VCN neurons. However, none of the neurons recorded in the VCN showed the reduction of neuronal activity following a prepulse that was observed in the CRNs.

It has been well established that the PPI of the ASR involves the other major component of the ASR neuronal circuit, the caudal pontine reticular nucleus

(PnC, reviewed in Fendt et al. 2001). However, the auditory prepulse information is relayed at the level of PnC through a long-multimodal circuit (Fendt et al. 2001). Several studies proposed the existence of fast neuronal connections to relay auditory information to the primary acoustic startle circuitry (Yeomans et al. 2006, Gómez-Nieto et al. 2008a). Both the cochlear root nucleus and PnC appear to be similar in the electrophysiological properties of their neurons (Sinex et al. 2001; Lingenhohl and Friauf 1992), afferents inputs (Fendt et al. 2001; Gómez-Nieto et al. 2008a), and neurotransmitters machinery (Bosch and Schmid 2006; Gómez-Nieto et al. 2008b). These resemblances support the idea that the ASR might be inhibited in both nuclei of the acoustic startle circuit. Our study showed that auditory prepulses presented at short ISIs reduced more effectively the CRNs' responses than those presented at longer ISIs. Therefore, we proposed a neuronal circuit for the mediation of the PPI based on ISIs (Fig. 8.3f). In agreement with the current models of PPI (Fendt et al. 2001; Yeomans et al. 2006), we proposed that CRNs might be inhibited at short ISIs through a short descending auditory pathway (Gómez-Nieto et al. 2008a) and PnC neurons might receive auditory prepulse information at long ISIs through the long multistep circuit.

Acknowledgements The authors thank #BFU2007-65210 and JCyL #GR221 to D.E.L; DC00341 from NIDCD to D.G.S.

References

Bosch D, Schmid S (2006) Activation of muscarinic cholinergic receptors inhibits giant neurones in the caudal pontine reticular nucleus. Eur J Neurosci 24:1967–1975

Fendt M, Koch M (1999) Cholinergic modulation of the acoustic startle response in the caudal pontine reticular nucleus of the rat. Eur J Pharmacol 370(2):101–107

Fendt M, Li L, Yeomans JS (2001) Brain stem circuits mediating prepulse inhibition of the startle reflex. Psychopharmacology 156:216–224

Gómez-Nieto R, Rubio ME, López DE (2008a) Cholinergic input from the ventral nucleus of the trapezoid body to cochlear root neurons in rats. J Comp Neurol 506:452–468

Gómez-Nieto R, Horta-Junior JA, Castellano O, Herrero-Turrión MJ, Rubio ME, López DE (2008b) Neurochemistry of the afferents to the rat cochlear root nucleus: possible synaptic modulation of the acoustic startle. Neuroscience 154:51–64

Harris DM, Dallos P (1979) Forward masking of auditory nerve fiber responses. J Neurophysiol 42:1083–1107

Hoffman HS, Ison JR (1980) Reflex modification in the domain of startle. I. Some empirical findings and their implications for how the nervous system processes sensory input. Psychol Rev 87:175–189

Horta-Júnior JdeA, López DE, Alvarez-Morujo AJ, Bittencourt JC (2008) Direct and indirect connections between cochlear root neurons and facial motor neurons: pathways underlying the acoustic pinna reflex in the albino rat. J Comp Neurol 507:1763–1779

Koch M, Kungel M, Herbert H (1993) Cholinergic neurons in the pedunculopontine tegmental nucleus are involved in the mediation of prepulse inhibition of the acoustic startle response in the rat. Exp Brain Res 97:71–82

Landis C, Hunt WA (1939) The startle pattern. Ferrar and Rinehart, New York

Lee Y, López DE, Meloni EG, Davis M (1996) A primary acoustic startle pathway: obligatory role of cochlear root neurons and the nucleus reticularis pontis caudalis. J Neurosci 16: 3775–3789

Li L, Du Y, Li N, Wu X, Wu Y (2009) Top-down modulation of prepulse inhibition of the startle reflex in humans and rats. Neurosci Biobehav Rev 33:1157–1167

Lingenhohl K, Friauf E (1992) Giant neurons in the caudal pontine reticular formation receive short latency acoustic input: an intracellular recording and HRP-study in the rat. J Comp Neurol 325:473–492

López DE, Merchán MA, Bajo VM, Saldaña E (1993) The cochlear root neurons in the rat, mouse and gerbil. In: Merchán MA, Juiz JM, Godfrey DA, Mugnaini E (eds) The mammalian cochlear nuclei: organization and function. Plenum Press, New York, pp 291–301

López DE, Saldana E, Nodal FR, Merchán MA, Warr WB (1999) Projections of cochlear root neurons, sentinels of the auditory pathway in the rat. J Comp Neurol 415:160–174

Merchán MA, Collía F, López DE, Saldana E (1988) Morphology of cochlear root neurons in the rat. J Neurocytol 17:711–725

Nodal FR, López DE (2003) Direct input from cochlear root neurons to pontine reticulospinal neurons in albino rat. J Comp Neurol 460:80–93

Osen KK, López DE, Slyngstad TA, Ottersen OP, Storm-Mathisen J (1991) GABA-like and glycine-like immunoreactivities of the cochlear root nucleus in rat. J Neurocytol 20:17–25

Pilz PK, Schnitzler HU, Menne D (1987) Acoustic startle threshold of the albino rat (*Rattus norvegicus*). J Comp Psychol 101(1):67–72

Shore SE (1995) Recovery of forward-masked responses in ventral cochlear nucleus Neurons. Hear Res 82:31–43

Sinex DG, López DE, Warr WB (2001) Electrophysiological responses of cochlear root neurons. Hear Res 158:28–38

Summers V, de Boer E, Nuttal AL (2003) Basilar-membrane responses to multicomponent (Schroeder-phase) signals: understanding intensity effects. J Acoust Soc Am 114:294–306

Swerdlow NR, Geyer MA (1993) Prepulse inhibition of acoustic startle in rats after lesions of the pedunculopontine tegmental nucleus. Behav Neurosci 107:104–117

Yeomans JS, Lee J, Yeomans MH, Steidl S, Li L (2006) Midbrain pathways for prepulse inhibition and startle activation in rat. Neuroscience 142(4):921–929

Part II
Masking

Chapter 9
FM Forward Masking: Implications for FM Processing

Neal Viemeister, Andrew Byrne, Magdalena Wojtczak, and Mark Stellmack

Abstract It has been argued and has been demonstrated in several species that there is specialized neural circuitry for processing frequency changes (FM). We have examined this possibility psychophysically in humans by examining the aftereffect of a brief sinusoidal FM masker on subsequent detection of FM. Using this FM forward masking paradigm, we have shown that: (1) there can be substantial masking; (2) this is not due to FM to AM conversion; (3) there is broad tuning for the modulation frequency of the FM; (4) the recovery period is approximately 300 ms; (5) similar results were seen at different carrier frequencies (4 kHz, 500 Hz). These results are qualitatively similar to those shown for AM forward masking (J Acoust Soc Am 118:3198–3210, 2005) and, as suggested in that paper for AM, are consistent with an adapting modulation filterbank for FM. More specifically, the notion is that there are hardwired neural processes selectively tuned to FM and that these channels adapt and recover over a relatively brief time scale (300 ms). An alternative model involves higher level, plastic processes such as noisy template matching and short-term memory. In this paper, we will highlight the basic psychophysical data and discuss the theoretical implications for models of FM processing.

Keywords FM • Adaptation • AM • Tuning • Masking • Perception

9.1 Introduction

Frequency changes (FM) that occur over time (spectrotemporal changes) are a primary basis for auditory perception and present a daunting theoretical and empirical challenge for understanding hearing. The general approach has been to treat these aspects separately, for example, to examine auditory frequency resolution when temporal factors are minimally involved (e.g., "tuning curves"), or temporal resolution when spectral factors are minimally involved (e.g., TMTFs).

N. Viemeister (✉)
Department of Psychology, University of Minnesota, 75 E. River Road, Minneapolis, MN, 55455, USA
e-mail: nfv@umn.edu

E.A. Lopez-Poveda et al. (eds.), *The Neurophysiological Bases of Auditory Perception*,
DOI 10.1007/978-1-4419-5686-6_9, © Springer Science+Business Media, LLC 2010

This approach, although limited, has yielded a wealth of basic, useful data but clearly only begins to capture the complexities of auditory perception. The focus in this presentation is on the temporal aftereffects of FM and thus, in a crude way, addresses this fundamental issue from a somewhat different perspective. More specifically, the question we begin to address is whether there are temporal aftereffects that are specific to FM.

This presentation highlights some new empirical findings about the perception and processing of FM, particularly those resulting from sinusoidal frequency modulation (SFM). We examine three related primary issues: (1) the magnitude of the effect; (2) its possible interpretation based on FM to AM conversion; (3) the frequency selectivity for SFM. In general, the rationale and procedure closely follow those from previous work on AM (Wojtczak and Viemeister 2005) and represent an extension to FM.

9.2 Procedure and Methods

Detection thresholds for SFM were measured using an adaptive 3IFC procedure. The nominal waveform of the signal was: $x_{FM}(t) = \cos[2\pi f_c t + (\Delta f / f_m)\sin(2\pi f_m t)]$ where f_c is the carrier frequency (4 kHz), f_m is the modulation frequency (20 Hz), and Δf is the maximum deviation of the instantaneous frequency from the carrier frequency (2 Δf is the total frequency excursion of the FM and $\Delta f/f_m$ is β as typically used in the engineering literature.)

Across signal intervals, Δf was varied with multiplicative increments/decrements to estimate the level of performance corresponding to 79.4% correct. For the two nonsignal intervals, $\Delta f = 0$. Trial-by-trial stimuli and responses were recorded to permit reconstruction of psychometric functions. Correct answer feedback was provided in all conditions.

The temporal sequence during a signal interval is schematized in Fig. 9.1 as a spectrographic plot. This illustrates the SFM forward masker condition and is the

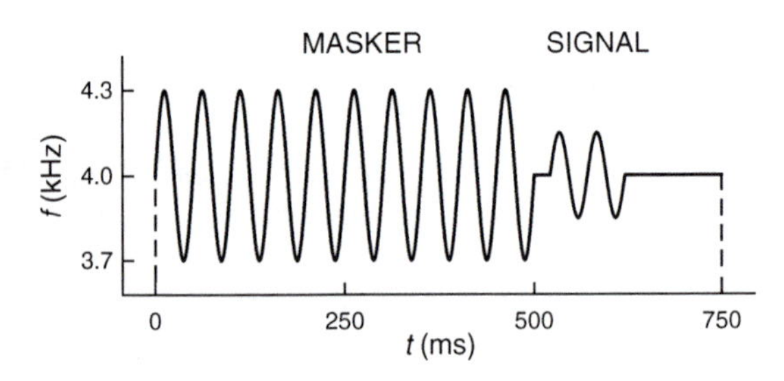

Fig. 9.1 The instantaneous frequency of the on-frequency, 20-Hz, masked stimuli. The total duration of the interval was 750 ms with 500-ms masker and 100-ms signal modulation durations. The full interval was gated, but a continuous carrier was used for 20-ms between the masker and signal and also for 130 ms following the signal

primary condition for this paper. The comparison conditions use an AM forward masker ($m = 1.0$) or an unmodulated forward masker.

The stimuli were presented monaurally via headphones to individual listeners ($n = 5$) all of whom had normal hearing according to our laboratory norms (see Wojtczak and Viemeister 2005 for further details). The data to be presented are based on asymptotic performance, which required up to 12 h of training.

9.3 Results

Figure 9.2 shows psychometric functions for one listener. These are typical of the reconstructed functions for the other listeners and show the magnitude of the effect of an SFM masker. The two functions with filled symbols indicate that modulating the masker produces a very large decrement in performance: The psychometric functions are far apart. To a rough approximation d' is proportional to Δf squared for both the masked and unmasked conditions. Thus, to assess the magnitude of the effect, it is more meaningful to present data as a ratio of thresholds (masked vs. unmasked). In those terms, the increase in threshold (for $P(C) = 79.4$) produced by the SFM masker is about a factor of 4. This corresponds to a change in performance from essentially perfect to chance. Clearly there is a very large effect, even by typical psychoacoustic measures.

A plausible explanation is that the FM is being detected as AM and that the masking is another manifestation of the AM forward masking shown by Wojtczak

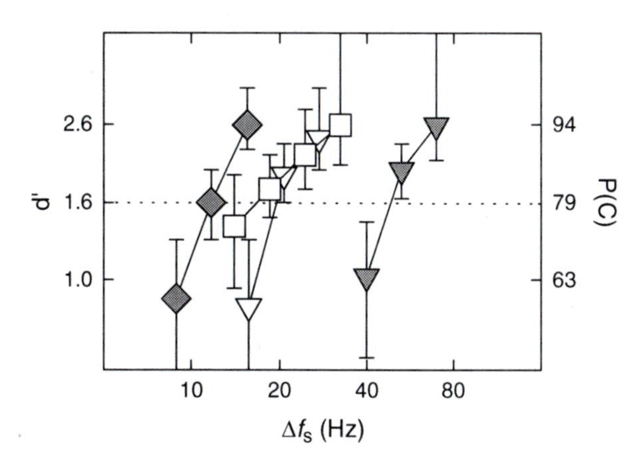

Fig. 9.2 Data from one subject was used to reconstruct psychometric functions from the adaptive tracks. On a 4-kHz carrier with a f_s of 20 Hz, the unmasked detection function (*shaded diamonds*), the on-frequency FM (*shaded triangles*) and AM (*open triangles*) masked functions, and the function with an AM masker at twice the f_s (*open squares*) are plotted. *Error bars* represent the 95% confidence intervals around the P(C) value

and Viemeister (2005). The notion is that as the instantaneous frequency changes due to the FM, the output of an auditory filter tuned to the carrier frequency will fluctuate as the signal passes in and out of the passband of the filter. There is an extensive literature that strongly suggests that for threshold detection of FM, this is the basis for detection (see Edwards and Viemeister 1994 for a review). Wojtczak and Viemeister (2005) have shown that there is a large effect in forward masking for detection of AM. Thus, the masking observed here may reflect FM-to-AM conversion: If both the masker and probe undergo FM-to-AM conversion, the masking we observe may reflect masking of AM by AM. This raises the question of whether the FM masking we observe is specific to FM.

In Fig. 9.2, the open symbols show FM detection performance in the presence of an AM masker. The modulation frequency of the masker is equal to that of the FM signal (triangles) or twice that frequency (squares). (The rationale for the 2*fm condition is that the FM masker may, depending on the auditory filter shape, produce AM at twice the modulation frequency.) The data shown in Fig. 9.2 indicate that an AM masker causes a decrease in performance (filled diamonds vs. open symbols). There are several interpretations of this effect. The more plausible is that the FM signal is being detected as AM (FM-to-AM conversion) and that the AM masker produces an aftereffect such as shown by Wojtczak and Viemeister (2005). Another interpretation is that the masking reflects simple "spectral" forward masking: When the (amplitude) spectra of the masker and signal are similar (as for the AM and FM maskers used here), the detectability of the sidebands will be reduced, and as a consequence, the threshold for detecting FM will be increased.

Although these explanations for the effects of an AM masker are plausible, it should be emphasized that an AM masker produces a change in performance that is small relative to that produced by an FM masker. At this point, we think it is appropriate to focus on the FM-specific effect: An FM masker produces a much larger decrement in performance than that produced by AM.

The immediate question then becomes that of the specificity of the FM forward masking effect. Figure 9.3 shows a masking pattern for detection of a 20-Hz FM signal. The *x*-axis is the modulation frequency of the masker. The *y*-axis is the elevation in threshold, expressed as a ratio, produced by modulating the masker. At 20 Hz, the average modulation threshold is raised by a factor of over 5. Consistent with the discussion regarding the psychometric functions for S2 (Fig. 9.2), this average effect is very large – an increase in threshold by a factor of 5 is equivalent to a change in performance from perfect to chance: Modulating the masker with FM will render highly detectable FM inaudible.

As shown in Fig. 9.3, there is substantial but broad frequency selectivity for FM detection with an FM forward masker. The major point of Fig. 9.3 is that there is a large effect of prior exposure to FM that depends on modulation frequency. This raises the question of whether there are FM detectors that are tuned to FM modulation frequency. Is there a modulation filterbank for FM as has been proposed for AM (Dau and Puschel 1996)?

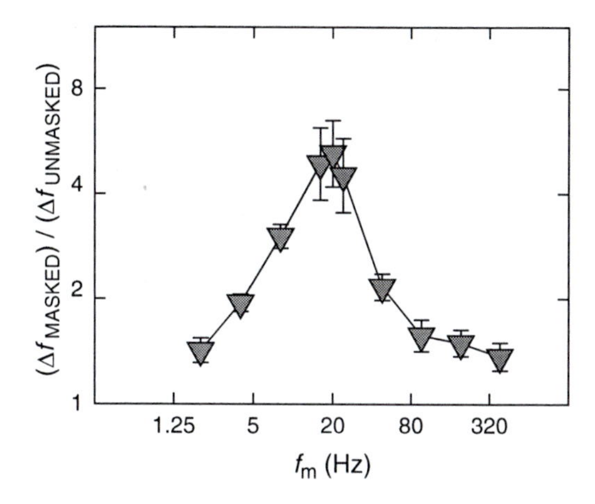

Fig. 9.3 The mean 20-Hz tuning curve from five subjects using a 4-kHz carrier is shown. The y-axis scale is the ratio of the masked Δf (Hz) to the unmasked Δf (Hz). *Error bars* represent standard deviation of the mean

9.4 Discussion

The data shown in Figs. 9.2 and 9.3 indicate that that there is very large effect of an FM masker on subsequent detection of FM. This does not appear to be due to FM-to-AM conversion. An AM masker does mask FM (Fig. 9.2), but the amount of masking is far less than that produced by an FM masker. Thus it appears that that the masking is dominated by the FM.

An issue that merits discussion is the relationship between the present findings and those obtained using a "selective adaptation" paradigm. Those studies typically used long duration adaptors, of the order of minutes, and long duration signals, much longer than those used in the present experiments. The adaptation effect shown in those experiments is much smaller than that shown here, it may disappear with training (Moody et al. 1984), and may result from the use of an improper template for decisions based on correlation (Wakefield and Viemeister 1984). We do not think that the basis for the effects shown here is the same as that shown in these earlier studies. The time scale of the effect is much different – the maskers are 500 ms (vs. min), the signal is relatively brief (100 ms), the masker–signal delay is brief (20 ms), and the effect is negligible after a masker–signal delay of 300 ms (not shown here). Furthermore, the effect does not disappear with training – our six subjects had extensive experience with the task and showed no significant change in performance over the many hours of testing. Although we cannot rule out the possibility that the adaptation effect shown in the earlier studies is related to the effect seen here, it is our opinion that because of the vast differences in its temporal characteristics and in the magnitude of the effect that the effect results from a fundamentally different process.

There is a striking similarity between the FM-specific effects we have shown and those for AM shown by Wojtczak and Viemeister (2005): the size of the effect, temporal characteristics, and the modulation frequency selectivity are similar. A possible account for the present data is similar to that we suggested for AM forward masking: there are modulation frequency-selective neural circuits/modules that adapt in the presence of FM and recover their sensitivity after a relatively brief postexposure interval. There are physiological data suggesting that in cortex there is such a phenomenon that occurs for AM (Bartlett and Wang 2007). We know of no evidence that tuned, adaptable "detectors" exist for FM.

Acknowledgments This research was supported by grant DC00683 from NIH/NIDCD.

References

Bartlett EL, Wang X (2007) Neural representations of temporally modulated signals in the auditory thalamus of awake primates. J Neurophysiol 97:1005–1017

Dau T, Puschel D (1996) A quantitative model of the "effective" signal processing in the auditory system. I. Model structure. J Acoust Soc Am 99:3615–3622

Edwards BW, Viemeister NF (1994) Frequency modulation versus amplitude modulation discrimination: evidence for a second frequency modulation encoding mechanism. J Acoust Soc Am 96:733–740

Moody DB, Cole D, Davidson LC, Stebbins WC (1984) Evidence for a reappraisal of the psychophysical selective adaptation paradigm. J Acoust Soc Am 76:1076–1079

Wakefield GH, Viemeister NF (1984) Selective adaptation of linear frequency modulated sweeps: evidence for direction-specific FM channels? J Acoust Soc Am 75:1588–1592

Wojtczak M, Viemeister NF (2005) Forward masking of amplitude modulation: basic characteristics. J Acoust Soc Am 118:3198–3210

Chapter 10
Electrophysiological Correlates of Intensity Resolution Under Forward Masking

Daniel Oberfeld

Abstract Nonsimultaneous masking can severely impair auditory intensity resolution, but the effect strongly depends on the stimulus configuration. For example, an intense forward masker causes a pronounced impairment in intensity resolution for standards presented at intermediate levels, but not for standards presented at low and high levels, resulting in a midlevel hump pattern (Zeng et al., Hear Res 55:223–230, 1991). Several aspects of the phenomenon cannot be explained by mechanisms in the auditory periphery. For instance, backward maskers cause midlevel humps at least as large as the humps caused by forward maskers. The present experiment was aimed at studying the relation between the effects of forward maskers on intensity resolution and on the slow components N1 and P2 of the auditory evoked potential. The EEG was recorded while listeners performed a one-interval intensity discrimination task in quiet and under forward masking. The 90-dB SPL masker caused a stronger reduction in sensitivity for a 60-dB SPL than for a 30-dB SPL standard, reflecting the midlevel hump. The effect of the masker on the N1 and the P2 amplitude paralleled the behavioral effects. The amplitude reduction caused by the masker was stronger for the 60-dB SPL than for the 30-dB SPL standard, thus also following a midlevel hump pattern. Listeners who showed a strong N1 midlevel hump tended to also exhibit a strong midlevel hump in sensitivity.

Keywords Auditory intensity discrimination • Forward masking • Signal detection theory • Auditory evoked potentials • N1 • P2

10.1 Introduction

Zeng et al. (1991) were the first to demonstrate that an intense forward masker (e.g., 90 dB SPL) causes strongly elevated intensity-difference limens (DLs) for a midlevel pure-tone standard, relative to the DL in quiet. On the other hand, there is

D. Oberfeld (✉)
Department of Psychology, Johannes Gutenberg-Universität, 55099 Mainz, Germany
e-mail: oberfeld@uni-mainz.de

E.A. Lopez-Poveda et al. (eds.), *The Neurophysiological Bases of Auditory Perception*,
DOI 10.1007/978-1-4419-5686-6_10, © Springer Science+Business Media, LLC 2010

only a small effect of the masker on the DLs for standards presented at low and high levels, resulting in the so-called *midlevel hump in intensity discrimination*, which is observed for masker-target intervals up to 400 ms (Zeng and Turner 1992). Although several explanations for the effects have been suggested, none of them is capable of accounting for the complete range of findings (Oberfeld 2008, 2009). Zeng et al. (1991) proposed that the effect was due to adaptation of low spontaneous-rate (SR) auditory nerve neurons showing slower recovery from prior stimulation than high-SR neurons (Relkin and Doucet 1991). However, subsequent experiments provided evidence for a contribution of more central mechanisms. In this context, two important findings are the midlevel hump caused by backward maskers, and the influence of the masker-target similarity on the masker-induced reduction in intensity resolution.

The fact that the midlevel hump is observed with backward maskers (Plack and Viemeister 1992) places a rather strong constraint on the potential physiological origins. For instance, if the neural response to the target persisted longer than 100 ms, then a backward masker presented 100 ms after target offset could interfere with this neural activity by terminating or reducing it. At the level of the auditory nerve (Harris and Dallos 1979), the cochlear nuclei (Rhode 1991), or the inferior colliculus (Nuding et al. 1999), no evidence for persistence over an interval of 100 ms has been reported, however. On the other hand, neuronal activity has been observed at up to 500 ms after signal offset in the medial geniculate body (Aitkin and Dunlop 1969) and in the auditory cortex (Brosch et al. 1999). A second potential explanation for the effect of a backward masker on intensity resolution would be that the masker presented in the first observation interval produces adaptation or inhibition, and therefore reduces the neural response to the target presented in the second interval. However, the two observation intervals were separated by more than 500 ms in the relevant experiments. Thus, to account for a midlevel hump at least as strong with backward maskers as with forward maskers (Plack and Viemeister 1992), it would be necessary to identify neuronal elements for which the adaptation by preceding stimulation is equally strong at a 100 ms masker-target interval (as with forward maskers) and at a 500-ms interval (as with backward maskers). Such a characteristic has not been observed at early processing stages (e.g., Aitkin and Dunlop 1969; Shore 1995). However, recovery times of several hundred milliseconds were found for neurons in the auditory cortex (e.g., Schreiner et al. 1997; Wehr and Zador 2005). Thus, both in terms of persistence and inhibition, the primary auditory cortex seems to be the first structure in the ascending auditory pathway that has physiological properties compatible with the midlevel hump caused by backward maskers (cf. Brosch et al. 1998).

Another finding suggesting central processing stages as the origin of the midlevel hump is *similarity effects* (for a review see Oberfeld 2008). For instance, Schlauch et al. (1997) found that adding a 4.133-kHz "cue tone" to a 1-kHz forward masker strongly reduced the size of the midlevel hump for a 1-kHz standard, presumably by helping the listeners to differentiate between the masker and the standard. A related finding is that a 10-ms forward masker causes a stronger DL elevation for a midlevel 10-ms standard than does a 250-ms masker (Schlauch et al. 1997).

The latter result is incompatible with adaptation in the auditory nerve that *increases* with the duration of the masker (e.g., Harris and Dallos 1979). Instead, the two findings suggest an effect of the *perceptual similarity* of masker and standard, akin to the well-established effects of the target-distractor similarity in auditory perception and other domains (e.g., Baddeley 1966; Duncan and Humphreys 1989; Kidd et al. 2002). Oberfeld (2008) also discussed whether the midlevel hump *per se* could be a similarity effect. In earlier experiments, a fixed-level, intense masker had been combined with various standard levels, so that the *masker-standard level difference* and the standard level were correlated. For a low-level standard, the level difference was always larger than for a medium-level standard. Thus, the different DL elevations at different standard levels could have been due to the variation in the masker-standard level difference rather than to the variation in standard level, as previous studies assumed. The data by Oberfeld (2008) showed, however, that the effect of a forward-masker is stronger at midlevels even if the masker-standard level difference is controlled, thus providing an even stronger definition of the midlevel hump. To account for the similarity effects, Oberfeld (2008) proposed a model based on the loudness enhancement hypothesis (Carlyon and Beveridge 1993), which assumes a relation between the masker-induced impairment in intensity resolution and masker-induced changes in target loudness (Oberfeld 2007).

Motivated by the evidence for central origins of the midlevel hump, the present combined psychoacoustic and event-related potentials (ERP) experiment for the first time studied the effect of forward maskers on sensitivity in an intensity discrimination task and on the long-latency component waveforms N1 and P2 of the auditory evoked potential (AEP; cf. Picton et al. 1974). The EEG was recorded while the listeners actively performed a one-interval intensity discrimination task (Oberfeld 2006). The N1 is evoked by a relatively abrupt change in acoustic energy at a given frequency and peaks approximately 100 ms after stimulus onset (cf. Näätänen and Picton 1987). It is in part generated by an auditory-specific supratemporal source located in primary auditory cortex. There are at least two other sources contributing to the N1, one located in the lateral temporal lobe, and one unknown source probably located in the frontal lobe (e.g., Giard et al. 1994). The supratemporal subcomponent appears as a positive deflection at the mastoids if the nose is used as the reference (Näätänen and Picton 1987). According to the concept by Näätänen and Winkler (1999), the N1 indexes the storage and the processing of a stimulus in auditory sensory memory and represents a *sensory feature trace* (Durlach and Braida 1969; Massaro 1975; Cowan 1984). At this point, it should be noted that in more general terms, the effects of backward maskers and the influence of the masker-target similarity can be viewed as effects on the *memory representation* of target intensity (Plack and Viemeister 1992; Carlyon and Beveridge 1993; Oberfeld 2008). In this line of thinking, we expected the N1 to reflect the behavioral consequences of a forward masker in an intensity discrimination task. Note, however, that for our stimuli, the N1 amplitude is unlikely to be a direct correlate of the target level presented on a given trial, that is, to be higher if the standard-plus-increment rather than the standard is presented. This is because the N1 is a response to sound onset and is correlated with the overall sound pressure level (e.g., Mulert et al. 2005), so that the rather small level difference between the standard and the standard-plus-increment (e.g., 60 dB SPL versus 65 dB SPL) should have only a very small effect on the N1 amplitude (Tanis 1971). Rather, we assume that the N1 amplitude reflects the precision of the sensory trace representation of intensity.

Although the P2 has often been treated as unitary with the N1 (e.g., Davis and Zerlin 1966), there is growing evidence that the P2 represents an independent component (Crowley and Colrain 2004). The functional significance is less clear than for N1, but it has been suggested that positive deflections in the AEP occurring around 200 ms after sound onset might be related to stimulus classification and discrimination (Novak et al. 1992).

10.2 Method

Eleven normal-hearing listeners participated in the experiment. One of them was the author, the others were volunteers who received partial course credit or payment and provided written informed consent according to the Declaration of Helsinki. Due to a poor EEG data quality, the data of two subjects were excluded from the analyses. The remaining nine participants (two men) were between 21- and 35-year-old and right-handed.

The stimuli were generated digitally, played back via an RME ADI/S digital-to-analog converter ($f_S = 44.1$ kHz), attenuated (TDT PA5), buffered (TDT HB7), and presented to both ears via Sennheiser HDA 200 headphones. EEG was recorded with a NeuroScan SynAmps system.

A one-interval, two-alternative forced-choice intensity discrimination task was used (absolute identification; Braida and Durlach 1972). On each trial, a pure-tone standard with a frequency of 1 kHz and a duration of 50 ms (including 5-ms ramps) was presented. A level increment was added to the standard with an a-priori probability of 0.5. The task was to decide whether the softer tone (standard) or the louder tone (standard-plus-increment) had been presented. The level increment was fixed within each block. A 30-dB SPL and a 60-dB SPL standard were presented in quiet, and combined with a 90-dB SPL forward masker. The forward masker was a 1-kHz sinusoid with a duration of 100 ms (including ramps). The silent interval between masker offset and standard onset was 120 ms.

The experiment consisted of six sessions in which only behavioral data were collected (termed *psychoacoustics sessions* in the following), followed by one session in which EEG was recorded while the listeners performed the same identification task. Visual trial-by-trial feedback was provided in the psychoacoustics sessions but not in the EEG session, in order to avoid visually evoked potentials.

Sessions 1 and 2 were practice sessions. In session 3, an individual level increment was selected, which was used in the main experiment. Intensity resolution was measured for a 60-dB standard in quiet, and for a 30-dB SPL and 60-dB SPL standard combined with a 90-dB SPL forward masker. On the basis of the resolution-per-dB ($\delta = d'/\Delta L$) per listener and condition (Durlach and Braida 1969), one individual level increment ΔL_i was selected, so that the arithmetic mean of the sensitivity in the easiest and in the most difficult condition could be expected to be $d' = 1.6$. This increment was used in the subsequent sessions constituting the main experiment. ΔL_i ranged from 2.9 to 9.0 dB ($M = 5.0$ dB, $SD = 1.97$ dB). In each of the sessions 4–6, three 120-trial blocks were presented for each of the four masker/standard level combinations, with ΔL fixed to the individually selected value. In session 7, in which EEG was recorded, the same stimuli and the same task as in sessions 4–6 were used. Three 46-trial blocks were run for each masker/standard level combination.

For each block, the sensitivity (d') was calculated on the basis of a signal detection theory (SDT) model assuming equal-variance Gaussian distributions (Green and Swets 1966). The "log-linear correction" for extreme proportions was used (cf. Hautus 1995).

The EEG was recorded at 21 electrode sites compatible with the 10–20 system (Fp1, Fp2, F7, F3, Fz, F4, F8, T7, C3, Cz, C4, T8, P7, P3, Pz, P4, P8, O1, O2, and mastoids LM and RM), using the nose as reference. Vertical and horizontal electrooculograms (EOGs) were also recorded. Impedances were kept below 5 kΩ. The sampling frequency was 500 Hz. The EEG was filtered online with a 70-Hz low pass, and offline with a 1–20 Hz band pass. For artifact rejection, a standard deviation criterion of 30 μV in a 200-ms window for electrode Fz and the vertical and horizontal EOG was used. AEPs were analyzed in 750-ms epochs with a prestimulus baseline of 200 ms. The valid epochs were averaged for each subject and experimental condition separately, namely for each combination of standard level (30, 60 dB SPL), masker level (in quiet, 90 dB SPL), and target type (standard, standard-plus-increment), and then averaged across all subjects to obtain the grand mean waveforms. As can be seen in Fig. 10.2, the 100-ms masker started at 0 ms, followed by the 50-ms target after a silent interval of 120 ms. The N1 and P2 amplitudes were calculated as the mean voltage in the 40-ms period centered at the peak in the grand average waveform. In quiet and in forward masking, the N1-component peaked at 90 ms and 104 ms after target onset, respectively. The peak of the P2-component occurred at 176 ms and 190 ms after target onset in quiet and in forward masking, respectively. The channels Fz and Cz, where the largest responses were obtained (Näätänen and Picton 1987; Crowley and Colrain 2004), were pooled for the statistical analysis.

10.3 Results and Discussion

10.3.1 Sensitivity

Mean sensitivity in the psychoacoustics sessions is shown in panel A of Fig. 10.1. In quiet, the sensitivity was higher for the 60-dB SPL than for the 30-dB SPL standard, $t(8) = 4.79$, $p = 0.001$ (two-tailed), reflecting the near-miss to Weber's law.

Paired-samples t-tests indicated that the masker-induced decrease in sensitivity was significant at the 30-dB SPL standard level, $t(8)=5.01$, $p=0.001$ (two-tailed), as well as at the 60-dB SPL standard level, $t(8)=8.38$, $p=0.001$ (two-tailed). In previous experiments using an adaptive procedure, the midlevel hump was defined as a stronger masker-induced elevation of the intensity DL (relative to the DL in quiet) for a midlevel standard than for a low-level or high-level standard (Zeng and Shannon 1995). The DLs measured via an adaptive procedure correspond to a fixed level of sensitivity, d' can be assumed to be proportional to ΔL (e.g., Jesteadt et al. 2003), and in the present experiment, a constant level increment was used for a given listener in all conditions. Therefore, we expected the 90-dB SPL masker to cause a stronger decrease in d' at the 60-dB SPL than at the 30-dB SPL standard level, compatible with the midlevel hump. The sensitivity was analyzed via a repeated-measures analysis-of-variance (ANOVA) based on a univariate approach. The two within-subjects factors were masker level and standard level. Partial η^2 is reported as a measure of effect. There was a significant Masker Level × Standard Level interaction, $F(1, 8)=49.46$, $p<0.001$, $\eta^2=0.86$, compatible with a midlevel hump. The effect of standard level and the effect of masker level was also significant [$F(1, 8)=10.46$, $p=0.012$, $\eta^2=0.57$ and $F(1, 8)=48.89$, $p<0.001$, $\eta^2=0.86$, respectively].

In order to check whether the stronger masker-induced reduction in d' at the 60-dB SPL standard level could be due to a floor effect, a test by Marascuilo (1970) was used to determine whether d' was significantly higher than 0 for a given listener and masker/standard level combination. The hits and false alarms were pooled across the three blocks obtained per condition. For two listeners, d' was not significantly higher than 0 ($p>0.05$, one-tailed) in the forward masking conditions. With the data from these two listeners excluded, a Masker Level × Standard Level repeated-measures ANOVA again showed a significant Masker Level × Standard Level interaction, $F(1, 6)=57.91$, $p<0.001$, $\eta^2=0.91$. Thus, the stronger masker-induced reduction in sensitivity at the 60-dB SPL than at the 30-dB SPL standard level cannot be attributed to a floor effect.

Sensitivity in the EEG session, which is not displayed due to space limitations, showed a pattern very similar to sensitivity in the psychoacoustic sessions, apart from a general reduction in sensitivity that was likely due to the absence of trial-by-trial feedback.

10.3.2 Auditory-Evoked Potentials

The grand-mean AEPs in quiet are shown by the dashed lines in Fig. 10.2, averaged for standard and standard-plus-increment (see below). The AEPs are depicted for the most informative electrodes Fz and Cz, and the left mastoid (LM). As can be seen by the confidence intervals in panels B and C of Fig. 10.1, all tones elicited a significant N1 and P2 component. At the mastoids, the characteristic polarity inversion was observed (Näätänen and Picton 1987). A Standard Level × Target Type

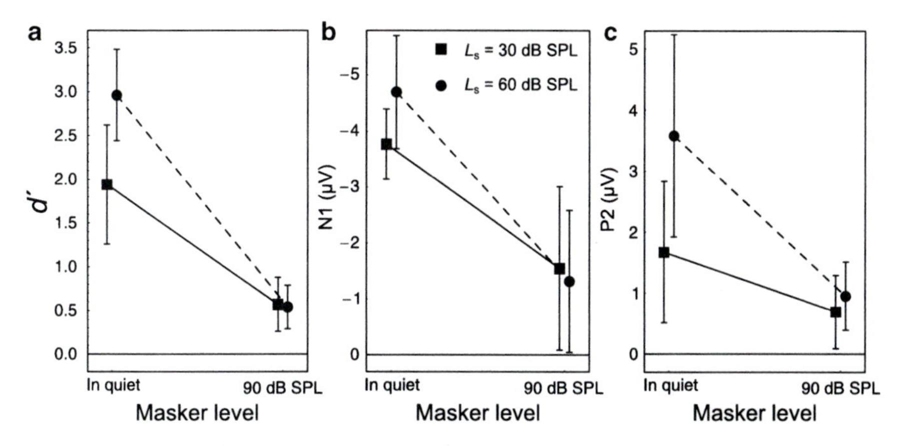

Fig. 10.1 Panel **a**: Mean sensitivity (d') in the absolute intensity identification task as a function of masker level and standard level (L_S). The level increment was individually selected and constant across all masker/standard level combinations. *Squares*: 30-dB SPL standard. *Circles*: 60-dB SPL standard. Panels **b** and **c**: Mean N1 and P2 amplitudes in response to the target. Pooled channels Fz and Cz; responses to standard and standard-plus-increment averaged. *Error bars* show 95% confidence intervals

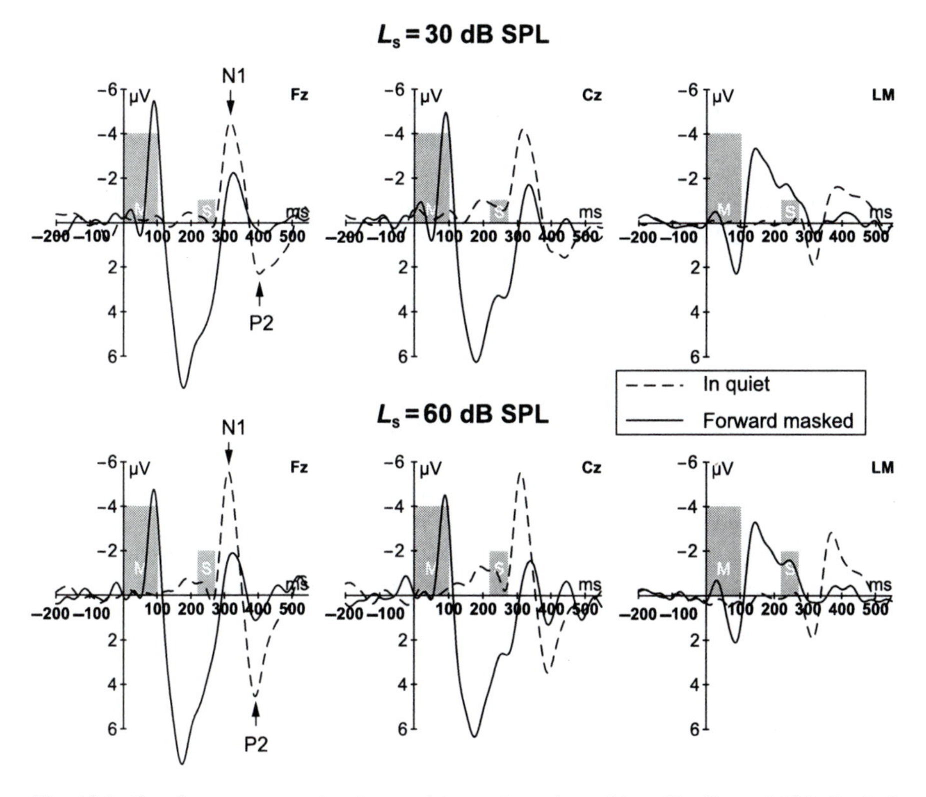

Fig. 10.2 Grand mean event-related potentials at electrode positions Fz, Cz, and LM. *Dashed curves*: target tone presented in quiet. *Solid curves*: with 90-dB SPL forward masker. *Upper row*: 30-dB SPL standard. *Lower row*: 60-dB SPL standard. The *gray rectangles* indicate the temporal positions of masker (*M*) and standard (*S*). In the in quiet condition, the masker was omitted

(standard, standard-plus-increment) repeated-measures ANOVA conducted on the N1 amplitude at pooled electrodes Fz and Cz showed a marginally significant increase with standard level, $F(1, 8)=4.81$, $p=0.060$, $\eta^2=0.39$, compatible with previous findings concerning the intensity dependence of the N1 (e.g., Adler and Adler 1989; Mulert et al. 2005). The N1 amplitude in response to the standard-plus-increment was not significantly higher than in response to the standard, $F(1, 8)=0.17$, $p=0.70$. This result is compatible with data by Tanis (1971); see above. The Standard Level × Target Type interaction was not significant, either, $F(1, 8)=0.40$, $p=0.55$.

The N1 was followed in time by the P2. There was a significant effect of standard level on the P2 amplitude, $F(1, 8)=45.1$, $p<0.001$, $\eta^2=0.85$. The effect of target type and the Standard Level × Target Type interaction failed to reach significance, $F(1, 8)=1.17$, $p=0.31$, and $F(1, 8)=3.17$, $p=0.113$, respectively. Because there were no effects involving target type, neither for N1 nor for P2, the amplitudes of the responses to the standard and to the standard-plus-increment were averaged in all further analyses.

The solid lines in Fig. 10.2 show the AEPs in the masking conditions. As can be seen in Fig. 10.1, panels B and C, the forward-masked N1 and P2 amplitudes to the target tones were small but significantly different from 0 μV. The N1 and P2 amplitudes followed a similar pattern as the sensitivity (Fig. 10.1). The forward maskers caused a decrease in both the N1 and the P2 amplitude, at the 30-dB SPL standard level [N1: $t(8)=2.83$, $p=0.022$ (two-tailed); P2: $t(8)=2.11$, $p=0.068$ (two-tailed)] as well as at the 60-dB SPL standard level [N1: $t(8)=4.41$, $p=0.002$ (two-tailed); P2: $t(8)=3.60$, $p=0.007$ (two-tailed)]. The reduction in the N1 and in the P2 amplitude caused by the forward masker was stronger for the 60-dB SPL than for the 30-dB SPL standard, just as the reduction in sensitivity was stronger for the midlevel standard. Thus, the N1 and P2 amplitudes showed a pattern compatible with the midlevel hump. This observation was partially confirmed by two Masker Level × Standard Level repeated-measures ANOVAs conducted separately for the N1 and the P2 amplitudes. For the N1, the Masker Level × Standard Level interaction was marginally significant, $F(1, 8)=3.74$, $p=0.089$, $\eta^2=0.32$. The effect of masker level was also significant, $F(1, 8)=15.41$, $p=0.004$, $\eta^2=0.66$. The standard level had no significant effect on the N1-amplitude, $F(1, 8)=1.53$, $p=0.25$, $\eta^2=0.16$. In the P2 time window, there was a significant Masker Level × Standard Level interaction, $F(1, 8)=7.76$, $p=0.024$, $\eta^2=0.49$. The effects of standard level and masker level were also significant, $F(1, 8)=58.51$, $p<0.001$, $\eta^2=0.88$, and $F(1, 8)=11.30$, $p=0.01$, $\eta^2=0.59$, respectively.

At this point, it should be noted that contamination of the response to the target by residual activity resulting from the forward masker presents a potential problem for the interpretation of the data. However, exactly the same 90-dB SPL masker preceded the 30-dB SPL and the 60-dB SPL target tone at exactly the same ISI. Therefore, under the usual assumption of linear additivity of the EEG responses (e.g., Hansen 1983), it is valid to compare the two conditions with respect to the masker-induced amplitude reduction. Consequently, the conclusion that the amplitude reduction caused by the 90-dB SPL masker was stronger at the intermediate standard level is unchallenged by potential residual activation. Could the different amounts of reduction in the N1 amplitude caused by the masker at the two different standard levels be due to the refractoriness of the N1 (e.g., Budd et al. 1998)? As the N1 subcomponents have refractory periods of 3–10 s (Näätänen and Picton 1987), a forward masker identical in frequency to the target can reduce the amplitude of the N1 to the target because the time interval between the target and the sound preceding it is considerably shorter than in quiet. However, in the present experiment, the temporal configuration was identical at the two standard levels. Therefore, it is unlikely that refractoriness caused the stronger N1 amplitude reduction at the intermediate standard level.

10.3.3 Relation Between the Behavioral and Electrophysiological Consequences of Forward Masking

The mean data displayed in Fig. 10.1 show a similar pattern for sensitivity and for the N1 and P2 amplitudes, and the statistical analyses confirmed this similarity. But was the relation between the behavioral and the electrophysiological consequences of a forward masker suggested by these data also present at the level of the individual? Imagine a listener for whom the masker had a stronger effect on sensitivity than for the 30-dB SPL standard. Does such a listener showing a strong "behavioral" midlevel hump also exhibit a strong "AEP" midlevel hump in the sense that the masker has a stronger effect on the N1 amplitude at the 60-dB SPL than at the 30-dB SPL standard level? To answer this question, we first computed the masker-induced reduction in sensitivity, $\Delta d' = d'(\text{quiet}) - d'(\text{masked})$, for each subject and standard level. The difference between the reduction in d' at the 60-dB SPL and the 30-dB SPL standard level ($\Delta d'_{60} - \Delta d'_{30}$) is a measure for the behavioral midlevel hump. Similarly, the difference between the N1 amplitude reduction at the two standard levels ($\Delta \text{N1}_{60} - \Delta \text{N1}_{30}$) is a measure for the AEP midlevel hump. As can be seen in Fig. 10.3, these two differences were positively correlated, Spearman rank correlation coefficient $r_S = 0.63$, $p = 0.034$ (one-tailed), $N = 9$. For the P2, the correlations between the difference in the d' reduction between the two standard levels and the difference in the amplitude reduction were not significant ($r_S = -0.017$). As Fig. 10.3 shows, the masker had a stronger effect on d' at the 60-dB SPL than at the 30-dB SPL standard level for all listeners, while for the N1, there were three listeners showing the opposite pattern, evident by negative values on the x-axis. These data are of course at odds with the assumed correlation between the d' and the N1 midlevel hump. Inspection of the individual data indicated, however, that two of the negative values on the x-axis seem to be due to problems with the EEG data, because for one of the three listeners, the N1 amplitude was *positive* at the 30-dB SPL standard level under masking, and for another listener, the N1 amplitude at the 60-dB SPL standard level was higher under masking than in quiet.

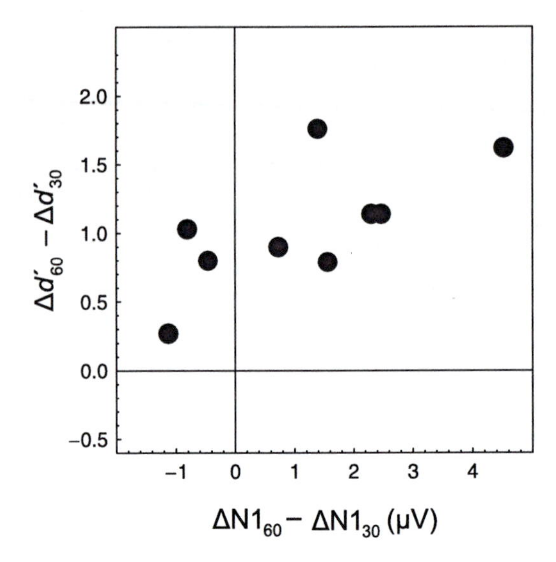

Fig. 10.3 Relation between the midlevel humps in sensitivity and in the N1 amplitudes. *Horizontal axis*: difference between the reduction in the N1 amplitude at the 60-dB SPL and the 30-dB SPL standard level ($\Delta N1_{60} - \Delta N1_{30}$), which is a measure for the N1 midlevel hump. *Vertical axis*: difference between the reduction in d' at the 60-dB SPL and the 30-dB SPL standard level ($\Delta d'_{60} - \Delta d'_{30}$), which is a measure for the behavioral midlevel hump. Each *data point* represents one listener

10.4 Summary

The experiment studied the effects of forward masking on intensity resolution and on the cortical auditory evoked potentials N1 and P2 to the target. The effects of the forward maskers on the N1 and P2 amplitudes paralleled the behavioral reduction in sensitivity because all showed a midlevel hump pattern. At the level of the individual, we found a relation between the masker-induced sensitivity reduction and the reduction in the N1 amplitude. Thus, the N1 represents an electrophysiological correlate of the effects of nonsimultaneous masking on intensity resolution. This finding is compatible with suggestions that the masker-induced reduction in sensitivity can be understood in terms of effects on the memory representation of target intensity (cf. Oberfeld 2008) because the N1 indexes the processing in auditory sensory memory (Näätänen and Winkler 1999). For the P2, the relation to the behavioral effects was less clear cut.

It would be interesting to study electrophysiological responses from processing stages preceding or following the processing stage indexed by N1, in order to further narrow down the locus of the mechanisms causing the midlevel hump in intensity discrimination. For example, Näätänen and Winkler (1999) proposed that the auditory feature trace is transformed into a long lasting and partially analyzed auditory stimulus representation, which is indexed by the *mismatch negativity* (MMN; for a recent review see Näätänen et al. 2007). Many studies found that the MMN is closely related to psychophysical performance (cf. Näätänen and Alho 1997), to a greater extent than the N1. Thus, the MMN amplitude and latency can be expected to exhibit an even stronger correlation with intensity resolution than the N1.

References

Adler G, Adler J (1989) Influence of stimulus intensity on AEP components in the 80- to 200-millisecond latency range. Audiology 28:316–324

Aitkin LM, Dunlop CW (1969) Inhibition in the medial geniculate body of the cat. Exp Brain Res 7:68–83

Baddeley AD (1966) Short-term memory for word sequences as a function of acoustic semantic and formal similarity. Q J Exp Psychol 18:362–365

Braida LD, Durlach NI (1972) Intensity perception: II. Resolution in one-interval paradigms. J Acoust Soc Am 51:483–502

Brosch M, Schulz A, Scheich H (1998) Neuronal mechanisms of auditory backward recognition masking in macaque auditory cortex. Neuroreport 9:2551–2555

Brosch M, Schulz A, Scheich H (1999) Processing of sound sequences in macaque auditory cortex: response enhancement. J Neurophysiol 82:1542–1559

Budd TW, Barry RJ, Gordon E, Rennie C, Michie PT (1998) Decrement of the N1 auditory event-related potential with stimulus repetition: habituation vs. refractoriness. Int J Psychophysiol 31:51–68

Carlyon RP, Beveridge HA (1993) Effects of forward masking on intensity discrimination, frequency discrimination, and the detection of tones in noise. J Acoust Soc Am 93:2886–2895

Cowan N (1984) On short and long auditory stores. Psychol Bull 96:341–370

Crowley KE, Colrain IM (2004) A review of the evidence for P2 being an independent component process: age, sleep and modality. Clin Neurophysiol 115:732–744

Davis H, Zerlin S (1966) Acoustic relations of the human vertex potential. J Acoust Soc Am 39:109–116

Duncan J, Humphreys GW (1989) Visual search and stimulus similarity. Psychol Rev 96:433–458

Durlach NI, Braida LD (1969) Intensity perception: I. Preliminary theory of intensity resolution. J Acoust Soc Am 46:372–383

Giard MH, Perrin F, Echallier JF, Thevenet M, Froment JC, Pernier J (1994) Dissociation of temporal and frontal components in the human auditory N1 wave: a scalp current density and dipole model analysis. Electroencephalogr Clin Neurophysiol 92:238–252

Green DM, Swets JA (1966) Signal detection theory and psychophysics. Wiley, New York

Hansen JC (1983) Separation of overlapping waveforms having known temporal distributions. J Neurosci Methods 9:127–139

Harris DM, Dallos P (1979) Forward masking of auditory-nerve fiber responses. J Neurophysiol 42:1083–1107

Hautus MJ (1995) Corrections for extreme proportions and their biasing effects on estimated values of *d'*. Behav Res Methods Instrum Comput 27:46–51

Jesteadt W, Nizami L, Schairer KS (2003) A measure of internal noise based on sample discrimination. J Acoust Soc Am 114:2147–2157

Kidd G Jr, Mason CR, Arbogast TL (2002) Similarity, uncertainty, and masking in the identification of nonspeech auditory patterns. J Acoust Soc Am 111:1367–1376

Marascuilo LA (1970) Extensions of the significance test for one-parameter signal detection hypotheses. Psychometrika 35:237–243

Massaro DW (1975) Experimental psychology and information processing. Rand McNally College Pub, Chicago

Mulert C, Jäger L, Propp S, Karch S, Störmann S, Pogarell O, Möller HJ, Juckel G, Hegerl U (2005) Sound level dependence of the primary auditory cortex: simultaneous measurement with 61-channel EEG and fMRI. Neuroimage 28:49–58

Näätänen R, Alho K (1997) Mismatch negativity – the measure for central sound representation accuracy. Audiol Neurootol 2:341–353

Näätänen R, Paavilainen P, Rinne T, Alho K (2007) The mismatch negativity (MMN) in basic research of central auditory processing: a review. Clin Neurophysiol 118:2544–2590

Näätänen R, Picton T (1987) The N1 wave of the human electric and magnetic response to sound: a review and an analysis of the component structure. Psychophysiology 24:375–425

Näätänen R, Winkler I (1999) The concept of auditory stimulus representation in cognitive neuroscience. Psychol Bull 125:826–859

Novak G, Ritter W, Vaughan HG (1992) Mismatch detection and the latency of temporal judgments. Psychophysiology 29:398–411

Nuding SC, Chen GD, Sinex DG (1999) Monaural response properties of single neurons in the chinchilla inferior colliculus. Hear Res 131:89–106

Oberfeld D (2006) Forward-masked intensity discrimination: evidence from one-interval and two-interval tasks. In: Langer S, Scholl W, Wittstock V (eds) Fortschritte der Akustik: Plenarvorträge und Fachbeiträge der 32. Deutschen Jahrestagung für Akustik DAGA '06, Braunschweig. Deutsche Gesellschaft für Akustik, Berlin, pp 309–310

Oberfeld D (2007) Loudness changes induced by a proximal sound: loudness enhancement, loudness recalibration, or both? J Acoust Soc Am 121:2137–2148

Oberfeld D (2008) The mid-difference hump in forward-masked intensity discrimination. J Acoust Soc Am 123:1571–1581

Oberfeld D (2009) The decision process in forward-masked intensity discrimination: evidence from molecular analyses. J Acoust Soc Am 125:294–303

Picton TW, Hillyard SA, Krausz HI, Galambos R (1974) Human auditory evoked-potentials. I. Evaluation of components. Electroencephalogr Clin Neurophysiol 36:179–190

Plack CJ, Viemeister NF (1992) Intensity discrimination under backward masking. J Acoust Soc Am 92:3097–3101

Relkin EM, Doucet JR (1991) Recovery from prior stimulation. I: Relationship to spontaneous firing rates of primary auditory neurons. Hear Res 55:215–222

Rhode WS (1991) Physiological–morphological properties of the cochlear nucleus. In: Altschuler RA, Bobbin RP, Hoffman DW, Clopton BM (eds) Neurobiology of hearing: the central auditory system. Raven Press, New York, pp 47–77

Schlauch RS, Lanthier N, Neve J (1997) Forward-masked intensity discrimination: duration effects and spectral effects. J Acoust Soc Am 102:461–467

Schreiner CE, Mendelson J, Raggio MW, Brosch M, Krueger K (1997) Temporal processing in cat primary auditory cortex. Acta Otolaryngol (Stockh) 117:54–60

Shore SE (1995) Recovery of forward-masked responses in ventral cochlear nucleus neurons. Hear Res 82:31–43

Tanis DC (1971) A signal-detection analysis of auditory evoked potentials in an intensity discrimination task. Unpublished PhD thesis, Washington University, Saint Louis, MO

Wehr M, Zador AM (2005) Synaptic mechanisms of forward suppression in rat auditory cortex. Neuron 47:437–445

Zeng FG, Shannon RV (1995) Possible origins of the non-monotonic intensity discrimination function in forward masking. J Acoust Soc Am 82:216–224

Zeng FG, Turner CW (1992) Intensity discrimination in forward masking. J Acoust Soc Am 92:782–787

Zeng FG, Turner CW, Relkin EM (1991) Recovery from prior stimulation II: effects upon intensity discrimination. Hear Res 55:223–230

Chapter 11
Neuronal Measures of Threshold and Magnitude of Forward Masking in Primary Auditory Cortex

Ana Alves-Pinto, Sylvie Baudoux, Alan Palmer, and Christian J. Sumner

Abstract Psychophysical forward masking is an increase in the threshold of detection of a brief sound (probe) when preceded by another sound (masker). These effects are reminiscent of the reduction in physiological responses following prior stimulation. However, previous studies of the response of auditory nerve fibers (Relkin and Turner, 1988) found probe threshold shifts following stimulation that were considerably smaller than those found perceptually. Although such threshold shifts are larger in some units of the cochlear nucleus (Ingham et al., 2006), these are either inhibitory interneurons or project to inhibitory neurons. A better account is obtained at the level of the IC in the awake marmoset (Nelson et al., 2009).

In the present study, we measure responses of neurons in the primary auditory cortex of the anesthetised guinea pig to forward masked pure tones. Signal detection theory methods are used to infer probe detection thresholds. The objective is to determine whether forward masked thresholds in cortical neurons are higher than those of sub-cortical neurons.

Changes in the neurometric function (the computed % correct against probe level) due to prior stimulation are diverse: for some units the function is shifted towards higher probe levels; for others the slope of the function becomes shallower. Thresholds shifts (e.g., 50 dB for a 60-dB SPL masker) calculated for individual units are on average much larger than seen in sub-cortical nuclei. Across the population, the minimum thresholds are also larger than the thresholds observed psychophysically. There is little evidence that persistent activity in response to the masker is contributing to masking.

Keywords Forward masking • Primary auditory cortex • Guinea pig • Signal detection theory

A. Alves-Pinto (✉)
MRC Institute of Hearing Research, Science Road, University Park, Nottingham, NG7 2RD, UK
e-mail: ana@ihr.mrc.ac.uk

E.A. Lopez-Poveda et al. (eds.), *The Neurophysiological Bases of Auditory Perception*,
DOI 10.1007/978-1-4419-5686-6_11, © Springer Science+Business Media, LLC 2010

11.1 Introduction

Psychophysical forward masking is an increase in the threshold of detection of a sound (probe) when this is preceded by another sound (masker). Although several physiological mechanisms have been proposed (for a review see Oxenham 2001), it remains to be demonstrated which physiological responses can account for the magnitude of threshold shifts seen in psychophysical forward masking.

Physiologically, the effect of prior stimulation can be observed in the reduction of the response of auditory nerve (AN) fibers to the probe stimulus (Harris and Dallos 1979; Smith 1979). Similar effects are observable throughout the auditory pathway (Bleeck et al. 2006; Brosch and Schreiner 1997; Calford and Semple 1995; Shore 1995). However, probe detection thresholds inferred from AN responses, using signal detection theory techniques that simulate the methods used to measure psychophysical masking, showed only modest increases in threshold (Relkin and Turner 1988; Turner et al. 1994). Threshold shifts vary across AN fibers between 4 and 20 dB, for an 80-dB increase in masker level, with larger shifts observed for low spontaneous rate fibers. The most sensitive fibers yield much smaller threshold shifts than the 14 dB increase observed psychophysically for an 80-dB SPL narrow band masker (Turner et al. 1994), or the 18-42 dB observed for a 60-dB increase in masker level, in a tone-on-tone masking experiment (Figs. 5 and 6 in Plack and Oxenham 1998). The mismatch between physiology and psychophysics suggests that information carried by AN fibers must be processed in a suboptimal way and that central mechanisms are likely to introduce further masking effects. This interpretation is supported by observations of forward masking in cochlear-implant patients (Chatterjee 1999; Chatterjee et al. 2006; Shannon 1990), who show quite normal forward masking.

A recent study (Ingham et al. 2006) measured threshold shifts in the cochlear nucleus, using signal detection theory techniques. Although onset and onset-chopper neurons were observed to produce thresholds shifts of similar magnitude to the shifts measured psychophysically, the authors concluded that due to the inhibitory nature of these units, they are unlikely to contribute directly to forward masking. Other cell types produced lower threshold shifts.

Another recent study measured forward masking threshold shifts from units in the inferior colliculus (IC) of the awake marmoset (Nelson et al. 2009). The average amount of masking observed across the population was 18.8 ± 9.6 dB, for a 40-dB increase in masker level. Also, threshold shifts increased with masker level with a slope of 0.5 dB/dB. Overall, these results provide a better match with the psychophysics (Jesteadt et al. 1982; Plack and Oxenham 1998).

Nevertheless, the extent to which the amount of forward masking increases, or does not, at a higher stage in the auditory system remains to be verified. Previous studies of forward masking in primary auditory cortex have provided evidence that suppression following a stimulus is generally larger and long lasting in cortical neurons (Brosch and Schreiner 1997; Calford and Semple 1995) than in sub-cortical regions. These studies investigated the general properties of forward masking in cortical neurons, but did not quantify probe detection thresholds using signal detection theory methods. Their results nevertheless suggest an increase in the magnitude of psychophysical forward masking in the auditory cortex.

The present study investigated the extent to which detection thresholds inferred from responses of neurons in primary auditory cortex using signal detection theory methods are affected by a preceding masker. The stimuli and method of analysis of cortical responses were similar to those employed by Relkin and Turner (1988) and Ingham et al. (2006) to study forward masking in the AN and cochlear nucleus.

11.2 Methods

11.2.1 Physiological Recordings

The data were collected from 12 healthy, adult, pigmented guinea pigs (*Cavia porcellus*), weighing between 410 g and 638 g. The animals were anesthetized with an initial injection of urethane (1.3 g/kg, in 20% solution in 0.9% saline; IP) and maintained with Hypnorm (0.2 ml I/M; fentanyl citrate 0.315 mg/ml and fluanisone 10 mg/ml). Atropine sulfate (0.06 mg/kg) was administered to reduce bronchial secretion. Heart rate and respiratory rate were monitored, and the oxygen supply was regulated to maintain normal end-tidal CO_2 levels. Core temperature was maintained in the range of 37–38°C with a thermostatically controlled heating blanket. The animal was placed in a stereotaxic frame in a double walled sound attenuating room. A polythene tube was inserted into an opening in each bulla to equalize middle ear pressure, and the hole resealed with petroleum jelly. The membrane overlying the foramen magnum was opened to release the pressure of the cerebrospinal fluid. A craniotomy with a diameter of around 5 mm was performed to expose the primary auditory cortex, the dura was removed and the brain was covered with a layer of agar. All procedures were performed in accordance with the United Kingdom (Scientific Procedures) Act of 1986. Neural responses were recorded with glass-insulated tungsten multi-electrodes, the outputs of which were fed into a Tucker Davis Technologies (TDT) System 3 recording setup (Alachua, FL). Electrodes were advanced together into AI by a piezoelectric motor (Burleigh Inchworm IW-700/710). Spikes were recorded and analyzed on-line using Brainware (developed by J. Schnupp, University of Oxford, UK). They were further analyzed off-line with Plexon (Dallas, TX) software to isolate action potentials from single and/or multi-units. Auditory stimuli were delivered through modified Radio Shack 40–1,377 tweeters coupled to probe tubes that fitted in to hollow specula, which replaced the stereotax ear bars.

11.2.2 Stimuli

Units were isolated using a wideband search noise, and then their frequency response area and characteristic frequency (CF) were determined by measuring the response to 50 ms pure tones of varying frequency and level. Subsequently, the response of the unit to forward masked tones was measured. The stimuli used to study forward masking were similar to Relkin and Turner (1988), except that whilst they employed a tracking algorithm to find thresholds, we collected complete neurometric functions (see below). Stimuli consisted of a 102-ms pure tone masker, followed by a 25-ms probe, both with a frequency near to CF. All tones were gated with 2 ms on and off cosine-squared ramps. There was no time interval separating masker and probe. The masker and probe levels were varied independently, and randomly interleaved. The probe level was varied from levels below threshold up to suprathreshold levels (>50 dB in 5 dB steps). Maskers were usually presented at four or more levels, including one condition when the masker was maximally attenuated (i.e., to well below threshold). Stimuli were presented diotically with interstimulus intervals longer than 1 ms.

11.2.3 Estimation of Probe Thresholds

For each unit, the spike count measured in response to the probe was compared with that measured within an equivalent temporal window when no probe was presented. Detection was considered to occur when the probe condition elicited more spikes than the no-probe condition. If spike counts were equal, a guess would be made. Responses to the probe and no-probe conditions were paired and compared across all the 50 repetitions for each condition, and the number of "detections" was counted. The probability of detecting the probe was calculated as the proportion of pairs for which the response to the probe was higher than the response to the no-probe condition. This procedure was repeated for the different probe levels to derive the neurometric function (probability of detection as a function of probe level) for each unit and for each masker level. The threshold of detection of the probe was finally estimated from the neurometric function as the smallest probe level for which detection occurred in at least 75% of the comparisons.

11.3 Results

For each unit, the data consisted of a set of detailed rate-level functions (RLF) for probe tones preceded by maskers of equal frequency at various sound levels, and an accompanying "neurometric function" plotting the proportion of trials in a 2-interval 2-alternative forced choice task on which the probe tone was detected.

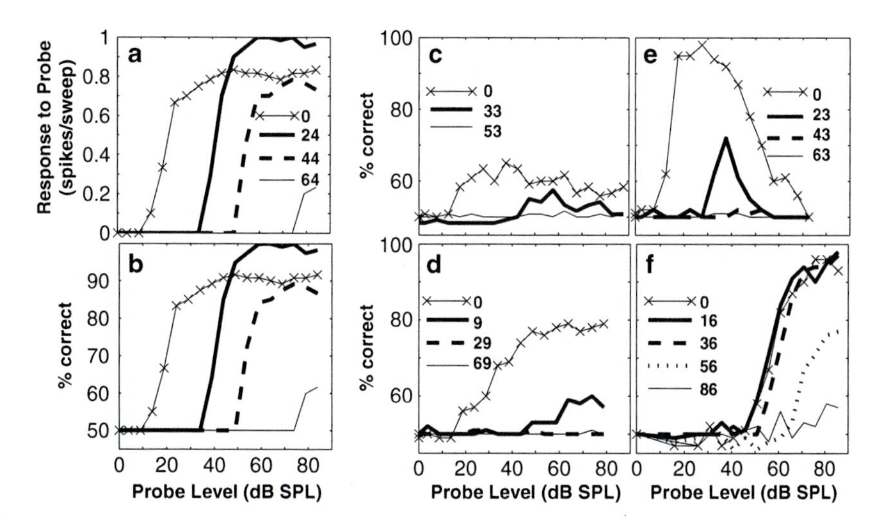

Fig. 11.1 Example responses of cortical neurons to forward masked tones. Panel **a** illustrates the mean spike count in response to the probe as a function of level and panel **b** illustrates the corresponding neurometric function for the same unit. Panels **c–f**, illustrate neurometric functions for other four units. *Different lines* correspond to different masker levels, as indicated in the legends

The results described here are based on the analysis of 48 units, 19 single units and 29 multi-units, for which there were reliable neurometric functions (not noisy and >30 repetitions).

The effect of the masker on the probe RLF, and the corresponding neurometric function, varied across units. Figure 11.1 illustrates five example responses. For the unit in panel a, the mean spike count vs. level function shifted along the *x*-axis towards higher levels as the level of the masker increased. This shift was also observed in the corresponding neurometric functions, illustrated in Fig. 11.1b. In general, neurometric and RLF were similar. Neurometric functions in some units reached 100% correct (Fig. 11.1b), indicating that spike count distributions for the masked and unmasked conditions were completely separated and that, at those probe levels, the presence of the probe was signaled unambiguously. Such neurometric functions tended to differ from their RLFs, as spike count could continue to grow at higher probe levels. Figure 11.1c–f shows examples of neurometric functions from four additional units. Some units (5/48) also showed a change in the slope of the neurometric function (Fig. 11.1d) with masker level, while others (7/48) showed non-monotonic neurometric functions (Fig. 11.1e).

Most units did not reach 100% correct responses, even when the neurometric function saturated (Fig. 11.1c). Failing to reach this maximum indicates that the spike count distributions to the probe and no-probe conditions overlap each other partially even at high probe levels. Furthermore, for most of the units, there were conditions at which the neurometric function did not reach the 75% criterion necessary to estimate a detection threshold. This occurred especially at high masker levels and is illustrated in several of the panels in Fig. 11.1, for example in panel f for the

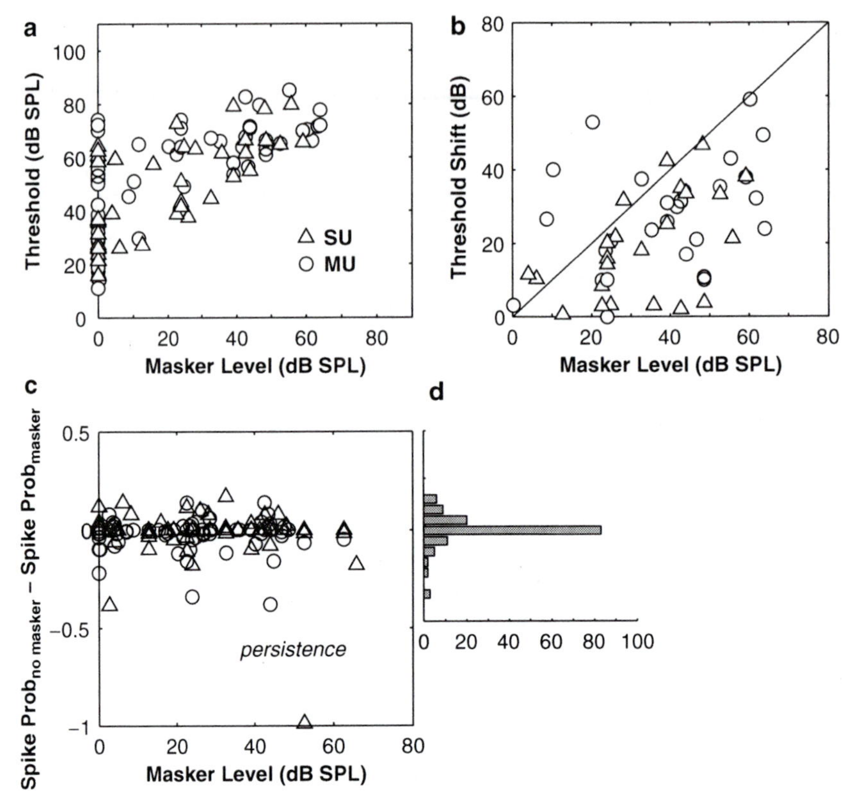

Fig. 11.2 (**a**) Probe detection threshold as a function of masker level. Thresholds correspond to the probe level at 75% correct. (**b**) Threshold shift as a function of masker level. *Squares* illustrate results obtained with multi-units and *triangles* with single units. (**c**) Difference between the mean spike count when no masker was presented and after the presentation of the masker. The probe was maximally attenuated in order to evaluate only the effect of the masker on the spike count. The spike count was computed in the time window [102,147]-ms, that includes the time window of presentation of the probe plus an interval of 20 ms that takes into account the delay of the response onset relative to the stimulus onset. The spike count difference was calculated individually for different units and masker levels and plotted as a function of masker level. (**d**) Histogram of the distribution of the spike count differences plotted in panel **c**

masker levels of 56 and 86 dB SPL. Finally, units which had high unmasked thresholds (6/48) typically showed very little threshold shift (Fig. 11.1f).

Probe detection thresholds were derived as the probe tone level at which the neurometric function crossed a 75% criterion. Figure 11.2a illustrates probe detection thresholds across the population as a function of masker level. These thresholds were derived from units with a wide range of CFs. At each masker level, thresholds varied across units. Thresholds as high as 70 dB were found for a masker level of about 68 dB SPL. At very high masker levels, there was complete masking for most of the units, that is, it was not possible to estimate thresholds following the 75% criterion for threshold used (see above).

If we were to assume that performance is determined by the most sensitive neurons for a given stimulus condition, then the line joining the minimum probe thresholds would represent the psychophysical growth of masking predicted by this population of neurons. This yielded a growth of masking slope of approximately 0.87 dB/dB. The largest slope obtained for individual units was 1.2 dB/dB, and the average slope was 0.6 dB/dB. No differences were apparent in threshold and threshold shift for single and multi-units.

The increase in probe threshold relative to the unmasked threshold in the same unit is plotted in Fig. 11.2b as a function of masker level. The magnitude of the shifts varied across units, with maximum changes increasing with masker level. Though not illustrated in the figure, the largest shifts occurred for the most sensitive units (i.e., with lower unmasked thresholds).

The increase in probe threshold reflects changes in the activity of cortical neurons when the probe is preceded by the masker. Psychophysical theories and models have suggested that masking might be due to persistent neural activity that outlasts the duration of the stimulus and "swamps" the response to the probe (Moore 1988; Plomp 1964; Zwicker 1984). To evaluate this possibility in our data, the mean spike count in the absence of the probe was calculated in a given time window after the presentation of the masker, and it was compared to the spike count in the same time window when the masker was absent (spontaneous activity). The difference in spike count was calculated separately for each unit and for each probe and masker level and is illustrated in Fig. 11.2c, as a function of masker level. The distribution of the spike count differences is plotted in Fig. 11.2d. In 30% of the cases the presence of the masker produced an increase in the mean spike count relative to the spontaneous activity. This effect did not increase with masker level.

11.4 Discussion

The largest thresholds observed for AI neurons (Fig. 11.2b) were about 70 dB, for a masker level of about 68 dB SPL. Considering the minimum thresholds obtained for an 0-dB SPL masker across the population, this corresponds to a threshold shift of about 55 dB. This change is much larger than the maximum 20 dB shift estimated from the responses of individual AN fibers for an 80-dB increase in masker level (Relkin and Turner 1988). It is also larger than the shift estimated from the N_2 component of the CAP (Relkin and Smith 1991), larger than the shifts observed for onset units in the cochlear nucleus (Ingham et al. 2006), and larger than the shifts observed for some units in the IC (e.g., Fig. 5 in Nelson et al. 2009). Growth of masking functions in the cortex also have higher slopes than in the AN (0.62 dB/dB; e.g., Fig. 3 in Relkin and Smith 1991) and than the average 0.5 dB/dB observed in the IC (Nelson et al. 2009). However, in the earlier studies in the AN and cochlear nucleus, the shortest masker–probe intervals were 2 (Relkin and Smith 1991) and 2.8 ms (Ingham et al. 2006). In the default condition in the study of Nelson et al. (2009), a 10-ms gap was used. In the experiments described here, there was no silent interval between the masker and the probe. This difference in the gap might have affected the difference in magnitude of the threshold shifts observed, especially when compared with the results in the IC.

The mean number of spikes evoked by the probe was generally small, often less than one, even in the unmasked conditions. The activity of cortical neurons is therefore better expressed in terms of firing probability (Calford and Semple 1995), rather than firing rate. The small number of spikes produced by AI neurons may have contributed to the amount of masking observed in AI, in addition to forward suppressive processes. This sparsity in the activity of cortical neurons is likely to have been augmented by the anesthetic. Anesthetics are known to depress the activity of cortical neurons (Gaese and Ostwald 2001; Syka et al. 2005; Zurita et al. 1994). In particular, urethane leads to a decrease in spontaneous activity (Capsius and Leppelsack 1996) and in the stimulus-related response amplitude (Albrecht and Davidowa 1989; Capsius and Leppelsack 1996), amongst other effects.

The present data provides little evidence for a relevant contribution of persistence of the response to the masker to the effect of masking (Fig. 11.2d). Furthermore, the observation of non-monotonic rate-level and neurometric functions (Fig. 11.1d) suggests, in agreement with previous studies, that inhibition processes are also likely to contribute to masking, at least at this level of the auditory pathway. These kind of responses have also been reported previously (e.g., Calford and Semple 1995). A non-monotonic variation in the response level for a stimulus within the excitatory region of the neuron may reflect the existence of inhibitory inputs that overtake the excitatory inputs above a certain stimulus level, dominating the response of the unit. Taking into account that onset units of the cochlear nucleus are inhibitory in nature and more likely contribute indirectly to forward masking (Ingham et al. 2006), it is possible that inhibitory mechanisms involved in physiological masking take place beyond the cochlear nucleus, maybe at the IC or at the level of the cortex. This follows what has been suggested previously (e.g., Calford and Semple 1995), that masking inhibition is generated at high levels in the auditory pathway.

The stimuli used here were chosen to facilitate comparison with previous physiological studies (Relkin and Turner 1988; Ingham et al. 2006), rather than to do a direct comparison with human psychophysical data. Nonetheless, the present results seem to show larger amount of masking in AI than has been observed psychophysically. For example, probe detection thresholds of about 65 dB SPL were measured in the present study for masker levels of approximately 60 dB SPL (Fig. 11.2a). This is larger than the 40-dB SPL measured by Plack and Oxenham (1998) for the same masker level, though in that study a longer probe (30 ms) and gap (2 ms) were used. The approximate 50 dB increase in threshold measured with the present data, at this masker level, is also much larger than the 10 dB obtained by Turner et al. (1994) for stimuli with similar durations and a gap of 2 ms, but using a narrowband noise as a masker. The growth of masking slopes obtained from thresholds determined by the most sensitive neurons in the present study is nevertheless, within the range of slopes measured psychophysically (Plack and Oxenham 1998).

Psychophysical thresholds depend on several stimulus features. One is stimulus onset and offset ramps (Turner et al. 1994), with longer ramps producing larger thresholds. Masker–probe interval is another of the influencing factors, with steeper functions occurring at shorter intervals (Jesteadt et al. 1982). Species may also play a role. Forward masking in animal models is similar to that seen in humans, but less extensively investigated (Halpern and Dallos 1986; Salvi et al. 1982). The comparison is further limited here by the use of anesthesia in our experiments. These factors mean that any quantitative comparisons of our data with psychophysical threshold shifts must be made with caution and is not conclusive. Nevertheless, it is clear that the growth of forward masking in the auditory cortex is considerably stronger than that found in more peripheral nuclei.

Acknowledgments We would like to thank Chris Scholes, who helped in some of the data collection for this work. Work supported by the Medical Research Council, UK.

References

Albrecht D, Davidowa H (1989) Action of urethane on dorsal lateral geniculate neurons. Brain Res Bull 22:923–927

Bleeck S, Sayles M, Ingham NJ, Winter IM (2006) The time course of recovery from suppression and facilitation from single units in the mammalian cochlear nucleus. Hear Res 212:176–184

Brosch M, Schreiner CE (1997) Time course of forward masking tuning curves in cat primary auditory cortex. J Neurophysiol 77:923–943

Calford MB, Semple MN (1995) Monaural inhibition in cat auditory cortex. J Neurophysiol 73:1876–1891

Capsius B, Leppelsack HJ (1996) Influence of urethane anesthesia on neural processing in the auditory cortex analogue of a songbird. Hear Res 96:59–70

Chatterjee M (1999) Temporal mechanisms underlying recovery from forward masking in multi-electrode-implant listeners. J Acoust Soc Am 105:1853–1863

Chatterjee M, Galvin JJ 3rd, Fu QJ, Shannon RV (2006) Effects of stimulation mode, level and location on forward-masked excitation patterns in cochlear implant patients. J Assoc Res Otolaryngol 7:15–25

Gaese BH, Ostwald J (2001) Anesthesia changes frequency tuning of neurons in the rat primary auditory cortex. J Neurophysiol 86:1062–1066

Halpern DL, Dallos P (1986) Auditory filter shapes in the chinchilla. J Acoust Soc Am 80:765–775

Harris DM, Dallos P (1979) Forward masking of auditory nerve fiber responses. J Neurophysiol 42:1083–1107

Ingham NJ, Bleeck S, Winter IM (2006) The magnitude of forward masking and the time course of its recovery as a function of unit type in the ventral cochlear nucleus. In: 29th Annual Midwinter Research Meeting of the Association for Research in Otolaryngology, Baltimore, USA, p 36

Jesteadt W, Bacon SP, Lehman JR (1982) Forward masking as a function of frequency, masker level, and signal delay. J Acoust Soc Am 71:950–962

Moore BC, Glasberg BR, Plack CJ, Biswas AK (1988) The shape of the ear's temporal window. J Acoust Soc Am 83:1102–1116

Nelson PC, Zachary MS, Young ED (2009) Wide-dynamic-range forward masking suppression in marmoset inferiour colliculus neurons is generated centrally and accounts for perceptual masking. J Neurosci 29:2553–2562

Oxenham AJ (2001) Forward masking: adaptation or integration? J Acoust Soc Am 109:732–741
Plack CJ, Oxenham AJ (1998) Basilar-membrane nonlinearity and the growth of forward masking. J Acoust Soc Am 103:1598–1608
Plomp R (1964) Rate of decay of auditory sensation. J Acoust Sco Am 36:277–282
Relkin EM, Smith RL (1991) Forward masking of the compound action potential: thresholds for the detection of the N1 peak. Hear Res 53:131–140
Relkin EM, Turner CW (1988) A reexamination of forward masking in the auditory nerve. J Acoust Soc Am 84:584–591
Salvi RJ, Ahroon WA, Perry JW, Gunnarson AD, Henderson D (1982) Comparison of psychophysical and evoked-potential tuning curves in the chinchilla. Am J Otolaryngol 3:408–416
Shannon RV (1990) Forward masking in patients with cochlear implants. J Acoust Soc Am 88:741–744
Shore SE (1995) Recovery of forward-masked responses in ventral cochlear nucleus neurons. Hear Res 82:31–43
Smith RL (1979) Adaptation, saturation, and physiological masking in single auditory-nerve fibers. J Acoust Soc Am 65:166–178
Syka J, Suta D, Popelar J (2005) Responses to species-specific vocalizations in the auditory cortex of awake and anesthetized guinea pigs. Hear Res 206:177–184
Turner CW, Relkin EM, Doucet J (1994) Psychophysical and physiological forward masking studies: probe duration and rise-time effects. J Acoust Soc Am 96:795–800
Zurita P, Villa AE, de Ribaupierre Y, de Ribaupierre F, Rouiller EM (1994) Changes of single unit activity in the cat's auditory thalamus and cortex associated to different anesthetic conditions. Neurosci Res 19:303–316
Zwicker E (1984) Dependence of post-masking on masker duration and its ralation to temporal effects in loudness. J Acoust Sco Am 75:219–223

Chapter 12
Effect of Presence of Cue Tone on Tuning of Auditory Filter Derived from Simultaneous Masking

Shunsuke Kidani and Masashi Unoki

Abstract We investigated whether the presence of cue tone affected the psychophysical frequency selectivity. We measured masked thresholds with/without the presentation of cue tone in notched-noise in a simultaneous masking experiment or band-pass noise in a simultaneous masking experiment, and then estimated the shapes of auditory filters from these data. Filter-Qs were calculated from the derived filters to compare the tuning of the filters with/without the cue tone. As a result, we found that the tip of the derived filter was sharpened by the presence of cue tone as signal levels were low and the range of improvements was limited to around the signal frequency. This suggested that the presence of cue tone may affect improvements in the psychophysical frequency selectivity.

Keywords Frequency selectivity • Auditory filter • Simultaneous masking • Tuning • Cue tone

12.1 Introduction

The psychophysical frequency selectivity has been investigated in studies on estimates of tuning of auditory filters. The tuning of auditory filters has been estimated from masked thresholds obtained from both simultaneous and forward masking experiments using notched-noise maskers (Patterson 1976) under various conditions. We know that the auditory-filter shapes are varied by signal frequencies, signal levels, and temporal placements of probe and masker (e.g., Moore and Glasberg 1981; Oxenham and Shera 2003; Unoki et al. 2007). It is also known that the tuning of auditory filter can be improved by eliminating the effects of suppression, especially when the signal levels are low (Heinz et al. 2002).

S. Kidani (✉)
School of Information Science, 1-1 Asahidai, Nomi, Ishikawa 923-1292, Japan
e-mail: kidani@jaist.ac.jp

E.A. Lopez-Poveda et al. (eds.), *The Neurophysiological Bases of Auditory Perception*,
DOI 10.1007/978-1-4419-5686-6_12,

However, it has been reported that hearing ability in signal detection can be improved by the presence of cue tone by using the probe-signal method (Ebata et al. 1968; Greenberg and Larkin 1968). They suggested that the auditory filter used in signal detection is focused on the cue tone, that is, auditory attention. However, the effects of off-frequency listening and the tuning of the auditory filter at low-signal levels were not considered in their studies.

The tuning of the auditory filter may be sharpened through the presence of cue tone. What effects the presence of cue tone has on the tuning of auditory filters have not been considered in signal detection, especially in the masking experiments. Our question is whether the presence of cue tone affects estimates of the tuning of the auditory filter. If the presence of cue tone affects the tuning of the auditory filter, this effect could be related to hearing ability in signal detection.

In this study, we attempt to investigate what effect the presence of cue tone has on the tuning of the auditory filter by using simultaneous notched-noise masking or band-pass noise masking (Fletcher 1940), similar to their effects on signal detection. We measure masked thresholds with/without the presentation of cue tone in symmetrical/asymmetrical notched-noise or band-pass noise in simultaneous masking experiments. Tuning of the auditory filter is estimated from the slope of the masked threshold function at different notch width.

12.2 Simultaneous Masking with Notched-Noise Masker

12.2.1 Methods

The probe frequency (f_c) was 1.0 kHz. The notched-noise masker consisted of two bands of white noise where each bandwidth was fixed at $0.4 \times f_c$. The values of the notch width ($\Delta f_c \,/\, f_c$) under seven symmetric conditions were 0.0, 0.05, 0.1, 0.15, 0.2, 0.3, and 0.4. The notched-noise under six asymmetric conditions was shifted $f_c \times 0.2$ from f_c to either lower or upper frequency, and the values of $\Delta f_c \,/\, f_c$ were 0.05, 0.1, and 0.2. The probe level (P_s) was fixed to 10 or 20 dB SL. At a fixed probe level, the masker levels (N_0) at the masked thresholds were measured in simultaneous masking. The cue tone was also fixed to the same probe frequency and probe level as P_s. The masker, probe, and cue tone were presented at 300 ms (15 ms raised-cosine ramps and a 270 ms steady state).

The layout for the time conditions used in our masking experiments is shown in Fig. 12.1. Figure 12.1a has the conditions without the cue tone, and Fig. 12.1b has the conditions with the cue tone. The masked thresholds with/without presentation of the cue tone in symmetrical/asymmetrical notched-noise were measured using a three-alternative forced-choice (3AFC) three-down one-up procedure that tracked the 79.4% point on the psychometric function. Three intervals of stimuli were presented sequentially using a 1,300 ms inter-stimulus interval in each trial. When the cue tone was presented, the interval for 500 ms was set and presented before each noise. The cue tone was spaced from the

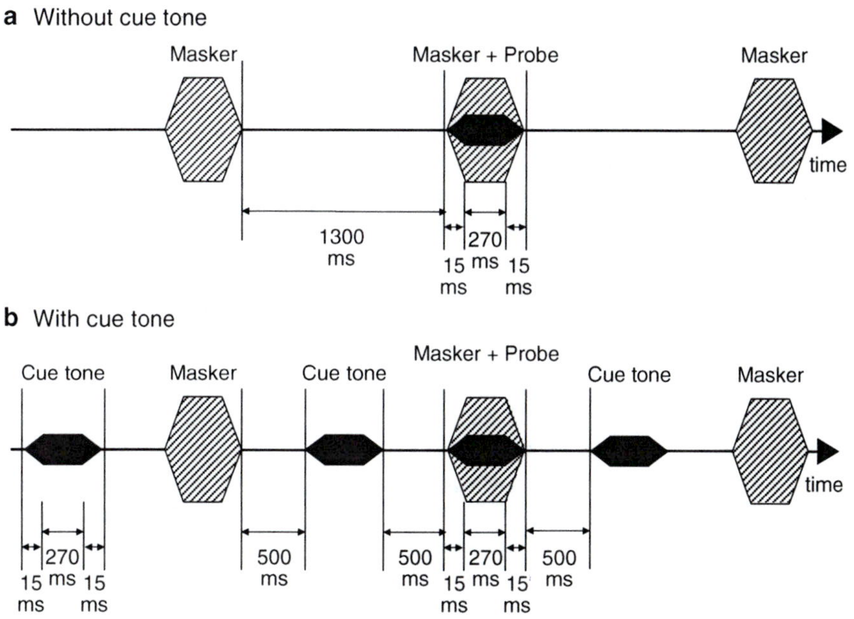

a Without cue tone

b With cue tone

Fig. 12.1 Position of stimuli used in (**a**) simultaneous masking without presentation of cue tone and (**b**) with presentation of cue tone

masker, so that excitation by the presence of the cue tone did not affect the masking experiment.

All stimuli were regenerated digitally at a sampling frequency of 48 kHz and presented via a Tucker-Davis Technologies (TDT) system III. The stimuli were presented monaurally to the subjects in a double-walled sound-attenuating booth via an Etymotic Research ER2 insert earphone. The levels of the stimuli were calibrated using an Artificial Ear Simulator (B&K 4152) with a 2-cm^3 coupler (B&K DB 0138) and a Modular Precision Sound Level Meter (B&K 2231).

Four normal-hearing listeners, aged 23–25, participated in the experiments. The absolute thresholds for all listeners, measured through a standard audiometric tone test using a RION AA-72B audiometer, were 15 dB HL or less for both ears at octave frequencies between 0.125 and 8.0 kHz. All listeners were given at least 2 h of practice.

The listeners were required to identify the intervals that carried the probe signals using numbered push buttons on a response box. Feedback was provided by LEDs that lighted up corresponding to the correct interval on the response box after each trial. A run was terminated after 12 reversals. The step size was 5 dB for the first four reversals and 2 dB after that. The threshold was defined as the mean probe level at the last eight reversals.

12.2.2 Results

The masking thresholds at the masker level were measured for four subjects and are shown in Fig. 12.2. The circles denote the masked thresholds under symmetric notched-noise conditions. The triangles pointing to the right denote the masked thresholds under the asymmetric conditions to the upper notched-noise condition. The triangles pointing to the left denote the masked thresholds under the asymmetric conditions to the lower notched-noise conditions. The open circles and triangles pointing show the case in which cue tone were not presented. The closed circles and triangles indicate cases in which cue tone were presented.

When cue tone was not presented, we found that the masked thresholds increased as the notch width was increased. We also found that the asymmetric notched-noise conditions for the lower noise band were higher than the asymmetric notched-noise conditions for the higher noise band. This suggests that the auditory-filter shapes were asymmetric, with a steeper high frequency slope. These results are the same as those from traditional research, i.e., those from simultaneous masking experiments (e.g., Unoki et al. 2007).

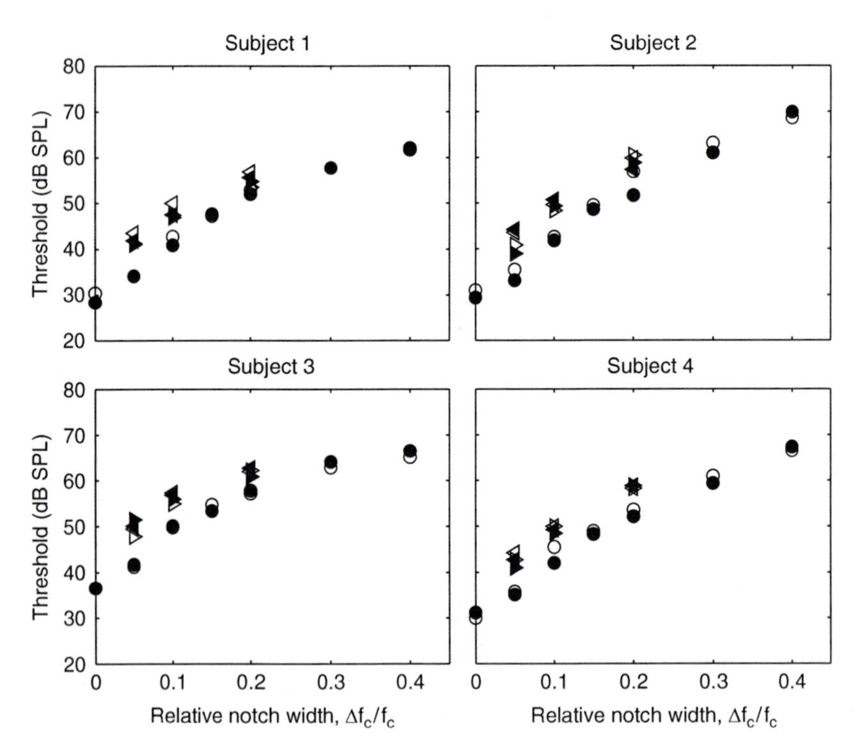

Fig. 12.2 Masked thresholds at masker level measured from notched-noise masking experiments with/without presenting cue tone when the probe level was 10 dB SL

When cue tone was presented, we found also the same tendency as in the results of the simultaneous notched-noise masking experiments. In this case, we found that the masked thresholds decreased as the notch width narrowed. This suggested that the presence of cue tone affected the slope of the masked threshold function at different notch width.

12.3 Estimates of Tuning of Auditory Filter Derived from Notched-Noise Masking Data

We did not use the PolyFit procedure, and attempted to estimate the auditory-filter shapes from the slope of the masked threshold function at different notch width as Patterson and Moore (1986) had done. When the weighting function applied by the auditory filter is *W(f)*, the power-spectrum model of masking is

$$P_s = KN_0 \int_0^{(f_c - \Delta f)} W(f) df \quad (12.1)$$

where K is efficiency, f_c is the probe frequency, and Δf is the separation between f_c and the noise edge, i.e., $fc - \Delta f$ is the notch width. If we differentiate both sides of (12.1) for notch width, the auditory-filter shape can be estimated from the slope of the masked threshold at different notch width.

$$W(f_c - \Delta f) = P_s K^{-1} \left[\frac{\mathrm{d}(1/N_0)}{\mathrm{d}(f_c - \Delta f)} \right] \quad (12.2)$$

The auditory-filter shapes that were estimated by using (12.2) are shown in Fig. 12.3. The open circles denote that no cue tone was presented, and the asterisks denote that cue tone was presented. The solid line plots the probe levels at 20 dB SL, and the dashed line plots the probe levels at 10 dB SL. We can see that the derived auditory-filter shapes were sharpened by the presence of cue tone. Moreover, we found the tuning of the auditory filter when the probe level was 10 dB SL sharpened in response to the presence of cue tone that affected the auditory-filter shapes compared with when the probe level was 20 dB SL. Therefore, we calculated filter-Qs, indicating the sharpness of a filter, from the derived filter to compare the tuning of the derived filter with/without the presentation of cue tone. Three filter-Qs ($Q_{3\,\mathrm{dB}}$, $Q_{10\,\mathrm{dB}}$, and $QERB_N$) were measured to investigate the derived auditory filters near their tip, middle, and throughout the entire filters. The measured filter-Qs when the probe level was 10 dB SL are listed in Table 12.1. Filter-Qs were increased when the tuning of the auditory filters sharpened.

When the probe level was 10 dB SL, the filter-Qs increased because of the presence of cue tone except for Subject 4. This suggested that the tuning of the auditory filter was affected by the presence of cue tone. This tendency was also more prevalent in

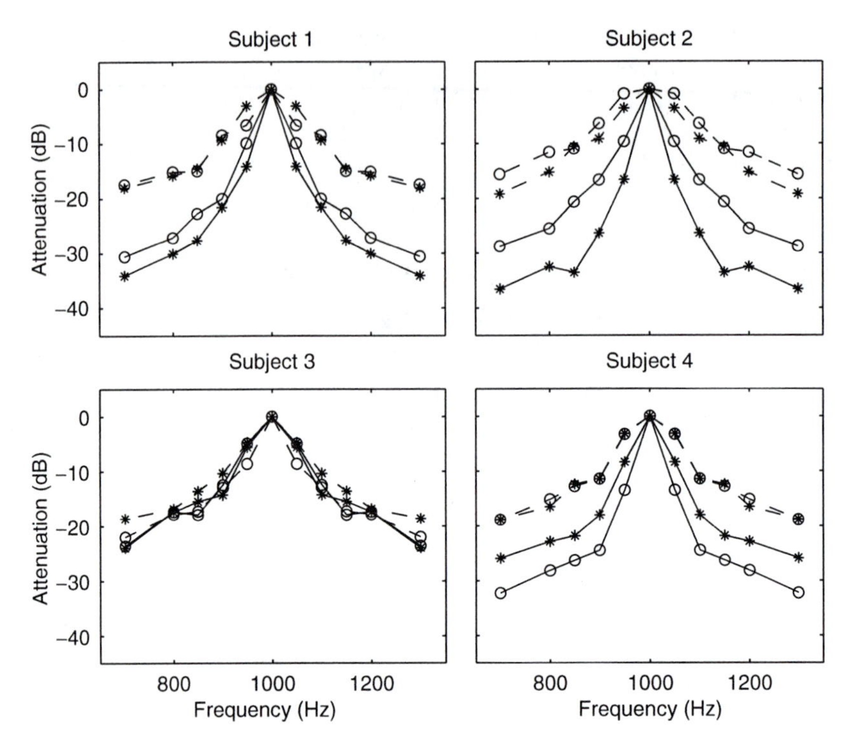

Fig. 12.3 Auditory-filter shapes derived from masked threshold data measured in notched-noise masking experiments with/without presentation of cue tone

Table 12.1 Filter-Qs were estimated from the measured notched-noise masking data

	Subject 1	Subject 2	Subject 3	Subject 4
$Q_{3\,dB}$	45.00/30.00	54.00/33.75	16.87/15.00	27.00/45.00
$Q_{10\,dB}$	14.21/10.00	16.87/9.64	6.43/5.87	8.44/13.50
$QERB_N$	5.49/5.59	6.34/4.91	4.09/4.16	5.45/5.90

The values listed in the table indicate the presentation/nonpresentation of cue tone

$Q_{3\,dB}$ than in $Q_{10\,dB}$ or $QERB_N$. This suggested that the presence of the cue tone increased the tuning of the auditory filter around the signal frequency.

No differences in predominance due to the presence of cue tone appeared under asymmetrical notched-noise conditions in the slope of the masked threshold, or filter-Qs. The presence of cue tone did not affect asymmetrical notched-noise conditions. This means that the range for the presence of cue tone to affect the tuning of the auditory filter was less than $\Delta fc/fc = 0.20$.

The effects of the presence of cue tone were increased where there was a lower signal level and around the signal frequency. Therefore, we attempted to investigate the range for the effect of the presence of cue tone by using band-pass noise.

12.4 Simultaneous Masking with Band-Pass Noise Masker

12.4.1 Method

We experimented by changing the notched-noise in the previous experiment into band-pass noise. The band-pass noise was symmetrically placed at f_c under ten conditions. The bandwidths of the band-pass noise were 12, 16, 20, 40, 60, 100, 150, 200, 250, and 300 Hz. The probe level (P_s) was only 10 dB SL; this is because the results of the previous experiment revealed that the presence of cue tone had an increased effect when the signal levels were low.

Four normal-hearing listeners who were different from the subjects in the previous experiment, aged 22–24, participated in the experiments. The absolute thresholds for all listeners, measured by using the same method as in the previous experiment, were 15 dB HL or less for both ears at octave frequencies between 0.125 and 8.0 kHz. All subjects were given at least 2 h of practice.

The time conditions with/without the presentation of cue tone and the experimental procedure were the same as in the previous experiment as shown in Fig. 12.1.

12.4.2 Results

The masking thresholds at masker levels measured using the four subjects are shown in Fig. 12.4. The open circles plot cases when cue tone was not presented, and the closed circles plot cases when cue tone was presented.

Irrespective of the presence of cue tone, we found that the masked thresholds decreased as bandwidth increased. This can be explained by the increased amount of noise that passed through inside the auditory filter. Moreover, when the bandwidth of the masker was wider than 150 Hz, the masked thresholds almost became fixed values. This is because the bandwidth of the masker reached the critical band.

The difference in the masked thresholds due to the presence of cue tone mostly disappeared from the critical or wider bands. When the tuning of the auditory filter increased because of the presence of cue tone, the increase in the masked thresholds could be explained based on the power-spectrum model of masking. Therefore, the tuning of the auditory filter was sharpened by the presence of cue tone in the critical band. We also attempted to estimate the auditory filter in this experiment to investigate what effects the presence of cue tone had on the tuning of the auditory filter.

12.5 Estimates of Tuning of Auditory Filter Derived from Band-Pass Noise Masking Data

The shapes of the auditory filters in this experiment are expressed as a function of the bandwidth of noise. The shapes of the auditory filters that were derived from the masked thresholds were estimated and measured in a symmetrical band-pass

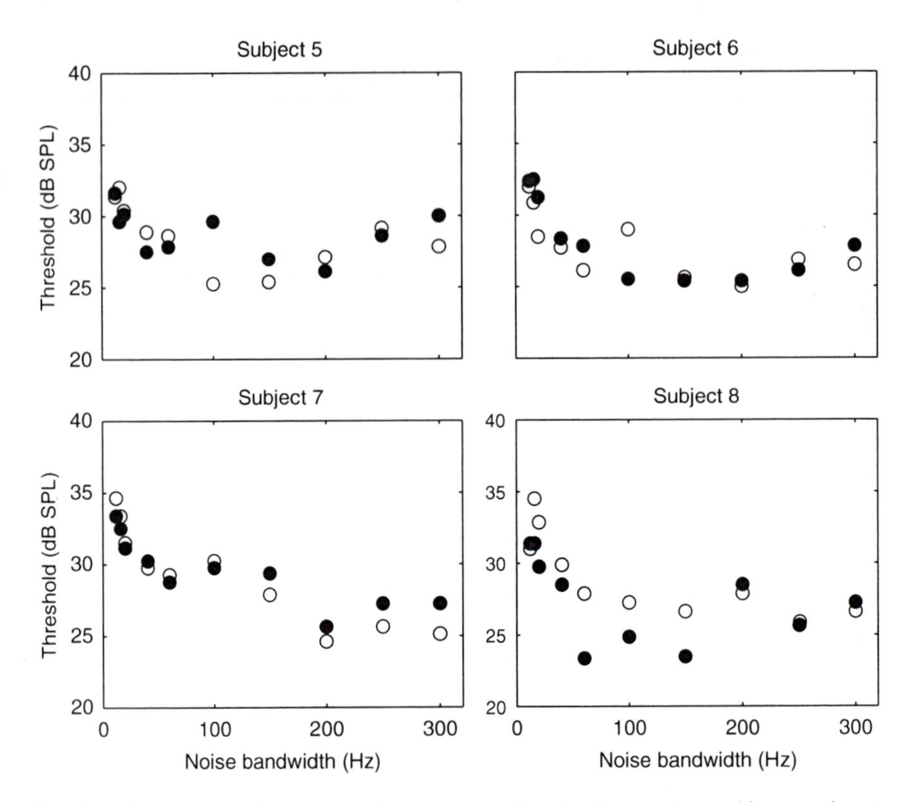

Fig. 12.4 Masked thresholds at masker level measured from band-pass noise masking experiments with/without presentation of cue tone

noise experiment. The filter-Qs were calculated from the derived filter to evaluate the tuning of the auditory filter. $Q_{3\ dB}$ was only calculated to investigate the tip of the derived filter. The measured filter-Qs are listed in Table 12.2. The filter-Qs were increased by two for the four subjects.

The tuning of the auditory filter estimated from the masked thresholds was measured under conditions where the narrow bandwidth of noise became sharp. This suggested that tuning of the auditory filter was none the less increased by the presence of cue tone around the signal frequency.

12.6 Summary

We considered what effect the presence of cue tone had on the tuning of auditory filters. Three main findings resulted from this study:

- The tuning of auditory filter was affected by the presence of cue tone.
- These effects increased when the signal levels were low.
- These effects improved around the signal frequency.

Table 12.2 Filter-Qs estimated from measured band-pass noise masking data

	Subject 5	Subject 6	Subject 7	Subject 8
$Q_{3\ dB}$	23.44/20.69	24.40/29.70	22.06/23.15	35.82/20.98

The values listed in the table indicate the presentation/nonpresentation of cue tone

These suggested that the presence of cue tone may improve psychophysical frequency selectivity, and the range of improvements is limited to around the signal frequency.

12.7 Comment by Andrew Oxenham

Cue tones have primarily been used in the past to cue the expected frequency of signal. For such a cue to be effective, the frequency of the following signal should not be known in the absence of the cue. However, it seems that in this experiment, the signal frequency remained the same throughout each run and was therefore expected by the subject with or without the cue tone. This may explain why the cue tone had only a small and unreliable effect on overall thresholds and filter shapes.

12.8 Reply Shunsuke Kidani

Thank you for your suggestions. Our purpose in these experiments was to find whether the tuning of the auditory filter that derived from the masked thresholds was affected by the presence of the cue tone or not. Therefore, we measured the masked threshold using the same method as that in the previous notched-noise masking experiments to derive the auditory-filter shape (e.g., Unoki et al. 2007), in which the cue tone had the same signal frequency and levels with the signal. We then used the effects on frequency selectivity by estimating the tuning of the auditory filter derived from masked thresholds. This is why we used notched-noise masking in the 3AFC three-down one-up adaptive procedure with/without cue tone. However, as you mentioned, we agree that the frequency of the following signal should not be known by subjects in the absence of the cue. We think that this problem can be resolved by conducting a constant method with 2AFC in which cue tone frequencies are not only the signal frequency but also other frequencies in a run. It might be that the different signal frequencies should also be used in the same run. Moreover, *d′* can be calculated using this method so that we can discuss the reliable effect of the presence of the cue tone for frequency selectivity. We intend to investigate this further in future studies.

Acknowledgments This work was supported by a Grant-in-Aid for Scientific Research from the Ministry of Education, Culture, Sports, Science, and Technology of Japan (No. 20300064).

References

Ebata M, Sone T, Nimura T (1968) Improvement of hearing ability by directional information. J Acoust Soc Am 43:289–297

Fletcher H (1940) Auditory patterns. Rev Mod Phys 12:47–65

Greenberg GZ, Larkin WD (1968) Frequency response characteristic of auditory observers detecting signals of a signal frequency in noise: the probe-signal method. J Acoust Soc Am 44:1513–1523

Heinz MG, Colburn HS, Carney LH (2002) Quantifying the implications of nonlinear cochlear tuning for auditory filter estimates. J Acoust Soc Am 111:996–1101

Moore BCJ, Glasberg BR (1981) Auditory filter shapes derived in simultaneous and forward masking. J Acoust Soc Am 70:1003–1014

Oxenham AJ, Shera CA (2003) Estimates of human cochlear tuning at low levels using forward and simultaneous masking. J Assoc Res Otolaryngol 4:541–554

Patterson RD (1976) Auditory filter shapes derived with noise stimuli. J Acoust Soc Am 59:640–654

Patterson RD, Moore BCJ (1986) Auditory filters and excitation patterns as representations of frequency resolution. In: Moore BCJ (ed) Frequency selectivity in hearing. Academic, London, pp 123–177

Unoki M, Miyauchi R, Tan C-T (2007) Estimates of tuning of auditory filter using simultaneous and forward notched-noise masking. In: Kollmeier B, Klump G, Hohmann V, Langemann U, Mauermann M, Uppenkamp S, Verhey J (eds) Hearing from sensory processing to perception. Heidelberg, Springer Verlag, pp 19–26

Part III
Spectral Processing and Coding

Part III
Spectral Processing and Coding

Chapter 13
Tone-in-Noise Detection: Observed Discrepancies in Spectral Integration

Nicolas Le Goff, Armin Kohlrausch, Jeroen Breebaart, and Steven van de Par

Abstract Spectral integration for several tone-in-noise conditions is discussed and an experiment is conducted for both monaural (NoSo) and binaural conditions (NoSπ). Monaural detection thresholds for running-noise maskers increased for masker bandwidths up to the critical bandwidth and remained constant for larger bandwidths. Binaural conditions and monaural conditions with a frozen-noise masker revealed different spectral integration patterns that are monotonic for masker bandwidth below and beyond the critical bandwidth, an effect known as the apparently wider binaural critical band. Finally, we show a different type of spectral integration obtained for binaural conditions with a reduced masker correlation (NρSπ) or for NoSπ with an overall interaural level difference. In these cases, the integration patterns are nonmonotonic with a maximum for masker bandwidths around the critical bandwidth.

Keywords Binaural masking • Spectral intregration • Masker correlation

13.1 Introduction

In the context of tone-in-noise detection, spectral integration refers to the dependency of the detection thresholds on the bandwidth of the noise masker. In order to simplify the description, we consider situations where the power spectral density (spectral level) of the masker is kept constant for all bandwidths.

Spectral integration was first formalized by the concept of the critical band proposed by Fletcher (1940). In this concept, thresholds for tones spectrally centered in a noise masker were increasing with increasing masker bandwidth

N. Le Goff (✉)
Technische Universiteit Eindhoven,
P.O. Box 513, 5600 MB Eindhoven, The Netherlands
e-mail: n.legoff@tue.nl

E.A. Lopez-Poveda et al. (eds.), *The Neurophysiological Bases of Auditory Perception*,
DOI 10.1007/978-1-4419-5686-6_13, © Springer Science+Business Media, LLC 2010

up to a specific (the critical) bandwidth and remained constant for larger masker bandwidths. In subcritical situations, thresholds were increasing as the total energy of the masker. In supracritical situations, thresholds remained constant because the filtering occurring on the basilar membrane removed all masker components outside the critical band. This view was later modified by Bos and de Boer (1966). They proposed a refinement for subcritical situations, where it was found that besides energy principles that lead to integration rates of about 3 dB/octave, the external masker variability could also be the limiting factor for running-noise maskers resulting in integration rates of about 1.5 dB/octave. This approach was then extended to account for a phenomenon that was primarily observed for binaural conditions where the auditory filter appeared to be wider than that measured in monaural conditions (Sever and Small 1979). In these conditions, a monotonic increase of the thresholds is observed for bandwidths below and beyond the critical band. The phenomenon was explained by taking into account the contribution of information contained in off-frequency channels for conditions, where the limiting factor for the detection process are the internal errors of the auditory system (van de Par and Kohlrausch 1999; Breebaart et al. 2001c).

In addition to those known spectral integration patterns, we will present experimental data that reveal a third pattern. In this type of spectral integration, thresholds are increasing with increasing masker bandwidths for subcritical situations and are decreasing for further increases of the masker bandwidths. This type of nonmonotonic spectral integration was observed in binaural conditions, where the binaural stimuli were presented with an overall interaural level difference (ILD) and in conditions with a reduced interaural correlation of the noise masker.

13.2 Experiment

Spectral integration was measured for several conditions of tone-in-noise detection. The signal had a frequency of 500 Hz, and the noise masker was centered on this frequency. It had a bandwidth between 10 and 1,000 Hz and was presented as either a running or frozen noise. The experiment included monaural and binaural conditions. In addition to these common conditions, we also included binaural conditions with an overall ILD of 30 dB and with a reduced interaural masker correlation.

13.2.1 Method and Stimuli

A three-interval forced-choice procedure with adaptive signal-level adjustment was used to determine the thresholds. The three intervals of 300-ms duration were separated by pauses of 200 ms. A 200-ms sinusoid was added to the temporal center of one of the masker intervals. Feedback was provided to the subjects after each trial. The signal level was adjusted according to a two-down one-up rule tracking the 70.7% correct response level (Levitt 1971). The initial step size for adjusting the level was 8 dB. After each second reversal of the level track, the step size was halved until a step size of 1 dB was reached. The run was then continued for another eight reversals. The median of the levels at these last eight reversals was calculated and used as a thresholds value. At least three threshold values were obtained and averaged for each parameter value and subject. Three subjects participated in this experiment, among them were two of the authors. All subjects had normal hearing.

The noise masker was, unless stated otherwise, presented diotically (No), and the signal (500 Hz) was either presented diotically (So) or out-of-phase between the two ears (Sπ). The masker bandwidth was either 10, 20, 50, 100, 200, 500, or 1,000 Hz. Bandwidths of 20, 50, 200, and 500 Hz were not measured for all conditions. For each masker bandwidth, the masker level was set to an overall sound pressure level of 65 dB. For running-noise conditions, the noise samples for each interval were obtained by randomly selecting 300-ms segments from a two-channel, 2,000-ms bandpass-noise buffer. The 2,000-ms noise buffer was created as a white Gaussian noise in the time domain that was filtered to the desired bandwidth in the frequency domain. For frozen-noise conditions, only one fixed 300-ms noise sample was used in all three intervals of an entire run. These bandlimited noise samples were generated in the same way as the noise buffers for random noise conditions followed by a normalization of their rms value. To exclude the possibility that the frozen-noise thresholds would depend on the specific waveform token, a different frozen-noise sample was used for each run. Not all conditions were measured with frozen-noise maskers. The partially interaurally correlated ($\rho = 0.93$) noise maskers were generated by combining two independent noise samples. In order to avoid spectral splatter, the signals and the maskers were gated with Hanning windows that had 50-ms onset and offset ramps.

13.3 Results

The graphical pattern of spectral integration or its absence depends on the experimental conditions and how the detection thresholds are represented. The present experiment was conducted with noise maskers that had a constant overall power regardless of their bandwidth. Consequently, a variation of the masker bandwidth is in fact a redistribution of the energy of the noise masker in the frequency domain, which will therefore lead to variation in the spectral density of the noise. In order to solely study the effect of spectral expansion of the masker and not its level variation, thresholds are represented in terms of signal to noise spectral density ratios, which will give the same integration pattern as a representation of the thresholds in dB SPL for an experiment conducted with a constant spectral level of the noise masker.

Average thresholds for three subjects are shown in Fig. 13.1 as a function of the masker bandwidth. Running-noise conditions are represented by the open symbols

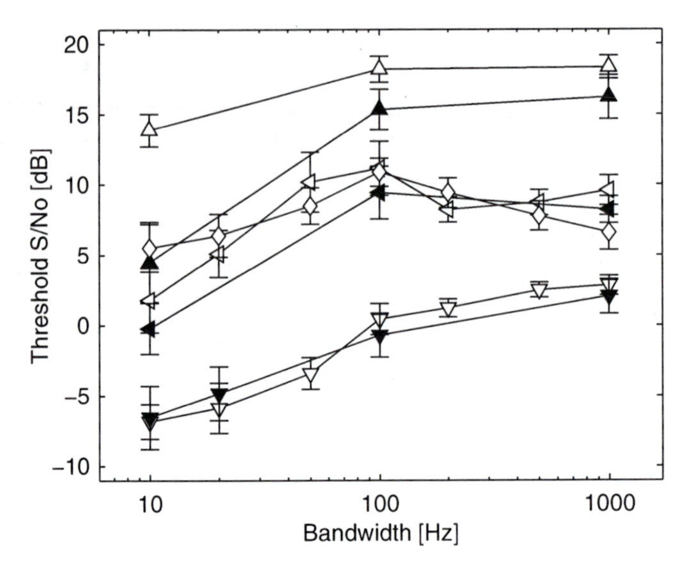

Fig. 13.1 Monaural and binaural masked thresholds expressed as signal to spectral density ratios are shown as a function of the masker bandwidth. *Filled symbols* represent thresholds obtained for frozen-noise maskers and *open symbols* represent thresholds obtained with running-noise maskers. *Upward triangles* represent NoSo thresholds. *Downward triangles* represent NoSπ thresholds. *Left pointing triangles* represent thresholds obtained for NoSπ conditions with an ILD of 30 dB. *Diamonds* represent NρSπ thresholds with an interaural masker correlation of 0.93

and frozen-noise conditions by the filled symbols. The error bars denote the standard deviation across the complete data set for each condition. For masker bandwidths smaller than 1 ERB, about 78 Hz at 500 Hz (Glasberg and Moore 1990), thresholds in all conditions are increasing with increasing masker bandwidth. On the contrary, for masker bandwidths wider than 1 ERB, three different behaviors are observed; detection at constant signal to noise spectral density ratio, thresholds still increasing with increasing masker bandwidths and an uncommon behavior of thresholds decreasing with increasing masker bandwidths.

Regarding the monaural conditions (upward triangles), we observe a threshold change between masker bandwidths of 10 and 100 Hz of 1.3 dB/octave and 3.2 dB/octave for the running-noise (open symbols) and the frozen-noise masker (filled symbols) respectively. In the case of the frozen-noise masker, the spectral integration corresponds to the variation of energy of the noise masker within the auditory filter, where a doubling of the bandwidth results in a 3 dB increase of the detection thresholds for bandwidths up to 1 ERB. In the case of the running-noise masker, the integration rate is close to 1.5 dB/octave which fits with the assumption that detection is limited in this case by the variability of the noise masker on a per sample basis (Bos and de Boer 1966). For masker bandwidths wider than 1 ERB, we observe no further spectral integration, which is in line with the energy masking principles and the concept of the auditory filter. Some minor spectral integration can arguably be seen for the frozen-noise masker thresholds. This phenomenon has been previously reported (Breebaart et al. 2001c) and related to the apparently wider auditory filter known from binaural conditions (van de Par and Kohlrausch 1999).

The phenomenon of an apparently wider critical band is particularly visible for binaural conditions (NoSπ, downward triangles) where one can clearly see that for both running- and frozen-noise maskers, the increase in thresholds is monotonic below and beyond the auditory filter bandwidth. The thresholds for these binaural conditions are very similar for both types of noise masker. The spectral integration appears to be stronger for bandwidths smaller than 1 ERB, about 2.5 dB/octave, and weaker for bandwidths larger than 1 ERB and amounts to about 0.7 dB/octave. The subcritical value is in good agreement with the hypothesis that detection requires a constant change in the stimulus correlation. This happens at a constant signal-to-noise ratio and would ideally give a gain of 3 dB/octave. Such a behavior is expected if detection is not limited by external variability but internal noise in the auditory system. The monotonic increase of the thresholds beyond the critical bandwidth or in other words, the apparent wider critical bandwidth has been explained by assuming the contribution of off-frequency auditory filters to the detection for conditions in which detection is limited by internal errors (van de Par and Kohlrausch 1999; Breebaart et al. 2001c). When the masker is narrower than 1 ERB, the phenomenon is not particularly visible in the spectral integration pattern, but as the masker bandwidth slowly increases beyond 1 ERB, the information in the off-frequency auditory filters becomes gradually unusable (covered by the masker), resulting in minor spectral integration beyond 1 ERB. The phenomenon also occurs to some extent for monaural conditions with a frozen-noise masker but not with a running-noise masker. Effectively, as seen previously, in monaural conditions with running-noise, the detection process is limited by the external variability of the masker waveform, while in conditions with frozen-noise the process is, similarly to binaural conditions (NoSπ), limited by internal limitations.

Detection thresholds for binaural conditions with additional overall ILD (left-pointing triangles) or a reduced masker correlation (open diamonds) lie between those obtained for monaural conditions (NoSo) and the binaural condition (NoSπ). Likewise, they elicit spectral integration that also lies between patterns given by a detection process limited by external variability (increase of 1.5 dB/octave) and internal errors (increase of 3 dB/octave) for bandwidths smaller than 1 ERB. The thresholds obtained for the NoSπ condition with an overall ILD of 30 dB are clearly higher than those obtained for the plain NoSπ condition and are below the thresholds obtained for the monaural conditions. The thresholds are similar for both types of noise maskers. The integration rate is about 2.9 dB/octave. The spectral integration obtained for the NoSπ condition with a reduced masker correlation is about 1.7 dB/octave for masker bandwidths smaller than 1 ERB. In both cases, an extension of the masker bandwidth beyond 1 ERB leads to a decrease in thresholds. Conditions with an overall ILD show a small decrease of about 1.6 dB for an increase of the bandwidth from 100 to 1,000 Hz. For the same bandwidths, the decrease is about 4.3 dB for the thresholds obtained for the NoSΠ condition with a reduced masker correlation.

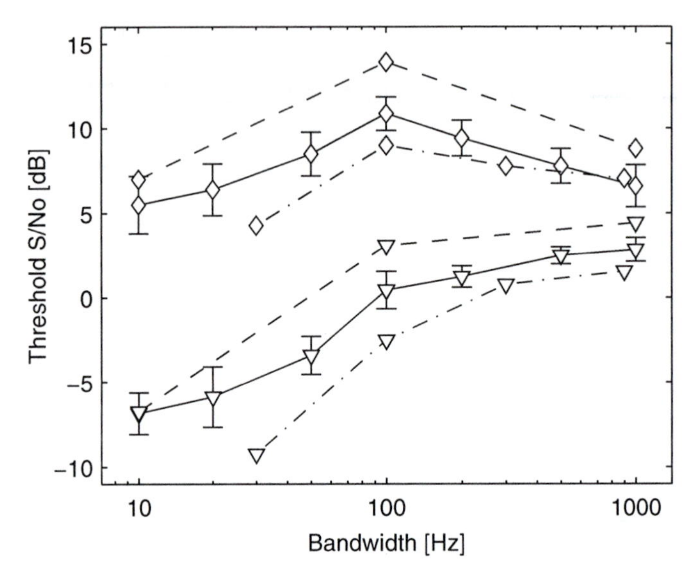

Fig. 13.2 Binaural masked thresholds expressed as signal to spectral density ratios are shown as a function of the masker bandwidth for running-noise maskers. Data are represented with the same convention of symbol types as in Fig. 13.1. The *continuous lines* show our data, the *dashed lines* represent data from Breebaart and Kohlrausch (2001); data adapted from van der Heijden and Trahiotis (1998) are shown as *dot-dashed lines*

A comparison of our data with other running-noise measurements adapted from the literature is shown in Fig. 13.2. Regarding the NoSπ conditions (downward triangles), our data (continuous lines), the data from Breebaart and Kohlrausch (2001) (dashed lines), and the data from van der Heijden and Trahiotis (1998) (dot-dashed lines) all show a monotonic increase in thresholds, including masker bandwidths beyond 1 ERB. Regarding the NoSπ conditions with a reduced masker correlation (all diamonds), our data and those from Breebaart and Kohlrausch (2001) are obtained for similar experimental conditions and the same masker correlation of 0.93, those from van der Heijden and Trahiotis (1998) are taken from a data set as the best fit to our own data set, which was found for a masker correlation of 0.87. For this condition, we also see a common behavior across the three data sets: thresholds are increasing with increasing masker bandwidth up to about 1 ERB and decrease for further extension of the bandwidth.

13.4 Discussion

While the various spectral integration patterns that we observed for the monaural (NoSo) and the binaural conditions (NoSπ) have been reported and modeled previously, that of the binaural condition with either an overall ILD or a reduced masker

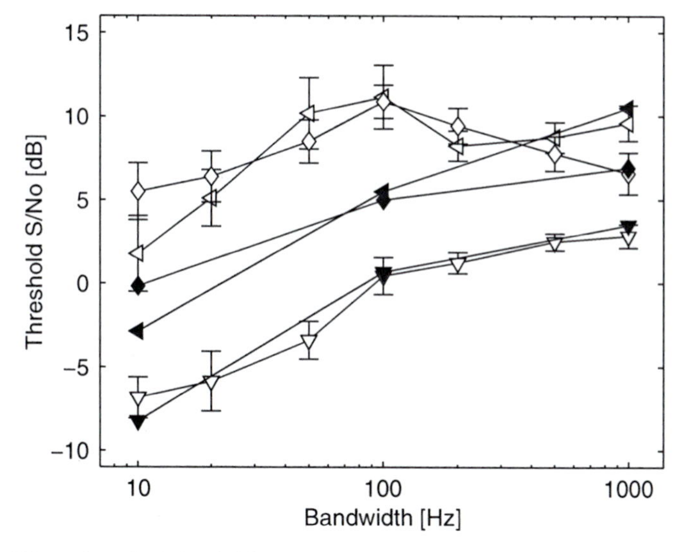

Fig. 13.3 Binaural masked thresholds expressed as signal to spectral noise density ratios are shown as a function of the masker bandwidth for running-noise maskers. Symbol types are used with the same convention as for Fig. 13.1. *Open symbols* represent experimental data, and *filled symbols* represent the equivalent simulated thresholds

correlation still needs to be fitted in a model approach. Figure 13.3 shows a comparison of the behavior of the listeners and the model proposed by Breebaart et al. (2001a). The open symbols represent the experimental data, and the filled symbols represent the equivalent simulated thresholds. One can see that the model accurately predicts the binaural thresholds (NoSπ), shown by the downward triangles. Especially, the wider apparent auditory filter is predicted due to the capacity of the model to integrate information from both on-frequency and off-frequency auditory filters. However, the model is not able to predict the unusual spectral integration that is observed for the binaural condition with overall ILD (left-pointing triangles) or reduced masker correlation (diamonds). The prediction for these conditions are however correct for the widest masker bandwidth. Effectively for a masker bandwidth of 1,000 Hz, the model prediction for both cases rises by an amount from the base NoSπ condition that is in line with the experimental data. In the model, this dependence results from an increase of model activity in the internal representation of the reference intervals that is due to the overall ILD or the masker decorrelation that prevent a total cancellation of activity in the binaural processor. A reduction of the masker bandwidth to 1 ERB and smaller values makes the model following the same behavior as for the base NoSπ condition (wider apparent critical band), and therefore predicts the same type of spectral integration. Consequently, it is unlikely that the nonmonotonic spectral integration that is observed for the binaural condition with either an overall ILD or a reduction in the masker correlation could be explained in terms of off-frequency listening or across-frequency integration.

This sort of nonmonotonic spectral integration has been reported for several other experimental conditions. To stay in the field of tone-in-noise detection, three studies on the monaural detection of a short, high-frequency tone in a noise masker reported the same type of nonmonotonic spectral integration. Figure 13.3 in Oxenham (1998) shows (a) an increase of detection thresholds of a 6-kHz tone for masker bandwidths increasing from 70 to 1,200 Hz and (b) a decrease of thresholds for wider masker bandwidths. Thresholds are reported for three signal durations (2, 20, and 300 ms, masker duration 500 ms), and the effect is stronger for the shortest durations. The author comments that "it is not clear what mechanism should underlie this result" (Oxenham 1998, p. 1,037). For similar experimental conditions (4-kHz signal, 10-ms duration), Bacon and Smith (1991) also reported that detection thresholds decrease by about 2.5 dB for masker bandwidths wider than 1 ERB. Likewise, Wright (1997) reported that detection thresholds for a 20-ms (noise) signal centered around 2,500 Hz decreased by about 2 dB as the masker bandwidth increased from 1,000 to 8,000 Hz.

Another case of nonunderstood spectral integration for supracritical situations was reported by Gabriel and Colburn (1981). They conducted, among others, an experiment to measure interaural correlation detection as a function of the stimulus bandwidth at a reference correlation of 1. One can see in their Fig. 4 that thresholds increase with increasing bandwidth beyond the critical band. They comment that this phenomenon is not consistent with the concept of optimal processing. This remark is in line with simulations reported in Breebaart et al. (2001b) of these conditions. The model has a central processor which is an optimal detector and, as can be seen in their Fig. 13.3, it predicts a monotonic improvement of the performance with increasing bandwidth but certainly not a decrease of performance. The fact that both these conditions by Gabriel and Colburn (1981) and our NoSπ conditions with a reduced masker correlation involve a major role of correlation in the discrimination process could suggest that the unexpected dependence of the thresholds beyond 1 ERB is somehow related to the influence of interaural decorrelation.

13.5 Conclusion

To conclude this overview on spectral integration, we observe that for subcritical situations, spectral integration for all conditions varies between what one would predict based on a pure energy integration (frozen-noise NoSo and running- and frozen-noise NoSπ, gain of 3 dB/octave) and what one would predict on a more statistical approach (running-noise NoSo, gain of 1.5 dB/octave). For supracritical conditions, we observe several behaviors for thresholds as a function of masker bandwidth: (a) constant thresholds (running-noise NoSo), (b) minor spectral integration, reflecting the effect of the apparent wider auditory filter (frozen- and running-noise NoSπ and frozen-noise NoSo), (c) the nonmonotonic case of spectral integration where thresholds decrease again for bandwidths beyond 1 ERB (NoSπ with overall ILD or reduced masker correlation). This third situation can not be predicted with a model based on optimal detection and can also not be explained in terms of off- frequency listening. One possibility could be, as it has been suggested previously in a different approach by van der Heijden and Trahiotis (1998), to replace the internal noise that is currently defined in the model by Breebaart et al. (2001a) as independent in each individual auditory channel by a bandwidth-dependent internal noise.

References

Bacon SP, Smith MA (1991) Spectral, intensive, and temporal factors influencing overshoot. Q J Exp Psychol 43A:373–400

Bos CE, de Boer E (1966) Masking and discrimination. J Acoust Soc Am 39:708–715

Breebaart J, Kohlrausch A (2001) The influence of interaural stimulus uncertainty on binaural signal detection. J Acoust Soc Am 109:331–345

Breebaart J, van de Par S, Kohlrausch A (2001c) An explanation for the apparently wider critical bandwidth in binaural experiments. In: Breebaart J, Houtsma AJM, Kohlrausch A, Prijs VJ, Schoonhoven R (eds) Physiological and psychological bases of auditory function, Shaker publishing BV, Maastricht 153–160

Breebaart J, van de Par S, Kohlrausch A (2001a) Binaural processing model based on contralateral inhibition I. Model structure. J Acoust Soc Am 110:1074–1088

Breebaart J, van de Par S, Kohlrausch A (2001b) Binaural processing model based on contralateral inhibition III. Dependence on temporal parameters. J Acoust Soc Am 110:1105–1117

Fletcher H (1940) Auditory patterns. Rev Mod Phys 12:47–65

Gabriel KJ, Colburn H (1981) Interaural correlation discrimination: i. bandwidth and level dependence. J Acoust Soc Am 69:1394–1401

Glasberg BR, Moore BCJ (1990) Derivation of auditory filter shapes from notched noise data. Hear Res 47:103–138

Levitt H (1971) Transformed up-down methods in psychoacoustics. J Acoust Soc Am 49:467–477

Oxenham AJ (1998) Temporal integration at 6 kHz as a function of masker bandwidth. J Acoust Soc Am 103:1033–1042

Sever J, Small A (1979) Binaural critical masking bands. J Acoust Soc Am 66:1343–1350

van de Par S, Kohlrausch A (1999) Dependence of the binaural masking level differences on center frequency, masker bandwidth, and interaural parameters. J Acoust Soc Am 106:1940–1947

van der Heijden M, Trahiotis C (1998) Binaural detection as a function of interaural correlation and bandwidth of masking noise: implications for estimates of spectral resolution. J Acoust Soc Am 103:1609–1614

Wright BA (1997) Detectability of simultaneously masked signals as a function of masker bandwidth and configuration for different signal delays. J Acoust Soc Am 101:420–429

Chapter 14
Linear and Nonlinear Coding of Sound Spectra by Discharge Rate in Neurons Comprising the Ascending Pathway Through the Lateral Superior Olive

Daniel J. Tollin and Kanthaiah Koka

Abstract Sound source location is computed at central levels in the auditory system based on the neural representations of the spectral and temporal characteristics of the sounds arriving at the ears. Interaural level differences, an acoustical cue to location, are first encoded by neurons in the lateral superior olive (LSO). The sound spectra at the left and right ears must be encoded accurately by the afferents to the LSO in order for the LSO neurons to extract the ILD. Here, we use a systems approach to examine spectral coding capabilities of the afferent inputs to the LSO - globular (GBC) and spherical (SBC) bushy cells of the cochlear nucleus and the medial nucleus of the trapezoid body (MNTB) - and LSO neurons themselves. Extracellular recordings were made in the trapezoid body and were classified according to their responses to short tone bursts: primary-like (PL), PL-with-notch (PLN) and chopper, which correspond to SBC, GBC, and stellate cell types, respectively. The Random Spectral Shape (RSS) method was used to estimate how spectral level is weighted by neurons, both linearly and non- linearly, across a wide band of frequencies. The first (linear) and second (non-linear) order spectral weighting functions were measured for 22 PL, 21 PLN, 16 chopper, 42 MNTB, and 15 LSO neurons. The validity of the estimated weighting functions was tested for each neuron by predicting the rate responses to arbitrary RSS stimuli and head related transfer function-filtered broadband noise. The fraction of explained variance, ranging from 1 (perfect fit) to 0 or less (poor fit), increased significantly for all neuron types by including the second order, non-linear terms. The results demonstrate that the neural pathways through the LSO can accurately encode sound spectra based on discharge rate. The psychophysical implications (sound localization, spectral shape discrimination, etc.) of these findings will be discussed.

D.J. Tollin (✉)
Department of Physiology and Biophysics,
University of Colorado Health Sciences Center,
Aurora, CO, USA
e-mail: Daniel.tollin@ucdenver.edu

E.A. Lopez-Poveda et al. (eds.), *The Neurophysiological Bases of Auditory Perception*,
DOI 10.1007/978-1-4419-5686-6_14, © Springer Science+Business Media, LLC 2010

Keywords Sound localization • Spectrotemporal receptive field • Frequency selectivity • Auditory brainstem

14.1 Introduction

Sound localization requires that the acoustical cues to location be encoded by neurons accurately. Two cues, interaural level differences (ILDs) and spectral-shape cues, require that the shapes and magnitudes of sound spectra be represented. Spectral-shape cues arise from direction- and frequency-dependent reflection and diffraction of sounds by the head, torso, and pinna that result in broadband spectral patterns, or shapes, which change with location; one prominent feature is the deep notches that occur at some locations (Tollin and Koka 2009). ILD cues result from frequency- and direction-dependent modifications of sound by the head and pinnae and are defined as the difference in spectra of the signals at the two ears. The neurons comprising the lateral superior olive (LSO) are hypothesized to be the most peripheral to encode ILDs in a functionally meaningful way (Tollin et al. 2008). LSO neurons receive excitatory input from the spherical bushy cells (SBCs) of the ipsilateral cochlear nucleus (CN) and an inhibitory input from the contralateral ear via the medial nucleus of the trapezoid body (MNTB); the MNTB, in turn, receives excitatory input from the globular bushy cells (GBCs) of the contralateral CN. SBCs and GBCs receive excitatory inputs from the auditory nerve (AN). LSO neurons are sensitive to ILDs. Yet, in order to compute the frequency-dependent ILD, the spectra at the left and right ears must be encoded accurately by the afferents to LSO. Here, we test this hypothesis. The random spectral shape (RSS) technique (Yu and Young 2000) was used to unify our approach to understanding the functional organization in the ILD-coding pathways through the LSO.

14.2 Methods

Details of the general physiological methods can be found in Tollin et al. (2008). Briefly, extracellular responses of well-isolated neurons were recorded with tungsten microelectrodes in the brainstem of barbiturate-anesthetized cats. To study the CN neurons, extracellular recordings were made in the trapezoid body and were classified according to their responses to short tone bursts: primary-like (PL), PL-with-notch (PLN), and chopper, which correspond to the SBC, GBC, and stellate cell types, respectively (Smith et al. 1998). The criteria of Smith et al. (1998) were used to classify MNTB principal cells based on (1) monaural responses only to contralateral-ear stimulation, (2) the presence of a prepotential in the action potential waveform, and (3) a PL or PLN post stimulus time histogram (PSTH) to short tone burst stimuli. LSO units were characterized by low spontaneous rates, "chopping" PSTHs to tone pips, and "IE" binaural interaction, in that they were excited (E) by ipsilateral and inhibited (I) by contralateral acoustical stimulation. Most MNTB and LSO neurons were also confirmed via histology.

The RSS technique (see Young and Calhoun 2005) was used to study how these neurons linearly and nonlinearly weight sound spectra to increase or decrease their discharge rate. Here, 264 different pseudorandom noises were created, each consisting of the sum of 512 random-phase tones logarithmatically spaced in frequency and covering the range from 0.17 to 40 kHz in 1/8 octave steps. The tones were grouped into 64 frequency bins each containing 8 tones so that each bin spanned 1/8 octave. Relative to the mean overall spectral level of each stimulus, the amplitude of the tones in each of the 64 bins were chosen randomly from a distribution that had a mean and standard deviation of 0 and 10 dB, respectively, so that all eight components in a single bin have the same amplitude. 4/264 RSS stimuli had a flat spectrum. Across the 264 stimuli, the amplitudes in each frequency bin were specifically constructed to be uncorrelated with the amplitudes in all other bins. This constraint allows the use of linear least squares techniques to compute the spectral weights from the rate responses to the ensemble of RSS stimuli. Computation of the second-order weighting functions was done as described by Reiss et al. (2007). The ensemble of 264 RSS stimuli was presented at 4–10 overall levels spanning ~20 dB below to ~40 dB above the threshold level for the flat spectrum noise alone. The RSS stimuli were 100 ms in duration and were gated on/off with 10-ms linear ramps. Each of the 264 RSS stimuli were presented to the neuron only once at each stimulus level.

The RSS weight function model is based on the following equation

$$r = R_0 + \sum_{j=1}^{M} w_j S(f_i) + \sum_{j=1}^{M}\sum_{k=j}^{M} w_{jk} S(f_i) S(f_k) \tag{14.1}$$

where r is the rate, $S(f_i)$ the dB values of the RSS stimuli at different frequencies, w_j [in units of spikes/(s·dB)] the first-order weights, and w_{jk} [units of spikes/(s·dB^2)] the second order, or nonlinear weights. R_0 is the rate response to the flat spectrum stimulus. The model parameters for (14.1) were estimated using the discharge rates in responses to single presentations of each of the 264 RSS stimuli. The variable R_0 and the first- and second-order weights were estimated simultaneously by the method of normal equations by minimizing the chi-square error.

14.3 Results

Results are based on 22 SBC, 21 GBC, 16 stellate, 42 MNTB, and 15 LSO neurons (116 total). In 17/42 MNTB neurons, we also collected responses to 100-ms duration noise that was filtered by the directional transfer functions (DTFs) of an adult cat (Tollin and Koka 2009). The DTFs covered 625 locations in the frontal hemisphere, which contain spectral notch cues. Responses to DTF stimuli were predicted for each neuron using the computed RSS weighting functions.

14.3.1 General Properties of Spectral Weight Functions

Figure 14.1a–c shows an example neuron (MNTB) measured with RSS stimuli at 15 dB re: flat-spectrum threshold. Figure 14.1a shows the first-order weight function, which exhibits an excitatory area near BF (best frequency, 20 kHz). Like a tuning curve, the weight function has peak weight (spikes/s/dB) at BF and smaller weights at adjacent frequencies. The correlation between BF measured with RSS and the characteristic frequency measured with tones was nearly perfect for all neuron types ($R^2=0.992$, slope$=0.997$, $N=116$ neurons). The weights indicate how spectral levels in each frequency bin contribute to the discharge rate. As such, weight functions are more informative of the functional properties of the neuron than is a frequency tuning curve or response area. Positive weights indicate excitation – discharge rate increases with increases in signal energy at that frequency –

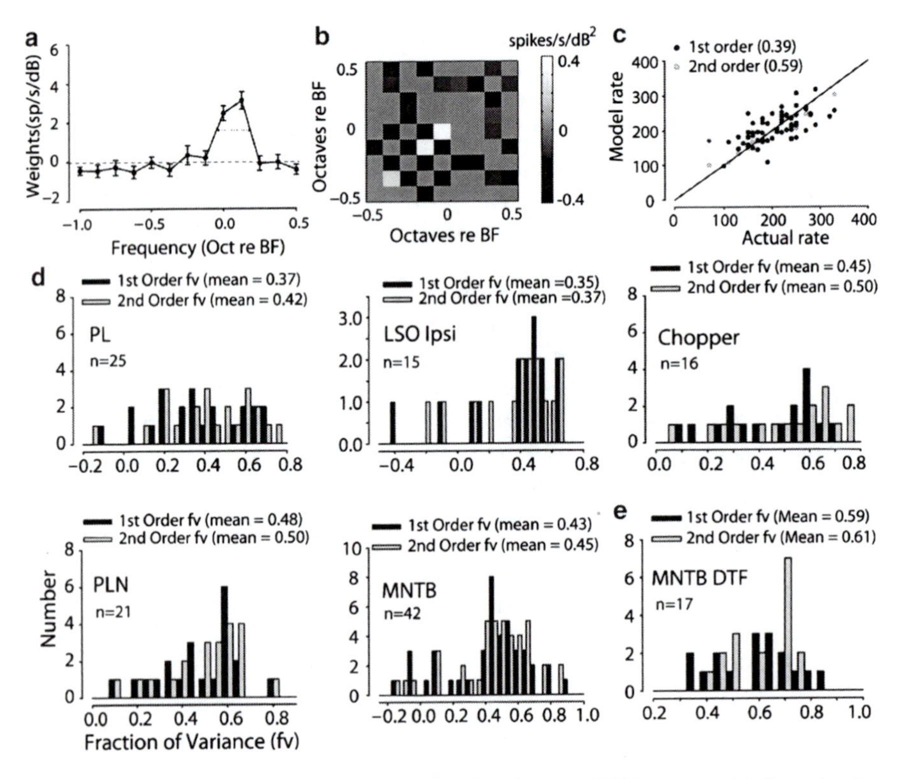

Fig. 14.1 (**a**) First-order spectral weighting function for one MNTB neuron. (**b**) Second-order weights for the neurons in (**a**). (**c**) Model rate based on weights in (**a**) and (**b**) as a function of actual empirical rate. (**d**) Histograms of model prediction quality (*fv*) for each of the neuron types – GBC (PLN), SBC (PL), stellate (chopper), ipsi-LSO, contra-LSO, and MNTB. (**e**) Model prediction quality (*fv*) for DTF-filtered noise in 17 MNTB neurons

while negative weights indicate inhibition – decreases in rate with increases in signal energy. Although not shown in Fig 14.1a, in many neurons, there was also significant sideband inhibition/suppression (weight<0). For example, in MNTB 13/42 neurons showed high-side while 5/42 showed low-side inhibition. Error bars show standard deviations of the weights estimated using statistical resampling techniques (Efron and Tibshirani 86). Weights ±1 SD away from 0 were considered significant and were subsequently used for predictions (see below). The second order weights show positive (white) values at BF and negative weights (black) at adjacent off diagonal frequencies (only significant terms are plotted). Second-order weights measure the contribution to the neuron rate response of quadratic terms like the energy-squared at a particular frequency [e.g., w_{jj} for $S^2(f_j)$ in (14.1)] or the product of the energy at two different frequencies. In summary, the weight function ((14.1) and Fig 14.1a, b) describes the transformation of spectral levels into discharge rate and indicates the frequency range over which these transformations occur (i.e., spectral selectivity).

14.3.2 Testing the Validity of Spectral Weight Functions

The validity of the model was tested by using the fitted weights and (14.1) to make predictions of rate responses to arbitrary RSS (and DTF-filtered) stimuli that were not used for the fitting. To test the predictive power of the model, rate responses to 264 RSS stimuli were divided in to two groups: responses to 200 RSS stimuli were used to estimate the weights using (14.1) and the responses to the remaining 64 RSS stimuli were used for prediction using (14.1) and the fitted weights. If DTF-filtered noise was used, then the prediction set consisted of responses at 625 different spatial locations. Model quality was quantified by the fraction of explained variance:

$$fv = 1 - \frac{\sum_j \left(r_j - \hat{r}_j\right)^2}{\sum \left(r_j - \bar{r}_j\right)^2} \tag{14.2}$$

Here, for the ith RSS (or DTF) stimulus, r_i is the empirical rate, $\hat{r}_i$ is the rate predicted by the model, and $\bar{\hat{r}}$ is the mean rate computed over all RSS stimuli. *fv* values vary from a maximum of 1 (perfect fit) and decrease with poorer predictions; *fv* can take values <0 when the fit is particularly poor. For the example neuron in Figs 14.1a, c plots the modeled rates, based on the first- and second-order weights in Fig 14.1a, b, respectively, using (14.1). *fv* was measured for the first-order linear model and again with the addition of the second-order nonlinear terms. In this example, addition of the second-order terms improves the quality from *fv* of 0.39 to 0.59. Thus, the full model explains 60% of the variance in the discharge rates. Note that $fv \cong R^2$, so *fv* of 0.6 corresponds to R of 0.77. *fv* is a stricter test of the model because it is sensitive to deviations of predictions from a slope of 1.0, whereas correlation coefficient is not.

For all neuron types and at all sound levels tested, the quality of the spectral weight model was assessed by the accuracy of the rate predictions. Figure 14.1d plots histograms of the first- and second-order prediction qualities, *fv*, for each of the neuron types. For PLN, PL, and Chopper neuron types (corresponding to GBC, SBC, and stellate neurons), the addition of the second-order, nonlinear terms significantly improved the predictions; however, the magnitude of the improvement was quite small. MNTB neurons were not improved significantly by second-order terms. That the second-order terms only marginally increased the accuracy suggests that linear weighting is a good model for how these neurons encode sound spectra. Across neuron types, the linear model explained 40–50% of the variance ($r \sim 0.63$–0.71). Although this value appears low (given the 1.0 maximum value for *fv*), this value is actually quite good considering the variance (nearly Poisson) of the empirical discharge rate data. Given the single repetition of each RSS stimulus used here, the large response variance ensures that the *fv* will always be less than 1.0; that is, some fraction of the responses are random and thus not predictable by the model. Correcting *fv* for this variance (assuming Poisson) reveals that the maximum *fv* we could theoretically obtain with our empirical data was ~0.6–0.7 (see also Young and Calhoun 2005). Finally, in 17 MNTB neurons, the ability of the model to predict responses to DTF-filtered stimuli was examined. *fv* for these stimuli (Fig 14.1e) was 0.59 for first-order, increasing to 0.61 with the second-order terms. In summary, the spectral weight models based on the RSS stimuli appears to capture the general spectral coding capabilities of the neuron types studied here.

14.3.3 Spectral Weight Function Properties for the Different Neuron Types

The spectral weight model for each neuron type produced relatively accurate predictions of responses to arbitrary stimuli. This fact suggests that the properties of the weight functions have functional meaning. Figure 14.2a shows the populations of the first-order spectral weight functions (grey lines) and their means (filled symbol with error bars) for each neuron type for three different frequency ranges. For each neuron, the weight function is plotted in Fig 14.2a for a sound level 10 dB re: threshold. We found that the magnitude of the maximum weights increased systematically with BF, being lower for low BF neurons (<3 kHz) and progressively larger for higher BFs (3–10 and 10–30 kHz groups). Interestingly, across neuron types that provide input directly or indirectly to LSO, there was little difference in either the shapes of the weight functions or the maximum spectral weight for a given frequency range.

What about weight functions of LSO neurons? Figure 14.2b shows an example of the first-order weight functions for one LSO neuron (CF = 18.35 kHz). Both the

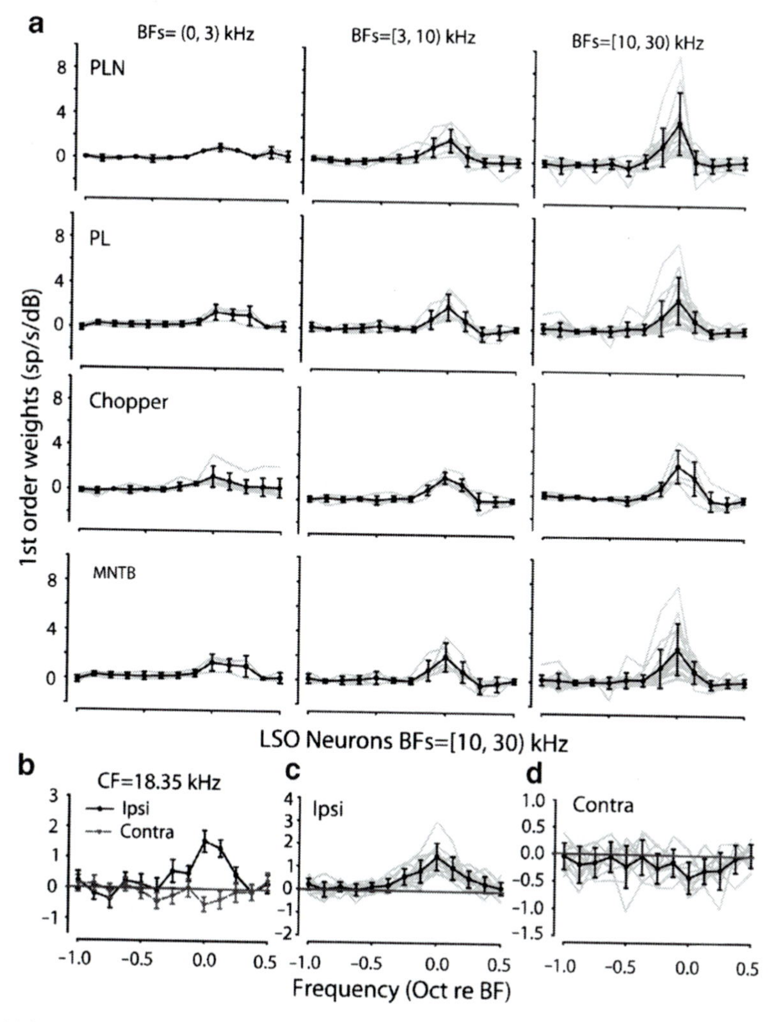

Fig. 14.2 (**a**) First-order spectral weighting functions for each of the neuron types. Functions have been grouped into three broad frequency ranges (*top*). In each panel, individual neuron weight functions are shown for each neuron, plus the across-neuron mean weight ±SD (*filled symbols* and *error bars*). (**b**) Ipsi- and contra-LSO weight functions for a single LSO neuron. (**c**, **d**) First-order spectral weight functions for ipsi-LSO (**c**) and contra-LSO (**d**). Across-neuron means are also shown in (**c**) and (**d**)

ipsilateral excitatory and contralateral inhibitory weight functions are shown. The contralateral function has negative weight because increases in spectral levels near BF at the contralateral ear led to reductions in rate. On the other hand, the weight functions for ipsilateral excitatory stimulation of LSO produced positive weights. The signs of the LSO weight functions are consistent with the IE binaural interaction exhibited by these neurons, which is responsible for their ILD sensitivity. For the LSO neuron in Fig 14.2b, the excitatory and inhibitory BFs were matched.

However, this was not the case for all LSO neurons. The across-neuron consequence of this can be seen in Fig 14.2d where the inhibitory weight functions are shown for 15 high-BF (>10 kHz) LSO neurons. Across neurons, the mean weight function re: excitatory BF was significantly negative at BF, but because there were some neurons that had their inhibitory BF away from the excitatory, the magnitude of the across-neuron mean weight was reduced. In general, the inhibitory weights were smaller than the comparable excitatory weights.

Because the general shapes of the spectral weight functions were quite similar across the different neuron types (over similar BF ranges, Fig 14.2a), the functions can be summarized by two main parameters: maximum weight at BF and the bandwidth of the weight function at half the maximal weight. The bandwidth at half-height gives an estimate of the frequency selectivity of the neuron (and can be directly related to Q_{10}, Young and Calhoun 2005) and the maximum weight at BF indicates the strength by which the neuron transforms spectral levels into discharge rate. Figure 14.3 summarizes the means ±SD of the bandwidth (Fig 14.3a, b) and the weight at BF (Fig 14.3c, d) for each neuron type as a function of the sound level re: threshold. Because we only had data for high-BF (>10 kHz) LSO neurons, the across-neuron type comparisons are made in Fig 14.3 only for neurons with BF >10 kHz. Data have been grouped into the relevant constituent parts of the neural

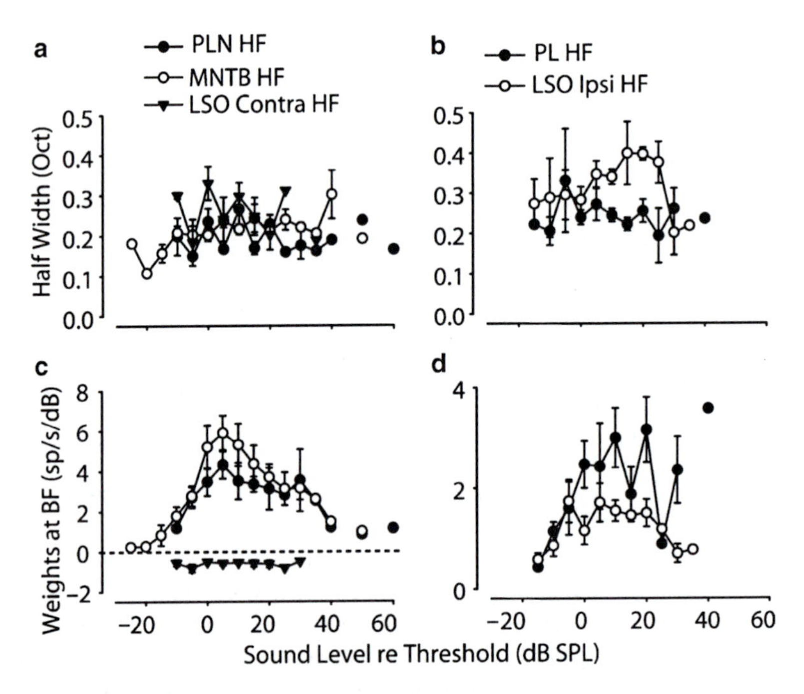

Fig. 14.3 (**a**, **b**) Across-neuron mean spectral weight function bandwidths for each neuron type as a function of sound level re: threshold for flat spectrum noise. (**c**, **d**) Across-neuron mean spectral weight at BF for each neuron type as a function of sound level as in (**a**) and (**b**). Only neurons with BF> 10 kHz are shown in this figure. *Error bars*, ±SD

circuits through the LSO. Weight function parameters for neuron types comprising the contralateral inhibitory input to LSO – GBC, MNTB, and contra-LSO – are plotted in Fig 14.3a, c while the ipsilateral excitatory neuron types – SBC and ipsi-LSO – are plotted in Fig 14.3b, d. For the ipsilateral excitatory pathway, the half-height bandwidths of the SBCs (PL) and the ipsi-LSO were comparable except for a small range of levels around 20 dB. On the other hand, the weights at BF were generally larger for SBCs (PL) than for the ipsi-LSO responses. For the contralateral-inhibitory pathway, the bandwidths were remarkably similar for the GBCs (PLN), MNTB, and contra-LSO responses. There was a tendency for the maximum MNTB weights at BF to be somewhat larger than the GBC, but only at mid sound levels. Both these neuron types had weight magnitudes that were much larger than the contra-LSO responses.

14.4 Discussion

Here, we used a system identification approach to examine how the LSO and its afferents process sound spectra. The ultimate goal of these studies is to determine how LSO neurons individually and collectively compute a neural correlate of the ILD cue to sound source location. To this end, we used the RSS technique, which is a model of the way these neurons linearly and nonlinearly weight sound spectra to produce a discharge rate. For each neuron, the validity of this model was tested by using the fitted model parameters to make predictions of the discharge rates to arbitrary stimuli that were not used in the fitting procedure. In general, spectral weight models of all neuron types here produced predictions that were broadly consistent with the model hypothesis. For the most part, these peripheral neurons were well modeled by a linear weighting of sound spectra, consistent with previous reports of CN neurons using RSS stimuli (Yu and Young 2000; Yu 2003). The second-order terms, when needed, reflect the need to model the curvature of the rate-level function near threshold and at high levels. Given this, we believe that the characteristics of the weight functions, like bandwidth and the magnitude of the weights, have functional meaning.

Here, we examined properties of the spectral weight functions of neurons comprising the pathways through LSO, looking for differences and similarities that might give insight into spectral processing mechanisms and transformations. For example, where there is likely a one-to-one synapse, like the GBC synapse with MNTB, to the extent to which the gain of the synapse is one, there would be expected to be little differences between the properties of the GBC (PLN) and the MNTB weight functions. Our data here support that hypothesis. Both the weights and the bandwidths of the GBC (PLN) and MNTB were virtually identical at comparable sound levels as would be expected from a one-to-one synapse with no loss (Fig 14.3a, c). If there was any transformation of spectral coding at the GBC–MNTB synapse, as has been suggested by Kopp-Scheinpflug et al. (2002), it was not apparent in these across-neuron averages; the latter finding is consistent with

McLaughlin et al. (2008). However, large differences were apparent in the inhibitory pathway from MNTB to LSO. While the contra-LSO (i.e., contralateral ear inputs via MNTB) bandwidths were comparable to the inputs from GBC (PLN) and MNTB, the magnitude of the spectral weight was substantially smaller by a factor of 2–4. The comparable bandwidth suggests little convergence of MNTB (at least by neurons of widely different BF) onto LSO, so the lower spectral weights for contra-LSO may be due to a reduced synaptic gain for inhibition at LSO (e.g., Sanes 1990). Also using RSS stimuli, Yu (2003) reported similar findings in one LSO and one MNTB neuron. In synapses through the excitatory half of the LSO circuit where there is likely to be some convergence, such as the SBC (PL) inputs to the LSO, there might be expected to be differences in the weight functions. In this latter projection, our data suggest that the ipsi-LSO bandwidth may be somewhat larger than the bandwidths of the presumptive inputs from the SBCs (PL), consistent with convergence. On the other hand, this convergence does not translate to stronger spectral weighting because the weights for ipsi-LSO were smaller than those for the inputs from the SBCs (PL). Thus, while there is likely convergence, the synaptic strength may be reduced for each input. Finally, comparing the ipsi-LSO weight functions to contra-LSO, we find that the contra-LSO bandwidths were somewhat narrower (for levels around 20 dB) but the spectral weights were much smaller. The latter suggests that the gain of the contra-inhibition relative to the ipsi-excitation is <1.0.

Finally, how might these data be used to interpret psychophysical findings related to encoding sound spectra necessary for localization? In one example, as discussed by Lopez-Poveda et al. (2007), it is known that the quality of the discharge rate-place representation of spectral notch cues to location is degraded at high SPLs (Rice et al. 1995). It has been hypothesized that the SPL-dependent neural degradation is due to the low threshold and narrow dynamic range of high-spontaneous rate AN fibers and to the broadening of their frequency selectivity with increasing SPL; the latter resulting from the broadening of basilar membrane (BM) tuning at high levels (Ruggero et al. 1997). These facts suggest that ANF rate-place responses would be unable to represent the spectral pattern of high-frequency spectral notch cues at high stimulus levels (Lopez-Poveda et al., 2007). Consistent with this, using the RSS technique, Young and Calhoun (2005) confirmed that the bandwidths of the spectral weight functions for ANFs broadened with increases in stimulus level, although not as much as that seen in the BM. It is possible that the signal representation in ANFs may not contain the information relevant for predictions of behavioral spectral selectivity. For example, our results in the CN and MNTB showed quite consistent spectral selectivity over a wide dynamic range and there was little evidence that bandwidths increased with increasing sound level in the way that ANFs do (Fig 14.3a, b). A potential mechanism for maintaining spectral selectivity with increasing level may be the sideband suppression/inhibition observed in many of the weight functions. The CN and MNTB neurons studied here, then, may be able to represent spectral patterns such as the spectral notches and ILDs over a much larger dynamic range and with better fidelity than their ANF inputs. (A similar transformation occurs from ANF to CN in the enhancement of phase locking to low-frequency stimuli.) Even so, the input neurons to LSO will still be restricted at very high SPLs by not only lower spectral weights (i.e., Fig 14.3c, d) but also increased response variance (assuming ~Poisson variability in spike count). Thus, there will be a compression of the across neuron rate-place representation of sound spectra (like notches) and the precision of the neural representation will also be degraded due to increased response variance. This is a potential explanation for the degradation in sound localization accuracy in humans with high SPLs, often called the negative-level effect (Hartmann and Rakerd 1993).

Acknowledgments Supported by NIH NIDCD grant DC006865 to DJT.

References

Hartmann WM, Rakerd B (1993) Auditory spectral discrimination and the localization of clicks in the sagittal plane. J Acoust Soc Am 94:2083–2092

Kopp-Scheinpflug C, Lippe WR, Dorrscheidt GJ, Rubsamen R (2002) The medial nucleus of the trapezoid body in the gerbil is more than a relay: comparison of pre- and post-synaptic activity. J Assoc Res Otolaryngol 4:1–23

Lopez-Poveda EA, Alves-Pinto A, Palmer AR (2007) Psychophysical and physiological assessment of the representation of high-frequency spectral notches in the auditory nerve. In: Kollmeir B, Klump G, Hohmann V, Langemann U, Mauermann M, Uppenkamp S, Verhey J (eds) Hearing-from sensory processing to perception. Springer, Berlin Heidelberg, pp 51–59

McLaughlin M, van der Heijden M, Joris PX (2008) How secure is in vivo synaptic transmission at the calyx of held? J Neurosci 28:10206–10219

Reiss LAJ, Bandyopadhyay S, Young ED (2007) Effects of stimuls spectral contrast on receptive fields of dorsal cochlear nucleus neurons. J Neurophysiol 98:2133–2143

Rice JJ, Young ED, Spirou GA (1995) Auditory-nerve encoding of pinna-based spectral cues: rate representation of high-frequency stimuli. J Acoust Soc Am 97:1764–1776

Ruggero MA, Rich NC, Recio A, Narayan SS, Robles L (1997) Basilar-membrane responses to tones at the base of the chinchilla cochlea. J Acoust Soc Am 101:2151–2163

Sanes DH (1990) An in vitro analysis of sound localization mechanisms in the gerbil lateral superior olive. J Neurosci 10:3494–3506

Smith PH, Joris PX, Yin TCT (1998) Anatomy and physiology of principal cells of the medial nucleus of the trapezoid body (MNTB) of the cat. J Neurophysiol 79:3127–3142

Tollin DJ, Koka K, Tsai JJ (2008) Interaural level difference discrimination thresholds for single neurons in the lateral superior olive. J Neurosci 28:4848–4860

Tollin DJ and Koka K (2009) Postnatal development of sound pressure transformations by the head and pinnae of the cat: monaural characteristics. J Acoust Soc Am 125(2):980–994

Young ED, Calhoun BM (2005) Nonlinear modeling of auditory-nerve rate responses to wideband stimuli. J Neurophysiol 94:4441–4454

Yu JJ (2003) Spectral information encoding in the cochlear nucleus and inferior colliculus: a study based on the random spectral shape method. Dissertation, Johns Hopkins University, Baltimore, MD

Yu JJ, Young ED (2000) Linear and nonlinear pathways of spectral information transmission in the cochlear nucleus. Proc Natl Acad Sci USA 97:11780–11786

Chapter 15
Enhancement in the Marmoset Inferior Colliculus: Neural Correlates of Perceptual "Pop-Out"

Paul C. Nelson and Eric D. Young

Abstract Although new information about external stimuli cannot be generated centrally, it is clear that the auditory system can selectively suppress or enhance different features of the peripheral response to acoustic stimulation. One example is the robust perceptual "pop out" of a single component within a broadband sound whose onset time is delayed relative to the remainder of the complex. Single auditory nerve fibers do not exhibit enhanced responses using such stimuli (J Acoust Soc Am 97:1786–1799, 1995); the percept is presumably derived from the amplification in the central auditory system of some set of features within the population peripheral response. The goals of this study were (1) to determine whether this neural integration occurs at or below the inferior colliculus (IC), and (2) to compare the effects of specific stimuli and parameter variations between physiological and psychophysical experiments. Single-unit activity was recorded in the IC of the awake marmoset in response to stimuli loosely modeled after previous behavioral and physiological studies of the enhancement effect. A 100-ms best-frequency (BF) tone was presented within a wideband sound with a spectral notch centered on BF. In many IC neurons, responses were significantly larger to this stimulus when a 500-ms preceding signal consisting of the band-reject complex was presented than when silence was presented prior to the probe.

Keywords Inferior colliculus • Marmoset • Enhancement • Context dependence

---

P.C. Nelson (✉)
Department of Biomedical Engineering, Johns Hopkins University, Baltimore, MD, USA
e-mail: pcnelson@jhu.edu

E.A. Lopez-Poveda et al. (eds.), *The Neurophysiological Bases of Auditory Perception*,
DOI 10.1007/978-1-4419-5686-6_15,

15.1 Introduction

One apparent function of the central auditory system is to transform distributed peripheral population codes into more explicit representations of stimulus features relevant for perception. In the process, sound attributes that arise from new or dynamic sources are often amplified (e.g., the oddball effect, Ulanovsky et al. 2004), while those likely to originate from a single unchanging source may be subject to active attenuation (e.g., forward masking, Nelson et al. 2009). The main purpose of this study was to determine whether a neural representation of the phenomenon referred to as enhancement in the psychophysical literature (e.g., Viemeister 1980; Viemeister and Bacon 1982; Wright et al. 1993) is emergent by the level of the inferior colliculus (IC).

When the onset of a single tonal component is delayed with respect to the remainder of a longer broadband signal (see the schematic spectra at the bottom of Fig. 15.1), the delayed signal can stand out perceptually from the rest of the complex. Psychoacoustic studies have confirmed this anecdotal observation (1) by measuring reduced detection thresholds for the introduced component, and (2) by showing that the introduced component is a more effective forward masker (Viemeister and Bacon 1982; Thibodeau 1991; Wright et al. 1993). The latter finding is an important one because it strongly suggests that enhancement reflects a true increase in the effective (internal) gain at the target frequency as opposed to a decrease in the relative magnitude of the response elicited by the conditioner.

Auditory nerve (AN) fibers only exhibit relative enhancement, in the sense that firing rate in response to the target is reduced at all best frequencies (BFs) by the presence of the conditioner, but is reduced *less* in fibers with BFs near the target frequency (Palmer et al. 1995). It appears that central stages act to amplify these relative across-BF contrasts to generate absolute or "true" enhancement. There is a small amount of absolute enhancement in ventral cochlear nucleus neurons (Scutt 2000), but the effect is often limited to the initial portion of the response to the introduced target component (within the first 25 ms). Both the AN and VCN data sets are from anesthetized guinea pigs.

Here, we show that single IC neurons in an awake primate often exhibit robust enhancement of their rate responses. Typically, the optimal preceding conditioner to generate maximal target enhancement is a medium-to-high-SPL stimulus with a spectral gap of ~½ octave surrounding the component of interest. The low frequency portion of the conditioner seems to be responsible for the effect more so than the high frequency portion. Based on responses to the conditioner in isolation, we also conclude that the mechanism underlying the increased response to the target is more complex than a burst of spikes caused by rebound from inhibition. Details of the results are qualitatively consistent with an idea put forth in the initial reports of the psychophysical phenomena (adaptation of inhibition; Viemeister and Bacon 1982), although multiple neural mechanisms are likely involved in IC enhancement.

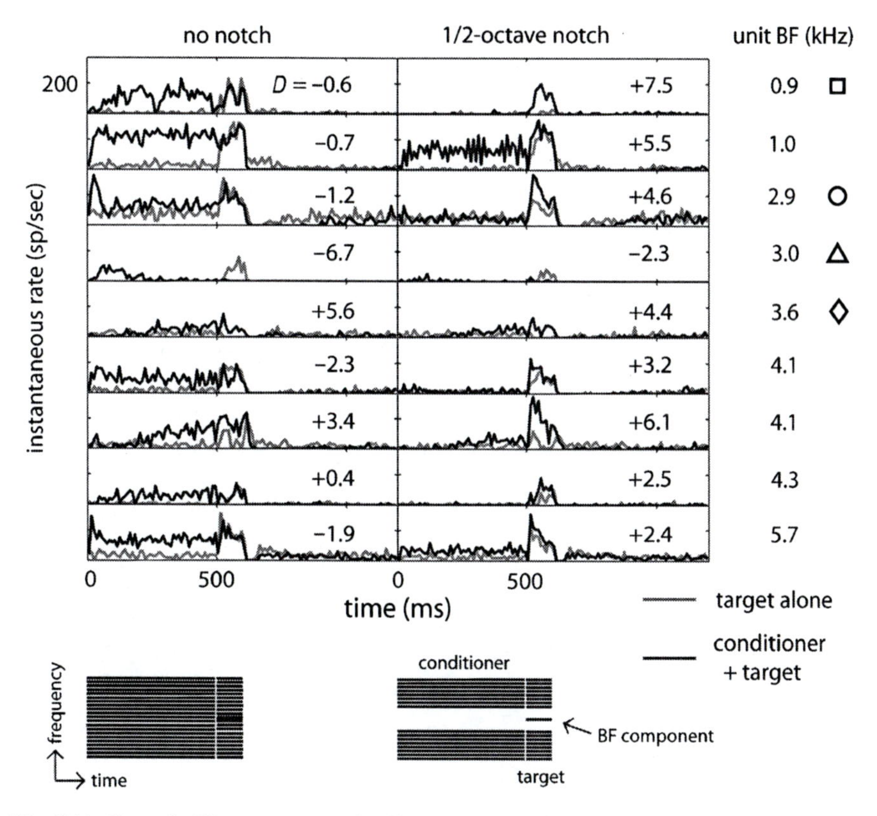

Fig. 15.1 Example IC responses to stimuli that cause psychophysical enhancement. PST histograms (10-ms bins) are shown for conditioners and targets with no notch (*left column*) and for stimuli with a ½-octave spectral notch (*right column*). Target-alone responses (*gray*) are shown along with conditioner followed by target responses (*black*) for each conditioner type and neuron. Firing-rate response-differences during the target interval are quantified with a standard separation index (*D*, see text), included in each panel. Schematic spectrograms (logarithmic frequency axis) of the stimuli are shown below the data. Symbols to the right of the panels correspond to the neurons with further data shown in Fig. 15.2

15.2 Methods

Single-neuron responses were recorded in the inferior colliculus of unanesthetized common marmosets. Methodological details for chronic recording of single-neuron extracellular activity in the awake marmoset IC are described elsewhere (Nelson et al. 2009) and are similar to the methods used for cortical recordings in the same animal (e.g., Lu et al. 2001). Head-fixed animals sat comfortably upright in a plastic chair while a tungsten electrode was advanced medially and ventrally through a craniotomy positioned on the lateral surface of the skull. Sounds were presented in the free field from a speaker located ~1 m directly in front of the animal's head.

When a neuron was encountered, tone bursts (200 ms) were presented over a range of frequencies and sound levels to characterize the BF, receptive field, and rate response with sound level. The frequency of the tone to be tested for enhancement was always set to BF. Three configurations were presented: (1) conditioner-alone (500 ms duration), (2) target-alone (100 ms duration, with the same spectrum as the conditioner except the addition of a component at BF), and (3) conditioner-then-target (corresponding to the potentially enhanced case). Both target and conditioner included 10-ms on/off ramps; there was no additional delay between the end of the conditioner offset and the target onset. The conditioner's spectrum was made up of sine-phase, equal-amplitude tonal components, logarithmically spaced (1/10th-octave spacing), over a wide range (200 Hz–25.6 kHz) and presented at approximately 50 dB SPL per component. A spectral notch of varying width was produced by omitting the appropriate components surrounding BF. Each stimulus was repeated ten times; to quantify the amount of enhancement (or suppression), a standard separation metric (*D*, Sakitt 1973) was used: $D_{(\mathrm{CT,TA})} = \frac{R_{\mathrm{CT}} - R_{\mathrm{TA}}}{\sqrt{SD_{\mathrm{CT}} SD_{\mathrm{TA}}}}$, where *R* represents the average rate in the 100-ms window containing the BF tone, either with the conditioner preceding the target (R_{CT}), or with the target presented alone (R_{TA}), and the SD terms are corresponding across-trial standard deviations of the rate estimates. Positive values of *D* reflect absolute enhancement, while negative values indicate suppression of the target response when it is preceded by the conditioner.

15.3 Results

15.3.1 Examples of Conditioner Influence on Target Response

Data from nine example IC neurons are shown in Fig. 15.1. Response PSTHs for two types of conditioners are shown: in the left column, the conditioner included all of the components in the original sound; in the right column, all components within ½ octave of the neuron's BF were removed from the conditioner spectrum. Within each panel, responses are shown for the target-alone (gray lines) and for the conditioner followed by the target sound (black lines). Values of *D* for each comparison are also shown in each panel.

With no notch imposed on the conditioner, IC neurons typically responded less in the target interval when the conditioner was present. In other words, a conditioner with energy at BF usually led to suppression of the subsequent target. However, it was not uncommon to observe robust absolute enhancement in single IC neurons when a conditioner with a ½-octave spectral notch (logarithmically centered at BF) preceded the target complex. This behavior is in stark contrast to AN fibers (Palmer et al. 1995), which almost always yield absolute suppression even with a notch in the signal (i.e., $D<0$, see left column of their Fig. 7). There did not appear to be a strong dependence on BF: diverse enhancement effects with respect to their magnitudes and time courses were observed across the range of BFs tested. The details of the enhancement were not clearly related to whether the conditioner itself elicited spikes. However, the presence of a build-up in the neuron's response to the conditioner did seem to coincide with enhancement (especially evident in the no-notch responses that resulted in target facilitation, rows five and seven).

15.3.2 Response Dependence on Notch Width and Isolation of Conditioner Components

The ½-octave notch condition (Fig. 15.1, right column) was the narrowest spectral gap that was tested. Standard separation (*D*) values reflecting the influence of a preceding conditioner over a range of stimulus notch widths from 0 to 2 octaves are shown in Fig. 15.2a for the four neurons marked by symbols in Fig. 15.1 (the same symbols are used in the panels of Fig. 15.2a). When enhancement was present (in three of the four example neurons), it was typically maximal for a notch width of ½ or 1 octave, and

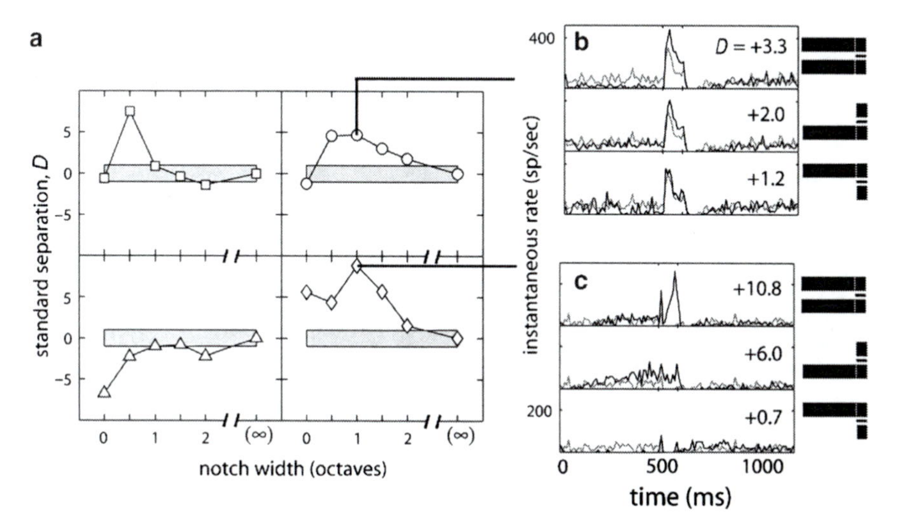

Fig. 15.2 Characterization of the effects of manipulating the conditioner spectrum. (**a**) Example functions showing the relationship between enhancement (or suppression) and the width of the spectral notch carved from the conditioner spectrum for 4 IC neurons. *Gray areas* span a range of *D* values from −1 to +1 (i.e., an insignificant effect of the conditioner on the rate response to the target). The rightmost point in each plot is a trivial comparison between two identical responses ($D=0$), representing the case where the conditioner is silence (the target is presented in isolation). (**b**, **c**) PSTHs from the two neurons in the *right panels* of **a** for different conditioner configurations using a 1-octave notch width. *Top panels*: both high and low frequency portions of the conditioner spectrum included; *middle panels*: low frequency side presented alone; *bottom panels*: high frequency side presented alone. As in Fig. 15.1, *black lines* show responses to the conditioner-then-target stimulus, and *gray lines* represent target-alone responses

declined at broader notch widths. The neuron that was suppressed in both conditions in Fig. 15.1 (fourth example from the top) also did not exhibit enhancement for any of the broader notch widths (Fig. 15.2a, triangles). The functions in the left panels of Fig. 15.2a do not show an effect for notch widths greater than 0.5 octaves, while those plotted in the right panels exhibit enhancement that persists up to a notch width of 1.5 octaves. This bandwidth matches up well with the bandwidth of inhibition revealed in pure-tone frequency response maps for these two neurons at the SPL per component used in the enhancement stimuli (data not shown).

The low-frequency portion of the conditioner stimulus seemed to drive much of the enhancement in our population of IC neurons. In the eight neurons exhibiting the most enhancement with the conditioner both above and below BF that were tested with isolated high- and low-frequency components, seven showed larger enhancement with the conditioner below BF only and one showed more enhancement from the high-frequency side of the conditioner. More commonly, we observed responses like those in Fig. 15.2a–c. The neuron depicted in Fig. 15.2b was modestly inhibited by all three of the conditioner spectra, but exhibited clear enhancement only when the low-frequency side was presented (alone or in combination with the high-frequency side). The effect in Fig. 15.2b is roughly linear, while the combination of high- and low-frequency conditioner segments led to a strongly nonlinear effect for the target response in Fig. 15.2c: clearly, neither high nor low frequencies alone were sufficient to account for the large enhancement generated when both were presented together. The PSTHs during the conditioner suggest an interaction between sustained high-frequency-driven inhibition, low-frequency excitation, and adapting below-BF inhibition. When all of these inputs were engaged with the appropriate stimulus spectrum and time course, the dramatic enhancement shown in the upper panel of Fig. 15.2c was produced.

15.3.3 Enhancement is Not Coupled to the Presence of Postinhibitory Rebound Spikes

Perhaps the simplest possible mechanism to account for the observed neural enhancement is an excitatory postinhibitory rebound (PIR) driven by the band-reject conditioner. Intracellular studies have shown both subthreshold and suprathreshold (i.e., spiking) depolarization following sound-driven inhibition in IC neurons (e.g., Covey et al. 1996; Kuwada et al. 1997). If spiking PIR were responsible for enhancement in the IC, then the response to the conditioner alone should include an excitatory component locked to sound offset.

Figure 15.3 shows that this does not seem to be the case, as robust enhancement was observed even though most neurons exhibited suppression of activity following the conditioner when the target was omitted. Conditioner-alone responses (½-octave notch width) from the same nine neurons as Fig. 15.1 are shown in Fig. 15.3a. The PSTHs are ordered by the magnitude of the target enhancement, with the most robust enhancement in the top panels and the single incident of target suppression in the bottom panel (i.e., a different order from that presented in Fig. 15.1). None

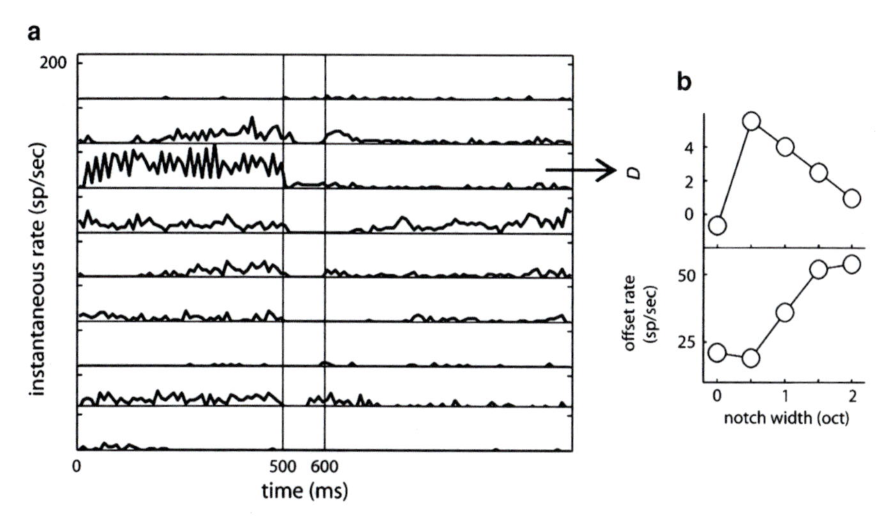

Fig. 15.3 Conditioner-alone offset spiking responses do not correlate well with enhancement. (**a**) Responses to the ½-octave notch width conditioner alone for the same nine neurons shown in Fig. 15.1. From *top* to *bottom*, the neurons are ordered in terms of decreasing amounts of enhancement observed in the response to the target. Note the conspicuous lack of spikes in the window marked by the *vertical lines* (between 500 and 600 ms, occupying the position of the target where it presented). (**b**) Comparison of one neuron's (BF= 1.0 kHz, marked with *arrow*) target enhancement (*upper panel*) and its corresponding offset response to the conditioner (*bottom panel*, with rates measured in the 100-ms window starting immediately after conditioner offset) over a range of stimulus notch widths

of the postconditioner responses included a clear burst of offset-locked spikes, even though all but one showed enhancement (see Fig. 15.1). The single example in which the rate was not at least transiently driven to zero following the conditioner (third row of Fig. 15.3a) is characterized in more detail in Fig. 15.3b. This neuron did have an offset response to the conditioner, but its presence did not correlate with the enhancement effects observed when the target was presented. The strongest offset response was observed for the widest tested notch width (2 octaves, lower panel of Fig. 15.3b). In contrast, the strongest target enhancement effect was measured when the narrowest imposed notch width of 0.5 octaves was used, and larger stimulus notches resulted in systematic relaxation of the response enhancement (Fig. 15.3b, upper panel).

15.4 Discussion

15.4.1 Neural Mechanisms Underlying Enhancement

We speculate that enhancement in IC neurons is driven by interactions between two or more separate processes. One likely candidate for target response facilitation involves the adaptation of off-BF inhibition over the course of the 500-ms conditioner, allowing for a stronger (i.e., less simultaneously suppressed) response to the BF component when it is introduced to the complex; this effect is shown in the data by the buildup of spiking during the conditioner. In three of the nine examples shown here, pure-tone response maps revealed firing patterns at different frequencies that may be consistent with such an explanation: at BF, the response was short-latency and sustained, while below-BF tones could elicit a longer-latency buildup pattern. Above-BF tones did not elicit such a response; instead, a sustained suppression of spikes below spontaneous rate was usually observed throughout the duration of the pure tone stimulus (at frequencies sufficiently close to BF). These features emerged only at higher SPLs, near the levels per component used to generate the enhancement-generating stimuli (~50 dB SPL and above). The presence of buildup on the low-frequency side but not the high-frequency side of the response map lines up well with the findings in Fig. 15.2b–c that point to the low-frequency components as important contributors to the neural phenomenon.

The fact that only a subset of the neurons exhibited clear buildup in the conditioner response or in the pure-tone response maps suggests that adaptation of inhibition may not be entirely responsible for the observed effects (or that it may not always manifest itself in the form of buildup). It remains possible that an additional mechanism actively locked to the offset of the conditioner is also playing a role. In Fig. 15.3, we ruled out the possibility that a simple PIR spike burst was causing the enhanced target response. Subthreshold PIR could still be effectively priming the neurons' responsiveness to the BF component of the target; although based on intracellular work (Covey et al. 1996; Kuwada et al. 1997), it seems unlikely that PIR would never be accompanied by spiking activity. One intriguing set of inputs to the IC that could potentially be involved in enhancement is from the superior paraolivary nucleus. These neurons provide inhibitory (GABA-ergic) projections to the ipsilateral IC (Kulesza and Berrebi 2000) and respond reliably to sound offsets and temporal gaps interrupting BF tones (rabbit: Kuwada and Batra 1999; rat: Kadner and Berrebi 2008). It would be interesting to know the response properties of these neurons when stimulated with notched noise, as used here and in psychophysical studies of enhancement.

15.4.2 Comparison with Perception

Several fundamental features of psychophysical enhancement of forward masking appear to be reflected directly in the responses of single IC neurons in the unanesthetized marmoset. For example, in both psychoacoustics and IC physiology, enhancement is usually greatest for intermediate notch widths (Wright et al. 1993). At the same time, it can also be readily observed in some subjects and in some IC neurons without a notch and also with very wide notches. Higher overall signal SPL gives rise to stronger psychophysical enhancement (Viemeister 1980; Thibodeau 1991); we also observed a systematic dependence on overall SPL in IC neurons consistent with this trend (data not shown). Our comparison of high-side versus low-side conditioners in isolation is connected to more controversial aspects of the behavioral phenomena. Viemeister (1980) concluded that the high-frequency components were most critical to the generation of signal enhancement (i.e., lower thresholds for a delayed-onset component) using an incomplete harmonic series, but Carlyon (1989) found that the low-frequency portion of the masker was more important in a study using a slightly different stimulus configuration. One direction for future work is to test human listeners in the task using stimuli matched to those used in the physiological experiments to better understand the origin(s) of this discrepancy.

Interestingly, listeners with hearing impairment show considerably less enhancement than normal-hearing listeners (Thibodeau 1991). Given our tendency to attribute much of the neural effect to dynamic inhibitory mechanisms, it is important to note that central inhibitory neurotransmission can be severely compromised in animals with hearing loss (for a review, see Caspary et al. 2008). In this sense, psychoacoustic measures of enhancement may provide a window to central processing disorders. Any observed differences between normal-hearing and hearing-impaired listeners may be caused by a reduced ability to emphasize changes in sound spectra caused by an effective loss of adaptive inhibition.

Enhancement and pure-tone forward masking can be cast as fundamentally different phenomena: in one, subsequent sound responses are facilitated, and in the other, they are suppressed. This apparent dichotomy may instead reflect two specific realizations of a general mechanism whose purpose is to emphasize features of sound that are new or likely to arise from a new source, at the expense of preserving details of sounds from a static source. We would argue that findings from the two perceptual tasks are heavily influenced by central processing, especially at higher SPLs (also see Nelson et al. 2009), and that the behaviorally relevant neural modifications are largely complete by the level of the IC for both tasks.

15.4.3 Additional Considerations

Although we show clearly facilitated responses to delayed-onset components within a broadband sound, it is important to note that we can only indirectly comment on the exact relationship between our findings and psychophysical enhancement of forward masking. This is because we did not explicitly measure neural detection thresholds for an additional short probe sound following the target. Nonlinearities accumulated along the auditory pathway make it impossible to infer the effectiveness of a forward masker, given only the magnitude of the spiking response it elicits (e.g., Nelson et al. 2009). Nonetheless, the findings are consistent with anecdotal observations of the "pop-out" of a tonal component that is introduced with a delayed onset, if it is assumed that a neural correlate of the effect is a local increase in the firing rates of IC neurons with BFs at or near the signal frequency. To estimate the extent of the tonotopic spread of neural enhancement for a given stimulus configuration, a different paradigm would be required in which a fixed stimulus was presented to a population of neurons regardless of their BF (as in Palmer et al. 1995).

15.5 Comment by Skyler Jennings

Viemeister and Bacon (1982) proposed "adaptation of suppression" as an explanation of enhancement. In the discussion of the same paper, they suggest this adaptation may exist in the auditory periphery when they state: "Although adaptation of suppression is not seen peripherally, preliminary psychophysical data from our laboratory are not incompatible with this notion..."

The straightforward interpretation of the data presented in the current study does not support the idea that adaptation of suppression occurred in the auditory periphery. Presumably, this is because previous studies in the auditory nerve (AN) and cochlear nucleus (CN) did not show enhancement. However, one notable difference between your study and the studies in the AN and CN is the fact that your data come from awake animals. Is it possible that adaptation of suppression is mediated peripherally by the olivocochlear efferents, but enhancement was not seen in the AN and CN due to the effects of anesthesia on the efferent system?

15.6 Reply by Paul Nelson

While we cannot rule out the possible influence of efferent activity, it seems unlikely that the entire effect is generated via an anesthetic-sensitive component of the olivocochlear system. First, the unpublished CN data (Scutt 2000) that were collected using the same animal and anesthesia regiment as the AN study (Palmer et al. 1995) did show modest enhancement in some neurons. This suggests that low-level central processing can produce the effect, but it does not directly address the question of how the ascending representation might differ in our awake preparation. The only way to unequivocally rule out a direct peripheral influence on IC enhancement would be to record from AN fibers in unanesthetized marmosets.

Attentional effects are also commonly linked to olivocochlear influences. The animals in this study were passively listening to the stimuli (i.e., their attention was not controlled). If the olivocochlear system influence is attention-modulated and responsible for the response facilitation, one might expect that an active listening task would be required before enhancement would be reflected in responses of single neurons. We speculate that a more automatic set of central processes underlies the phenomenon at the level of the IC.

References

Carlyon RP (1989) Changes in the masked thresholds of brief tones produced by prior bursts of noise. Hear Res 41:223–236

Caspary DM, Ling L, Turner JH, Hughes LF (2008) Inhibitory neurotransmission, plasticity and aging in the mammalian central auditory system. J Exp Biol 211:1781–1791

Covey E, Kauer JA, Casseday JH (1996) Whole-cell patch-clamp recording reveals subthreshold sound-evoked postsynaptic currents in the inferior colliculus of awake bats. J Neurosci 16: 3009–3018

Kadner A, Berrebi AS (2008) Encoding of temporal features of auditory stimuli in the medial nucleus of the trapezoid body and superior paraolivary nucleus of the rat. Neuroscience 151:868–887

Kulesza RJ Jr, Berrebi AS (2000) Superior paraolivary nucleus of the rat is a GABAergic nucleus. J Assoc Res Otolaryngol 1:255–269

Kuwada S, Batra R, Yin TC, Oliver DL, Haberly LB, Stanford TR (1997) Intracellular recordings in response to monaural and binaural stimulation of neurons in the inferior colliculus of the cat. J Neurosci 17:7565–7581

Kuwada S, Batra R (1999) Coding of sound envelopes by inhibitory rebound in neurons of the superior olivary complex in the unanesthetized rabbit. J Neurosci 19:2273–2287

Lu T, Liang L, Wang W (2001) Neural representations of temporally asymmetric stimuli in the auditory cortex of awake primates. J Neurophysiol 85:2364–2380

Nelson PC, Smith ZM, Young ED (2009) Wide dynamic range forward suppression in marmoset inferior colliculus neurons is generated centrally and accounts for perceptual masking. J Neurosci 29(8):2553–2562

Palmer AR, Summerfield Q, Fantini DA (1995) Responses of auditory-nerve fibers to stimuli producing psychophysical enhancement. J Acoust Soc Am 97:1786–1799

Sakitt B (1973) Indices of discriminability. Nature 241:133–134

Scutt MJ (2000) Temporal enhancement in the cochlear nucleus of the guinea pig. Dissertation, University of Nottingham

Thibodeau LM (1991) Performance of hearing-impaired persons on auditory enhancement tasks. J Acoust Soc Am 89:2843–2850

Ulanovsky N, Las L, Farkas D, Nelken I (2004) Multiple time scales of adaptation in auditory cortex neurons. J Neurosci 24:10440–10453

Viemeister NF (1980) Adaptation of masking. In: van den Brink G, Bilsen FA (eds) Psychophysical, physiological and behavioral studies in hearing. Delft UP, Delft, The Netherlands

Viemeister NF, Bacon SP (1982) Forward masking by enhanced components in harmonic complexes. J Acoust Soc Am 71:1502–1507

Wright BA, McFadden D, Champlin CA (1993) Adaptation of suppression as an explanation of enhancement effects. J Acous Soc Am 94:72–82

Chapter 16
Auditory Temporal Integration at Threshold: Evidence of a Cortical Origin

B. Lütkenhöner

Abstract A reanalysis of a previous study of wave N100m of the auditory evoked field (AEF) shows that the level dependence of the response *amplitude* is, near the perceptual threshold, consistent with a model recently developed for the compound action potential of the auditory nerve. The response *latency*, decremented by a constant transmission delay, is about inversely proportional to the response amplitude. This result, although valid only for very low stimulus levels, supports the view that the considerable increase of the N100m latency with decreasing stimulus level arises, at least in part, from temporal integration. To allow conclusions as to where the temporal integration takes place, first results of a new experiment are presented, in which the auditory evoked potential elicited by a series of eight near-threshold tone-pulses was recorded. Neither wave V of the auditory brainstem response nor the middle-latency response component P_a differed between first and last pulse of the series, suggesting that temporal integration occurs more centrally than primary auditory cortex.

Keywords Auditory evoked field • Auditory evoked potential • Auditory cortex • Temporal integration • Subthreshold activity • Auditory brainstem response

16.1 Introduction

Numerous experiments have shown that the perceptual threshold decreases with increasing stimulus duration, but it is still not clear where this kind of temporal integration occurs. Moreover, there is an ongoing debate as to whether the quantity being integrated is proportional to sound intensity or sound pressure (see e.g. the discussion triggered by Heil and Neubauer 2004). Own results appear to support

B. Lütkenhöner (✉)
Section of Experimental Audiology, Münster University Hospital,
Münster, Germany
e-mail: lutkenh@uni-muenster.de

E.A. Lopez-Poveda et al. (eds.), *The Neurophysiological Bases of Auditory Perception*,
DOI 10.1007/978-1-4419-5686-6_16, © Springer Science+Business Media, LLC 2010

the latter hypothesis. In a recent study of the auditory evoked field (AEF), the latency of component N100m showed a level dependence that was consistent with the assumption of perfect sound-pressure integration (Lütkenhöner and Klein 2007). Moreover, a model developed for the compound action potential (CAP) of the auditory nerve suggests that, at the perceptual threshold, the CAP amplitude can be approximated very well by a linear function of sound pressure (Lütkenhöner 2008). This theoretical prediction was experimentally confirmed by studying wave V of the auditory brainstem response (ABR) to brief 4-kHz tone pulses (Lütkenhöner and Seither-Preisler 2008). The results of that study suggest that a hypothetical temporal integrator located centrally to the generator of wave V would receive an input that linearly increases with sound pressure. Temporal integration of such activity would consequently appear as sound-pressure integration.

This chapter has two main concerns. First, the data of Lütkenhöner and Klein (2007) are reanalyzed. The conclusions of that study are based on empirical curve fitting, and it remains to be checked to what extent the data are consistent with the predictions of the later modeling study (Lütkenhöner 2008). Second, first results of a follow-up study of Lütkenhöner and Seither-Preisler (2008) are shown – designed to allow a clear-cut distinction between peripheral and central integration.

16.2 Theory

To provide a basis for the reanalysis of the N100m data, some theoretical ideas (mostly based on Lütkenhöner 2008) shall be presented in a nutshell. Starting from well-founded assumptions about the rate-intensity functions of single auditory-nerve fibers (Yates et al. 1990) and the pattern of cochlear excitation caused by a tone (e.g., Russell and Nilsen 1997), formulas for the intensity dependence of the auditory-nerve response near threshold were derived. In the simplest model, response amplitude A and intensity I are related by the equation

$$A(I) = \alpha \log_{10}\left(1+\frac{I}{I_{\text{ref}}}\right), \tag{16.1}$$

where α is an amplitude factor and I_{ref} is a reference intensity. At very low intensities, (16.1) can be approximated as

$$A(I) \approx \frac{\alpha}{\ln(10)} \cdot \frac{I}{I_{\text{ref}}} \tag{16.2}$$

Response amplitude and intensity are proportional in this *low-intensity approximation* – a property inherited from the rate-intensity functions of single auditory-nerve fibers (at extremely low intensities, the gross response of the nerve may be assumed to be determined by the most sensitive fibers). At high intensities, (16.1) can be approximated as

$$A \approx \frac{\alpha\,(L - L_{\mathrm{ref}})}{10}. \qquad (16.3)$$

where

$$L = 10\,\log_{10}(I) \qquad (16.4)$$

is the sound level corresponding to the intensity I, and L_{ref} is, accordingly, the sound level corresponding to the intensity I_{ref}. In this *high-level approximation*, the response amplitude is a linear function of sound level.

The black solid curve in Fig. 16.1a shows the function $A(I)$ as defined in (16.1). The parameters α and I_{ref} were set to one, for the sake of simplicity. This means that both the response amplitude A and the intensity I are considered to be normalized. The low-intensity approximation (16.2) and the high-level approximation (16.3) are represented by a dashed and a dotted curve, respectively. By transforming the abscissa, the curves plotted in Fig. 16.1a turn into the curves shown in Fig. 1b and 1c, respectively. In Fig. 16.1b, the abscissa represents the normalized sound pressure

$$p = \left(\frac{I}{I_{\mathrm{ref}}}\right)^{1/2}. \qquad (16.5)$$

In this representation, the black solid curve is almost linear around $p=1$. A series expansion yields

$$\mathrm{A} \approx \alpha/\ln(10)(p - p_0) \qquad (16.6)$$

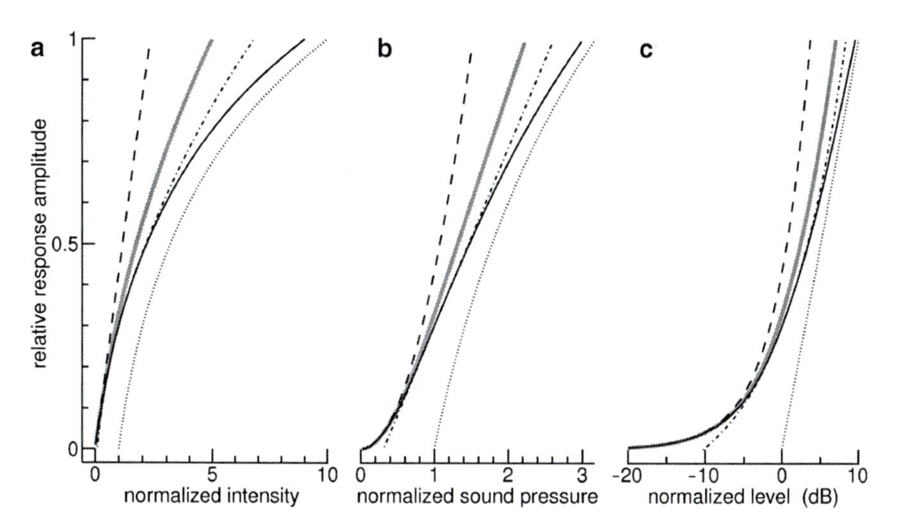

Fig. 16.1 Relative amplitude of an auditory evoked response as a function of (**a**) intensity, (**b**) sound pressure, and (**c**) sound level. The exact model (16.1) is represented by a *black solid curve*; the low-intensity approximation, the high-level approximation, and the pressure approximation are displayed as a *dashed*, a *dotted*, and a *dash-dotted curve*, respectively. A *gray curve* shows a transformed version of Zwislocki's (1965) loudness function

with $p_0 = 1-\ln(2) \approx 0.307$. This approximation, shown as a dash-dotted line, shall be called the *pressure approximation*. In Fig. 16.1c, the abscissa represents the normalized sound level. The high-level approximation appears as a straight line (dotted) now. The pressure approximation (dash-dotted curve) becomes zero at −10.3 dB, which is the normalized sound level corresponding to the normalized sound pressure p_0.

In contrast to the pressure approximation and the high-level approximation, the exact model (16.1) denies the existence of an absolute auditory threshold (in the sense that the response amplitude becomes zero if the sound intensity falls below a certain threshold value). Regarding the detection of a significant stimulus-related effect, this implies that any reduction in sound intensity can theoretically be compensated for by increasing the measurement time. In reality, the measurement time is, of course, always limited, and owing to this limitation there is a detection threshold. When recording auditory evoked responses, a response is detectable at a specific sound intensity whenever the measurement time is long enough to ensure a sufficient signal-to-noise ratio. For sounds with an intensity near the perceptual threshold or below, measurement times of the order of hours may be required (Lütkenhöner and Seither-Preisler 2008). As to the perceptual threshold itself, "measurement time" has to be understood as the time period over which neural activity can be accumulated by the auditory cortex. This *integration time* appears to be below 1 s. The consideration suggests that a perfect match between perceptual threshold and detection threshold for auditory evoked responses cannot be expected. Thus, the excellent agreement found by Lütkenhöner and Klein (2007) has to be interpreted as being coincidental.

For intensities near the perceptual threshold, it appears plausible to assume that the amplitudes of auditory evoked responses reflect aspects of loudness coding, at least qualitatively (Lütkenhöner et al. 2007). This view is corroborated by the fact that $A(I)$, as specified in (16.1), qualitatively resembles Zwislocki's (1965) loudness function. After appropriate transformations, that function may be written as

$$A(I) = \alpha / \ln(10)\ ((1+I/I_{\mathrm{ref}})^m - 1) / m\ , \tag{16.7}$$

with $m=0.27$. The function is shown as a gray curve in Fig. 16.1. Equations (16.1) and (16.7) share the same low-intensity approximation (16.2).

16.3 Auditory Evoked Field at Threshold Revisited

The AEF is dominated, under a great variety of conditions, by a strong response with a typical latency of 100 ms: the N100m. This response is generated in the supratemporal auditory cortex and is considered to be a reliable index of cortical function (Mäkelä 2007). The amplitude of the N100m increases with increasing stimulus duration up to a duration of about 20–40 ms (Joutsiniemi et al. 1989; Forss et al. 1993; Gage and Roberts 2000; Gage et al. 2006). These studies were done at higher sound levels though. Thus, it is not unreasonable to assume that the integration time window is somewhat longer at threshold. A recent study by Lütkenhöner and Klein (2007) focused on very low sound levels. The data of this study shall be reexamined now in the light of the theory outlined above.

16.3.1 Data

Starting point for the reanalysis are the curves shown in Fig. 16.2a (based on Fig. 16.3a of Lütkenhöner and Klein 2007). They show, as a function of time for sound levels between 2 and 40 dB sensation level (SL), the moment of a dipole representing the N100m generators in the auditory cortices of the left hemisphere (grand-average of five subjects). The absolute value of the dipole moment may be

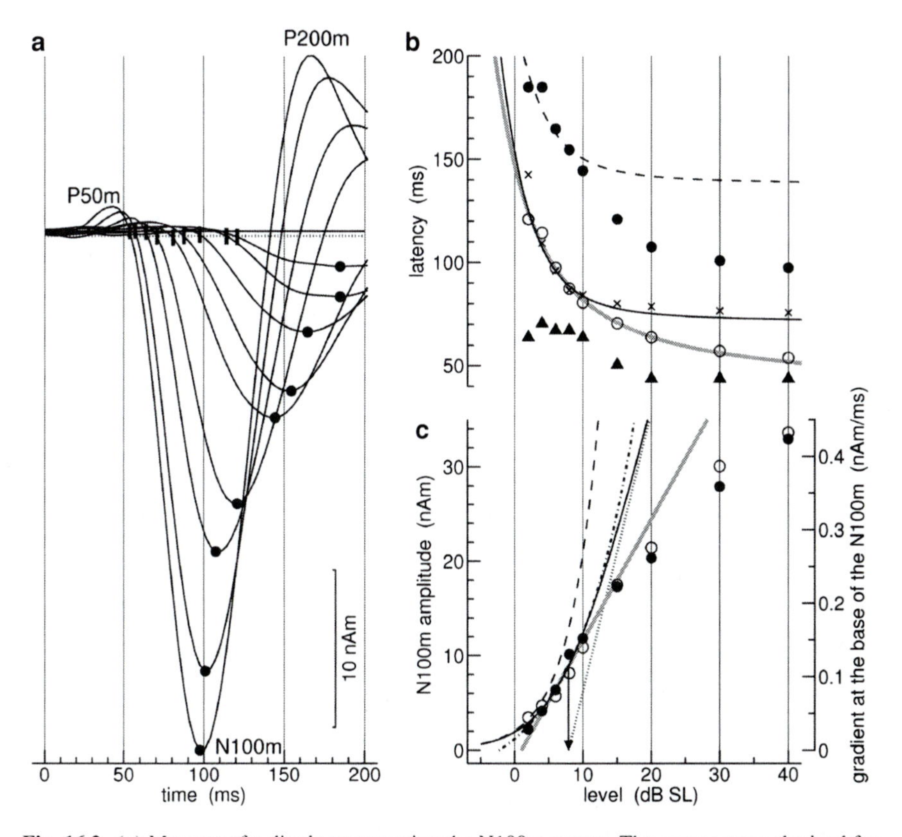

Fig. 16.2 (**a**) Moment of a dipole representing the N100m source. The *curves* were obtained for sound levels between 2 and 40 dB SL. Base and peak of the N100m are marked by a *bar* and a *filled circle*, respectively. (**b**) Latencies as a function of level. N100m peak: *filled circles*; N100m base: *open circles*; latency difference between peak and base: triangles. *Curves* discussed in text. (**c**) Amplitude measures as a function of level. N100m amplitude: *filled circles*; N100m gradient at base: *open circles*; low-intensity approximation: *dashed curve*; high-level approximation: *dotted line*; pressure approximation: *dash-dotted curve*; reference-level approximation: *gray line*

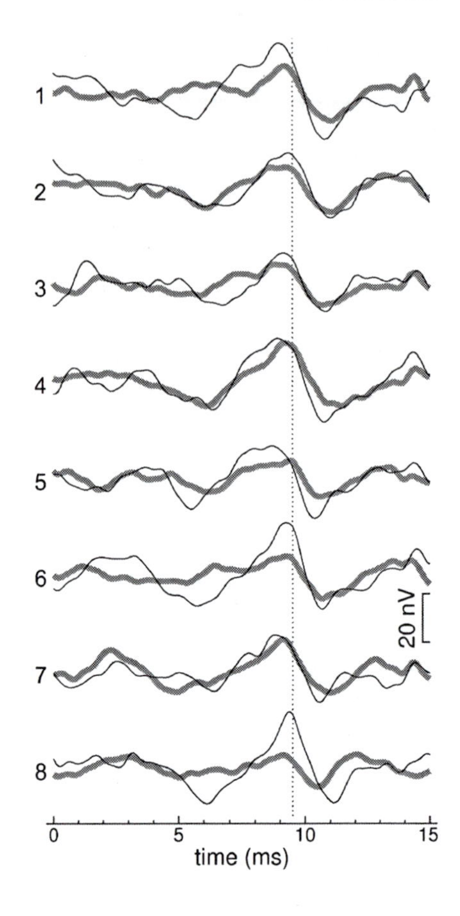

Fig. 16.3 Auditory brainstem responses to the eight pulses of the tonepulse series. The *gray curves* were obtained 1 dB below the threshold for a single pulse, the *black curves* 2 dB above that threshold

considered as a measure of the strength of cortical activation. The N100m peak is marked by a filled circle in each curve. The filled circles in Fig. 2b, c show how latency and amplitude of the peak depend on sound level.

Vertical bars in Fig. 16.2a mark the base of the N100m peak, which was defined as the sampling point closest to the (somewhat arbitrary) criterion level of –0.3 nAm (dotted horizontal line). The latencies of the marked points are shown as open circles in Fig. 16.2b. The latency difference between peak and base (filled triangles) is roughly constant for levels up to 10 dB SL (mean: 66.5 ms), but it decreases at higher levels. The gradient at the base (difference between marked and subsequent sampling point divided by the sampling interval) served as an amplitude measure (open circles in Fig. 16.2c; corresponding scale on the right). As the open circles are qualitatively in good agreement with the solid circles, only the latter will be analyzed below.

16.3.2 Amplitude Analysis

In the original article (Lütkenhöner and Klein 2007), the N100m amplitude near threshold was reported to be a roughly linear function of level with an extrapolated threshold of 0.22 dB SL. At first glance, this interpretation of the data appears to correspond to the high-level approximation given in (16.3). But things are not what they seem. The black solid curve in Fig. 16.2c represents (16.1). The function was fitted to the data using a least-squares approach. The reference level obtained by this means was $L_{ref} = 7.90$ dB (indicated by an arrow). The fit accounted for levels up to 10 dB SL, and for that level range, a good agreement with the data was achieved. By contrast, the high-level approximation (dotted line) fails to explain the data. Thus, this approximation cannot correspond to the linear function of sound level reported in the original article.

However, this does not mean that there is an inconsistency between the original analysis and the present reanalysis. To show this, (16.1) has to be rewritten as a function of sound level. Expanding the resulting function around L_{ref} yields

$$A \approx \alpha / \ln(10)\,(L - L_0)/10 \tag{16.8}$$

with

$$L_0 = L_{ref} - 10\ln(2)\,. \tag{16.9}$$

This approximation shall be called the *reference-level approximation*. As the level L_0 is $10 \ln(2) \approx 6.93$ dB lower than the reference level L_{ref}, L_0 is 0.97 dB SL in the present case. The reference-level approximation is shown as a gray line in Fig. 16.2c. The approximation basically agrees with the linear function found in the original analysis: Both functions are consistent with the data up to a level of 15 dB SL, and the level L_0 is close to the extrapolated threshold reported in the original article.

16.3.3 Latency Analysis

The N100m latency is about 100 ms at the highest sound levels and approaches 200 ms at the lowest levels. Lütkenhöner and Klein (2007) described this latency increase by an exponential term. The time constant of this term turned out to be consistent with the assumption of temporal integration by an almost perfectly operating sound-pressure integrator. In the reanalysis presented here, it is postulated that the input to the temporal integrator has the amplitude $A(I)$ so that the integration time is proportional to $1/A(I)$. As in the original article, it is assumed that, in addition to an intensity-dependent integration time, there is a constant term T_∞, which represents the response latency at the highest possible intensity. Thus, the model for the observed latencies is

$$T(I) = T_\infty + \beta / A(I) \tag{16.10}$$

where β is a constant factor.

The function $A(I)$ in (16.10) shall be tentatively identified now with the function fitted to the amplitude data (black solid curve in Fig. 16.2c). The resulting function $T(I)$ is shown as a solid curve in Fig. 16.2b. The two parameters T_∞ and β were determined by fitting the model to the latency of the N100m base (open circles), taking into account the levels between 2 and 10 dB SL. Shifting this curve by the mean latency difference between peak and base (again considering the levels up to 10 dB SL) resulted in the dashed curve. Crosses represent a transformation of the filled circles in Fig. 16.2c. Model and data are evidently consistent up to 8 dB SL. But with further increasing sound level, the observed latencies decrease much faster than predicted by the model, especially in the case of the N100m peak (filled circles).

An almost perfect fit to the latency data can be obtained by adjusting the reference intensity I_{ref}. The gray curve in Fig. 2b was obtained after multiplying the reference intensity of the amplitude fit by the factor 0.05, which results in a reference level L_{ref} of about –5 dB SL.

16.3.4 Discussion

The experimental finding of a roughly linear relationship between sound level and response amplitude (Lütkenhöner and Klein 2007) could be related to the reference-level approximation of the model (16.8). This means that the original analysis and the present reanalysis are entirely consistent. The lowest level studied corresponded to the upper limit of the range where response amplitude and intensity may be considered proportional. This follows from the fact that the dashed curve in Fig. 2c, representing the low-intensity approximation of the model (16.2), begins to diverge from the solid curve at about 2 dB SL.

Supposed that the level dependence of the N100m latency reflects temporal integration, the observed amplitudes and latencies must be somehow related. The simplest model conceivable is represented by (16.10). Although the model appears to be consistent with the data at low sound levels, there is a serious problem: When applying (16.10), $A(I)$ refers to the *input* of the hypothetical integrator, whereas the observed amplitudes are supposed to be dependent on the *output* of the integrator. This inconsistency may be one of the reasons why the model failed at sound levels greater than 8 dB SL. Fitting the latencies regardless of the amplitudes resulted in an almost perfect match between model and data, for all levels studied. But this success may simply be a consequence of the increased number of free parameters and does not necessarily indicate that the model is adequate from a physiological point of view.

16.4 Auditory Evoked Response to a Series of Tone Pulses

Lütkenhöner and Seither-Preisler (2008) showed that ABR wave V can be measured at sound levels below the perceptual threshold for a single stimulus. But the interpretation of the results is complicated by the fact that, during the electrophysiological measurements, the stimulus was presented in relatively rapid succession. The repetitive stimulation caused a faint auditory sensation, indicating some kind of temporal integration. The question arises as to where in the auditory pathway this integration takes place and to what extent it alters the response. This question gave rise to a follow-up study.

16.4.1 Methods

A series of eight Gaussian-shaped tone pulses (4-kHz carrier frequency; full width at half maximum of 0.5 ms; pulses presented at 16-ms intervals) recurred at a rate of 4/s. It was clear from the outset that recording a response to this stimulus at the perceptual threshold would require an exorbitant measuring time, rendering the experiment almost impossible. This is why the author was the only subject. The data are based on 61 measuring sessions during which the author was doing deskwork. It took nearly 10 months to complete the experiment. The experimental procedure corresponded, in most respects, to the previous study (Lütkenhöner and Seither-Preisler 2008). Only two sound levels will be considered here: −1 and 2 dB SL_1 (the SL_1 scale refers to the perceptual threshold for a *single* stimulus). The number of stimulus repetitions for these levels were 652,000 and 284,000, respectively. The data shown here were passed through a fourth-order Butterworth filter in both the forward and the reverse direction (Matlab routine FILTFILT). The passband was 100–1,500 Hz.

16.4.2 Results

Figure 16.3 shows, in separate panels, the ABR elicited by the eight pulses of the tone-pulse series. The gray and black curves represent the responses at −1 and 2 dB SL_1, respectively. At both stimulus levels, there is a clear positive deflection with a latency of about 9.5 ms (dotted vertical line). This is ABR wave V. The curves are relatively noisy (especially the curves at 2 dB SL_1, owing to the much smaller number of averaged epochs), but the pulse number (position of the pulse within the series) appears to have no obvious effect. Similar conclusions were obtained for the cortical wave P_a, with a latency of about 30 ms (data not shown here).

16.4.3 Discussion

Assuming that the hypothesized temporal integrator is located peripheral to the site of the P_a generation, the response to the last stimulus in the series should be higher in amplitude than the response to the first stimulus. This is not the case. Thus, at least for the type of stimulus considered in this study, temporal integration appears to have a cortical origin.

16.5 Conclusions

Lütkenhöner and Klein (2007) speculated that the level dependence of the N100m latency might reflect sound-pressure integration. The idea implies that the input to the hypothetical integrator is roughly a linear function of sound pressure. But according to the theory outlined above, this is to be expected only for a limited sound-pressure range around the perceptual threshold. Equation (16.10) represents an attempt to overcome this problem. However, the N100m latency data support the equation only at very low sound levels so that the underlying idea may still be too simplistic.

In the pulse-series experiment, the cortical image of the stimulus showed aspects that were roughly independent of sound level. What is most remarkable is that precise information about the time structure of the stimulus was conveyed up to the cortex even though each individual pulse was below the perceptual threshold. The following consideration attempts to explain this observation. Without a stimulus, the various levels of the auditory pathway may be assumed to be stochastically coupled, in the sense that random fluctuations in the spontaneous activity at level n cause correlated fluctuations at level $n+1$. If a stimulus is presented only once at an intensity well below the perceptual threshold, its effect on the spontaneous activity of the auditory-nerve fibers will be hardly noticeable, and the idea of stochastic coupling suggests that this will also be true for higher levels of the auditory pathway. Notwithstanding that, even the faintest stimulus may be assumed to affect the firing *probabilities* of the neurons. But this effect can be detected only on the basis of a sufficient number of stimulus repetitions. The latter was huge in the pulse-series experiment, and this explains why the data showed clear responses to sub-threshold pulses.

The idea that subthreshold stimuli are able to affect the firing probabilities of neurons along the whole auditory pathway may also serve as a basis for explaining temporal integration at threshold. Multiple integration steps, also at subcortical levels, are conceivable. But for the stimulus used in the pulse-series experiment, it appears that temporal integration takes place more centrally than the P_a generation – this means more centrally than primary auditory cortex (Lütkenhöner et al. 2003).

References

Forss N, Mäkelä JP, McEvoy L, Hari R (1993) Temporal integration and oscillatory responses of the human auditory cortex revealed by evoked magnetic fields to click trains. Hear Res 68:89–96

Gage NM, Roberts TP (2000) Temporal integration: reflections in the M100 of the auditory evoked field. Neuroreport 11:2723–2726

Gage N, Roberts TP, Hickok G (2006) Temporal resolution properties of human auditory cortex: reflections in the neuromagnetic auditory evoked M100 component. Brain Res 1069:166–171

Heil P, Neubauer H (2004) Auditory thresholds re-visited. In: Pressnitzer D, de Cheveigné A, McAdams S, Collet L (eds) Auditory signal processing: psychophysics, physiology and modeling. Springer, New York, pp 454–470
Joutsiniemi SL, Hari R, Vilkman V (1989) Cerebral magnetic responses to noise bursts and pauses of different durations. Audiology 28:325–333
Lütkenhöner B (2008) Threshold and beyond: modeling the intensity dependence of auditory responses. J Assoc Res Otolaryngol 9:102–121
Lütkenhöner B, Klein JS (2007) Auditory evoked field at threshold. Hear Res 228:188–200
Lütkenhöner B, Seither-Preisler A (2008) Auditory brainstem response at the detection limit. J Assoc Res Otolaryngol 9:521–531
Lütkenhöner B, Krumbholz K, Lammertmann C, Seither-Preisler A, Steinsträter O, Patterson RD (2003) Localization of primary auditory cortex in humans by magnetoencephalography. Neuroimage 18:58–66
Lütkenhöner B, Klein JS, Seither-Preisler A (2007) Near-threshold auditory evoked fields and potentials are in line with the Weber-Fechner law. In: Kollmeier B, Klump G, Hohmann V, Langemann U, Mauermann M, Uppenkamp S, Verhey J (eds) Hearing – from sensory processing to perception. Springer, Berlin, pp 215–225
Mäkelä JP (2007) Magnetoencephalography. Auditory evoked fields. In: Burkard RF, Eggermont JJ, Don M (eds) Auditory evoked potentials. Basic principles and clinical applications. Lippincott Williams & Wilkins, Philadelphia, pp 525–545
Russell IJ, Nilsen KE (1997) The location of the cochlear amplifier: spatial representation of a single tone on the guinea pig basilar membrane. Proc Natl Acad Sci U S A 94:2660–2664
Yates GK, Winter IM, Robertson D (1990) Basilar membrane nonlinearity determines auditory nerve rate-intensity functions and cochlear dynamic range. Hear Res 45:203–219
Zwislocki J (1965) Analysis of some auditory characteristics. In: Luce RD, Bush RR, Galanter E (eds) Handbook of mathematical psychology, vol III. Wiley, New York, pp 1–97

Part IV
Pitch and Timbre

Chapter 17
Spatiotemporal Characteristics of Cortical Responses to a New Dichotic Pitch Stimulus

Caroline Witton, Arjan Hillebrand, and G. Bruce Henning

Abstract This study used behavioural measures and magnetoencephalography (MEG) to explore a binaural pitch, which is perceived when dichotic frequency modulation (FM) is applied to a band-limited noise stimulus. Although the stimulus contains no spectral, temporal or binaural components expected to produce a pitch, listeners were able to consistently match it to a pure tone. The MEG measures explored spatiotemporal characteristics of the cortical response evoked by the onset of this pitch within a noise. This had a latency of approximately 170 ms and was located in auditory cortex posterior to Heschl's gyrus. In terms of spectral content, pitch onset was associated with a burst of activity in the theta and alpha frequency bands. Listeners' ability to perceptually match the stimulus to a pure tone, and the similarity of the evoked response to previously described pitch-onset responses suggest that it does elicit a true pitch.

Keywords Pitch • Binaural • Magnetoencephalography

17.1 Introduction

A previous psychophysical study (Witton et al. 2005) explored sensitivity to diotic and dichotic frequency modulations (FM) of band-limited noise stimuli. In discriminating dichotic from diotic FM at modulation rates above about 40 Hz, listeners reported using faint, lateralised pitch cues. However, the stimulus contained none of the spectral, temporal or binaural characteristics expected a priori to elicit a percept of pitch. In this study, we used behavioural pitch-matching and magnetoencephalography (MEG) to explore whether a true

C. Witton (✉)
Wellcome Trust Laboratory for MEG Studies,
Aston University, Birmingham B4 7ET, UK
e-mail: c.witton@aston.ac.uk

E.A. Lopez-Poveda et al. (eds.), *The Neurophysiological Bases of Auditory Perception*,
DOI 10.1007/978-1-4419-5686-6_17,

pitch is elicited by dichotic FM of band-limited noise. If our stimulus elicits a binaural pitch, then listeners should be able to match it reliably to a pure tone, and for it to elicit a cortical response similar to that evoked by other pitch stimuli.

Three tonal binaural pitches have been well described in the literature: Huggins pitch (Cramer and Huggins 1958) is created when an interaural phase delay is introduced in a narrow frequency within a noise band; binaural edge pitch (Klein and Hartmann 1981) and binaural coherence-edge pitch (Hartmann and McMillon 2001) result from the introduction of an interaural phase difference above or below a certain frequency "edge" within a noise. Our dichotic FM noise stimulus differs slightly from those described above because it contains a dynamic interaural phase difference, rather than a static one; the interaural phase difference is modulated at the rate of frequency modulation, and the mean interaural phase over the period of modulation is zero. If considered in terms of a simple reduction in phase coherence in the dichotic compared to the diotic condition, it is not unlike a binaural coherence edge pitch stimulus, except that the reduced coherence is across the whole bandwidth.

Huggins pitch has also been used in neuroimaging experiments to explore auditory cortical responses to pitch (Hertrich et al. 2005; Chait et al. 2006). In common with some other broad-band stimuli, such as iterated rippled noise, the introduction of a Huggins pitch does not alter the overall energy content of the noise and therefore, makes a well-controlled neuroimaging stimulus. Krumbholz et al. (2003) used IRN in an MEG experiment to describe a "pitch-onset response", a neuromagnetic field evoked by the pitch onset. Its neuronal source was different from the source of the onset of sound itself, and its latency was dependent on the pitch elicited by the IRN. Huggins pitch elicits a similar pitch-onset response (Hertrich et al. 2005; Chait et al. 2006), although fMRI investigations suggest that the source of peak responses to Huggins pitch and IRN differ, with cortical responses to pitch stimuli originating from areas posterior to Heschl's gyrus, into planum temporale (Hall and Plack 2008). In this study, we used MEG to explore the spatiotemporal characteristics of cortical responses to our dichotic FM noise stimulus enabling us to compare both time series and localisation data for our stimulus with previous work on responses to pitch.

17.2 Methods

17.2.1 Stimuli

Each stimulus was a 1 s, 250-Hz band-limited noise centred at 500 Hz. Components were spaced at 1-Hz intervals and had random (approximately Gaussian) phase and amplitude. During the first 500 ms, the components were unmodulated, but at 500 ms, a 60-Hz sinusoidal FM was introduced on each component. The same complex was presented to each ear, but the FM was either diotic or dichotic.

In the diotic condition, the FM applied to the signal for each ear was identical, but in the dichotic condition there was a 180° phase delay between the FM applied to the signals for the left and right ears, so the dichotic stimuli effectively contained a 60-Hz interaural phase modulation. The band-limited noise was generated afresh for each presentation of the stimuli, and the stimuli were gated with 1-ms raised cosine rise and fall times, as was the onset of the FM. Modulation of the noise did not add energy to the stimulus, except for possible spectral splatter at the onset.

17.2.2 *Experiment 1: Pitch Matching*

Two experienced listeners matched the pitch of the dichotic FM to a pure tone. The depth of FM was set just above detection threshold for each participant, so that the modulation could be comfortably perceived; the modulation index was 0.3 for Listener 1 and 0.2 for Listener 2. Stimuli were generated on a laptop computer and presented through Sennheiser HD-50 earphones. An exemplar of the dichotic FM complex was followed after a 500-ms pause by a 500-ms pure tone. The frequency of the pure tone was first set to 1 kHz and the listener used a joystick to adjust its frequency until the pitches of the alternating FM-noise and pure tone matched. Initial adjustments were made with minimum steps of 10 Hz and a maximum of about 90 Hz was achievable with maximum displacement of the joystick. Listeners were then able to change the adjustment scale to a minimum step size of 1 Hz. This experimental protocol was based on similar work by Roberts and Brunstrom (2001). The listeners usually achieved a match with 20–30 stimulus pair presentations.

17.2.3 *Experiment 2: MEG*

MEG recordings were made using a 275 channel CTF MEG system with third-order axial gradiometers, while participants were seated with their eyes open. Stimuli were presented through echoless plastic tubing with foam ear inserts. One hundred exemplars of each of the diotic and dichotic FM noise stimuli were presented in a pseudorandom order with a 1-s silent interstimulus interval, and data were segmented into 1.5-s epochs starting at 500 ms before stimulus onset. Each epoch was baseline-corrected using the pre-stimulus period and comb-filtered to remove powerline artefacts (50 Hz).

Data for seven listeners were analysed at the sensor level by averaging the epochs for each stimulus condition. Volumetric source analysis was performed using a beamformer method, synthetic aperture magnetometry (SAM; Robinson and Vrba 1999), which uses the covariance matrix of the data to compute a set of weights, which effectively apply a spatial filter to the data. SAM is most commonly used to localise non-stimulus-locked or "induced" activity with particular time–frequency characteristics (see Hillebrand et al. 2005 for review), but more recent implementations have developed this approach to localise evoked activity by calculating the average evoked power in the SAM time series for a particular latency (Cheyne et al. 2007). The benefit of using SAM is that it provides excellent artefact rejection (e.g. Adjamian et al. 2009), obviating the need to remove trials with eye movements, for example, thus providing a strong signal-to-noise ratio in reconstructed time series. Additionally, no priori knowledge about the number of active sources is required, as is the case with dipole modelling.

The MEG dataset for each participant was spatially coregistered with a structural MRI using a modification of the surface-matching approach described in Adjamian et al. (2004). A source space with 5 mm voxels was derived from the scalp outline for each individual, and a multisphere head model was used to model the volume conduction. Beamformer methods tend to minimise sources where the time course of activity is tightly correlated (Hadjipapas et al. 2005; Brookes et al. 2007). For an evoked response resulting from binaural stimulation where activity in the left and right hemispheres may be tightly stimulus-locked, at least some proportion of the activity is likely to be correlated. We overcame this potential problem by analysing the data for each side of the head (i.e. left- and right-hemisphere channels) separately (Herdman et al. 2003). For the initial source model, the SAM weights were computed over the 5–25-Hz frequency band (chosen based on a pilot spectral analysis), and over the two time periods chosen to capture: first, the evoked responses to the onset of the sound (50–250 ms) and second, the onset of the frequency modulation (550–750 ms). Evoked SAM activations were computed using the method described by Cheyne et al. (2007). For each individual, the latency of the peak response evoked by the onset of sound and the onset of dichotic FM was recorded, and evoked SAM images were computed for these latencies using the SAM weights corresponding to that time period. "Virtual electrode" time series were then reconstructed for the peak voxels of the evoked SAM images for each individual. Averaging these data across epochs provided a spatially filtered evoked response for the cortical location. Time-frequency spectrograms were also produced for these averaged time series using Morlet wavelet analysis.

17.3 Results

17.3.1 Experiment 1: Pitch Matching

Listeners reliably matched the pitch of the dichotic FM stimulus to a pure tone: Listener 1's matching frequency was 331 Hz (SD: 1.89 Hz) and for Listener 2 the frequency was also 331 Hz (SD: 1.98). The small standard-deviations of the pitch matches are within the bounds suggested by frequency resolution for discrimination of pure tones within this frequency range (Henning 1966).

17.3.2 Experiment 2: MEG Data

All seven participants were able to report a percept of pitch and reported that it was strongly lateralised. Figure 17.1a shows time series data from a single MEG sensor in the left and right hemisphere, for one participant. Each plot shows a clear N1–P2 complex in response to the onset of sound, which is the same in the dichotic and diotic condition. The trace for the dichotic condition shows a clear evoked response in the right hemisphere with a positive peak at 670 ms, 170 ms after the onset of FM, and followed by a slower negative wave. This evoked response is less evident in the dichotic response for the left hemisphere, and is not present in the traces for the diotic condition in either hemisphere. Figure 17.1b shows field patterns for the same participant, for the time point at 120 ms post-stimulus (the peak of the N1m response for this participant) and at 170 ms post-stimulus onset at the peak of the response to the FM. The field pattern for the response to the onset of sound is roughly symmetrical, but the field pattern for the onset of the dichotic pitch is stronger in the right. This listener, a right-handed male, had reported the percept of pitch to be lateralised in his left ear.

Figure 17.2 shows volumetric images of the peak evoked SAM activations for the "P2" component of the response to the onset of sound, and to the response to

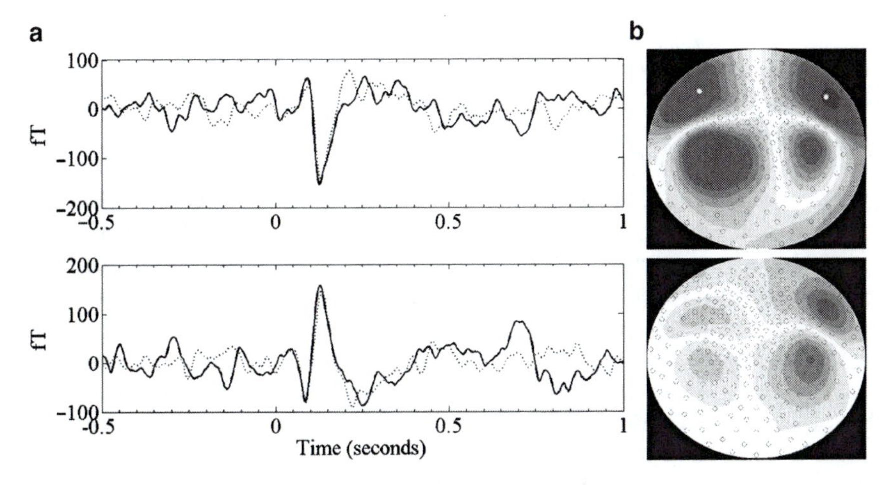

Fig. 17.1 (**a**) Shows averaged traces from a single sensor in each of the left (*upper panel*) and right (*lower panel*) hemispheres for a single participant. Each sensor was located at the peak of the anterior portion of the field pattern to the onset of sound (see below). The *dotted lines* indicate responses evoked by the diotic FM noise stimulus and *unbroken lines* of the dichotic stimulus. Sound onset was at 0 ms and FM onset was at 500 ms. Data have been band-pass filtered between 1 and 30 Hz. (**b**) Shows field patterns from the average from the dichotic condition taken at the largest peak of the onset response (210 ms in this participant) and at the peak of the response to the onset of FM (170 ms). The black-and-white scale obscures the fact that the dipolar patterns were in mirror-image across hemispheres, as expected for an auditory response. The *white dots* indicate the locations of the sensors in (**a**)

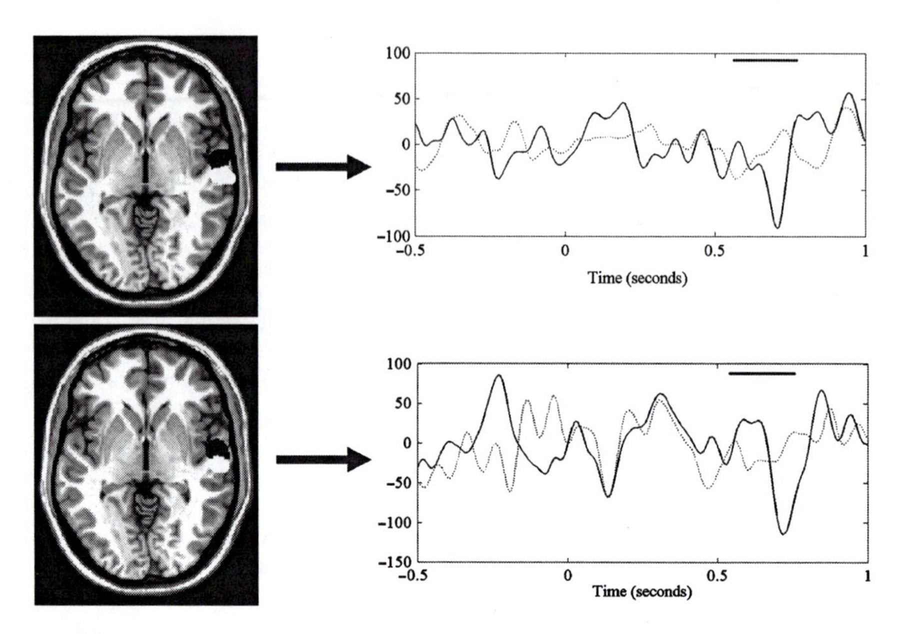

Fig. 17.2 The brain images show the peak of the evoked-SAM response at the latency of the "P2" response (*black mark*) and the latency of the response to the onset of dichotic FM (*white mark*) for the right hemisphere only in two participants. The *right-hand panels* show the normalised averaged virtual electrode time series data for the voxel indicated in *white* (peak dichotic FM response). *Unbroken lines* indicate the dichotic condition and *dotted lines* the diotic condition; *horizontal bars* indicate the time period for which SAM weights were computed. Data have been band-pass filtered from 1 to 10 Hz

the onset of dichotic FM for two participants in their right hemisphere. Previous research (Lütkenhöner and Steinsträter 1998) has shown that the P2 component of the onset response originates from primary auditory cortex in Heschl's gyrus and is replicated in the data presented here. The response to the onset of dichotic FM is clearly adjacent, but posterior to the P2 response. The evoked virtual electrode time series data for the right hemisphere voxel with the peak response to the onset of dichotic FM is shown adjacent to the brain images. A negative evoked response following the onset of dichotic FM is observed for each participant, confirming that this voxel is a source of an evoked waveform like that presented in the sensor level analysis in Fig. 17.1. Note that the response to the onset of sound is not expected to be observed in this voxel.

Figure 17.3 shows wavelet spectrograms of the evoked activity in the diotic (left) and dichotic (right) conditions for the same two participants. The white box in the right-hand spectrograms highlights the spectral power change on which the evoked-SAM source localisation was based. The dichotic FM evokes a burst of activity predominantly in the theta and alpha bands, which is relatively well sustained for the duration of the dichotic FM; such a pattern of activity is not observed in the diotic condition.

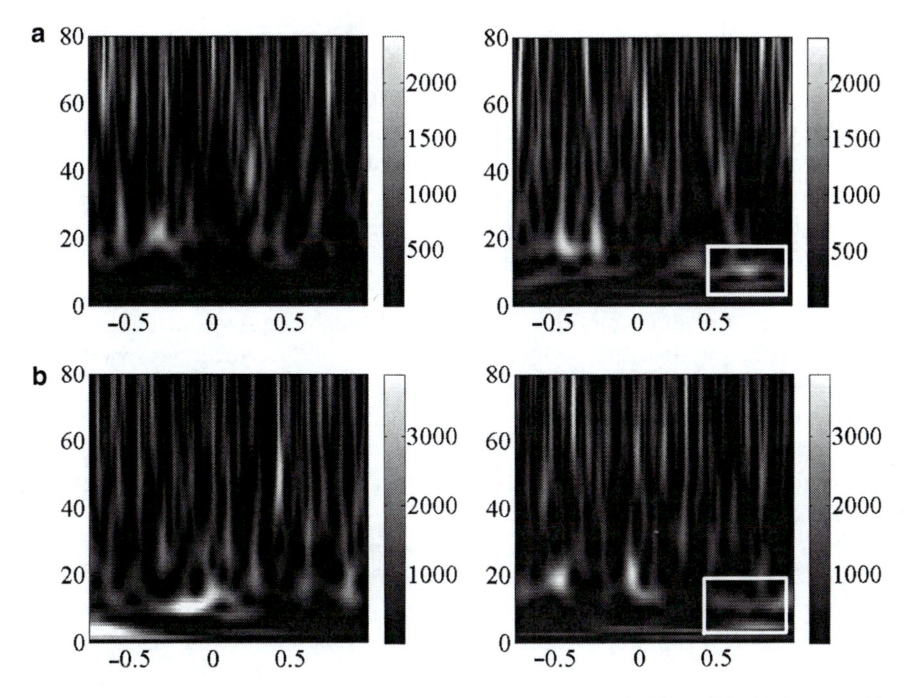

Fig. 17.3 Wavelet spectrograms of the average traces are shown in Fig. 17.2. (**a**) (*top panels*) shows the diotic (*left*) and dichotic (*right*) conditions for one participant and (**b**) shows the same conditions for the second participant. The *white box* indicates the spectral power change associated with the evoked-SAM localisation

17.3.3 Discussion

The observation that both listeners were able to reliably match the percept associated with the dichotic FM to a pure tone supports our report (Witton et al. 2005) that the dichotic FM noise stimuli induce a strong pitch. Both listeners consistently matched it to the same frequency, 331 Hz; however, it is not clear how this frequency relates to the binaural properties of the stimulus.

For Huggins pitch and the binaural edge pitches, the height of the pitch is systematically related to the frequency at which the interaural delay is introduced: the centre frequency for a Huggins pitch and the edge frequency for the edge pitches. With frequency modulation, the energy in the carrier is distributed through spectral sidebands placed at multiples of the modulation frequency above and below the centre frequency, so additional spectral components are added to the noise components on the introduction of diotic or dichotic FM, with the distribution of energy between each sideband and carrier determined by the modulation index (Goldman 1948). At relatively low modulation depths, such as those used in this study, only a single pair of spectral sidebands are likely to contain significant energy. Our noise stimuli were centred on 500 Hz with a 250 Hz bandwidth, so noise spectral components were present between 375 and 625 Hz. During the FM, a lower sideband of the 375-Hz component is present at 315 Hz, but our listeners consistently matched the pitch of the dichotic FM noise to 331 Hz. Pitch estimates of Huggins and binaural edge pitches are usually consistent and within 1–2% of the spectral frequency of interest (Klein and Hartmann 1981; Hartmann and McMillon 2001) and our pitch-match was 5% above the 315-Hz component. Although these data alone cannot test any hypothesis of how the pitch is generated, it seems unlikely that participants are simply detecting a lower spectral sideband of the FM noise.

A percept of pitch for diotic AM and FM, at modulation frequencies above the critical modulation frequency where sidebands exceed the critical bandwidth, has been described (Zwicker 1962). Also known as the "residue" (Schouten et al. 1962; Ritsma 1962), this has a pitch close to the modulation frequency. Although 60 Hz may have been beyond the critical modulation frequency for our listeners at our carrier frequency, the pitch of our tone is higher than expected if they were only detecting a binaural residue pitch.

Our MEG data can be compared with those of Hertrich et al. (2005) and Chait et al. (2006), who contrasted neuromagnetic response to Huggins pitch with responses to "non-binaural" pitches (iterated rippled noise and tone-in-noise, respectively). Huggins pitch evokes a pitch onset response with a latency of around 160 ms, dependent on pitch height. We observed a pitch-onset response of similar morphology and latency with our stimulus. Our evoked SAM analysis provided accurate localisation of the source of response to the onset of dichotic FM, posterior to Heschl's gyrus and to the source of the P2 component of the sound onset response. This localisation is in accordance with previous reports for the localisation of pitch-sensitive areas in auditory cortex; posterior to Heschl's gyrus and extending into planum temporale (Hall and Plack 2008).

The spectrograms in Fig. 17.3 show that our dichotic FM stimulus evokes a burst of oscillatory activity, predominantly in the theta and alpha frequency bands (4–12 Hz). Theta–alpha activity is frequently implicated in top–down processes such as attention (e.g. Gootjes et al. 2006). It has been suggested that pitch-onset responses, especially where pitch is extracted from noise, may reflect auditory "object" extraction (Chait et al. 2006; Hall and Plack 2008). A potential confounding factor in our comparison between diotic and dichotic FM is that the dichotic stimulus also induces a change in spatial "spread" between the ears due to the reduction in interaural correlation although this is very subtle at the modulation depths used here. It is possible that the response reflects a combination of the extraction of pitch and spatial characteristics of the sound; nevertheless, the data show a striking correspondence with previous measurements of pitch-onset responses in morphology and spatial origin.

In summary, this study shows that dichotic FM of a band-limited noise stimulus elicits a percept of pitch, which is strongly lateralised; we do not believe that this binaural pitch has been described previously. Cortical responses to the onset of this binaural pitch are similar to those previously described for Huggins pitch and other non-binaural pitches, providing further evidence that pitch is extracted by listeners presented with this stimulus. Further work is needed to explore how the binaural spectral and/or temporal characteristics of dichotic FM elicit a pitch.

References

Adjamian P, Barnes GR, Hillebrand A et al (2004) Co-registration of magnetoencephalography with magnetic resonance imaging using bite-bar-based fiducials and surface-matching. Clin Neurophysiol 115:691–698

Adjamian P et al (2009) Effective electromagnetic noise cancellation with beamformers and synthetic gradiometry in shielded and partly shielded environments. J Neurosci Methods. doi:10.1016/j.jneumeth.2008.12.006

Brookes MJ, Stevenson CM, Barnes GR et al (2007) Beamformer reconstruction of correlated sources using a modified source model. Neuroimage 34:1454–1465

Chait M, Poeppel D, Simon JZ (2006) Neural response correlates of detection of monaurally and binaurally created pitches in humans. Cereb Cortex 16:835–848

Cheyne D, Bostan AC, Gaetz W, Pang EW (2007) Event-related beamforming: a robust method for presurgical functional mapping using MEG. Clin Neurophysiol 118:1691–1704

Cramer EM, Huggins WH (1958) Creation of pitch through binaural interaction. J Acoust Soc Am 30:413–417

Goldman S (1948) Frequency analysis, modulation and noise. McGraw-Hill, New York

Gootjes L, Bouma A, van Strien JW et al (2006) Attention modulates hemispheric differences in functional connectivity: evidence from MEG recordings. Neuroimage 30(1):245–253

Hadjipapas A, Hillebrand A, Holliday IE et al (2005) Assessing interactions of linear and nonlinear neuronal sources using MEG beamformers: a proof of concept. Clin Neurophysiol 116:1300–1313

Hall DA, Plack CJ (2008) Pitch processing sites in the human auditory brain. Cereb Cortex 19(3):576–585. doi:10.1093/cerecor/bhn108

Hartmann WM, McMillon CD (2001) Binaural coherence edge pitch. J Acoust Soc Am 109:294–303

Henning GB (1966) Frequency discrimination of random amplitude tones. J Acoust Soc Am 39:336–339

Herdman AT, Wollbrink A, Chau W, Ishii R, Ross B, Pantev C (2003) Determination of activation areas in the human auditory cortex by means of synthetic aperture magnetometry. Neuroimage 20:995–1005

Hertrich I, Mathiak K, Menning H et al (2005) MEG responses to rippled noise and Huggins pitch reveal similar cortical representations. Neuroreport 16:193–196

Hillebrand A, Singh KD, Holliday IE et al (2005) A new approach to neuroimaging with magnetoencephalography. Hum Brain Mapp 25:199–211

Klein MA, Hartmann WM (1981) Binaural edge pitch. J Acoust Soc Am 70:51–61

Krumbholz K, Patterson R, Seither-Preisler A et al (2003) Neuromagneteic evidence for a pitch-processing center in Heschl's gyrus. Cereb Cortex 13:765–772

Lütkenhöner B, Steinsträter O (1998) High-precision neuromagnetic study of the functional organisation of the human auditory cortex. Audiol Neurootol 3(2–3):191–213

Ritsma RJ (1962) Existence region of the tonal residue. J Acoust Soc Am 34:1224–1229

Roberts B, Brunstrom JM (2001) Perceptual fusion and fragmentation of complex tones made inharmonic by applying different degrees of frequency shift and spectral stretch. J Acoust Soc Am 110:2479–2490

Robinson SE, Vrba J (1999) Functional neuroimaging by synthetic aperture magnetometry (SAM). In: Yoshimoto T, Kotani M, Kuriki S, Karibe H, Nakasato N (eds) Recent advances in biomagnetism. Tohoku University Press, Sendai, pp 302–305

Schouten JF, Ritsma RJ, Lopes Cardozo B (1962) Pitch of the residue. J Acoust Soc Am 34:1418–1424

Witton C, Green GGR, Henning GB (2005) Binaural 'sluggishness' as a function of stimulus bandwidth. In: Pressnitzer D, de Cheveigné A, McAdams S et al (eds) Auditory signal processing: physiology, psychophysics, and models. Springer, New York

Zwicker E (1962) Direct comparison between the sensations produced by frequency modulation and amplitude modulation. J Acoust Soc Am 34:1425–1430

Chapter 18
A Temporal Code for Huggins Pitch?

Christopher J. Plack, Suzanne Fitzpatrick, Robert P. Carlyon, and Hedwig E. Gockel

Abstract Periodic sound waves produce periodic patterns of phase-locked activity in the auditory nerve and in nuclei throughout the auditory brainstem. It has been suggested that this temporal code is the basis for our sensation of pitch. However, some stimuli evoke a pitch without monaural pitch information (temporal or otherwise). Huggins pitch (HP) is produced by presenting the same wideband noise to both ears except for a narrow frequency band which is interaurally decorrelated. "Complex" HP (CHP) can be produced by generating HP components at harmonic frequencies. The frequency-following response (FFR) is an electrophysiological measure of phase locking in the upper brainstem. The FFR was measured for a 300-Hz CHP in a 0–2 kHz noise, and a perceptually similar stimulus comprising a series of narrowband noise (NBN) harmonics of a 300-Hz fundamental presented in a 0–2 kHz background noise at different relative levels of NBN and background. The FFR measurements revealed a phase-locked response to the NBN harmonics, even for NBN stimuli with pitch salience below that of the CHP. Little evidence for phase locking to the CHP stimulus was found, although there was a weak component in the FFR at 300 Hz relative to neighboring frequencies. The results suggest that HP is not associated with an enhanced temporal response to the decorrelated frequency band at the level of the upper brainstem.

Keywords Huggins pitch • Phase locking • Frequency-following response

18.1 Introduction

Pitch can be defined as the sensation whose variation is associated with musical melodies (Plack et al. 2005). Periodic sounds are composed of a harmonic series: a set of sinusoidal partials whose frequencies are integer multiples of the repetition

---

C.J. Plack (✉)
Division of Human Communication and Deafness,
University of Manchester, Manchester, UK
e-mail: chris.plack@manchester.ac.uk

E.A. Lopez-Poveda et al. (eds.), *The Neurophysiological Bases of Auditory Perception*,
DOI 10.1007/978-1-4419-5686-6_18, © Springer Science+Business Media, LLC 2010

rate, or fundamental frequency (F0). These complex tones give rise to a "residue" pitch that corresponds to F0. Each place on the basilar membrane (BM) in the cochlea is tuned to a particular characteristic frequency (CF), and neurons in the auditory nerve which innervate a particular region of the BM will tend to fire in phase with the vibration in that region, a phenomenon called "phase locking." Hence, periodic sound waves produce periodic patterns of phase-locked activity in the auditory nerve, and in nuclei throughout the auditory brainstem, providing a potential temporal code for pitch. Changes in the F0 of a sound will be reflected by changes in the pattern of phase locking. In addition, as the frequencies of low-numbered harmonics change, so do the regions of the BM that vibrate maximally. This will, in turn, affect which auditory-nerve fibers fire most, producing a potential "place-of-excitation" or "rate-place" code for pitch.

There is some evidence that the temporal code is the basis for our sensation of musical pitch. For example, the perception of musical melody is limited to frequencies (less than about 5 kHz) that have been shown to generate phase locking (Attneave and Olson 1971). Several successful neurally based models of pitch perception are based on the temporal code (Licklider 1951; Patterson 1994; Meddis and O'Mard 1997; de Cheveigné 1998). Although there is still some debate as to whether or not the temporal code is *sufficient*, by itself, to explain pitch perception for complex tones (Oxenham et al. 2004), it is a reasonable hypothesis that the temporal code is *necessary* for the perception of musical pitch.

A potential problem with the temporal code hypothesis is that there are some stimuli that produce a clear musical pitch, but for which there is no physiological evidence that the pitch is coded temporally. Huggins pitch (HP) is produced by presenting the same wideband noise to both ears, except for a narrow frequency band which is interaurally decorrelated (Cramer and Huggins 1958). Complex HP (CHP) can be produced by generating HP components at harmonic frequencies (Bilsen 1977; Gockel et al. in press). HP stimuli produce a clear musical pitch, supporting melody recognition (Akeroyd et al. 2001). However, although there is phase locking to the activity at each place on the BM excited by the noise, there is no monaural information (temporal or otherwise) concerning the frequency of the decorrelated band, since the input to each ear is just a random noise.

It is thought that the medial superior olive (MSO) in the brainstem combines the information from the two ears and enhances or extracts the decorrelated band of an HP stimulus. In response to a wideband noise, the outputs of auditory-nerve fibers in the two ears with a given CF will resemble the outputs to narrowband noises (NBNs) centered on that CF. For an HP stimulus, the difference between these two auditory-nerve outputs (as could be derived in the MSO) will also resemble the response to a NBN centered on the decorrelated band. The amplitude of this "difference response" noise will be greater in neurons tuned to CFs where the noise is interaurally decorrelated, giving rise to a potential place code (cf. Durlach 1963), but the pitch mechanism could alternatively work on the temporal structure in the difference response (reflected in the pattern of phase locking), which would also resemble that of an NBN. A crucial question, therefore, is how the frequency of the decorrelated band is represented: is there phase locking to the band at the level of MSO and above, or is the frequency coded by a rate-place representation? If the latter, then a temporal code is not necessary for pitch.

The frequency-following response (FFR) is an electrophysiological measure of phase locking at the level of the upper brainstem, using electrodes attached to the scalp. The FFR is thought to reflect phase-locked activity at the level of the lateral lemniscus (LL) and/or inferior colliculus (IC) (Krishnan 2006). Periodic sounds produce periodic electric potentials, and hence the FFR provides a potential measure of temporal pitch coding in the human brainstem (Greenberg et al. 1987). Furthermore, there is evidence that the FFR reflects the perceptual enhancement of a pure tone when presented at a different interaural phase to a masking noise in a "binaural masking level difference" condition (Wilson and Krishnan 2005), suggesting that the measure is sensitive to binaural processing.

In the present experiment, the FFR was used to test the hypothesis that CHP produces a temporal pitch code. The FFR to CHP was compared to the FFR to a perceptually similar complex-tone stimulus containing NBN harmonics presented in a noise background. The level of the harmonics relative to the background was varied to produce a range of NBN stimuli with a pitch salience less than, and greater than, that evoked by the CHP, estimated using behavioral F0 difference limens (F0DLs). If the FFR revealed that CHP is associated with enhanced phase locking to the regions of decorrelation, then that would be evidence for a temporal code. If no such activity were observed, yet there was an FFR to the NBN stimulus matched in pitch salience, this would be evidence against the generation of a temporal code for CHP.

18.2 Methods

18.2.1 Listeners

Three normally hearing listeners (<20 dB HL between 250 Hz and 8 kHz) were tested in the behavioral experiment. All were highly experienced in psychoacoustic testing. Their ages were 25, 27, and 35 years. Nine different normally hearing listeners were tested in the FFR experiment. Their mean age was 23.3 years old, ranging from 19 to 34 years.

18.2.2 Stimuli

The CHP stimuli consisted of a Gaussian noise lowpass filtered at 2 kHz and presented diotically, except for five 30-Hz wide frequency regions in which a π phase shift was generated between the ears. The center frequencies of the phase-shift bands were 300, 600, 900, 1,200, and 1,500 Hz (i.e., the first five harmonics of a 300-Hz F0).

The NBN stimuli consisted of a background diotic Gaussian noise lowpass filtered at 2 kHz, to which was added five 30-Hz wide bands of diotic rectangular NBN. The center frequencies of the bands were the same as for the phase-shift bands in the CHP stimulus. The spectrum level of the NBN harmonics was 3 ("NBN+3"), 6 ("NBN+6"), 9 ("NBN+9"), or 12 ("NBN+12") dB greater than the spectrum level of the background noise. For the FFR, a diotic reference stimulus (REF) was generated using either the left or the right channel (alternated between runs) of the CHP stimulus. The overall level for all the stimuli was fixed at 80 dB SPL. Total duration was 150 ms including 10-ms raised-cosine onset and offset ramps. The spectral characteristics of the stimuli are illustrated in Fig. 18.1. For

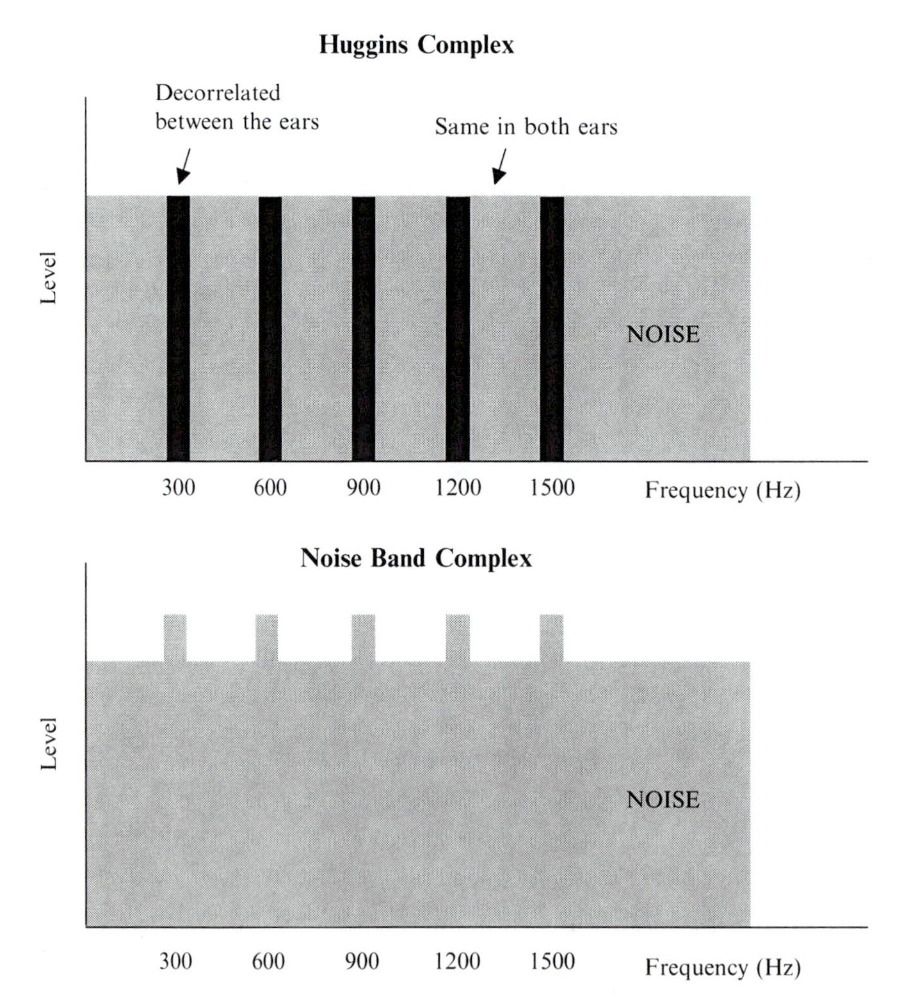

Fig. 18.1 Schematic spectra of the stimuli used in the experiment

both the CHP and NBN stimuli, an independent sample of noise was used for each presentation for the behavioral measurements, but the noise was frozen (i.e., same waveform on each presentation) for the FFR measurements.

18.2.3 F0DL Measurement Procedure

For each stimulus (except the REF), the F0DL was measured using a three-interval, three-alternative forced choice task. The interstimulus interval was 500 ms. Two intervals contained the standard stimulus with a 300-Hz F0. One interval (chosen at random) contained a comparison stimulus with a lower F0. The listeners' task was to select the interval with the lower F0 by pressing a button on a computer keyboard. The percentage difference between the F0s of standard and comparison was varied adaptively using a geometric track with a two-down, one-up rule. A run consisted of 16 reversals (changes in track direction). The step size was a factor of two for the first four reversals, and a factor of 1.414 for the remaining 12 reversals. The F0DL was taken as the geometric mean of the F0 difference at the last 12 reversals. For each block all five conditions were presented in a random order. At least five such blocks were run for each listener, and the geometric mean F0DL was taken as the final estimate.

Listeners were tested in an IAC double-walled sound-attenuating room. Stimuli were presented via Sennheiser HD580 headphones and a computer display provided feedback after each trial.

18.2.4 FFR Recording Procedure

Listeners reclined comfortably in an IAC double-walled sound-attenuating room. Stimuli were presented, and FFR waveforms recorded, using Intelligent Hearing Systems hardware and software. Mu-metal shielded etymotic ER2 insert earphones with E-A-R LINK foam ear-tips were used to present the stimuli. Stimuli were presented at a rate of 3.93 per second, and each run consisted of 1,024 stimulus presentations. A batch consisted of runs of CHP, REF, NBN+3, NBN+6, NBN+9, NBN+12, in this order. This was repeated two or three times for each listener.

A single-channel recording system was used. Evoked potentials were recorded differentially between electrodes attached to the midline of the forehead just below the hairline and to the C7 vertebra at the base of the neck. A common-ground electrode was positioned on the midline at the lower forehead (Fpz). Inter-electrode impedances were maintained below 3,000 Ohms. The electroencephalogram inputs were amplified by 150,000, filtered between 100 and 1,500 Hz, and recorded digitally at a sampling rate of 5,714 Hz.

FFR recordings for each stimulus run and each listener were gated from 5 to 155 ms relative to stimulus onset. A discrete Fourier transform was used to extract the spectral power in 30-Hz wide bands centered at, and midway between, the harmonic frequencies. The spectral power from individual runs was averaged for each listener.

18.3 Results

18.3.1 F0DLs

The pattern of behavioral results was similar across the three listeners, and geometric mean results are shown in Fig. 18.2. The NBN thresholds show a progressive decline (increasing salience) as the level of the noise bands increases relative to the background. The results suggest that the CHP and NBN+6 stimuli are roughly equally salient, and that the CHP stimulus is more salient than the NBN+3 stimulus.

18.3.2 FFR

Figure 18.3 shows the mean level of each 30-Hz wide band in the FFR for the test stimuli relative to the mean level of the same band in the FFR for the REF stimulus. For the NBN stimuli, peaks are evident at harmonic frequencies, particularly the first, third, and fifth harmonic. Recordings from three listeners with the earphones

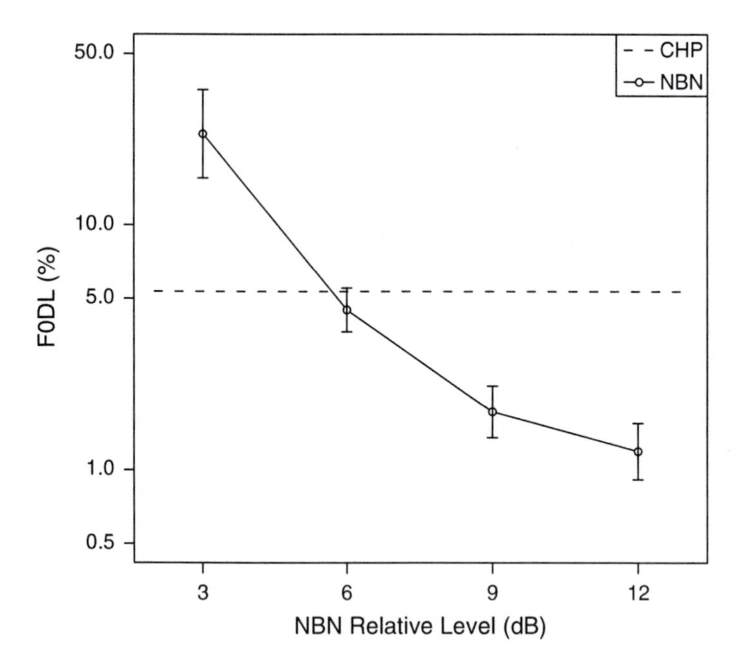

Fig. 18.2 The results of the behavioral experiment. The geometric mean F0DLs are plotted as a function of NBN level relative to the background. The *horizontal dashed line* shows the F0DL for the CHP stimulus. *Error bars* show standard errors across listeners for the NBN thresholds normalized to the CHP threshold for each listener (this shows to what extent the *pattern* of thresholds was consistent across listeners)

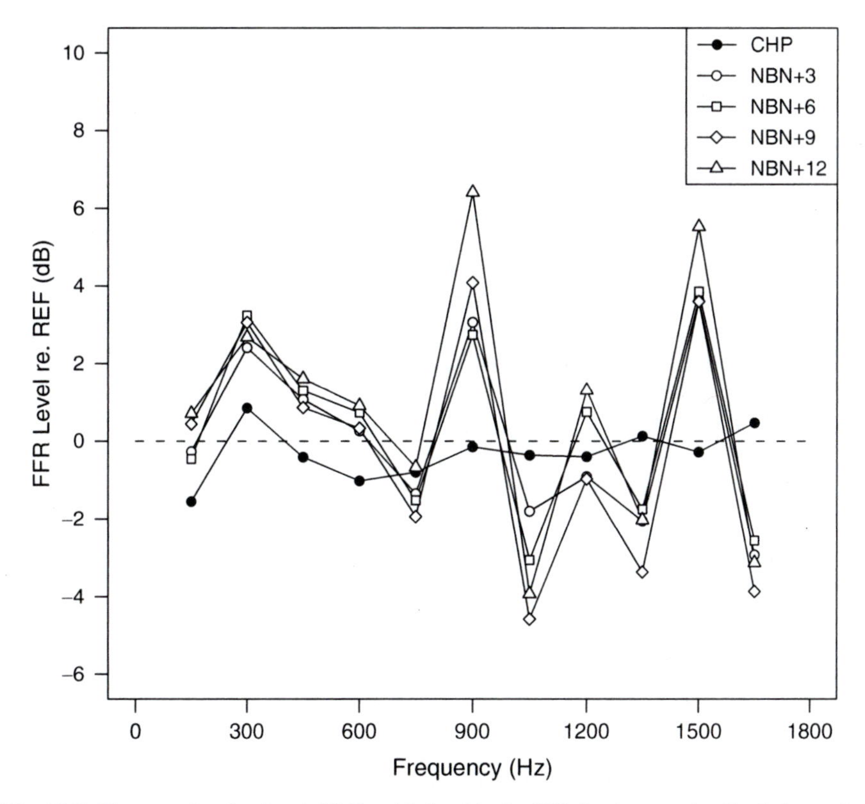

Fig. 18.3 The mean levels of each 30-Hz wide band in the FFR for the test stimuli relative to the mean level of the same band in the FFR for the REF stimulus. The *x*-axis shows the center frequency of the 30-Hz band

disconnected from the ears revealed that these peaks were not produced by stimulus artifact. For example, for these three listeners, the mean response to the NBN+12 stimulus with the earphones disconnected was 3.9, 3.2, 10.1, 4.7, and 6.0 dB below the mean response when the earphones were inserted, for harmonic frequencies of 300, 600, 900, 1,200, and 1,500 Hz respectively. Furthermore, the response to the REF stimulus with earphones inserted was greater than the disconnected response for all NBN stimuli at all measurement frequencies. In contrast to the results for the NBN stimuli, for the CHP stimulus there is little evidence for any enhanced response at harmonic frequencies, although there is a small peak at 300 Hz.

Planned comparisons of FFR level at 300 Hz revealed no significant difference between the CHP and REF ($t=0.70$, $df=8$, $p=0.25$, one tailed), but a nearly significant difference between NBN+3 and REF ($t=1.69$, $df=8$, $p=0.065$, one tailed), and a significant difference between NBN+6 and REF ($t=2.61$, $df=8$, $p=0.016$, one tailed). There was no significant difference between NBN+3 and CHP, but a significant difference between NBN+6 and CHP ($t=3.25$, $df=8$, $p=0.012$, two tailed). Hence, the NBN stimulus produces a significantly greater FFR response at 300 Hz than the CHP stimulus when presented with roughly equal perceptual salience.

Arguably, a more sensitive measure of the phase-locked response is to compare the response at the harmonic frequency with the response at intermediate frequencies above and below. This allows for the possibility that a response at the harmonic frequencies might be accompanied by a suppression of the response at intermediate frequencies. In a post-hoc comparison, the CHP response at 300 Hz expressed relative to REF, was significantly greater than the average CHP response at 150 and 450 Hz, both expressed relative to REF ($t=2.79$, $df=8$, $p=0.012$, one tailed). For this particular measure, there was no significant difference between CHP and NBN+3, or between CHP and NBN+6. So there is tentative evidence for an enhanced response at 300 Hz in the CHP stimulus compared to neighboring frequencies, although this test was not a planned comparison.

Finally, a 2-way repeated-measures ANOVA (NBN level X harmonic frequency) was conducted on the FFR level for the NBN stimuli relative to REF. The effects of both level [$F(1,8)=8.92$, $p=0.017$] and frequency [$F(4,32)=5.81$, $p=0.0012$] were significant with no significant interaction. There was a weak but significant linear increase in relative FFR strength with NBN level [$R(178)=0.14$, $p=0.028$, one tailed].

18.4 Discussion

The results do not provide convincing evidence for a temporal response to CHP. The FFR to CHP was less than that for an NBN stimulus with similar salience, and was only significantly different from the REF when the response at 300 Hz was compared to that at 150 and 450 Hz. The results provide only very tentative support for an *enhanced* phase-locked response to the decorrelated band relative to the background, and only at one frequency. However, it could be argued that the phase-locked responses to the decorrelated bands were tagged by the binaural system in another way, such as a separate place code indicating which frequency channels were uncorrelated. This might still allow for the frequencies of the decorrelated bands (or the CHP F0) to be extracted by a subsequent temporal mechanism if the channels were tagged by a place code at the level of the LL/IC. In this scenario, since the spectrum of the CHP stimulus was flat up to 2,000 Hz, the FFR would not reveal any enhanced phase locking to the HP bands relative to neighboring frequency bands. Alternatively, it is possible that the binaural information from the MSO was swamped by the monaural response to the noise. This may not have occurred for the NBN stimuli, since there are clear spectral peaks in the monaural representation. Finally, it is conceivable that the binaural temporal information has a different dipole orientation compared to the monaural information, thereby generating only a weak response. At the very least, however, we can say that the representations of a CHP stimulus and an NBN stimulus with equal pitch salience differ in the upper brainstem.

A much stronger FFR was seen for the NBN stimuli. A response was observed even for NBN+3, which has a very low pitch salience (much less than that of the

CHP stimulus). Although there was considerable variability in the data, the results showed an effect of pitch salience on FFR strength in the expected direction, largely driven by the NBN+12 condition. The results show that phase locking to noise stimuli is measurable in the human brainstem, even for stimuli whose pitch is only just detectable.

Acknowledgments The authors are most grateful to Dr. Ravi Krishnan for his expert advice and support regarding the recording equipment, the experimental design, and the interpretation of the FFR.

References

Akeroyd MA, Moore BCJ, Moore GA (2001) Melody recognition using three types of dichotic-pitch stimulus. J Acoust Soc Am 110:1498–1504
Attneave F, Olson RK (1971) Pitch as a medium: a new approach to psychophysical scaling. Am J Psychol 84:147–166
Bilsen FA (1977) Pitch of noise signals: evidence for a central spectrum. J Acoust Soc Am 61:150–161
Cramer EM, Huggins WH (1958) Creation of pitch through binaural interaction. J Acoust Soc Am 30:413–417
de Cheveigné A (1998) Cancellation model of pitch perception. J Acoust Soc Am 103:1261–1271
Durlach NI (1963) Equalization and cancellation theory of binaural masking-level differences. J Acoust Soc Am 35:1206–1218
Gockel HE, Carlyon RP, Plack CJ (2009) Pitch discrimination interference between binaural and monaural or diotic pitches. J Acoust Soc Am 126:281–290
Greenberg S, Marsh JT, Brown WS, Smith JC (1987) Neural temporal coding of low pitch. I. Human frequency-following responses to complex tones. Hear Res 25:91–114
Krishnan A (2006) Frequency-following Response. In: Burkhard RF, Don M, Eggermont J (eds) Auditory evoked potentials: basic principles and clinical application. Lipincott Williams and Wilkins, Philadelphia
Licklider JCR (1951) A duplex theory of pitch perception. Experientia 7:128–133
Meddis R, O'Mard L (1997) A unitary model of pitch perception. J Acoust Soc Am 102:1811–1820
Oxenham AJ, Bernstein JGW, Penagos H (2004) Correct tonotopic representation is necessary for complex pitch perception. Proc Nat Acad Sci USA 101:1421–1425
Patterson RD (1994) The sound of a sinusoid: time-interval models. J Acoust Soc Am 96:1419–1428
Plack CJ, Oxenham AJ, Fay RR, Popper AN (eds) (2005) Pitch: neural coding and perception. Springer, New York
Wilson JR, Krishnan A (2005) Human frequency-following responses to binaural masking level difference stimuli. J Am Acad Audiol 16:184–195

Chapter 19
Understanding Pitch Perception as a Hierarchical Process with Top-Down Modulation

Emili Balaguer-Ballester, Nicholas R. Clark, Martin Coath, Katrin Krumbholz, and Susan L. Denham

Abstract Previous studies suggest that the auditory system uses a wide range of time scales to integrate pitch-related information, and that the effective integration time is both task- and stimulus-dependent. None of the existing models of pitch processing can account for such task- and stimulus-dependent variations in processing time scales. This study presents an idealized neurocomputational model, which provides a unified account of the multiple time scales observed in pitch perception. The model is evaluated using a range of perceptual studies and a neurophysiological experiment. In contrast to other approaches, the current model contains a hierarchy of integration stages and uses feedback to adapt the effective time scales of processing at each stage in response to changes in the input stimulus. The model suggests a key role for efferent connections from central to sub-cortical areas in controlling the temporal dynamics of pitch processing.

Keywords Pitch perception processing • Top-down modulation • Temporal dynamics of pitch

19.1 Introduction

Pitch, one of the most important features of auditory perception, is usually associated with periodicities in sounds. Hence, a number of models of pitch perception are based upon a temporal analysis of the neural activity evoked by the stimulus (Licklider 1951). Most of these models compute a form of short-term autocorrelation of the simulated auditory nerve activity using an exponentially weighted integration

E. Balaguer-Ballester (✉)
Computational Neuroscience group, Central Institute for Mental Health (ZI), Ruprecht-Karls University of Heidelberg, J5, Mannheim, Germany
e-mail: emili.balaguer@zi-mannheim.de

E.A. Lopez-Poveda et al. (eds.), *The Neurophysiological Bases of Auditory Perception*,
DOI 10.1007/978-1-4419-5686-6_19, © Springer Science+Business Media, LLC 2010

time window (Meddis and O'Mard 1997; Balaguer-Ballester et al. 2008). Autocorrelation models have been able to predict the reported pitches of a wide range of complex stimuli. However, choosing an appropriate integration time window has been problematic, and none of the previous models has been able to explain the wide range of time scales encountered in perceptual data in a unified fashion. These data show that, in certain conditions, the auditory system is capable of integrating pitch-related information over time scales of several hundred milliseconds (Hall and Peters 1981; Plack and White 2000), while at the same time being able to follow changes in pitch or pitch strength with a resolution of only a few milliseconds (Bregman et al. 1994; Wiegrebe 2001).

The trade-off between temporal integration and resolution is not exclusive to pitch perception, but is a general characteristic of auditory temporal processing (de Boer 1985); therefore, it appears that the integration time of auditory processing varies with the stimulus and task. To our knowledge, no model has yet quantitatively explained the stimulus- and task-dependency of integration time constants.

Another major challenge for pitch modeling is to relate perceptual phenomena to neurophysiological data. Functional brain-imaging studies strongly suggest that pitch is processed in a hierarchical manner (Kumar et al. 2007), starting in sub-cortical structures and continuing up through Heschl's Gyrus on to the *planum polare* and *planum temporale*. Within this processing hierarchy, there is an increasing dispersion in response latency, with lower pitches eliciting longer response latencies than higher pitches (Krumbholz et al. 2003). This suggests that the time window over which the auditory system integrates pitch-related information depends on the pitch itself. However, no attempt has yet been made to explain this latency dispersion.

In this study, we present a unified account of the multiple time scales involved in pitch processing (Balaguer-Ballester et al. 2009). We suggest that top-down modulation within a hierarchical processing structure is important for explaining the stimulus-dependency of the effective integration time for extracting pitch information. A highly idealized model, formulated in terms of interacting neural ensembles, is presented. The model represents a natural extension of previous autocorrelation models of pitch (Balaguer-Ballester et al. 2008) in a form resembling a *hierarchical generative* process (Friston 2003), in which higher (e.g., cortical) levels modulate the responses in lower (e.g., sub-cortical) levels via feedback connections. The model can account not only for a wide range of perceptual data, but also for neurophysiological data on pitch processing (Krumbholz et al. 2003).

19.2 Methods

The model consists of a feed-forward process, as well as a feedback process, which modifies the parameters of feed-forward processing. A schematic diagram of the model is shown in Fig. 19.1.

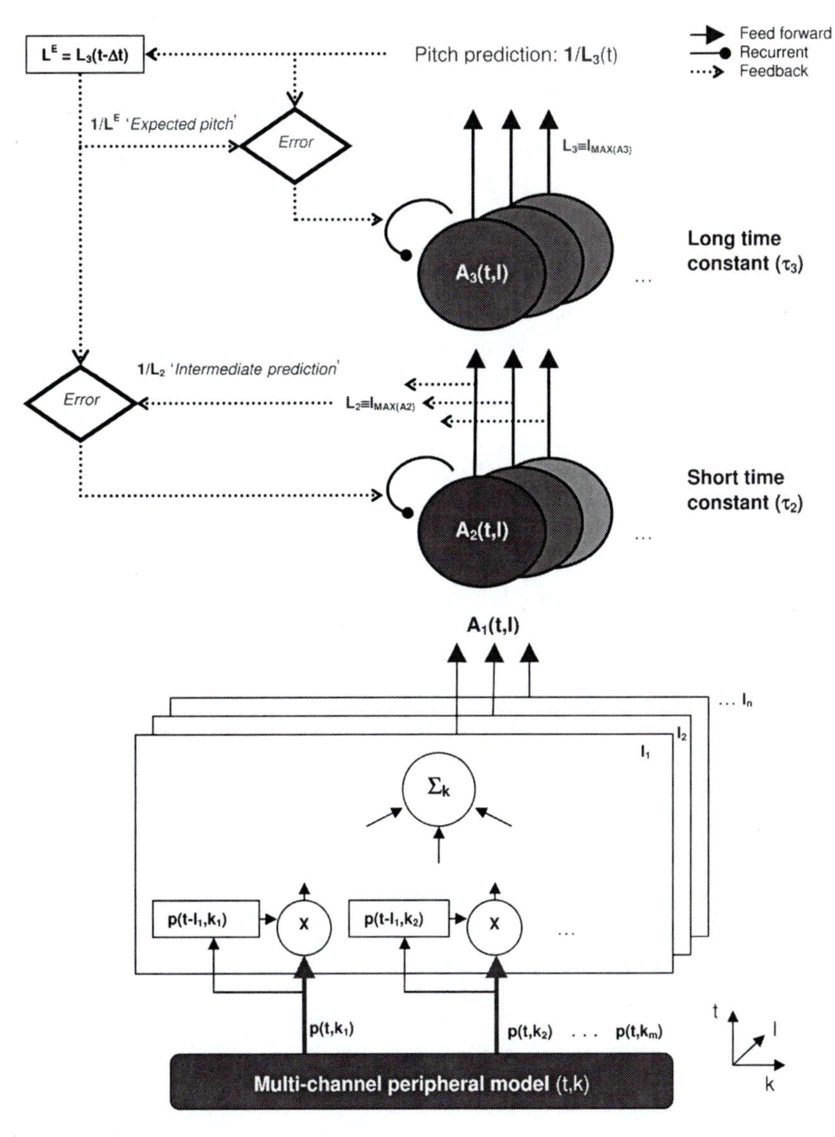

Fig. 19.1 Schematic outline of the model. The model consists of: (1) a simulation of auditory nerve spiking probabilities, $p(t,k)$, in response to a sound for each cochlear frequency channel, k; (2) a cross-product of the auditory nerve activity with a time-delayed version of itself for a range of different time lags, l (in the diagram, processing relating to different lags is represented by *stacked boxes*); (3) two integration stages, A_2 *and* A_3, shown by *ellipses*, which represent highly idealized models of collective neuronal responses using a shorter (τ_2) and a longer (τ_3) time constant, respectively. $L_2(t)$ is the lag yielding the maximum response at the second processing stage, $A_2(t,l)$; its inverse, $1/L_2(t)$ represents an intermediate pitch prediction of the model. Similarly, $1/L_3(t)$ represents the ultimate pitch estimate predicted by the model. When the pitch estimate changes over time, a mismatch between the previous pitch estimate at level 3 (labeled "expected pitch" or $1/L^E$) and the current prediction at the first integration stage, $1/L_2$, feeds back to modulate the recurrent processes (*curved lines*) at both integration stages. See text for details

19.2.1 Feed-Forward Processing

The role of the feed-forward process (solid lines in Fig. 19.1) is to predict the pitch of the incoming stimulus. Like in autocorrelation models of pitch perception, this model analyzes the periodicities of the signal within the auditory-nerve channels, and then uses these periodicities to derive a pitch estimate by computing the reciprocal of the periodicity that is most prevalent across frequency channels. In the current model, each cochlear filter was implemented as a *dual resonant nonlinear gammatone filter* and then passed through a hair cell transduction model (Lopez-Poveda and Meddis 2001; Sumner et al. 2002). The model was implemented twice using MAP (Matlab Auditory Peripheral, http://www.essex.ac.uk/psychology/psy/PEOPLE/meddis/webFolder08/WebIntro.htm) (Meddis and O'Mard 2006) and DSAM (Development System for Auditory Modelling http://www.pdn.cam.ac.uk/groups/dsam/).

The hair cell transduction model generates auditory-nerve spike probabilities, $p(t,k)$, as a function of time, t, in each frequency channel, k. The first processing stage (open boxes in Fig. 19.1) computes the joint probability that a given auditory nerve fiber produces two spikes, one at time t and another at $t-l$, where l is a time delay or lag. These joint probabilities are generated by computing the cross-product of the auditory-nerve firing probability, $p(t,k)$, with time-delayed versions of itself for a range of time delays. The cross-products are then summed across all frequency channels, k, to generate the output of the first stage of the model $A_1(t,l)$:

$$A_1(t,l) = \sum_k p(t,k)p(t-l,k) \tag{19.1}$$

The activity at the second processing stage, $A_2(t,l)$ (circles in Fig. 19.1), is computed as a leaky integration (i.e., a low-pass filter using an exponentially decaying function) of the input activity, $A_1(t,l)$, using relatively short time constants, τ_2. It may therefore be assumed to represent subthalamic neural populations. In the third stage, $A_3(t,l)$ (circles in Fig. 19.1), the output of the second stage is integrated over a longer time scale, τ_3. This stage is assumed to be located more centrally. Both integration stages can be simply described as time-varying exponential averages,

$$A_n(t,l) = A_n(t-\Delta t,l)e^{-\Delta t/E_n(t)} + \frac{\Delta t}{\tau_n}\frac{A_{n-1}(t,l)}{g_n(t)}, \quad n = 2,3. \tag{19.2}$$

In (19.2), Δt is the time step of the integration, and $E_n(t)$ is the instantaneous exponential decay rate of the response at each integration stage ($E_n(t) \leq \tau_n$), which will henceforth be referred to as the *effective integration window*. Establishing an appropriate time constant is, as has been mentioned, one of the major difficulties in formulating a general model of pitch perception. Hence, the value of $E_n(t)$ in the model proposed here is not constant but is controlled by changes in the properties of the stimulus.

The factors $g_n(t)$ normalize the input to each stage by the corresponding integration window ($g_2 \equiv 1$; $g_3(t) = E_2(t)/\tau_2$). The control of $E_n(t)$ will be discussed below.

At each time step, $A_n(t,l)$ will have a maximum at some value of l, which we will write as L_n. The inverse of this lag for the output of stage 2, $1/L_2(t)$, represents the intermediate pitch prediction of the model (see Fig. 19.1). Similarly, the inverse of the lag corresponding to the maximum response in stage 3, $1/L_3(t)$ is the final pitch prediction. For convenience, we refer to the final pitch prediction from the preceding time step $1/L_3(t-\Delta t)$ as the pitch *expectation,* $1/L^E$.

As a trivial example, Fig. 19.2 shows the model response to a sequence of pure tones (Fig. 19.2a) with random frequencies and durations. Figure 19.2b shows the first stage of the model $A_1(t,l)$ and Fig. 19.2c the effective integration windows. Figure 19.2d shows the final model output; the red color highlights the lag-channels with strong responses. The lag of the channel with the maximum response at a given time corresponds to the reciprocal of the pitch predicted by the model. Note that the response $A_3(t,l)$ in Fig. 19.2d was normalized to a maximum of unity after each time step and mapped exponentially onto the color scale to make the plot clearer.

19.2.2 Feed-Back Processing

The necessity for stimulus-driven modulation of the effective integration time, $E_n(t)$, becomes clear from a consideration of existing autocorrelation models. If $E_2(t)$ were constant over time, i.e., $E_2(t) \equiv \tau_2$, then $A_2(t,l)$ would correspond to the *summary autocorrelation function* (SACF) proposed by Meddis and O'Mard (1997). If, in addition, $E_3(t) \equiv \tau_3$ then $A_3(t,l)$ would represent an additional leaky integrator with a longer time constant. This is equivalent to the *cascade autocorrelation model* proposed by Balaguer-Ballester et al. (2008). The right panel in Fig. 19.3a illustrates the success of this purely feed-forward model in response to a click train stimulus with alternating inter-click intervals (Carlyon et al. 2008). The arrow indicates the average pitch reported by listeners. The pitch of such alternating click train stimuli has been difficult to predict with autocorrelation models consisting of only one integration stage with a short time constant (see right panel in Fig. 19.3b).

However, the longer time scale used in the second stage of the cascade autocorrelation model prevents the detection of rapid pitch changes such as in the sequence of pure tones shown in Fig. 19.2. The left panel in Fig. 19.3a clearly shows that the cascade autocorrelation model fails to distinguish the pitches of individual tones in the tone sequence used in Fig. 19.2, while the left panel in Fig. 19.3b shows that the SACF model does so fairly well. Therefore, stimulus-dependent changes in the effective integration windows are required.

A detailed description of the stimulus-dependent algorithm for $E_n(t)$ and of the parallels with more formal population models and Hierarchical Generative Bayesian models will appear in Balaguer-Ballester et al. (2009). A Matlab©-based software implementation of the model is freely available from the first author or in http://www.zi-mannheim.de/1944.html.

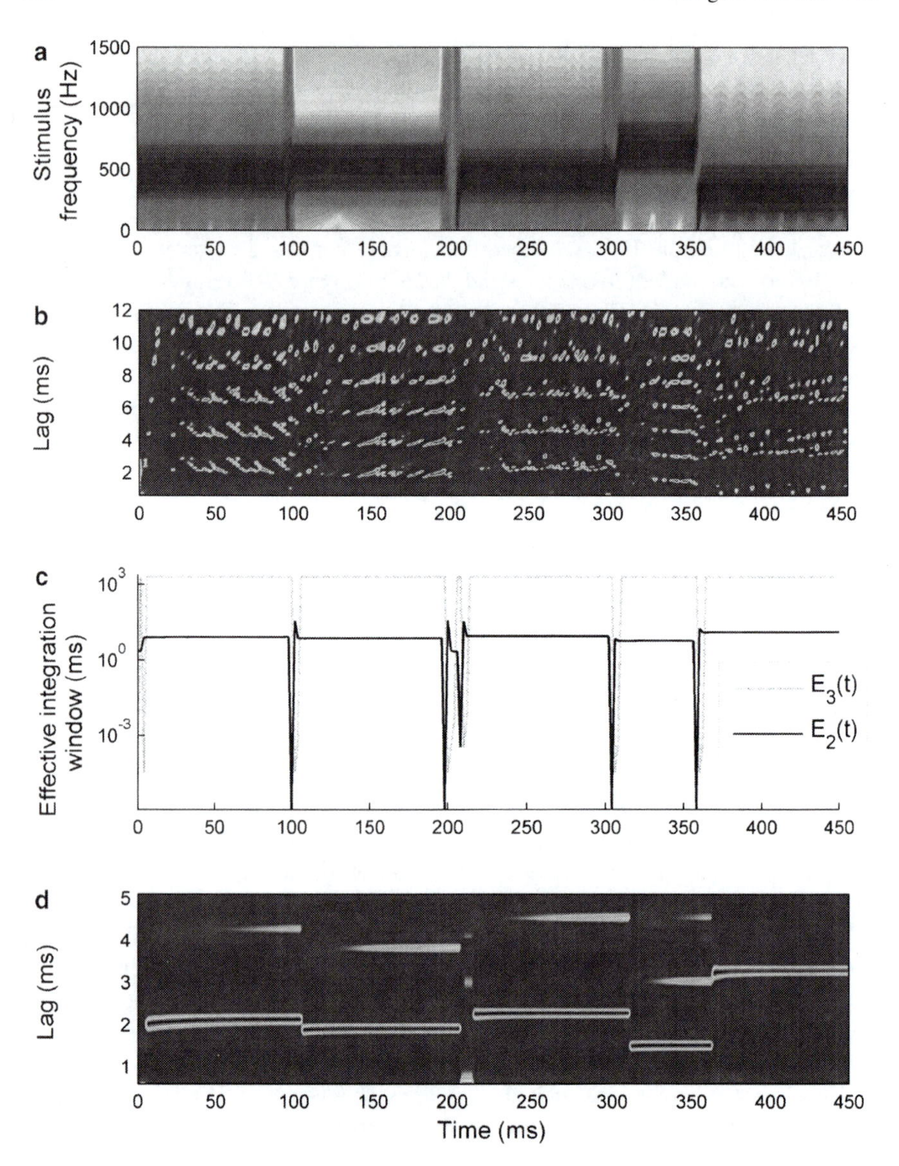

Fig. 19.2 Example of the model output in response to an arbitrary sequence of pure tones with random frequencies and durations. (**a**) Spectrogram of the stimulus as a function of time. (**b**) Response of the second processing stage, $A_1(t,l)$, plotted as a function of time, t (abscissa), and time lag, l (ordinate). (**c**) Effective integration window of the second and third processing stages, $A_2(t,l)$ ($E_2(t)$, *solid line*) and $A_3(t,l)$ ($E_3(t)$, *dotted line*). $E_2(t)$ represents the integration time at the lag corresponding to the maximum response in the second stage. (**d**) Response of the third processing stage, $A_3(t,l)$. $A_3(t,l)$ was normalized to a maximum response of unity and exponentially enhanced after each time step for illustrative purposes. Plots (**b**) and (**d**) represent the activation strength as a percentage of the maximum response at that time. Thus, the lag channel corresponding to the current pitch estimate appears dark

19.3 Results and Discussion

The model was evaluated using a representative set of psychophysical experiments (Hall and Peters 1981; Plack and White 2000; Carlyon et al. 2008), which illustrate the different time scales of temporal integration and resolution in pitch perception; and further experiment was conducted specifically for this study. Finally, the proposed model can also replicate neurophysiological data (Krumbholz et al. 2003). We refer to Balaguer-Ballester et al. (2009) for details on those evaluations.

We present here a simple example of the model capabilities (Fig. 19.3, panels c and d). Hall and Peters' (1981) experiment highlighted an unsolved problem concerning the balance between synthetic and analytic listening in response to a sequence of pure tones. The stimuli consisted of three tones played sequentially either in quiet (Fig. 19.3c, left panel) or against a background of white noise (Fig. 19.3c, right panel). Each tone lasted 40 ms and was separated from the following tone by a gap of 10 ms. Tone frequencies were 650, 850 and 1,050 Hz (similar results were obtained with a harmonic sequence). The overall level of the noise was about 15 dB above the level of the tones. The individual tones in the sequence were perceived in both conditions. In the experiment, listeners were instructed to match the *lowest* pitch that they perceived, and in the quiet condition, this was the first of the tones (650 Hz). However, in the noise condition, the non-simultaneous tones combine to create a lower global pitch of about 213 Hz, which is not perceived in the quiet condition.

Figure 19.3d shows the responses $A_3(t,l)$ over time. As in Fig. 19.2, the responses after each time step have been normalized for visualization purposes (however, it should be noted that their real magnitudes, which are close to zero during the silent gaps, are not evident in the figure). The maximum of $A_3(t,l)$ correctly predicts the pitches perceived in quiet, which correspond approximately to the frequencies of the individual tones at each moment in time (Fig. 19.3d, left plot). However, when background noise is present (Fig. 19.3d, right plot), a global pitch gradually emerges (horizontal arrow in the right plot), and the peak in the final response occurs at the reciprocal of the perceived pitch of 213 Hz (4.7 ms). The above results match precisely with the listeners' responses in this study (Hall and Peters 1981). Many other model evaluations will be presented in Balaguer-Ballester et al. 2009).

In summary, we propose a neurocomputational model to explain the observed paradox between temporal integration and temporal resolution in the auditory processing of pitch information. This model is an extension of the autocorrelation theory of pitch perception formulated in terms of equations describing the activity of neural ensembles (Balaguer-Ballester et al. 2009). The principal novelty of the model is the suggestion that top-down connections to sub-cortical areas determine the temporal dynamics of auditory perception, and that this influence is mediated through feedback modulation of recurrent inhibitory circuits. As a result, the responses at each stage adapt to recent and relevant changes in the input stimulus; i.e., feedback in the model essentially determines the dynamics of the "effective" integration window used at each stage. This approach is consistent with recently available

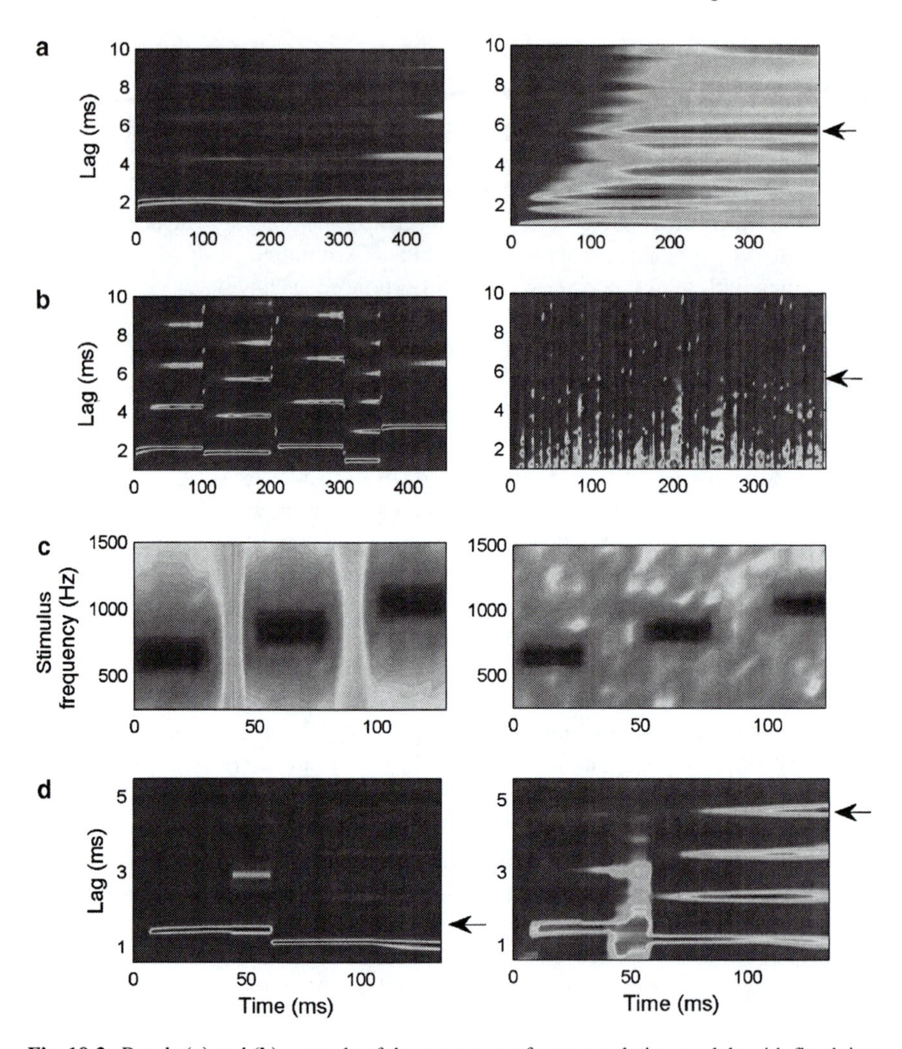

Fig. 19.3 Panels (**a**) and (**b**) example of the responses of autocorrelation models with fixed time constants. (**a**) Response of the cascade autocorrelation model (Balaguer-Ballester et al. 2008); *left plot*: to the sequence of random tones shown in Fig. 19.2a, and *right plot*: to the alternating click train (Carlyon et al. 2008). (**b**) Response of short-term integration stage of the cascade autocorrelation model. As in panel (**a**), the left panel shows the response to the tone sequence and the right panel shows the response to the click train (*horizontal arrows* mark the reported pitch of the click train). Panels (**c**) and (**d**) response of the model to stimuli used in the Hall and Peters' (1981) experiment. (**c**) Spectrogram of a rapid sequence of three 40-ms tones presented in quiet (*left panel*) and after the addition of white noise (*right panel*). (**d**) Response of the third stage of the model, $A_3(t,l)$, for the stimulus in quiet (*left panel*) and in noise (*right panel*). In the noise condition, the response represents the average over three different random realizations of the noise background. *Horizontal arrows* indicate the lowest pitch reported by listeners in each condition

neuroimaging data (Gutschalk et al. 2007; Kiebel et al. 2008; Kadner and Berrebi 2008; von von Kriegstein et al. 2008). A prediction of the model is that blocking the feedback circuits would impair the ability to separate sounds over time.

Acknowledgments This work was supported by EmCAP (Emergent Cognition through Active Perception, 2005–2008); a research project in the field of Music Cognition funded by the European Commission (FP6-IST, contract 013123) and by the EPSRC grant EP/C010841/1 (COLAMN). EBB wants to thank Ray Meddis for his very generous support, which was critical to the completion of this study.

References

Balaguer-Ballester E, Clark N, Coath M, Krumholz K, Denham SL (2009) Understanding pitch perception as a hierarchical generative with top-down modulation. PLoS Comput Biol 5(3):e1000301

Balaguer-Ballester E, Denham SL, Meddis R (2008) A cascade autocorrelation model of pitch perception. J Acoust Soc Am 124:2186–2195

Bregman AS, Ahad PA, Kim J, Melnerich L (1994) Resetting the pitch-analysis system: 1. Effects of rise times of tones in noise backgrounds or of harmonics in a complex tone. Percept Psychophys 56:155–162

Carlyon RP, Mahendran S, Deeks JM, Long CJ, Axon P, Baguley D, Bleeck S, Winter IM (2008) Behavioural and physiological correlates of temporal pitch perception in electric and acoustic hearing. J Acoust Soc Am 123:973–985

de Boer E (1985) Auditory time constants: a paradox? In: Michelsen A (ed) Time resolution in auditory systems. Springer, Berlin, pp 141–158

Friston K (2003) Learning and inference in the brain. Neural Netw 16:1325–1352

Gutschalk A, Patterson RD, Scherg M, Uppenkamp S, Rupp A (2007) The effect of temporal context on the sustained pitch response in human auditory cortex. Cereb Cortex 17:552–561

Hall JW, Peters RW (1981) Pitch for nonsimultaneous successive harmonics in quiet and noise. J Acoust Soc Am 69:509–513

Kiebel SJ, Daunizeau J, Friston KJ (2008) A hierarchy of time-scales and the brain. PLoS Comput Biol 4(11):e1000209

Krumbholz K, Patterson RD, Seither-Preisler A, Lammertmann C, Lutkenhoner B (2003) Neuromagnetic evidence for a pitch processing center in Heschl's gyrus. Cereb Cortex 13:765–772

Kumar S, Stephan KE, Warren JD, Friston KJ, Griffiths TD (2007) Hierarchical processing of auditory objects in humans. PLoS Comput Biol 3(6):e100

Licklider JCR (1951) A duplex theory of pitch perception. Experientia 7:128–134

Lopez-Poveda EA, Meddis R (2001) A human nonlinear cochlear filter bank. J Acoust Soc Am 110:3107–3118

Meddis R, O'Mard L (1997) A unitary model of pitch perception. J Acoust Soc Am 102:1811–1820

Meddis R, O'Mard L (2006) Virtual pitch in a computational physiological model. J Acoust Soc Am 120:3861–3868

Plack CJ, White LJ (2000) Perceived continuity and pitch perception. J Acoust Soc Am 108:1162–1169

Sumner CJ, O'Mard LP, Lopez-Poveda EA, Meddis R (2002) A revised model of the inner-hair cell and auditory nerve complex. J Acoust Soc Am 111:2178–2189

Wiegrebe L (2001) Searching for the time constant in of neural pitch extraction. J Acoust Soc Am 107:1082–1091

Kadner A, Berrebi S (2008) Encoding of the temporal features of auditory stimuli in the medial nucleous of the trapezoid body and superior paraolivary nucleous of the rat. Neuroscience 151:868–887

von Kriegstein K, Patterson RD, Griffiths TD (2008) Task-dependent modulation of medial geniculate body is behaviourally relevant for speech recognition. Curr Biol 18(23-2):1855–1859

Chapter 20
The Harmonic Organization of Auditory Cortex

Xiaoqin Wang

Abstract Harmonicity is a unique feature of human speech, animal vocalizations, music, and many other natural or man-made sounds. Experimental evidence has shown that many neurons in the primary auditory cortex exhibit characteristic responses to harmonically related frequencies, in the form of facilitation and inhibition. Recent neurophysiology and imaging experiments have identified regions of nonprimary auditory cortex in humans and nonhuman primates that have selective responses to harmonic pitches. In this chapter, I propose that a fundamental organizational principle of auditory cortex is based on the harmonic structure of sounds.

Keywords Harmonicity • Auditory cortex • Marmoset

A fundamental property of acoustic signals encountered in daily life is the harmonicity. Harmonicity is a unique feature of human speech and animal vocalizations. Harmonic structures are found not only in vocal communication sounds but also in echolocating signals of species like bats. Harmonicity is also an essential component of music. Harmonic sounds are produced by vocal apparatuses of many species and music instruments of many types whose design results in resonances at harmonic frequencies. Harmonic sounds are also produced as a result of non-linear characteristics of acoustic generators or reflectors. It is not an exaggeration to say that we live in a world of harmonicity. It is nearly in every aspect of our hearing environment. The perception of harmonicity reflects the need of the auditory system to discriminate between vocal signals and environmental sounds (e.g., wind blowing). An important difference between these two classes of sounds is that environmental sounds are generally inharmonic, whereas most vocalizations contain harmonic structures.

X. Wang (✉)
Laboratory of Auditory Neurophysiology, Department of Biomedical Engineering,
Johns Hopkins University, Baltimore, MD 21205, USA
e-mail: Xiaoqin.wang@jhu.edu

E.A. Lopez-Poveda et al. (eds.), *The Neurophysiological Bases of Auditory Perception*,
DOI 10.1007/978-1-4419-5686-6_20, © Springer Science+Business Media, LLC 2010

In auditory system, the cochlea generates nonlinear distortion products that contain harmonics of frequencies heard. The nonlinear processing in auditory nerve, cochlear nucleus and other structures leading to auditory cortex also generate harmonic by-products. Therefore, the auditory cortex is flooded from the onset of hearing with exogenous harmonics from the acoustic environment and endogenous harmonics generated by subcortical auditory systems. Given what we know about developmental plasticity, auditory cortex must be imprinted with harmonic signatures. Such an "imprinting processing" is likely reinforced through evolutional history of a species. Cortical neural circuitry thus may have evolved to accommodate the sensory environments with the prevalence of harmonicity in natural and man-made sounds. In this chapter, I propose that a fundamental organizational principle of auditory cortex is based on the harmonic structure of sounds. I review available evidence on harmonic processing in auditory cortex that supports such an organizational principle and suggest future experiments to further validate it.

20.1 Harmonic Inputs to Auditory Cortex

A hallmark of neurons throughout the ascending auditory systems is frequency tuning, first established in the cochlea. A typical auditory neuron is tuned (with excitatory responses) to one particular frequency within the hearing range of a species. In the auditory cortex, many neurons are tuned to more than one frequency, often referred to as "multi-peak neurons". Such neurons are found in various mammalian species, from bats (Suga et al. 1983), cats (Abeles and Goldstein 1970; Phillips and Irvine 1981; Sutter and Schreiner 1991) to primates (Kadia and Wang 2003). In marmoset primary auditory cortex (A1), about 20% of neurons were found to have a "multi-peak" frequency response area when tested with pure tones. Figure 20.1a shows an example of such neurons. This neuron had one excitatory peak at 10 dB SPL (CF = 14.5 kHz) and three excitatory peaks at 40 dB SPL (21.9, 28.1 and 35.3 kHz, corresponding approximately to 1.5CF, 2CF and 2.5CF, respectively). In echolocating bats, neurons usually show multiple excitatory peaks at harmonics of CF

Fig. 20.1 (continued) tone frequency for two sound levels (10, 40 dB SPL). Several frequency peaks were identified. Note that the frequency peaks at both sound levels were harmonically related to the neuron's CF of 14.5 kHz. (**b**) An example of two-tone facilitation in a single-peak A1 neuron. A two-tone pair is defined by S1 (Stimulus 1, fixed tone) and S2 (Stimulus 2, variable tone). S1 is at CF (1.47 kHz), 50 dB SPL and S2 at 0.12–5.88 kHz, 70 dB SPL (in linear steps of ~122 Hz). This neuron has a monotonic rate-level function at CF with the threshold (Th) at 0 dB SPL. Percent change in discharge rate is plotted versus S2 frequency. The asterisk and horizontal dotted line indicate the discharge rate of responses to the S1 tone alone. (**c**) An example of a single-peak A1 neuron showing distant off-CF inhibitions. This neuron has a nonmonotonic rate-level function with a preferred sound level of 50 dB and threshold of 20 dB. S1 is at CF (12.25 kHz), 50 dB SPL and S2 at 1–39.6 kHz, 40 dB SPL (in linear steps of ~1,000 Hz). The strongest distant off-CF inhibition is at 3*CF (indicated by an arrow). The asterisk and horizontal dotted line indicate the discharge rate of responses to the S1 tone alone

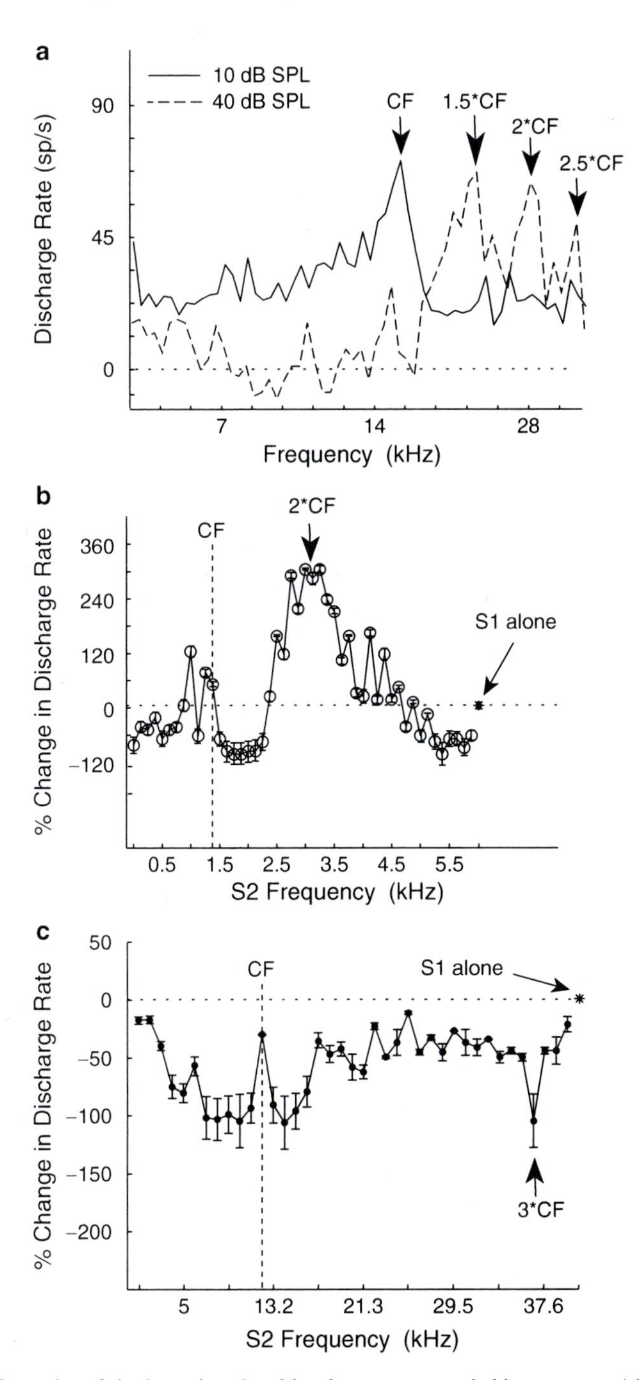

Fig. 20.1 Examples of single-peak and multipeak neurons recorded in marmoset A1 (from Kadia and Wang 2003). (**a**) Example of a multipeaked A1 neuron with different excitatory frequency peaks at different sound levels revealed by single-tone stimulation. Discharge rate is plotted versus

(e.g., 2CF, 3CF, 4CF, etc.) and the spectral locations of the excitatory peaks correspond to spectral components of ultrasonic calls emitted by the bats (Suga et al. 1983). In marmoset auditory cortex, in contrast, multipeak neurons show excitatory peaks not only at harmonics of CF, but also at integer ratios of CF such as 1/2, 3/2, etc. as illustrated by the example neuron in Fig. 20.1a. Figure 20.2a shows a population summary of the excitatory frequencies of multipeak neurons in relationship to CF. There is a dominant peak at 2CF as well as a prominent peak at 1.5CF. The frequencies of the excitatory peaks of these multipeak neurons fall into the frequency range of marmoset vocalizations (~4–16 kHz) in some but not all cases (Kadia and Wang 2003). In many cases, harmonically related excitatory peaks are outside the vocalization range of marmosets. Like bats, spectral components of marmoset vocalizations contain only multiples of a fundamental frequency (between 4 and 8 kHz for marmosets, ~30 kHz for mustache bats), which is a constraint of the vocal apparatus. The fact that neurons in marmoset auditory cortex exhibit spectral tuning to multiple frequencies that do not obey octave ratios suggests that they are designed to detect sounds other than their species-specific vocalizations.

In species other than echolocating bats, the proportions of multipeak neurons with harmonically related excitatory peaks is small when tested by single pure tome. However, when neurons tuned to only one frequency ("single-peak neurons") were tested with two simultaneously presented tones, a much larger portion of them showed harmonically related peaks in their responses. Figure 20.1b shows two-tone response from such a single-peak neuron. In this case, in addition to a CF-tone (S1), the frequency of a second tone (S2) was varied. As one can see from Fig. 20.1b, the two-tone combination generated a much stronger facilitatory response when S2 frequency was near 2CF. Using the two-tone method, we found many more single-peak neurons in marmoset auditory cortex that exhibit harmonically related responses (Kadia and Wang 2003). Single-peak neurons also showed harmonically related inhibitions, as shown by the example in Fig. 20.1c. This neuron's response was nearly completely suppressed when the S2 tone was placed at 3CF, far away from its CF of 12.25 kHz. We refer to this type of inhibition as "distant inhibition" because it is distinctly different from the inhibition flanking the excitatory region near CF. What makes this distant inhibition interesting and important is that it is often harmonically related to CF. There are a larger proportion of single-peak neurons that show harmonically related distant inhibition than harmonically related facilitation. In contrast to multi-peak neurons (Fig. 20.2a), we did not observe in single-peaked neurons a strong preference for one-octave separation between two tones in facilitatory modulation, but found a clear preference in inhibitory modulation when the second tone was one octave higher or lower in frequency than the first tone. Figure 20.2b summarizes population data of facilitatory and inhibitory interactions between two tones in single-peak neurons in marmoset auditory cortex. The data in this figure show prominent peaks centered on 1/2CF, 2CF, and 3CF, indicating widespread, harmonically related inputs to A1. Distant inhibitory influences in A1 have also been observed in the auditory cortex of other nonprimate species (e.g., bat: Kanwal et al. 1999; cat: Sutter et al. 1999).

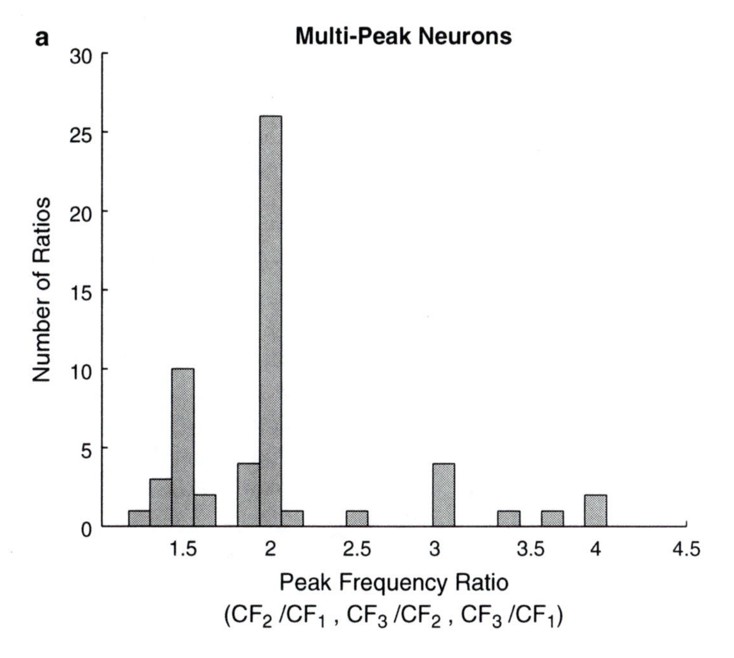

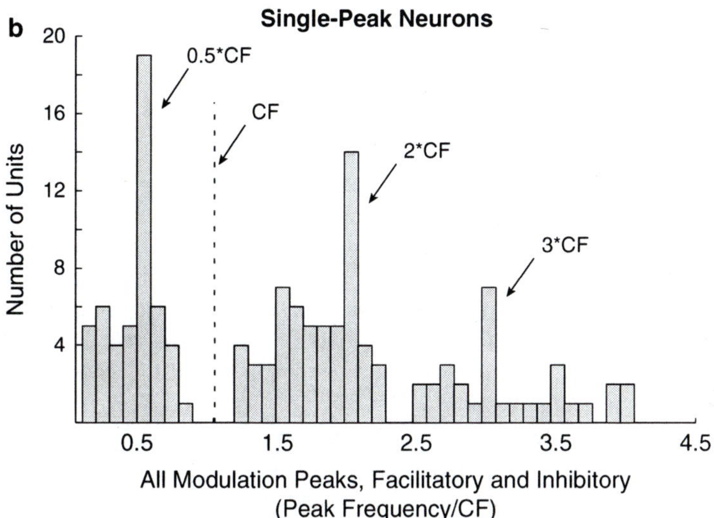

Fig. 20.2 Population properties showing harmonic interactions in marmoset A1 (from Kadia and Wang 2003). (**a**) Distribution of all peak frequency ratios (CF_2/CF_1, CF_3/CF_1, CF_3/CF_2) of multipeak neurons. CF_1, CF_2 or CF_3 are the frequency peaks of multipeak neurons. Since a multipeak neuron could have more than two peaks, it may be represented by more than one peak frequency ratio in this plot. A total of 56 peak frequency ratios were obtained from 38 multipeak neurons (from Kadia and Wang 2003). (**b**) Response modulations in the population of single-peak neurons in marmoset A1. Distribution of all modulatory peaks (facilitatory or inhibitory) measured from 76 of 113 neurons that showed facilitation and/or inhibition in their two-tone responses. There were a total of 139 measured peaks

Neurons with facilitative interactions between harmonically related frequency peaks are optimally stimulated by specific combinations of spectral elements. Therefore, these neurons can function to extract harmonic components embedded in complex sounds as a unitary object. It is also conceivable that such neural mechanisms can facilitate detection of signals in noisy environments by binding harmonic spectrotemporal features of an acoustic object. The frequency-specificity of the distant inhibition in single-peaked neurons suggests that the distant off-CF inhibition may have a different functional role than the distant off-CF facilitation. It further suggests that single-peaked neurons, which represent the majority of A1 neurons, may process harmonically related spectral components in a different manner than multipeaked A1 neurons. Just as harmonics are useful in assembling components of a complex acoustic object into one entity, they can also introduce confusion regarding the identification of the fundamental frequency. The strong inhibitory influences by harmonically related frequencies might enhance the perception and identification of the fundamental frequency. The harmonically related inhibition could also serve as a mechanism to remove unwanted harmonic artifacts in natural environment. It is also conceivable that, by combining harmonically related facilitation and inhibition, the auditory cortex could determine whether a sound is harmonic, a function that is important for a wide range of auditory perception.

20.2 Harmonic Pitch Processing

Pitch perception is crucial for speech perception, music perception, and auditory object recognition in a complex acoustic environment. Our auditory system relies on pitch, among other things, to enable us to accomplish such tasks. Changes in pitch are used to convey information in speech (for example, prosodic information in European languages, and semantic information in tonal languages like Chinese, Vietnamese and Thai). Pitch is closely associated with the perception of harmonically structured or periodic sounds. The perception of fundamental frequency of complex sounds with harmonic structures, and its corresponding neural mechanisms in cortex is therefore of particular importance.

A recent study has identified a region in nonprimary auditory cortex of marmosets, where neurons with pitch-selective responses were found (Bendor and Wang 2005). These pitch-selective neurons not only are tuned to low frequency pure tones, but also responded to missing fundamental harmonic complex sounds with a pitch near a neuron's CF. They did not respond to individual components in a harmonic complex tone that were outside its tone-derived excitatory frequency response area. An example of a pitch-selective neuron is shown in Figure 20.3a. The location of pitch-selective neurons is in the low-frequency region of auditory cortex near the anterolateral border of A1, similar to a pitch-region found in human auditory cortex (Bendor and Wang 2006). A typical pitch-selective neuron also responded to an array of spectrally dissimilar pitch-evoking sounds (harmonic complex tones, click trains, iterated ripple noise) when the pitch was near the neuron's preferred fundamental frequency.

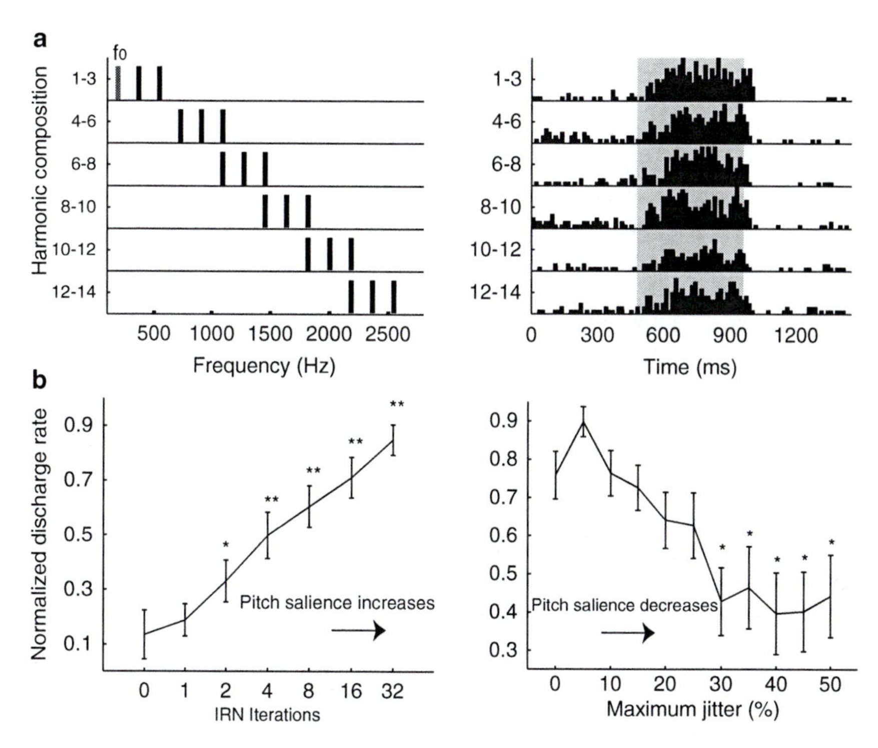

Fig. 20.3 Examples of pitch-selective responses recorded in marmoset auditory cortex (from Bendor and Wang 2005). (**a**) *Left*: Frequency spectra of a series of harmonic complex stimuli. The fundamental frequency component (f_0) and its higher harmonics have equal amplitudes of 50 dB SPL. *Right:* Peristimulus time histograms of a pitch-selective neuron's responses to the harmonic complex stimuli. Stimuli were presented from 500 to 1,000 ms (indicated by the shaded region on the plot). (**b**) Responses of pitch-selective neurons increases with increasing pitch salience. *Left*: Averaged population response of pitch-selective neurons as a function of the iterations of IRN stimuli. The response to IRN stimuli with 0 iterations is used as a reference for statistical comparison at other iterations. *Right:* Averaged population response of pitch-selective neurons to irregular click trains as a function of maximum jitter. The response to a regular click train is used as a reference for statistical comparison at other jitter values

It was also found that pitch-selective neurons increased their firing rates as the pitch salience increased and preferred temporally regular sounds (Fig. 20.3b), in agreement with the imaging studies by (Patterson et al. 2002) and (Penagos et al. 2004). Because of technical limitations in single-unit recordings, not all auditory cortex areas in marmosets have been investigated for their possible pitch-processing functions. The similarity in the anatomic structures of auditory cortex between New World and Old World monkeys and humans suggests that the pitch center identified in marmosets may exist in other nonhuman primate species and that this pitch center shares similar functions as the lateral Heschl's gyrus in humans (Bendor and Wang 2006).

A recent human imaging study using melodies composed of iterated rippled noise (IRN) stimuli localized a "pitch center" to lateral Heschl's gyrus and regions anterior to lateral Heschl's gyrus in the right hemisphere (Patterson et al. 2002). In another study, Penagos and colleagues (Penagos et al. 2004) used four harmonic complex sounds that had either a low or high pitch and occupied either a low or high spectral range. Bilaterally, a restricted region of nonprimary auditory cortex in the lateral Heschl's gyrus anterolateral to A1 was found more strongly responsive to the stimuli with high pitch salience than the stimulus with low pitch salience. Furthermore, the activity in this region was not significantly different in the high pitch salience stimuli when comparing across different fundamental frequencies or frequency ranges. All together, a number of human imaging studies published in recent years have confirmed the location of a pitch center in lateral Heschl's gyrus (Patterson et al. 2002; Penagos et al. 2004; Gutschalk et al. 2004; Schneider et al. 2005; Hall et al. 2005; Hall et al. 2006; Ritter et al. 2005; Chait et al. 2006).

However, two recent studies have suggested that there might be more than one region in auditory cortex for pitch processing (Hall and Plack 2007; Hall and Plack 2008). In Hall and Plack 2007 study, the authors used a dichotic pitch stimulus known as a Huggin's pitch. A weak pitch percept is created by playing a broadband noise to each ear that is identical, except for a narrow frequency band (190–210 Hz in this study) in which a phase shift is introduced in one of the ears relative to the other. The Huggins stimulus was found to evoke a weak activation of cortex and no significant activity in lateral Heschl's gyrus (Hall and Plack 2007). These results suggest that lateral Heschl's gyrus is only involved in processing diotic rather than diochotic pitch. In the second study, (Hall and Plack 2008) played to the same listeners a variety of pitch-evoking stimuli and failed to show that all pitch is processed in a single locus in auditory cortex. Their results suggest that parts of the planum temporale are more relevant for pitch processing than lateral Heschl's gyrus. In some listeners, pitch responses occurred elsewhere, such as the temporo-parieto-occipital junction or prefrontal cortex. Taken together, the available human imaging data have shown that pitch-evoking stimuli are selectively processed in particular regions of auditory cortex and that there may exist more than one pitch-processing centers in auditory cortex, including but not limited to the lateral Heschl's gyrus. The notion of multiple pitch processing regions in auditory cortex is not all that surprising given the complexity of pitch perception (Plack and Oxenham 2005). Further studies need to investigate specific roles played by each cortical region that has been implicated in pitch processing.

20.3 Temporal Periodicity Processing

In addition to studies of harmonically related spectral inputs, studies have also shown neural selectivity for temporal periodicity. Periodic sounds have a "regular" temporal structure as opposed to noises or random temporal sequences that have "irregular" temporal structures. The coding of periodic sounds in auditory cortex is often related to the neural processing of harmonics. Neural recordings in gerbils showed that A1 neurons could respond to the periodicity of amplitude-modulated tones with the spectral components located outside neuron's excitatory frequency response area, but with the periodicity ranging much higher than pitch range found in humans (Schulze and Langner 1997). Schulze et al. (2002) has also reported a semi-circularly shaped map of best fundamental frequency in gerbil auditory cortex using optical imaging techniques. Langner et al. 2002 reported a topographic arrangement of periodicity information in chinchilla auditory cortex that was orthogonal to the tonotopic axis.

Neurons that are preferentially driven by temporally modulated periodic sounds have been observed throughout auditory cortex in many species (Joris et al. 2004), including several nonhuman primates (Bieser and Muller-Preuss 1996; Steinschneider et al. 1998; Liang et al. 2002; Malone et al. 2007). A large proportion of neurons in marmoset A1 showed preferential responses to amplitude- or frequency-modulated tones and, interestingly, some of these neurons could only be driven by temporally modulated tones but not by unmodulated pure tones (Liang et al. 2002). Neurons in marmoset auditory cortex are also found responsive to periodic click train stimuli, by either stimulus synchronized or unsynchronized discharges in both A1 (Lu et al. 2001) and rostral fields (Bendor and Wang 2007; Bendor and Wang 2008).

However, the response to temporally modulated periodic sounds alone can not differentiate whether a neuron or cortical region responds to temporal regularity or average repetition rate. This issue was not examined in the studies referred earlier. Using MEG imaging method, (Gutschalk et al. 2002) identified two separate sources adjacent to primary auditory cortex in humans that exhibit differential sensitivity to temporal regularity tested by click trains. One source, located in lateral Heschl's gyrus, was particularly sensitive to regularity and largely insensitive to sound level. The second, located just posterior to the first in planum temporale, was particularly sensitive to sound level and largely insensitive to regularity. In marmosets, except within the pitch-processing region, we find that neurons in auditory cortex respond to temporally varying acoustic signals with repetition rates in the range of pitch, yet they do not show sensitivity to temporal regularity, i.e., they have similar responses to periodic and aperiodic acoustic pulse trains with the same repetition rate (Bendor and Wang 2009). In contrast, neurons within the pitch-processing region were found to be sensitive to temporal irregularity and decrease their firing rates for aperiodic acoustic signals.

20.4 Harmonic Organizations of Auditory Cortex

From the evidence reviewed in this article, we can conclude that harmonicity (spectrally) and periodicity (temporally) processing is a unique feature of auditory cortex. Harmonicity and periodicity are two closely related characteristics of sounds. A sound with harmonic structure spectrally, usually has a periodic waveform temporally, although the periodicity depends on the phase relationships between the spectral components within the sound. Likewise, a periodic sound like a click train or a sequence of tone or noise pulses has a harmonic spectrum. Although a sinusoidal amplitude-modulated sound has only two sideband components besides the carrier component, its spectrum resembles a harmonic spectrum more than inharmonic ones. In a broader sense, neural coding of harmonicity and periodicity can be considered the coding of spectral or temporal regularity. From the information-coding standpoint, a spectrally or temporally regular sound can encode specific acoustic information, whereas a truly irregular sound (spectrally and temporally) like a white noise or random click train codes little specific information other than parameters associated with its statistical structure. Given the mirror relationship between the harmonicity and periodicity, we may consider their neural representations being governed by a unified framework.

In A1, neurons show characteristic responses to harmonic spectral structures and periodic temporal modulations. In the pitch-region of nonprimary auditory cortex in primates, pitch-selective neurons exhibit unique properties for processing harmonic pitch and temporal regularity that are not observed in A1. Much remains unknown on harmonic processing outsides A1 and the pitch-region. Here, I propose that a fundamental organizational principle of auditory cortex is based on and shaped by the harmonic nature of sounds. Thus, various cortical areas may be specialized in processing particular harmonic structures. In this scenario, the pitch-region anterolateral to A1 (Bendor and Wang 2006) is one of such cortical areas of a harmonics-based functional organization. Other cortical areas outside A1 may be specialized in processing, for example, harmonic structures outside the pitch domain, spectral and temporal structures that contain preferentially used music intervals, etc. The diversity of harmonic interactions in A1 responses revealed by two-tones (Figs. 20.1, 20.2) suggest possible readouts by different higher auditory cortical areas. Recent imaging studies (Hall and Plack 2007; Hall and Plack 2008) demonstrating activations in multiple cortical areas in humans by different types of pitch evoking stimuli adds further support to the proposed harmonic organization of auditory cortical processing.

Acknowledgment This work has been supported by NIH grant DC–03180.

References

Abeles M, Goldstein MH Jr (1970) Functional architecture in cat primary auditory cortex: columnar organization according to depth. J Neurophysiol 33:172–187

Bendor D, Wang X (2009) Neural processing of temporal regularity in auditory cortex, unpublished data

Bendor D, Wang X (2005) The neuronal representation of pitch in primate auditory cortex. Nature 436(7054):1161–1165

Bendor DA, Wang X (2006) Cortical representations of pitch in monkeys and humans. Curr Opin Neurobiol 16:391–399

Bendor D, Wang X (2007) Differential neural coding of acoustic flutter within primate auditory cortex. Nat Neurosci 10:763–771

Bendor D, Wang X (2008) Neural response properties of core fields AI, R, and RT in the auditory cortex of marmoset monkeys. J Neurophysiol 100:888–906

Bieser A, Muller-Preuss P (1996) Auditory responsive cortex in the squirrel monkey: neural responses to amplitude-modulated sounds. Exp Brain Res 108:273–284

Chait M, Poeppel D, Simon JZ (2006) Neural response correlates of detection of monaurally and binaurally created pitches in humans. Cereb Cortex 16(6):835–848

Gutschalk A, Patterson RD, Rupp A, Uppenkamp S, Scherg M (2002) Sustained magnetic fields reveal separate sites for sound level and temporal regularity in human auditory cortex. Neuroimage 15:207–216

Gutschalk A, Patterson RD, Scherg M, Uppenkamp S, Rupp A (2004) Temporal dynamics of pitch in human auditory cortex. Neuroimage 22(2):755–766

Hall DA, Plack CJ (2007) The human 'pitch center' responds differently to iterated noise and Huggins pitch. Neuroreport 18(4):323–327

Hall DA, Plack CJ (2008) Pitch processing sites in the human auditory brain. Cereb Cortex 19(3):576–585

Hall DA, Barrett DJ, Akeroyd MA, Summerfield AQ (2005) Cortical representations of temporal structure in sound. J Neurophysiol 94(5):3181–3191

Hall DA, Edmondson-Jones AM, Fridriksson J (2006) Periodicity and frequency coding in human auditory cortex. Eur J Neurosci 24(12):3601–3610

Joris PX, Schreiner CE, Rees A (2004) Neural processing of amplitude-modulated sounds. Physiol Rev 84(2):541–577

Kadia SC, Wang X (2003) Spectral integration in A1 of awake primates: neurons with single- and multipeaked tuning characteristics. J Neurophysiol 89(3):1603–1622

Kanwal JS, Fitzpatrick DC, Suga N (1999) Facilitatory and inhibitory frequency tuning of combination-sensitive neurons in the primary auditory cortex of mustached bats. J Neurophysiol 82:2327–2345

Langner G, Albert M, Briede T (2002) Temporal and spatial coding of periodicity information in the inferior colliculus of awake chinchilla (Chinchilla laniger). Hear Res 168:110–130

Liang L, Lu T, Wang X (2002) Neural representations of sinusoidal amplitude and frequency modulations in the auditory cortex of awake primates. J Neurophysiol 87:2237–2261

Lu T, Liang L, Wang X (2001) Temporal and rate representations of time-varying signals in the auditory cortex of awake primates. Nat Neurosci 4:1131–1138

Malone BJ, Scott BH, Semple MN (2007) Dynamic amplitude coding in the auditory cortex of awake rhesus macaques. J Neurophysiol 98:1451–1474

Patterson RD, Uppenkamp S, Johnsrude IS, Griffiths TD (2002) The processing of temporal pitch and melody information in auditory cortex. Neuron 36(4):767–776

Penagos H, Melcher JR, Oxenham AJ (2004) A neural representation of pitch salience in nonprimary human auditory cortex revealed with functional magnetic resonance imaging. J Neurosci 24(30):6810–6815

Phillips DE, Irvine DRF (1981) Responses of single units in physiologically defined primary auditory cortex (A1) of the cat: frequency tuning and response to intensity. J Neurophysiol 45:48–58

Plack CJ, Oxenham AJ (2005) The psychophysics of pitch. In: Plack CJ, Oxenham AJ, Fay RR, Popper AN (eds) Pitch: neural coding and perception. Springer, New York, pp 7–55

Ritter S, Gunter Dosch H, Specht HJ, Rupp A (2005) Neuromagnetic responses reflect the temporal pitch change of regular interval sounds. Neuroimage 3:533–543

Schneider P, Sluming V, Roberts N, Scherg M, Goebel R, Specht HJ, Dosch HG, Bleeck S, Stippich C, Rupp A (2005) Structural and functional asymmetry of lateral Heschl's gyrus reflects pitch perception preference. Nat Neurosci 8(9):1241–1247

Schulze H, Langner G (1997) Periodicity coding in the primary auditory cortex of the Mongolian gerbil (Meriones unguiculatus): two different coding strategies for pitch and rhythm? J Comp Physiol (A) 181:651–663

Schulze H, Hess A, Ohl FW, Scheich H (2002) Superposition of horseshoe-like periodicity and linear tonotopic maps in auditory cortex of the Mongolian gerbil. Eur J Neurosci 15:1077–1084

Steinschneider M, Reser DH, Fishman YI, Schroeder CE, Arezzo JC (1998) Click train encoding in primary auditory cortex of the awake monkey: evidence for two mechanisms subserving pitch perception. J Acoust Soc Am 104(5):2935–2955
Suga N, O'Neill WE, Kujirai K, Manabe T (1983) Specialization of "combination-sensitive" neurons for processing of complex biosonar signals in the auditory cortex of the mustached bat. J Neurophysiol 49:1573–1626
Sutter ML, Schreiner CE (1991) Physiology and topography of neurons with multipeaked tuning curves in cat primary auditory cortex. J Neurophysiol 65:1207–1226
Sutter ML, Schreiner CE, McLean M, O'Connor KN, Loftus WC (1999) Organization of inhibitory frequency receptive fields in cat primary auditory cortex. J Neurophysiol 82:2358–2371

Chapter 21
Reviewing the Definition of Timbre as it Pertains to the Perception of Speech and Musical Sounds

Roy D. Patterson, Thomas C. Walters, Jessica J.M. Monaghan, and Etienne Gaudrain

Abstract The purpose of this paper is to draw attention to the definition of timbre as it pertains to the vowels of speech. There are two forms of size information in these "source-filter" sounds, information about the size of the excitation mechanism (the vocal folds), and information about the size of the resonators in the vocal tract that filter the excitation before it is projected into the air. The current definitions of pitch and timbre treat the two forms of size information differently. In this paper, we argue that the perception of speech sounds by humans suggests that the definition of timbre would be more useful if it grouped the size variables together and separated the pair of them from the remaining properties of these sounds.

Keywords Musical pitch • Voice pitch • Vocal timbre

21.1 Timbre, Speech Sounds and Acoustical Scale

> "Timbre is that attribute of auditory sensation in terms of which a listener can judge that two sounds similarly presented and having the same loudness and pitch are dissimilar." ["American national standard acoustical terminology" (1994). American National Standards Institute, ANSI S1.1-1994 (R1999)]

Informally, the standard definition of timbre is regarded with considerable amusement. You might expect the definition of timbre to tell you something about what timbre is, but all the definition tells you is that there are a few things that timbre is not. It is not pitch, it is not loudness, and it is not duration. It is everything else. Despite the poverty of the definition, it appears in most popular introductory books on hearing and auditory perception, and aside from the definition, these books actually have

R.D. Patterson (✉)
Centre for the Neural Basis of Hearing, University of Cambridge, Downing Site, Cambridge, CB2 3EG, UK
e-mail: rdp1@cam.ac.uk

E.A. Lopez-Poveda et al. (eds.), *The Neurophysiological Bases of Auditory Perception*,
DOI 10.1007/978-1-4419-5686-6_21,

rather little to say on the topic of timbre. This is somewhat surprising given how important the concept is in music and speech perception. Timbre is what distinguishes a trumpet from a violin when they are playing the same sustained note at the same loudness and for the same duration, and timbre is what distinguishes vowels spoken by a person on the same note at the same loudness and for the same duration. It is a very important concept in hearing and it is, perhaps, time to consider revising the definition of timbre, at least as it pertains to vocal sounds, to make it conform more with what we hear.

The aspects of timbre that are important in this paper are readily illustrated with the sustained sounds of speech, that is, sustained vowels. The waveform and spectrum of a short segment of a synthetic /a/ vowel, like that spoken by a child, are presented in the upper and lower panels of Fig. 21.1, respectively. The waveform shows that a vowel is a stream of glottal pulses, each of which is accompanied by a decaying resonance that reflects the filtering of the vocal tract above the larynx. In this case, the glottal pulse rate (GPR) is 200 pulses per second (pps), so the time between glottal pulses (the period of the wave) is 5 ms. The set of vertical lines in the lower panel of Fig. 21.1 shows the long-term magnitude spectrum of the sound, and the dashed line connecting the tops of the vertical lines shows the *spectral envelope* of the vowel. The peaky structures in the spectral envelope are the formants of the vowel;

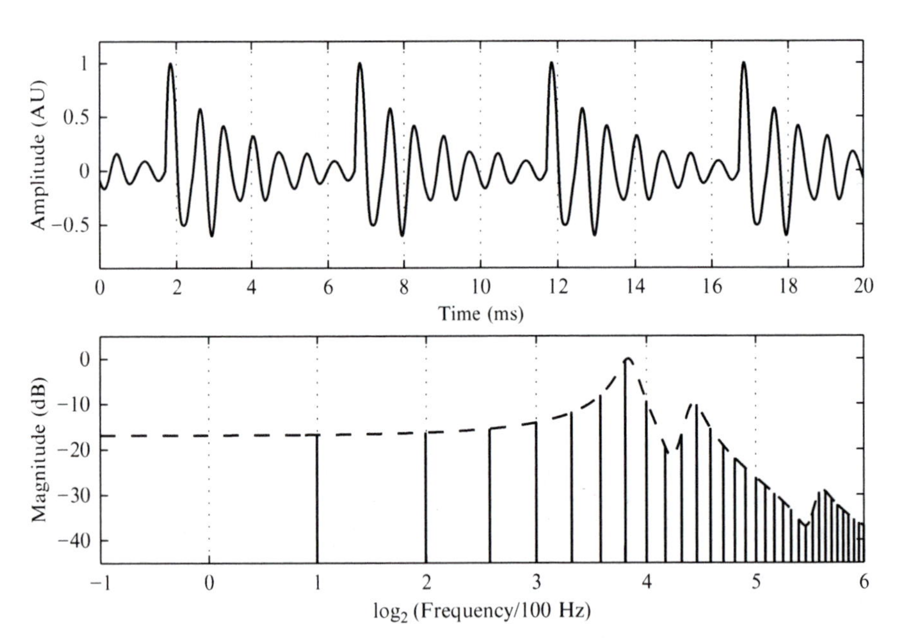

Fig. 21.1 Waveform (*upper panel*) and magnitude spectrum of a synthetic three-formant /a/ vowel, as might be spoken by a small child. The glottal pulse rate is 200 pulses per second. The solid lines in the lower panel present the magnitude spectrum of the waveform; the dashed line is the spectral envelope

the shape of the envelope in the spectral domain corresponds to the shape of the damped resonance in the time domain.

When children begin to speak, they are about 0.85 m tall and as they mature their height increases by about a factor of two. Vocal tract length increases in proportion with height (Fitch and Giedd 1999; Turner et al. 2009, Fig. 4), and so the formant frequencies of children's vowels decrease by about an octave as they mature (Lee et al. 1999; Turner et al. 2009). It is also the case that the GPR of the voice decreases by, on the order of, an octave as the vocal cords become longer and more massive, although the effect is more pronounced in males than in females. These effects are simple examples of the fact that larger objects vibrate more slowly than smaller objects; see Patterson et al. (2008) for examples from speech and music. In speech, the pattern of formants that defines a given vowel type remains largely unchanged as people grow up (Peterson and Barney 1952; Lee et al. 1999; Turner et al. 2009). The effect of growth on the spectrum of a vowel is quite simple to characterize, provided the spectrum is plotted on a logarithmic frequency scale. In this case, the set of harmonics that define the fine structure of the spectrum simply moves, as a unit, toward the origin as the child matures into an adult, by about one octave. Similarly, the spectral envelope shifts toward the origin without changing shape, and it shifts during maturation by about an octave.

In acoustical terms, "the position of the fine-structure of the spectrum on a logarithmic frequency scale" is the *acoustic scale* of the sound produced by the source of excitation; it is a property of the sound as it occurs in the air (Cohen 1993) – a property that in this case is determined by the Glottal Pulse Rate (GPR) of the speaker. For brevity, it will be referred to as "the scale (***S***) of the source (***s***)" and designated S_s. Similarly, in acoustical terms, "the position of the envelope on a logarithmic frequency scale" is the acoustic scale of the filter in the vocal tract that produces the formants in the spectrum; it is also a property of the sound but in this case it is largely determined by the speaker's vocal tract length (VTL). For brevity, it will be referred to as "the scale (***S***) of the filter (***f***)" and designated S_f. Turner et al. (2009) have recently reanalyzed several large databases of spoken vowels and shown that almost all of the variability in formant frequency data is either attributable to vowel type (80%) or S_f (16%). The vocoder STRAIGHT (Kawahara and Irino 2004) can be used to manipulate the acoustic scale variables independently and produce versions of notes with a wide range of combinations of S_s and S_f.

21.2 Timbre and the Perception of Speech Sounds

The logic of the definition of timbre involves taking variables of auditory perception that appear to be associated with physical properties of sound, and separating them from other variables, which by default remain part of timbre; the perceptual variables that the definition encompasses are duration, loudness and pitch. Duration is the variable that is most obviously separable from timbre, and it illustrates the logic underlying the definition of timbre. If a singer holds a note for a longer rather than a shorter period, it produces a discriminable change in the sound but it is not a change in timbre. Duration has no effect on the magnitude spectrum of a sound, once the duration is well beyond that of the temporal window used to produce the magnitude spectrum. Since the current example involves sustained vowels and the window used to produce the magnitude spectrum in Fig. 21.1 is on the order of 50 ms, duration has no effect on timbre in this example. In general, the perceptual change associated with a change in the duration of a sustained note is separable from changes in the timbre of the note.

Loudness is also largely separable from timbre. If we turn up the volume control when playing a recording, the change will be perceived as an increase solely in loudness. The pitch of any given vowel and the timbre of that vowel will be essentially unaffected by the manipulation. The increase in intensity produces a change in the magnitude spectrum of the vowel – both the fine structure and the envelope shift vertically upwards – but there is no change in the frequencies of the components of the fine structure and there is no change in the relative amplitudes of the harmonics. Nor is there any change in the shape of the spectral envelope. So, loudness is also separable from timbre.

This illustrates the logic behind the definition of timbre; acoustic variables that do not affect either the fine structure or the envelope of the magnitude spectrum do not affect the timbre of the sound. The question is "What happens when a simple shift is applied to the position of the fine structure or the envelope of a sound, that is, when we change S_s, S_f, or both?" The current definition of timbre suggests that a change in S_s, which is heard as a change in pitch, does not affect the timbre of the sound, whereas a change in S_f, which is heard as a change in speaker size, does affect the timbre. This is where the current definition of timbre becomes problematic.

Note, in passing, that shifting the fine structure of the magnitude spectrum while holding the envelope fixed produces large changes in the relative amplitudes of the harmonics as they move through the region of a formant peak. So the relative magnitude of the components in the spectrum can change substantially without producing a change in timbre, by the current definition. Note, also, that shifting the envelope of the magnitude spectrum while holding the fine-structure frequencies fixed produces similarly large changes in the relative amplitudes of the components as they move through formant regions. Shifting the envelope does not change the perceived vowel-type; nevertheless, these spectral changes are considered to produce timbre changes, by the current definition.

21.2.1 Timbre in the Perception of "Acoustic Scale Melodies"

The timbre issues in the remainder of the paper are more readily understood when presented in terms of "melodies" in which the acoustic scale values of the notes (S_s and S_f) vary according to the diatonic scale of Western music. The melodies are shown in Fig. 21.2; all of them have four bars containing a total of eight notes. The melodies are in 3/4 time, with the fourth and eight notes extended to give the sequence a musical feel. The black and grey notes show the progression of intervals for S_s and S_f,

Fig. 21.2 Musical notation for Melodies 1 to 4: The black notes represent the acoustic scale of the source (S_s), i.e., the pitch. The grey notes represent the acoustic scale of the filter (S_f). The original speaker's voice defines the note E for both dimensions

respectively, as the melody proceeds. The S_s component of the melodies is presented in the key of C major as it is the simplest to read. The melodies in the demonstration waves are actually in the key of G major. The sound files for the melodies are available at http://www.pdn.cam.ac.uk/groups/cnbh/teaching/sounds_movies/.

Melody 1. The first example simulates a normal melody in which the VTL of the singer, and thus S_f, is fixed (grey notes), and the GPR, and thus S_s, varies (black notes). The singer is an adult male and the pitch of the voice drops by an octave over the course of the melody from about 200 to 100 pps [ISH09_PWMG_Melody_1; Fig. 21.2, Staff (1)]. This descending melody is within the normal range for a tenor, and the melody sounds natural. The melody is presented as a sequence of syllables to emphasize the speech-like qualities of the source; the "libretto" is "pi, pe, ko, kuuu; ni, ne, mo, muuu." In auditory terms, this phonological song is a complex sequence of distinctive timbres produced by a sequence of different spectral envelope shapes. The timbre changes engage the phonological system and emphasize the role of envelope shape in producing the libretto of the melody. As the melody proceeds, the fine-structure of the spectrum (S_s) shifts, as a unit, with each change in GPR, and over the course of the melody, it shifts an octave towards the origin. The definition of timbre indicates that these relatively large GPR changes, which produce large pitch changes, do not produce timbre changes, and this seems correct in this case. This suggests that pitch is largely separable from timbre, much as duration and loudness are, and much as the definition of timbre implies.

Melody 2. But problems arise when we extend the example with the help of STRAIGHT and synthesize a version of the same melody but with a singer that has a much shorter VTL, like that of a small child [ISH09_PWMG_Melody_2; Fig. 21.2, Staff (2)]. There is no problem at the start of the melody; it just sounds like a child singing the melody. The starting pulse rate is low for the voice of a small child but not impossibly so. As the melody proceeds, however, the pitch decreases by a full octave, which is beyond the normal range for a child. In this case, the voice quality seems to change and the child comes to sound rather more like a dwarf. The ANSI definition of timbre suggests that the voice quality change from a child to a dwarf is not a timbre change, it is just a pitch change. But traditionally, voice quality changes are thought to be timbre changes. This is the first form of problem with the standard definition of timbre – changes that are nominally pitch changes producing what would normally be classified as a timbre change.

Melody 3. The next example [ISH09_PWMG_Melody_3; Fig. 21.2, Staff (3)] involves reversing the roles of the variables $\boldsymbol{S_s}$ and $\boldsymbol{S_f}$, and using STRAIGHT to manipulate the position of the spectral envelope, $\boldsymbol{S_f}$, while holding $\boldsymbol{S_s}$, and thus the pitch, fixed. Over the course of the melody, the position of the envelope, $\boldsymbol{S_f}$, shifts by an octave towards the origin. This simulates a doubling of the singer's VTL, from about 10 to 20 cm. As with the previous $\boldsymbol{S_s}$ melodies, the specific values of $\boldsymbol{S_f}$ are determined by the diatonic musical scale of Western music. In other words, the sequence of $\boldsymbol{S_f}$ ratios have the same numerical values as the sequence of $\boldsymbol{S_s}$ ratios used to produce the first two melodies. This effectively extends the domain of notes from a diatonic musical scale to a diatonic musical plane, like that illustrated in Fig. 21.3. The abscissa of the plane is $\boldsymbol{S_s}$ or GPR; the ordinate is $\boldsymbol{S_f}$ or VTL. Both of the axes are logarithmic.

The syllables of the libretto were originally spoken by an adult male (author RP) with a VTL of about 16.5 cm and an average GPR of about 120 pps. Then STRAIGHT was used to generate versions of each syllable for each combination of $\boldsymbol{S_s}$ and $\boldsymbol{S_f}$ in the musical plane. The note corresponding to the original singer is [E, E] on this version of the $\boldsymbol{S_s}$-$\boldsymbol{S_f}$ plane; so what we refer to as "C" is acoustically "G" (123 pps). Melody 1 was synthesized with the VTL of the original singer, that is, with notes from the E row of the plane. Melody 2 was synthesized with the VTL of a child, that is, with notes from the upper C row of the plane. Melody 3 was synthesized with a fixed pitch, upper C, and notes from the upper C column of the matrix.

Perceptually, as Melody 3 proceeds and the envelope shifts down by an octave, the child seems to get larger and the voice comes to sound something like that of a counter tenor, that is, a tall person with an inordinately high pitch. The definition of timbre does not say anything specific about how changes in the spectral envelope affect timbre; the acoustic scale variable, $\boldsymbol{S_f}$, was not recognized when the standard was written. Nevertheless, the definition gives the impression that any change in the spectrum that produces an audible change in the perception of the sound produces a change in timbre, provided it is not simply a change in duration, loudness or pitch, which are all fixed in the current example. Experiments with scaled vowels and syllables show that the just noticeable change in $\boldsymbol{S_f}$ is about 7% for vowels (Smith et al. 2005) and 5% for syllables (Ives et al. 2005), so all but the smallest intervals

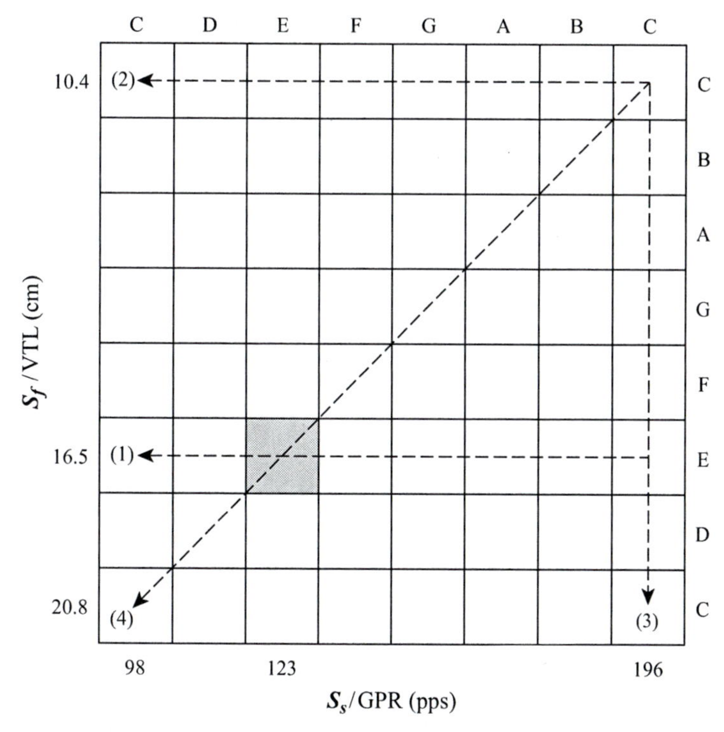

Fig. 21.3 The S_s–S_f plane, or GPR–VTL plane. The abscissa is the acoustic scale of the source (S_s), or the GPR, increasing from left to right over an octave. The ordinate is the acoustic scale of the filter (S_f), or the VTL, doubling from top to bottom. The plane is partitioned in squares that represent the musical intervals. The *square* associated with the original speaker is highlighted in grey. The *dashed lines* show the progression of the notes in the melodies

in the melody would be expected to produce perceptible S_f changes. Since traditionally, voice quality changes are thought to be timbre changes, the perception of the S_f melody in this example seems entirely compatible with the definition of timbre; both voice quality and timbre are changing. However, we are left with the problem that large changes in S_s and S_f both seem to produce changes in voice quality, but whereas the perceptual changes associated with large shifts of the *fine-structure* along the log-frequency axis *are not* timbre changes, the perceptual changes associated with large shifts of the *envelope* along the same log-frequency axis *are* timbre changes according to the standard. They both produce changes in the relative amplitudes of the spectral components, but neither changes the spectral envelope pattern *and* neither form of shift alters the libretto.

Melody 4. The problems involved in attempting to unify the perception of voice quality with the definition of timbre become more complex when we consider melodies where both S_s and S_f change as the melody proceeds. Consider the melody produced by co-varying S_s and S_f to produce the notes along the diagonal of the matrix [ISH09_PWMG_Melody_4]. The musical notation for the melody is shown in Fig. 21.2, Staff (4). The melody is perceived to descend an octave as the sequence proceeds, and there is a progressive increase in the perceived size of the singer from a child to an adult (with one momentary reversal at the start of the second phrase). It is as if we had a set of singers varying in age from 4 to 18 in a row on stage, and we had them each sing their assigned syllable in order, and in time, to produce the melody. The example makes it clear that there is an entire plane of singers with different vocal qualities defined by combinations of the acoustic scale variables, $\boldsymbol{S}_s$ and $\boldsymbol{S}_f$. The realization that there is a whole plane of voice qualities makes it clear just how difficult it would be to produce a clean definition of timbre that excludes one of the acoustic scale variables, $\boldsymbol{S}_s$, and not the other, $\boldsymbol{S}_f$. If changes in voice quality are changes in timbre, then changes in pitch ($\boldsymbol{S}_s$) can produce changes in timbre. This would seem to undermine the utility of the current definitions of pitch and timbre.

21.2.2 The Second Dimension of Pitch Hypothesis

At first glance, there would appear to be a fairly simple way to solve the problem, which is to consider the acoustic scale variable associated with the filter, $\boldsymbol{S}_f$, to be a second dimension of pitch, which could then be excluded from the definition of timbre along with the first dimension of pitch, $\boldsymbol{S}_s$. In this case, manipulation of the second dimension of pitch on its own would sound like the change in perception produced by Melody 3 where $\boldsymbol{S}_s$ is fixed at C and $\boldsymbol{S}_f$ decreases by a factor of two over the course of the melody. Semitone changes in the second dimension of pitch, $\boldsymbol{S}_f$, are unlikely to be sufficiently salient to support accurate reporting of novel melodies (e.g., Krumbholz et al. 2000; Pressnitzer et al. 2001) for which pitch discrimination has to better than about 3%. This second form of pitch is more like the weak pitch associated with a set of unresolved harmonics where pitch discrimination is possible if the changes are relatively large, say 4 semitones. Nevertheless, the second form of pitch would, in some sense, satisfy the ANSI definition of pitch, which says that "Pitch is that attribute of auditory sensation in terms of which sounds may be ordered on a scale extending from low to high." Moreover, it does not seem unreasonable to say that the notes at the start of Melody 3 are higher than the notes at the end, which would support the "second dimension of pitch" hypothesis.

It does, however, lead to a problem. To determine the pitch of a sound, it is traditional to match the pitch to the pitch or either a sinusoid or a click train. It seems likely that if listeners were asked to pitch match each of the notes in Melody 3, among a larger set of sounds that diverted attention from the orderly progression of $\boldsymbol{S}_f$ in the melody, they would probably match all of the notes to the same sinusoid or the same click train, and the pitch of the matching stimulus would be the upper C. This would leave us with the problem that the second form of pitch changes the perception of the sound but it does not change the pitch (as matched). However, by the current definition, a change in perception that is not a change in pitch (or loudness, or duration) is a change in timbre. Thus, the "second dimension of pitch" hypothesis would appear to lead us back to the position that changes in $\boldsymbol{S}_f$ produce changes in the timbre of the sound.

It is also the case that the "second dimension of pitch" hypothesis implies that voice quality changes like those produced by moving around on the musical plane of Fig. 21.3, would be pitch changes rather than timbre changes. And there is one further problem. Many people hear the perceptual change in Melody 3 as a change in speaker size, and they hear a more pronounced change in speaker size when changes in S_f are combined with changes in S_s, as in Melody 4. To ignore the perception of speaker size, is another problem inherent in the "second dimension of pitch" hypothesis; source size is an important aspect of perception, and pretending that changes in the perception of source size are just pitch changes does not seem like a good idea.

21.2.3 The Scale of the Filter, S_f, as a Dimension of Timbre

Rather than co-opting the acoustic scale of the filter, S_f, to be a second dimension of pitch, it might make more sense to consider it as a separable, but nevertheless, internal dimension of timbre – a dimension of timbre that is associated with the perception of voice quality and speaker size. This, however, leads to a problem which is, in some sense, the inverse of the "second dimension of pitch" problem. Once it is recognized that shifting the position of the fine structure of the spectrum is rather similar to shifting the position of the envelope of the spectrum, and that the two position variables are different aspects of the same property of sound (acoustic scale), then it somehow seems unreasonable to have one of these variables, S_f, within the realm of timbre and the other, S_s, outside the realm of timbre. For example, consider the issue of voice quality; both of the acoustic scale dimensions affect voice quality and they interact in the production of a specific voice quality. Moreover, the scale of the source, S_s, affects the perception of the singer's size, in a way that is similar to the effect of the scale of the filter, S_f. Thus, if we define the scale of the filter, S_f, to be a dimension of timbre, then we need to consider that the scale of the source, S_s, may also need to be a dimension of timbre. After all, large changes in S_s affect voice quality, which is normally considered an aspect of timbre. The problem, of course, is that these seemingly reasonable suggestions lead to the conclusion that pitch is a dimension of timbre.

21.2.4 The Independence of Spectral Envelope Shape

There is one further aspect of the perception of these melodies that should be emphasized, which is that neither of the acoustic scale manipulations causes a change in the libretto; it is always "pi, pe, ko, kuuu; ni, ne, mo, muuu." That is, the changes in timbre that give rise to the perception of a sequence of syllables are unaffected by changes in S_s and S_f, even when those changes are large (Smith et al. 2005; Ives et al. 2005). The changes in timbre that define the libretto are associated with changes in the shape of the envelope, as opposed to the position of the envelope or the position of the fine structure. Changes in the shape of the envelope produce changes in vowel type in speech and changes in instrument family in music. Changing the position of the envelope and changing the position of the fine structure both produce substantial changes in the relative amplitudes of the components of the magnitude spectrum, but they do not change the timbre category of these sounds, that is, they do not change the vowel type in speech or the instrument family in music.

21.3 Conclusions

Cohen (1993)'s hypothesis that acoustic scale is a basic property of sound leads to the conclusion that the major categories of timbre (vowel type and instrument family) are determined by spectral envelope *shape*, and that these categories of timbre are relatively independent of both the acoustic scale of the excitation source *and* the acoustic scale of the resonant filter. In speech, the acoustic scale variables, S_s and S_f, largely determine the voice quality of the speaker, and thus our perception of their sex and size (e.g., Smith and Patterson 2005). With regard to timbre, this suggests that when dealing with tonal sounds that have pronounced resonances like the vowels of speech, it would be useful to distinguish between aspects of timbre associated with the shape of the spectral envelope, on the one hand, and aspects of timbre associated with the acoustic scale variables, S_s and S_f, on the other hand. This would lead to a distinction between the "what" and "who" of timbre, that is, what is being said, and who is saying it.

Acknowledgments Research supported by the UK Medical Research Council [G0500221, G9900369].

References

Cohen L (1993) The scale transform. IEEE Trans Acoust 41:3275–3292

Fitch WT, Giedd J (1999) Morphology and development of the human vocal tract: a study using magnetic resonance imaging. J Acoust Soc Am 106:1511–1522

Irino T, Patterson RD (2002) Segregating information about the size and shape of the vocal tract using a time-domain auditory model: the stabilized wavelet-Mellin transform. Speech Commun 36:181–203

Ives DT, Smith DRR, Patterson RD (2005) Discrimination of speaker size from syllable phrases. J Acoust Soc Am 118:3816–3822

Kawahara H, Irino T (2004) Underlying principles of a high-quality speech manipulation system STRAIGHT and its application to speech segregation. In: Divenyi PL (ed) Speech separation by humans and machines. Kluwer Academic, MA

Krumbholz K, Patterson RD, Pressnitzer D (2000) The lower limit of pitch as determined by rate discrimination. J Acoust Soc Am 108:1170–1180

Lee S, Potamianos A, Narayanan S (1999) Acoustics of children's speech: developmental changes and spectral parameters. J Acoust Soc Am 105:1455–1468

Patterson RD, van Dinther R, Irino T (2007) The robustness of bio-acoustic communication and the role of normalization. In: Proceedings of 19th international congress on acoustics, Madrid, September 2007, pp 7–11

Patterson RD, Smith DRR, van Dinther R, Walters TC (2008) Size information in the production and perception of communication sounds. In: Yost WA, Popper AN, Fay RR (eds) Auditory perception of sound sources. Springer Science/Business Media, LLC, New York

Peterson GE, Barney HI (1952) Control methods used in the study of vowels. J Acoust Soc Am 24:75–184

Pressnitzer D, Patterson RD, Krumbholtz K (2001) The lower limit of melodic pitch. J Acoust Soc Am 109:2074–2084

Smith DRR, Patterson RD (2005) The interaction of glottal-pulse rate and vocal-tract length in judgements of speaker size, sex, and age. J Acoust Soc Am 118:3177–3186

Smith DRR, Patterson RD, Turner RE, Kawahara H, Irino T (2005) The processing and perception of size information in speech sounds. J Acoust Soc Am 117:305–318

Turner RE, Walters TC, Monaghan JJM, Patterson RD (2009) A statistical, formant-pattern model for segregating vowel type and vocal-tract length in developmental formant data. J Acoust Soc Am 125:2374–2386

Chapter 22
Size Perception for Acoustically Scaled Sounds of Naturally Pronounced and Whispered Words

Toshio Irino, Yoshie Aoki, Hideki Kawahara, and Roy D. Patterson

Abstract Recent studies have demonstrated that listeners can extract information about the length of a speaker's vocal tract from voiced vowels (Smith et al., 2005) and syllables (Ives et al. 2005). Smith and Patterson (2005) have also reported that they can extract size information from unvoiced vowels. In this paper, we extend the observations to words recorded in natural conversation mode to demonstrate that size perception is robust to changes in the mode of vocal excitation. We used a high-quality vocoder, STRAIGHT to produce scaled versions of voiced words with natural F0-contours, and scaled versions of unvoiced words that sounded like they were whispered. Size discrimination performance was measured for five reference speakers, using a two-alternative, forced-choice paradigm (2AFC) with the method of constant stimuli. The listener was asked to choose the interval with the word, or words, spoken by the smaller person. The just-noticeable-difference (JND) discrimination of speaker size was found to be about 5 % independent to the mode of vocal excitation. This value is a little greater than that reported in the previous syllable experiments and a little smaller than that reported in the vowel experiments. Moreover, the natural F0-contour in the voiced words and noise excitation in the whispered words did not affect the JND value. The results support the hypothesis that the auditory system segregates size and shape information at an early stage in the processing (Irino and Patterson, 2002).

Keywords Size-shape information • Scale • Vocoder • Timbre • Fundamental frequency

T. Irino (✉)
Faculty of Systems Engineering, Wakayama University, 930 Sakaedani,
Wakayama, 640-8510, Japan
e-mail: irino@sys.wakayama-u.ac.jp

E.A. Lopez-Poveda et al. (eds.), *The Neurophysiological Bases of Auditory Perception*,
DOI 10.1007/978-1-4419-5686-6_22, © Springer Science+Business Media, LLC 2010

22.1 Introduction

The sounds that convey words from a speaker to a listener contain information about the length of the speaker's vocal tract as well as its shape (the message). Humans can extract the message from the voices of men, women, and children without being confused by the size information, and they can extract the size information without being confused by the message. Irino and Patterson (2002) have argued that the auditory system segregates the acoustic features in speech sounds associated with the shape of the vocal tract from those associated with its length at an early stage in auditory processing. Smith et al. (2005) and Turner et al. (2006) argued that good recognition performance outside the normal speech range favors the normalization hypothesis of Irino and Patterson (2002). Smith et al. (2005) performed discrimination and recognition experiments using vowel sounds scaled to simulate speakers with a large range of vocal tract length, and showed that the ability to discriminate speaker size extends beyond the normal range of speaker sizes. Ives et al. (2005) performed size discrimination experiments using a much larger set of speech sounds (180 syllables) and found similar results. Van Dinther and Patterson (2006) reported discrimination and recognition experiments on size perception using musical instrument sounds. Together the experiments suggest that information about the size of a source is segregated from information about the shape and structure of the source, automatically, at an early stage in the processing.

Smith and Patterson (2005) also reported that performance of the size discrimination and recognition tasks in these experiments is largely unaffected when the voicing cue is removed; that is, when noise is used to resynthesize the vowels. Similarly, recognition performance for unvoiced vowels is remarkably resilient to changes in vocal tract length (VTL) – listeners can accurately recognize unvoiced vowels scaled to VTL values well beyond the range encountered in everyday life. These experiments suggest that size information can be extracted from unvoiced speech to inform perceptual decisions, in much the same way as with voice speech.

In this chapter, we extend the size discrimination experiments from vowels and syllables to naturally spoken and whispered words to determine whether the earlier observations apply to the sounds of everyday conversation. We used a database of Japanese words to produce scaled versions of voiced words with natural F0-contours, and scaled versions of unvoiced words that sounded like they were whispered. The experimental methods are similar to those of previous studies (Smith et al. 2005; Ives et al. 2005; Irino et al. 2007) to facilitate the comparison of performance, and to clarify the effect of the mode of vocal excitation on size perception.

22.2 Experiment

The JND for speaker size was measured in a discrimination experiment using naturally voiced and whispered words.

22.2.1 Stimuli

We extended the domain of size perception from vowels and syllable phrases to more naturally spoken words. We used recorded speech sounds from a database of Japanese four-mora words (FW03). The words in this database are controlled with respect to both word familiarity and phonetic balance. The words were pronounced naturally and collected from four native speakers (two males and two females). The words for the experiment were selected from those of one of the male speakers (mya) who had an average fundamental frequency of about 150 Hz.

The stimuli in the experiment consisted of words that were analyzed, manipulated, and resynthesized by STRAIGHT, which is a high-quality vocoder developed by Kawahara et al. (1999) (Kawahara 2006). STRAIGHT was used to generate all the different versions of each word with the specific combinations of VTL and GPR values required for the experiment; the specific combinations are presented in the next section. It is possible to resynthesize unvoiced words by exciting the vocal tract filter function with noise instead of a burst of glottal pulses. We produced two types of unvoiced words to investigate whether the spectral tilt affects the JND for size discrimination. The first type was produced with white noise excitation and no modification to the STRAIGHT spectrogram. It is referred to as "unvoiced" hereafter. The second type that was produced with white noise excitation STRAIGHT was modified to lift the spectrograms by 6 dB per octave, as in Smith and Patterson (2005). This type is referred to as "whispered" since it simulates whispered speech, which has relatively more energy at higher frequencies than voiced speech due to the turbulence of noise excitation in the human vocal tract (Fujimura and Lindqvist 1971). Thus, there were three forms of each word extracted from database, FW03: naturally pronounced (referred to as "voiced" hereafter), unvoiced, and whispered. The average fundamental frequency (F0) of the words was approximately 150 Hz and the standard deviation was 8.0 Hz.

Figure 22.1a shows the five combinations of VTL ratio and GPR ratio used as reference combinations for the voiced words; they were chosen to characterize the same five speaker types as in Ives et al. (2005). The crosses show the ratios for reference sounds and the circles show the ratios for the test sounds. The GPR ratios were 0.5, 1, and 2, relative to GPR of the original word. Since the words in the database were produced in conversational mode, the GPR varies within a word and the GPR contours are different for different words. The GPR-VTL ratios for the reference stimuli were (1) [0.5, 1.19], (2) [2, 1.19], (3) [1, 0.89], (4) [0.5, 0.67], (5) [2, 0.67]. The VTL ratios of the test sounds were $2^{-5/12}$, $2^{-3/12}$, $2^{-1/12}$, $2^{1/12}$, $2^{3/12}$, and $2^{5/12}$ (0.75, 0.84, 0.94, 1.06, 1.19, and 1.33) relative to the reference speaker. The reference speakers for voiced words cover a range of VTL and GPR values that extends beyond that normally encountered in everyday life.

The scaled sounds for unvoiced and whispered words were shown in Fig. 22.1b. Since there is no voicing, there are only three reference speakers with the same VTLs

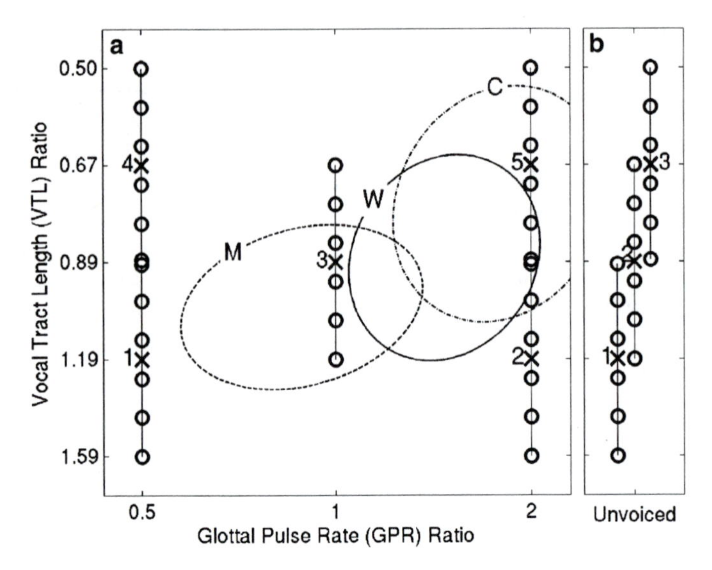

Fig. 22.1 (**a**) The GPR-VTL combinations for the voiced (naturally spoken) words. A GPR ratio of 1 corresponds to the GPR of the original words (i.e., 150 Hz on average). A VTL ratio of 1 corresponds to a VTL of 15 cm (i.e., a rough estimate for the speaker). Five reference speakers are shown by the numbered crosses; the test speakers are shown by the circles. The three ellipses show approximate distributions of GPR and VTL values in normal speech for men (*dashed lines*), women (*solid lines*), and children (*dashed* and *dotted lines*). The three ellipses were derived from the classic data of Peterson and Barney (1952) and each ellipse encompasses 96% of the population (Turner et al. 2006, Turner et al. 2009; Smith et al. 2005). (**b**) The VTL combinations for the unvoiced and whispered words. Three reference speakers are shown by the *numbered crosses*; the test speakers are shown by the *circles*

as is the case with the reference speakers in the voice-word experiment, as shown by the crosses in Fig. 22.1b. The circles show the VTL ratios for the test speakers associated with each reference speaker. The VTL ratios for the reference speakers were (1) 1.19, (2) 0.89, and (3) 0.67 for each of the unvoiced and whispered words. So, there were six (=3 × 2) reference speakers in total. The VTL ratios for the test speakers were $2^{-5/12}$, $2^{-3/12}$, $2^{-1/12}$, $2^{1/12}$, $2^{3/12}$, and $2^{5/12}$ (0.75, 0.84, 0.94, 1.06, 1.19, and 1.33) relative to the reference speaker (i.e., the same as in the voiced-word stimuli.).

22.2.2 *Discrimination Procedures and Listeners*

The JND for speaker size was measured in a discrimination experiment. It employed a two-interval, two-alternative forced-choice (2AFC) paradigm with the method of constant stimuli. There were six Japanese listeners (three male and three female between 21 and 24 years of age). They all had normal hearing thresholds between 125 and 8,000 Hz. The listeners were seated in a sound attenuated room. The words were played by EDIROL FA-101, with a 48-kHz sampling rate over headphones

(Sennheiser HD-580) at an rms level of 70 dB SPL on average. The sound level for each interval was roved between words over a 6-dB range.

Discrimination performance was measured simultaneously for the 11 reference speakers (5 voiced words as in Fig. 22.1a, 3 unvoiced words and 3 whispered words as in Fig. 22.1b), using a two-alternative forced-choice (2AFC) paradigm with the method of constant stimuli. The listener was asked to choose the interval with the words spoken by the smaller person, and to indicate their choice by clicking on the appropriate interval box on a computer screen. On each trial, one interval contained a reference speaker (randomly chosen from the 11 shown in Fig. 22.1a and b), and the other interval contained a test speaker whose VTL was either larger or smaller than that of the reference speaker. The two intervals of each trial contained words of the same type; that is, voiced words were compared with voiced words; unvoiced words were compared with unvoiced words; whispered words were compared with whispered words. The data from speakers with six different VTL values were combined to produce a psychometric function for each reference speaker.

Informal listening suggested that it might prove difficult to judge speaker size from a single word because of the variation in pitch. So, we also prepared an experiment with two words in each interval. It produced a more natural simulation of everyday listening than the single word experiment. The two words were from the same speaker (i.e., they had the same combination of VTL and average GPR). The only consistent difference between the two intervals of a trial was the VTL ratio. We also introduced training sessions in which the listener was initially presented with the VTL-GPR combination associated with one reference speaker and was finally presented with all of the VTL-GPR combinations as the same structure as the runs in the main experiment. The training session included feedback; during the main experiment there was no feedback. The duration of the training session was 3.5 h on average.

In the main experiment, the total number of trials per listener was 1,320, i.e., 600 voiced words (5 reference speakers × 6 comparison speakers × 2 orders × 10 sessions), 360 unvoiced words, and 360 whispered words (3 reference speakers × 6 comparison speakers × 2 orders × 10 sessions). The total number of words used in the experiment was 5,280, i.e., 2,400 voiced words, 1,440 unvoiced words, and 1,440 whispered words. They were selected at random from the list in the FW03 database with the highest familiarity (list No.4), which contained 1,000 words. No word was allowed to be selected more than six times in the experiment. Within individual trials, the word selection algorithm included the restriction that no word appears more than once. Accordingly, there was little chance of selecting the same sequence of words in different trails.

22.2.3 Results on Voiced Words

The results of the voiced words for each of the five reference speakers are shown in Fig. 22.2, with one psychometric function for each of the five reference

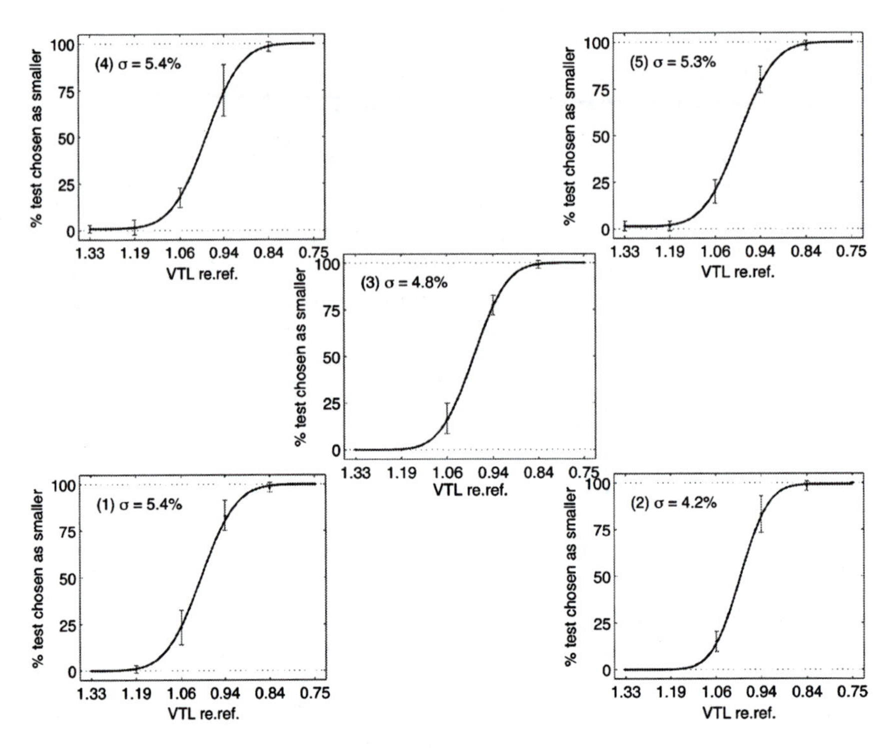

Fig. 22.2 Psychometric functions for the voiced words for the five reference speakers (shown in Fig. 22.1a) averaged over all listeners. The error bars represent ±1 standard deviations

speakers. The layout of the figure mirrors that for the reference conditions presented on the GPR-VTL plane in Fig. 22.1a. For example, reference speaker 1 is at the bottom left, and reference speaker 2 is at the bottom right. The abscissa for the psychometric function is VTL expressed as a ratio of the reference VTL; the ordinate is the percentage of trials on which the test interval was identified as having the smaller speaker. The data have been averaged across listeners because there was a little difference between listeners. The error bars represent ±1 standard deviation over listeners. A cumulative Gaussian function has been fitted to the data for each psychometric function (Wichmann and Hill 2001) and used to calculate the JND. The JND was defined as the difference in VTL for a 26% increase in performance from 50 to 76% performance (d′ = 1 in this 2AFC task); it is shown on each graph at the top left corner in percent. The cumulative Gaussians in Fig. 22.2 are all quite steep indicating that the listeners had no difficulty with the size discrimination task, even when the voice of the reference speaker has an unusual combination of GPR and VTL. The JND values range from 4.2 to 5.4%; there is no consistent trend in the JND values across the GPR-VTL plane; the average JND value is 5.0%.

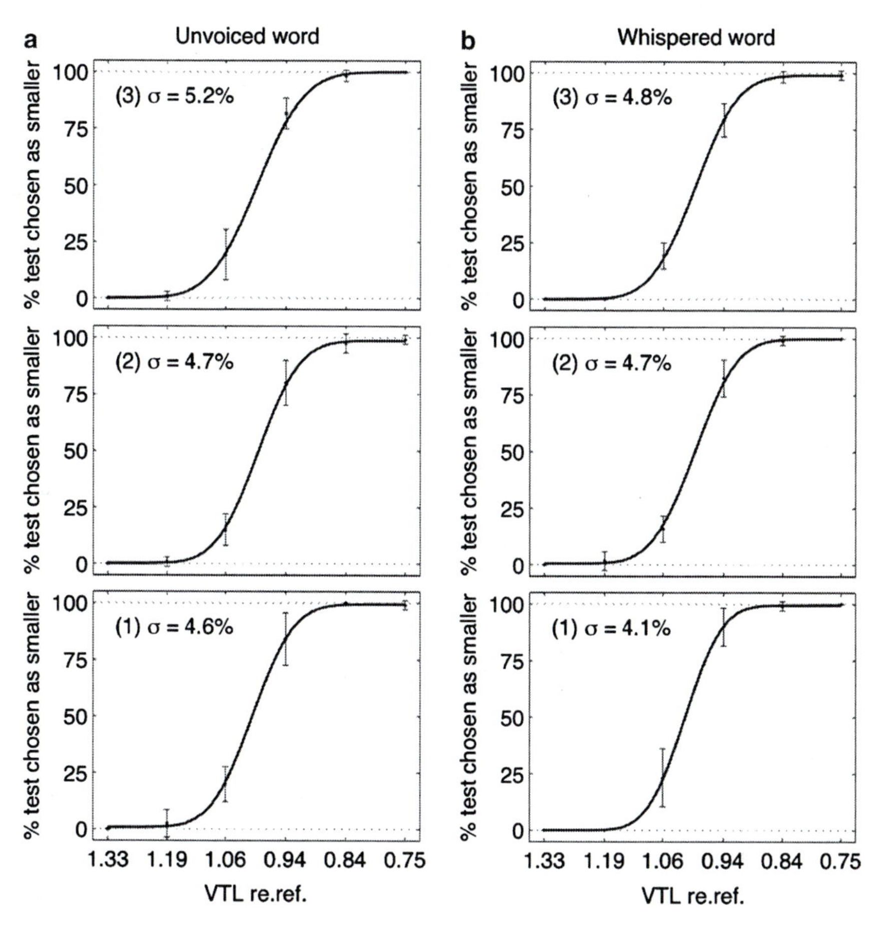

Fig. 22.3 Psychometric functions for the unvoiced and whispered words for the three reference speakers (shown in Fig. 22.1b) averaged over all listeners. (**a**) Unvoiced words and (**b**) whispered words. The error bars represent ±1 standard deviations

22.2.4 Results on Unvoiced and Whispered Words

The results of the unvoiced and whispered words are presented as two sets of three psychometric functions in Fig. 22.3a and b. The layout mirrors that of the reference-speaker conditions presented in Fig. 22.1b. The figure shows that discrimination with unvoiced and whispered words is essentially the same as with voiced words in Fig. 22.2. The JND values for the unvoiced word experiment (Fig. 22.3a) range from 4.7 to 5.2%; the average value is 4.8%. The JND values for the whispered word experiment (Fig. 22.3b) range from 4.1 to 4.8%; the average value is 4.5%. So, the average values are comparable to the average for the voiced words. Since there is a very little difference between the unvoiced, and whispered words, it can be concluded that the spectral tilt (i.e., the 6 dB/oct lift) did not affect the listeners'

Table 22.1 The average JND values in percent (%) for voiced, unvoiced and whispered words for comparison with the average JND values reported for vowels (Smith et al. 2005; Smith and Patterson 2005), and syllables (Ives et al. 2005)

	Voiced	Unvoiced	Whispered
Word	5.0	4.8	4.5
Vowel	8.3	–	9.4
Syllable	5.1	–	–

judgments about speaker size. The size judgment is more likely made on the basis of fine detail in the spectrum.

22.2.5 Summary and Comparison

Table 22.1 shows the summary of the results. The average JND for VTL discrimination in this experiment is essentially 5.0%, and there is little variation with either the mode of vocal excitation or the presence or absence of a spectral tilt. The JND for words (5%) does appear to be somewhat smaller than the JND reported by Smith et al. (2005) for voiced vowels, 8.1%, and by Smith and Patterson (2005) for voiced vowels, 8.3%, and for whispered vowels, 9.4%. This implies that there is more, or better, information about the acoustic scale of the speaker's voice in words, but it is not clear why this should be the case. The JND for words (5%) is almost the same as the JND reported by Ives et al. (2005) for voiced syllables, 5.0%. The inter-subject variability is greater with words (Fig. 22.2 of the current experiment) than syllables [Fig. 2 of Ives et al. (2005)], which might also be due to the variability of the GPR contours in the word experiment. This size discrimination experiment was performed with two words in each interval to make the listeners' experience more speech like. Previous studies (Aoki et al. 2008a, b) have shown that the size JND for one word per interval is roughly the same as the size JND for two words per interval. It appears that it is possible to judge the size of a speaker for discrimination purposes using a single word with a few syllables.

Finally, it should be noted that the size JND for words and syllables is considerably smaller than the JNDs of most other perceptual dimension, such as loudness, 10%; brightness, 15%; or odor, 25%.

22.3 Conclusions

The results show that listeners can make fine judgments about the relative size of speakers from voiced and unvoiced speech sounds, scaled well beyond the normal range. The results show that the auditory system can extract size information from whispered sounds as well as voiced sounds, and size discrimination performance is just about as good with whispered sounds as it is with voiced sounds. In everyday life, we encounter very few environments where the GPR and VTL values of the speakers are beyond the normal range, so normal experience provides very little data that would assist the brain to learn how to generalize from normal speech to the extreme combinations encountered in the current experiments. The robustness of speech perception suggests that the auditory system has an automatic mechanism to segregate vocal tract length information from vocal tract shape information *prior to* the application of discrimination and recognition processing (Irino and Patterson 2002).

Acknowledgments This research was supported by the Japan Society for the Promotion of Science Grant-in-Aid (B18300060) and the UK Medical Research Council (G0500221).

References

Aoki Y, Irino T, Kawahara H, Patterson RD (2008a) "Speaker size discrimination for acoustically scaled versions of naturally spoken words," in Abstracts of ARO 31th Midwinter meeting, Phoenix, AZ, USA

Aoki Y, Irino T, Kawahara H, Patterson RD (2008b) Speaker size discrimination for acoustically scaled versions of whispered words. J Acoust Soc Am 123(5 Pt 2):3718

van Dinther R, Patterson RD (2006) Perception of acoustic scale and size in musical instruments sounds. J Acoust Soc Am 120:2158–2177

Fujimura O, Lindqvist J (1971) Sweep-tone measurements of vocal-tract characteristics. J Acoust Soc Am 49:541–558

Irino T, Patterson RD (2002) Segregating information about the size and shape of vocal tract using a time-domain auditory model: the stabilized wavelet-Mellin transform. Speech Commun 36:181–203

Irino T, Aoki Y, Hayashi Y, Kawahara H, Patterson RD (2007) "Discrimination and recognition of scaled word sounds." In: Proc Interspeech 2007, pp 378–381, Antwerp, Belgium

Ives DT, Smith DRR, Patterson RD (2005) Discrimination of speaker size from syllable phrases. J Acoust Soc Am 118(6):3816–3822

Kawahara H (2006) STRAIGHT, exploitation of the other aspect of VOCODER: perceptually isomorphic decomposition of speech sounds. Acoust Sci Tech 27(6):349–353

Kawahara H, Masuda-Kasuse I, de Cheveigne A (1999) Restructuring speech representations using pitch-adaptive time-frequency smoothing and instantaneous-frequency-based F0 extraction: possible role of repetitive structure in sounds. Speech Commun 27(3–4):187–207

Peterson GE, Barney HL (1952) Control methods used in the study of vowels. J Acoust Soc Am 24:175–184

Smith DRR, Patterson RD, Turner R, Kawahara H, Irino T (2005) The processing and perception of size information in speech sounds. J Acoust Soc Am 117:305–318

Smith DRR, Patterson RD (2005) "The perception of scale in whispered vowels," Meeting of the British Society of Audiology, Cardiff, Wales. http://www.pdn.cam.ac.uk/groups/cnbh/research/posters_talks/BSA2005/SPbsa05.pdf

Turner RE, Al-Hames MA, Smith DRR, Kawahara H, Irino T, Patterson RD (2006) Vowel normalization: time-domain processing of the internal dynamics of speech. In: Divenyi P, Greenberg S, Meyer G (eds) Dynamics of speech production and perception, NATO science series, series A: Life sciences. IOS Press, Amsterdam, pp 153–170

Turner RE, Walters TC, Mongaghan JJM, Patterson RD (2009) A statistical, formant-pattern model for segregating vowel type and vocal-tract length in developmental formant data, J Acoust Soc Am 125:2374–2386

Wichmann FA, Hill NJ (2001) The psychometric function: I. Fitting, sampling, and goodness of fit. Perception and Psychophysics, 63:1293–1313

Part V
Binaural Hearing

Chapter 23
Subcomponent Cues in Binaural Unmasking

John F. Culling

Abstract The roles of fluctuations in interaural amplitude and phase in binaural unmasking were separated experimentally and examined as a function of frequency. Narrow bands of noise (1 ERB wide) with a range of centre frequencies (250–1,500 Hz) and of 500-ms duration were sinusoidally modulated at 20 Hz using either amplitude or quasi-frequency modulation (AM or QFM). In a two-interval, forced-choice task, modulation was applied in both intervals, interaurally out of phase in the signal interval and in phase in the non-signal interval. To emulate a narrow-band binaural unmasking task, the bands were used in isolation. To emulate a broadband unmasking task, flanking bands of diotic noise were added, but separated from the target band by 1-ERB-wide notches. Discrimination thresholds for narrowband AM were roughly constant as a function of frequency. However, AM with flanking bands displayed higher thresholds that increased with frequency, suggesting the presence of a cross-frequency interference effect. QFM produced a quite different data pattern, suggesting the operation of a different mechanism. Thresholds were lower than for AM up to 500 Hz, but increased sharply with frequency, and there was no effect of the flanking bands.

Keywords Unmasking • Modulation • Non-linearity • Binaural interference

23.1 Introduction

The binaural system can assist in the detection of signals in background noise. This binaural unmasking effect is typified by the binaural masking level difference (BMLD) in the N0Sπ binaural configuration. Here, a low-frequency signal is presented out of phase to the two ears against an interfering noise that is in phase. Detection threshold is substantially lower in this case than when both signal and

J.F. Culling (✉)
School of Psychology, Cardiff University, Tower Building, Park Place, Cardiff, CF10 3AT, UK
e-mail: cullingj@Cardiff.ac.uk

E.A. Lopez-Poveda et al. (eds.), *The Neurophysiological Bases of Auditory Perception*,
DOI 10.1007/978-1-4419-5686-6_23, © Springer Science+Business Media, LLC 2010

noise are in phase (N0S0). In the frequency channel that bears the signal, fluctuations in interaural phase and amplitude result from the differing vector summation of signal and noise at each ear. At other frequencies, no such fluctuations occur. A number of studies have attempted to separate the potential roles of interaural phase and amplitude in the unmasking effect, by creating either static (Hafter and Carrier 1970; McFadden et al. 1971; Yost 1972) or fluctuating (van de Par and Kohlrausch 1998; Breebaart et al. 1999) interaural differences of each type in separate conditions. As van de Par and Kohlrausch (1998) pointed out, since the addition of a tonal signal to a Gaussian noise gives rise to fluctuating cues, the latter studies are more ecologically valid. The present study builds upon their work, but using a different technique, and extends upon the ecological validity by simulating masking by broadband as well as narrowband noises. The pattern of results corresponds well with much of what we know about binaural unmasking, but at the same time suggests the presence of two underlying mechanisms.

Van de Par and Kohlrausch (1998) and Breebaart et al. (1999) employed a multiplied noise technique in order to control the way signal and noise sum at each ear. This technique allowed them to measure signal detection thresholds much like those of a normal BMLD measurement. In the present study, the interaural phase and amplitude differences of a band of Gaussian noise (1 ERB wide, cf. Moore and Glasberg 1983) were controlled directly, without the addition of a signal. This control was achieved by using interaurally out-of-phase amplitude modulation (AM) or quasi-frequency modulation (QFM) to create interaural amplitude and phase modulation, respectively. This technique has the following features. Interaural amplitude and phase modulations can be independently controlled, with zero mean. No systematic cues are introduced in the long-term power spectrum. The modulation index, m, of AM and of QFM are equivalent, facilitating their direct comparison. These modulation indices relate to interaural correlation, r, in the same way. Variability across stimuli of r, for a given m, is minimal (considerably smaller than for a given signal-to-noise ratio in N0Sπ). Interaural modulation frequency is controlled.

When presented in isolation, these narrowband, interaurally modulated noisebands thus produced cues similar to, but more closely controlled than, those of an N0Sπ stimulus with a narrowband masker. In order to produce results that were also comparable with detection thresholds for broadband maskers, these "target bands" were flanked by simultaneous bands of diotic noise. This noise filled out the spectrum from 0 to 3 kHz, save for a 1-ERB-wide notch on either side of the target band. The notches were designed to prevent interference between the flanking bands and the modulation sidebands of the target band.

23.2 Experiment

The experiment separately examined listeners' sensitivity to modulations in interaural phase and interaural amplitude, as a function of the centre frequency of the target band, and for both narrowband and broadband stimuli. A modulation rate of 20 Hz was used in order to modulate the cues at a higher rate than can be temporally resolved by the binaural system (Culling and Summerfield 1998), whilst creating the least possible spectral spread due to the modulation sidebands.

23.2.1 Method

All stimuli were generated fresh for each trial using MatLab. For each 500-ms stimulus, a Gaussian noise was generated and filtered in the frequency domain to a 1-ERB bandwidth around a centre frequency. The Hilbert transform was taken, and the complex analytic signal was multiplied by two different modulation functions (one for each ear) with a π phase difference between them. Inverse transformation then created the target band for each ear. In order to apply AM, 20-Hz sinusoidal modulation functions of amplitude m were real and added to a mean level of 1. To apply QFM, the same functions were imaginary and added to a real part set to 1. Thus, the only difference between the ways the two types of modulation were implemented was a factor of i applied to the sinusoidal modulation function. To create Broadband stimuli, flanking bands of equal spectrum level were created by generating an additional Gaussian noise, filtering it to contain a 3-ERB-wide notch centred on the centre frequency of the target band and adding it to the stimulus for each ear. All stimuli were gated by 10-ms raised-cosine onset and offset ramps.

Two such stimuli were presented on each trial separated by 500 ms of silence. In the non-signal interval, zero modulation was applied. In the signal interval, m was adapted according to a 2-down/1-up adaptive track (Levitt 1971). At the beginning of each adaptive track, m was set to 1. For the first two reversals, its size was increased or decreased by a factor of 1.2 and on the subsequent ten reversals by a factor of 1.1. The last ten reversals were averaged (geometric mean) to give the interaural modulation detection threshold.

Target bands were centred at 250, 500, 750, 1,000, 1,250, and 1,500 Hz. To cover these six modulation frequencies and the two types of modulation (AM and QFM), there were 12 measurements, presented in a random order, in each 45-min session. Broadband and narrowband stimuli were tested in separate sessions. Three listeners, including the author, practiced using the full experimental procedure until performance reached asymptote. Three thresholds were then recorded from each participant in each condition. A fourth listener began the experiment, but proved unable to achieve comparable thresholds to the other three listeners in the narrowband conditions.

23.2.2 Results

Figure 23.1 shows the results of the experiment for each individual listener and for the mean. Broadly, thresholds for QFM stimuli are lowest at low frequency, but increase sharply as frequency increases, while thresholds for AM stimuli are somewhat higher at the lowest frequencies, but, being roughly constant with increasing

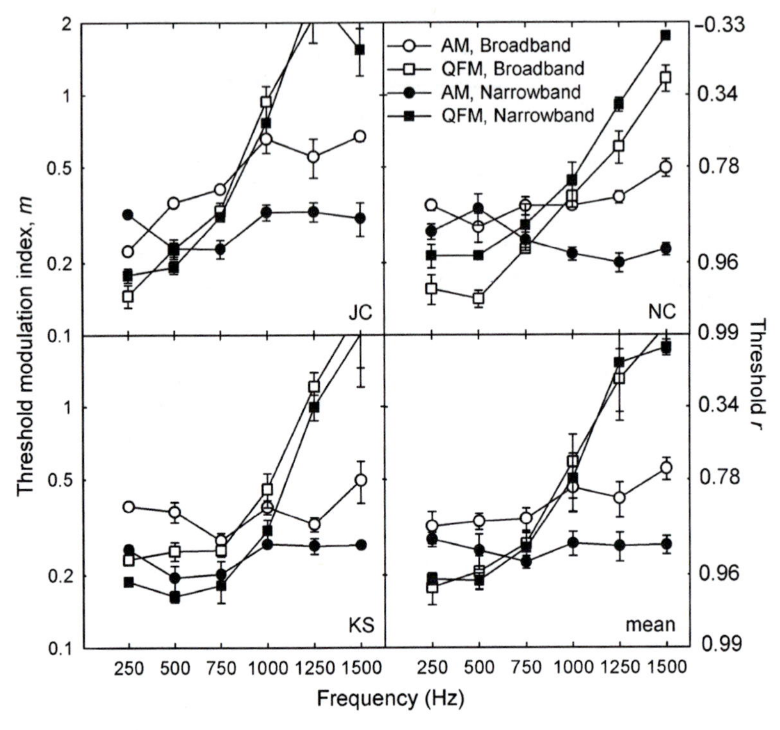

Fig. 23.1 Results of the experiment. Thresholds for the detection of interaurally out-of-phase AM (*circles*) and QFM (*squares*) as a function of target-band centre frequency. *Filled symbols* are "narrowband" thresholds, with no flanking bands. *Open symbols* are "broadband" thresholds taken with simultaneous flanking bands present. The *first three panels* show thresholds for individual listeners averaged (arithmetic mean) across three measurements. The *bottom-right panel* shows the same thresholds averaged (arithmetic mean) across the three listeners. Error bars are one standard error of the mean

frequency, are lower than the QFM thresholds at 1,000 Hz and above. Some individual variation in the pattern of data is evident. In particular, listener NC displays consistently lower thresholds for broadband QFM stimuli than for narrowband QFM stimuli, while listener KS shows the opposite pattern. Listener JC and the mean data show no overall difference between the two types. In contrast, all listeners show markedly lower thresholds for narrowband than for broadband AM stimuli, and this difference increases with centre frequency.

23.2.3 Discussion

In a real N0Sπ stimulus, both interaural phase and amplitude fluctuations will be present simultaneously and in approximately equal proportion in terms of m. Consequently, we may expect both cues to make some contribution to detection in this condition, but the cue to which the binaural system is most sensitive might be expected to dominate the resulting detection threshold. Monaural cues are unlikely to play any detectable role in N0Sπ thresholds, at least at low frequencies (Culling 2007). So, one can initially make the crude approximation that the N0Sπ threshold will correspond to the lowest of the observed thresholds in the present data from either type of cue. No correspondence can be drawn, on the other hand between these data and N0S0 thresholds, since no monaural cues are present. There is also, therefore, no correspondence with the BMLD, which depends upon N0S0 thresholds as a baseline. Given these caveats, the pattern of mean data is consistent with a wide range of data on binaural unmasking.

Detection thresholds in N0Sπ for broadband noise increase with frequency by about 16 dB between 250 and 1,400 Hz (Hirsh and Burgeat 1958). Correspondingly, we see that the lowest threshold from either of the two cues in the broadband condition increases from $m = 0.18$ to $m = 0.55$ between 250 and 1,500 Hz. Detection threshold for N0Sπ for approximately 1-ERB-wide bands of noise (equivalent to the narrowband condition) can be selected from the data of van de Par and Kohlrausch (1999). Since their maskers were presented at a constant overall level, a 3-dB/oct. correction is required to make them comparable with the present, constant-spectrum-level data. Once this is done, their threshold signal-to-noise ratios are seen to be roughly constant between 125 and 4,000 Hz. Correspondingly, lowest threshold for m, though not constant, increases relatively modestly from 0.18 to 0.26 between 250 and 1,500 Hz.

Zurek and Durlach (1987) found that, at 250 Hz, thresholds for N0Sπ are independent of masker bandwidth for bandwidths over about 100 Hz (again, a 3-dB/oct. correction is needed). At 4 kHz, however, their reported thresholds increase steeply with bandwidth for a constant spectrum level. In the present data, one correspondingly observes that the lowest broadband and narrowband thresholds (i.e. for QFM) are equal at low frequency, but, at 1,000 Hz and above, the lowest broadband thresholds (i.e. for AM) are markedly higher than the lowest narrowband thresholds.

One can think of this pattern also in terms of binaural interference. In N0Sπ detection, the presence of an additional band of diotic noise at a remote frequency can increase detection threshold for a tone in narrowband noise (Bernstein and Trahiotis 1993, 1995). However, this effect is asymmetric in frequency with high-frequency targets being affected by low-frequency interferers, but not vice versa. In the present data, one can regard the flanking bands as interferers. Interference is seen at high frequencies, particularly 1,000 Hz and above, but not at low frequencies. In N0Sπ, frequencies as low as 800 Hz can be subject to interference (Bernstein and Trahiotis 1993). The present data correspondingly indicates that interference in N0Sπ should begin somewhere between 750 and 1,000 Hz. Interestingly, the data also show that in the AM conditions, where modulation of interaural phase is absent, interference occurs at all frequencies.

Finally, we can compare the data with van de Par and Kohlrausch's (1998) experiment with multiplied noise maskers. Their stimuli are comparable with the present narrowband data and show an almost identical pattern. They also observed a steep increase in threshold with centre frequency when only interaural phase modulations were present. They also observed that thresholds are roughly constant with centre frequency when only interaural amplitude modulations were present. They also found that there was a cross-over point at about 750–1,000 Hz with interaural phase modulation producing slightly lower thresholds at lower frequencies and interaural amplitude modulation producing lower thresholds at 1,000 Hz and above. They attributed the small advantage for phase modulations at the lowest frequencies to a product of cochlear compression (Ruggero and Rich 1991), which lowers the salience of amplitude modulations once they have been encoded by the auditory periphery. Similarly, the advantage of amplitude modulations at higher frequency was attributed to a progressive of loss of peripheral phase locking to waveform fine structure with increasing frequency (e.g. Weiss and Rose 1988). Van der Heijden and Joris (this volume) have also confirmed the dominance of phase modulations at low frequencies.

The present data point to the same conclusions, but the differences in pattern between the narrowband and broadband data point to an additional conclusion. The fact that sensitivity to interference appears to be a feature of detection of interaural amplitude modulation (AM), but not of interaural phase modulation (QFM) suggests that these two cues rely on qualitatively distinct detection mechanisms. Possibly interaural phase modulation is detected by a within-channel mechanism, whilst interaural amplitude modulation is detected by a mechanism that features obligatory across-frequency integration. In this interpretation, the interaction between stimulus frequency and masker bandwidth and the frequency asymmetry of binaural interference are complimentary products of two different mechanisms operating in different circumstances.

23.3 Modelling

As noted above, van de Par and Kohlrausch (1998) interpreted the apparent advantage for the detection of interaural phase compared to amplitude modulation at low frequencies to the action of cochlear compression. They suggested that the two cues might be detected by a common mechanism in this low frequency region, despite the disparity in detection threshold. They suggested further that this mechanism is sensitive to interaural correlation, r, a statistic that is affected by both interaural phase and amplitude modulation. All disparities in threshold are explained in this scheme through the action of peripheral non-linearities that transform the stimulus presented to the ears and make the internal r between the transduced stimuli at each ear, r_i, different from the r between external stimulus wave forms, r_e. At threshold, r_i should thus be constant in all stimulus conditions.

These ideas have also been pursued by Bernstein and Trahiotis (1996, 2003). The latter article contains a specific package of operations for evaluating r_i. In the present modelling, this scheme has been adopted and the model parameters adjusted in order to produce a fit to the current data. The results indicate that the scheme works well for three out of the four conditions used in the experiment, but it is unable to account for the AM broadband condition.

23.3.1 Method

The data was fitted using the fminsearch function in MatLab, which implements the Simplex technique for fitting multi-parameter models (Nelder and Mead 1965). The Bernstein and Trahiotis model (2003) consists of the following processes (1) the waveforms at each ear are passed though a gammatone filter tuned to the signal frequency; (2) the envelopes of the waveforms are compressed through manipulation of their complex analytic waveforms; (3) the waveforms are half-wave rectified and squared; (4) the waveforms are low-pass filtered using a fourth-order Butterworth filter; (5) the internal interaural correlation is evaluated using the normalised correlation (as opposed to the normalised covariance). The free parameters in the model are the power-law compression factor and the cut-off frequency of the low-pass filter.

Since the resulting r_i should be identical at threshold in all conditions, the quality of fit with the data was evaluated by calculating the variance of r_i for threshold stimuli in all conditions. The Simplex algorithm minimised this variance.

23.3.2 Results

The results of the modelling can be seen in Fig. 23.2. The left-hand panel illustrates the r_i values at threshold, when minimised across all 24 conditions. Ideally, they should all be equal. In order to produce this fit, the model used a compression factor of 0.12 and a low-pass cut-off frequency of 674 Hz. It is noteworthy that the compression factor is about half the size of that generally reported in the literature and also of that fitted to the data of Bernstein and Trahiotis (2003). In addition, there remains a systematic mismatch between the internal correlations at threshold for AM and QFM stimuli across the low frequencies and between the two AM conditions across all frequencies. These mismatches result in a residual variance of 7.06×10^{-5}. The limitations in the quality and plausibility of this fit led to an alternative approach.

The right-hand panel of Fig. 23.2 shows the results of allowing the model to ignore the AM broadband data and fit the other three conditions. Now, the fitted parameters were a more plausible compression ratio of 0.272 and a slightly higher low-pass cut-off of 716 Hz. The residual variance was reduced to 4.17×10^{-5}. The internal correlation, r_i, for these best-fitting parameters is around 0.98 at threshold.

23.3.3 Discussion

The modelling results showed that the Bernstein et al. modelling approach works well for three of the four conditions, but has difficulty accommodating the AM broadband condition. This outcome is not surprising, because the

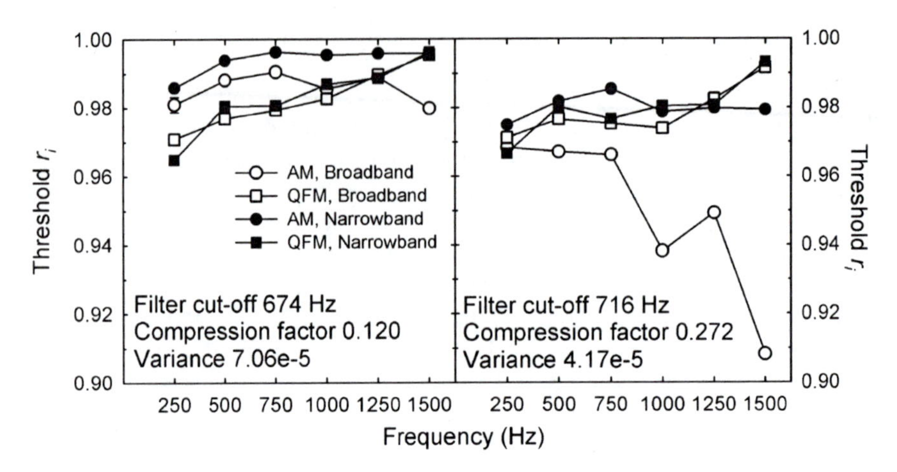

Fig. 23.2 Results of modelling peripheral non-linearities. Internal interaural correlation, r_i, for stimuli at mean threshold after the application of a model of peripheral transduction. The model parameters are optimised to minimise the variance of r_i at threshold. The *left panel* shows the results for parameters optimised using all 24 stimulus conditions. The *right panel* shows results obtained by ignoring the AM broadband condition and optimising for the remaining 18 stimulus conditions

difference in the measured thresholds between this condition and AM narrowband appears to be a form of binaural interference and Bernstein and Trahiotis' model includes no interference process. The fact that the model works well for the other three conditions indicates that interaural phase and amplitude modulation may be detected by the same r_i-sensitive mechanism at low frequencies. This mechanism is immune to binaural interference. At higher frequencies, this mechanism becomes very insensitive, but here a second interference-susceptible mechanism can still detect interaural amplitude modulations.

23.4 Conclusions

The results of Experiment 1 suggest that detection of tones in N0Sπ may be dependent upon separate mechanisms in different frequency ranges. Up to 750 Hz, thresholds are probably determined by interaural phase modulations, but given what we know about peripheral transduction, the mechanism may be equally sensitive to either cue in the internal representation of the stimulus. Although sensitive to interaural correlation, the mechanism is not necessarily a correlation detector per se and may be based upon an interaural cancellation mechanism (Durlach 1963, 1972; Culling 2007). This mechanism is not subject to across-frequency interference. From 1,000 Hz upwards, N0Sπ thresholds are determined by detection of interaural amplitude modulations, which are subject to across-frequency interference, suggesting a distinct detection mechanism. Although this picture may seem complex, it also appears to have explanatory power.

Acknowledgment I am indebted to Marcel van der Heijden for introducing me to the QFM technique.

References

Bernstein LR, Trahiotis C (1993) Spectral interference in binaural detection tasks: effects of masker bandwidth and temporal fringe. J Acoust Soc Am 98:155–163

Bernstein LR, Trahiotis C (1995) Binaural interference effects measured with masking-level difference and with ITD- and ILD-discrimination paradigms. J Acoust Soc Am 94:735–742

Bernstein LR, Trahiotis C (1996) The normalized correlation: accounting for binaural detection across center frequency. J Acoust Soc Am 100:3774–3784

Bernstein LR, Trahiotis C (2003) Enhancing interaural-delay-based extents of laterality at high frequencies using "transposed stimuli." J Acoust Soc Am 113:3335–3347

Breebaart J, van de Par S, Kohlrausch A (1999) The contribution of static and dynamically varying ITDs and IIDs to binaural detection. J Acoust Soc Am 106:979–992

Culling JF (2007) Evidence specifically favoring the equalization-cancellation theory of binaural unmasking. J Acoust Soc Am 122:2803–2813

Culling JF, Summerfield Q (1998) Measurements of the binaural temporal window using a detection task. J Acoust Soc Am 103:3540–3553

Durlach NI (1963) Equalization and cancellation theory of binaural masking level differences. J Acoust Soc Am 35:1206–1218

Durlach NI (1972) Binaural signal detection: equalization and cancellation theory. In: Tobias JV (ed) Foundations of modern auditory theory, vol II. Academic, New York

Hafter ER, Carrier SC (1970) Masking level differences obtained with a pulsed tonal masker. J Acoust Soc Am 47:1041–1047

Hirsh IJ Burgeat M (1958) Binaural effects in remote masking. J Acoust Soc Am 30:827–832

Levitt H (1971) Transformed up-down methods in psychoacoustics. J Acoust Soc Am 49:467–477

McFadden D, Jeffress, LA, Ermey, HL (1971) Differences of interaural phase and level in detection and lateralization. J Acoust Soc Am 50:1484–1493

Moore, BCJ, Glasgerg BR (1983) Suggested formulae for calculating auditory-filter bandwidths and excitation patterns. J Acoust Soc Am 74:750–753

Nelder JA, Mead R (1965) A simplex method for function minimization. Comput J 7:308

van de Par S, Kohlrausch A (1998) Diotic and dichotic detection using multiplied-noise maskers. J Acoust Soc Am 103:2100–2110

van de Par S, Kohlrausch A (1999) Dependence of binaural masking level differences on center frequency, masker bandwidth, and interaural parameters. J Acoust Soc Am 106:1940–1947

Ruggero MA, Rich NC (1991) Furosemide alters organ of Corti mechanics: evidence for feedback of outer hair cells on the basilar membrane. J Neurosci 11:1057–1067

Weiss TF, Rose C (1988) A comparison of synchronization filters in different auditory receptor organs. Hear Res 33:175–180

Yost (1972) Tone-on-tone masking for three binaural listening conditions. J Acoust Soc Am 52:1234–1237

Zurek, PM, Durlach NI (1987) Masker-bandwidth dependence in homophasic and antiphasic tone detection. J Acoust Soc Am 81:459–464

Chapter 24
Interaural Correlations Between +1 and −1 on a Thurstone Scale: Psychometric Functions and a Two-Parameter Model

Helge Lüddemann, Helmut Riedel, and Andre Rupp

Abstract This contribution investigates how the interaural correlation (IAC/ρ) is represented in terms of perceptual units, to understand how the binaural auditory system separates meaningful sound sources from diffuse ambient noise. Psychometric functions for the discriminability of narrowband noise signals with different IAC were obtained from pairwise comparisons in a novel 2-pair-AFC-paradigm. The nine reference correlations covered the whole parameter range between −1 and +1.

The shapes and slopes of the psychometric functions critically depended on the reference IAC, and IAC discrimination thresholds were inconsistent with the predictions of signal detection theory (SDT) if the normalized IAC was used as a decision variable.

Therefore, a Thurstone model was applied to transform the stimulus IAC into a decision variable $T(\rho)$ which is, in contrast to the normalized IAC itself, characterized by standard normal distributions in the entire parameter range and thus compatible with SDT. In good approximation, $T(\rho)$ is proportional to the difference between the dB-scaled energies in the correlated vs. the anticorrelated signal components after noisy peripheral preprocessing. $T(\rho)$ allows one to predict the percentage of correct discriminations for arbitrary pairs of IAC with a simple analytical two-parameter-formula.

Keywords Binaural listening • Decision variable • Thurstone model case V

H. Lüddemann (✉)
Medizinische Physik, Institut für Physik, Universität Oldenburg,
26111 Oldenburg, Germany
e-mail: h.lueddemann@uni-oldenburg.de

E.A. Lopez-Poveda et al. (eds.), *The Neurophysiological Bases of Auditory Perception*,
DOI 10.1007/978-1-4419-5686-6_24,

24.1 Introduction

Binaural listening can facilitate the identification of auditory objects and improve speech intelligibility in complex acoustical situations. However, it is well known that temporal and spectral fluctuations of binaural cues have a tremendous impact on the auditory system's ability to localize – and thereby separate – relevant signals and concurrent auditory events, e.g., reverb or ambient noise. Many studies support the idea that the interaural correlation (IAC) between the two ear signals $l(t)$ and $r(t)$ plays a key role in the weighting of binaural information during spatial scene analysis (see Faller and Merimaa (2004) for an overview).

The IAC is usually specified as the normalized scalar product of the time vectors $l(t)$ and $r(t)$:

$$\rho = \int l(t)r(t)dt \left[\int l^2(t)dt \int r^2(t)dt \right]^{-1/2} \tag{24.1}$$

As an integrative measure of coincidence, the positive range of ρ is often associated with the perceptual continuum between compact ($\rho=+1$) and diffuse ($\rho=0$) spatial listening impressions. However, most research on the topic concentrates on IAC discrimination thresholds in the vicinity of integer ρ, but little is known about the decision statistics, which describes the perceptual distance between intermediate IAC and the respective discrimination probability. The present study tackles this issue by fitting a "Thurstone model case V" to pairwise comparison data, including the values of ρ in the entire range between -1 and $+1$.

24.1.1 *Reasons Against the Use of the Normalized IAC*

The discriminability of two signals with different interaural correlations critically depends on the IAC of the reference stimulus, ρ_{ref}: the just noticeable difference (ρ-JND) between ρ_{ref} and a deviant IAC ρ_{dev} is about 10 times smaller for $\rho_{ref}=+1$ than for $\rho_{ref}=0$. This relation is found throughout the literature and is illustrated in Fig. 24.1, using data by Pollack and Trittipoe (1959a).

It has often been argued (see, e.g., Gabriel and Colburn 1981; Breebaart and Kohlrausch 2001) that this dependence on ρ_{ref} could be due to the variability of single observations, since the actual correlation ρ in temporal or spectral portions of a signal with an overall IAC of ρ_0 is a random variable with a variance being roughly proportional to $(1-\rho_0^2)$.

Indeed, the ρ-JNDs by Pollack and Trittipoe (1959a) are well approximated by the fit function $f(\rho_{ref})=0.41(1-\rho_{ref}^2)$, and thus proportional to the variance of single observations (least square fit, solid line in Fig. 24.1). In contrast, the assumption that JNDs were a multiple of the respective standard deviation, i.e., proportional to $(1-\rho_{ref}^2)^{1/2}$, results in a poorer fit (dashed line). Signal detection theory (SDT),

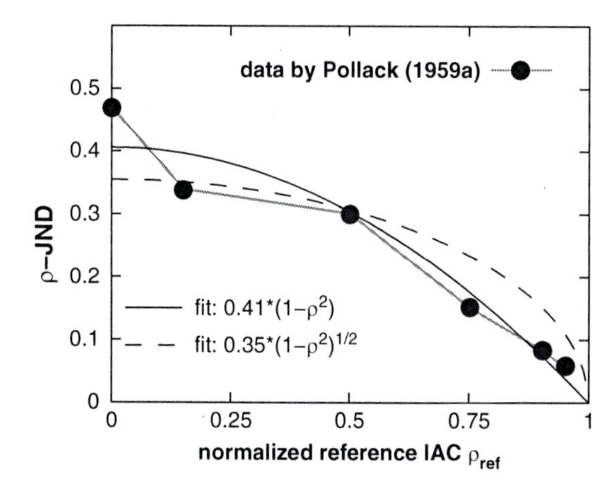

Fig. 24.1 Just noticeable IAC difference (ρ-JND) as function of the reference IAC ρ_{ref}. Data by Pollack and Trittipoe (1959a) are shown as *dots*. The *fitted curves* indicate that ρ-JNDs are proportional to the variance (*solid line*) rather than to the standard deviation (*dashed line*) of sample correlations

however, predicts that ρ-JNDs should be proportional to the standard deviation rather than to the variance of the decision statistics.

Hence, it is implausible that the auditory system uses the normalized ρ itself as a decision variable, that would allow for a perceptually adequate description of IAC differences by means of equal variance Gaussian signal detection theory (SDT).

24.1.2 Alternative Hypothesis: Spatial Percept Represented by the dB Scaled Ratio of N_0- and N_π-Components

One alternative way to describe the coincidence between the two ear signals l and r is to calculate the energies of the correlated (N_0, $\rho=+1$) and anticorrelated (N_π, $\rho=-1$) signal components and to specify their ratio in dB:

$$\rho_{0\pi,\alpha}\ [\mathrm{dB}(N_0/N_\pi)_\alpha] = 10 \cdot \log\,(1+\rho+\alpha^{-2})\,/\,(1-\rho+\alpha^{-2}) \qquad (24.2)$$

In the following, we denote $\rho_{0\pi,\alpha}$ as the "effective dB(N_0/N_π) scaled IAC" because the scale parameter α simulates the degrading effect of additive uncorrelated noise (e.g., peripheral transmission errors) on the effective interaural correlation of a signal prior to binaural processing (van der Heijden and Trahiotis 1997). Furthermore, finite ratios between signal- and noise power (SNR$=\alpha^2$) also avoid that $\rho_{0\pi,\alpha}$ takes infinite values.

24.2 Methods

24.2.1 Psychoacoustical Experiments

Psychometric functions for the discriminability of signals with different IAC were measured in a 2-AFC paradigm, with two pairs of narrow band stimuli (500 ms duration, 1 ERB centered at 500 Hz, and 65 dB SPL) in each trial. The two stimuli in the target pair had different correlations, ρ_{ref} and ρ_{dev}, while the stimuli in the control pair both had an IAC of ρ_{ref}. Stimuli were generated in Matlab, using the technique described by Culling et al. (2001)

The 2-Pair-2-AFC paradigm avoids any preference or misconception of particular IAC values (e.g., selection of the signal with more compact/broader percept or with higher/lower IAC). It further minimizes any subjective bias and ensures a chance level of 50% in the case of indistinguishable ρ_{ref} and ρ_{dev}.

The 94 combinations of reference and deviant IAC were selected after pilot testing, so that psychometric functions (1) can be constructed with respect to nine reference correlations, including $\rho_{ref}=+1$, 0, −1 and six intermediate values, (2) have all roughly the same number of samplings points, (3) cover the whole range of possible discrimination rates except for the vicinity of 100% and (4) have a maximal intersect of absolute IAC levels, $\rho_{dev}=\rho_{ref}+\Delta\rho$.

For each combination of ρ_{ref} and ρ_{dev}, a total number of 275 trials was acquired, including data from six normal hearing subjects (three males/three females, aged between 25 and 32).

24.2.2 Thurstone Scaling

A Thurstone model case V was used to investigate the relation between the external stimulus parameter ρ and the internal decision variable in an IAC discrimination task. This approach implies the apriori assumption that IAC discrimination is based on a single decision variable, which is monotonically related to ρ by the scale transform $T(\rho)$.

There are no further constraints about the functional character of $T(\rho)$. Instead, the Thurstone model generates $T(\rho)$ as a "lookup table" and adjusts the difference between any two scale values, $\Delta T=T(\rho+\Delta\rho)-T(\rho)$, so that the percentage of correctly discriminated stimuli can be predicted by one unique Gaussian cumulative density function (CDF) for all ρ_{ref} and ρ_{dev}, i.e., $p_{correct}=\Phi(|\Delta T|/s)$, with s being a constant global scale parameter independent of ρ.

24.3 Results

The psychometric functions in Fig. 24.2a are characterized by very different shapes and slopes, depending on both ρ_{ref} and the sign of the IAC deviation. These dependencies are also apparent in the thresholds for the just noticeable IAC increase or

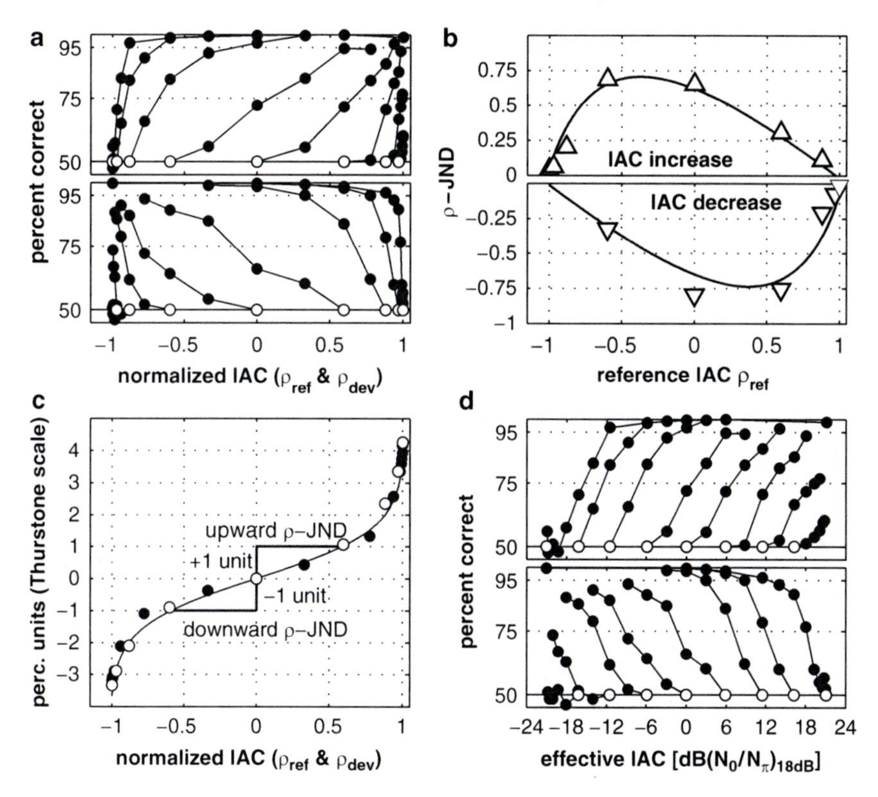

Fig. 24.2 Discriminability of narrowband noise stimuli with different interaural correlations (500 ms, 1 ERB centered at 500 Hz). (**a**) Psychometric functions for the discriminability of signals with ρ_{dev} (*filled dots*) with respect to ρ_{ref} (*open dots* at 50% chance level). (**b**) ρ-JND as a function of ρ_{ref}: JNDs for an increase/decrease of IAC were derived from the 75%-level in the upper/lower part of (**a**). The *solid lines* were calculated assuming that the JND corresponds to a constant difference of 6.5 dB(N_0/N_π) between reference and deviant IAC. (**c**) Relation between normalized ρ and Thurstone scale values $T(\rho)$: the fit function $f_{\alpha,s}(\rho)=\rho_{0\pi,\alpha}(\rho)/s$ [*solid line*, $\rho_{0\pi,\alpha}$ according to (24.2)] approximates $T(\rho)$ best for $\alpha=18$ dB and $s=5.73$ dB$(N_0/N_\pi)_{18\text{ dB}}$ per perceptual unit. (**d**) Same data as in (**a**) but with the IAC expressed as the effective dB(N_0/N_π) scaled IAC, assuming an SNR of 18 dB in the monaural periphery [cf. (24.2)]

decrease (Fig. 24.2b): ρ-JNDs are best for $\rho_{ref}=+1$ (0.015), slightly larger for $\rho_{ref}=-1$ (0.042) and largest in the vicinity of zero (0.65/−0.80 for positive/negative deviations). Hence, the ρ-JND strongly depends on $|\rho_{ref}|$. But also the sign of the IAC deviation, and the issue, if the task is performed in the positive/negative range of ρ have (interacting) effects on the ρ-JND.

Figure 24.2c shows that the Thurstone scale values (dots) are roughly proportional to the effective dB(N_0/N_π)-transformed IAC after noisy monaural preprocessing with an SNR of 18 dB (solid line). The proportionality factor of the fit function indicates that one unit on the Thurstone scale corresponds to 5.73 dB$(N_0/N_\pi)_{18\text{ dB}}$. However, in contrast to the dB(N_0/N_π) transform, the Thurstone scale is not exactly symmetric in the sense $T(\rho)=-T(-\rho)$. Instead, positive correlations are mapped to scale values between 0 and 4.3, whereas the negative range of ρ is represented by only 3.3 units.

24.4 Discussion

The dependence of the ρ-JND on ρ_{ref} observed in the present work is in qualitative agreement with previous literature, e.g., Pollack and Trittipoe (1959a) or Boehnke et al. (2002). The quantitative differences to their data can be understood as an effect of stimulus bandwidth (Gabriel and Colburn 1981).

In addition, the comparison of ρ-JNDs for corresponding positive and negative ρ_{ref} and for positive and negative IAC deviations (Fig. 24.2b) shows clearly that the ρ-JND depends not only on ρ_{ref}, but also on the sign of ρ and $\Delta\rho$. In particular, the JND is not simply a multiple of the variance of normalized sample correlations (cf. Sect. 24.1.1) because thresholds proportional to $(1-\rho_{ref}^2)$ would show the same dependence on $|\rho_{ref}|$ for positive and negative ρ_{ref}, in contradiction to the data. The worst ρ-JND is presumably not found at $\rho_{ref}=0$, but instead at small positive/negative ρ_{ref} for negative/positive IAC deviations.

Although the main trends in the data can be described by a comparatively simple analytical formula with two parameters [cf. Fig. 24.2c, (24.2)], the outcome of the Thurstone scaling procedure suggests that IAC sensitivity is generally better in the positive range of ρ than in the negative range. This minor asymmetry might be due to hair cell transduction or different weighting of positively and negatively correlated signal components.

Nevertheless, it is evident that the effective $\mathrm{dB}(N_0/N_\pi)$ scaled IAC after noisy monaural preprocessing resembles the decision variable of the auditory system in an IAC discrimination task by means of equal variance Gaussian SDT far better than the normalized ρ. After replotting the data from Fig. 24.2a on the $\mathrm{dB}(N_0/N_\pi)_{18\,\mathrm{dB}}$ scale, all psychometric functions in Fig. 24.2d appear as shifted (for different ρ_{ref}) or mirrored (for IAC increase/decrease) copies of one unique Gaussian CDF. Consequently, the JND between $\mathrm{dB}(N_0/N_\pi)$ scaled reference and deviant correlations corresponds to roughly 6.5 $\mathrm{dB}(N_0/N_\pi)_{18\,\mathrm{dB}}$, independent of ρ_{ref} or the sign of ρ or $\Delta\rho$ (solid lines in Fig. 24.2b), in accordance with the proportionality factor of the fit in Fig. 24.2c. Furthermore, the effective $\mathrm{dB}(N_0/N_\pi)$ scaled IAC is a neurophysiologically plausible measure of coincidence: it is linearly related to the amplitude of LAEP in response to an IAC transition (Lüddemann et al. 2007, 2009), and it is qualitatively compatible with fMRI data by Budd et al. (2003) and with single cell recordings by Coffey et al. (2006).

We, therefore, suggest that binaural models should not use the normalized ρ, but instead use the effective dB scaled ratio of correlated and anticorrelated signal components.

Acknowledgments This work was supported by the Deutsche Forschungsgemeinschaft (DFG, project KO 942/19-1).

References

Boehnke SE, Hall SE, Marquardt T (2002) Detection of static and dynamic changes in interaural correlation. J Acoust Soc Am 112(4):1617–1626

Breebaart J, Kohlrausch A (2001) The influence of interaural stimulus uncertainty on binaural signal detection. J Acoust Soc Am 109(1):331–345

Budd TW, Hall DA, Goncalves MS, Akeroyd MA, Foster JR, Palmer AR, Head K, Summerfield AQ (2003) Binaural specialisation in human auditory cortex: an fMRI investigation of interaural correlation sensitivity. Neuroimage 20(3):1783–1794

Coffey CS, Ebert CS Jr, Marshall AF, Skaggs JD, Falk SE, Crocker WD, Pearson JM, Fitzpatrick DC (2006) Detection of interaural correlation by neurons in the superior olivary complex, inferior colliculus and auditory cortex of the unanesthetized rabbit. Hear Res 221(1–2):1–16

Culling JF, Colburn HS, Spurchise M (2001) Interaural correlation sensitivity. J Acoust Soc Am 110(2):1020–1029

Faller C, Merimaa J (2004) Source localization in complex listening situations: selection of binaural cues based on interaural coherence. J Acoust Soc Am 116(5):3075–3089

Gabriel KJ, Colburn HS (1981) Interaural correlation discrimination: I. bandwidth and level dependence. J Acoust Soc Am 69(5):1394–1401

Lüddemann H, Riedel H, Kollmeier B (2007) Logarithmic scaling of interaural cross-correlation: a model based on evidence from psychophysics and EEG. In: Kollmeier B, Klump G, Hohmann V, Langemann U, Mauermann M, Uppenkamp S, Verhey J (eds) Hearing: from sensory processing to perception – 14th International Symposium on Hearing. Springer, Berlin

Lüddemann H, Riedel H, Kollmeier B (2009) Asymmetries in electro-physiological and psychophysical sensitivity to interaural cross-correlation steps. Hear Res 256(1–2):39–57

Pollack I, Trittipoe WJ (1959) Binaural listening and interaural noise cross correlation. J Acoust Soc Am 31(9):1250–1252

Van der Heijden M, Trahiotis C (1997) A new way to account for binaural detection as a function of interaural noise correlation. J Acoust Soc Am 101(2):1019–1022

Chapter 25
Dynamic ITDs, Not ILDs, Underlie Binaural Detection of a Tone in Wideband Noise

Marcel van der Heijden and Philip X. Joris

Abstract To examine the relative contributions of dynamic ITDs and ILDs in binaural detection, we developed a novel signal processing technique that selectively degrades different aspects of binaural stimuli. We applied this selective scrambling technique to the stimuli of a classic N0Sπ task: detecting an antiphasic 500-Hz signal in a diotic wideband noise. Data from five listeners revealed (1) selective scrambling of ILDs had little effect on binaural detection; (2) selective scrambling of ITDs significantly degraded detection; and (3) combined scrambling of ILDs and ITDs had the same effect as scrambling of ITDs only. We conclude that (1) dynamic ITDs dominate detection performance; (2) ILDs are largely irrelevant; and (3) interaural correlation is a poor predictor of detection. We describe two stimulus-based models, each reproduces all binaural aspects of the data quite well: (1) a single-parameter detection model using ITD variance as detection criterion; and (2) a crosscorrelator preceded by waveform compression. We propose that the observed sensitivity to ITDs and insensitivity to ILDs reflect enhanced temporal coding and limited dynamic range found in bushy cells, the monaural inputs to the binaural processor.

Keywords MLD • Binaural • Masking • ITD

25.1 Introduction

In an N0Sπ detection task, an interaurally phase-reversed tone Sπ is masked by an interaurally identical noise N0. For wideband maskers and low-frequency (<1 kHz) tones, the masking level difference (MLD) *re* the N0S0 condition is ~12 dB.

M. van der Heijden (✉)
Laboratory of Auditory Neurophysiology, Medical School, Campus Gasthuisberg, K.U. Leuven, B-3000 Leuven, Belgium
e-mail: m.vanderheyden@erasmusmc.nl

E.A. Lopez-Poveda et al. (eds.), *The Neurophysiological Bases of Auditory Perception*,
DOI 10.1007/978-1-4419-5686-6_25,

The MLD shows listeners' sensitivity to interaural disparities caused by adding the Sπ signal to the N0 masker, but does not reveal which aspects of the stimulus ("cues") are used for detection and which are not. Any quantitative model or theory of binaural detection must quantify interaural disparity of auditory stimuli (Colburn and Durlach 1978). In the classic stimulus-oriented theories of binaural detection, two approaches prevail. In the first, the disparity is quantified in terms of interaural correlation of the stimulus, and detection is associated with a critical change in correlation caused by the signal (Dolan and Robinson 1967). The second approach distinguishes two types of binaural disparity: interaural time differences (ITDs) and interaural level differences (ILDs). Only a limited spectral noise band around the signal frequency is effective in masking the tone, and this band is equivalent to a tone with slowly varying amplitude and phase. In N0 noise, these modulations are identical in the two ears, but the addition of an Sπ signal introduces slowly varying, random ITDs and ILDs. It is the listeners' sensitivity to these dynamical interaural disparities that is assumed to underlie binaural detection in the second type of models. Webster (1951) hypothesized that binaural detection of tones in broadband noise requires a certain magnitude of dynamic ITD. Later work also considered dynamic ILDs (Hafter and Carrier 1970). Alternatively, in many physiologically inspired models of binaural detection (e.g., Colburn 1973), contrasting roles of ILDs and ITDs are not described explicitly but implicitly in the form of assumptions concerning peripheral processing (van de Par and Kohlrausch 1998), sources of internal noise (Durlach 1963), and/or decision criteria.

It is difficult to assess the relative importance of ITDs and ILDs because they are inextricably linked in most detection tasks. In the N0Sπ case, the magnitudes of both ITDs and ILDs grow with signal level, leading to an effective equivalence between correlation-based models, in which ILDs and ITDs merge into the single correlation metric, and models that do discriminate between dynamic ITDs and ILDs (Domnitz and Colburn 1976). Here, we present a novel approach, based on imposing "jitter" on the stimulus and assessing its disruptive effect on binaural detection. Starting with a classic wideband N0Sπ task, the stimuli are manipulated in different ways that selectively scramble the different cues: ITDs, ILDs, or both. The decline in performance with the different types of scrambling then reveals the relative importance of the scrambled cues.

25.2 Methods

25.2.1 Binaural Modulation

The general idea is as follows. The stimulus to each ear is modulated in a manner that affects both the amplitude and the phase of the waveform. The amplitude modulation (AM) introduces maxima and minima in the envelope; the phase modulation (QFM) introduces phase leads and phase lags *re* the unmodulated waveform. As long as the modulation is identical in the two ears, no binaural differences occur, but when the modulation is interaurally antiphasic, envelope maxima in one ear coincide with envelope minima in the other, and phase leads in one ear coincide with lags in the other. This opponency leads to ILDs and ITDs. Interestingly, it is also possible to choose modulation phases in the two ears in such a way that the envelopes are interaurally reversed, but the phases are identical: this yields ILDs but almost no ITDs (only at large modulation depth is there a second order effect on ITDs, called "spurious ITDs" below). Yet another choice of modulation phases yields identical envelopes and antiphasic phase shifts: ITDs, no ILDs. Figure 25.1 illustrates how three metrics of binaural disparity (interaural correlation; RMS[ITD]; RMS[ILD]) are affected by the different types of binaural modulation.

We imposed binaural modulation on the *complete waveforms* comprising the N0Sπ (or N0S0) stimuli: both masker and signal (if present) were subjected to the same modulation, leaving their mutual interaction unaltered. The unmodulated condition was either N0Sπ or N0S0. Modulation type is indicated by subscripts, e.g., $(\mathrm{N0S}\pi)_\mathrm{d}$ is diotic modulation applied to a N0Sπ. The stimulus types are listed in Table 25.1.

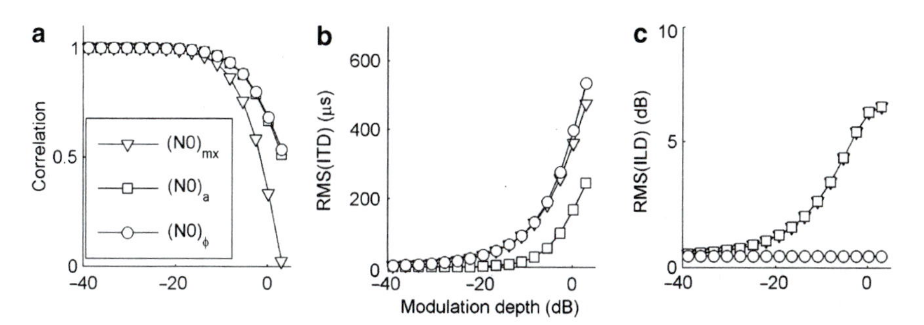

Fig. 25.1 The effect of binaural modulation on three metrics of binaural disparity. (**a**) normalized interaural correlation; (**b**) RMS of dynamic ITD; (**c**). RMS of dynamic ILD. Symbols indicate the type of binaural modulation imposed on the stimulus: mixed (mx: *triangles*), binaural AM (*a*: *squares*), and binaural QFM (ϕ: *circles*). The binaural modulation was applied to a diotic noise stimulus

Table 25.1 Stimulus configurations

Stimulus	Modulation type	Binaural cues affected
$(\mathrm{N0S0})_\mathrm{d}$	Diotic	None
$(\mathrm{N0S}\pi)_\mathrm{d}$	Diotic	None
$(\mathrm{N0S}\pi)_\mathrm{mx}$	Binaural mixed	Dynamic ITDs and ILDs
$(\mathrm{N0S}\pi)_\phi$	Binaural QFM	Dynamic ITDs
$(\mathrm{N0S}\pi)_\mathrm{a}$	Binaural AM	Dynamic ILDs (spurious ITDs)

25.2.2 Stimuli and Data Collection

Detection of a 500-Hz signal was examined in the presence of a Gaussian noise (100–3,000 Hz, total level of 70 dB SPL). The 280-ms signals were temporally centered within the 300-ms maskers. Durations include 10-ms ramps. Independent noise tokens were generated for each presentation. Mixed modulation described above was imposed on both noise-alone ("reference") and noise-plus-tone ("target") intervals. The modulation frequency was 20 Hz, a value high enough to prevent the binaural system from "following" the fluctuations in binaural parameters (Grantham and Wightman 1978), and low enough for the type of modulation to be uncorrupted by peripheral filtering. Modulation depth is expressed in dB and equals 20 log *m*. Computational details of the stimulus generation will be published elsewhere.

For each stimulus configuration, modulation depths of −∞, −12, −7, −2 and 3 dB were tested, where −∞ dB means no modulation at all ($m=0$). Conditions were interleaved and presented in random order. Thresholds are the average of four estimates. Stimulus generation and presentation are described in Van der Heijden and Trahiotis (1999). Five female students with normal hearing, 21–25 years old, served as listeners. They received 4 h of training prior to data collection.

25.3 Results

Figure 25.2a–e show the thresholds of the individual listeners. Stimulus configurations (see Table 25.1) are indicated in the graph. The data obtained with the unmodulated stimuli (−∞dB modulation depth) serve as a reference: they are conventional N0Sπ and N0S0 thresholds. MLDs ranged from 8 to 13 dB across listeners.

The $(N0S0)_d$ and $(N0S\pi)_d$ thresholds (filled symbols) do not show a systematic variation with modulation depth; their curves are basically flat. Because these are the conditions in which the modulation is diotic, this implies that modulation per se, although clearly audible, does not affect the monaural and binaural thresholds of the 500 Hz tone. This strongly suggests that the cues underlying monaural and binaural detection of the tone detection are unaffected by the modulation as long as the modulation does not introduce any binaural disparities. This justifies an interpretation of any effects in the remaining configurations solely in terms of the *binaural* aspects of the modulation.

In contrast to the diotically modulated conditions, the $(N0S\pi)_{mx}$ thresholds (open triangles) show a clear effect of modulation. With increasing *m*, thresholds grow from a minimal, N0Sπ, value and approach the N0S0 threshold at the highest value of *m*, 3 dB. Thus, the 20-Hz mixed binaural modulation, when strong enough, causes a substantial reduction of the binaural advantage. The modulation in this case is antiphasic in both its QFM and AM constituents, resulting in a decorrelation of the waveform that is mediated by both dynamic ITDs and ILDs (Fig. 25.1, triangles).

We now turn to the cases where binaural modulation was imposed either in an ILD-specific way or an ITD-specific way. The $(N0S\pi)_a$ thresholds (open squares)

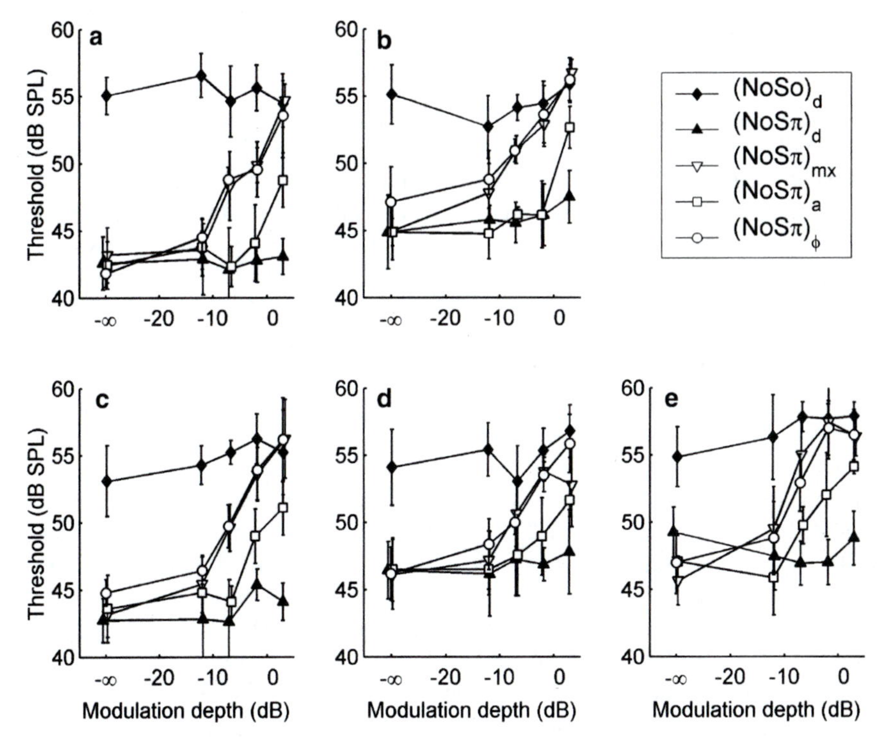

Fig. 25.2 Thresholds for detecting a 500-Hz tone in the presence of wideband noise. Each *symbol* is the average of four thresholds; error bars indicate ± one standard deviation. Each *panel* displays the data from one listener. Different *symbols* indicate different conditions. *Closed symbols* are used for the control conditions, in which diotic modulation was applied to N0S0 (*closed diamonds*) and N0Sπ (*closed triangles*). *Open symbols* indicate different types of binaural modulation as indicated in the graph

show a weak effect of modulation: only at the two highest values of modulation depth are the thresholds elevated by binaural AM. In contrast, the $(\mathrm{N0S}\pi)_\phi$ curves (open circles) practically coincide with the $(\mathrm{N0S}\pi)_{\mathrm{mx}}$ curves. Thus, binaural QFM has a much more pronounced effect on detection than binaural AM, implying that dynamic ITDs are more important than dynamic ILDs in the N0Sπ task. The overlap of $(\mathrm{N0S}\pi)_{\mathrm{mx}}$ thresholds (in which ITDs and ILDs are equally manipulated) and $(\mathrm{N0S}\pi)_\phi$ thresholds (in which only ITDs are manipulated) strongly suggests that dynamic ITDs completely determine the 500-Hz N0Sπ condition used in the present study, and that ILDs are irrelevant.

The occurrence of spurious ITD modulation in the $(\mathrm{N0S}\pi)_{\mathrm{a}}$ condition (cf. Fig. 25.3b) leaves open the possibility that the effect of modulation in this condition is in fact caused by spurious ITD modulation, not by the "proper" ILD modulation. If that is true, the binaural amplitude modulation itself has no effect at all, not even at the highest values of m, and the dominance of dynamic ITDs in the N0Sπ detection task would be complete.

25.4 Modeling the Data

Models based on interaural correlation of the (effective) stimulus predict identical thresholds for $(N0S\pi)_a$ and $(N0S\pi)_\phi$ because these conditions are equivalent in terms of correlation. Such models fail to reproduce the clear ILD/ITD asymmetry in our data. Figure 25.3a shows predictions of such a correlation model (Van der Heijden and Trahiotis 1999). Symbols are data from listener 1. The $(N0S\pi)_{mx}$ predictions are quite good, but, as anticipated, $(N0S\pi)_a$ and $(N0S\pi)_\phi$ predictions show large and systematic deviations from the data. We next adapted this model by inserting a stage of monaural waveform compression preceding the crosscorrelator (Van de Par and Kohlrausch 1998). Using a power law with an I/O slope of 0.25 dB/dB, predictions are much improved; the model now correctly describes the essential features of the data as described above.

We finally consider a very different approach: a variation of Webster's (1951) lateralization model of binaural detection. It considers the magnitude of ITD fluctuations, expressed as RMS(ITD) over the stimulus duration. The detection criterion is an increase of at least 130 μs in RMS(ITD) upon adding the signal. We need to consider the *relative* magnitudes of ITD fluctuations because most of our reference ("noise alone") conditions contain baseline ITDs. The modified Webster model predicts the data quite well (Fig. 25.3c). It is exclusively based on ITD, and yet correctly predicts the increase of $(N0S\pi)_a$ thresholds at large values of m. There is no need to consider ILDs at all!

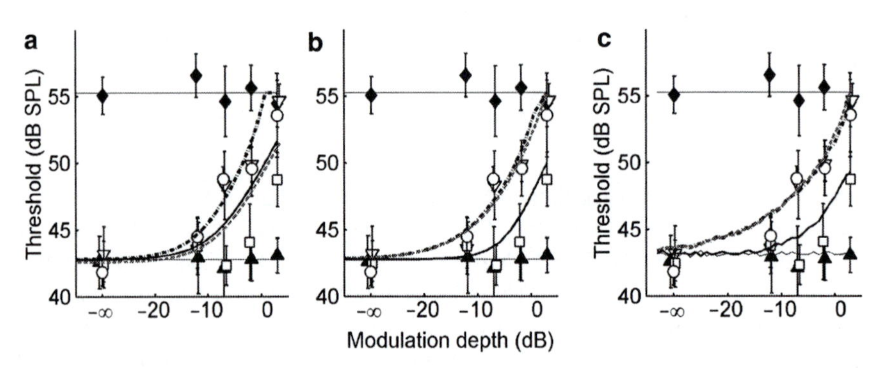

Fig. 25.3 Threshold predictions by three different models. (**a**) Predictions based on interaural correlation (see text). Symbols are the data replotted from Fig. 25.2a. *Line styles* and *symbols* indicate the type of binaural modulation: mixed (*thick dash-dotted line*, *open triangles*); binaural AM (*thin solid line*, *open squares*); binaural QFM (*dashed line*, *open circles*). *Closed symbols* and *horizontal lines* are the control conditions. Interaural correlation was computed directly from the filtered waveforms. The predictions for $(N0S\pi)_a$ and $(N0S\pi)_\varphi$ are identical. (**b**) Compression + correlation model. Same as (**a**), except the filtered waveforms were subjected to a compressive power law (0.25 dB/dB) prior to computing the correlation. (**c**) Webster-type lateralization model, requiring a 130-μs increment of RMS(ITD) for binaural detection (see text)

25.5 Discussion

In this study, we applied a binaural modulation technique that introduces binaural disparities consisting of dynamic ITDs, ILDs, or both. We used this technique to impose specific types of "jitter" on the stimuli of a binaural detection task. We found that binaural detection of a 500-Hz tone in wideband noise is much more sensitive to the introduction of ITD-specific jitter of the stimulus than to ILD-specific jitter. The asymmetry between ITDs and ILDs reveals that interaural correlation is a poor predictor of binaural detection in this N0Sπ task. The data are in agreement with the hypothesis that detection is solely based on the analysis of dynamic ITDs, and that ILDs do not play a role. This can either be understood in terms of a dynamical tracking of fluctuating ITDs, or in terms of interaural correlation computed from strongly compressed monaural inputs. The former view corresponds to a lateralization model of MLD, in which the fluctuating ITD is subject to rapid dynamical tracking. The latter view explains the dominance of ITDs in terms of monaural inputs that are somehow "compressed." The combination of such a compressive stage and a traditional cross-correlator explains our findings without invoking fast dynamical tracking of binaural parameters.

While our findings unambiguously identify those aspects of the stimulus that underlie binaural detection, the simultaneous success of two contrasting models shows that our data alone cannot identify the physiological mechanisms behind the processing of these stimulus parameters. That question requires specific physiological work.

Regarding the "dynamic ITD tracking" model (Fig. 25.3c), it is important that at the inferior colliculus is well able to accurately "follow" fast changes in ITD (Joris et al. 2006). Thus, from binaural physiology there is no a priori reason to exclude lateralization models of detection. Psychoacoustical data by Grantham and Wightman (1978), however, show that dynamical processing of binaural stimulus parameters is severely limited. Regarding the "compression + correlation" model (Fig. 25.3b), it is unlikely that the compression has a cochlear-mechanical origin, as suggested in (Van de Par and Kohlrausch 1998). Basilar membrane responses to wideband stimuli is known to be virtually linear, its Rayleigh-distributed envelope revealing the complete absence of cochlear envelope compression for wideband stimuli (Recio et al. 2009). It is more realistic to assume that the effective compression of the monaural inputs is a consequence of the *neural* representation of wideband stimuli. This would mean that the monaural inputs are highly specialized in coding stimulus phase, and are less sensitive to amplitude fluctuations. Binaurally, this bias toward temporal acuity would then be reflected in a high sensitivity to dynamic ITDs and a poor sensitivity to ILDs. It is interesting that bushy cells in the cochlear nucleus, which provide the inputs to the binaural crosscorrelator, show exactly such a bias when compared to their own inputs (auditory nerve fibers): improved phase locking and reduced dynamic range (Joris et al. 1994).

References

Colburn HS (1973) Theory of binaural interaction based on auditory-nerve data. I. General strategy and preliminary results on interaural discrimination. J Acoust Soc Am 54:1458–1470
Colburn HS, Durlach NI (1978) Models of binaural interaction. In: Carterette EC (ed) Handbook of perception, vol 4, Hearing. Academic Press, New York
Dolan RE, Robinson DE (1967) Explanation of masking-level differences that result from interaural intensive disparities of noise. J Acoust Soc Am 42:977–981
Domnitz RH, Colburn HS (1976) Analysis of binaural detection models for dependence on interaural target parameters. J Acoust Soc Am 59:598–601
Durlach NI (1963) Equalization and cancellation theory of binaural masking-level differences. J Acoust Soc Am 35:1206–1218
Grantham DW, Wightman FL (1978) Detectability of varying interaural temporal differences. J Acoust Soc Am 63:511–523
Hafter ER, Carrier SC (1970) Masking level differences obtained with pulsed tonal maskers. J Acoust Soc Am 47:1041–1047
Joris PX, Carney LH, Smith PH, Yin TC (1994) Enhancement of neural synchronization in the anteroventral cochlear nucleus. I. Responses to tones at the characteristic frequency. J Neurophysiol 71:1022–1036
Joris PX, Van de Sande B, Recio-Spinoso A, Van der Heijden M (2006) Auditory midbrain and nerve responses to varying interaural correlation. J Neurosci 26:279–289
Recio A, Narayan SS, Ruggero MA (2009) Basilar-membrane responses to noise at a basal site of the chinchilla cochlea: quasi-linear filtering. J Assoc Res Otolaryngol 10(4):471–484
Van de Par S, Kohlrausch A (1998) Diotic and dichotic detection using multiplied-noise maskers. J Acoust Soc Am 103:2100–2110
Van der Heijden M, Trahiotis C (1999) Masking with interaurally delayed stimuli: The use of "internal" delays in binaural detection. J Acoust Soc Am 105:388–399
Webster FA (1951) The influence of interaural phase on masked thresholds I. the role of interaural time-deviation. J Acoust Soc Am 23:452–462

Chapter 26
Effect of Reverberation on Directional Sensitivity of Auditory Neurons: Central and Peripheral Factors

Sasha Devore, Andrew Schwartz, and Bertrand Delgutte

Abstract In reverberant environments, acoustic reflections interfere with the direct sound arriving at a listener's ears, distorting the binaural cues for sound localization. Using virtual auditory space simulation techniques, we investigated the effects of reverberation on the directional rate responses of single neurons in the inferior colliculus (IC) of unanesthetized rabbits. We find that reverberation degrades the directional sensitivity of single neurons, although the amount of degradation depends on the characteristic frequency (CF) and the type of binaural cues available. To investigate the extent to which these midbrain results reflect peripheral processing of the monaural input signals, we extracted directional information from spike trains recorded from auditory nerve (AN) in anesthetized cat for the same VAS stimuli. Our results suggest that the frequency-dependent degradation in ITD-based directional sensitivity in reverberation originates in the auditory periphery.

Keywords Auditory nerve • Binaural • Inferior colliculus • Reverberation • Sound localization

26.1 Introduction

Indoors and in nature alike, the auditory scenes that we perceive unfold in reverberant environments. In a reverberant sound field, reflected acoustic waves reach the listener from all directions, interfering with the direct sound and distorting the binaural cues for sound localization such as interaural time and level differences (ITD and ILD). In previous work (Devore et al. 2009), we showed that reverberation degrades the directional sensitivity of low frequency, ITD-sensitive neurons in the inferior colliculus (IC) of anesthetized cats, although

S. Devore (✉)
Eaton-Peabody Laboratory, Massachusetts Eye & Ear Infirmary, Boston, MA, USA
e-mail: sashad@alum.mit.edu

E.A. Lopez-Poveda et al. (eds.), *The Neurophysiological Bases of Auditory Perception*,
DOI 10.1007/978-1-4419-5686-6_26,

not as much as predicted by an interaural cross-correlation model. Here, we extend this work by characterizing directional sensitivity in neurons across a wide range of the tonotopic axis in an awake rabbit preparation, while maintaining our focus on neurons that are sensitive to ITD.

Low frequency IC neurons are typically sensitive to ITD in the waveform fine structure, while high frequency IC neurons are sensitive to ITD in the amplitude envelopes (Batra et al. 1993; Joris 2003; Griffin et al. 2005; Yin et al. 1984). At all characteristic frequencies (CFs), the rate responses of IC neurons can be altered by imposing ILDs (Batra et al. 1993; Palmer et al. 2007); however, for stimuli with naturally co-occurring binaural cues, ILDs may be a more potent directional cue than envelope ITDs in high frequency neurons (Delgutte et al. 1995).

We investigated the effects of reverberation on directional sensitivity of ITD-sensitive neurons in the IC of awake rabbits using virtual auditory space (VAS) stimuli containing different binaural cues (ITD-only and ITD + ILD). We find that reverberation degrades the directional sensitivity of single neurons, although the amount of degradation depends on both the CF and the type of binaural cues available. We also compared results from IC neurons with measures of directional information extracted from coincidence analysis of spike trains recorded from auditory nerve (AN) fibers in anesthetized cats. Together our results suggest that the frequency-dependent degradation in ITD-based directional sensitivity partly originates in the auditory periphery and can be attributed to differential degradation of interaural envelopes and fine-time structure in reverberation.

26.2 Methods

VAS stimuli. Two sets of reverberant binaural room impulse responses (BRIRs) were simulated using the room-image method (Allen and Berkley 1979; Shinn-Cunningham et al. 2001) for azimuths spanning ±90° at a distance of 1 m with respect to the center of the "head" in a room with dimensions 11 × 13 × 3 m. Anechoic impulse responses were created by time-windowing the direct wavefront from the reverberant BRIRs. For one set of simulations, we used two sensors separated by 12 cm to represent the ears, with no explicit model of the head, so that the resulting BRIRs contained ITD but essentially no ILD cues. In the second set of simulations, we modeled the rabbit head as a rigid sphere (12 cm diameter), so that the BRIRs contained both ITD and ILD cues. VAS stimuli were created by convolving BRIRs with bursts of reproducible 400-ms broadband noise. We refer to simulations using the two sets of BRIRs as ITD-only and ITD+ILD stimuli.

Single-unit recordings in IC of awake rabbit. Methods for recording from single neurons in the IC of unanesthetized Dutch-Belted rabbits (*Oryctolagus cuniculus*) were based on those of Kuwada et al. (1987). Our protocol for characterizing the directional sensitivity of IC neurons is similar to that described previously (Devore et al. 2009). Responses were measured as a function of azimuth for each virtual room condition (anechoic, reverberant) and stimulus type (ITD-only, ITD + ILD) in pseudorandom order. We typically used 13 azimuths (15° spacing) or, occasionally, 7 azimuths (30° spacing). Stimuli were presented at 15–20 dB above broadband noise threshold.

Single-unit recordings in AN of anesthetized cat. Methods for recording from single fibers in the of anesthetized cats were as described by Dreyer and Delgutte (2006). We used the same ITD-only stimuli as in the rabbit IC experiments. Neural responses were measured separately for the 0°-azimuth left and right ITD-only VAS stimuli for both anechoic and reverberant conditions, in pseudorandom order. Stimuli were presented at 5–24 dB above broadband noise threshold. Shuffled cross-correlograms (SXC; Louage et al. 2004) between the responses to left-ear and right-ear stimuli were computed from the AN spike trains.

26.3 Results

Our results are based on recording from 192 ITD-sensitive neurons in the IC of four awake rabbits. CFs of our neurons spanned 255–9,600 Hz, with 53% lying below 2 kHz.

26.3.1 *Sensitivity to ITD in the Envelope and Fine Structure of Noise in the Awake Rabbit IC*

Similar to results from anesthetized cat (Joris 2003), we find that low-CF neurons are primarily sensitive to ITD in the fine-time structure of broadband noise (ITD_{fs}), while high-CF neurons are sensitive to ITD in the envelope induced by cochlear filtering (ITD_{env}). ITD_{fs}- and ITD_{env}-sensitivity were distinguished by comparing noise delay functions obtained with homophasic and antiphasic noise (Fig. 26.1a, b). Inverting the stimulus at one ear causes a 180° phase shift in fine-time structure but

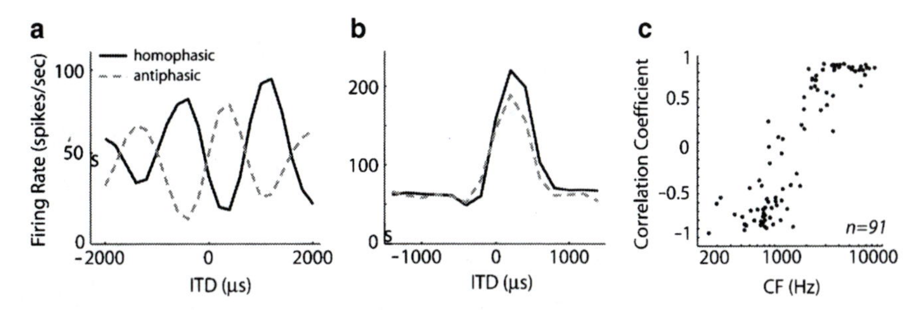

Fig. 26.1 (**a**) Homophasic (*black line*) and antiphasic (*gray line*) noise delay functions for a low-CF IC neuron (CF = 642 Hz) and (**b**) high-CF neuron (CF = 3,427 Hz). Spontaneous rate indicated by "s" at the left of each panel. (**c**) Correlation coefficient for homophasic and antiphasic noise delay functions versus CF for the IC neuron population

does not alter the envelope. Thus, in ITD_{fs}-sensitive cells (Fig. 26.1a), noise delay functions for antiphasic stimuli show peaks and valleys that interleave with those of noise delay functions for homophasic stimuli, while the two noise delay functions are similar in ITD_{env}-sensitive cells (Fig. 26.1b). Figure 26.1c shows the correlation coefficient between noise delay functions for homophasic and antiphasic noise as a function of CF, demonstrating the transition from ITD_{fs}- to ITD_{env}-sensitivity as CF increases from 0.8 to 2 kHz, consistent with the Joris (2003) results in cat.

26.3.2 *Characterization of Directional Sensitivity Using ITD-Only in the Awake Rabbit IC*

We measured rate responses as a function of azimuth using ITD-only stimuli in 192 neurons. Because these stimuli contain no ILD cues, the anechoic directional response functions (DRFs) are essentially equivalent to noise delay functions sampled with high resolution within the naturally occurring range of ITD (Fig. 26.2a, b, d, e, solid black lines). In general, neurons tended to respond more vigorously to sources in the hemifield contralateral to the recording site (positive azimuths, Fig. 26.2a, b), although some neurons had nonmonotonic DRF (Fig. 26.2d, e).

In those units where we obtained a sufficient number of stimulus trials at each azimuth, we quantified directional sensitivity by computing the information

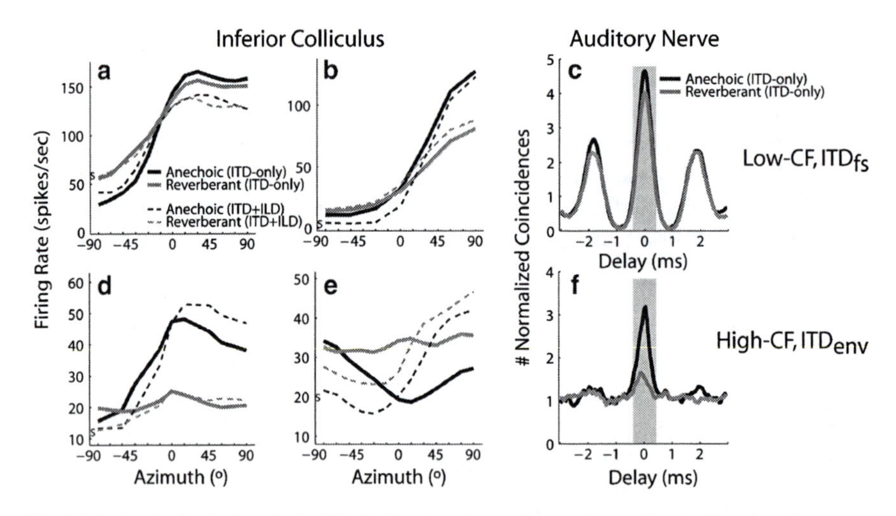

Fig. 26.2 (**a**, **b**, **d**, **e**) Anechoic (*black lines*) and reverberant (*gray lines*) directional response functions (DRFs) obtained using ITD-only (*solid lines*) and ITD + ILD (*dashed lines*) virtual space simulations for four IC neurons with CFs of (**a**) 597 Hz, (**b**) 791 Hz, (**d**) 5,215 Hz, (**e**) 7,200 Hz. Spontaneous rate indicated by "s" at the left of each panel. (**c**, **f**) Anechoic (*black line*) and reverberant (*gray line*) shuffled cross-correlograms (SXC) computed from spike trains obtained from (**c**), low-CF (588 Hz) and (**f**), high CF (10.7 kHz) AN fibers in anesthetized cat. *Gray region* represents range of ITDs corresponding to ±90° in the anechoic ITD-only stimuli

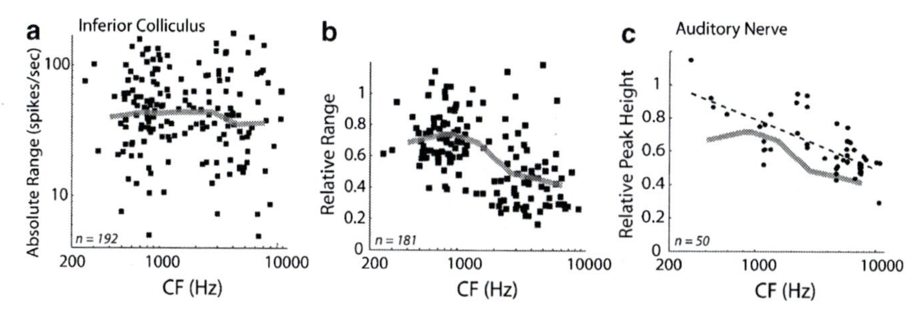

Fig. 26.3 (**a**) Absolute range of anechoic DRF (obtained using ITD-only) versus CF for IC neuron population. *Solid line* depicts mean absolute range within each of six CF bins (edge frequencies = 250, 700, 950, 2,000, 3,500, 5,000, 10,000). (**b**) Relative range of reverberant DRF (obtained using ITD-only) versus CF for IC neuron population. (**c**) Relative peak height of reverberant SXC versus CF for AN fiber population. *Dashed line* depicts mean relative peak height within each CF bin; *solid line* has been superimposed from panel (**b**)

transfer between stimulus azimuth and spike count. Information transfer measures the extent to which stimulus azimuth is unambiguously encoded by spike count and is sensitive to both mean rate as well as trial-to-trial variability. Consistent with previous observations in anesthetized cat IC (Devore et al. 2009), we found that information transfer is highly correlated with the absolute range of firing rates across azimuth (data not shown, $r^2=0.83$, $p<0.001$). Because we did not obtain enough stimulus trials to reliably estimate information transfer in many of the present units, we use the absolute range of firing rates to quantify directional sensitivity.

Figure 26.3a shows absolute rate range against CF for our neuron population. The gray line represents the mean absolute range within each of six CF bins containing approximately equal numbers of data points. Units in the lower three CF bins are primarily sensitive to ITD_{fs}, while units in the upper three bins are primarily sensitive to ITD_{env} (see Fig. 26.1c). Absolute range is highly variable within each bin, and there is no significant dependence of the mean absolute range on CF [ANOVA, $F(5,186)=0.66$, $p=0.679$]. This finding suggests that, for anechoic stimuli, comparable directional information is available in the rate responses of ITD_{fs}-sensitive and ITD_{env}-sensitive neurons.

Differences between high-CF ITD_{env}-sensitive neurons and low-CF ITD_{fs}-sensitive neurons emerge when we evaluate directional sensitivity in reverberation. We obtained DRFs in 181 neurons using reverberant ITD-only stimuli that had a direct-to-reverberant energy ratio of 0 dB, typical of a medium-sized classroom. In general, reverberation caused a compression of the range of firing rates (Fig. 26.2a, b, d, e, solid gray lines), such that peak firing rates were reduced and minimum firing rates were increased. The compression was more pronounced in high-CF, ITD_{env}-sensitive neurons (Fig. 26.2d, e), than in low-CF ITD_{fs}-sensitive neurons (Fig. 26.2a, b).

To quantify the effects of reverberation on directional sensitivity in individual neurons, we computed the *relative range*, the ratio of the range of firing rates across azimuths in reverberation to the range of firing rates in the anechoic condition.

Figure 26.3b shows relative range versus CF for our neuron population; the solid line depicts the mean relative range in each of the six CF bins defined above. Although relative range is somewhat variable within each bin, there is a significant dependence of relative range on CF [ANOVA: $F(5,172)=10.06$, $p<0.001$). Post hoc analysis (*t*-test with Bonferroni corrections) reveals that the relative range is significantly smaller in each of the three higher CF bins than in each of the three lower CF bins ($p<0.002$, all pairwise comparisons). However, the differences between bins *within* the lower three and upper three CF bins are not significant, suggesting that the decrease in relative range with increasing CF may reflect the transition from ITD_{fs}- to ITD_{env}- sensitivity.

26.3.3 Peripheral Factors Determining ITD-Only Sensitivity in Reverberation

In previous work, we showed that, at low CFs, compression of DRF in reverberation can be qualitatively attributed to decorrelation of the ear input signals after taking into account peripheral filtering (Devore et al. 2009). Here, we use recordings from the AN to examine whether the disparity in the effect of reverberation for low-CF (ITD_{fs}-sensitive) and high-CF (ITD_{env}-sensitive) neurons can be explained by a similar, common type of binaural processing across all frequencies.

We analyzed spike trains recorded from AN fibers in anesthetized cat in response to the ITD-only stimuli using the shuffled-correlation technique (Louage et al. 2004). This technique implements processing by a bank of ideal coincidence detectors receiving delayed inputs from left and right AN fibers. Figure 26.2c, f shows SXC for two AN fibers computed from responses to anechoic (black lines) and reverberant (gray lines) ITD-only stimuli at 0° azimuth. The SXC are normalized, so that 1 corresponds to the expected normalized coincidence count for uncorrelated random spike trains. As shown by Joris (2003), SXC resemble noise delay functions from IC neurons (Fig. 26.1a, b), and reflect the phase locking of AN spikes to the quasi-periodic fine-time structure at low CFs (Fig. 26.2c) and phase locking to the aperiodic cochlear-induced envelope at high CFs (Fig. 26.2f). Similar to our observations in the IC, reverberation causes a compression of SXC that is more severe at high CFs (Fig. 26.2f) than at low CFs (Fig. 26.2c).

To quantify the effects of reverberation on the SXC, we computed the relative peak height i.e., the ratio of the SXC peak height in the reverberant condition to that in the anechoic condition. We subtracted 1 from the peak height in each condition *before* computing the ratio, in order to remove the baseline coincidence count representing random firings. Figure 26.3c shows the relative peak height versus CF for the AN fiber population. The dashed line depicts the least-squares regression line of the relative peak height on CF, with the solid line from Fig. 26.3b superimposed to facilitate comparison. As with the relative range in IC neurons (Fig. 26.3b), the relative peak height decreases with increasing CF (Fig. 26.3c), although it appears that the main drop may occur at somewhat higher CFs in the AN than in the IC (Fig. 26.3c, compare solid and dashed lines). These results suggest a partial peripheral origin for the frequency-dependent effects of reverberation on ITD sensitivity in the IC.

26.3.4 Directional Sensitivity in the IC with ITD and ILD Cues

To examine the effects of reverberation in more realistic conditions, we characterized directional sensitivity using the ITD + ILD stimuli that had an azimuth-dependent ILD as well as ITD. Figure 26.2a, b, d, e shows DRFs for ITD + ILD stimuli as dashed lines along with responses to ITD-only stimuli (solid lines). For high-CF, ITD_{env}-sensitive neurons (Fig. 26.2d, e), there is a systematic, azimuth-dependent difference between anechoic DRF for ITD + ILD and ITD-only stimuli: Firing rates for ITD + ILD stimuli tend to be lower than ITD-only firing rates at ipsilateral azimuths and higher at contralateral azimuths, often resulting in a larger absolute range. The differences between directional responses to ITD-only, and ITD + ILD stimuli were less systematic and less prominent in low-CF, ITD_{fs}-sensitive neurons (Fig. 26.2a, b).

To quantify the relative influence of ILD on anechoic DRFs, we defined an *ILD influence index* as,
$$\frac{\text{Absolute range for VAS}_{\text{both}} - \text{Absolute range for VAS}_{\text{ITD}}}{\text{Absolute range for VAS}_{\text{both}} + \text{Absolute range for VAS}_{\text{ITD}}}.$$

This metric takes on values between ±1 and is equal to 0 when the ranges of firing rates are the same for both VAS stimuli. Figure 26.4a shows the ILD influence index versus CF for the 44 neurons, in which responses to both types of VAS stimuli were measured. We split the population into low (<2,000 Hz) and high (>2,000 Hz) CF groups, essentially dividing the population on the basis of ITD_{fs}- and ITD_{env}-sensitivity (see Fig. 26.1c). The ILD influence index is significantly

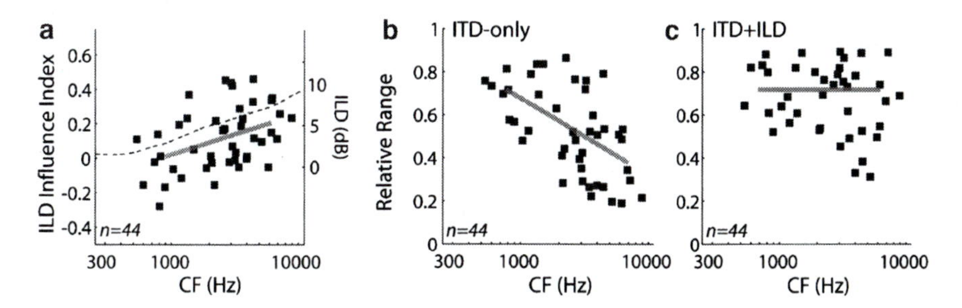

Fig. 26.4 (**a**) ILD influence index versus CF for IC neuron population. *Solid gray line* depicts mean ILD influence index within each CF bin. *Dashed line* depicts across-azimuth average ILD in 1/3-octave bands, with values corresponding to axis at the right edge of the panel. (**b**) Relative range of reverberant DRF obtained using ITD-only versus CF for IC neuron population tested in both VAS conditions. (**c**) Relative range of DRF obtained using ITD + ILD for IC neuron population

larger in the high-CF group (one-sided *t*-test, $p<0.001$), indicating that ILDs that are consistent with ITDs improve directional sensitivity at high CFs more so than at low CFs. The ILD influence index parallels the magnitude of ILD in 1/3-octave bands (Fig. 26.4a, dashed line) with increasing frequency, suggesting an acoustic origin for the increasing influence of ILD on directional rate responses at high frequencies.

Consistent with the results for ITD-only stimuli, reverberation generally causes a compression of DRF for ITD+ILD stimuli. In low-CF, ITD_{fs}-sensitive neurons, reverberation tends to similarly compress the DRFs for both stimuli (Fig. 26.2a, b). In contrast, in high-CF neurons, DRF can show much less compression for ITD+ILD stimuli than for ITD-only stimuli (Fig. 26.2e), although in some units, directional sensitivity is impoverished in both conditions (Fig. 26.2d).

Figure 26.4b, c show the relative range plotted against CF for ITD-only and ITD+ILD stimuli, respectively. The gray lines depict the average relative range for CF bins above and below 2 kHz. There is a significant interaction ($p=0.011$) between CF (low versus high) and VAS condition in a two-way ANOVA, resulting from the fact that there is a substantial improvement in directional sensitivity at high CFs, but not low CFs, when the VAS stimuli contain both binaural cues. Consequently, the directional sensitivity in reverberation is stable across the tonotopic axis for the more realistic ITD+ILD stimuli, whereas in the ITD-only condition, less information is available from high-CF neurons.

26.4 Discussion

We found that, for anechoic stimuli, directional sensitivity of high-CF ITD_{env}-sensitive neurons is comparable to that of low-CF ITD_{fs}-sensitive neurons. Consistent with previous observations (Devore et al. 2009), reverberation degrades directional sensitivity in ITD-sensitive IC neurons by compressing the range of firing rates across azimuths, although the amount of degradation depends on the CF and the types of binaural cues available. When VAS stimuli contain only ITD cues, directional sensitivity in reverberation is significantly worse in high-CF ITD_{env}-sensitive neurons than in their low-CF ITD_{fs}-sensitive counterparts. However, directional sensitivity at high CFs can be significantly improved when VAS stimuli contain both ILD and ITD cues, suggesting that, at high CFs, ILDs contribute importantly to directional sensitivity in reverberation. Moreover, given that the ILD influence index parallels the magnitude of ILD in the virtual space stimuli (Fig. 26.4a, dashed line), we would expect ILD to be of even greater importance for more realistic head-related transfer functions, which can exhibit ILDs up to ±30 dB for species with directional pinnae.

To examine the contribution of peripheral processing to the IC results, we used ideal coincidence detectors to extract ITD-based directional information from spike trains recorded from cat AN fibers with the ITD-only stimuli. Similar to the trend observed in the IC, we found that reverberation caused an increasing reduction in the normalized peak height of the shuffled cross-correlation with increasing CF, suggesting a partial peripheral origin for the IC results. Consistent with Joris' (2003) observation that the transition from ITD_{fs}-sensitivity to ITD_{env}-sensitivity occurs at higher CFs in the AN than in the IC, we found that the degradation in directional sensitivity produced by reverberation seemed to occur at higher CFs in the AN than in the IC, although more AN data are needed to reach a definitive conclusion (compare Fig. 26.3b, c). The fact that the effect of reverberation covaries with the transition from ITD_{fs} to ITD_{env} in both structures is evidence that it is not an effect of frequency, per se; rather, it suggests that reverberation degrades directional coding more so for envelopes than fine-time structure. The difference in transition frequency between the AN and the IC may be accounted for by additional degradation in fine structure coding between the AN and the binaural processor in the auditory brainstem (Joris 2003).

Although differences in species (cat versus rabbit) and state (anesthetized versus awake) need to be considered when interpreting the two sets of results, the similarity of AN fiber tuning properties in cat and rabbit (Borg et al. 1988) helps validate the cross-species comparison. Moreover, our observations are consistent with Sayles and Winter's (2008) finding that reverberation degrades neural sensitivity to envelope pitch cues more severely than fine structure cues in the cochlear nucleus of anesthetized guinea pig.

The present neurophysiological results parallel those of human psychophysical studies of spatial hearing in reverberant environments. Rakerd et al. (2006) found that ITD discrimination thresholds for 1/3-octave noise bands degrade more rapidly with increasing reverberation at high frequencies (2,850 Hz) than at low frequencies (715 Hz), consistent with our observation that reverberation causes more severe compression of DRF in high-CF, ITD_{env}-sensitive IC neurons than in low-CF ITD_{fs}-sensitive neurons. Rakerd and Hartmann also found that ITD-discrimination for reverberant noise bands degraded further when they introduced a small ILD in opposition to the ITD, suggesting that ILD has an important influence on localization in reverberation. Rakerd and Hartmann observed effects of ILD at both low and high frequencies, while the influence of ILD on neural responses was most prominent in high-CF, ITD_{env}-sensitive neurons. The two sets of results may be reconciled if, as suggested by Fig. 26.4a, the influence of ILD on neural responses is determined by the magnitude of the ILD. Since Rakerd and Hartmann used the same ILDs at low and high frequencies, similar effects are expected for the two frequency bands in their experiment. In contrast, the ILD in our experiments were determined by a rigid sphere model for the head and therefore increased with frequency. Nevertheless, single-unit experiments using stimuli with opposing ITD and ILD are needed to better understand the relationships between our results and those of Rakerd and Hartmann.

The duplex theory of sound localization (Rayleigh 1907) stipulates that listeners use ITD at low frequencies and ILD at high frequencies. The present results, which represent one of the few investigations of ITD sensitivity across a wide range of CFs using a common set of stimuli, suggest that, in the anechoic condition, ITD_{env}-sensitivity at high CFs can rival that based on fine-time structure at low CFs. However, the comparatively poor reliability of ITD_{env} cues in reverberant listening conditions may explain why these cues receive a low perceptual weight for sound localization by most human listeners (Macpherson and Middlebrooks 2002).

Acknowledgments The authors thank Laurel Carney and Shig Kuwada for their help with the awake rabbit preparation, Ken Hancock for software support, and Melissa Wood and Connie Miller for technical assistance. Supported by NIH grants RO1 DC002258 and P30 DC005209 and a Helen Carr Peake Research Assistantship to SD.

References

Allen JB, Berkley DA (1979) Image method for efficiently simulating small-room acoustics. J Acoust Soc Am 65:943–950

Batra R, Kuwada S, Stanford TR (1993) High-frequency neurons in the inferior colliculus that are sensitive to interaural delays of amplitude-modulated tones: evidence for dual binaural influences. J Neurophysiol 70:64–80

Borg E, Engstrom B, Linde G, Marklund K (1988) Eighth nerve fiber firing features in normal-hearing rabbits. Hear Res 36:191–201

Delgutte B, Joris PX, Litovsky RY, Yin TC (1995) Relative importance of different acoustic cues to the directional sensitivity of inferior colliculus neurons. In: Manley GA, Klump GM, Koeppl C, Fastl H, Oeckinghaus H (eds) Advances in hearing research. World Scientific, Singapore, pp 288–299

Devore S, Ihlefeld A, Hancock K, Shinn-Cunningham B, Delgutte B (2009) Accurate sound localization in reverberant environments is mediated by robust encoding of spatial cues in the auditory midbrain. Neuron 62:123–134

Dreyer A, Delgutte B (2006) Phase locking of auditory-nerve fibers to the envelopes of high-frequency sounds: implications for sound localization. J Neurophysiol 96:2327–2341

Griffin SJ, Bernstein LJ, Ingham NJ, McAlpine DM (2005) Neural sensitivity to envelope delays in the inferior colliculus of guinea pigs. J Neurophysiol 93:3463–3478

Joris PX (2003) Interaural time sensitivity dominated by cochlea-induced envelope patterns. J Neurosci 23:6345–6350

Kuwada S, Stanford TR, Batra R (1987) Interaural phase sensitive units in the inferior colliculus of the unanesthetized rabbit. Effects of changing frequency. J Neurophysiol 57:1338–1360

Louage DH, van der Heijden M, Joris PX (2004) Temporal properties of responses to broadband noise in the auditory nerve. J Neurophysiol 91:2051–2065

Macpherson EA, Middlebrooks JC (2002) Listener weighting of cues for lateral angle: the duplex theory of sound localization revisited. J Acoust Soc Am 111:2219–2236

Palmer AR, Liu LF, Shackleton TM (2007) Changes in interaural time sensitivity with interaural level differences in the inferior colliculus. Hear Res 223:105–113

Rakerd B, Hartmann WM, Pepin E (2006) Localizing noise in rooms via stead state interaural time differences. J Acoust Soc Am 120:3082–3083

Rayleigh L (1907) On our perception of sound direction. Philos Mag 13:214–232

Sayles M, Winter IM (2008) Reverberation challenges the temporal representation of the pitch of complex sounds. Neuron 58:789–801

Shinn-Cunningham BG, Desloge JG, Kopco N (2001) Empirical and modeled acoustic transfer functions in a simple room: effects of distance and direction. In: 2001 IEEE workshop on applications of signal processing to audio and acoustics. New Pfaltz, New York, pp 419–423

Yin TCT, Kuwada S, Sujaku Y (1984) Interaural time sensitivity of high frequency neurons in the inferior colliculus. J Acoust Soc Amer 76:1401–1410

Chapter 27
New Experiments Employing Raised-Sine Stimuli Suggest an Unknown Factor Affects Sensitivity to Envelope-Based ITDs for Stimuli Having Low Depths of Modulation

Leslie R. Bernstein and Constantine Trahiotis

Abstract This chapter reports an extension of work reported at the previous ISH meeting wherein threshold interaural temporal disparities (ITDs) were measured with high-frequency, raised-sine stimuli. The threshold ITDs reported here were obtained while varying, independently and parametrically, the depth of modulation, the frequency of modulation, and the value of the exponent of the raised-sine stimuli (which affects the relative "peakedness/dead-time" of their envelopes). Graded *increases* in the exponent led to graded *decreases* in threshold ITD for frequencies of modulation ranging from 32 to 256 Hz. Thresholds also generally increased with decreases in modulation depth. Unexpectedly, however, variations of the exponent of the raised sine interacted with variations in depth of modulation such that smaller increases in threshold ITD occurred for decreases in the depth of modulation when the exponent was large. This interaction proved interesting because an interaural correlation-based model that is generally able to account for changes in threshold ITD produced by separate changes in exponent, depth of modulation, and frequency of modulation of raised-sine stimuli could not account for the interaction. The magnitudes of the psychophysically measured effects suggest that it would be beneficial and important that auditory physiologists conduct parallel investigations while employing similar raised-sine stimuli.

Keywords Raised-cosine stimuli • Inter-aural time difference • Binaural hearing • Modulation • Envelope information

L.R. Bernstein (✉)
Deptartments Of Neuroscience and Surgery (Otolaryngology), University of Connecticut Health Center, Farmington, CT 06030, USA
e-mail: les@neuron.uchc.edu

E.A. Lopez-Poveda et al. (eds.), *The Neurophysiological Bases of Auditory Perception*,
DOI 10.1007/978-1-4419-5686-6_27, © Springer Science+Business Media, LLC 2010

27.1 Introduction

In 1997, van de Par and Kohlrausch reported data showing that "transposed" stimuli enhanced high-frequency binaural processing in that they yielded NoSπ thresholds of detection that were comparable to the much lower thresholds of binaural detection routinely obtained at low vs. high center frequencies. Following that, Bernstein and Trahiotis reported that transposed stimuli yielded smaller threshold interaural temporal disparities (ITDs) (Bernstein and Trahiotis 2002), larger extents of ITD-based laterality (Bernstein and Trahiotis 2003), and were resistant to binaural interference (Bernstein and Trahiotis 2004, 2005). In addition, physiological studies have also revealed that high-frequency transposed stimuli enhance envelope-based neural timing (Griffin et al. 2005; Dreyer and Delgutte 2006).

In order to determine which aspect(s) of the envelopes of high-frequency stimuli lead to efficient processing of ongoing ITDs, we have been conducting a series of experiments employing "raised-sine" stimuli. As reported at the last ISH meeting, the algorithm used to generate raised-sine stimuli (see John et al. 2002) allows one to vary independently the frequency of modulation, the depth of modulation, and the "peakedness" or "sharpness" of the envelope of a high-frequency waveform. These are features that cannot be varied independently with conventional stimuli such as SAM tones, repeated Gaussian clicks (e.g., Buell and Hafter 1988; Stecker and Hafter 2002) or the transposed tones used in our previous studies. The threshold ITDs reported here were obtained while varying, independently and parametrically, the depth of modulation, the frequency of modulation, and the value of the exponent of the raised-sine stimuli (which affects the relative "peakedness/dead-time" of their envelopes).

27.2 Generating Raised-Sine Stimuli

The equation used to generate such stimuli is:

$$y(t) = (\sin(2\pi f_c t))(2m(((1+\sin(2\pi f_m t))/2)^n - 0.5)+1) \qquad (27.1)$$

where f_c is the frequency of the carrier, f_m is the frequency of the modulator, m is the modulation index, and n is the exponent denoting the power to which the DC-shifted modulator is raised.

The left side of Fig. 27.1 (taken from our previous ISH chapter) shows the time-waveforms for cases in which a 128-Hz modulating tone was raised using exponents of 1, 2, 4, or 8 prior to multiplication with a 4-kHz carrier. The bottom row of the figure shows a 128-Hz tone transposed to 4 kHz. An exponent of 1.0 yields a conventional SAM waveform. This figure makes clear that the peakedness of the envelope increases directly with the value of the exponent to which the modulator is raised. Note that, concomitantly, the "dead-time" or "off-time" between individual

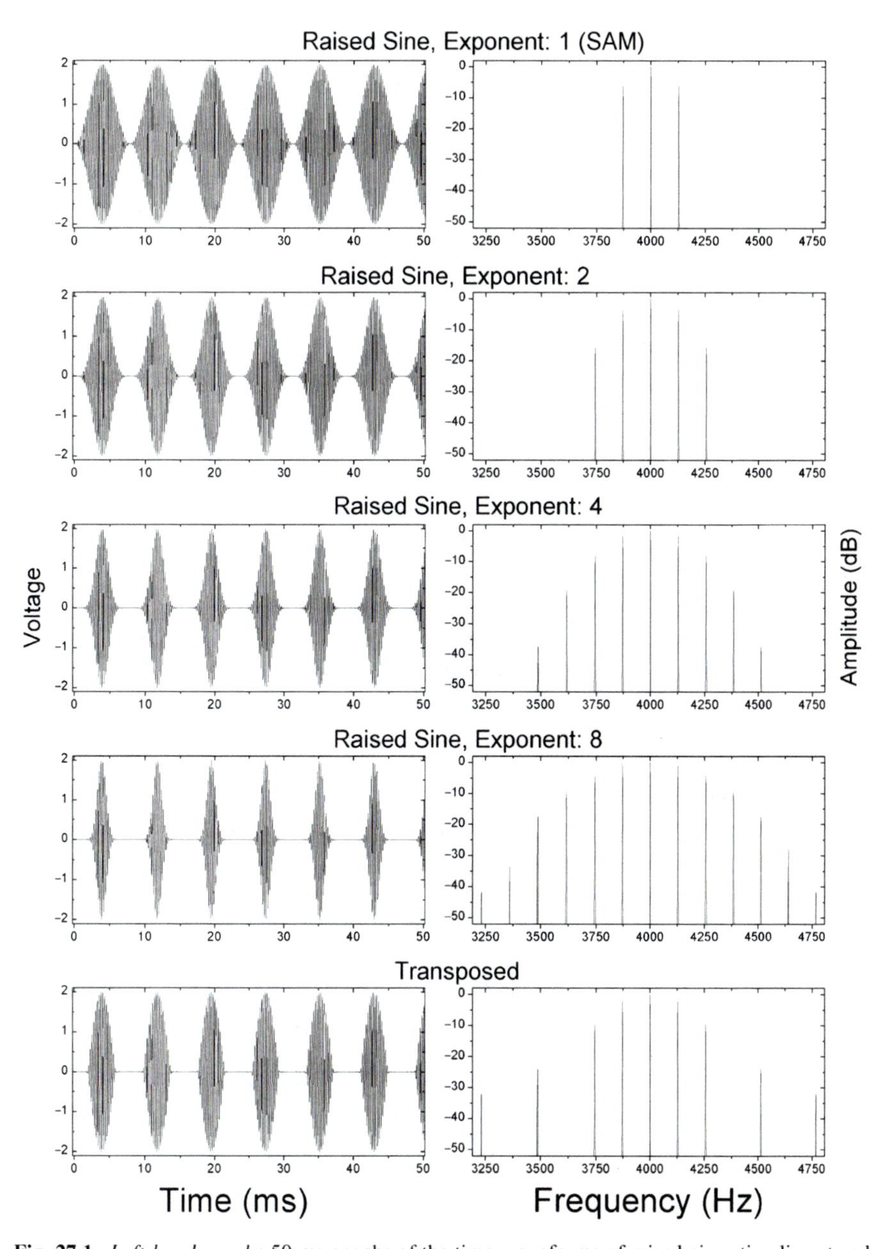

Fig. 27.1 *Left-hand panels*: 50-ms epochs of the time-waveforms of raised-sine stimuli centered at 4 kHz, modulated at 128 Hz, and having exponents of 1, 2, 4, and 8 (rows 1–4, respectively). The bottom row depicts a 4-kHz-centered transposed stimulus modulated at the same rate. *Right-hand panels*: Each row depicts the long-term power spectrum of the time-waveform shown immediately to its left

lobes of the envelope also increases with increasing values of the exponent. The right side of the figure shows the corresponding long-term spectrum of each stimulus. Note that increasing the value of the exponent also increases the number of sidebands and their spectral extent. Still, for each of the stimuli depicted, the vast majority of its energy falls within the approximately 500-Hz wide auditory filter centered at 4 kHz (see Moore 1997).

27.3 Procedure, Results, and Discussion

Threshold ITDs were measured using a two-cue, two-interval, adaptive forced-choice task for raised-sine stimuli having exponents of 1.0 (equivalent to a SAM tone), 1.5, 2.0, 4.0, and 8.0 and for transposed-tones. For both types of stimuli, rates of modulation ranged between 32 and 256 Hz. All stimuli were centered at 4 kHz, were 300 ms-long, and were presented at 70 dB SPL. A continuous diotic noise, low-passed at 1.3 kHz (spectrum level equivalent to 30 dB SPL) was presented to preclude the listener's use of low-frequency distortion products arising from normal, nonlinear peripheral auditory processing (e.g., Nuetzel and Hafter 1976; Bernstein and Trahiotis 1994). Four normal-hearing adults served as listeners.

The solid squares within each plot of the upper panel of Fig. 27.2 represent mean "normalized" threshold ITDs, calculated across the four listeners. Normalized thresholds are shown in order to remove differences in absolute sensitivity to ITD across listeners commonly found with high-frequency, complex stimuli (e.g., Bernstein et al. 1998). The normalization was accomplished by dividing an individual listener's threshold ITDs by that listener's threshold ITD obtained with a SAM tone (raised-sine exponent equal to 1.0) having a frequency of modulation of 128 Hz. The error bars represent ±1 standard error of the mean.

Note that, for all rates of modulation, threshold ITDs decrease with increases in the exponent of the raised-sine and approximate threshold ITDs obtained with transposed stimuli when the exponent is 8.0. This suggests that graded changes in the amounts of peakedness of the envelope lead to graded changes in sensitivity to ongoing envelope-based ITDs. Those changes appear to be largest where threshold ITDs are generally largest (32-Hz modulation rate) and smallest where threshold

Fig. 27.2 (continued) three sections of the figure contains mean normalized thresholds (*bars*) obtained with a single value of the exponent (1.0, 1.5, or 8.0) and four values of depth of modulation (m=0.25, 0.5, 0.75, or 1.0). Predictions represented by the *closed squares* were calculated using the interaural correlation-based model and employing the same criterion change in interaural correlation that provided the best-fitting (*dashed-line*) predictions shown in the figure. The time-waveforms corresponding to four of the stimuli are depicted atop their corresponding bars

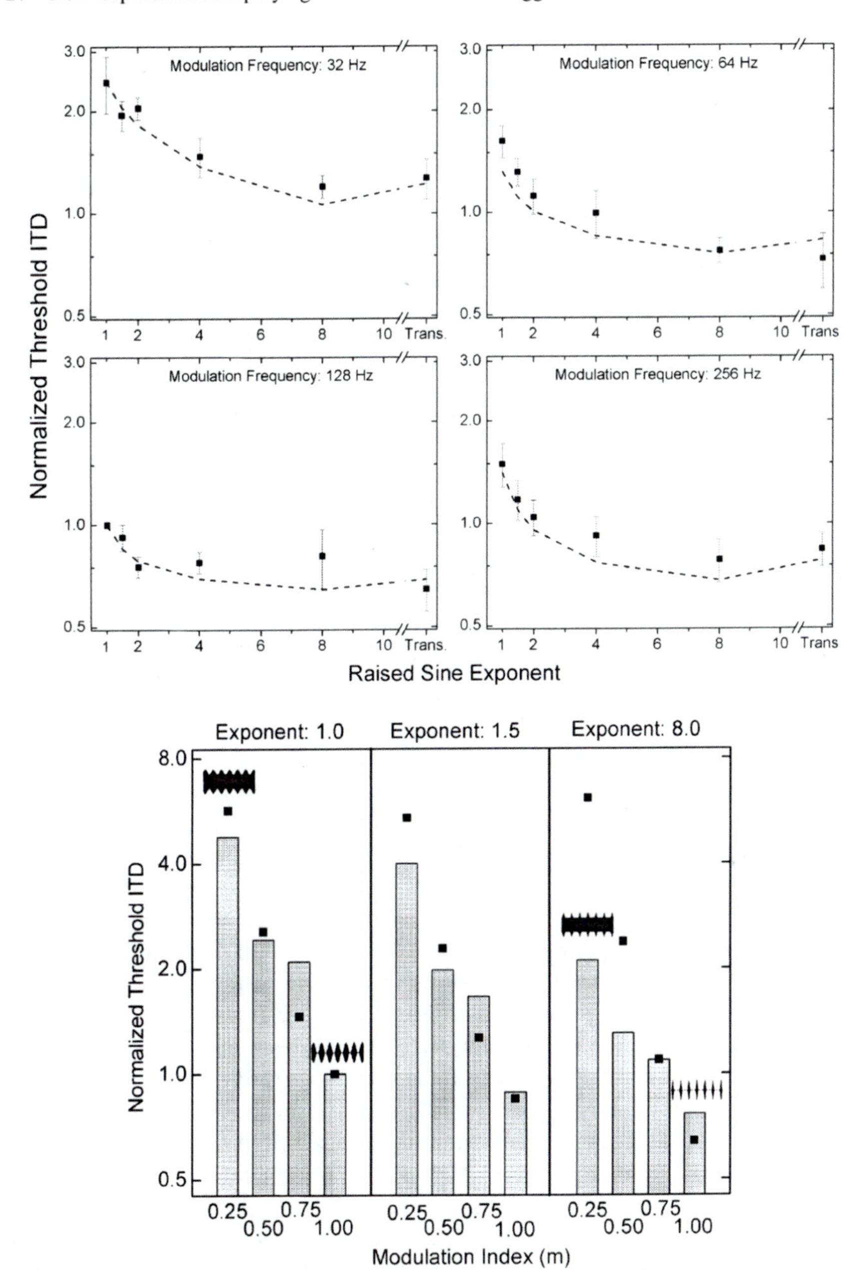

Fig. 27.2 *Upper panel*: Normalized threshold ITDs represented by the means calculated across the four listeners (*squares*). The normalization was achieved by dividing an individual listener's threshold ITDs by that listener's threshold ITD obtained with a SAM tone (raised-sine exponent equal to 1.0) having a frequency of modulation of 128 Hz. Each of the four panels of the figure contains data obtained with a single depth of modulation and raised-sine exponents having values of 1.0, 1.5, or 8.0. The error bars represent ±1 standard error of the mean. The *dashed lines* represent predictions obtained from an interaural correlation-based model. *Lower panel*: Each of the

ITDs are generally smallest (128-Hz modulation rate). Second, threshold ITDs decrease with increases in rate of modulation from 32 to 128 Hz and then increase slightly when the rate of modulation is increased to 256 Hz. The latter trend was found previously with SAM and transposed tones by Bernstein and Trahiotis (2002). These trends in the patterning of the data were verified statistically via analysis of variance.

The dashed lines within each plot of the upper panel of Fig. 27.2 represent predictions of the threshold ITDs obtained via our correlation-based model that incorporates bandpass filtering, envelope compression, and half-wave, square-law rectification and single-pole low-pass filtering of the envelope as suggested by Kohlrausch et al. (2000) and Ewert and Dau (2000). The predictions account for 91% of the data.

Additional data were gathered for a rate of modulation of 128 Hz while varying, in a parametric fashion, both the exponent and the depth of modulation of the raised-sine stimuli. This was done to assess the relative influences on ITD-discrimination of the peakedness of the envelope and its magnitude between those peaks (and any interactive effects between the two variables).

In the same manner as before, threshold ITDs were obtained for 4-kHz-centered raised-sine stimuli having exponents of 1.0, 1.5, and 8.0 at a rate of modulation of 128 Hz. For each of the three raised-sine exponents, thresholds were measured at indices of modulation (m) of 0.25, 0.5, 0.75, and 1.0. The bars in the lower panel of Fig. 27.2 display the mean normalized threshold ITDs, calculated across four listeners, three of whom participated in experiment 1. One of the four listeners was unable to consistently perform the task when the raised-sine exponent was 1.0 and the index of modulation was 0.25. For purposes of computing the normalized threshold ITDs, this listener's threshold for that condition was coded as 1 ms.

Each of the three sections of the plot contains data obtained with a single value of the exponent (1.0, 1.5, or 8.0) and four values of depth of modulation (m=0.25, 0.5, 0.75, or 1.0). The time-waveforms corresponding to four of the stimuli are depicted atop their corresponding bars. The data in the lower panel of Fig. 27.2 show that the decreases in threshold produced by increases in the value of the exponent of raised-sine stimuli occur for all four values of depth of modulation. It is also the case that, for all values of the exponent of the raised sine, threshold ITDs increase as the index of modulation is decreased from 1.00 to 0.25. They do not, however, do so equally. Decreasing the depth of modulation led to larger increases in threshold when the value of the exponent was 1.0 or 1.5, as compared to when the exponent was 8.0. In retrospect, it is clear that parametric variation of the two variables was advantageous in that it revealed an interaction that, a priori, we had no reason to suspect would occur. All of the trends in the data discussed above were verified via an analysis of variance.

The squares in the lower panel of Fig. 27.2 represent predictions obtained from the same correlation-based model that accounted successfully for the data in the upper panel of Fig. 27.2. The criterion amount of change in the normalized interaural correlation to reach threshold ITD was assumed to be the same as that which provided the best fit to the data in the upper panel of Fig. 27.2. Note that the model captures fairly accurately the changes in normalized threshold ITD that occur with changes in depth of modulation for raised-sine stimuli having exponents of 1.0 or 1.5. This is logically consistent with the analysis of Bernstein and Trahiotis (1996). They showed that a quantitative account based on interaural correlation could account for the threshold ITDs, taken as a function of the depth of modulation, for high-frequency SAM tones (Nuetzel and Hafter 1981) and high-frequency two-tone complexes (McFadden and Pasanen 1976). Nonetheless, the model does not capture the fact that, decreasing the depth of modulation from 1.00 to 0.25 when the exponent of the raised-sine is 8.0 results in smaller degradations in sensitivity to ITD than was the case for the other two values of the exponent. In fact, looking across the three sections of the lower panel of Fig. 27.2 reveals that the model predicts that the effects of decreasing the depth of modulation would be nearly the same independent of the exponent of the raised-sine. More specifically, the model predicts that when the depth of modulation was 0.25 or 0.5 there would be no improvement in sensitivity to ITD, were the exponent of the raised-sine to be increased from 1.0 to 8.0.

In order to gain an intuitive understanding of why the predictions of the model take the form that they do, it is helpful to consider the schematized representations of the interaural correlation functions of the envelopes of stimuli having exponents of either 1.0 or 8.0. These are shown in the upper and lower panels, respectively, of Fig. 27.3. In each panel, the heavy solid and dashed lines represent nondelayed and delayed versions of the envelope, respectively, for a depth of modulation of 1.0. The lighter solid and dashed lines represent nondelayed and delayed versions of the envelope, respectively, for a depth of modulation of 0.25. In both panels, the interaural delay for a depth of modulation of 1.0 (heavier lines) was chosen such that the normalized interaural correlation (the value along the ordinate at "lag zero") was 0.5. This large a change was chosen to make it easy to see the effects of interest. Note that a smaller value of ITD was required to produce the same decrease in normalized interaural correlation for the raised-sine stimulus with an exponent of 8.0. This is consistent with: (1) the smaller values of threshold ITD measured with that stimulus and (2) the fact that a constant reduction in interaural correlation accounts for the thresholds obtained with both types of stimuli.

When the depth of modulation is reduced to 0.25 (lighter lines in both panels), the same values of interaural delay result in a normalized interaural correlation of 0.8 for both types of stimuli. This suggests that, within the experiment, in order to overcome the reduction in depth of modulation, the same relative increase in interaural delay would be required to reach threshold for both exponents of 1.0 (SAM) and 8.0. This helps to explain why the predictions in the lower panel of Fig. 27.2 for stimuli having low depths of modulation are so similar regardless of the value of the exponent. Thus, the discrepancies between the predictions and the data for those stimulus conditions represent a clear failure of a correlation model that only takes into account activity at lag zero.

Prompted by a suggestion made by Dr. Wes Grantham, we wondered whether better predictions could be made by considering the displacements and patterning

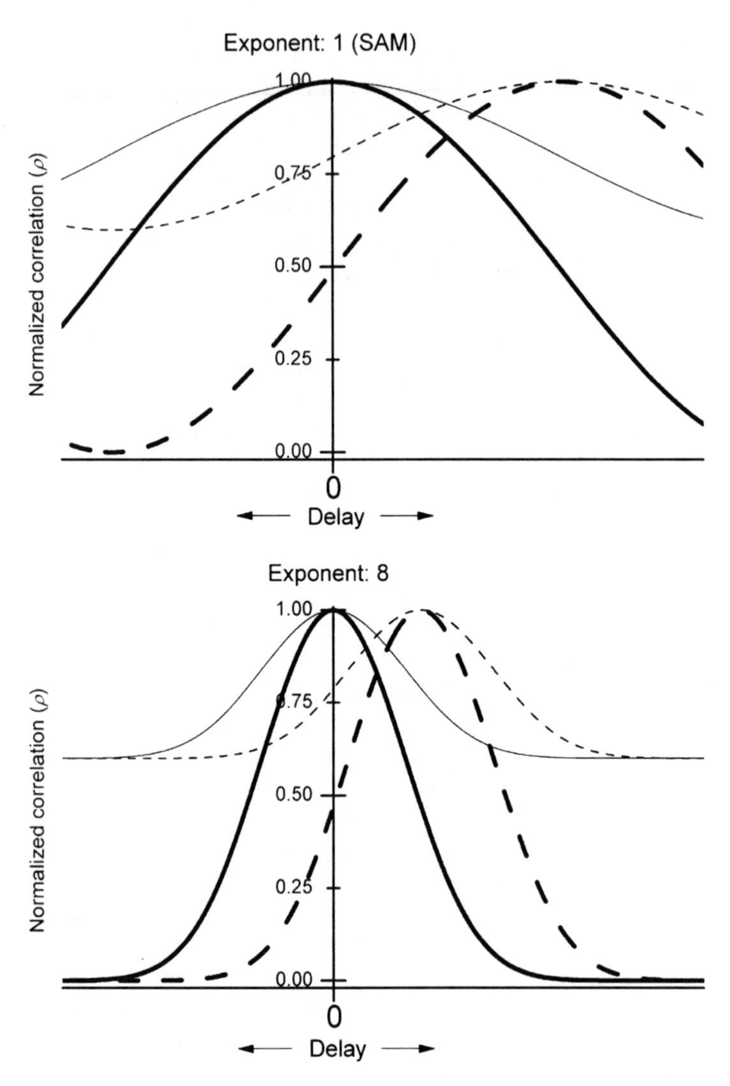

Fig. 27.3 Schematized representations of the interaural correlation functions of the envelopes of stimuli having exponents of either 1.0 (*top panel*) or 8.0 (*bottom panel*). In each panel, the *heavy solid and dashed lines* represent nondelayed and delayed versions of the envelope, respectively for a depth of modulation of 1.0. The *lighter solid and dashed lines* represent nondelayed and delayed versions of the envelope, respectively, for a depth of modulation of 0.25. In both panels, the interaural delay for a depth of modulation of 1.0 (*heavier lines*) was chosen such that the normalized interaural correlation (the value along the ordinate at "lag zero") was 0.5

of the interaural cross-correlations functions produced by the same stimuli. To do so, we evaluated the changes in the "mean-to-sigma" properties of the stimuli along the delay axis that were produced by reducing the depth of modulation from 1.0 to 0.25. The reader is reminded that the relative displacements of the heavier lines (depth of modulation of 1.0) in each panel were chosen to represent the relative magnitudes of the interaural delays required to reach threshold in each case. From this mean-to-sigma point of view, one can easily appreciate that a greater displacement of the function along the delay axis would be required for an exponent of 1.0 in order to differentiate its activity from its nondelayed counterpart. This is so because the variance of that function (it's "width" along the delay axis) is substantially larger than that for the envelope of a stimulus having an exponent of 8.0.

In order to evaluate whether this type of "mean-to-sigma approach" would improve the accuracy of the predictions for these types of stimuli, one need only: (1) determine the relative increase in the width (variance) of each function that results from decreasing the depth of modulation from 1.0 to 0.25 and (2) compare those relative increases across the two stimuli. When this was done, we found that, the differences in the shapes of the functions for two values of exponent notwithstanding, the relative increases in the widths of the functions were essentially identical when the depth of modulation was reduced. This once again suggests that, within the experiment, in order to overcome the reduction in depth of modulation, the same relative increase in interaural delay would be required to reach threshold for both exponents of 1.0 (SAM) and 8.0. Consequently, within the context of these stimuli, the interaural correlation-based model fails to predict the interaction between the value of the exponent and the depth of modulation of raised-sine stimuli independent of whether one considers only activity at lag zero or mean-to-sigma displacements of the correlation function along the delay axis.

Several analyses were conducted in attempts to understand why the model failed to predict threshold ITDs obtained with low indices of modulation and a high value of the raised-sine exponent. These included generating predictions after altering the form of the model in the following ways: (1) removing peripheral auditory processing and considering only the Hilbert transforms of the stimuli; (2) increasing the frequency of the envelope low-pass filter; (3) considering the summed outputs of gammatone filters surrounding and including 4 kHz; (4) replacing the gammatone filters by gammachirp filters (e.g., Irino and Patterson 1997; Unoki et al. 2006; Irino and Patterson 2006); (5) incorporating physiologically measured values of "modulation gain" reported by Joris and Yin (1992); (6) evaluating effects stemming from linear half-wave vs. square-law rectification and amount of compression.

None of these manipulations redressed the inability of the model to account for the interaction. At this writing, we are continuing our efforts to discover how the model could be changed or augmented so that it captures the interaction in the data between changes in index of modulation and raised-sine exponent while still providing satisfactory accounts of the host of data to which it has been successfully applied.

Acknowledgments This research was supported by research grants NIH DC-04147 and DC-04073 from the National Institute on Deafness and Other Communication Disorders, National Institutes of Health.

References

Bernstein LR, Trahiotis C (1994) Detection of interaural delay in high-frequency SAM tones, two-tone complexes, and bands of noise. J Acoust Soc Am 95:3561–3567

Bernstein LR, Trahiotis C (1996) The normalized correlation: accounting for binaural detection across center frequency. J Acoust Soc Am 100:3774–3784

Bernstein LR, Trahiotis C (2002) Enhancing sensitivity to interaural delays at high frequencies by using "transposed stimuli". J Acoust Soc Am 112:1026–1036

Bernstein LR, Trahiotis C (2003) Enhancing interaural-delay-based extents of laterality at high frequencies by using "transposed stimuli". J Acoust Soc Am 113:3335–3347

Bernstein LR, Trahiotis C (2004) The apparent immunity of high-frequency "transposed" stimuli to low-frequency binaural interference. J Acoust Soc Am 116:3062–3069

Bernstein LR, Trahiotis C (2005) Measures of extents of laterality for high-frequency "transposed" stimuli under conditions of binaural interference. J Acoust Soc Am 118:1626–1635

Bernstein LR, Trahiotis C, Hyde EL (1998) Inter-individual differences in binaural detection of low-frequency or high-frequency tonal signals masked by narrow-band or broadband noise. J Acoust Soc Am 103:2069–2078

Buell TN, Hafter ER (1988) Discrimination of interaural differences of time in the envelopes of high-frequency signals: integration times. J Acoust Soc Am 84:2063–2066

Dreyer A, Delgutte B (2006) Phase locking of auditory-nerve fibers to the envelopes of high-frequency sounds: implications for sound localization. J Neurophysiol 96:2327–2341

Ewert SD, Dau T (2000) Characterizing frequency selectivity for envelope fluctuations. J Acoust Soc Am 108:1181–1196

Griffin SJ, Bernstein LR, Ingham NJ, McAlpine D (2005) Neural sensitivity to interaural envelope delays in the inferior colliculus of the guinea pig. J Neurophysiol 93:3463–3478

Irino T, Patterson RD (1997) A time-domain, level-dependent auditory filter: the gammachirp. J Acoust Soc Am 101:412–419

Irino T, Patterson RD (2006) A dynamic compressive gammachirp auditory filterbank. IEEE Trans Audio Speech Lang Processing 14:2222–2232

John MS, Dimitrijevic A, Picton T (2002) Auditory steady-state responses to exponential modulation envelopes. Ear Hear 23:106–117

Joris PX, Yin TC (1992) Responses to amplitude-modulated tones in the auditory nerve of the cat. J Acoust Soc Am 91:215–232

Kohlrausch A, Fassel R, Dau T (2000) The influence of carrier level and frequency on modulation and beat-detection thresholds for sinusoidal carriers. J Acoust Soc Am 108:723–734

McFadden D, Pasanen EG (1976) Lateralization at high frequencies based on interaural time differences. J Acoust Soc Am 59:634–639

Moore BCJ (1997) Frequency analysis and pitch perception. In: Crocker M (ed) Handbook of acoustics. Wiley, New York

Nuetzel JM, Hafter ER (1976) Lateralization of complex waveforms: effects of fine-structure, amplitude, and duration. J Acoust Soc Am 60:1339–1346

Nuetzel JM, Hafter ER (1981) Discrimination of interaural delays in complex waveforms: spectral effects. J Acoust Soc Am 69:1112–1118

Stecker GC, Hafter ER (2002) Temporal weighting in sound localization. J Acoust Soc Am 112:1046–1057

Unoki M, Irino T, Glasberg B, Moore BCJ, Patterson RD (2006) Comparison of the roex and gammachirp filters asrepresentations of the auditory filter. J Acoust Soc Am 120:1474–1492

van de Par S, Kohlrausch A (1997) A new approach to comparing binaural masking level differences at low and high frequencies. J Acoust Soc Am 101:1671–1680

Chapter 28
Modeling Physiological and Psychophysical Responses to Precedence Effect Stimuli

Jing Xia, Andrew Brughera, H. Steven Colburn, and Barbara Shinn-Cunningham

Abstract Many perceptual and physiological studies of sound localization have explored the precedence effect (PE), whereby two dichotic clicks coming from different directions and arriving at the ears close together in time are perceived as one event coming from the location of the click arriving first. Here, we used a computational model of low-frequency inferior colliculus (IC) neurons to account for both physiological and psychophysical responses to PE stimuli. In the model, physiological suppression of the ITD-tuned lagging response depends on the inter-stimulus delay (ISD) between the leading and lagging sound as well as the ITD of the lead. Psychophysical predictions generated from a population of model IC neurons estimate the perceived location of the lagging click as near that of the lead click at short ISDs, consistent with subjects perceiving both lead and lag as coming from the lead location. As ISD increases, the estimated location of the lag becomes closer to the true lag location, consistent with listeners perceiving two sounds coming from separate locations. Together, these physiological and perceptual simulations suggest that ITD-dependent suppression in IC neurons can explain the behavioral phenomenon known as the PE.

Keywords Computational model • Inferior colliculus • Localization dominance • Sound localization model

---

B. Shinn-Cunningham (✉)
Department of Biomedical Engineering, Boston University, 677 Beacon Street, Boston, MA 02215, USA
e-mail: shinn@cns.bu.edu

E.A. Lopez-Poveda et al. (eds.), *The Neurophysiological Bases of Auditory Perception*,
DOI 10.1007/978-1-4419-5686-6_28, © Springer Science+Business Media, LLC 2010

28.1 Introduction

Listeners are remarkably accurate at localizing sound despite the reflections that are present in most settings. It is thought that the auditory system accomplishes this feat by attributing greater perceptual weight to the location cues at sound onsets and suppressing cues from later-arriving reflections. Wallach et al. (1949) introduced the term "precedence effect" (PE) to describe the phenomenon whereby a pair of dichotic clicks presented with a brief inter-stimulus delay (ISD) are typically heard as a single "fused" sound image whose perceived direction is near the location of the leading click.

Psychophysically, the PE can be described by three phases. *Summing localization*, in which listeners perceive one single fused auditory image located about halfway between the lead and lag sources, occurs for ISDs from 0 to about 1 ms. *Localization dominance*, in which the paired sounds are localized near the lead, occurs for ISDs ranging from about 1 to 10 ms. Finally, for ISDs more than about 10 ms, two separate auditory images can be heard, one near the location of the lead and one near the lag location.

Neural correlates of the PE have also been observed in physiological studies of extracellular responses, especially in inferior colliculus (IC; Fitzpatrick et al. 1995; Tollin et al. 2004). For small ISDs, neural responses to the lag are reduced or eliminated, while at longer ISDs, responses to the lagging source recover, mirroring psychophysical results (Litovsky and Yin 1998a, b).

The objective of this study is to build a computational model that simulates the three phases of the PE as ISD changes, as well as the way these phenomena depend on the relative directions of the leading and lagging stimuli. In contrast to previous models that simulate the behavioral aspects of the PE (Lindemann 1986; Hartung and Trahiotis 2001), the modeling approach in this study emphasizes the physiological basis of PE found in the binaural responses of IC neurons.

28.2 Methods

28.2.1 Stimuli

PE stimuli were generated by presenting a pair of binaural clicks (the lead and the lag) separated by an ISD (defined as the time difference between the onsets of the lead and lag delivered to the right ear). ITDs were imposed separately on the lead and lag. We assumed that the left and right ICs were mirror symmetric and generated models of individual left IC neurons. By convention, positive ITDs were generated by advancing the stimulus in the ear contralateral to the model cell (right) and delaying the stimulus in the ear ipsilateral to the model cell (left).

28.2.2 Model Structure

The model (Fig. 28.1) consists of a hierarchy of processing stages, mimicking the stages of the auditory periphery. Our IC model neurons, which are based on the previous IC model (Cai et al. 1998), are innervated by medial superior olive (MSO) model neurons (Brughera et al. 1996), with inhibitory inputs passed through the dorsal nuclei of the lateral lemniscus (DNLL). The MSO model neurons receive bilateral inputs from models of bushy cells in the cochlear nucleus (Rothman et al. 1993), which receive convergent inputs from existing models of auditory-nerve fibers (Carney 1993).

Connections from model MSO cells to a model IC cell are based on anatomical and physiological evidence. Excitatory inputs from MSO to ipsilateral IC lead to ITD sensitivity in low-frequency cells of the IC (Kuwada and Yin 1983; Carney and Yin 1989; Loftus et al. 2004). Different model MSO neurons are characterized by different "best" ITDs, the ITDs that lead to the maximal firing rates (e.g., the ITD that best compensates for, among other things, any difference in the neural transmission delays to the neuron from the ipsi- and contralateral ears). With our conventions, model cells with positive best ITD are in the left IC.

Our IC model is based on the hypothesis that directionally tuned parallel inhibition from ipsilateral and contralateral MSOs (via the corresponding DNLL) produces

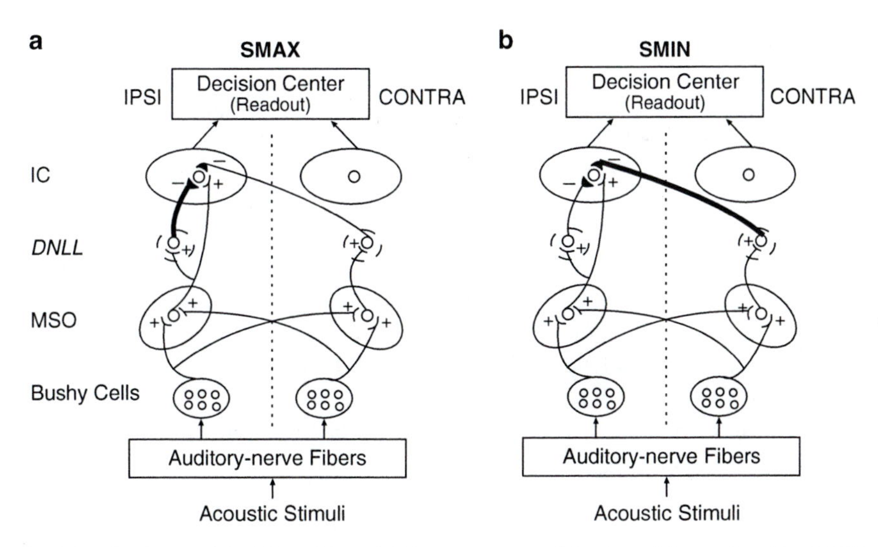

Fig. 28.1 Structure of the IC model, which incorporates models of medial superior olive (MSO) neurons, bushy cells in cochlear nucleus, and auditory-nerve fibers. The dorsal nucleus of the lateral lemniscus (DNLL) is included in the model only as a relay mechanism for generating delayed inhibitory input to the IC from the MSO. The details of DNLL cell behaviors are not included. Excitatory synapses are marked by "+" and inhibitory synapses by "–." (**a**) The structure of an SMAX model neuron, which has strong ispilateral inhibition. (**b**) The structure of an SMIN model neuron, which has strong contralateral inhibition

suppression of the lagging response (Fitzpatrick et al. 1995). The differential equations describing the membrane potential of the model IC neurons are the same as those used by Cai et al. (1998). Inhibitory inputs to the IC are delayed, presumably because of the extra synapse at the DNLL. As a result, inhibition suppresses responses after sound onsets, but has little effect on onset responses. Some model IC neurons receive more inhibition from the ipsilateral DNLL than from the contralateral DNLL (Fig. 28.1a). These neurons, known as suppression-at-maximum (SMAX) neurons (Litovsky and Yin 1998b), should show the greatest suppression of the lagging response when the leading sound comes from the neuron's best ITD. In contrast, suppression-at-minimum (SMIN) neurons (Litovsky and Yin 1998b) receive stronger inhibition from the contralateral DNLL (Fig. 28.1b), and should show the greatest suppression of the lag when the lead comes from positions that elicit minimal response.

28.2.3 Data Analysis

The techniques used here to compute the discharge rate in response to the leading and lagging stimuli were the same as those used in physiological studies (Fitzpatrick et al. 1995; Litovsky and Yin 1998a; Tollin et al. 2004). Specifically, to quantify the response to a single binaural click, the number of the spikes in an *analysis window* was counted. The window began when the number of discharges first was greater than two standard deviations above the mean spontaneous rate, which was computed over the 10 ms prior to the start of the stimulus. The window ended when the response fell below the mean spontaneous rate. Response latency was taken as the time of the onset of the analysis window. For a pair of binaural clicks, two windows (the *leading window* and the *lagging window*) were used to calculate responses. The start of the leading window was taken to be the latency of the response. The end of the leading window was the end the analysis window in response to the single binaural click. The lagging window started at the time of the response latency plus the ISD and had a duration equal to the duration of the leading window. For small ISDs, the two windows overlapped with each other; the response to the lag was estimated by subtracting the number of spikes in response to a single leading stimulus from the number of spikes counted from the onset of the leading window to the offset of the lagging window.

28.2.4 The Readout

The absolute discharge rate of a single IC neuron cannot account for the perceived location of acoustic inputs. Readout of the responses of a population of IC neurons with the same frequency tuning was accomplished by forming an estimate of the perceived ITD from the population response. In the current work, we concentrated on the low characteristic frequency $f=500$ Hz and assumed that the population consisted of neurons whose best ITDs were equally distributed over a symmetrical range from $-T$ to T ms, where T is some integer multiple of half the period of the characteristic frequency. For this neural distribution, the maximum likelihood estimate (MLE) of the input IPD (φ) can be computed as the complex angle of a weighted sum of complex values (Colburn and Isabelle 1992). The magnitude of each complex value representing one neuron's response is given by the firing rate (L) integrated across the time window corresponding to the leading and lagging stimulus (see data analysis). The phase depends on the best ITD (τ_m) of the neuron.

$$\hat{\varphi}_i(f) = \angle \sum_{\tau_m=-T}^{\tau_m=T} L_{i,\tau_m}(f) e^{j2\pi f \tau_m}, \tag{28.1}$$

where $i=1$, 2, corresponds the response to the leading and lagging stimulus in the paired source, respectively. While the phase of the complex-valued sum estimates the source IPD, the normalized magnitude of the sum (r) estimates the reliability of the observed population response estimate (Shinn-Cunningham and Kawakyu 2003)

$$\hat{r}_i(f) = \left| \sum_{\tau_m=-T}^{\tau_m=T} L_{i,\tau_m}(f) e^{j2\pi f \tau_m} \right|, \tag{28.2}$$

For the PE stimuli used here, we assumed that the perceived ITD (α) to the lead ($i=1$) and to the lag ($i=2$) is a weighted sum of the lead and lag IPDs estimated by (28.1).

$$\hat{\alpha}_i = \frac{c_i \hat{\varphi}_1 + (1-c_i)\hat{\varphi}_2}{f}, \tag{28.3}$$

where the weight (c_i) depends on the population response to the lead or lag, relative to the population response to the lead or lag alone. If the instruction is to match the location of the leading stimulus, $i=1$

$$c_1 = \frac{\hat{r}_1}{\hat{r}_{S1}} \tag{28.4}$$

If the instruction is to match the location of the lagging stimulus, $i=2$.

$$c_2 = 1 - \frac{\hat{r}_2}{\hat{r}_{S2}} \tag{28.5}$$

where r_{S1} and r_{S2} are the population response estimated using (28.2) to a single binaural click presented in isolation at the leading and lagging location, respectively.

The weight c is restricted to fall between 0 and 1. Consistent with the weight c described by Shinn-Cunningham et al. (1993), $c_i=1$ indicates that the lead dominates lateralization entirely ($\alpha_i=\varphi_1/f$), whereas $c_i=0$ indicates that the lag dominates lateralization completely ($\alpha_i=\varphi_2/f$). In the current study, we assume that the response to the leading stimulus is not affected by the presence of the lag ($c_1 \approx 1$, $\alpha_1 \approx \varphi_1/f$). If the instruction is to match the lagging ITD, the PE is strong when the population response to the lagging stimulus is substantially suppressed ($c_2 \approx 1$, $\alpha_2 \approx \varphi_1/f$). When the lagging stimulus begins to evoke a population response that closely resembles the neural response to the lagging source presented in isolation, PE is weak ($c_2 \approx 0$, $\alpha_2 \approx \varphi_2/f$).

28.3 Results

28.3.1 Simulations of Physiological Data

A model SMAX and SMIN neuron with the best ITD of 300 μs was built to simulate the response properties observed in physiological studies of PE. The ITD of the lagging click was fixed at the best ITD of the model IC neuron (300 μs). The ITD of the leading click varied from −900 to +900 μs, and ISDs ranged from 1 to 20 ms.

For both 10- and 20-ms ISDs, the responses to the leading sound were similar to those to a single stimulus, being strongest on the contralateral side for an ITD of 300 μs; the model cell responded minimally at the negative ITDs on the ipsilateral side (see Fig. 28.2b). The similarity of the curves indicates that the response to the leading click is not affected by the presence of the lagging click. Figure 28.2a displays corresponding physiological data (Litovsky and Yin 1998b), which is in good agreement with the model.

Figure 28.2c and d show the response of a SMAX cell observed by Litovsky and Yin (1998b) and the model simulation, respectively. The amount of suppression of the lagging response depends on the stimulus location of the leading source as well as on the ISD. At 20-ms ISD, the response dips at about 300 μs, when the leading click is near the neuron's best ITD. The model SMAX cell received more inhibition from an ipsilateral MSO cell tuned to the same ITD as its excitatory projection; therefore, the suppression of the lagging response was the strongest when the leading source produced maximal response of the neuron. At 10 ms, the suppression increased and the dip broadened. Finally at 5 ms, the response was suppressed completely at nearly all positions of the leading sound. This suppression of the lagging response is similar to the behavioral phase of localization dominance in that the perceived position of the paired sources appears to be governed by the leading source only.

Litovsky and Yin (1998b) also found a small number of units for which the lagging responses are most reduced when the lead produces the lowest responses. The model SMIN cell was able to simulate such response properties (compare Fig. 28.2e, f). The model SMIN cell received more inhibition from a contralateral MSO cell tuned to the ITD to which its excitatory projection would minimally respond. Therefore, suppression was strongest at the trough of the leading response.

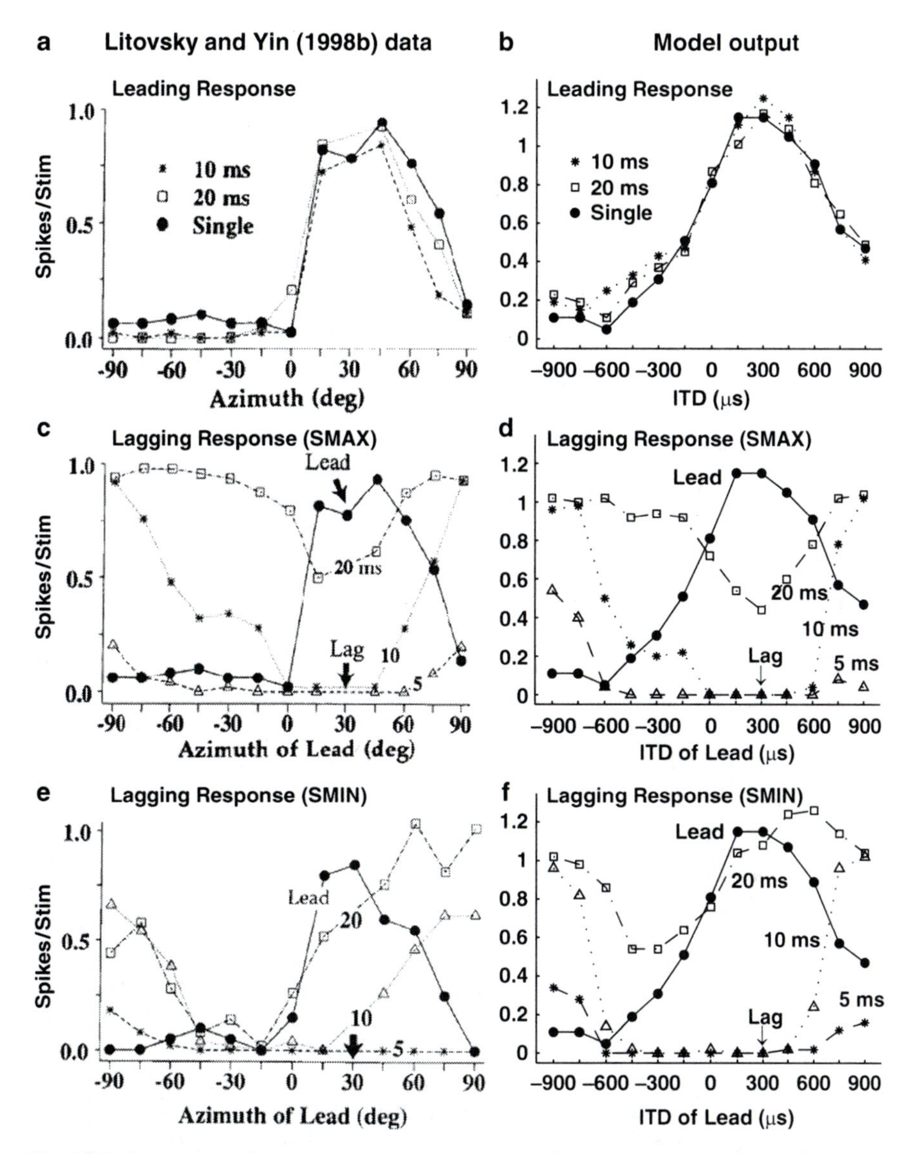

Fig. 28.2 Comparison of neuron response patterns to model response patterns. (**a**, **c**, **e**) Responses of neurons. Reproduced from Litovsky and Yin (1998b) with permission from J Neurophysiol. (**b**, **d**, **f**) Responses of model IC cells. *Top*: responses to the leading stimulus with the lag delayed by 10 and 20 ms, as well as to a single click as a function of the location of the lead. *Middle*: responses of SMAX neurons to the lagging stimulus as a function of lead location. *Bottom*: responses of SMIN neurons. Responses to single clicks are shown as well. *Down arrow* indicates the location of the lagging click, which evokes the maximal possible response from the neuron if presented in isolation. Responses to the lagging click at ISDs of 5, 10, and 20 ms are shown

28.3.2 Simulations of Psychophysical Data

For the generation of psychophysical predictions, the IC neuron population only consisted of SMAX units because SMAX units are thought to be more prevalent in the auditory pathway (Litovsky and Yin 1998b). We used this population to simulate the results of behavioral experiments in which subjects were asked to indicate the perceived location(s) of the lead/lag target by adjusting the ITDs of a pointer stimulus (Litovsky and Shinn-Cunningham 2001). For each stimulus, two matches were made: one in which subjects matched the "right-most" image, and one in which they matched the "left-most" image. In the model simulation, the ITDs of the lead were chosen between 0 and –400 μs, and the lagging ITD was held constant at 400 μs. To allow a direct comparison with the conditions measured, all the data obtained when the lagging stimulus was located on the left side (with the ITD of –400 μs) were generated by assuming the left/right symmetry of the model.

We found that lagging responses were suppressed, in some cases for ISDs as long as 20 ms. Clearly, a second sound image can be heard at ISDs for which some suppression remains. On the other hand, at 10-ms ISD, some individual neurons had already recovered completely. Although the responses to the lagging stimulus had recovered for some neurons at moderate ISDs, the population activity was spread out across neurons with different best ITDs. Therefore, the strength of the resulting perceived location was reduced, corresponding to a relatively low population response. It seems that perception of a second image occurs for ISDs that lead to two distinct neural responses across a population of neurons (one due to the lead and one due to the lag), both of them containing sufficient information to convey perception of a source at the corresponding location.

As in the behavioral measures, our model predictions show that the matched ITD was near the lead ITD at all ISDs when listeners matched the lead image (compare Fig. 28.3a, b, filled symbols). At short ISDs, the listener matched the lead ITD regardless of instructions, suggesting that the localization cues of the lead dominates (see open symbols in Fig. 28.3). When instructions were to match the lag, the matched position approached the lag ITD for ISDs longer than 10 ms. One minor discrepancy between model and behavioral results is that in the model, localization dominance is slightly stronger at long ISDs. When lead and lag were spatially near to one another, the likelihood of perceiving two distinct images at their respective sources was lower than when lead and lag were spatial far apart in both model and behavior. However, this effect is greater in the simulations than in the behavioral results, reflecting the fact that SMAX model cells showed stronger suppression of the lag when the lead and lag locations were close together.

28.4 Conclusions

A model was developed that simulated the responses of IC neurons in response to a pair of binaural clicks that could evoke the PE. The model used existing models for auditory-nerve fibers, bushy cells in the cochlear nucleus, and cells in MSO.

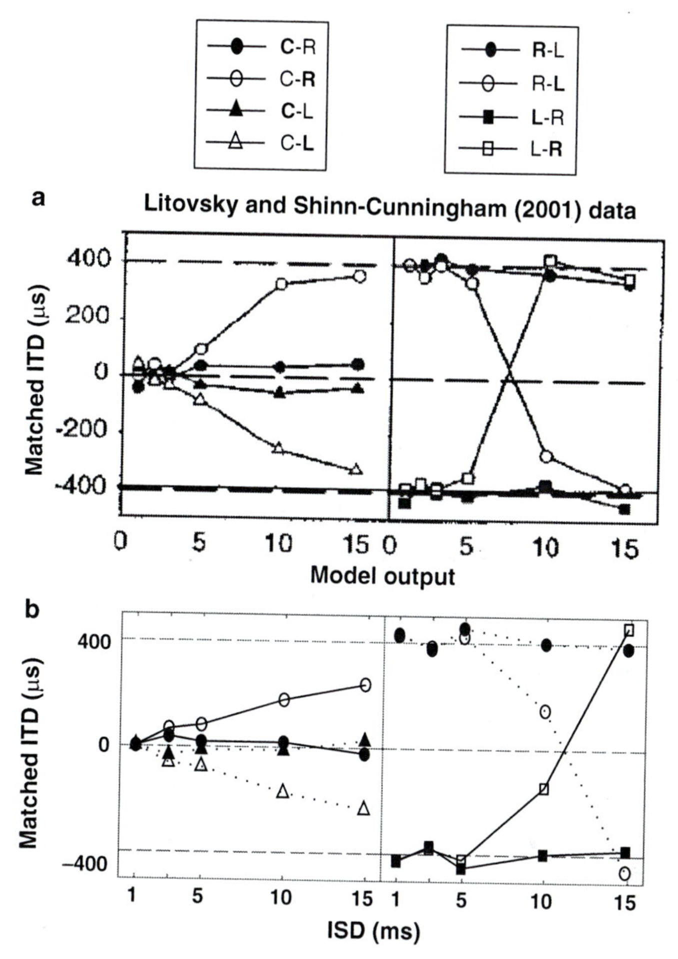

Fig. 28.3 The matched ITD α under various conditions, at ISDs of 1–15 ms. In the legend, lead–lag positions are denoted by order. The *bold letter* indicates whether instructions are consistent with matching the lead or lag. *Left column*: the lead is always at center (0 μs), and the lag is either on the right or left. *Right column*: lead and lag are on opposite sides. (**a**) Matching results for subject S1. Reproduced from Litovsky and Shinn-Cunningham (2001) with permission from J Acoust Soc Am. (**b**) Estimates of the model

The IC model neuron received excitatory inputs from an ipsilateral MSO model cell, as well as inhibitory inputs from both ipsilateral and contralateral MSO model cells via the DNLL. Suppression of the lagging response arises because of the long-lasting inhibition evoked by the leading stimulus. This suppression is modulated by ITD because the inhibition comes from cells that are themselves sensitive to stimulus ITD. Consistent with previous data, the model neuron cells showed suppression of the lagging response at short ISDs, with greatest suppression at ISDs from 1 to 5 ms. By adjusting the relative strength of inhibition from both sides, some model neurons displayed the strongest suppression of the lagging response for a lead at the neuron's best ITD, whereas others had the strongest suppression for a lead placed in the hemifield opposite the best ITD. A model IC built from the first type of model neurons can explain localization dominance reported in psychophysical studies of PE, whereby at short ISDs, the perceived location of a pair of clicks was dominated by the leading source. The strength of dominance decreased and the lagging sound was more likely to be heard near its own true location as the spatiotemporal separation of the lead and lag increased.

References

Brughera AR, Stutman ER, Carney LH, Colburn HS (1996) A model with excitation and inhibition for cells in the medial superior olive. Aud Neurosci 2:219–233

Cai H, Carney LH, Colburn HS (1998) A model for binaural response properties of inferior colliculus neurons. I. A model with ITD-sensitive excitatory and inhibitory inputs. J Acoust Soc Am 103:475–493

Carney LH (1993) A model for the responses of low-frequency auditory-nerve fibers in cat. J Acoust Soc Am 93:401–417

Carney LH, Yin TCT (1989) Responses of low-frequency cells in the inferior colliculus to interaural time differences of clicks: excitatory and inhibitory components. J Neurophysiol 62:144–161

Colburn HS, Isabelle SK (1992) Models of binaural processing based on neural patterns in the medial superior olive. In: Cazals Y, Demaney L, Horner K (eds) Auditory physiology and perception. Pergamon, Oxford, pp 539–545

Fitzpatrick DC, Kuwada S, Batra R, Trahiotis C (1995) Neural responses to simple simulated echoes in the auditory brain stem of the unanesthetized rabbit. J Neurophysiol 74:2469–2486

Hartung K, Trahiotis C (2001) Peripheral auditory processing and investigations of the "precedence effect" which utilizes successive transient stimuli. J Acoust Soc Am 110:1505–1513

Kuwada S, Yin TCT (1983) Binaural interaction in low-frequency neurons in inferior colliculus of the cat. I. Effects of long interaural delays, intensity, and repetition rate on interaural delay function. J Neurophysiol 50:981–999

Lindemann W (1986) Extension of a binaural cross-correlation model by contralateral inhibition. II. The law of the first wave front. J Acoust Soc Am 80:1623–1630

Litovsky RY, Shinn-Cunningham BG (2001) Investigation of the relationship among three common measures of precedence: fusion, localization dominance, and discrimination suppression. J Acoust Soc Am 109:346–358

Litovsky RY, Yin TCT (1998a) Physiological studies of the precedence effect in the inferior colliculus of the cat. I. Correlates of psychophysics. J Neurophysiol 80:1285–1301

Litovsky RY, Yin TCT (1998b) Physiological studies of the precedence effect in the inferior colliculus of the cat. II. Neural mechanisms. J Neurophysiol 80:1302–1316

Loftus WC, Bishop DC, Saint Marie RL, Oliver DL (2004) Organization of binaural excitatory and inhibitory inputs to the inferior colliculus from the superior olive. J Comp Neurol 472(3):330–344

Rothman SR, Young ED, Manis PB (1993) Convergence of auditory nerve fibers onto bushy cells in the ventral cochlear nucleus: implications of a computational model. J Neurophysiol 70:2562–2583

Shinn-Cunningham BG, Kawakyu K (2003) Neural representation of source direction in reverberant space. In: Proceedings of the 2003 IEEE workshop on applications of signal processing to audio and acoustics, New Paltz, NY, October 2003, pp 79–82

Shinn-Cunningham BG, Zurek PM, Durlach NI (1993) Adjustment and discrimination measurements of the precedence effect. J Acoust Soc Am 93:2923–2932

Tollin DJ, Populin LC, Yin TCT (2004) Neural correlates of the precedence effect in the inferior colliculus of behaving cats. J Neurophysiol 92:3286–3297

Wallach H, Newman EB, Rosenzweig MR (1949) The precedence effect in sound localization. Am J Psychol 52:315–336

Chapter 29
Binaurally-Coherent Jitter Improves Neural and Perceptual ITD Sensitivity in Normal and Electric Hearing

M. Goupell, K. Hancock, P. Majdak, B. Laback, and B. Delgutte

Abstract The sensitivity of listeners to interaural time differences (ITDs) in pulse-train stimuli decreases with increasing rate beyond a few hundred pulses per second (pps). Hafter and Dye [*J. Acoust. Soc. Am.* 73: 644–651] showed that the decreasing sensitivity is related to the reduced effectiveness of the ongoing signal and called the effect "binaural adaptation." We hypothesized that binaurally-coherent temporally-random ("jittered") pulse trains would improve ITD sensitivity at high pulse rates by causing a recovery from binaural adaptation. Perceptual ITD sensitivity was measured in six normal-hearing listeners in a left-right discrimination task. For rates higher than 400 pps, we found that ITD sensitivity was much better for jittered pulse trains than for periodic pulse trains.

To investigate the neural basis for this effect, we recorded from single units in the inferior colliculus (IC) of bilaterally-implanted, anesthetized cats. Responses to pulse trains were measured as a function of pulse rate, jitter, and ITD. For pulse rates above 300 pps, ongoing neural firing rates were greater when the pulse trains were jittered. Action potentials tended to occur at sparse preferred times across repeated presentations of a jittered pulse train. Neurons stimulated with jittered high-rate pulse trains showed ITD tuning comparable to that produced by low-rate pulse trains. Thus, jitter appears to improve neural ITD sensitivity by restoring sustained firing in IC neurons.

In addition, the response to jittered pulse trains was observed in a physiologically-based model of auditory nerve. We found that the random nature of the pulse trains caused increased firing at specific times across repeated presentations, which is consistent with the firing pattern observed in the IC.

Keywords ITD • Cochlear implants • Binaurally-coherent jitter • Inferior colliculus • Temporal coding • Binaural hearing

M. Goupell (✉)
Acoustics Research Institute, Austrian Academy of Sciences, Wohllebengasse 12-14, A-1040 Vienna, Austria
e-mail: matt.goupell@gmail.com

E.A. Lopez-Poveda et al. (eds.), *The Neurophysiological Bases of Auditory Perception*,
DOI 10.1007/978-1-4419-5686-6_29,

29.1 Introduction

Most cochlear implants (CIs) encode information about sounds by amplitude modulating periodic electrical pulse trains whose rates are typically on the order of 1,000 pulses per second (pps). These high rates make it difficult to deliver binaural timing cues with bilateral CIs because the sensitivity of CI listeners to interaural time differences (ITDs) decreases with increasing pulse rate above 100 pps (Majdak et al. 2006; Laback et al. 2007; van Hoesel 2007). The cause of this insensitivity to ITD may be a form of adaptation that prevents the binaural pathway from responding to high-rate stimulation (Hafter and Dye 1983; Laback and Majdak 2008). For example, neurons in the inferior colliculus (IC) of implanted cats best respond to pulse rates below about 100 pps (Snyder et al. 1995), and the decrease in firing rate at higher pulse rates impairs the ability to code ITD in bilaterally-implanted animals (Smith and Delgutte 2007).

The ability of CI users to discriminate ITDs at high pulse rates is greatly improved by imposing binaurally-coherent jitter on the interpulse intervals (IPIs) (Laback and Majdak 2008). The work described here combines psychophysics, neurophysiology, and modeling techniques to further characterize the effects of temporal jitter on ITD coding and gain a better understanding of the underlying neural mechanisms. First, we used acoustic stimuli that closely mimic the electrical pulse trains used with CI listeners to show that ITD discrimination in normal-hearing (NH) subjects is also improved by jitter. Second, we studied neural correlates of the effect of jitter on ITD sensitivity by recording from IC neurons in implanted and NH cats. Lastly, we modeled neural responses to pulse trains to better understand the specific mechanism by which jitter improves ITD coding. Our results show clear parallels between the perceptual, neural, and modeling data.

29.2 Perceptual Experiment

29.2.1 Method

Six NH human subjects participated in the experiment. They had normal hearing according to standard audiometric tests.

Stimuli were 300-ms pulse trains with trapezoidal amplitude modulation (AM) (Fig. 29.1a). There were four trapezoids per stimulus, each with a rise-fall time of 20 ms, and a steady-state duration of 20 ms. The inter-trapezoid interval was 20 ms.

Monophasic rectangular pulse trains were passed through a digital sixth-order bandpass Butterworth filter centered at 4.6 kHz (1.5-kHz bandwidth). The A-weighted sound pressure level (SPL) of the stimuli was 70 dB re 20 μPa. Binaurally-uncorrelated, low-pass filtered, white noise was used to mask low-frequency components that might contain useful binaural cues. The noise corner frequency was 3.5 kHz, with a 24 dB/oct roll-off, and the A-weighted SPL of the noise was 60 dB.

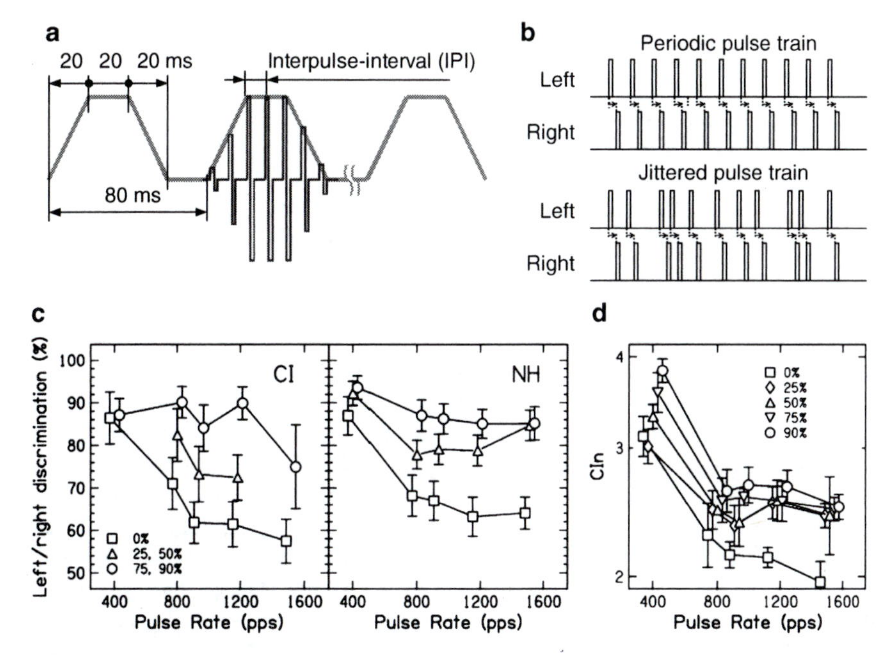

Fig. 29.1 Perceptual experiment and neural modeling. (**a**) The envelope of the experimental stimuli. For clarity, only three of four trapezoids are shown and the pulses are shown in one trapezoid only. (**b**) The stationary portion of a periodic pulse train (*upper*) and of a jittered pulse train (*lower*). Note that the binaurally-coherent jitter preserves the interaural time difference (marked with *arrows*). (**c**) Percent correct scores for left/right discrimination as a function of the pulse rate. The data are averaged over all subjects, interaural time difference, and jitter (when needed). The error bars represent 95% confidence intervals. (**d**) Correlation index (CIn) results. The data points show the mean ± standard deviation of the CIn measurements over five manifestations

The nominal IPI corresponded to the average IPI over the stimulus duration. To preserve the ITD information in the pulse timing, the jitter was coherent between the two ears (Fig. 29.1b). For each pulse train, the IPI was drawn from a rectangular distribution. The amount of jitter was represented by the width of the distribution relative to the nominal IPI in percent. For 0% jitter, the stimuli were periodic. For 100% jitter, the largest possible IPI was twice the nominal IPI and the smallest possible IPI was zero.

The independent variables were pulse rate (400, 800, 938, 1,182, and 1,515 pps), ITD (200, 400, and 600 μs), and jitter (0, 25, 50, 75, and 90%). Stimulus conditions corresponding to combinations of the independent variables were presented in a balanced design. Each condition was repeated 100 times. Each trial represented a new random jitter manifestation.

A two-interval, two-alternative forced-choice procedure was used in a lateralization discrimination test. The first interval contained a reference stimulus with zero ITD and no jitter, evoking a centralized auditory image. The second interval contained the target stimulus with nonzero ITD. The interstimulus interval was 400 ms. The listeners indicated whether the second stimulus was perceived to the left or to the right of the first stimulus. Visual response feedback was provided after each trial.

29.2.2 *Results*

The right side of Fig. 29.1c shows the present experimental results, and the left side shows a subset of the results with five CI listeners from Laback and Majdak (2008). For the periodic stimuli, ITD sensitivity decreases with increasing pulse rate. For both CI and NH listeners, jitter improves ITD sensitivity, thereby compensating for much of the performance decrease with increasing pulse rate. The larger jitter (75 and 90%) improved ITD discrimination more than smaller jitter (25 and 50%), with the exception of the NH listeners at 1,515 pps.

An analysis of variance (ANOVA) (factors: listener type, rate, ITD, and jitter) on comparable conditions across CI and NH listeners showed a significant difference between the types of listeners ($p=0.003$). The NH listeners had an average percent correct of 80.7%, whereas the CI listeners had an average percent correct of 76.6%. Therefore, two separate repeated-measures (RM) ANOVAs (factors: rate, ITD, and jitter) were performed, one for the CI listeners and one for the NH listeners. Both types of listeners showed mostly the same significant main effects and interactions. There was a significant effect of rate, ITD, and jitter ($p<0.0001$ for all). There was a significant interaction of rate × jitter (CI: $p=0.001$; NH: $p<0.0001$). For just the NH listeners, there was a significant interaction of ITD × jitter (CI: $p=0.67$; NH: $p=0.035$).

The data at 400 pps were examined with separate RM ANOVAs (factors: ITD and jitter) and subsequent Tukey HSD post hoc tests since the jitter effect is much less pronounced at this rate. For the CI listeners, the RM ANOVA showed no significant effect of jitter ($p>0.05$). For the NH listeners, the post hoc tests showed a significantly ($p<0.05$) higher performance for all jittered conditions than for periodic conditions.

29.2.3 *Discussion*

The results show that binaurally-coherent jitter can improve ITD sensitivity at high pulse rates. The ITD sensitivity for periodic pulse trains decreased with increasing pulse rate, consistent with the idea of rate limitations. The larger the amount of irregularity in the pulse timing, the larger the improvement that occurred when compared to the periodic pulse trains. For NH listeners, this even occurred for 400 pps, the lowest rate tested. In general, the trends were the same between the CI and NH listeners, but the NH listeners had better scores (4.1%) on average.

29.3 Neurophysiology

29.3.1 *Method*

Neural correlates of the effect of jitter on ITD discrimination were investigated by recording responses to pulse trains in the IC of anesthetized cats. We used seven deaf cats bilaterally implanted with 8-contact intracochlear electrode arrays (Cochlear Corp.) and two NH cats. The recording methods were as described previously (Smith and Delgutte 2007).

In the CI experiments, stimuli were trains of 50-μs biphasic (anodic/cathodic) current pulses presented at 2–6 dB above single-pulse threshold using a wide bipolar configuration. In the NH experiment, 50-μs monophasic rectangular pulses were presented binaurally through closed acoustic systems at 10–30 dB above threshold. In both types of experiments, pulse trains were 300 ms in duration, and were repeated once every 600 ms. Pulse rates were typically varied in half-octave steps from 20 to 640 pps in CI experiments and up to 2,560 pps in the NH experiment.[1] The NH data only include units with characteristic frequency (CF) > 3,000 Hz to avoid providing ITD cues in the fine structure near CF, consistent with the bandpass filtering used in the psychophysical experiments.

Jittered pulse trains were created as described for the perceptual experiments, except that there was no trapezoidal AM, no bandpass filtering, and that the jitter was "frozen": the same sequence of pulses was presented on every stimulus trial at a given pulse rate. Responses were obtained using 50 and 90% jitter; for clarity, the results for 50% jitter are omitted.

29.3.2 *Jitter Can Restore Ongoing Neural Firing at High Pulse Rates*

Figure 29.2a shows firing rate as a function of stimulus pulse rate for diotic stimulation, both with and without jitter, for one IC neuron from an implanted deaf cat. The corresponding temporal discharge patterns are shown as dot rasters in Fig. 29.2c. Without jitter, firing rate initially increases linearly with pulse rate as the neuron fires one spike per pulse (indicated by regularly spaced firing patterns). For pulse rates above 112 pps, the firing rate drops dramatically, and essentially goes to zero by 448 pps. This pattern of response to periodic pulse trains is typical of IC neurons (Snyder et al. 1995; Smith and Delgutte 2007). When the pulse train is jittered, the response at low pulse rates is similar to that in the periodic condition, but the decline at high pulse rates is smaller and the neuron fires robustly for pulse rates up to 640 pps. Overall, jitter significantly increased the firing rate of 29 out of 102 (28%) IC neurons tested in response to high-rate (≥320 pps) pulse trains. Interestingly, the temporal firing patterns evoked by high-rate jittered pulse trains are characterized by sparse preferred times of firing rather than randomly-occurring spikes (Fig. 29.2c, top).

Figure 29.2b shows a similar effect of jitter on the firing rate of a neuron from the IC of an NH cat (dot rasters not shown). Jitter significantly increased the firing rate of all 13 IC neurons studied in NH cats. Thus, in both implanted and NH cats, jitter can restore ongoing firing at high pulse rates.

[1] The stimulus artifact with electrical stimulation made it difficult to obtain responses to pulse rates above 640 pps.

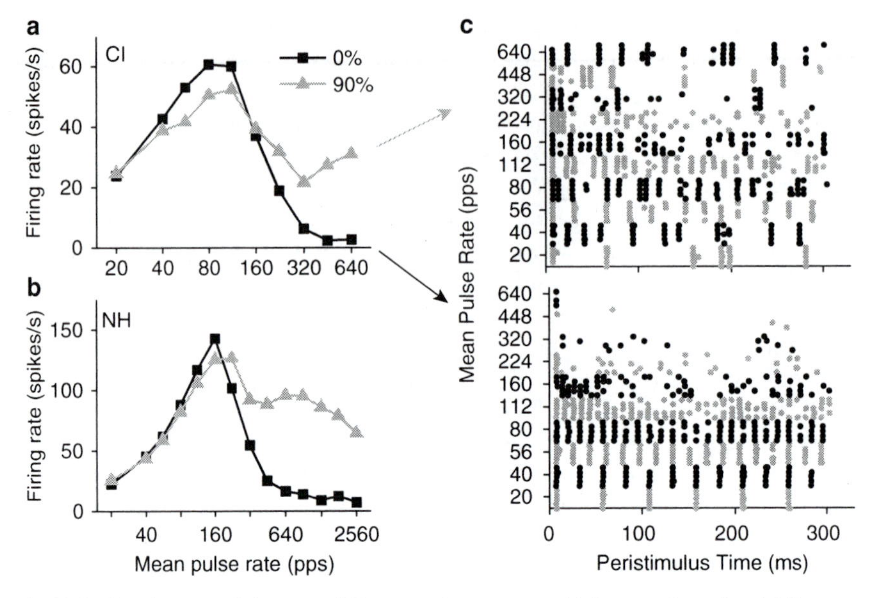

Fig. 29.2 Jitter increases firing rate of IC neurons in response to high-rate pulse trains. (**a**) Response of one neuron in a bilaterally-implanted cat. (**b**) Normal-hearing cat. (**c**) Dot rasters of the temporal response patterns corresponding to (**a**). Each *dot* indicates a single spike, and alternating *colors* distinguish blocks of responses to different pulse rates. The neuron tends to entrain for pulse rates less than about 160 pps, with (*top*) or without jitter (*bottom*). In the jittered case, for pulse rates greater than about 320 pps, the neuron does not entrain, but exhibits sparse preferred times of firing (*top*)

29.3.3 Restoration of Ongoing Firing Reveals ITD Sensitivity

A critical question is the extent to which jitter can improve ITD coding at high pulse rates. Figure 29.3a, b shows rate-ITD curves for the two neurons of Fig. 29.2, obtained using pulse rates of 640 (a) and 2,560 pps (b). In both cases, periodic pulse trains evoke weak activity that is not obviously influenced by changes in ITD. Jittered pulse trains, in contrast, produce much larger firing rates that are clearly modulated by ITD. Neural ITD sensitivity was quantified using a signal-to-noise ratio (SNR) metric which normalizes the across-ITD variance in spike rate by the total spike rate variance, including across-trial variance at each ITD. The SNR ranges between 0 and 1, meaning that ITD accounts for none and all of the total variance, respectively. ITD SNRs are shown to the right of the curves in Fig. 29.3a, b, and are much larger in response to jittered pulse trains in these two units.

Figure 29.3c, d shows the mean ITD SNR over the IC population as a function of pulse rate for the CI and NH experiments, respectively. ITD SNR is significantly greater for jittered pulse trains when compared to periodic pulse trains (one-tailed paired *t*-test) for both the CI ($p=0.02$, $n=108$ rate-ITD curves from 52 neurons)

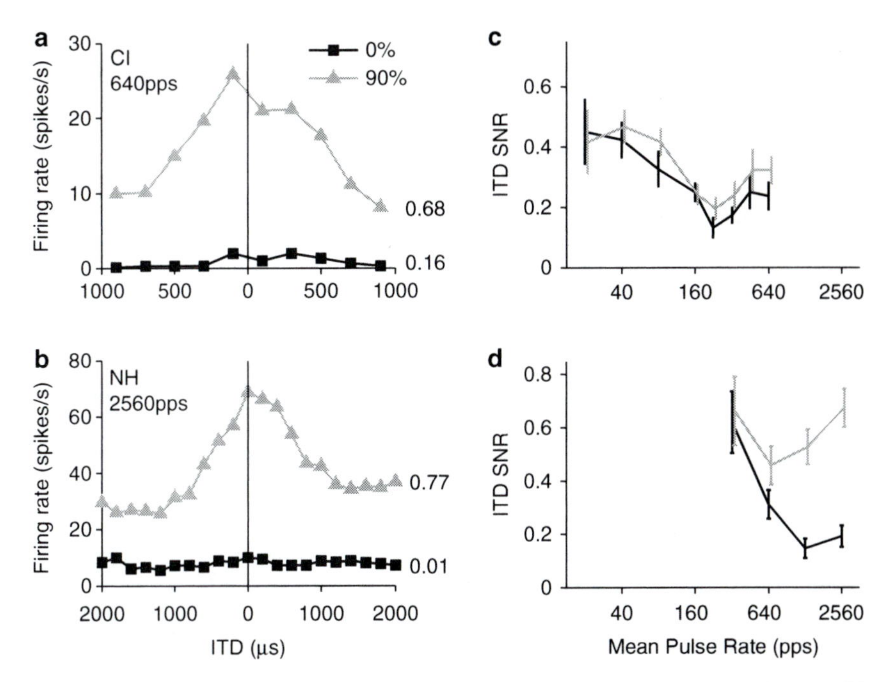

Fig. 29.3 Jitter restores neural ITD sensitivity in response to high-rate pulse trains. (**a, b**) Rate-ITD curves for the same neurons as in Fig. 29.2. Numbers to the right of each curve indicate ITD signal-to-noise ratio (SNR, see text). (**a**) Bilaterally-implanted cat. (**b**) Normal-hearing cat. (**c, d**) Corresponding population data. ITD SNR as a function of pulse rate (mean ± standard error, 3–21 data sets per pulse rate). The NH data are limited to units with CF > 3,000 Hz

and NH data ($p < 0.001$, $n = 33$ rate-ITD curves from 11 neurons). The effect of jitter on ITD sensitivity is clearly larger in the NH case (Fig. 29.3d), consistent with the psychophysical results, but a rigorous comparison is difficult because different ranges of pulse rates were used in the two groups of animals.

For the CI experiments, the effect of jitter on ITD sensitivity depends on pulse rate. It is significant for pulse rates greater than or equal to 320 pps ($p = 0.03$, paired t-test) but not for pulse rates below 320 pps ($p = 0.11$). This is consistent with the notion that jitter improves ITD coding primarily through its effect on firing rate: low-rate periodic pulse trains evoke robust responses that jitter can do little to enhance. A comparable analysis for the NH experiments is not possible because very few rate-ITD curves were measured with pulse rates below 320 pps.

In summary, the physiological data are consistent with the hypothesis of Laback and Majdak (2008) regarding the mechanism by which jitter improves ITD discrimination. Specifically, jitter restores ongoing neural firing at high pulse rates in 28% of IC neurons in CI cats, and all IC neurons in NH cats. Moreover, the responses evoked under such conditions are often ITD-sensitive. However, the overall effect on neural ITD sensitivity is relatively modest in CI experiments, and involves a minority of neurons. Detection-theoretic analyses are needed to quantitatively compare the effect of jitter on neural activity with the improvement in psychophysical performance (Hancock and Delgutte 2004).

29.4 Neural Modeling

To explore possible reasons why binaurally-coherent jitter improves ITD sensitivity, we examined the response of a model of the NH human auditory nerve (AN) (Meddis 2006) to the jittered stimuli used in the perceptual experiments. Figure 29.4 shows the firing probability of a model AN fiber (4.6-kHz center frequency) for 400, 800, and 1,515 pps pulse trains with either jitter of 0 or 90%. After a strong onset response, the firing probability is fairly flat throughout the periodic pulse train. In contrast, the jittered pulse train shows much greater fluctuations in firing probability after the onset. The instances of increased AN firing during the ongoing portion of the signal resulting from the jitter might be utilized by the binaural system to improve ITD sensitivity at high pulse rates. For acoustic stimulation, the pronounced fluctuations with jitter may be partly due to the temporal overlap of the physical pulses at such a high rate and large jitter, which can introduce amplitude modulations. Additional filtering by the cochlea and the hair-cell synapse are also likely to contribute. These fluctuations are consistent with the neural data from the IC, where firings with jittered pulse trains tended to occur at precise sparse times that may correspond to the peaks in the firing probability for model AN fibers.

We quantitatively measured the change in synchrony of the model neural response when jitter was applied to a pulse train. We used the correlation index (CIn) derived from shuffled correlations (Joris et al. 2006), which provides a general measure of the tendency of a neuron to fire at specific times. Unlike the synchronization coefficient, this metric is applicable to both periodic and aperiodic stimuli. Because binaural processing of ITD in the medial superior olive (MSO) is

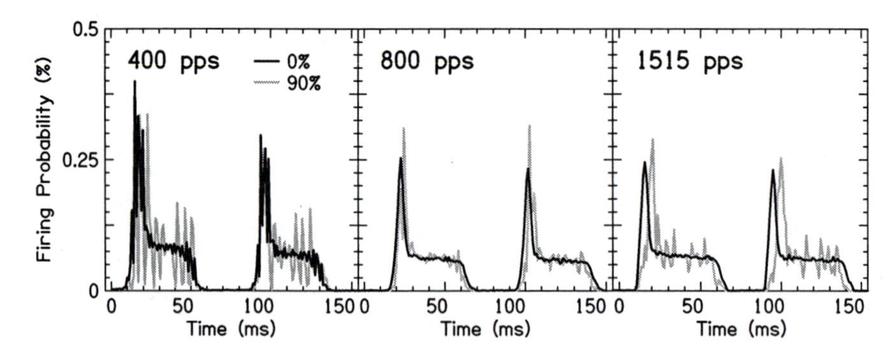

Fig. 29.4 Firing probability of an auditory nerve fiber in response to 400-, 800-, and 1,515-pps trapezoidally-modulated periodic and jittered (90%) pulse trains. The bin width was 1 ms. For clarity, two trapezoids are shown for each pulse rate

based on a form of coincidence detection, CIn represents an appropriate measure of the strength of ITD coding. The CIn is based on the counting of neural spike coincidences from multiple presentations of the same stimulus. We used a bin width of 10 μs and 100 stimulus presentations for each CIn calculation. CIn has a value of one for random uncorrelated spike trains, and a value greater than one for a correlated response.

Figure 29.1d shows that, as expected, the synchrony (CIn) decreases with increasing pulse rate. Introducing jitter increases the synchrony over that for the periodic condition. The AN firing is more synchronized at the relatively low rate of 400 pps when compared to the higher rates. Interestingly, the CIn largely parallels the overall trends in the NH psychophysical data.

29.5 General Discussion

We found that binaurally-coherent jitter improves ITD perception for high-rate pulse trains in NH humans, and can also improve ITD sensitivity in the IC of implanted and NH cats by restoring ongoing firing at high pulse rates. We further found that jitter produces prominent fluctuations in the responses of an AN model to acoustic bandpass-filtered pulse trains, consistent with the sparse, precisely timed firings observed in IC neurons in response to jittered pulse trains.

These response fluctuations produced by jitter were observed specifically in a model of normal hearing, and hence partly reflect AM produced by the overlap of filtered pulses in the physical stimulus as well as the overlap of cochlear filter impulse responses. Nevertheless, such fluctuations are likely robust to exact model assumptions because most filters will transform random frequency modulations of pulse timing into AM. Thus, similar AM could be created in the central auditory system of both NH and CI subjects by neural membrane time constants and synaptic transmission.

Hafter and Dye (1983) proposed the concept of binaural adaptation to account for the observation that only the first few pulses in a high-rate periodic pulse train contribute to ITD perception. This concept is consistent with the low ITD sensitivity for high-rate periodic pulse trains in both CI and NH listeners, and with the onset-only response of most ITD-sensitive IC neurons to high-rate periodic pulse trains in NH and CI animals. Two key questions are: At what site(s) does binaural adaptation occur, and how does jitter overcome the effects of binaural adaptation?

Although firing rate adaptation has been observed with high-rate pulse trains in AN fibers of both NH (Wickesberg and Stevens 1998) and CI (Miller et al. 2008) animals, the effect appears too small and not sufficiently specific to account for binaural adaptation, so that the primary site of adaptation is likely to be central (Wickesberg and Stevens 1998). One possibility is that, for periodic pulse trains at high pulse rates, decreased phase locking makes synaptic inputs too asynchronous to trigger spikes in the coincidence detectors of the binaural pathway. Neurons in the MSO are sensitive to ITD via a binaural coincidence detection mechanism (Yin and Chan 1990). There is also evidence that bushy cells in the ventral cochlear nucleus (VCN), which provide the excitatory inputs to MSO, act as monaural coincidence detectors (Carney 1990). Consistent with this idea, the responses of model AN fibers show minimal phase locking to bandpass filtered pulse trains for rates above 400 pps. Both MSO principal neurons and VCN bushy cells contain low-threshold potassium (K^+) channels thought to play a role in coincidence detection (Manis and Marx 1991; Smith 1995). The relatively steady synaptic inputs to these neurons at high pulse rates may produce a sustained activation of the low-threshold K^+ channels, thereby suppressing subsequent repolarization and blocking firing (Colburn et al. 2008). In this view, jitter improves ITD sensitivity by increasing synchrony in the inputs to the coincidence detectors (Fig. 29.1d), thereby overcoming the depolarization block and allowing MSO neurons to fire more often.

Another possible mechanism underlying binaural adaptation is inhibition at the level of the IC. Smith and Delgutte (2008) showed that a simple model in which IC neurons receive both a brief excitatory input and a delayed long-lasting inhibitory input can qualitatively account for both the onset responses observed in the IC at high pulse rates, and the restoration of sustained firing and ITD sensitivity for amplitude-modulated pulse trains. Presumably, the fluctuations in firing rate introduced by jitter (Fig. 29.4) could overcome the inhibition in the same way as externally-imposed AM. Interestingly, our perceptual stimuli had a low-rate trapezoidal AM, suggesting that they may have induced substantial firing rates in IC neurons, even without jitter. Nevertheless, the improvement in ITD sensitivity suggests that the addition of jitter may have increased IC responses further. Consistent with this idea is that the jitter effect also persists in perceptual experiments for unmodulated high-rate pulse trains (Goupell et al. 2008).

The ITD discrimination performance is better for the NH subjects than for the CI subjects (who were all postlingually deafened). This finding seems paradoxical because phase-locking to sinusoids in the AN is better with electric stimulation than with acoustic stimulation (Dynes and Delgutte 1992), and additional degradation in phase locking is expected with pulse trains in the NH case because of basilar membrane filtering. Perhaps, additional factors like decreased neuronal survival caused the overall worse performance of the implanted subjects. Alternatively, prolonged deprivation of sound inputs prior to cochlear implantation may have induced changes in binaural processing pathways. For example, the balance between excitation and inhibition is known to be altered in deafened animals (Vale and Sanes 2002).

In summary, we have shown that introducing binaurally-coherent jitter to high rate pulse trains improves both psychophysical and neural ITD sensitivity in CI and NH subjects. Although these improvements may be produced by multiple neural mechanisms operating at different processing stages, these mechanisms are likely triggered by the increased synchrony of neural firings produced by jitter in the AN. Regardless of the mechanism, the beneficial effect of jitter offers hope for more effective coding of binaural timing cues in bilateral cochlear implants.

Acknowledgments Supported by NIH Grants RO1 DC005775 and P30 DC005209, and FWF Grant P18401-B15. KEH and BD thank Dr. David Ryugo for providing congenitally deaf cats from his colony.

References

Carney LH (1990) Sensitivities of cells in anteroventral cochlear nucleus of cat to spatiotemporal discharge patterns across primary afferents. J Neurophysiol 64:437–456

Colburn HS, Chung Y, Zhou Y, Brughera A (2008) Models of brainstem responses to bilateral electrical stimulation. J Assoc Res Otolaryngol 10:91–110

Dynes SB, Delgutte B (1992) Phase-locking of auditory-nerve discharges to sinusoidal electric stimulation of the cochlea. Hear Res 58:79–90

Goupell MJ, Majdak P, Laback B (2008) Interaural-time-difference sensitivity to acoustic temporally-jittered pulse trains. J Acoust Soc Am 123:3562

Hafter ER, Dye RH Jr (1983) Detection of interaural differences of time in trains of high-frequency clicks as a function of interclick interval and number. J Acoust Soc Am 73:644–651

Hancock KE, Delgutte B (2004) A physiologically based model of interaural time difference discrimination. J Neurosci 24:7110–7117

Joris PX, Louage DH, Cardoen L, van der Heijden M (2006) Correlation index: a new metric to quantify temporal coding. Hear Res 216–217:19–30

Laback B, Majdak P (2008) Binaural jitter improves interaural time-difference sensitivity of cochlear implantees at high pulse rates. Proc Natl Acad Sci U S A 105:814–817

Laback B, Majdak P, Baumgartner WD (2007) Lateralization discrimination of interaural time delays in four-pulse sequences in electric and acoustic hearing. J Acoust Soc Am 121:2182–2191

Majdak P, Laback B, Baumgartner WD (2006) Effects of interaural time differences in fine structure and envelope on lateral discrimination in electric hearing. J Acoust Soc Am 120:2190–2201

Manis PB, Marx SO (1991) Outward currents in isolated ventral cochlear nucleus neurons. J Neurosci 11:2865–2880

Meddis R (2006) Auditory-nerve first-spike latency and auditory absolute threshold: a computer model. J Acoust Soc Am 119:406–417

Miller CA, Hu N, Zhang F, Robinson BK, Abbas PJ (2008) Changes across time in the temporal responses of auditory nerve fibers stimulated by electric pulse trains. J Assoc Res Otolaryngol 9:122–137

Smith PH (1995) Structural and functional differences distinguish principal from nonprincipal cells in the guinea pig MSO slice. J Neurophysiol 73:1653–1667

Smith ZM, Delgutte B (2007) Sensitivity to interaural time differences in the inferior colliculus with bilateral cochlear implants. J Neurosci 27:6740–6750

Smith ZM, Delgutte B (2008) Sensitivity of inferior colliculus neurons to interaural time differences in the envelope versus the fine structure with bilateral cochlear implants. J Neurophysiol 99:2390–2407

Snyder R, Leake P, Rebscher S, Beitel R (1995) Temporal resolution of neurons in cat inferior colliculus to intracochlear electrical stimulation: effects of neonatal deafening and chronic stimulation. J Neurophysiol 73:449–467

Vale C, Sanes DH (2002) The effect of bilateral deafness on excitatory and inhibitory synaptic strength in the inferior colliculus. Eur J NeuroSci 16:2394–2404

van Hoesel RJ (2007) Sensitivity to binaural timing in bilateral cochlear implant users. J Acoust Soc Am 121:2192–2206

Wickesberg RE, Stevens HE (1998) Responses of auditory nerve fibers to trains of clicks. J Acoust Soc Am 103:1990–1999

Yin TC, Chan JC (1990) Interaural time sensitivity in medial superior olive of cat. J Neurophysiol 64:465–488

Chapter 30
Lateralization of Tone Complexes in Noise: The Role of Monaural Envelope Processing in Binaural Hearing

Steven van de Par, Armin Kohlrausch, and Nicolas Le Goff

Abstract The lateralization of tone complexes presented against a diotic band-pass noise background was measured as a function of level and Interaural Time Difference (ITD) of the tone complex. The bandwidths of the sine-phase tone complex and the noise were 600 Hz, centered at 550 Hz, the component spacing of the tones was 20 Hz. Due to this spectral configuration, none of the tonal components is spectrally resolved, and the binaural cues of the combined stimulus vary with the 20 Hz cycle of the stimulus. Results show that listeners can judge the lateral position of the tone complex down to levels of −5 dB SNR. A second experiment determined masked thresholds of the tone complex in the diotic noise as a function of ITD. Detection thresholds are as low as −20 dB SNR for ITDs of 0.7 ms, and about −7 dB for ITDs of 0 ms, a condition in which only monaural information could be used by the listeners. It seems that despite the *audibility* of the tone complex at rather low levels (−20 dB SNR) based on binaural cues, the *ability to lateralize* requires the audibility of monaural envelope information, which is restricted to levels above about −7 dB SNR. Thus, monaural envelope information may be a prerequisite for the readout of lateralization information corresponding to the tone complex from the binaural display.

Keywords Binaural Hearing • Lateralization

30.1 Introduction

The binaural system is an important component of our hearing function. It is well known that in the presence of noise, a target stimulus is heard better when it has a different interaural configuration than the noise (e.g., Hirsh 1948); a phenomenon

S. van de Par (✉)
Philips Research Laboratories, High Tech Campus 36, 5656 AE, Eindhoven, The Netherlands
e-mail: Steven.van.de.Par@philips.com

E.A. Lopez-Poveda et al. (eds.), *The Neurophysiological Bases of Auditory Perception*,
DOI 10.1007/978-1-4419-5686-6_30, © Springer Science+Business Media, LLC 2010

referred to as binaural *detection*. Binaural hearing allows us to determine the lateral direction of a sound source (e.g., Yost 1981) even when it is presented against a background of noise (Lorenzi et al. 1999), referred to as binaural *lateralization*. Finally, binaural hearing allows us to improve discrimination between different stimuli that are presented against a background of noise provided that stimuli and noise have different interaural configurations (van de Par et al. 2004), a condition referred to as binaural *discrimination*.

Binaural detection is believed to be mediated by changes in the interaural properties of the composite stimulus when the target signal is added to the noise masker. For example, in the case of a diotic noise masker, an interaurally time-delayed target signal that spectrally and temporally overlap with the masker will result in Interaural Time Delays (ITDs) and Interaural Level Differences (ILDs) that provide cues for detecting the presence of the target signal. It is important to realize that detection of the target signal based on these cues does not imply that any of the stimulus properties of the target signal are heard such as its lateral position or its temporal structure. Van de Par et al. (2007b) demonstrated that a tone complex with an ITD of 0.6 ms presented in diotic noise can be detected at a level that is about 7 dB lower than the level needed for determining the lateral orientation of the noise plus tone complex stimulus.

Since the usefulness of the hearing function in everyday life probably does not depend so much on hearing the presence of a certain signal per se, but rather on being able to determine the position and nature of the stimulus, it is of interest to study the lateralization of a target stimulus in background noise. In our previous ISH contribution (van de Par et al. 2007a; see also Schimmel et al. 2008), we presented results that suggest that listeners can determine the lateral orientation of a tone complex that is mixed with noise of the same level even when both signals are temporally and spectrally fully overlapping.

In this study, the main experiment will aim at measuring the perceived lateral position of an interaurally time-delayed tone-complex signal that is presented in noise at several signal-to-noise ratios. Whereas in our previous study, listeners only had to indicate the lateral orientation (left, right), now also the perceived "magnitude" of the laterality will be determined. If listeners were to depend only on analyzing the short-term cross correlation pattern, one would have to assume that the binaural processing stage can somehow infer the lateral position of the tone-complex from this pattern. However, when it is assumed that monaural envelope information is driving the readout of the short-term cross-correlation pattern, perceived laterality of the tone-complex would be based on those time instances, where the tone-complex dominates in the composite stimulus. An implication would be, however, that when the tone complex cannot be heard anymore in the monaural envelope of the composite stimulus, the ability to hear the lateral position of the tone complex would be lost. Thus, the main question of this study is whether binaural processing alone is sufficient for determining the lateral position of a tone complex presented simultaneously with a noise masker.

As a reference for the lateralization data, in an additional experiment, *detection* thresholds of the tone complex in diotic noise are measured for various ITDs of the tone-complex. For an ITD of zero, the lowest level at which the monaural cues of the tone complex can be detected will be known. For ITDs different from zero, it will be known at what level the interaural difference cues resulting from adding the tone complex to the noise can still be heard.

Besides monaural envelope cues that can drive the readout of the binaural display, the temporally changing pattern of binaural cues may also drive the selection of the best moments to readout the short-term cross-correlation pattern for obtaining information about the lateral position of the tone complex. This assumption obtains some credibility because in an earlier ISH contribution (van de Par et al. 2004), we showed that listeners can discriminate an antiphasic tone complex from an antiphasic noise that was presented in diotic noise at levels sufficiently low to assume that no monaural cues could contribute to this discriminability. Such a cue may also be available in our main lateralization experiment to drive the readout of the short-term cross correlation pattern. Therefore, another experiment is conducted to determine the threshold levels at which listeners were able to *discriminate* between a target tone-complex and a noise with a certain ITD presented in diotic noise. This experiment will indicate whether in this stimulus configuration, the binaural display alone may provide enough temporal information to drive lateralization of interaurally time delayed tone-complexes in noise.

30.2 Detection Experiment

In this first experiment, thresholds for detecting a tone-complex signal presented simultaneously with a diotic masker will be measured. The tone-complex signal is presented with various ITDs.

In order to determine the detection thresholds of the tone-complex, a 3-interval-forced choice 2-down 1-up adaptive staircase method was used to adjust the level of the tone complex. The initial step size of the adaptive track was 8 dB, which was halved after each second reversal of the track until a step size of 1 dB was reached. The median value of the following six reversals served as the threshold value. Each condition was measured three times.

All three intervals contained the diotic noise masker. One randomly selected interval in addition contained the tone-complex signal. Listeners had to indicate which interval contained the tone complex and were subsequently informed whether their answer was correct or not.

The tone complex that was used consisted of 30 components in sine phase spanning a range from 260 to 840 Hz with 20-Hz spacing. The diotic Gaussian masking noise covered a very similar spectral range from 250 to 850 Hz and was always presented at a level of 65 dB SPL. Both the tone-complex signal and the noise masker had a duration of 400 ms and were gated with 30-ms raised cosine on and off-set ramps. Thresholds were measured for various values of the tone-complex ITD ranging from 0 to 1 ms. Three normal hearing subjects participated in this experiment of which two were authors of this paper.

30.3 Results and Discussion

In Fig. 30.1, detection thresholds of the tone-complex are shown as a function of its ITD. As can be seen, thresholds decrease when the ITD of the tone complex increases. When the ITD of the tone complex is zero, only monaural detection cues are present, and thresholds are about 7 dB lower than the level of the diotic noise masker (65 dB SPL). For ITDs of 0.7 ms and larger, thresholds saturate at a level about 20 dB below the level of the masker.

These results are in line with the general observation that differences in the interaural parameters of the masker and target signal lead to lower thresholds. The improvement in detection of the tone-complex can be assumed to be the result of interaural disparities of the composite stimulus and can therefore not be interpreted to indicate that the listeners can hear the presence of the tone-complex as such, but rather as a change in the spatial image due to the presence of the tone complex. The next experiment is designed to find out at what levels the tone-complex can be discriminated from a noise with the same level and ITD.

30.4 Discrimination Experiment

In this experiment, thresholds for discriminating between a target tone-complex signal and a target noise signal were measured which were presented simultaneously with a diotic masker. The two target signals always had the same level and were presented with the same ITD for each condition. As a result, the interaural cues that were generated by the addition of the target tone-complex or target noise in itself were not enough to perform the discrimination task. Listeners had to be able to also determine the difference in temporal structure of the tone complex and the noise targets. The same three subjects of the previous experiment participated in the discrimination experiment.

The same 3-interval-forced choice 2-down 1-up adaptive staircase method was used as in the previous experiment to determine the thresholds for discriminating

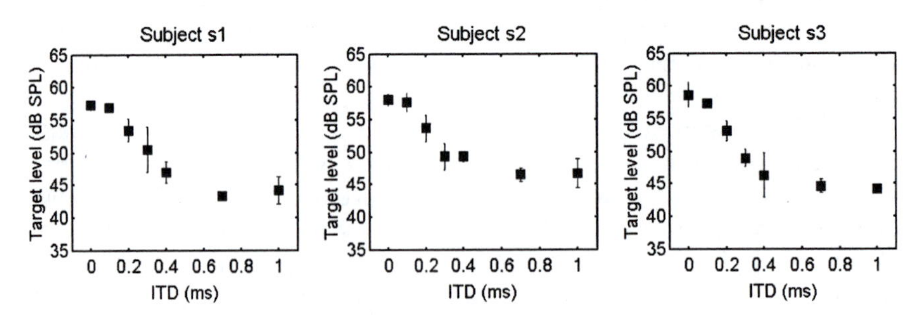

Fig. 30.1 Detection thresholds for the tone complex presented in diotic noise masker as a function of the ITD of the tone complex. The noise masker had a level of 65 dB SPL

between the target tone complex and target noise. All three intervals contained the diotic noise masker at a constant level of 65 dB SPL. One randomly selected interval in addition contained the tone-complex signal, the remaining two intervals contained the target-noise signal, both with the same adaptively varied level. Listeners had to indicate which interval contained the target tone complex and were subsequently informed whether their answer was correct or not.

The same tone complex and noise masker were used as in the previous experiments. The diotic Gaussian masking noise was always presented at a level of 65 dB SPL. Target ITDs of the tone complex ranged from −1 to +1 ms.

30.5 Results and Discussion

In Fig. 30.2, thresholds for discriminating between the target tone complex and target noise are shown as a function of the ITD of the targets. As can be seen, there is some decrease in the discrimination thresholds toward ITDs larger than zero. The discrimination thresholds of this experiment, however, tend to be up to 7 dB higher as compared to the detection thresholds of the previous experiment. Since for the discrimination task, listeners were presented with targets that had the same level and ITD in each interval, interaural disparities by themselves were not sufficient to discriminate between intervals containing the target tone complex and the target noise. In addition, the listeners needed to be able to hear the difference in the temporal envelope structure of the tone complex and the noise. Thus, the discrimination thresholds reported here can be interpreted as reflecting the tone-complex level that is needed to recognize some of the temporal envelope characteristics of the tone complex in the presence of background noise.

The results also indicate that the presence of binaural cues improves discriminability of the two target signals. This observation is in line with results of our earlier ISH contribution (van de Par et al. 2004), where we used a similar discrimination paradigm for NoSπ stimuli instead of the NoSτ stimuli of the present study.

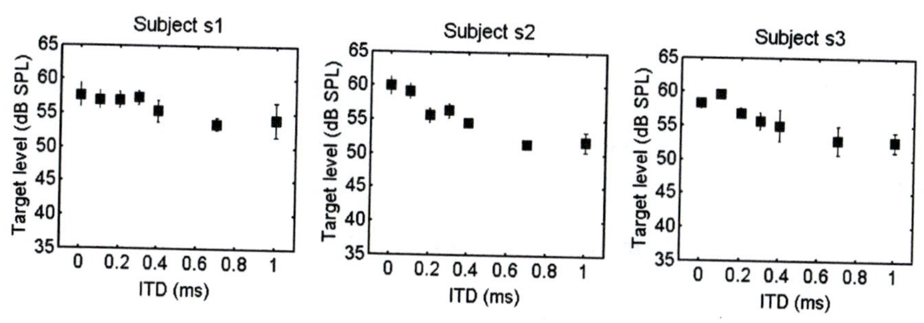

Fig. 30.2 Thresholds for discriminating between a target tone complex and a target noise presented against a diotic noise masker as a function of ITD of the targets signals. The noise masker had a level of 65 dB SPL

Interestingly, discrimination thresholds for conditions with zero ITD are close to the detection thresholds for the same ITD reported in the previous experiment. This may indicate that the primary cues for detecting the *presence* of the tone complex in noise correspond to the cues used for *discriminating* the tone complex from the noise target.

30.6 Lateralization Experiment

The main purpose of the lateralization experiment was to investigate whether listeners were able to determine the lateral position of an interaurally time-delayed tone complex signal presented in background noise, and how this ability depends on the signal-to-noise ratio. The same three subjects of the previous experiments participated in the lateralization experiment.

In order to determine the perceived lateral position of an interaurally time-delayed tone-complex signal in diotic noise, listeners had to match the adjustable ITD of a tone complex presented in isolation with the tone complex presented in noise. Each trial consisted of three intervals; the first and third interval contained the diotic noise plus the tone-complex signal with a certain fixed target ITD. The second interval contained a tone complex presented in isolation with an adjustable ITD. After each trial, listeners had to indicate whether the adjustable tone complex in isolation (second interval) sounded more to the left or right as compared to the target tone complex presented together with the noise (first and third interval). The tone complex had the same level in all three intervals. No Feedback was provided to the listeners.

Using a 1-up 1-down adaptive staircase method, the ITD was determined that gave the same perceived lateralization for the tone-complex signal presented in isolation and for the target tone complex presented together with diotic noise. The initial step size of the adaptive staircase procedure was 80 μs, and it was reduced with a factor two after each second reversal of the adaptive track until a step size of 10 μs was reached. The median value of the following six reversals then served as the measured adjusted ITD value. Adjusted ITDs were measured three times for each condition.

The same tone complex and noise masker were used as in the previous experiments. The diotic Gaussian masking noise was always presented at a level of 65 dB SPL, the tone complex was presented at levels of 65, 60, and 55 dB SPL. Target ITDs of the tone complex ranged from −1 to +1 ms.

Besides using a diotic noise masker, an additional set of conditions containing noise maskers with a roving ITD was used. The ITD was roved from trial to trial, always being identical in the first and third interval of a trial. The roved ITD was drawn from a uniform distribution ranging from −0.7 to 0.7 ms. For the roved condition, measurements were restricted to a target level of 65 dB SPL.

30.7 Results and Discussion

Results of the lateralization experiment are shown in Fig. 30.3 for three listeners. When we concentrate first on the condition with the highest level of the tone complex (65 dB SPL) using a diotic masker noise (open squares), we see a reasonable correspondence between the target and adjusted ITD, especially for ITDs in the range from –0.4 to 0.4 ms. This indicates that despite the presence of the diotic masking noise, listeners were able to recognize the lateral position of the tone complex. When we consider the comparable condition with roving ITD masking noise (filled squares), a very similar result is observed. Apparently, the large variability in the ITD of the masking noise that was presented simultaneously with the tone complex did not affect the ability to hear the lateral position of the tone complex, which would be in line with the idea that the tone complex is heard as a separate and localized auditory object.

When the tone complex was at a level of 60 dB SPL (circles), a similar result is obtained for the first two subjects, the third subject shows considerably more misalignment between the target and adjusted ITD, also the first subject did not perceive the correct lateral position for the largest ITDs. For the lowest target level of 55 dB SPL (crosses), all listeners show a systematic deviation from the target ITD, where the adjusted ITD is always smaller than the target ITD. This latter result should be seen with reference to the previous two experiments. At a level of 55 dB SPL, monaural cues are not sufficient for detecting the presence of the tone-complex signal, monaural detection thresholds are about 58 dB. When binaural cues are present, considerably lower detection thresholds are observed. However, these cannot be interpreted as if listeners were able to hear the nature of the tone complex at such low levels. As indicated by the discrimination experiment, depending on ITD, levels of at least 52 dB SPL are needed for listeners to hear that a tone complex target is present instead of a noise target. This low level is only achieved

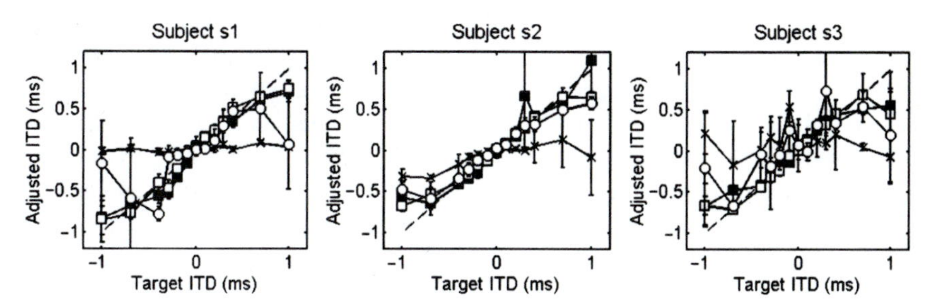

Fig. 30.3 Adjusted ITD (*y*-axis) as a function of target ITD (*x*-axis) for tone complexes presented at 65 dB SPL (*squares*), 60 dB SPL (*circles*), and 55 dB SPL (*crosses*). *Open symbols* represent conditions with a diotic noise masker, while *filled squares* represent roving ITD noise maskers. The *dashed lines* represent the situation where there would be a perfect match between target and adjusted ITD

for the largest ITDs suggesting that this discrimination at least partly was based on binaural cues. Apparently, these binaural cues were not sufficient to reliably determine the lateral position of the tone complex.

30.8 General Discussion

Results of the lateralization experiment showed that listeners can correctly determine the lateral position of an interaurally time delayed tone complex in a diotic noise masker even when the tone complex is 5 dB lower in level than the noise. For lower signal levels, the lateral position of the tone complex could not be determined anymore, despite the fact that for large ITDs, the temporal characteristics of the target tone complex could be heard. This can be concluded from the ability of listeners to discriminate a target tone complex from a target noise within the diotic noise masker, suggesting that some aspects of the temporal structure of the tone complex are reflected and retrieved for these conditions. Apparently, the binaural advantage that is obtained in the discrimination experiment is not a sufficient basis for perceiving the lateral position of the tone complex. However, the level range in which the tone complex could be correctly lateralized coincides with the range in which the tone complex can be detected based on monaural envelope cues which points to the role of monaural envelope cues in determining the lateralized position of the tone complex.

The ability to hear the correct lateral position of the tone complex did not seem to be affected by roving the ITD of the masking noise demonstrating the robustness of auditory localization in noisy conditions also when the properties of the noise are not known a priori. The ability to determine the lateral position of the tone complex in these conditions with an unpredictable masker ITD suggests that listeners do segregate the tone complex from the noise and attribute the correct lateral position to the segregated signal.

Since the tone complex and the masking noise covered the same spectral range, there are no auditory channels that are dominated by the tone complex which could allow an estimate of the lateral position of the tone complex based on the average interaural cues in this auditory channel. However, since the tone complex had a rather low periodicity of 20 Hz, temporal fluctuations in the binaural cues may provide cues for lateralizing the tone complex as suggested in our earlier work (van de Par et al. 2007a; Schimmel et al. 2008). According to this line of reasoning, the monaural envelope peaks of the sine-phase tone complex will dominate the composite stimulus every 50 ms when an envelope peak occurs in the tone complex. At these time instances, the binaural cues of the tone complex will dominate the composite stimulus and may accurately reflect the lateral position of the tone complex.

It is an interesting question whether the binaural display alone would provide enough information to extract the lateral position of the tone complex at higher target levels (60 and 65 dB SPL) were monaural cues of the tone complex are also perceived. When we consider the condition with a diotic masker, the pattern of changing binaural cues may be selectively read at the moments where the largest interaural differences occur, i.e., when the envelopes of the tone complex peaks. At these moments, the composite stimulus is dominated by the tone complex, and therefore it will mostly reflect the ITD of the tone complex. Such a purely binaural processing strategy that selects the moments with the largest ITDs would not be expected to be successful when the masker ITD is roved. When the masker ITD exceeds the ITD of the target tone complex, the largest interaural differences would reflect those of the noise. Thus, it seems that another means is needed to select the ITDs that correspond to the tone complex from the composite stimulus. It is interesting to note in this perspective that determining the lateral position of the tone complex was only possible when monaural envelope cues were sufficiently strong to allow detection of the tone complex suggesting that monaural envelope cues are used to select the binaural cues corresponding to the tone complex.

The temporal resolution that is required for resolving the dynamic pattern of changing ITDs in the current experiment seems considerably higher than those reported in some studies (e.g., Grantham and Wightman 1978). There, a rate of change of maximally a few Hz allowed the listeners to hear movement of the source. It was found, however, that rather than resolving an ongoing pattern of changes such as discussed in the study of Grantham and Wightman (1978), the detection of brief changes in ITDs seems to be governed by a better temporal resolution (Bernstein et al. 2001) more in line with the rate of change present in the stimuli of our experiments.

The comparison of the detection and discrimination experiments demonstrates that the classical binaural detection experiments do not necessarily determine the level at which essential properties of the signal such as its temporal envelope structure and lateral position can be determined. Such properties are related to important auditory functions such as recognition and localization of sound sources, where it seems that both are needed to interpret an auditory scene. For both these properties to be heard, considerably higher levels are needed than the levels that are sufficient for the detection of a signal. In the current stimulus configuration, these perceptual properties arise only at levels where monaural envelope cues of the target signal already start to play a role (cf. detection experiment for ITD = 0), and the current experiments provide no evidence that localization of sound sources in noisy backgrounds can be based on binaural cues alone and are more in line with the notion that monaural information steers the readout of binaural cues.

References

Bernstein LR, Trahiotis C, Akeroyd MA, Hartung K (2001) Sensitivity to brief changes of interaural time and interaural intensity. J Acoust Soc Am 109:1604–1615

Grantham DW, Wightman FL (1978) Detectability of varying interaural temporal differences. J Acoust Soc Am 63:511–523

Hirsh IJ (1948) The influence of interaural phase on interaural summation and inhibition. J Acoust Soc Am 20:11–19

Lorenzi C, Gatehouse S, Lever C (1999) Sound localization in noise in normal-hearing listeners. J Acoust Soc Am 105:1810–1820

Schimmel O, van de Par S, Kohlrausch A, Breebaart J (2008) Sound segregation based on temporal envelope structure and binaural cues. J Acoust Soc Am 124:1130–1145

van de Par S, Kohlrausch A, Breebaart J, McKinney M (2004) Discrimination of different temporal envelope structures of diotic and dichotic target signals within diotic wide-band noise. In: Pressnitzer D, de Cheveigné A, McAdams S, Collet L (eds) Auditory signal processing: physiology, psychoacoustics, and models. Springer, New York

van de Par S, Schimmel O, Kohlrausch A, Breebaart J (2007a) Source segregation based on temporal envelope structure and binaural cues. In: Kollmeier B, Klump G, Hohmann V, Langemann U, Mauermann M, Uppenkamp S, Verhey J (eds) Hearing – from basic research to applications. Springer, Heidelberg

van de Par S, Kohlrausch A, Schimmel O (2007b) A comparison of detection of lateralization of interaurally time-delayed signals in the presence of diotic masking noise. In: Calvo-Manzano A, Pérez-Lopéz A (eds) Proceedings of the 19th international Congress on acoustics, ICA'07, Madrid, Spain, 6 p

Yost WA (1981) Lateralization position of sinusoids presented with interaural intensive and temporal differences. J Acoust Soc Am 70:397–409

Chapter 31
Adjustment of Interaural-Time-Difference Analysis to Sound Level

Ida Siveke, Christian Leibold, Katharina Kaiser, Benedikt Grothe, and Lutz Wiegrebe

Abstract To localize low-frequency sound sources in azimuth, the binaural system compares the timing of sound waves at the two ears with microsecond precision. A similarly high precision is also seen in the binaural processing of the envelopes of high-frequency complex sounds. Both for low- and high-frequency sounds, interaural time difference (ITD) acuity is to a large extent independent of sound level. The mechanisms underlying this level-invariant extraction of ITDs by the binaural system are, however, only poorly understood. We use high-frequency pip trains with asymmetric and dichotic pip envelopes in a combined psychophysical, electrophysiological, and modeling approach. Although the dichotic envelopes cannot be physically matched in terms of ITD, the match produced perceptually by humans is very reliable, and it depends systematically on the overall sound level. These data are reflected in neural responses from the gerbil lateral superior olive and lateral lemniscus. The results are predicted in an existing temporal-integration model extended with a level-dependent threshold criterion. These data provide a very sensitive quantification of how the peripheral temporal code is conditioned for binaural analysis.

Keywords Binaural hearing • Loudness • Envelope • Spike timing

31.1 Introduction

Precise temporal coding is the hallmark of the auditory system. Like no other sensory modality, the auditory system relies on the neural analysis of spike timing for both object localization and identification. On the other hand, the peripheral auditory system

---

L. Wiegrebe (✉)
Biocenter, University of Munich, Munich, Germany
e-mail: wiegrebe@zi.biologie.uni-muenchen.de

E.A. Lopez-Poveda et al. (eds.), *The Neurophysiological Bases of Auditory Perception*, DOI 10.1007/978-1-4419-5686-6_31,

has to cope with a huge variability in the loudness of sounds; the dynamic range of natural acoustic input spans at least five orders of magnitude. Psychophysically, it has been shown that sound localization is stable across a wide range of stimulus levels (Blauert 1997). The neural processing underlying this level-invariant encoding of temporal stimulus properties is still poorly understood, in particular since even phase locking, as a basic measure of precise temporal coding, is sensitive to changes in overall sound level in the auditory nerve (Joris et al. 2004). Such level dependence is also seen in the temporal encoding of the envelopes of high-frequency complex tones (Dreyer and Delgutte 2006).

Recent studies on temporal integration of transient sounds have revealed that integration of the stimulus pressure envelope, rather than its intensity can explain both perceptual temporal integration and neural first-spike latency in the auditory nerve at threshold intensity (Heil and Neubauer 2003). It is unclear, however, whether this pressure-envelope integration is sufficient to explain neural integration and how the binaural system uses the temporal response characteristics to form a spatial percept.

In this study, we recruit the binaural system's exquisite temporal sensitivity to quantify temporal integration preceding binaural analysis. We exploit recent findings that interaural time differences (ITDs) of envelopes of high-frequency carriers, "transposed tones," are binaurally analyzed with a precision similar to that of low-frequency tones (Bernstein and Trahiotis 2002, 2003, 2007). At high carrier frequencies, transient stimuli can be constructed which carry different binaural temporal properties although they occupy the same frequency region in the two ears. If the envelopes of these stimuli are binaurally incongruent, the pressure waves cannot be physically matched by simple interaural time shifts. Nevertheless, the binaural system can produce a reliable perceptual match that in turn reflects the way the stimulus envelopes are processed preceding binaural analysis.

31.2 Methods

31.2.1 Psychophysics

31.2.1.1 Stimuli

The current experiments recruit temporally asymmetric pips as illustrated in Fig. 31.1. These pips consist of a high-frequency, pure-tone carrier modulated with a temporally asymmetric envelope. If such a pip is presented binaurally but temporally reversed in one ear, (Fig. 31.1), the pips cannot be physically matched by an ITD. If the sound level is low (Fig. 31.1a), only the tips of the stimulus envelope exceed threshold and consequently, one would expect that the stimulus in the ear with the steeper rise time has to be delayed by an interaural onset difference (IOD) to compensate for the rise-time difference between the ears. If the sound level is high (Fig. 31.1b), the envelope exceeds threshold very quickly and thus, a much smaller IOD would be needed to compensate for the rise-time difference.

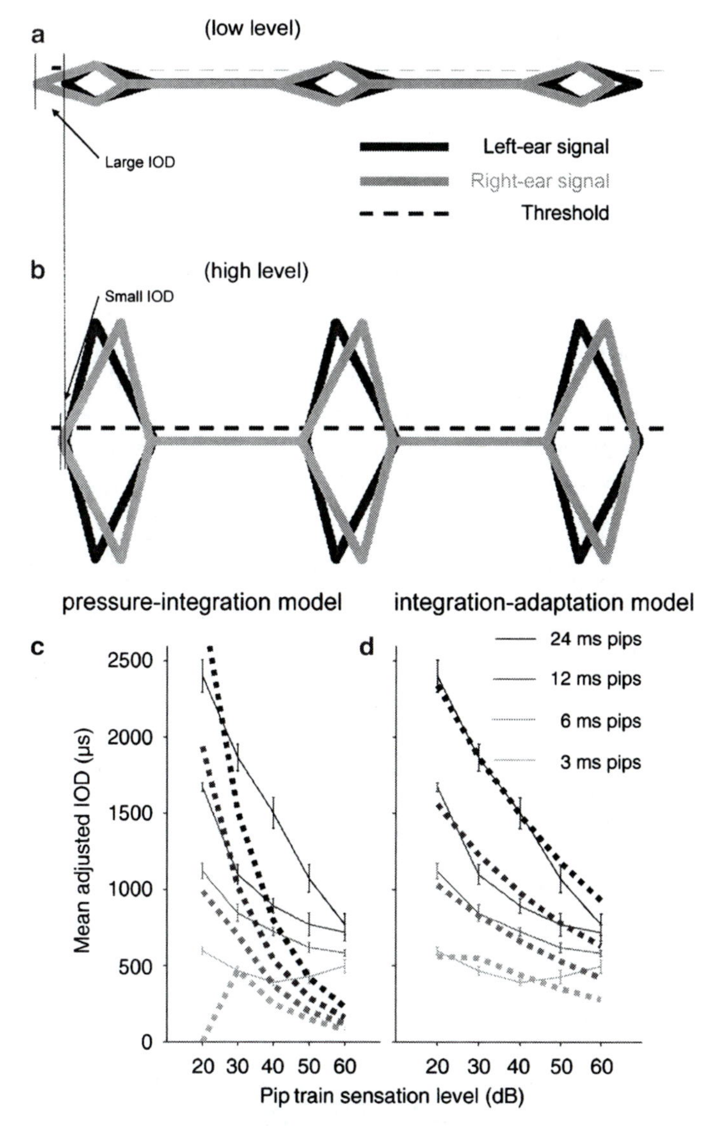

Fig. 31.1 Illustration of the experimental paradigm (**a**, **b**) and psychophysical and modeling results (**c**, **d**). Listeners were asked to adjust the IOD of binaural pip train to compensate for the interaural differences in the rise- and fall times of the pip envelopes (*black* left ear; *grey* right ear). It is assumed that at low stimulus levels, (**a**), a large IOD is required for the compensation, whereas at higher levels, (**b**), smaller IODs are sufficient. The psychophysically adjusted IODs (**c**, **d**) confirm this assumption: adjusted IODs systematically decrease with increasing sensation level and decreasing overall pip duration

The used stimuli were trains of such pips with a carrier frequency of 5 kHz. The interval between the pips was fixed at 10 ms. For the left-ear stimuli, the rise time was 1, 2, 4, or 8 ms, and the decay time was always double the rise time, resulting in tone pips with a duration of 3, 6, 12, or 24 ms. The pip trains in the right ear were either identical to the left-ear trains, or they were temporally reversed. Experiments were run with a continuous background noise (8 dB SPL/Hz) low-pass filtered at 1,500 Hz to preclude the binaural analysis of low-frequency aural distortion products.

31.2.1.2 Procedure

In a two-alternative, forced-choice paradigm without feedback, listeners were asked to judge the lateralization of a test stimulus when compared with a reference stimulus. The reference stimulus was always presented first. It consisted of a diotic pip train with identical rise and decay times in both ears and no IOD. In the standard condition, the test stimulus consisted also of a pip train with identical rise and decay times in the two ears but with an initially randomized IOD between ±500 μs. In the test condition, the pip train in the right ear was temporally reversed when compared to the left ear. As in the standard condition, listeners were asked to adjust the IOD of these pip trains, which cannot be physically matched to the perceived position of the diotic reference stimulus. In each trial, listeners judged whether the test stimulus was lateralized left or right to the reference stimulus. In an adaptive procedure, the IOD of the test stimulus was changed to compensate for the lateralization of the test stimulus. For the first and second reversal (a change in lateralization direction), IOD was changed in steps of 160 μs, after the second reversal, the step size was reduced to 80 μs, and after the fifth reversal, it was reduced to 40 μs. The adjusted IOD in a given experimental run is given as the mean IOD across reversals six to eleven. Individual data are based on at least six runs per experimental condition. An experimental session consisted of six runs in randomized order, three with temporally reversed test stimuli in the right ear, the other three without reversion. Lateralization was measured as a function of the individually measured pip-train sensation level.

31.2.1.3 Listeners

At least four normal hearing listeners, aged between 21 and 35 of both genders took part in each experiment. Listeners for a pip duration of 6 ms were different from those that acquired all the other data.

31.2.2 Electrophysiology

31.2.2.1 Animals

Recordings were obtained from 18 adult Mongolian gerbils (Meriones unguiculatus). Single cells were recorded in two different brainstem nuclei, the LSO ($n=32$) and the DNLL ($n=66$). All experiments were approved according to the German Tierschutzgesetz (AZ 55.2-1-54-2531-57-05). The detailed methods in terms of surgical preparation, acoustic stimulus delivery, stimulus calibration, and recording techniques have been described previously (Siveke et al. 2006).

31.2.2.2 Recording Procedure and General Neural Characterization

Extracellular single-cell responses were recorded as described earlier (Siveke et al. 2006). The recording sites of 24 of the 32 LSO neurons that we analyzed were histologically verified. Stimuli were generated at 48 kHz sampling rate in Matlab and delivered to the ear-phones (Sony MDR-EX70 LP). Using pure tones, we first determined audio-visually the neuron's characteristic frequency (CF) as that frequency which elicited a response at the lowest intensity, neuronal threshold (thr). For the pip-train experiments, it was first ensured that LSO neurons were IID sensitive, that DNLL neurons were ITD sensitive and that neither phase locked to CF tones.

31.2.2.3 Pip-Train Stimulation

The pips consisted of the same envelopes as in the psychophysical experiments. The carrier frequency was set to the cell's CF. The pips were presented in a 4-s train at a repetition rate of 40 Hz. For the LSO, five repetitions of the pip trains and temporally reversed pip trains were presented monaurally on the ipsilateral ear in a randomized sequence. Pip trains were presented with different pip durations (3, 6, 12, and 24 ms) and different sound levels (18–50 dB above threshold in 8 dB steps).

For the DNLL, the pip trains were presented binaurally: In the contralateral ear the temporal structure of the envelope was constant; in the ipsilateral ear, the stimulation was identical to the LSO monaural stimulation, i.e., the temporal envelope of the pip train was equal to the contralateral side or temporally reversed. Furthermore in the ipsilateral ear, the inter-pip interval was manipulated to generate time-variant envelope ITDs: Within 1 s of pip-train stimulation, the onset of the ipsilateral pip was varied across a range spanning ± one quarter of the pip duration. This onset range was centered on the best ITD of the neuron. For a 40 Hz pip rate, the ITD range thus consisted of 40 different ITDs. As for the LSO stimulation, pip trains were presented at different pip durations and different sound levels above threshold.

31.2.2.4 Analysis

To calculate the time shift in the period histograms recorded with the standard and time-reversed pips in the LSO, we performed a cross-correlation between the two histograms. A Gaussian was fitted to the cross-correlation function, and the delay at the maximum of the fit was taken as the time shift. For the binaural DNLL responses, the response rates of the neurons as a function of the IOD were fitted by Gaussians for both standard and ipsilaterally reversed pip trains. The differences between the IODs at which the two Gaussians had their maxima were taken as the IOD change. From the population of recorded neurons ($N=50$), only the neuronal data with a correlation coefficient >0.7 between the data and the Gaussian fits were used for further analysis ($N=30$).

31.3 Results

The IODs adjusted by the listeners to centralize the asymmetric pip trains for the 6-ms pip duration (2 ms rise and 4 ms decay time, or vice versa) are shown with solid lines in Fig. 31.1c. At a sensation level of 20 dB and a pip duration of 24 ms, the listeners adjusted the temporally reversed 24-ms pip train in the right ear (with a 16 ms rise and 8 ms decay time) to start about 2,400 μs earlier than the pip train in the left ear (with a 8 ms rise and 16 ms decay time). With increasing sensation level, the adjusted IOD decreased monotonically but, even at a high sensation level of 60 dB, the listeners still required an IOD of about 800 μs to compensate for the temporal reversion and the resulting different rise times in the two ears.

Not surprisingly, the adjusted IOD systematically depended on the overall pip duration. The adjusted IOD decreases with decreasing pip duration for all sensation levels. In all conditions even at the highest tested sensation levels, the adjusted IOD remained above 500 μs.

To study, whether the IOD shifts can be explained by the latency difference $L_l - L_r$, between the left and right-ear evoked neuronal activity, we employed two model variants of the pressure-integration model (Heil and Neubauer 2003). In contrast to the original model, in which latencies are predicted based on the sound pressure wave, we assume integration of the envelope $e(t)$ of the sound pressure wave. Then, the binaural latencies L_l and L_r of action potentials are assumed to be determined by the threshold crossings of the respective integrated envelopes S. Mathematically, this maps to the implicit equation

$$\Theta = S(L_x) = \int_0^{Lx} dt\, [e_x(t)]^k,$$

in which we allowed an additional exponent k as a fit parameter. In the left model variant (left), the threshold Θ is a constant and independent of sound level. This model thus has two fit parameters, k and Θ. In the second model variant (right), the threshold is a function of the signal level. Specifically, we modeled the threshold as

$$\Theta = \Theta_0 + \Theta_1\, \alpha,$$

where a denotes the amplitude of the envelope $e(t)$ in units of Pascal. The second model thus has three fit parameters, k, Θ_0 and Θ_1.

The model variant with a level independent threshold cannot reproduce the large perceived IODs at high sound levels, although the model qualitatively fits the result in that longer pips produce larger IODs (dashed lines in Fig. 31.1c). In contrast, the model variant that includes a level-dependent threshold can account for the saturation of perceived IODs at high sound levels (dashed lines in Fig. 31.1d).

Neural responses to ipsilateral monaural pip trains from 32 cells with CFs above 2 kHz were obtained from the gerbil LSO. The raster plot (Fig. 31.2b) shows the accurate locking of the cell to the 6 ms pip envelope. The period histograms (2C) show that the spike timing depends on the direction of the asymmetric envelope in that the cell fires earlier when the pip has a 2-ms rise and a 4-ms decay (black). With increasing stimulation level, the response rate increases, but the phase difference between the period histograms of the two envelope conditions decreases. This is quantified in the cross-correlograms of the period histograms in Fig. 31.2d. The peak in the cross-correlogram shifts from 810 μs at a sound level of 26 dB above threshold to 357 μs at 50 dB. These main features are preserved in the LSO population average, shown with filled symbols in Fig. 31.2e. Comparable to the psychophysical results, the time shifts extracted from the period histograms decrease with decreasing pip duration and increasing sound level.

Recordings in the DNLL were obtained to get an estimate of how the gerbil's binaural system evaluates the time shifts in the peripheral representation of the pip trains. We systematically searched for high-CF binaural units which showed sensitivity to envelope ITDs in response to Gaussian noise. We used binaural pip trains and introduced a 1-Hz binaural beat into the envelopes of the 4-s pip trains (see Methods). The interaction of this beat with the envelope ITD tuning of a DNLL cell is seen in the raster plot of Fig. 31.3b.

As in all other experiments, we recorded responses both with standard pips in both ears and with ipsilaterally reversed pips (Fig. 31.3a). This stimulation paradigm allows for the evaluation of both phase locking to the pip envelope period (25 ms) and of the binaural locking to the envelope beat period (1,000 ms). As in the LSO, also the DNLL cells locked reliably to the 25 ms period of the pip trains (Fig. 31.3b). Again, the response rates increase with increasing sound level, but this increase was more pronounced for the time-reversed pips (grey), which produced particularly weak responses at low levels. The binaural envelope beat was constructed to cover a range of ± one quarter of the pip duration within 1 s. Period histograms constructed with the envelope beat period of 1 s are shown in Fig. 31.3c. As within each 1 s period, the envelope IOD changes linearly, the abscissa can be relabeled as an IOD axis. A quantitative analysis of the IOD changes introduced by time reversion of the ipsilateral pips is obtained by fitting a Gaussian to the period histograms in the two conditions and extracting the difference between the fit maxima. The extracted time shifts decrease with increasing sound level (Fig. 31.3d). DNLL population data of the neurally extracted time shifts are shown in Fig. 31.3e. While the DNLL data show qualitatively the same trends as the psychophysical and LSO data, time shifts are significantly smaller than estimated both psychophysically and from the LSO data.

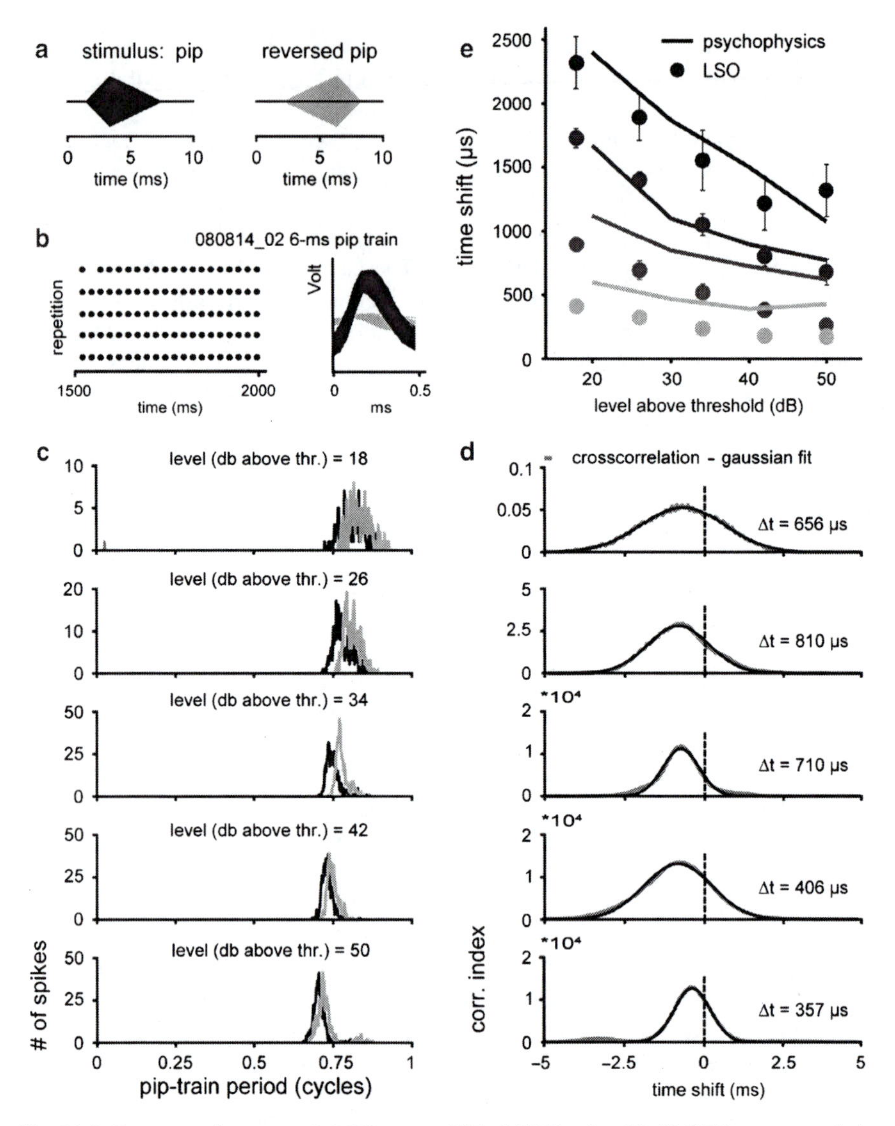

Fig. 31.2 Response of an example LSO neuron (CF=3.5 kHz; thr=20 dB SPL) to monaural pip train stimuli. (**a**) Schematic diagram of the presented standard and the time reversed 6-ms pip. (**b**) Example reaster plot to show the reliable locking of the LSO neuron to the pip-train envelope. (**c**) Period histograms for the standard (*black*) and the time reversed (*grey*) 6-ms pip trains are plotted for five different sound pressure levels ranging from 18 to 50 dB above neuronal threshold. (**d**) cross-correlations of the period histograms are plotted as correlation index versus time shift. The time shift is derived from the time delay (Δt) of the correlation maximum as obtained from a Gaussian fit (*black line*). (**e**) Estimated time shifts from the LSO neural population (N=32; *filled symbols*) compared to the psychophysical data (*solid lines*). Estimated time shifts are in good agreement with perceptually adjusted IODs

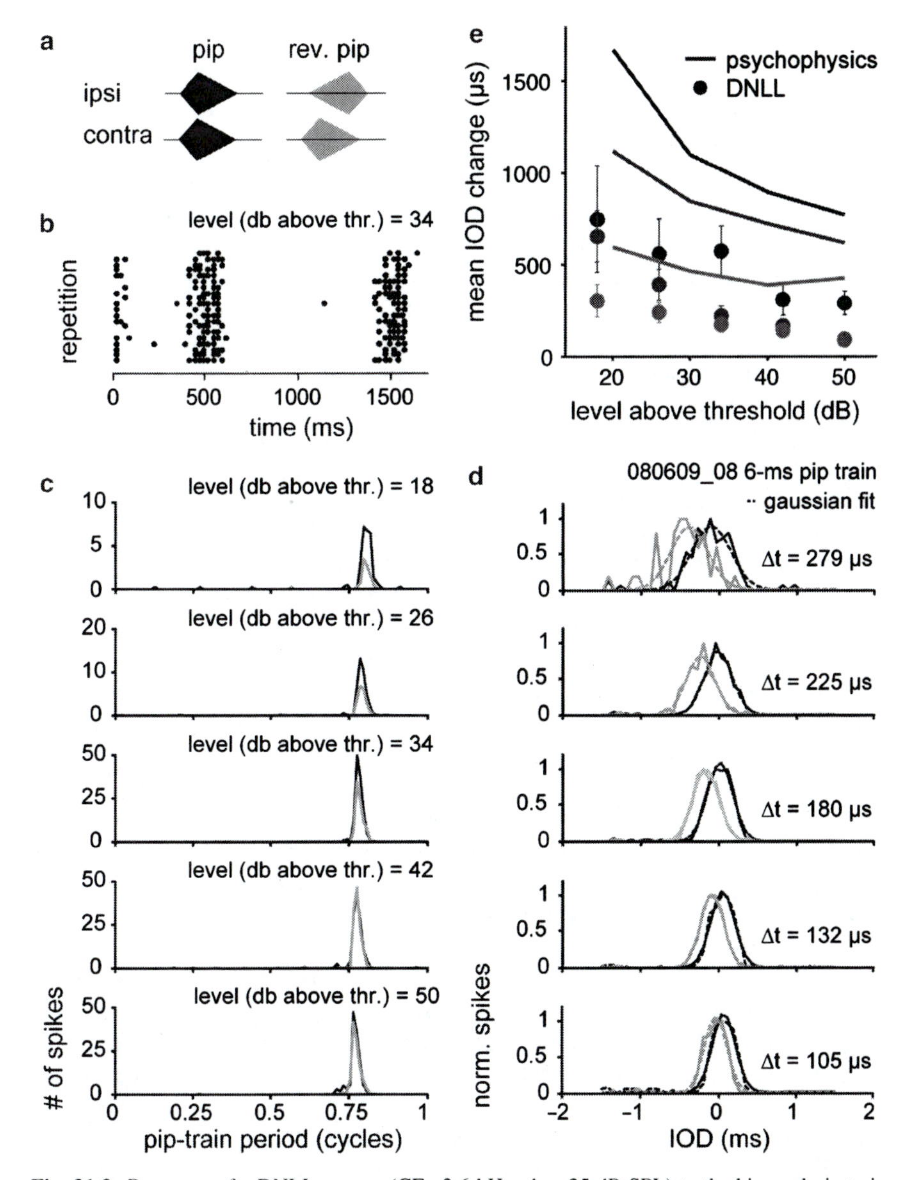

Fig. 31.3 Response of a DNLL neuron (CF = 3.6 kHz; thr = 25 dB SPL) to the binaural pip train stimuli. The schematic diagram of the presented binaural standard and the time-reversed pips are shown as insets are shown in (**a**). The raster plot (**b**) shows a defined acoustical response to a particular envelope IOD of the 6-ms pip train stimuli, which is, according to the "1 Hz beat" stimulation, repeated every second. Period histograms for the standard (*black*) and the time reversed (*grey*) 6-ms binaural pips are plotted for five different sound pressure levels ranging from 18 to 50 dB above neuronal threshold in (**c**). The tuning of the neuron to a particular envelope IOD is revealed by the period histograms constructed with the envelope beat period of 1 s (**d**). Within each second of the envelope period, the IOD changes linearly, and the abscissa is relabeled as IOD axis. The IOD change (Δt) between the envelope period histograms obtained from the standard (*black*) and the time reversed pip stimulation is obtained by the difference between the shifts of the two histogram maxima obtained from Gaussian fits (*dotted lines*). Population data for the time shifts estimated from the binaural stimulation data are shown in (**e**)

31.4 Discussion

Here, we have presented a binaurally asymmetric stimulus paradigm to investigate the level dependence of ITD processing. The paradigm exploits two key features of auditory processing: the precise encoding of the envelopes of high-frequency carriers and the exquisite precision of ITD analysis. The used pip stimuli are composed of linear up- and downward ramps that constitute the envelope for a sinusoidal high-frequency carrier. The stimuli were investigated with respect to human binaural perception as well as their neuronal representations that are revealed by electrophysiological recordings from the LSO and DNLL of anesthetized gerbils. These three sets of experimental data are in qualitative agreement in that an IOD change induced by the binaural asymmetry of the envelopes increases with signal duration and decreases with signal level.

While the monaurally recorded gerbil LSO data match the human psychophysical data reasonably well, the DNLL data predict much smaller time shifts. Several possible reasons can be envisaged: First, it is conceivable that the recorded monaural LSO responses may not be identical to the inputs of the binaural MSO cells whose output we recorded in the DNLL. Second, to accommodate the large stimulus protocol for the binaural stimulation, a dynamic stimulation paradigm was implemented. Although the acoustic motion introduced by this paradigm is relatively slow, it is conceivable that this dynamic component may interfere with the ITD extraction. Note, however, that time-variant ITDs are encoded with high fidelity in the gerbil DNLL (Siveke et al. 2008). Third, we only inverted the ipsilateral tone pips, but we did not invert the contralateral pips and leave the ipsilateral pips unchanged. These two paradigms would yield the same results if the ipsi- and contralateral inputs converging at the binaural processing stage are identical. If, however, ipsi- and contralateral inputs are not identical (Brand et al. 2002), it is possible that the binaurally observed time shifts depended on whether the ipsi- or contralateral rise time was shallower. This hypothesis remains to be tested.

To check whether the observed IODs can be explained by binaural latency differences, we employed two variants of a temporal integration model by Heil and Neubauer (Heil and Neubauer, 2003; Heil et al. 2007) that was originally proposed to explain the level dependence of the first spike latencies in the auditory nerve. Our experimental data are inconsistent with a simple temporal integration model (Heil and Neubauer 2003). However, including a level dependent threshold into such a model gives a reasonable match between data and model predictions. Heil and Neubauer proposed a neural correlate of their pressure-envelope integration to occur at the inner-hair cell, auditory-nerve synapse. Recent work has addressed the physiological basis of the pressure-envelope integration in the Heil model. In a constructive discourse, it was shown that the interplay between the stochasticity of synaptic events and a short time constant of synaptic transmission can explain the apparently long time constant of temporal integration at threshold (Meddis 2006a, b; Krishna 2006).

The physiological basis for the proposed level-dependent threshold cannot be assessed in the framework of the present study. Our data reveal that this level of adjustment must already be present in the input to the SOC. Dreyer and Delgutte

(2006) showed that at the level of the cat auditory nerve, the temporal representation of transposed tones, with an envelope modulation comparable to the current pip trains, deteriorates with increasing sound level. In contrast, the current data show that at the level of the gerbil LSO, the temporal representation of the pip-train envelopes do not deteriorate with increasing sound level. The current data are thus also in agreement with the psychophysical data showing that sensitivity to envelope ITDs elicited by transposed tones is stable over a wide range of sound levels (Dreyer and Oxenham 2008). These findings argue for a refinement of temporal envelope encoding at the level of the cochlear nucleus, similar to the refinement of phase locking to pure tones (Joris et al. 2004).

Any ITD of the envelope of a high-frequency sound is accompanied by a short-term IID. While it has been shown that IID sensitivity improves with increasing duration (Blauert 1997), the time constant of perceptual IID analysis, i.e., its temporal resolution is entirely unclear. If we assume a long time constant of IID extraction (>20 ms), the IIDs generated by the current stimulation would be negligible. If however, the IID time constant is short, time variant IIDs are generated by the current stimuli. This would be the case for all stimuli with an ITD. To address the question of the time scale of IID processing, it is conceivable to replicate the current psychophysical experiment asking listeners not to compensate the rise-time differences with an ITD but with an IID. Thus, the potential interaction of the current paradigm with IID extraction opens new opportunities to study the dynamics of binaural processing.

Taken together, our experimental paradigm provides insights into the level dependence of binaural processing. Both the perceptual and electrophysiological data can be explained in an existing temporal-integration model extended with a level dependent threshold criterion. These data provide a very sensitive quantification of how the peripheral temporal code is conditioned for binaural analysis.

Acknowledgments The authors would like to thank Alain de Cheveigne and Peter Heil for fruitful suggestions and discussions on the topic. Supported by the DFG and the Bernstein Center for Computational Neuroscience.

References

Bernstein LR, Trahiotis C (2002) Enhancing sensitivity to interaural delays at high frequencies by using “transposed stimuli”. J Acoust Soc Am 112:1026–1036

Bernstein LR, Trahiotis C (2003) Enhancing interaural-delay-based extents of laterality at high frequencies by using “transposed stimuli”. J Acoust Soc Am 113:3335–3347

Bernstein LR, Trahiotis C (2007) Why do transposed stimuli enhance binaural processing? Interaural envelope correlation vs envelope normalized fourth moment. J Acoust Soc Am 121:EL23–EL28

Blauert J (1997) Spatial hearing: the psychophysics of human sound localization. MIT Press, Cambridge MA

Brand A, Behrend O, Marquardt T, McAlpine D, Grothe B (2002) Precise inhibition is essential for microsecond interaural time difference coding. Nature 417:543–547

Dreyer A, Delgutte B (2006) Phase locking of auditory-nerve fibers to the envelopes of high-frequency sounds: implications for sound localization. J Neurophysiol 96:2327–2341

Dreyer AA, Oxenham AJ (2008) Effects of level and background noise on interaural time difference discrimination for transposed stimuli. J Acoust Soc Am 123:EL1–EL7
Heil P, Neubauer H (2003) A unifying basis of auditory thresholds based on temporal summation. Proc Natl Acad Sci U S A 100:6151–6156
Heil P, Neubauer H, Irvine DR, Brown M (2007) Spontaneous activity of auditory-nerve fibers: insights into stochastic processes at ribbon synapses. J Neurosci 27:8457–8474
Joris PX, Schreiner CE, Rees A (2004) Neural processing of amplitude-modulated sounds. Physiol Rev 84:541–577
Krishna BS (2006) Comment on "Auditory-nerve first-spike latency and auditory absolute threshold: a computer model" [J Acoust Soc Am 119:406–417 (2006)]. J Acoust Soc Am 120:591–593
Meddis R (2006a) Auditory-nerve first-spike latency and auditory absolute threshold: a computer model. J Acoust Soc Am 119:406–417
Meddis R (2006b) Reply to comment on "Auditory-nerve first-spike latency and auditory absolute threshold: a computer model". J Acoust Soc Am 120:1192–1193
Siveke I, Ewert SD, Grothe B, Wiegrebe L (2008) Psychophysical and physiological evidence for fast binaural processing. J Neurosci 28:2043–2052
Siveke I, Pecka M, Seidl AH, Baudoux S, Grothe B (2006) Binaural response properties of low-frequency neurons in the gerbil dorsal nucleus of the lateral lemniscus. J Neurophysiol 96: 1425–1440

Chapter 32
The Role of Envelope Waveform, Adaptation, and Attacks in Binaural Perception

Stephan D. Ewert, Mathias Dietz, Martin Klein-Hennig, and Volker Hohmann

Abstract The human binaural system is capable of using interaural temporal disparities in the fine structure and the envelope of sounds. At high frequencies, the lack of phase-locking to the fine structure restricts the use of interaural timing disparities to envelope cues only. However, it is still unclear which specific envelope waveform features ranging from "classical" onsets (and offsets) to periodic variations such as sinusoidal amplitude modulations play a role as dominant cues. Moreover, a major drawback of these classic features is that a variation of parameters such as overall level, bandwidth, and modulation depth or modulation rate necessarily cause a covariation of "secondary" parameters such as, e.g., attack and decay times, steepness, pause, and hold durations (off-/on-times). Here, psychophysical measurements and auditory model predictions to assess the role of these secondary parameters are presented. Just noticeable differences in the interaural time difference were measured for amplitude-modulated 4-kHz tones with systematic modifications of the envelope. The results indicate that the attack times and the pause times prior to the attack are the most important features. A model including neuronal adaptation prior to the binaural stage is suggested. The model was also tested with the dichotic pip trains from the concurrent study by Siveke et al. presented at this ISH.

Keywords Envelope waveform • Auditory models • Interaural time difference • Discrimination • Binaural • Amplitude modulation • Adaptation

32.1 Introduction

Interaural time differences (ITDs) are an important binaural cue for the human auditory system, helping to localize sound sources in space. At low frequencies, the fine structure of a sound is preserved and the ITD can be used for a variety of stimuli

---

S.D. Ewert (✉)
Medizinische Physik, Universität Oldenburg, 26111, Oldenburg, Germany
e-mail: stephan.ewert@uni-oldenburg.de

E.A. Lopez-Poveda et al. (eds.), *The Neurophysiological Bases of Auditory Perception*,
DOI 10.1007/978-1-4419-5686-6_32,

including pure tones. At high frequencies, above about 1,500 Hz, however, the lack of phase-locking causes the effective extraction of the stimulus' envelope in the auditory system. The envelope can convey binaural information in the form of onsets (and offsets) of a sound (referred to as "onset" or "transient" interaural delay, e.g., McFadden and Pasanen 1976) and in the form of "ongoing" ITDs which occur for sounds with time-varying temporal envelopes. In the latter case, ITD sensitivity on the basis of interaural delays of the envelope is observed (e.g., Klumpp and Eady 1956; Henning 1974; McFadden and Pasanen 1976; Nuetzel and Hafter 1976, 1981).

A number of recent studies have demonstrated that the processing of ongoing envelope ITDs for high-frequency stimuli depends on the envelope waveform and the degree of fluctuations in the envelope. Certain envelope waveforms as produced by "transposed" stimuli (van de Par and Kohlrausch 1997), which provide high-frequency auditory channels with envelope-based information mimicking the fine-structure based information typically available in low-frequency channels, show enhanced processing of the interaural disparities in the high-frequency region (e.g., van de Par and Kohlrausch 1997; Bernstein 2001; Bernstein and Trahiotis 2002). Earlier, the degree to which the envelopes of the stimuli fluctuate was suggested to describe binaural processing performance of ITDs conveyed by high-frequency stimuli (Dye et al. 1994). They quantified the degree of envelope fluctuations by the normalized fourth moment of the envelope (Hartmann 1987; Hartmann and Pumplin 1988). Bernstein and Trahiotis (2007), however, demonstrated that for transposed stimuli, the "internal" interaural envelope correlation accounts for binaural processing performance rather than the normalized fourth moment of the envelope as a measure of the degree of envelope fluctuations.

For arbitrary high-frequency stimuli, it still remains an open question which specific envelope features determine binaural processing performance for ongoing envelope ITDs. The use of "classical" features provided by periodic and/or deterministic envelope variations like sinusoidal amplitude modulations (SAMs) or transposed tones have the drawback that a modification of, e.g., overall level, bandwidth, and modulation depth or modulation rate necessarily cause a covariation of other "secondary" parameters like, e.g., attack and decay times, steepness, pause and hold durations (off-/on-times). These parameters also appear to be potentially suited features for the characterization of binaural auditory function and may additionally close the gap between onsets (and offsets) and ongoing temporal disparities in a generalized view of envelope features.

In the present study, psychoacoustic measurements were conducted employing 4-kHz tones with systematic modifications to their envelope waveform. The "secondary" parameters of the envelope were varied separately, and the just noticeable difference (JND) in the ITD was measured for each condition.

Simulations were performed with different model approaches: The normalized cross-correlation (NCC) model (Bernstein and Trahiotis 2002, 2007) and three models that determine the maximum difference between the left and right channel after different stages of neural adaptation.

32.2 Experiments

32.2.1 Method

32.2.1.1 Listeners

Five normal-hearing listeners ranging in age from 25 to 29 years participated in the experiments. Two of the subjects were the authors MD and MKH. All subjects received several hours of listening experience prior to the final data collection.

32.2.1.2 Apparatus and Stimuli

Subjects were seated in a double-walled sound attenuating booth in front of a computer keyboard and monitor. Subjects listened via Sennheiser HD580 headphones driven by a Tucker Davis HB7 headphone buffer. Signal generation and presentation during the experiments were computer controlled using the alternative forced-choice (AFC) software package for MATLAB, developed at the University of Oldenburg. The stimuli were digitally generated at a sampling rate of 48 kHz and converted to analog signals by a high-quality 24-bit sound card and external digital-to-analog converter (RME DIGI96/8 PAD, RME ADI-8 PRO).

Amplitude modulated 4-kHz pure-tone stimuli were used to measure the just noticeable ITD. The envelopes were periodic at a modulation rate of 35–100 Hz. The lower waveform in Fig. 32.1a represents of a single period of the envelopes used in the present study. Each envelope period was comprised of a pause segment, an attack segment (raised-cosine ramp), a hold segment, and a decay segment (raised-cosine ramp). The durations of the segments were varied independently to study their individual importance. A SAM resulted for pause and hold duration of 0 ms. Otherwise, a "square-wave" modulation (SWM) with variable duty cycle and attack/decay times resulted. The shortest attack/decay durations used in this study were 1.25 ms to control for spectral broadening of the stimuli. The ITD was applied to the envelope waveform as a whole (indicated by the dotted line) or to the attack only as indicated by the dashed line in Fig. 32.1a. Control experiments with ITDs in the decay segment only indicated that the ITD of the decay had virtually no influence on the lateralization. Stimuli were 500 ms in duration and were gated with 125-ms raised-cosine ramps to minimize the salience of onset cues. A low-pass filtered noise (fifth-order Butterworth at 1,000 Hz, white spectrum up to 200 Hz, −3 dB per octave slope up to 1,000 Hz) was simultaneously presented to preclude the listeners' use of any information at low frequencies (e.g., Nuetzel and Hafter 1976, 1981; Bernstein and Trahiotis 2007). The low-pass noise was 600 ms in duration, resulting in 50-ms temporal fringes during which it was gated with 50-ms raised-cosine ramps and had a level of 45 dB SPL. The level of the SAM stimuli was 60 dB SPL. All other stimuli had the same peak amplitude as the SAM stimuli, except for one

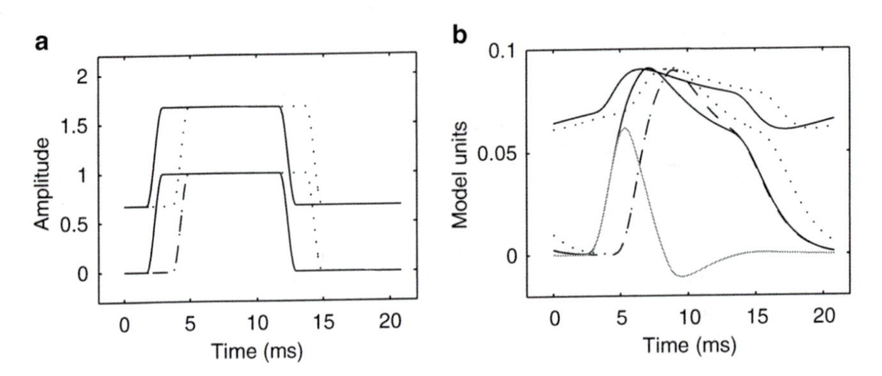

Fig. 32.1 (**a**) Schematic representation of a single stimulus envelope period (*solid line*) with an ITD applied to the attack only (*dashed line*) and the whole waveform (*dotted line*). Note that for the waveform ITD, the *dotted line* coincides with the *dashed line* during the attack. The upper traces are for the offset condition. (**b**) Adapted signals (AD model) of the respective envelope conditions in (**a**). The *gray line* at the bottom indicates the interaural difference which is used in the detector

stimulus with a doubled peak level (66 dB SPL). Additionally, in the "offset" condition (reduced modulation index of 0.43), the peak amplitude was scaled by a factor of 1.67 as shown by the upper waveform trace in Fig. 32.1a.

32.2.1.3 Procedure

An adaptive, two-interval, two-AFC procedure was used in conjunction with a one-up three-down tracking rule to estimate the 79.4% correct point on the psychometric function (Levitt 1971). Feedback was provided after each trial. Listeners responded via the computer keyboard or mouse. The ITD started at 2 ms, and the initial factor by which the ITD was reduced was 2. After the second and fourth reversals, the factors were 1.4 and 1.1, respectively. The adaptive run then continued for a further six reversals, and the threshold was defined as the geometric mean of the ITDs at those last six reversals. Four threshold estimates were obtained and geometrically averaged from each listener in each condition. Final thresholds reported here are the geometric means across listeners.

32.2.2 Results

The experimental data are shown in Fig. 32.2 as closed triangles. Figure 32.2a shows the just noticeable ITD as a function of the attack duration of the SWM (with the pause, hold, and decay duration fixed at 8.75, 8.75, and 1.25 ms, respectively). An almost linear dependence between the JND and the attack duration is

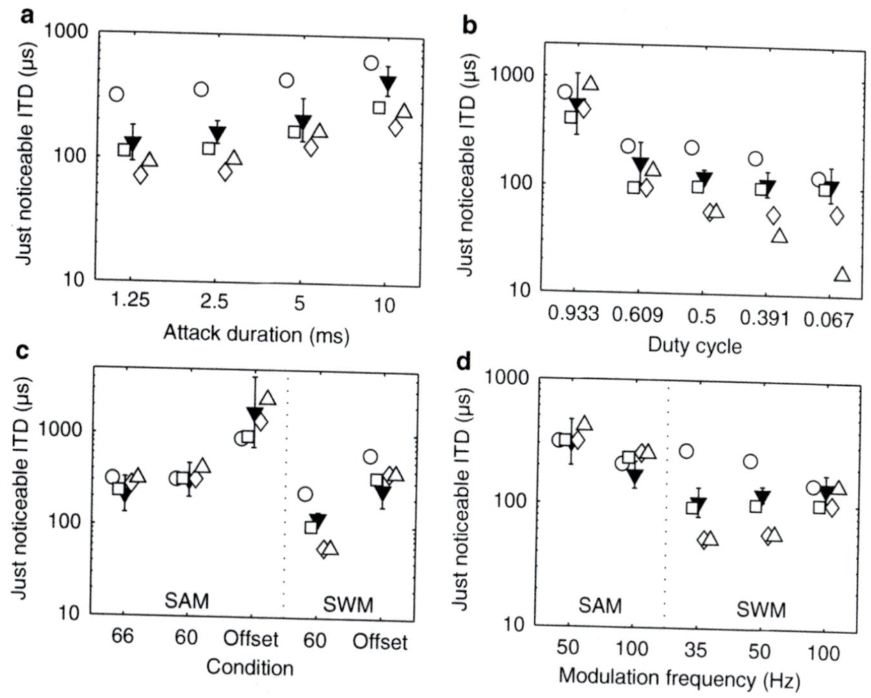

Fig. 32.2 Just noticeable ITDs for different stimulus conditions. (**a**) Effect of attack duration for SWM. (**b**) Effect of duty cycle for SWM. (**c**) Level and envelope offset variations for SAM and SWM. (**d**) Effect of modulation frequency for SAM and SWM. Experimental data are shown as closed triangles. Model predictions (*open symbols*) for the NCC, ADL, SADL, and AD model are indicated as *circles*, *triangles*, *diamonds*, and *squares*, respectively

observed, indicating that the ITD sensitivity of the binaural system is highly correlated to the steepness of the rising flank. Figure 32.2b shows data for different duty cycles of the SWM. The modulation rate was 50 Hz and the attack and decay duration was 1.25 ms. The highest JND is observed for the condition with the highest duty-cycle, in which the decay is directly followed by the attack without additional pause duration. Otherwise, the data are nearly independent of the duty cycle. Figure 32.2c shows the effect of level (60 and 66 dB, indicated as condition 60 and 66) and envelope offset for 50-Hz SAM (left) and 50-Hz SWM (right) with a duty cycle of 0.5 and attack and decay durations of 1.25 ms. A decreased sensitivity is observed for the offset condition. Particularly for the SAM condition, the JND increases by a factor of five in the offset condition while the steepness of the ramps is identical to condition 60. Additionally, it can be seen that the 6-dB level increase (doubled steepness) in condition 66 causes a slightly increased ITD sensitivity.

In Fig. 32.2d, the effect of modulation frequency is shown SAM (right) and SWM (left) with a duty cycle of 0.5 and attack and decay durations of 1.25 ms.

It can be seen that the JND is decreased by a factor of about 2 when the SAM rate is increased from 50 to 100 Hz. For the SAM stimuli, the increased modulation frequency results in an increased steepness of the flanks which was shown to influence the JND in Fig. 32.2a. In the case of the SWM, the steepness is constant and no effect on the JNDs is observed.

32.3 Model Predictions

32.3.1 Models

Four models were used to predict the experimental data. All models shared the (monaural) preprocessing consisting of middle-ear filtering (first order bandpass with cutoff frequencies of 500 and 8,500 Hz), auditory filtering (4-kHz, fourth-order Gammatone filter, Patterson et al. 1987), half-wave rectification and low-pass filtering (770-Hz fifth-order, Breebaart et al. 2001), and peripheral power-law compression with an exponent of 0.4 (e.g., Dietz et al. 2008).

The NCC model (Bernstein and Trahiotis 2002, 2007) operated directly on the output of the preprocessed signals at both ears. The model was fit to correctly predict psychoacoustic JND for a reference condition (50-Hz SAM).

In the adaptation loop (ADL) model, the preprocessed signals were passed to a series of five ADLs (Püschel 1988) as used in, e.g., Dau et al. (1996) and Jepsen et al. (2008). JNDs were determined by the ITD, at which the maximum difference of the adapted signals in the left and right channel exceeded a threshold value. Figure 32.1b shows a schematic plot of the adapted signals of the corresponding envelope conditions in Fig. 32.1a. The upper traces are for the offset condition. The grey line in Fig. 32.1b indicates the interaural difference. The maximum difference was calculated from a 100-ms steady-state part of the adapted signals (300–400 ms after stimulus onset). The threshold value was set individually for the ADL model, and two variations of the model given below, to match the psychoacoustic JND for the 50-Hz SAM condition.

The single ADL (SADL) model used only the first (and fastest) ADL of the ADL model with a 5-ms time constant of the low-pass filter stage in the loop. The same detector stage as in the ADL model was used.

The adaptation (AD) model employed a feed-forward mechanism to simulate adaptation instead of a feedback loop. The output of the preprocessing stage was divided by an rms-normalized and first-order low-pass filtered version of it (time constant of 5 ms). The output of the normalized low-pass stage was set to not be less than a threshold value of 0.9. The detector stage was again identical to the ADL and SADL model.

All four models include a first-order, 150-Hz low-pass filter prior to the binaural stage in order to account for monaural processing limitations of high-frequency envelopes (Kohlrausch et al. 2000; Ewert and Dau 2000).

32.3.2 Model Predictions

Model predictions are shown in Fig. 32.2 (open symbols) together with the psychoacoustic data (closed triangles). The NCC, ADL, SADL, and AD model are indicated by the circles, triangles, diamonds, and squares, respectively. The dependence on the attack duration in Fig. 32.2a is generally well described by all models. The ADL and SADL models, however, show generally too low JNDs, while the NCC model shows too high JNDs. Figure 32.2b shows that the data of the duty-cycle experiment are best described by the AD model and with minor deviations by the SADL model. In contrast to the data, the NCC model predicts a strong increase in sensitivity when the hold duration is reduced to zero (right-most condition) while particularly the ADL model overestimates the JND decrease with decreasing duty cycle (increased pause duration). In Fig. 32.2c, it can be seen that all models account quite well for the increased JNDs at the offset conditions with reduced modulation depth. However, the NCC model is per definition independent of the overall level. This is in contrast to the data of this study and in contrast to the data of Siveke et al. (2009). The ADL and SADL models also tend to underestimate the level dependence. The predictions of all models describe the trends in the data for different SAM frequencies in the left part of Fig. 32.2d well. As mentioned in the psychoacoustic results, the JNDs decrease with increasing steepness of the attacks. However, for the SWM stimuli in the right part of Fig. 32.2d, the NCC model does also predict decreasing JNDs with increasing modulation frequency which is not observed in the data. The AD model shows slightly increasing JNDs as in the data, caused by the slightly shorter pause duration at higher modulation frequencies. Again, the ADL and SADL models overestimate the importance of the pause duration: with decreasing modulation frequency, the JND predictions of these models decrease too much.

Figure 32.3 shows model predictions along with the data for the pip-train stimuli given in Fig. 31.1 of Siveke et al. (2009). The different shades of gray ranging from

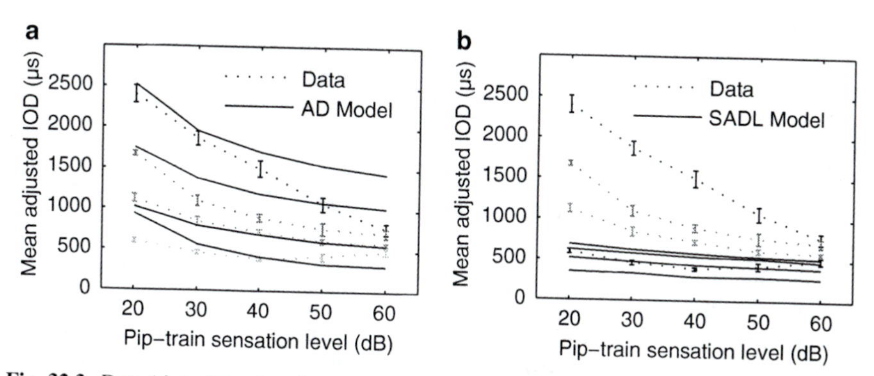

Fig. 32.3 Data (*dotted lines*) and model predictions (*solid lines*) for the pip-train experiment of Siveke et al. (2009, see their Fig. 31.1). The adjusted interaural onset difference (IOD) for a perceived mid-line image is plotted as a function of sensation level for 24, 12, 6, and 3-ms pip trains indicated by the *black* to *light grey* shading. (**a**) Shows the AD model predictions and (**b**) the predictions of the SADL model

black to light gray are for the 24-, 12-, 6-, and 3-ms pip trains, respectively. For details, the reader is referred to Siveke et al. (2009). The AD model in Fig. 32.3a agrees well with the data for the 6-ms pip train. For longer pip train durations and high levels, the mean adjusted interaural onset difference is overestimated. Figure 32.3b shows that the SADL model generally fails to predict the data, the same holds for the ADL model (not shown).

32.4 Discussion

The psychoacoustic results in Fig. 32.2 demonstrate an important role of the attack durations (rise time) and pause durations preceding the attack for the sensitivity to envelope ITD. In case of the pause durations, the main effect is already reached for pause durations as short as about 5 ms (Fig. 32.2b). Beyond 5 ms, further increase of the pause duration has little effect on the just noticeable ITD. A reduction of the modulation index to $m=0.43$ by replacement of the pause (silence) with a region of reduced carrier intensity (offset condition) leads to an increase in JND (Fig. 32.2c). A comparable effect was found in Stellmack et al. (2005) for their 128-Hz condition, while no difference between $m=1$ and $m>0.3$ was found for 300 Hz in Stellmack et al. (2005) and Nuetzel and Hafter (1981). The slight increase in sensitivity with level observed in Fig. 32.2c is in line with psychoacoustic data by Kohlrausch et al. (2000), where lower detection thresholds for monaural SAM were found and with Dietz et al. (2009), where an increased salience of envelope ITDs with increasing level was observed. Physiologic data on level dependence are inhomogeneous. In line with the current psychoacoustic data, Siveke et al. (2009) found a more precise tuning of binaural LSO neurons with increasing level. However, Dreyer and Delgutte (2006) found that the synchronization index of the auditory-nerve discharge pattern decreased with increasing level for both SAM and transposed tones.

The fact that the increase of modulation frequency does only result in a reduction of the just noticeable ITD for SAM and not for SWM (fixed attack duration), indicates that it is not the period duration of the envelope which determines the just noticeable (ongoing) ITD as implied by, e.g., the NCC model (Bernstein and Trahiotis 2002). Only in case of the SAM, the potential rise-time feature is directly correlated to the period duration. The stimuli used here might also be useful to test potential limitations of the IPD model (Dietz et al. 2009), where only the width of the modulation filters can account for differences between SAM and SWM. Furthermore, the current data support the hypothesis that the high sensitivity to ITDs in transposed tones might be less related to their special design to mimic the auditory nerve pattern of low-frequency pure-tones but rather to their pronounced pause regions and attacks when compared to SAM. With longer pause durations and steeper attacks (e.g. Fig. 32.2b), JNDs are even lower than for transposed tones in this study (not shown). Taken together, the data indicate that the slew rate of the attacks, additionally pronounced by preceding gaps, is the dominant temporal envelope feature for ongoing ITDs. Such behavior is expected if the binaural system operates on an internal signal after a form of neural adaptation.

The model results suggest that adaptation prior to the binaural stage is generally suited to account for the data. The ADL model includes too long time constants (up to 500 ms) which are not appropriate to be assumed prior to the binaural stage (see Fig. 32.2b) as in, e.g., Breebaart et al. (2001). With respect to the integration time constant responsible for the time course of the adaptation, there is no dramatic difference between the AD (feed-forward) and the SADL (feedback) model which both use the same time constant of 5 ms. Differences between the models (particularly for the pip trains in Fig. 32.3) are related to the strong "overshoot" limitation integrated in the AD model (by means of a threshold value for the output of the low-pass filter stage). Here, the SADL model shows a much more pronounced overshoot resulting in an excessive slew-rate boost of the attacks in the internal representation. The slew rate is reduced after adaptation by the 150-Hz low-pass filter, however, it is also plausible that there is an absolute slew-rate limitation in the neural responses independent of overall level. Such a limitation is not accounted for by the current model approaches. The missing effect of modulation index for $m>0.3$ on just noticeable ITDs for the rather high envelope rates of 300 Hz in Stellmack et al. (2005) and Nuetzel and Hafter (1981) is in line with limitations for envelope processing beyond 150 Hz prior to the binaural stage.

The NCC model shows a modulation rate dependent prediction in case of the SWM (Fig. 32.1b) and is too insensitive in Fig. 32.1a. Additionally, the effect of duty cycle cannot be predicted correctly. The NCC model is by design capable to explain the results as long as the just noticeable ITD is inversely related to the spectral bandwidth of the stimuli. The model was not applied to the pip-train stimuli since it is level independent.

In general, the feed-forward adaptation (AD) model shows the best agreement with the data. The major difference to the ADL model is the more severe overshoot limitation in the AD model. All three adaptation models tested here share the same binaural difference detector which was developed as a functional concept. However, the concept is also physiologically plausible: The binaural difference detector can be realized with excitatory-inhibitory cells, which are typical cells in the lateral superior olive, the region which is assumed to encode temporal disparities in the stimulus envelope (e.g., Joris and Yin 1995).

Acknowledgments This work was supported by the German Research Foundation (Deutsche Forschungsgemeinschaft) SFB/TRR 31.

References

Bernstein LR (2001) Auditory processing of interaural timing information: new insights. J Neurosci Res 66:1035–1046

Bernstein LR, Trahiotis C (2002) Enhancing sensitivity to interaural delays at high frequencies by using transposed stimuli. J Acoust Soc Am 112:1026–1036

Bernstein LR, Trahiotis C (2007) Why do transposed stimuli enhance binaural processing?: interaural envelope correlation vs envelope normalized fourth moment. J Acoust Soc Am 121:EL23

Breebaart J, van de Par S, Kohlrausch A (2001) Binaural processing model based on contralateral inhibition. I. Model structure. J Acoust Soc Am 110:1074–1088

Dau T, Püschel D, Kohlrausch A (1996) A quantitative model of the effective signal processing in the auditory system. I. Model structure. J Acoust Soc Am 99:3615–3622

Dietz M, Ewert SD, Hohmann V, Kollmeier B (2008) Coding of temporally fluctuating interaural timing disparities in a binaural processing model based on phase differences. Brain Res 1220:234–245

Dietz M, Ewert SD, Hohmann V (2009) Lateralization of stimuli with independent fine-structure and envelope based temporal disparities. J Acoust Soc Am 125(3):1622–1635

Dreyer AA, Delgutte B (2006) Phase locking of auditory-nerve fibers to the envelopes of high-frequency sounds: implications for sound localization. J Neurophysiol 96:2327–2341

Dye RH Jr, Niemiec AJ, Stellmack MA (1994) Discrimination of interaural envelope delays: the effect of randomizing component starting phase. J Acoust Soc Am 95:463–470

Ewert SD, Dau T (2000) Characterizing frequency selectivity for envelope fluctuations. J Acoust Soc Am 108:1181–1196

Hartmann WM (1987) Temporal fluctuations and the discrimination of spectrally dense signals by human listeners. In: Yose WA, Watson CS (eds) Auditory processing of complex sound. Erlbaum, Hillsdale, NJ

Hartmann WM, Pumplin J (1988) Noise power fluctuations and the masking of sine signals. J Acoust Soc Am 83:2277–2289

Henning GB (1974) Detectability of interaural delay in high-frequency complex waveforms. J Acoust Soc Am 55:84–90

Jepsen ML, Ewert SD, Dau T (2008) Modeling spectral and temporal masking in the human auditory system. J Acoust Soc Am 124:422–438

Joris PX, Yin TC (1995) Envelope coding in the lateral superior olive. I. Sensitivity to interaural time differences. J Neurophysiol 73:1043–1062

Klumpp RG, Eady HR (1956) Some measurements of interaural time difference thresholds. J Acoust Soc Am 28:859–860

Kohlrausch A, Fassel R, Dau T (2000) The influence of carrier level and frequency on modulation and beat-detection thresholds for sinusoidal carriers. J Acoust Soc Am 108:723–734

Levitt H (1971) Transformed up-down methods in psychoacoustics. J Acoust Soc Am 49:467–477

McFadden D, Pasanen EG (1976) Lateralization at high frequencies based on interaural time differences. J Acoust Soc Am 59:634–639

Nuetzel JM, Hafter ER (1976) Lateralization of complex waveforms: effects of fine-structure, amplitude, and duration. J Acoust Soc Am 60:1339–1346

Nuetzel JM, Hafter ER (1981) Discrimination of interaural delays in complex waveforms: spectral effects. J Acoust Soc Am 69:1112–1118

Patterson RD, Nimmo-Smith I, Holdsworth J, Rice P (1987) An efficient auditory filterbank based on the gammatone function. Paper presented at a meeting of the IOC speech group on auditory modeling at RSRE, 14–15 December 1987

Püschel D (1988) Prinzipien der zeitlichen Analyse beim Hören (Principles of Temporal Processing in Hearing). PhD thesis, University of Göttingen

Stellmack MA, Viemeister NF, Byrne AJ (2005) Discrimination of interaural phase differences in the envelopes of sinusoidally amplitude-modulated 4-kHz tones as a function of modulation depth. J Acoust Soc Am 118:346–352

van de Par S, Kohlrausch A (1997) A new approach to comparing binaural masking level differences at low and high frequencies. J Acoust Soc Am 101:1671–1680

Siveke I, Leibold C, Kaiser K, Grothe B, Wiegrebe L (2009) Adaptation of interaural-time-difference analysis to sound level. ISH, Salamanca, Spain, 2009

Chapter 33
Short-Term Synaptic Plasticity and Adaptation Contribute to the Coding of Timing and Intensity Information

Katrina MacLeod, Go Ashida, Chris Glaze, and Catherine Carr

Abstract In birds, the cochlear nucleus angularis (NA) is an obligatory relay for intensity processing, and the nucleus magnocellularis (NM) serves the same function for temporal information. Our recent experimental and modeling studies have shown that short-term synaptic plasticity is a major player in this division into parallel pathways. Short-term synaptic plasticity, by dynamically modulating synaptic strength, filters information contained in the firing patterns.

In NA, excitatory inputs express a mixture of short-term facilitation and depression, unlike those in NM and its target, NL, that only show synaptic depression. In NA, facilitation and depression at NA synapses were balanced such that postsynaptic response amplitude was often maintained throughout the train at high firing rates. The steady-state input rate relationship of the balanced synapses linearly conveyed rate information and therefore transmits intensity information encoded as a rate code in the nerve.

In the time coding arm of this pathway, recordings from NM and NL in vivo show that spiking activity decreases with time after the onset of the sound stimulus. Furthermore, NL neurons show faster decay than NM, suggesting differences in short term synaptic depression between the two time coding nuclei. Numerical simulations indicate that the large transient component of the input after the onset lessens ITD modulation. Therefore suppressing the transient input into NL may serve to facilitate ITD detection.

Keywords Synaptic plasticity • Neural coding • ITD

C. Carr (✉)
Department of Biology, University of Maryland, College Park, MD, USA
e-mail: cecarr@umd.edu

E.A. Lopez-Poveda et al. (eds.), *The Neurophysiological Bases of Auditory Perception*,
DOI 10.1007/978-1-4419-5686-6_33, © Springer Science+Business Media, LLC 2010

33.1 Introduction

In birds, the cochlear nucleus angularis (NA) is an obligatory relay for intensity processing, and the nucleus magnocellularis (NM) serves the same function for temporal information (Sullivan and Konishi 1984). Recent experimental and modeling studies have shown that short-term synaptic plasticity is a major player in this division into parallel pathways (Kuba et al. 2002; Cook et al. 2003; MacLeod et al. 2007). Short-term synaptic plasticity, by dynamically modulating synaptic strength, filters information contained in the firing patterns. The short-term synaptic plasticity found in auditory brainstem circuits appears to be tuned according to the requirements of sound information processing (MacLeod and Carr 2007). Adaptation of the spiking response during short (tens of millisecond) tone stimuli is also a fundamental feature of auditory and other sensory responses (see Laughlin 1989 and Wark et al. 2007). Mechanisms responsible for these "primary-like" responses in auditory nerve fibers are thought to reside in the afferent synapse; one putative mechanism is a short-term synaptic depression (Spassova et al. 2004).

Here, we report how adaptation may affect coding for intensity and timing in the brainstem, and how short-term plasticity may contribute to this adaptation. In the intensity pathway, spiking responses have varying degrees of firing rate adaptation, reflected in their descriptive categorization of type by post-stimulus time histogram (PSTH) profile (Blackburn and Sachs 1989; Köppl and Carr 2003). The short-term plasticity at the auditory nerve to NA neuron synapses also expresses varied frequency-dependent plasticity (MacLeod et al. 2007). In this chapter, we use simulations of synaptic conductance and natural auditory nerve spike trains to address how variation in synaptic plasticity contributes to variation in the PSTH.

In the timing pathway, nucleus laminaris (NL) neurons, which receive inputs from NM, show strong firing rate adaptation. Simulation of an NL neuron model shows that firing adaptation may be important for coding interaural time differences (ITD). Results suggest that this must be due to network and synaptic factors such as inhibitory input from the superior olivary nucleus and short-term synaptic depression.

33.2 Methods

33.2.1 Chick NA Physiology and Simulation

Detailed methods of synaptic recordings in NA brainstem slices, quantitative modeling of synaptic transmission and simulation can be found in MacLeod et al. (2007). Briefly, extracellular electrodes were used to stimulate auditory nerve fibers with pulse trains (10–250 Hz). Excitatory synaptic currents were recorded for 30 neurons in NA. A quantitative model describing synaptic transmission was used to fit the synaptic data. The model calculated the synaptic current in response to each pulse in the trains according to the following assumptions: that transmitter release resulted in the depletion of synaptic vesicles; vesicles were replenished with a double exponential that had a fast (tens of ms) and slow (seconds) time constant. Facilitation was implemented by incrementally increasing the probability of release (Pr) with each pulse; Pr then relaxed to its baseline value with an exponential decay time constant. There were six free parameters in the model: two time constants for vesicle recovery (τ_{fast}, τ_{slow}), one time constant for facilitation decay (τ_F), probability of vesicle release ($P_{release}$), facilitation increment (δ_F), and a vesicle pool size ratio (α).

For the simulations, we used auditory nerve spike trains recorded in vivo from chicks (courtesy of James Saunders, University of Pennsylvania). Data consisted of spike trains recorded in response to a 40 ms tone at CF at seven sound intensities (−5 to 40 dB relative to threshold), with 200 trains per intensity. The simulation convolved an alpha function representing synaptic conductance with each spike train. The alpha function amplitude was scaled according to the plasticity model and the interspike intervals present in each spike train. The conductance trains were summed over each intensity to generate a conductance histogram (gPSTH). For each set of gPSTH, the average conductance was measured during an early (0–10 ms) and late (30–40 ms) time window at a each intensity. The average early or late conductance was plotted versus intensity, to give two conductance versus intensity curves (gI curves). Each raw gI curve was fit with a logistics function (see text). From this function, we calculated the maximum conductance ($g_{max} = r_0 + r_1$) and dynamic range ($DR = 2 \times \log(2 + \sqrt{3})/b$). The logistic fit also yielded a correlation coefficient, *R*.

33.2.1.1 Recording from Owls' NM and NL Neurons In Vivo

Nineteen adult barn owls (Tyto alba) were used in this study. Animal husbandry and experimental protocols were approved by the Animal Care and Use Committee of the California Institute of Technology and the Animal Care and Use Committee of the University of Maryland. We followed the same procedures for surgery, acoustic stimulation, location of NM and NL, and electrophysiological recording as previously described (Peña et al. 2001; Köppl and Carr 2003). In brief, anesthesia was induced by injections of ketamine and xylazine or diazepam and supplementary doses were administered to maintain an adequate level of anesthesia. Body temperature was maintained at 39–40°C. Owls were placed in a sound-attenuating chamber. Acoustic stimuli generated were delivered by earphones placed into the ear canals. Sound pressure levels were calibrated before recordings using built-in miniature microphones. Stimuli were tone bursts of 60 ms duration at 40 dB SPL. Custom-written software (xdphys, Caltech) was used for controlling acoustic stimuli and data collection. Recordings were made with glass electrodes filled with 3 M potassium acetate or tungsten electrodes. NM and NL are easily located by monitoring the field potential while delivering sound stimuli because they are the only nuclei in the brainstem where neurophonic (sound-evoked sinusoidally oscillating extracellular field potentials) are observed. Neurophonic in NL shows ITD-tuning, whereas that in NM does not. Data were analyzed with custom-written Matlab scripts.

33.2.1.2 Simulation of NL Neuron Responses

The model neuron consists of two compartments, inexcitable cell body and excitable axonal node (Ashida et al. 2007). Fast sodium, low- and high-voltage activated potassium, and leak conductances are included. Input current to the neuron is modeled as AC+DC+noise (Ashida et al. 2007). The amplitude of the AC component changes with ITD (Funabiki and Konishi 2005), whereas DC and noise were assumed to be ITD-independent. To model the initial transient after the sound stimulus onset, we modified the input model as $h(t) \times (\mathrm{AC}+\mathrm{DC}+\mathrm{noise})$, where $h(t)$ is an exponentially decaying function with a time constant t_{input}. We changed t_{input} and examined the PSTHs and ITD-rate curves.

33.3 Results

33.3.1 Short-Term Plasticity Affects Intensity Coding in Nucleus Angularis

Many analyses of short-term synaptic plasticity have focused on the frequency-dependent transfer function of the steady state responses to constant frequency stimulus trains. Sensory stimuli, however, are often brief, and the auditory nerve responses show time-dependent changes in firing rate, typically observable in the PSTH. We investigated, therefore, how short-term plasticity might influence the pattern of firing in the cochlear nucleus. We addressed this question previously using a sample of three synaptic recordings (MacLeod et al. 2007); here, we extend this analysis to an expanded data set ($n=30$). To analyze the effects of short-term plasticity at individual recorded synapses, a model of synaptic transmission was fit to the data set of synaptic responses recorded during stimulation with trains of electrical impulses. Using an optimization procedure, the model was able to reproduce the data to within a standard deviation of the data by finding the best values for six parameters. Three of these parameters described synaptic vesicle depletion (τ_{fast}, $\tau_{slow,}$ α); three described facilitation ($P_{release}$, $\tau_{F,}$ δ_{F}).

We then simulated the effects of the short-term plasticity on the input by using natural auditory nerve spike train recorded in response to tone of different intensities. For each auditory nerve spike train, a conductance response in the model postsynaptic neuron was calculated, and the conductance responses for all spike trains at a given sound intensity was summed to generate a conductance PSTH (“gPSTH”)

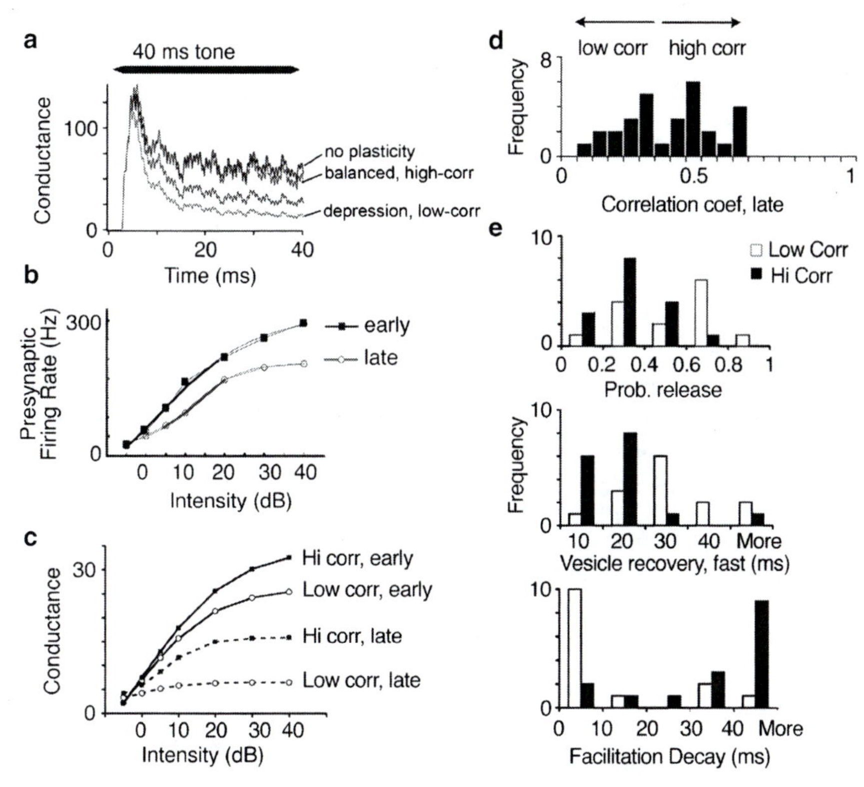

Fig. 33.1 (**a**) Short-term synaptic depression leads to adaptation of stimulus-generated input conductance, while "balanced" synaptic plasticity in NA prevents any further adaptation. Spike train input to the simulation was a 40 dB, 40 ms tone. *Top trace*: no plasticity. *Middle two traces*: NA synapses with nonmonotonic plasticity. *Bottom trace*: depressing NA synaptic plasticity. Adapted from MacLeod et al. 2007 with permission. (**b**) Logistics function fit of rate–intensity curve for presynaptic firing. Early (phasic, 0–10 ms) component, *closed symbols*; late (tonic, 30–40 ms) component, *open symbols*. Adaptation of the firing rate over time decreases the maximum firing rate, and reduces the dynamic range (*bold lines*). (**c**) Average logistics curves for high-correlation responses (*solid markers*) and low-correlation responses (*open markers*) (see text and (**d**)). Early components have higher maximum conductances and have wide dynamic ranges (*solid lines*); the late component for low-correlation synapses (*dashed lines*) has flattened considerably. (**d**) Histogram of correlation coefficients from logistic function fits for 30 NA synapses with differing plasticity properties. (**e**) Synaptic plasticity parameters differ for the high-correlation (*black bars*) and low-correlation groups (*white bars*). *Top graph*: $P_{release}$. *Middle graph*: τ_{fast}. *Bottom graph*: τ_F

(Fig. 33.1a). Here, we show example gPSTH for the original auditory nerve data ("no plasticity"), for a "balanced" synapse (one that expressed facilitation and relatively little depression), and for a "depressed" synapse.

These gPSTHs showed that a "balanced" synapse maintained the PSTH shape of the original presynaptic spike train, suggesting that balanced synapses may preserve whatever intensity information is encoded presynaptically over the duration of the stimulus. Intensity information was transmitted across the synapse not only during the higher frequency onset (phasic component) but also during the ongoing (tonic) activity. In contrast, while a depressing synapse similarly preserved intensity information during the phasic portion of the stimulus, it suppressed the tonic component of the gPSTH. The suppression was more severe with higher intensity, such that the gPSTH for different intensities converged by 10 ms into the response: the tonic conductances were independent of stimulus intensity. In other words, the depression mediated a gain control that eliminated intensity information late in the response.

Previously, we had found that the short-term plasticity could generally be divided into two qualitative categories based on their steady-state synaptic transfer function: simple, monotonic depression, and nonmonotonic, "balanced" synapses (MacLeod et al. 2007). Can the rate–intensity curves be similarly grouped? To better understand the effects of the plasticity in the population we observed, we repeated the simulation for a data set of 30 synapses and sought to characterize exactly how rate–intensity curves were transformed by short-term plasticity in both balanced and depressive synapses. To better define the intensity coding, we measured the total summed conductance during the phasic component (0–10 ms; "early" in Fig. 33.1) and the tonic component (30–40 ms; "late" in Fig. 33.1). These conductances were plotted versus the intensity (conductance–intensity curves, or "gI curves"), and then fitted with a logistics function:

$$r(s) = r_0 + r_1 / (1 + \exp(b(s_{\mathrm{mid}} - s))$$

where s is intensity in decibels, r is the response magnitude, r_0 is an offset, r_1 determines the ceiling for conductance (or firing rate), and s_{mid} is the decibel level at which response changes the most (zero-crossings in the second derivative). Coefficient b drives how quickly response changes as a function of decibel level and can be thought of as the slope over the dynamic range. From this function, we can define the maximum conductance (g_{max}, for synapse simulations; or maximum firing rate for presynaptic spike trains), dynamic range, midpoint (s_{mid}), and a correlation coefficient (R-value) (see Methods; Table 33.1). The early and late logistics curves for the presynaptic spike trains are plotted in Fig. 33.1b. For the presynaptic spike trains, adaptation resulted in several differences in the gI curves: decreased maximum firing rate, dynamic range, and R-value, and a shift of the midpoint to the right (higher intensity) (Fig. 33.1; Table 33.1, left two columns). Short-term plasticity always caused further "adaptation" in the late component

Table 33.1 Rate–intensity or conductance–intensity curve properties

	Presynaptic		Low-corr		High-corr	
	Early	Late	Early	Late	Early	Late
Max firing rate/ conductance	363.5	241	26.1	6.4	34.6	15.8
Correlation coeff.	0.79	0.66	0.79	0.27	0.81	0.52
Dynamic range (dB)	32.8	16.4	25.9	12.4	32.0	13.9
Midpoint (dB)	2.3	11.7	−0.3	1.6	0.3	5.5

gI curve, resulting in a decreased conductance, dynamic range, and *R*-value. However, the degree of this further adaptation had a large spread, depending on the plasticity.

To determine whether there was clustering in the synaptic plasticity effects, we analyzed the *R*-value, g_{max}, dynamic range, and s_{mid} values for groupings. The four descriptors defined by the logistics functions were generally continuous and covaried across the population of synapses. During the late component the distribution of the *R*-values of the logistic functions was bimodal, with a gap at 0.4 (Fig. 33.1d). We therefore divided the response profiles into two groups based on "late" *R*-values ("low-corr," *R*-values$<$0.4; "high-corr," *R*-values$>$0.4). Average gI curves for the early and late components are shown in Fig. 33.1c for these two groups. The flatter gI curves for the "low-corr" group (g_{max} low-corr$<g_{max}$ high-corr) implies that the postsynaptic firing rate–intensity curve will be flatter, assuming a given input–output spiking generation function for the postsynaptic neuron, which would result in less robust intensity coding. During the early component, there were small but significant differences between the two groups (g_{max}, dynamic range, and correlation; Table 33.1), but both had high *R*-values. During the late component, both groups had similar dynamic ranges.

The clustering of the low correlation and high correlation groups could be related to several significant differences in the plasticity parameters used in the model fit. The high-correlation responses correlated with synapses with significantly lower release probabilities ($P_{release}$), faster vesicle recovery (τ_{fast}), and facilitation (τ_F) (black bars in Fig. 33.1e). The low-correlation responses universally lacked facilitation (with facilitation decay times nearly instantaneous (τ_F~0; Fig. 33.1e, open bars)). These results suggest that short-term synaptic plasticity could contribute to adaptation of the intensity responses, depending on the type of plasticity present. The presence of two clusters suggests that there may be intensity "channels" within NA. Balanced synapses maintain ongoing intensity information better and minimize adaptation. In contrast, synaptic depression increases adaptation and reduces ongoing intensity information, but could signal the onset of a sound or changes in intensity.

33.3.2 Timing Pathway and Adaptation

Short-term plasticity also affects coding in the circuits involved in the computation of ITD (Carr and Konishi 1990). ITD sensitive neurons in the nucleus laminaris (NL) receive bilateral inputs from the axons in the nucleus magnocellularis (NM). NM neurons phase-lock to the auditory stimulus (Köppl 1997), while the NL neurons act as a coincidence detectors, and compares inputs from ipsi- and contralateral NM (Carr and Konishi 1990). In vitro experiments have shown that the NM–NL synapse is characterized by short-term synaptic depression, which would be predicted to enhance adaptation (Kuba et al. 2002; Cook et al. 2003). Here, we show more adaptation in NL than in NM, and suggest that the intrinsic properties of the NL neurons alone cannot account for the adaptation on this time scale.

33.3.2.1 Firing Rate Adaptation in NM and NL

NM recordings in vivo show gradual decrease in firing rate with a decay time constant of about 8 ms (Fig. 33.2a), while NL firing rates decay more rapidly, reaching steady state right after sound stimulus onset (Fig. 33.2b). The difference in the PSTH shapes between NM and NL suggest adaptation mechanisms that suppress the activity of NL after the stimulus onset, and suggest a role for rapid firing rate adaptation in ITD coding.

Behavioral studies have shown that owls readily lateralize brief correlated sound stimuli, and possess a neural temporal window for ITD detection (Wagner 1991). We examined the length of the time window over which spikes are counted to determine how short a time window was needed to encode changes in ITD. If the window is set as 60 ms after the sound stimulus onset, the spiking rate of NL neuron will show periodic modulation with ITD (Fig. 33.2c, thin line). Even when the spike-counting window was as short as 10 ms, however, the ITD-rate curve still shows clear peaks and troughs (Fig. 33.2c, thick line). This result corresponds to the known temporal precision of owls' behavior (Konishi 1973; Wagner 1991). The average spiking rate is higher when spike rate is calculated over a 10 ms interval, probably because of the onset spiking (Fig. 33.2a).

33.3.2.2 Simulation of NL Coding

In order to investigate the relationship between the fast decay in spiking activity and the efficiency of ITD coding in a very short time window, we performed numerical simulations using an NL neuron model. The model neuron PSTH shows a slower decay with increasing time constant of the input (Fig. 33.3a). If t_{input} is set equal to or smaller than 5 ms, the PSTH has a similar shape as the experimental PSTH (Fig. 33.2b). The ITD-rate curve of the model becomes flatter and the modulation depth, defined as the spike rate difference between the peak and the trough,

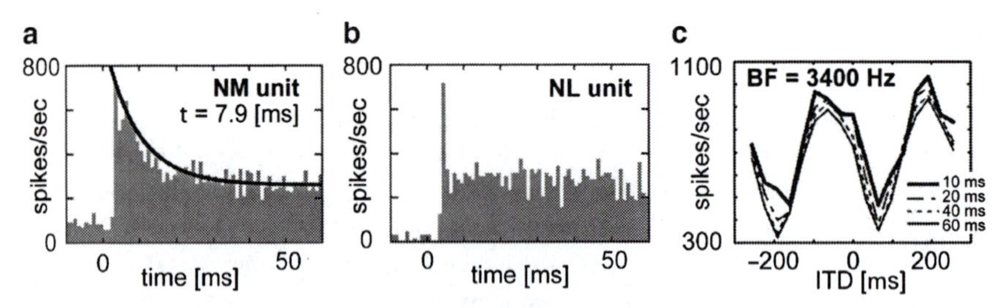

Fig. 33.2 (**a**) Peristimulus time histogram (PSTH) of an NM neuron. Decay time constant was calculated from the exponential fit. Sound stimulus of 60 ms duration was started at time 0. (**b**) PSTH of an NL neuron. Decrease in spiking rate was too rapid to calculate the decay time constant. (**c**) Spiking rate plotted against ITD (ITD-rate curve). Line types correspond to the length of time intervals over which spike numbers are counted

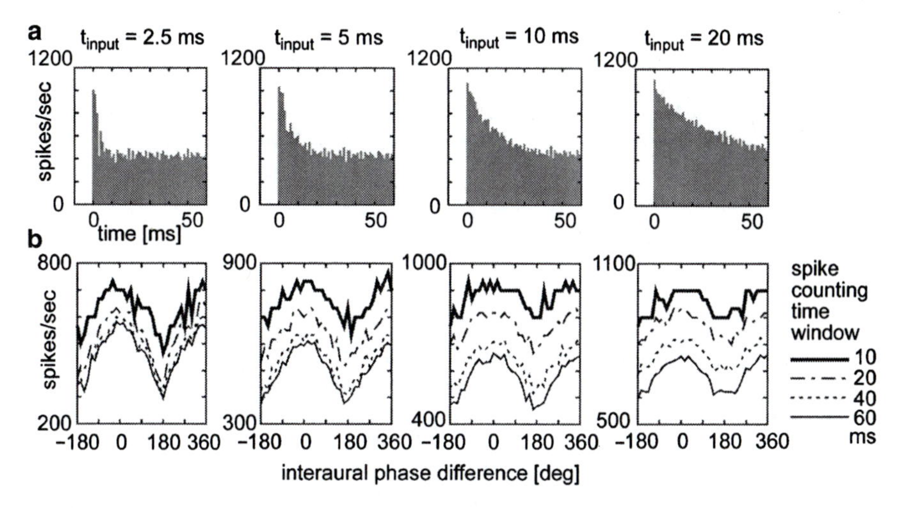

Fig. 33.3 (**a**) PSTH of the NL neuron model. Sound stimulus of 60 ms duration was started at time 0. (**b**) ITD-rate curves of the NL neuron model. Line types correspond to the length of time intervals over which spike numbers are counted. Input frequency is 4 kHz

becomes smaller with increasing t_{input} (Fig. 33.3b). If t_{input} is set small, average spiking rate calculated with the 10 ms time window is slightly higher than that with 60 ms window, and the ITD-rate curves have a similar shape with experimental results (Fig. 33.2c). These simulations show that the decay time constants of the input to NL need to be less than about 5 ms to reproduce the experimental results. However, the decay time constant of PSTH of NM observed in vivo was larger than 5 ms. Hence, additional mechanisms may suppress the transient component of the output of NM. Inhibitory synaptic input from the superior olivary nucleus and short-term synaptic depression, not yet included in our modeling, are possible candidates enabling ITD computation of brief sounds.

33.4 Conclusions

In this chapter, we present data and simulations that explore the relationship between the adaptation of neural responses to sound stimuli and short-term synaptic plasticity. Adaptation of neural activity over time may alter the transmission of auditory information. Short-term plasticity, by acting as a time- and frequency-dependent filter of presynaptic activity, can reinforce the adaptation, through short-depression, or minimize adaptation, through a balance of plasticity mechanisms. Adaptation reduces intensity information during ongoing activity. In the intensity pathway, this may allow emphasis of the transient, while in the timing pathway, it may allow intensity-independent computation of temporal information.

Acknowledgments The authors thank Kazuo Funabiki for kindly providing us with owl's NM and NL data; Timothy Horiuchi for assistance with the computational algorithms in the NA synaptic analysis; and James Saunders for the auditory nerve data. We acknowledge the support of the National Institutes of Health (DC000436 to CEC and DC007972 to KMM) and the Center for the Comparative and Evolutionary Biology Hearing (NIH grant DC04664).

References

Ashida G, Abe K, Funabiki K, Konishi M (2007) Passive soma facilitates submillisecond coincidence detection in the owl's auditory system. J Neurophysiol 97:2267–2282

Blackburn CC, Sachs MB (1989) Classification of unit types in the anteroventral cochlear nucleus: PST histograms and regularity analysis. J Neurophysiol 62(6):1303–1329

Carr CE, Konishi M (1990) A circuit for detection of interaural time differences in the brain stem of the barn owl. J Neurosci 10:3227–3246

Cook DL, Schwindt PC, Grande LA, Spain WJ (2003) Synaptic depression in the localization of sound. Nature 421:66–70

Funabiki K, Konishi M (2005) Intracellular study of auditory coincidence detector neurons in owls. In: 28th Assoc Res Otolaryngol, p 116

Konishi M (1973) How the owl tracks its prey. Am Sci 61:414–424

Köppl C (1997) Phase locking to high frequencies in the auditory nerve and cochlear nucleus magnocellularis of the barn owl, Tyto alba. J Neurosci 17:3312–3321

Köppl C, Carr CE (2003) Computational diversity in the cochlear nucleus angularis of the barn owl. J Neurophysiol 89:2313–2329

Kuba H, Koyano K, Ohmori H (2002) Synaptic depression improves coincidence detection in the nucleus laminaris in brainstem slices of the chick embryo. Eur J Neurosci 15:984–990

Laughlin SB (1989) The role of sensory adaptation in the retina. J Exp Biol 146:39–62

MacLeod KM, Carr CE (2007) Beyond timing in the auditory brainstem: intensity coding in the avian cochlear nucleus angularis. Prog Brain Res 165:123–133

MacLeod KM, Horiuchi TK, Carr CE (2007) A role for short-term synaptic facilitation and depression in the processing of intensity information in the auditory brain stem. J Neurophysiol 97:2863–2874

Peña JL, Viete S, Funabiki K, Saberi K, Konishi M (2001) Cochlear and neural delays for coincidence detection in owls. J Neurosci 21:9455–9459

Spassova MA, Avissar M, Furman AC, Crumling MA, Saunders JC, Parsons TD (2004) Evidence that rapid vesicle replenishment of the synaptic ribbon mediates recovery from short-term adaptation at the hair cell afferent synapse. J Assoc Res Otolaryngol 5:376–390

Sullivan WE, Konishi M (1984) Segregation of stimulus phase and intensity coding in the cochlear nucleus of the barn owl. J Neurosci 4:1787–1799

Wagner H (1991) A temporal window for lateralization of interaural time difference by barn owls. J Comp Physiol [A] 169:281–289

Wark B, Lundstrom BN, Fairhall A (2007) Sensory adaptation. Curr Opin Neurobiol 17:423–429

Chapter 34
Adaptive Coding for Auditory Spatial Cues

P. Hehrmann, J.K. Maier, N.S. Harper, D. McAlpine, and Maneesh Sahani

Abstract Adaptation of sensory neurons to the prevailing environment is thought to underlie improved coding of relevant stimulus features. Here, we report that neurons in the inferior colliculus (IC) of anesthetized guinea pigs adapt to statistical distributions of interaural time differences (ITDs – one of the binaural cues for sound-source localization). Each distribution contained all of the ITDs in the range of ±330 μs, but with a different high probability range (HPR) of ITDs in each from which ITDs were chosen 80% of the time. The center (mean) or the width (variance) of the HPR was altered systematically to assess the sensitivity of each neuron to the underlying distribution of ITDs. Neural rate-vs.-ITD functions shifted to accommodate the change in the underlying distribution when the mean of the HPR, but not the variance, was altered. Assuming a population of similarly tuned binaural neurons with Poisson spike statistics inputting to an IC neuron, we used an established model of synaptic depression to compute the IC postsynaptic current. By including a stochastic leaky integrate-and-fire mechanism in the IC spike generator, the model predicts the changes in neural gain and the shapes of rate-ITD functions as well as the shifts in Fisher information, a measure of coding accuracy, observed in the physiological data.

Keywords Binaural hearing • Interaural time differences • Fisher information • Synaptic depression

34.1 Introduction

Many sensory neurons appear to adjust their responses to take account of the statistical distribution of stimulus parameters. This ‘adaptive coding’ is suggested to underlie efficient coding of the current sensory environment (e.g., Ohzawa et al. 1982; Fairhall et al. 2001; Dean et al. 2005). Here, we examine adaptive coding

D. McAlpine (✉)
UCL Ear Institute, 332 Gray’s Inn Road, London WC1X 8EE, UK
e-mail: d.mcalpine@ucl.ac.uk

E.A. Lopez-Poveda et al. (eds.), *The Neurophysiological Bases of Auditory Perception*,
DOI 10.1007/978-1-4419-5686-6_34, © Springer Science+Business Media, LLC 2010

in inferior colliculus (IC) neuron to different distributions of interaural time differences (ITDs) – small differences in the timing of a sound at the two ears that underpin sound localization abilities in a wide range of species. ITDs are particularly suited to examining the site at which adaptive mechanisms arise in the central nervous system, as contributions from mechanisms below the level of binaural integration – the medial superior olive (MSO) of the brainstem – can be excluded. Further, since MSO neurons have been shown to encode instantaneous ITDs, independent of stimulus context, the observed adaptive effects can be assumed to arise at the level of IC. In experimental recordings, we find IC neurons to adapt to the most prevalent ITDs in a defined distribution – i.e., the same instantaneous ITD may evoke different neuronal responses depending on its context when the mean but not the variance of the distribution of ITDs is altered. We then modeled responses of IC neurons to the same stimuli investigated in the physiological recordings. Rate responses of MSO inputs to IC neurons were modeled with Gaussian-shaped ITD-functions. Assuming a population of such neurons similarly tuned for ITD and with independent Poisson spike generators, inputting to an IC neuron, we employed an established model of synaptic depression to compute the IC postsynaptic current. Like the physiological recordings, the model predicts small shifts in the slopes of neural rate-ITD functions when the center but not the width of the HPR was altered, especially when those ITDs were ipsilateral-leading. In addition, the gain of rate-ITD functions tended to be higher the more ipsilateral-leading were the commonly-occurring ITDs. Sensitivity to changes in the center but not the width of the HPR can be ascribed to the activation of a leaky integrate-and-fire mechanism in the former, but not the latter case. Taking into account both sides of the brain, these adaptive effects suggest a limited but clear capacity for neural coding to adjust to better represent the most common ITDs in a distribution.

34.2 Methods

Experiments were carried out in accordance with the guidelines of the UK Home Office, under control of the Animals (Scientific Procedures) Act 1986. For surgical preparation, sound generation, and presentation and spike collection, please refer to descriptions given in the previous publications (e.g., Dean et al. 2005).

34.2.1 Physiological Recordings

Single neurons were isolated by presenting 50-ms pure tones of variable intensities and frequencies at a rate of 5/s. Characteristic frequency (CF) and threshold were determined audio-visually and confirmed by plotting a frequency-vs.-level response area. To assess sensitivity to ITDs, a baseline function was obtained in response to 50-ms bursts of broadband noise with ITDs presented in a random order of 20–40 dB above threshold. Stimuli used to assess adaptive coding properties

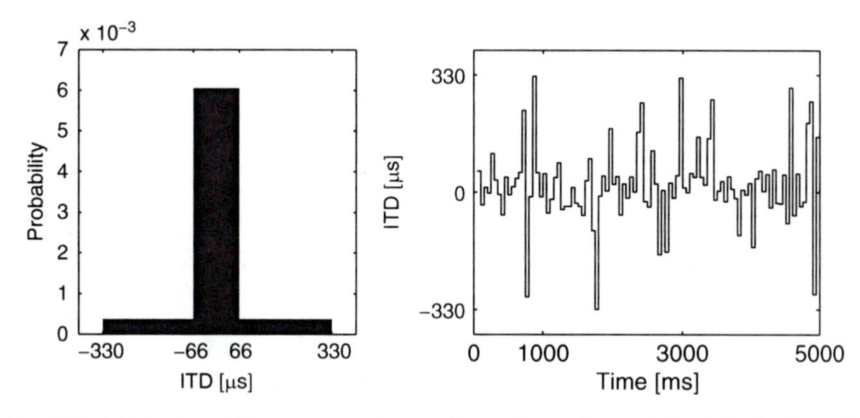

Fig. 34.1 (**a**) Number of 50-ms presentations randomly chosen from a defined distribution, with a high-probability region from which ITDs were selected with 80% probability, as a function of ITD. Exemplified is a distribution centered at 0 μs. (**b**) ITDs presented with a high probability region centered at 0 μs as a function of time

consisted of repeated presentations of interaurally-delayed white noise of 5-s duration, with the ITD sequence randomized across stimulus repetitions. The noise seed was also randomized across stimuli presented to the same neuron, and across neurons. ITDs were restricted to the maximum physiological range (±330 μs) suggested by Sterbing et al. (2003). ITDs were randomly chosen every 50 ms from a defined distribution, with a high-probability region, from which ITDs were selected with 80% probability (Fig. 34.1).

34.2.2 Data Analysis

Using TDT Brainware, action potentials were sorted according to waveform characteristics to ensure that data were collected from a single neuron. Spike times were then exported to MATLAB 7.0 (The MathWorks, Natick, MA) for off-line analysis. Taking neural latency into account, mean spike counts for each ITD were calculated from every presentation of a given ITD in a given distribution (Fig. 34.2). The probability $P_a[r|\tau]$ of neuron a giving r spikes to ITD τ was calculated. $P_a[r|\tau]$ was smoothed with a 2-D Gaussian, with standard deviations of 0.08 cycles (or 48 μs for the physiological range) and 0.5 spikes. Fisher information $f_a(\tau)$, for neuron a is given by

$$f_a(\tau) = \sum_r P_a[r \mid \tau]\left(\frac{\mathrm{d}\ln P_a[r \mid \tau]}{\mathrm{d}\tau}\right)^2$$

Assuming IC neurons generate spikes independently (Chechik et al. 2006; Popelář et al. 2003; Seshagiri and Delgutte 2003) the population Fisher information $F(\tau)$ for N neurons can be approximated by

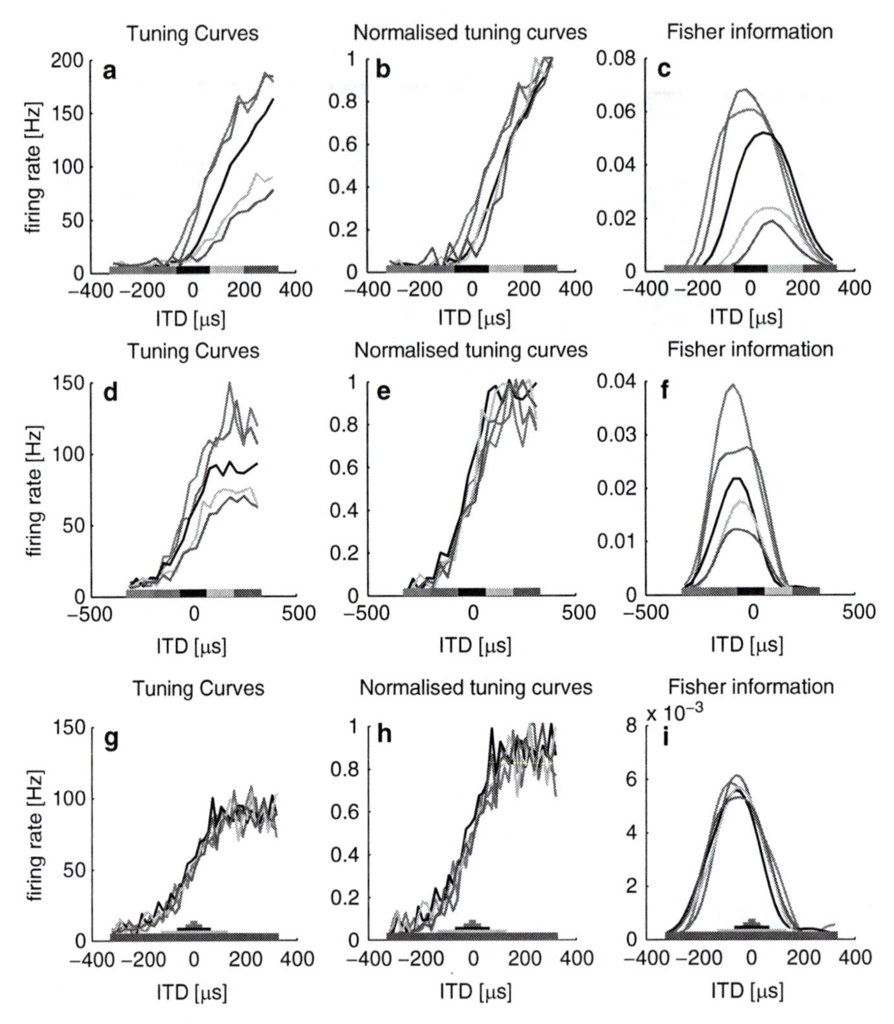

Fig. 34.2 (**a**, **d**) Spike rate-ITD functions, (**b**, **e**) normalized fits with an exponentiated sine based curve and (**c**, **f**) Fisher information of two neurons stimulated with five different high probability centers. (**g**) Spike rate-ITD functions, (**h**) normalized fits with an exponentiated sine based curve and (**i**) Fisher information of a with five different high probability widths. Neurons were stimulated over the maximum naturally-encountered range of ITDs (±330 μs) measured for the guinea pig. *Colors* indicate position of stimulus high probability region and corresponding neural response

$$F(\tau) = \sum_{a=1}^{N} f_a(\tau)$$

Peak Fisher information, indicating the highest coding accuracy, tends to occur over the slopes of a function – the region of the greatest change in neural activity. The spike count vs. ITD functions of the neurons were fitted with the equation:

$$g(\tau) = R\exp(A_1 \sin(2\pi(\theta - \phi_1)) + A_2 \sin(4\pi(\theta - \phi_2)) + A_3 \sin(6\pi(\theta - \phi_3))) + B$$

where θ is the interaural time difference τ expressed at a proportion of the period Q of the neurons characteristic frequency $\theta = \tau/Q$. All other parameters of the equation were varied to fit the data. The fitting was done by minimizing the least squared difference between the previous equation and the measured rate-ITD function using the standard matlab program fminsearch.m. The normalized curves fits were produced by means of the formula:

$$g_{norm}(\tau) = \frac{g(\tau) - g_{\min}}{g_{\max} - g_{\min}}$$

where $g_{\min}$ is the minimum value of the curve fit over the range of the data, and $g_{\max}$ the maximum value of the curve fit over the range of the data.

34.3 Results

34.3.1 Neurons Are Sensitive to Changes in the Mean of ITD Distributions

Figure 34.2a, d shows responses of two neurons to HPRs, where ITDs were drawn from distributions centered at −264, −132, 0, +132, or +264 μs, and the width of the HPR was fixed at ±132 μs. Rate-ITD functions showed highest gain when the HPR was positioned at ipsilateral-leading (negative) ITDs (Fig. 34.2, top row, magenta, and red curve), gain falling systematically as the HPR was shifted toward contralateral ITDs.

To test whether multiplicative gain changes constitute the only difference between the responses to different HPRs, the exponentiated sine-based fits to the rate-ITD functions were normalized (Fig. 34.2b, e). The normalized fits for ITDs show substantial differences in the shapes of the functions. For ipsilaterally-centered HPRs (magenta and red curves), the slope of the rate-ITD function appears to shift ipsilaterally as compared with the rate-ITD function centered at 0 μs (thick black curves). For HPRs centered contralaterally (green and blue curves), the gain is low, thus the spike counts involved are smaller and normalized rate-ITD functions more variable.

34.3.2 Coding Accuracy Shifts to Accommodate Shifts in HPR Mean

Fisher information was used to determine whether changes in neural sensitivity associated with changes in the HPR of the stimulus distribution are accompanied by adjustments in the accuracy with which neurons encode ITD. Assuming an optimal decoder, higher Fisher information reflects higher coding accuracy, i.e. a more accurate representation of the ITD presented and, thus, a higher capacity to discriminate nearby ITDs. Peak Fisher information evoked by stimulus distributions with HPRs centered at −264, −132, +132, and +264 μs (Fig. 34.2c, f), indicates that most neurons showed shifts in the position of peak Fisher information to accommodate the change in the HPR of the distribution, in good agreement with the shift in the position of the steepest slope. Note, however, that the shifts were not complete; peak Fisher information did not shift to align completely with the center of the HPR. Thus, although Fisher information for individual neurons follows, to some extent, the HPR, it does so relatively coarsely, and then only significantly so for HPRs centered at ipsilateral-leading (negative) ITDs.

34.3.3 Neurons Are Insensitive to the Changes in the Variance of ITD Distributions

In contrast to changes in the center of the HPR of ITDs, IC neurons are insensitive to changes in the stimulus variance of a distribution of ITDs. Figure 34.2g shows responses of one neuron in which the width of an HPR centered at zero was changed systematically. The width of each HPR ranged from ±6.5 to ±330 μs. Since the widest HPR covers 80% of the full range of ITDs explored, this distribution constitutes one in which all ITDs are equally likely (i.e. a flat distribution within the ecological range of ITDs). Normalizing the (largely overlapping) functions obtained to different HPRs widths indicates little change in the slope of the function to accommodate the increased variance in ITDs presented (Fig. 34.2h). Consistent with this, analysis of Fisher information indicated that widening the HPR from ±6.5 to ±330 μs around zero ITD had no effect on the width of Fisher information functions (Fig. 34.2i).

34.3.4 Neural Mechanisms Underlying Adaptive Coding of ITDs

Our data indicate that IC neurons show adaptive coding for ITD when the mean of the distribution of ITDs is changed, but not the variance. Assuming that firing rates of MSO neurons (the first stage of binaural processing) encode the instantaneous stimulus ITD independent of its context, but that those in the IC are sensitive to the context in which ITDs are presented (see Spitzer and Semple 1993, 1995), we modeled responses of IC neurons investigating the effect of short-term synaptic depression at afferent excitatory synapses on their target IC neurons. The rate response of a single MSO neuron to a stimulus with instantaneous ITD $s(t)$ was modeled by a Gaussian ITD-tuning function of width σ centered at a contralateral-leading ITD Δ_{max} outside the physiological range:

$$r(t) = r_{min} + c\exp\left(-\frac{(s(t)-\Delta_{max})^2}{2\sigma^2}\right).$$

Assuming a population of similarly tuned excitatory neurons with conditionally independent Poisson spiking behavior, we used the following model of synaptic depression (Tsodyks et al. 1998) to compute the postsynaptic current $I(t)$:

$$\frac{dx}{dt} = \frac{1-x(t)}{\tau_{rec}} - Ur(t)x(t) \quad \text{and} \quad \frac{dy}{dt} = -\frac{y(t)}{\tau_{in}} + Ur(t)x(t)$$

x quantifies the fraction of neurotransmitter that is available presynaptically, U is the fraction of available neurotransmitter that is released each time a spike occurs and τ_{rec} determines the rate at which the pool of available neurotransmitter recovers. y is the fraction of neurotransmitter that is active postsynaptically and τ_{in} determines its inactivation rate. Finally, $I(t)$ is assumed to be proportional to y:

$$T\frac{dV}{dt} = -V(t) + R_{mem}Wy(t) + \eta(t)$$

Model parameters were chosen so as to provide a good fit to the time course of response adaptation and recovery reported by Ingham and McAlpine (2004), where a binaural pure-tone stimulus was switched between the worst and best IPD of a neuron.

Stimulus-response functions for the postsynaptic current were calculated using the same stimuli as in the physiological recordings in order to investigate the effect of changing the stimulus HPR center and width (Fig. 34.3a–c and d–f, respectively). The model data, like the physiological data, suggest that the postsynaptic current is sensitive to changes in the HPR center, whereas changes in HPR width show no effect. More precisely, the gain of the stimulus-response functions decreases as the HPR center moves from the ipsilateral to the contralateral side, consistent with the experimental results. However, the lateral shifts observed in the ITD-functions of some neurons for changes of the HPR center, which alter the position of the peak Fisher information cannot be accounted for with a model using purely synaptic depression of the excitatory MSO-IC innervation (Fig. 34.3g, h). We suggest that by including a stimulus-distribution dependent leaky integrate-and-fire mechanism in the IC spiking generator, the model can predict the physiologically observed shifts of the ipsilateral rate-ITD- and Fisher information functions. A likely source of such a mechanism is the inhibitory input from the DNLL which in turn receives its input from the contralateral MSO and a slow conductance as produced by the neurotransmitter $GABA_B$ (~200 ms).

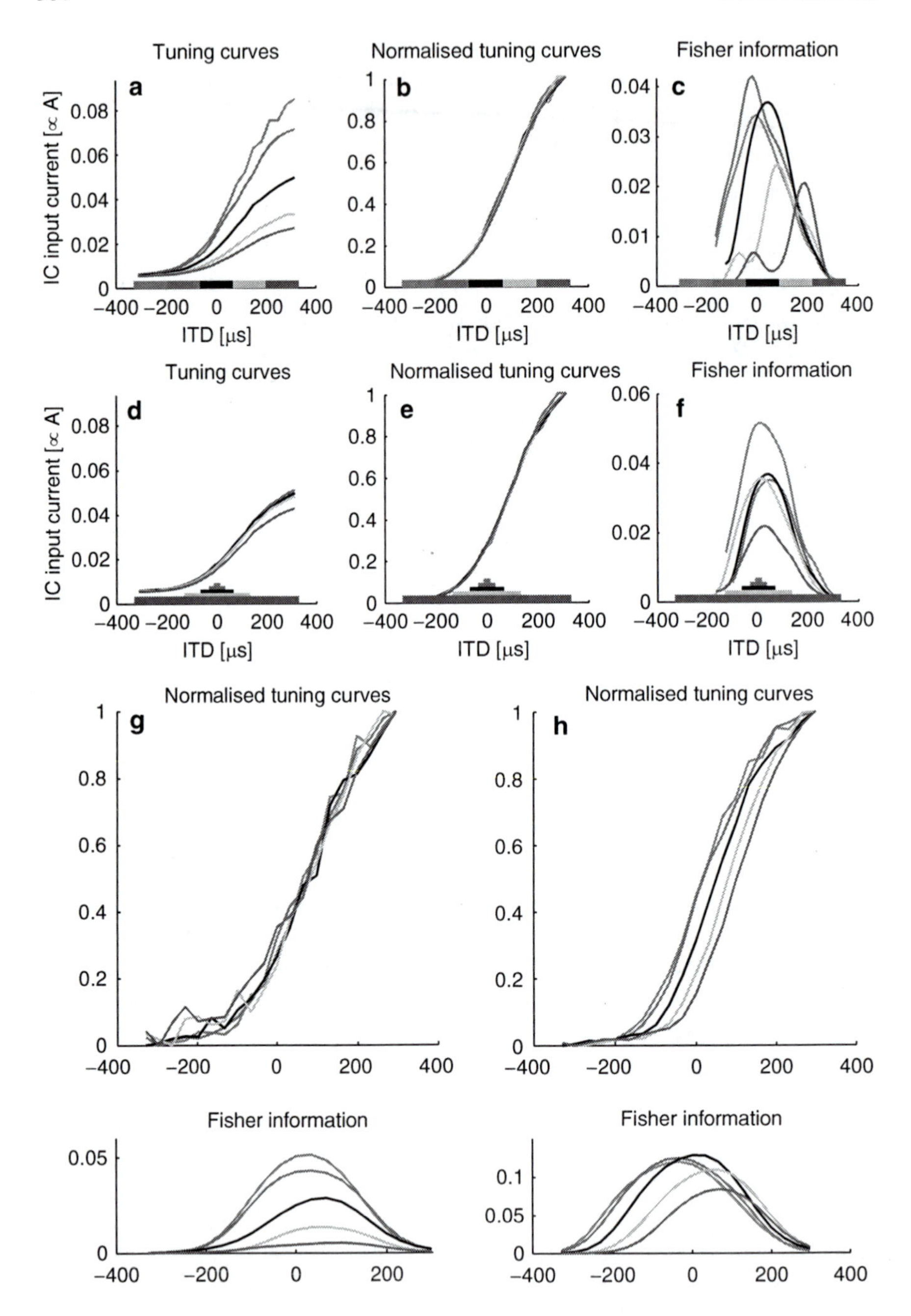

Fig. 34.3 (**a–f**) ITD tuning of the post-synaptic current (**g**, **h**) tuning curves of a leaky integrate-and-fire IC neuron with low (**g**) and high (**h**) membrane leakage

34.4 Discussion

The major finding of this study is that the gain and shape of rate-ITD functions of IC neurons adapt to take account of changes in the mean but not the variance of a statistical distribution of ITDs. For mean changes, this results in a shift in peak Fisher information toward the HPR in a stimulus distribution, at least for ipsilateral-leading HPRs. That coding accuracy that is greatest for ipsilateral-leading sounds is consistent with the reports of discrimination thresholds for static ITDs in the IC being smallest for ipsilateral-leading ITDs, with peaks of rate-ITD functions being situated contralaterally (Shackleton et al. 2003).

One significant contribution to changes in neural responses to different HPRs stems from neural gain, and our modeled responses that employ time constants previously reported for binaural adaptation to static interaural disparities (Ingham and McAlpine 2004) can account for these changes. However, although IC neurons scale multiplicatively in response to different HPRs, gain changes per se do not necessarily lead to informative changes in neural coding, as the neural responses to changes in HPR width demonstrates. In contrast, for changes of the HPR center, we also observed changes in the shape of the rate-ITD function which manifested as shifts in the position of the steepest slope in response to different HPRs, at least when the center of the HPR was changed. These shifts underlie the adjustments of the neural population's representation of ITD to encode most effectively the prevailing ITDs. Synaptic depression of the excitatory synaptic input models effectively the gain changes we observed when the center or width of the HPR was altered. If changes are to be informative of the HPR center, a shape change of the neural ITD-rate function is required. This can be modeled by a stimulus-distribution dependent leaky integrate and fire mechanism derived via inhibitory input from the DNLL. Usually, the DNLL is thought of as an instantaneous converter of excitation to inhibition of the activity from the contralateral MSO (e.g., Cai et al. 1998; Borisyuk et al. 2002). This however would lead to ITD-rate functions, shifted with respect to each other for ipsilateral ITDs (because for every 50 ms stimulus segment, an *instantaneous* inhibitory conductance is elicited from the DNLL), but show no shift for contralateral ITDs (because there is no conductance). This is true assuming that the inhibitory conductance is as rapid as that of the excitatory in tracking its presynaptic neural activity, which is reasonable if the receptors involved are $GABA_A$ or AMPA (~5 ms decay time constant for EPSPs and IPSPs). If, however, the inhibitory conductance is considerably slower than the 50-ms timescale of the stimulus (as would be the case for example with $GABA_B$, ~200 ms), then it will rather reflect the average activity in the DNLL – resulting in the physiologically observed shifts of the ITD-rate functions. The proposed mechanism does not contradict the notion that DNLL activity follows its MSO inputs instantaneously, but that its effect on IC neurons becomes sluggish by virtue of the lowpass filtering by $GABA_B$ receptors.

References

Borisyuk A, Semple MN, Rinzel J (2002) Adaptation and inhibition underlie responses to time-varying interaural phase cues in a model of inferior colliculus neurons. J Neurophysiol 88:2134–2146

Cai H, Carney LH, Colburn HS (1998) A model for binaural response properties of inferior colliculus neurons. II. A model with interaural time difference-sensitive excitatory and inhibitory inputs and an adaptation mechanism. J Acoust Soc Am 103:494–506

Chechik G, Anderson MJ, Bar-Yosef O, Young ED, Tishby N, Nelken I (2006) Reduction of information redundancy in the ascending auditory pathway. Neuron 51:359–368

Dean I, Harper NS, McAlpine D (2005) Neural population coding of sound level adapts to stimulus statistics. Nat Neurosci 8:1684–1689

Fairhall AL, Lewen GD, Bialek W, de Ruyter van Steveninck RR (2001) Efficiency and ambiguity in an adaptive neural code. Nature 412:787–792

Ingham NJ, McAlpine D (2004) Spike-frequency adaptation in the inferior Colliculus. J Neurophysiol 91:635–645

Ohzawa I, Sclar G, Freeman RD (1982) Contrast gain control in the cat visual cortex. Nature 298:266–268

Popelář J, Nwabueze-Ogbo FC, Syka J (2003) Changes in neural activity of the inferior colliculus in rat after temporal inactivation of the auditory cortex. Physiol Res 52:615–628

Seshagiri CV, Delgutte B (2003) Simultaneous single-unit recording from local populations in the inferior colliculus. Assoc Res Otolaryngol (Abstract 1280)

Shackleton TM, Skottun BC, Arnott RH, Palmer AR (2003) Interaural time difference discrimination thresholds for single neurons in the inferior colliculus of guinea pigs. J Neurosci 23:716–724

Spitzer MW, Semple MN (1993) Responses of inferior colliculus neurons to time-varying interaural phase disparity: effects of shifting the locus of virtual motion. J Neurophysiol 69:1245–1263

Spitzer MW, Semple MN (1995) Neurons sensitive to interaural phase disparity in gerbil superior olive: diverse monaural and temporal response properties. J Neurophysiol 73:1668–1690

Sterbing SJ, Hartung K, Hoffmann KP (2003) Spatial tuning to virtual sounds in the inferior colliculus of the guinea pig. J Neurophysiol 90:2648–2659

Tsodyks M, Pawelzik K, Markram H (1998) Neural networks with dynamic synapses. Neural Comput 10:821–835

Chapter 35
Phase Shifts in Monaural Field Potentials of the Medial Superior Olive

Myles Mc Laughlin, Marcel van der Heijden, and Philip X. Joris

Abstract Jeffress (J Comp Physiol Psychol 41:35–39, 1948) proposed that external interaural time differences (ITDs) are compensated by internal, axonal delays allowing ITD to be represented by a population of coincidence detectors in the medial superior olive (MSO). The MSO shows a strong extracellular field potential: the neurophonic. Studies in the barn owl reported a phase shift in the neurophonic along the nucleus laminaris and concluded that this phase shift is consistent with axonal delay lines as proposed by Jeffress. We recorded the neurophonic in the MSO of the cat at various locations along its short, dendritic axis. A phase shift of about 0.5 cycles was observed at depths close to the amplitude maxima, sometimes accompanied by localized amplitude minima. Current source density analysis for contralateral (ipsilateral) stimulation shows a current source close to a neurophonic amplitude maximum and a sink 100 μm ventromedially (dorsolaterally). These results indicate that some of the features of the neurophonic may be caused by a dipole field. Contralateral (ipsilateral) excitation causes a current sink at the ventromedial (dorsolateral) dendrites and a source at the soma and dorsolateral (ventromedial) dendrites. The difference in phase at the sink and source is 0.5 cycles, which closely resembles the phase shift that has been reported in the barn owl. Our interpretation in terms of a dipole field raises the question whether the neurophonic phase shift reported in the barn owl reflects axonal delays or simply a nucleus laminaris dipole configuration.

Keywords Neurophonic • Interaural time difference • Delay lines • Dipole field

P.X. Joris (✉)
Laboratory of Auditory Neurophysiology, Medical School, Campus Gasthuisberg, K.U.Leuven, B-3000, Leuven, Belgium
e-mail: Philip.Joris@med.kuleuven.be

E.A. Lopez-Poveda et al. (eds.), *The Neurophysiological Bases of Auditory Perception*, DOI 10.1007/978-1-4419-5686-6_35, © Springer Science+Business Media, LLC 2010

35.1 Introduction

In 1948, Jeffress proposed a model of binaural processing (Jeffress 1948) based on two central mechanisms: axonal delay lines and coincidence detection. Simply stated, external interaural time differences (ITDs) are compensated by internal, axonal delays so that signals from both ears arrive at the same time at the medial superior olive (MSO). MSO cells function as coincidence detectors, only firing a spike when input from both ears arrives at the same time. The model proposes that axonal delays are arranged in a systematic fashion, generating a range of internal delays. MSO neurons would thus form a spatial map, allowing external ITDs to be translated into an internal representation of auditory space.

Thus, a central prediction of the Jeffress model is that, due to axonal delay lines, signals arriving at the MSO should show a range of delays. Single cell recordings from a population of MSO neurons could be used to test this. However, isolating single neurons in vivo in the MSO is hampered by two main factors: (1) MSO action potentials are small (Scott et al. 2007; Yin and Chan 1990) and (2) a strong field potential termed the "neurophonic" (Biedenbach and Freeman 1964; Boudreau 1965) obscures these small action potentials.

Auditory neurophonics are AC signals that represent the phase coherent phase-locked activity of a local population of neurons. They can be measured in the auditory nerve (Snyder and Schreiner 1984), the cochlear nucleus (Marsh et al. 1970), and the superior olivary complex (in both the MSO and lateral superior olive) (Boudreau 1965; Galambos et al. 1959). In the MSO, neurons are arranged tonotopically (Guinan et al. 1972) and the dendrites project medially and laterally are oriented parallel to each other (Schwartz 1977; Smith 1995; Stotler 1953). Such an orderly, parallel arrangement gives rise to an "open field" (Biedenbach and Freeman 1964; Lorente De Nó 1947), causing postsynaptic potentials to summate and to create a large ensemble response here called the neurophonic.

Studies in the barn owl (Sullivan and Konishi 1986) reported a phase shift in the neurophonic along the nucleus laminaris and attribute this phase shift to axonal delay lines running along the dendritic (short) axis of the nucleus. Our goal is to examine whether the neurophonic in mammals provides clues regarding the origin of internal delays.

35.2 Methods

35.2.1 Surgical Preparation

Adult cats were anesthetized with a mixture of acepromazine (0.2 mg/kg) and ketamine (20 mg/kg). A venous cannula allowed infusion of Ringer's solution and sodium pentobarbital at doses sufficient to maintain an areflexic state. A laryngopharyngectomy was performed and a tracheal tube was inserted. The basioccipital bone was exposed after resection of the prevertebral muscles. The pinnas were removed, and the bullas were exposed and vented. The animal was placed in a double-walled soundproof room (Industrial Acoustics Company, Niederkrüchten, Germany). The trapezoid body (TB) was exposed by drilling a longitudinal slit as close as possible to the medial wall of the bulla and 3–5 mm rostral to the jugular foramen. A micromanipulator was used to support a five channel microdrive (TREC, Giessen, Germany) with quartz/platinum-tungsten electrodes (TREC, Giessen, Germany, 2–4 MΩ). The microdrive also contained a built-in broadband preamplifier (0.04 Hz to 20 kHz, 26 dB gain).The electrodes were positioned in the TB under visual control, just lateral or medial to the rootlets of the abducens nerve. The angle of penetration ranged from 15 to 30° mediolaterally relative to the mid-sagittal plane. The aim was to penetrate the low frequency region of the MSO along its dendritic (short) axis. After placing the electrodes in the TB, the basioccipital bone was covered with warm 3% agar.

35.2.2 Stimulus Generation and Signal Sampling

Dynamic speakers (supertweeter, Radio Shack, Forth Worth, TX) were connected to a hollow Teflon earpiece that fit in the transversely cut ear canals. Custom software, run on a personal computer, was used to calculate the stimuli and control the digital hardware (PD1 sys. 2 Tucker-Davis Technologies (TDT), Alachua, FL). The neural signal was amplified (×100), filtered (10–10,000 Hz) (EX1 and EX4-400, Dagan, Minneapolis, MN), and sampled (RX8 sys3, TDT, 50 kHz sampling frequency).

35.2.3 Stimuli and Data Collection

Monaural tone bursts (100–2,500 Hz, 50 or 100 Hz steps, 2,000 ms duration, 1 repetition, 2,500 ms repetition intervals, 50–70 dB) were played to the contralateral ear and then repeated to the ipsilateral ear. The electrodes were advanced through the MSO along its dendritic axis, and a coarse sampling of the neurophonic was performed at intervals of 100 μm. The electrodes were judged to be in the correct position if there was a strong neurophonic in response to both ipsi- and contralateral stimulation on at least three electrodes. Otherwise, the electrodes were retracted and repositioned. Once the electrodes were in the correct position, they were lowered to a depth of between 3,000 and 4,500 μm (depending on the strength of the neurophonic). They were then retracted in 50 μm steps and the response to the monaural ipsi- and contralateral stimulation was recorded.

35.3 Results

Data were collected from 27 electrode penetrations through six MSOs in three cats. Figure 35.1a shows the first 30 ms of a typical neurophonic response (thick line). The stimulus was an 800 Hz tone played at 70 dB to the contralateral ear and the penetration depth was 1,450 μm. The onset of the neurophonic is delayed by around 5 ms in relation to the stimulus. The neurophonic shows an onset response (positive deflection of ~10 ms) superimposed on top of a sustained stimulus-following response. A discrete Fourier transform (fft, Matlab, Mathworks, Natick, MA) was performed on the neurophonic response to the tone. The first 20 ms of the neurophonic were excluded to minimize the effect of the onset response. The resulting amplitude spectrum is shown is Fig. 35.1b. A strong component is clearly visible at the stimulus frequency (800 Hz). Smaller harmonic components are also visible at 1,600 and 2,400 Hz. At this depth, for contralateral stimulation, the neurophonic response to a range of stimulus frequencies was collected. An amplitude spectrum was calculated for each response and the amplitude of the component at the stimulus frequency is plotted in Fig. 35.1c. The neurophonic shows clear tuning to a broad range of frequencies between 500 and 2,500 Hz with a peak at 700 Hz. In this penetration, the neurophonic response to contralateral stimulation was collected at depth intervals of 50 μm between penetration depths of 600 and 3,750 μm. Figure 35.1d shows the same analysis as in Fig. 35.1c but repeated at different penetration depths (for clarity of illustration, only data at depth intervals of 100 μm are shown). The neurophonic response to ipsilateral stimulation was also collected at depth intervals of 50 μm and these data are shown in Fig. 35.1e (again only at intervals of 100 μm). Figure 35.1d, e shows that for both contralateral and ipsilateral stimulation, the neurophonic response is tuned in both frequency and depth. There are also differences in the response to contralateral stimulation and the response to ipsilateral stimulation. In the example shown here, the peak response to contralateral stimulation is reached at a depth of 1,450 μm (Fig. 35.1d, thick line) while the response to ipsilateral stimulation peaks at 1,550 μm (Fig. 35.1e, thick line). There are also differences in the frequency tuning of the contra- and ipsilateral neurophonic response. The contralateral response shows two peaks in frequency tuning, at 700 and 1,500 Hz. A similar pattern was seen for ipsilateral stimulation but the tuning, to both peaks, was much sharper and changed with depth.

The data shown in Fig. 35.1d, e can be replotted, for one specific frequency, as a function of depth. Data collected from a different cat are shown in such representation in Fig. 35.2a. The stimulus was a 1,400 Hz tone at 70 dB. The thick line shows the Fourier amplitude of the response to contralateral stimulation, and the thin line shows the response to ipsilateral stimulation. The response to contralateral stimulation increases from 0 to 600 μm. It then levels out before peaking at 1,200 μm. Ipsilateral stimulation with the same tone shows a similar response pattern but shifted to larger depths. There is also dip in the amplitude response around 1,100 μm. When performing Fourier analysis the phase spectrum was calculated in addition to the amplitude spectrum. Figure 35.2b shows the phase at the stimulus frequency for contralateral (thick line) and ipsilateral (thin line) stimulation at 1,400 Hz, for a range

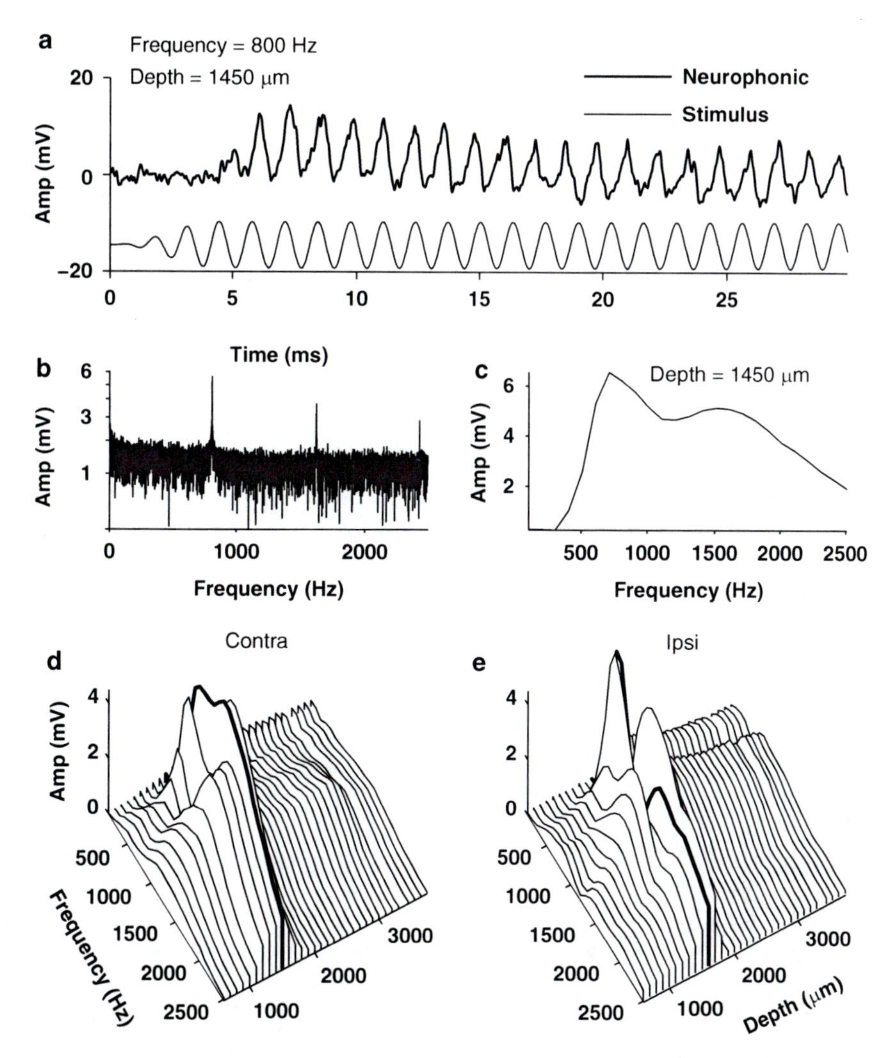

Fig. 35.1 Measurement and Fourier analysis of the neurophonic. (**a**) An 800 Hz tone played to the contralateral ear (*thin line*) and the resulting neurophonic response (*thick line*) at a penetration depth of 1,450 μm. (**b**) The amplitude spectrum of the neurophonic response shown in (**a**). There is a strong component at the stimulus frequency (800 Hz) and weaker harmonic components (1,600 and 2,400 Hz). (**c**) The Fourier amplitude at the stimulus frequency is plotted as a function of stimulus frequency for contralateral stimulation at 1,450 μm. There is broad frequency tuning in the neurophonic response. (**d**) The analysis shown in (**c**) is repeated for a range of penetration depths for contralateral stimulation, showing clear tuning with depth. The *thick line* indicates the depth at which the maximum amplitude was measured. (**e**) The analysis shown in (**d**) is repeated for ipsilateral stimulation

of penetration depths. The phase of the response to contralateral stimulation is relatively constant until 500 μm where it starts to decrease. Between 800 and 1,000 μm, it shows a sharp decrease in phase of almost half a cycle. For the rest of the penetration, the phase of the contralateral response remains relatively constant. The phase

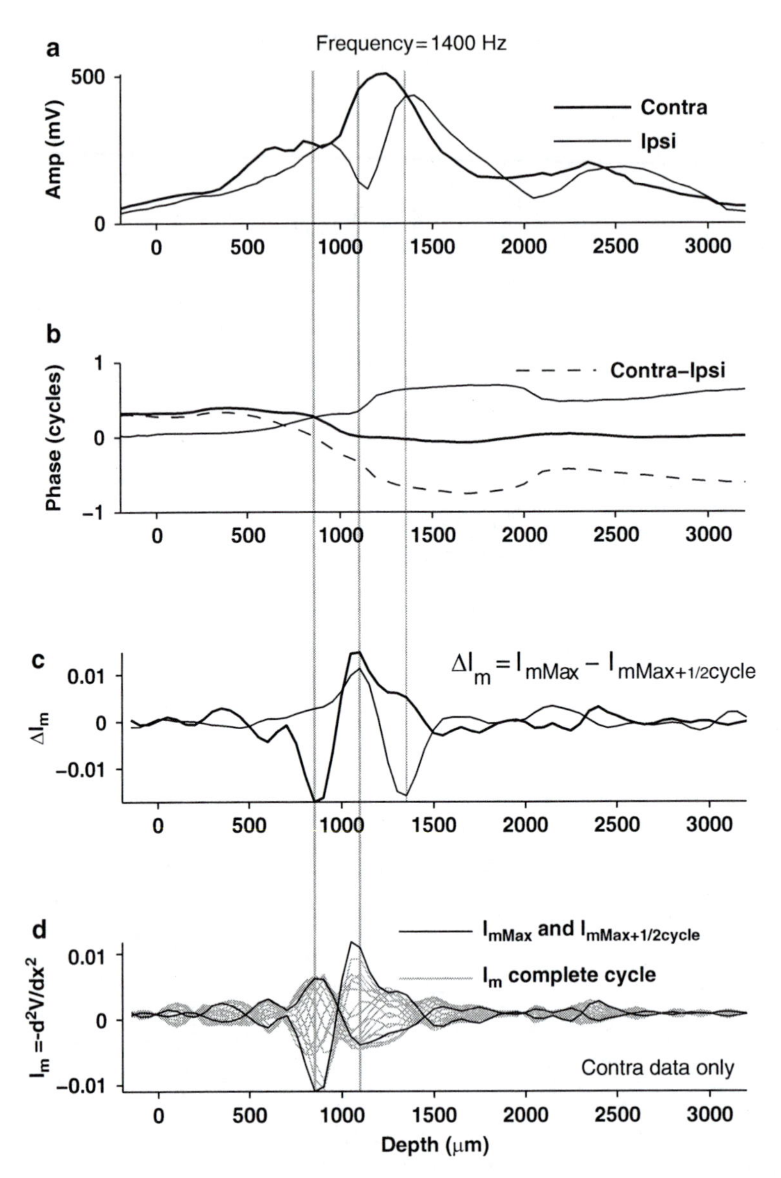

Fig. 35.2 Neurophonic amplitude, phase, and current source density analysis. (**a–c**) *Thick black line* shows contralateral data, *thin black line* shows ipsilateral data. (**a**) The analysis as shown in Fig. 35.1d, e is replotted for one frequency (1,400 Hz) as a function of depth (different dataset than Fig. 35.1). (**b**) Phase of the contralateral and ipsilateral neurophonic. The binaural phase (*dashed line*) is calculated by subtracting the contralateral from the ipsilateral phase. (**c**) A current source density analysis was performed on the neurophonic data (see text for details) to reveal the location of the contralateral current sources and sinks (*thick gray line* running across all panels) and ipsilateral current sources and sinks (*thin gray line*). ΔI_m shows the total change in transmembrane current at a given depth during one stimulus cycle. (**d**) The neurophonic was averaged over one stimulus cycle. Taking the second spatial derivative of the averaged neurophonic gives a measurement of the transmembrane current (I_m) during one complete stimulus cycle (*gray lines*). ΔI_m is calculated by finding the point in the cycle with the maximum transmembrane current (I_{mMax}, *black line*) and subtracting from this the transmembrane current half a cycle later ($I_{mMax+1/2cycle}$, *black line*)

of the ipsilateral response shows a more complex pattern. It shows an increase in phase of around half a cycle between 500 and 1,250 μm and a smaller decrease in phase at 2,100 μm. The dashed line in Fig. 35.2b shows the binaural phase: contralateral phase minus ipsilateral phase. The binaural phase shows a shift of around one complete cycle. Similar phase shifts have been reported before in both the cat (Bojanowski et al. 1989) and the barn owl (Sullivan and Konishi 1986).

35.3.1 Current Source Density Analysis

The half-cycle phase shift and amplitude minima suggest that the neurophonic can be interpreted in terms of a dipole field. A dipole field has a spatially segregated current sink and current source (Fig. 35.3b). Moving through the dipole field, a peak in the amplitude of the potential will be measured near the sink (Fig 35.3c). The amplitude of the potential will reach zero as the zero equipotential line is crossed before reaching a second peak close to the source. The phase of the potential will show a sharp phase shift of half a cycle when the zero equipotential line is crossed (Fig. 35.3d). Current source density (CSD) analysis can be used to gain temporal and spatial information about the location of current sources and sinks in the brain (Freeman and Nicholson 1975; Manis and Brownell 1983; Nicholson and Freeman 1975). If the region of interest contains a high degree of symmetry, as the MSO and its dendrites do, CSD can be used to derive information about the transmembrane currents. Current flow in three dimensions is related to the second spatial derivative of the extracellular field potential (expressed here in rectangular Cartesian coordinates),

$$I_{\mathrm{m}} = -K\left[\frac{\partial^2\phi}{\partial x^2}+\frac{\partial^2\phi}{\partial y^2}+\frac{\partial^2\phi}{\partial z^2}\right] \tag{35.1}$$

where I_{m} is the transmembrane current, ϕ is the field potential, and K is the conductivity constant. Assuming that potential only changes along one dimension (along the MSO dendritic axis) and the conductivity is constant,

$$I_{\mathrm{m}} \simeq -\frac{\partial^2\phi}{\partial x^2} \tag{35.2}$$

Therefore, an estimation of the transmembrane current can be obtained by taking the second spatial derivative of the measured potential, i.e., the neurophonic. Taking the derivative of a signal decreases its signal-to-noise ratio. Since the neurophonic is a periodic signal, this ratio can be improved by simple averaging: we cut the waveform up into segments of one cycle in length and took the average of these segments. This means that for each depth, there is an averaged neurophonic with a length equal to one cycle of the stimulus frequency. Taking the second spatial derivative along the depth dimension gives an estimate of transmembrane current, I_{m}, at each depth and at each point in the cycle. The gray lines in Fig. 35.2d show I_{m}; each line represents the transmembrane current at one point in the cycle,

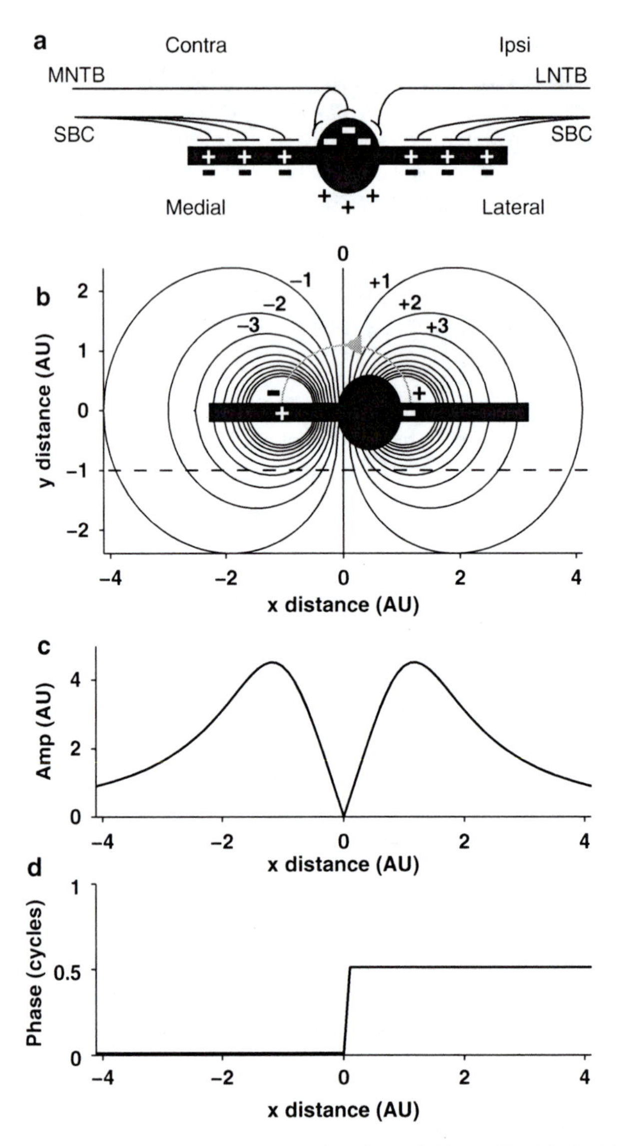

Fig. 35.3 Interpreting the neurophonic in terms of a dipole field. (**a**) The black object represents an MSO neuron with excitatory inputs from contralateral spherical bushy cells (SBC) on the medial dendrite and excitatory inputs from ipsilateral SBC on the lateral dendrite. Inhibitory input from the contralateral medial nucleus of the trapezoid body (MNTB) and the ipsilateral lateral nucleus of the trapezoid body (LNTB) is present on the soma. The *pluses* and *minuses* show the direction of change in intracellular (*white*) and extracellular (*black*) potential when an EPSP (SBC input) or IPSI (MNTB or LNTB input) arrives at the MSO. (**b**) With contralateral stimulation, the excitatory input on the medial dendrite causes a localized current sink and a corresponding current source near the soma. The current source and sink result in a dipole field. Note that the inhibitory input would also contribute to this source–sink configuration. The *black contours* show the equipotential lines of the dipole field. The direction of extracellular current flow is marked by the *gray line with the arrow*. (**c**) The amplitude of the dipole field at the location marked by the *dashed line* in (**b**). (**d**) The phase of the dipole field at the location marked by the *dashed line* in (**b**)

in steps of 1/18 of a cycle, across all depths. Two of these lines are shown in black: the depth profile yielding the maximal I_m value (I_{mMax}) and its counterpart separated by half a cycle. Subtraction of these two profiles captures the magnitude of changes in I_m during one cycle (ΔI_m). The thick line in Fig. 35.2c shows ΔI_m for contralateral stimulation, and the thin line shows ΔI_m for ipsilateral stimulation. For contralateral stimulation, a current sink–source pair can be identified and is marked across all panels by the thick gray lines. A current sink–source pair is defined as two local maxima in the transmembrane current which are half a cycle out of phase. Figure 35.2d shows that when I_m at 850 μm is at a minimum, I_m at 1,100 μm is at a maximum and visa versa. Interestingly, most of the phase change in the neurophonic to contralateral stimulation occurs between the current sink–source pair. For ipsilateral stimulation, a current sink–source pair can be seen at 1,100 and 1,350 μm, overlapping part of the contralateral sink–source pair. Again, the sharpest phase change in the neurophonic to ipsilateral stimulation occurs between the ipsilateral current sink–source pair.

35.4 Discussion

The CSD analysis (see Fig. 35.2c, d) shows current sink–source configurations in the neurophonic, which are spatially offset for contralateral and ipsilateral stimulation. We interpret these configurations on the basis of the known anatomy of MSO neurons and their inputs. The dendrites of MSO cells have a bipolar anatomy, with dendrites on the lateral side receiving ipsilateral input and dendrites on the medial side receiving contralateral input. For contralateral stimulation, strongly phased-locked excitation (Joris et al. 1994) on the medial dendrites causes phased locked EPSPs, which create a current sink on the medial dendrites. This gives rise to a current source located somewhere on the soma or possibly on the lateral dendrites. Note that glycinergic inputs from the MNTB and LNTB are predominantly placed near the soma (Adams and Mugnaini 1990; Cant and Hyson 1992; Helfert et al. 1989; Kapfer et al. 2002) and would also contribute a source near the soma and sink on the dendrites. Figure 35.3a shows a cartoon representation of both the excitatory and inhibitory inputs to the MSO. Ipsilateral stimulation would give rise to a current sink located on the lateral dendrites with a corresponding current source on the soma or medial dendrites. Figure 35.3b shows a cartoon of an MSO cell (black object) with a current source located on the medial dendrite and a corresponding current sink located near the soma. The equipotential lines of a cross-section through the resulting dipole field are shown as the concentric solid lines. Figure 35.3c shows the amplitude of the potential of the dipole field at the position indicated by the dashed line on Fig. 35.3b. This amplitude profile, with the two peaks separated by a dip, resembles the amplitude profile of the neurophonic shown in Fig. 35.2a. Figure 35.3d shows the phase of the dipole field at the position indicated by the dashed line on Fig. 35.3b. The dipole field shows a change in phase of half a cycle when moving from the sink to the source field. The amount of phase change resembles that seen in the neurophonic (Fig. 35.2b) but the phase change in the neurophonic is not as sharp, and its amplitude minima are not always pronounced (Fig. 35.2a, thick line). Initial results from a neurophonic modeling study suggest that the more gradual changes in phase and less pronounced amplitude minima may occur when a number of different dipoles, at different spatial locations and with different phases, contribute to the field.

The phase shifts associated with a dipole field complicate the use of neurophonic delays to draw conclusions about axonal delay lines. Sullivan and Konishi (1986) reported a phase change of around half a cycle for monaural stimulation in the barn owl, or a full cycle for the interaural phase. They concluded that this phase change was the result of a delay line, a key element of the Jeffress model. Phase and time shifts have also been reported in mammals. Biedenbach and Freeman (1964) described a reversal in polarity of the potential evoked by clicks. Bojanowski et al. (1989) found a phase shift in neurophonic to tonal signals and also concluded that this phase shift was the result of a delay line. In this study, we found a similar phase change. However, our conclusions as to the origins of the phase shift are markedly different. We used CSD analysis to show that there are spatially segregated current sources and sinks in the MSO. This configuration of sources and sinks results in a dipole field. This means that most of the phase change across the short axis of the cat MSO is probably a feature of that field and is not caused by a delay line. This raises the question as to whether the same dipole field is present in the nucleus laminaris of the owl and how much this field contributes to the reported phase shift. It should, however, be noted that there are differences in the anatomy of the owl and cat MSO, one important difference being that the cat MSO consists of a relatively thin sheet of neurons (three or four) (Stotler 1953) while nucleus laminaris in the barn owl is made up of a thick sheet of neurons (Carr and Konishi 1990).

35.5 Comment by Catherine Carr

The suggested difference between birds and mammals may rather be a difference between different cellular architectures. Both cat and chicken MSOs are characterized by a compact cell body layer, populated by bitufted coincidence detectors, with segregated inputs from each ear. Thus, one might expect similar dipole fields in cat and chicken since each would have spatially segregated sources and sinks. There is some published support for this in Schwarz (1992). We were able to confirm this well-localized nature in Koppl and Carr (2008). Using high-impedance electrodes, we could determine the location of maximal neurophonic amplitude to within 50 μm, and later showed this to be the cell body layer by analysis of small neurobiotin injections. Two other interesting points relate to comparisons to the barn owl's neurophonic. First, barn owl coincidence detectors do not have bitufted dendrites, except at very low best frequencies. Instead, they receive mixed ipsi/contra inputs onto very short dendrites that surround the cell body.

If the cell bodies act as current sources/sinks, then current density should be more uniformly distributed throughout NL, as appears to be the case. Second, Hermann Wagner and my recordings, so far only in abstract form, show that the half-cycle phase shift in Sullivan and Konishi (1986) is notable not for its phase shift, but for the time delay over the recorded depth, since the degree of monaural phase shift depends upon the best frequency of the recording. Higher best frequency recordings show more than 0.5 cycle, while lower best frequency recordings show less.

35.6 Reply Myles Mc Laughlin

As mentioned in the comment and as we point out in our discussion, there are important differences in the architectures of NL in the barn owl and MSO in the cat. In spite of these differences, both structures generate a strong extracellular field (the neurophonic) that share some surprisingly similar features, most notably the phase shift or delay along the short axis of the nucleus. In our paper, we argue that in the cat, two key features in the amplitude and phase data indicate that the delay across the short axis of the MSO is caused by a dipole and not a delay line: the dips in the amplitude depth profiles and the similarity of phase change across stimulus frequency. These data do not rule out the possibility that the phase shift in NL of the barn owl reflects systematic axonal delays: possibly the architectural differences between barn owl and cat are such that the neurophonics in the two species share phenomenological features but are generated by different mechanisms. Interestingly, a neurophonic is also recorded in the mammalian anteroventral cochlear nucleus (Marsh et al. 1970), which lacks the clear bipolar organization present in the MSO, and across the auditory nerve (Snyder and Schreiner1984). Further detailed study of the 3D distribution of amplitude and phase in these structures will illuminate the role of different cellular components in the generation of these potentials.

References

Adams JC, Mugnaini E (1990) Immunocytochemical evidence for inhibitory and disinhibitory circuits in the superior olive. Hear Res 49:281–298

Biedenbach MA, Freeman WJ (1964) Click-evoked potential map from the superior olivary nucleus. Am J Physiol 206:1408–1414

Bojanowski T, Hu K, Schwarz DW (1989) Analogue signal representation in the medial superior olive of the cat. J Otolaryngol 18:3–9

Boudreau JC (1965) Stimulus correlates of wave activity in the superior-olivary complex of the cat. J Acoust Soc Am 37:779–785

Cant NB, Hyson RL (1992) Projections from the lateral nucleus of the trapezoid body to the medial superior olivary nucleus in the gerbil. Hear Res 58:26–34

Carr CE, Konishi M (1990) A circuit for detection of interaural time differences in the brain stem of the barn owl. J Neurosci 10:3227–3246

Freeman JA, Nicholson C (1975) Experimental optimization of current source-density technique for anuran cerebellum. J Neurophysiol 38:369–382

Galambos R, Schwartzkopff J, Rupert A (1959) Microelectrode study of superior olivary nuclei. Am J Physiol 197:527–536

Guinan JJ, Norris BE, Guinan SS (1972) Single auditory units in the superior olivary complex. II: Locations of unit categories and tonotopic organization. Int J Neurosci 4:147–166

Helfert RH, Bonneau JM, Wenthold RJ, Altschuler RA (1989) GABA and glycine immunoreactivity in the guinea pig superior olivary complex. Brain Res 501:269–286

Jeffress LA (1948) A place theory of sound localization. J Comp Physiol Psychol 41:35–39

Joris PX, Carney LHC, Smith PH, Yin TCT (1994) Enhancement of synchronization in the anteroventral cochlear nucleus. I. Responses to tonebursts at characteristic frequency. J Neurophysiol 71:1022–1036

Kapfer C, Seidl AH, Schweizer H, Grothe B (2002) Experience-dependent refinement of inhibitory inputs to auditory coincidence-detector neurons. Nat Neurosci 5:247–253

Koppl C, Carr CE (2008) Maps of interaural time difference in the chicken's brainstem nucleus laminaris. Biol Cybern 98:541–559

Lorente De Nó R (1947) Action potential of the motoneurons of the hypoglossus nucleus. J Cell Comp Physiol 29:207–287

Manis PB, Brownell WE (1983) Synaptic organization of eighth nerve afferents to cat dorsal cochlear nucleus. J Neurophysiol 50:1156–1181

Marsh JT, Worden FG, Smith JC (1970) Auditory frequency-following response: neural or artifact? Science 169:1222–1223

Nicholson C, Freeman JA (1975) Theory of current source-density analysis and determination of conductivity tensor for anuran cerebellum. J Neurophysiol 38:356–368

Schwartz IR (1977) Dendritic arrangements in the cat medial superior olive. Neuroscience 2:81–101

Schwarz DW (1992) Can central neurons reproduce sound waveforms? An analysis of the neurophonic potential in the laminar nucleus of the chicken. J Otolaryngol 21:30–38

Scott LL, Hage TA, Golding NL (2007) Weak action potential backpropagation is associated with high-frequency axonal firing capability in principal neurons of the gerbil medial superior olive. J Physiol 583:647–661

Smith PH (1995) Structural and functional differences distinguish principal from nonprincipal cells in the guinea pig MSO slice. J Neurophysiol 73:1653–1667

Snyder RL, Schreiner CE (1984) The auditory neurophonic: basic properties. Hear Res 15:261–280

Stotler WA (1953) An experimental study of the cells and connections of the superior olivary complex of the cat. J Comp Neurol 98:401–431

Sullivan WE, Konishi M (1986) Neural map of interaural phase difference in the owl's brainstem. Proc Natl Acad Sci U S A 83:8400–8404

Yin TCT, Chan JK (1990) Interaural time sensitivity in medial superior olive of cat. J Neurophysiol 64:465–488

Part VI
Speech Processing and Perception

Chapter 36
Representation of Intelligible and Distorted Speech in Human Auditory Cortex

Stefan Uppenkamp and Hagen Wierstorf

Abstract The aim of the study is to identify cortical regions that process the meaning of spoken language, irrespective of acoustical realization, by means of functional MRI. Distorted speech material is used along with clean speech to allow for a controlled manipulation of intelligibility, depending on the presentation order. Speech distortion was performed by rotating spectral features. Psychophysical experiments were performed with 230 sentences of a speech intelligibility test in German language, to explore the effect of presentation order of clean and distorted speech. Results from 15 listeners indicate that distorted signals can reliably be presented either in a way well to understand, or completely unintelligible, depending on the exact presentation relative to the other speech material. The results were used to design a functional MRI experiment that allowed us to image brain activation for identical acoustic stimuli, only differing in their intelligibility. The fMRI data indicate that speech stimuli in general cause bilateral activation in the temporal lobes. The contrast between natural and distorted speech only comes up in regions beyond primary auditory areas, mainly in superior temporal gyrus. The contrast between acoustically identical stimuli only differing in their intelligibility reveals several areas, largely in the left hemisphere and clearly separate from the temporal lobes. These regions are interpreted to represent the processing of the meaning of intelligible speech, irrespective of their exact acoustical properties.

Keywords Speech intelligibility • Language • Auditory cortex • fMRI

S. Uppenkamp (✉)
Medizinische Physik, Universität Oldenburg,
Oldenburg 26111, Germany
e-mail: stefan.uppenkamp@uni-oldenburg.de

E.A. Lopez-Poveda et al. (eds.), *The Neurophysiological Bases of Auditory Perception*,
DOI 10.1007/978-1-4419-5686-6_36,

36.1 Introduction

Neuropsychology and brain imaging studies indicate a left hemispherical preference for intelligible speech in most listeners. However, the reason for this specialization, the starting point of hemispheric preference and the exact nature of the hierarchy of listening to spoken language, from the processing of basic speech sounds to the comprehension of complex sentences, are still not completely understood. There is no question that several localized brain regions such as Broca and Wernicke areas in the left hemisphere are representing language-specific skills, both for comprehension and articulation. Still, listening to spoken language is a fairly complex process, activating the complete auditory pathway from brainstem to cortex, and language processing in general is not restricted to the "classical" speech areas. Many separate structures and different processes are involved in the extraction, classification, and recombination of speech-specific features, which in the end transform an acoustic signal into a clear-cut utterance.

In a previous fMRI study (Uppenkamp et al. 2006), the initial stages of the processing of speech sounds were localized in the temporal lobes by comparing brain activation in response to synthesized human vowels and acoustically-matched control stimuli that were rated as completely nonspeech like. The results indicated that early cortical stages of processing, including primary auditory cortex, respond indiscriminately to speech and nonspeech sounds. Only regions in the superior temporal sulcus, lateral and inferior to anatomically defined auditory cortex, showed selectivity for speech sounds.

The current study is now aimed at identifying those brain regions that process the meaning of spoken language, irrespective of its acoustical properties, by means of functional MRI. Distorted speech material is used along with natural speech to allow for a controlled manipulation of intelligibility, depending on the presentation order of distorted and clean speech signals. The effect that distorted speech will be intelligible after a previous presentation of a clean version of the same material is utilized to record brain activation in response to physically identical acoustical stimuli, which only differ in their intelligibility.

36.2 Generation of Distorted Speech

Speech distortion was performed according to a procedure of rotating spectral features (Blesser 1972). These signals have recently been used in various neuroimaging studies to identify speech-specific brain activation (Scott et al. 2000; Narain et al. 2003). A matlab routine by Scott et al. (2000), which imitates the original analogue processing steps by Blesser (1972), was slightly modified for this study to adapt it to the own speech material, which is 230 sentences of a German speech intelligibility test (Kollmeier and Wesselkamp 1997). The sentences in this test are phrases from everyday life and vary in length from three to seven words.

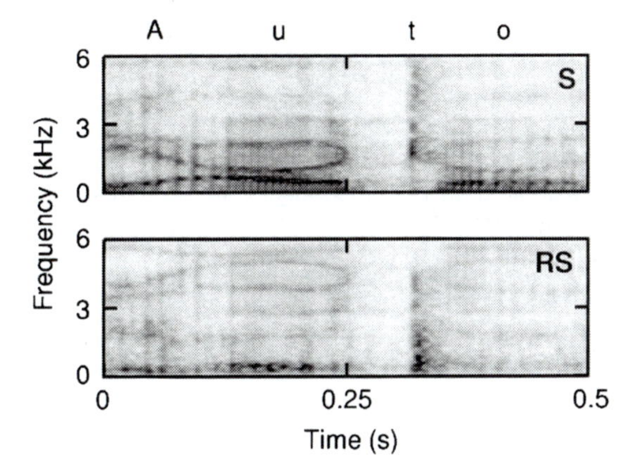

Fig. 36.1 Spectrograms of clean (*top*: *S*) and rotated (*bottom*: *RS*) versions of the German spoken word "Auto" (*car*)

The speech material is balanced for phonemic homogeneity. The recording is from a male speaker.

Spectral rotation around a centre frequency of 3 kHz is performed by the following steps: The speech material is first low-pass filtered at 5.7 kHz. The next step is a filter with an equalization curve according to the inverse of the long-term average speech-spectrum (Byrne et al. 1994) in order to achieve a speech-like long-term spectrum for the spectrally inverted signal. The third step is a multiplication with a sinusoid at 6 kHz to perform spectral rotation. Finally, the signal is again low-pass filtered at 5.7 kHz. One example is shown in the spectrograms in Fig. 36.1. The top panel shows the original spectrogram (speech signal *S*), low-pass filtered at 5.7 kHz, and the bottom panel shows the rotated spectrogram of the same word (*RS*). The figure illustrates that the spectral pattern is inverted along the frequency axis, while the envelope structure is not changed. Informal listening reveals that running speech is not intelligible after this manipulation, although most listeners realize that the signal has something to do with human speech.

36.3 Psychophysics of Spectrally Rotated Speech

In the original study by Blesser (1972), volunteers reached an intelligibility of rotated speech between 9% and 14% for single words and 5% for complete sentences. However, when the meaning of the sentence is known, either through a previous presentation of a clean version of the same sentence, or by simultaneous visual presentation, then the listeners have the impression that they can understand rotated speech, and recognition rates rise to more than 90%. A similar observation was demonstrated for sine-wave speech (Remez et al. 1981) and for noise-vocoded speech (Davis et al. 2005; Hervais-Adelman et al. 2008).

Psychophysical experiments were performed to quantify the effect of spectral distortion on the speech intelligibility. The main questions during this investigation were:

1. How much can the listeners understand of the rotated speech material in this study when they do not know the meaning of the presented sentences?
2. How much is understood when a clean version of the same sentence is presented immediately before the rotated version?
3. How long does it take for this effect (understanding distorted speech) to decrease over time, i.e. how does the intelligibility of rotated speech change when other speech material is presented between the clean and the rotated version of a particular sentence?

36.3.1 Subjects

Fifteen normal-hearing volunteers, aged between 21 and 39 years, participated in the speech intelligibility measurements. None of the volunteers had ever listened to spectrally rotated speech before, and the speech material itself was also unknown.

36.3.2 Stimulus Presentation

The canonical way to demonstrate the change in intelligibility of distorted speech, once the meaning of the sentence is known, would be the following presentation order: (1) Play a spectrally rotated sentence, which is initially unknown to the listener, *RS1*; (2) Play a clean version of the same sentence, *S*; (3) Repeat the spectrally distorted version, *RS2*. It is obvious that it will be easy to recognize the words in *RS2*, as long as the words of *S* are still preserved in memory. However, if some other speech material, either distorted or clean, is presented between *S* and *RS2*, the memory trace will decay rather quickly, and the intelligibility of *RS2* will decrease again as a function of distance between *S* and *RS2*. This effect is explored quantitatively in our paradigm.

Each volunteer was listening to about 200 different sentences, each sentence was presented three times, as *RS1*, *S*, and *RS2*, making a total of 600 presented items. The presentation order for the different items was interleaved. After a particular sentence $RS1_{(n)}$, between 0 and 30 different items were presented before the initial sentence was played as $S_{(n)}$. After that, either 1, 2, 3, 5, or 8 other items were played before $RS2_{(n)}$ was presented. After each item, the listeners were asked to repeat the words they had understood. At the end of the experiment, the percentage of correctly identified words is determined for each condition, *RS1*, *S*, and *RS2*. For the *RS2* conditions, this percentage is separately calculated for the five different distances between *S* and *RS2*.

In their original application for speech audiometry (Kollmeier and Wesselkamp 1997), the sentences are grouped in ten lists with 20 sentences each, which are balanced for equal intelligibility. These predefined test lists, however, were ignored throughout this study and the presentation order of the sentences was randomized for each listener to avoid an overall bias caused by a fixed sequence. All speech material was presented diotically via headphones Sennheiser HDA200 at a speech level of 65 dB SPL.

36.3.3 Results

All correctly identified words are counted for each *RS1* presentation and each listener, and plotted in relation to the overall number of presented words, to answer the first question, "How much can listeners understand during the initial presentation of rotated speech?" The results as a function of time, measured as total number of sentence already presented in the course of the experiment, are shown in Fig. 36.2a. The solid black line gives the accumulated percentage of correctly identified words. During the complete experiment, 198 sentences with three to seven words (1,011 words in total) were presented to each listener. The overall percentage of correct answers is less than 10% for most listeners, and the learning effect over time is comparatively small, indicating that spectrally rotated speech was completely unintelligible for nearly all participants, as long as the

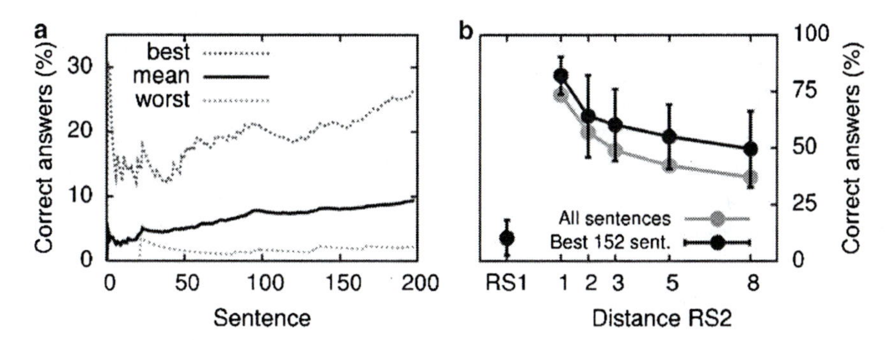

Fig. 36.2 (**a**) Mean learning effect for the intelligibility of spectrally rotated speech for 15 normal-hearing listeners. The *solid black line* shows the accumulated percentage of correctly identified words. During the complete experiment, 198 sentences with three to seven words (1,011 words in total) were presented to each listener, with a different presentation order for each individual. The *top dotted line* shows the individual learning curve for the most successful listener, the *bottom dotted line* the one for the least successful listener. (**b**) Mean intelligibility of spectrally rotated speech, after previous presentation of a clean version of the same sentences. The single dot marked as "*RS1*" illustrates the average intelligibility from (**a**), which is for the initial presentation of the rotated sentences. The other data points ("*RS2*") give the intelligibility as a function of the distance between the second presentation of the spectrally rotated sentence and an interim presentation of a clean version of the same sentence

sentences were not known beforehand. The top dotted line gives the individual learning curve for the most successful listener and the bottom dotted line for the least successful listener.

Figure 36.2b illustrates the effect of the interim presentation of a clean version of the same speech material, before spectrally rotated speech had to be identified for a second time. The single dot marked as "*RS1*" illustrates the average intelligibility (9.8%) and one standard deviation across subjects from Fig. 36.2a, which is for the initial presentation of the rotated sentences. The other data points ("*RS2*") give the intelligibility as a function of the distance between the second presentation of the spectrally rotated sentences and the interim presentation of the undistorted version. The overall intelligibility of this second presentation is strongly affected by the number of other sentences played in between. When rotated speech is immediately following an undistorted version of the same sentence (distance 0, not shown here), intelligibility is close to 100%. The effect of decreasing intelligibility with increasing distance is mainly one of a decay of memory traces. The grey line shows the data for all 230 sentences and the black line shows the results for those 152 sentences that exhibited the biggest difference between first and second presentation of the spectrally rotated version. These sentences were chosen as speech database for the following fMRI experiment, to maximize the perceptual contrast between conditions.

36.4 Brain Activation in Response to Distorted Speech Stimuli

The findings from the psychophysical tests were used to design a functional MR imaging experiment, aimed at identifying those regions in the brain that represent the intelligibility of spoken language, irrespective of acoustical features. Essentially, due to the effect of an interim presentation of an undistorted version of the same speech material, the intelligibility of physically identical distorted speech stimuli can be switched from less than 10% to more than 95%, or, depending on the number of additional items in between, to a value around 50%.

36.4.1 Stimulus Presentation

Seven stimulus conditions were chosen for the fMRI sessions, five different speech conditions and in addition *silence* (no acoustic stimulation) and *random noise* only as control conditions. An overview of all seven conditions is given in Table 36.1. Most stimulus conditions were repeated 38 times throughout the experiment. The conditions *RS1* and *S* occurred 76 times (in two groups of 38 presentations), which was required to have a complete set of control stimuli for the two separate *RS2* conditions, *RS2-0* and *RS2-5*. All stimuli were presented

Table 36.1 Overview of stimulus conditions in functional MRI experiment

Stimulus name	Description	Overall intelligibility (%)
Silence	No acoustic stimulus	–
Noise	Random noise only, 65 dB	–
SR	Speech in masking noise at –6 dB S/N	50
RS1	Initial presentation of rotated speech	10
S	Clean, undistorted speech	100
RS2-0	Second presentation of rotated speech, immediately following the presentation of a clean version	95
RS2-5	Second presentation of rotated speech, separated by five items from the previous clean version	50

diotically via MR compatible, electromagnetic headphones (Baumgart et al. 1998) at a presentation level of 65 dB. The presentation order of sentences was randomized for each participant.

36.4.2 Subjects and Task

A new group of 15 volunteers, aged between 18 and 35 years, was recruited to participate in the fMRI recordings. Twelve of the volunteers were right handed; for three listeners, the Edinburgh inventory (Oldeld 1971) exhibited an ambidextrous handedness. None of the participants had ever listened to spectrally rotated speech before, and nobody knew the speech material in advance. The task for the listeners was a simple subjective ad hoc assessment of overall intelligibility ("Did you understand the sentence?" yes/no buttons). This task was mainly introduced to maintain the attention throughout the experiment. However, the responses were compared with the results from the full speech intelligibility tests and essentially reflected the previous findings.

36.4.3 Scanning Procedure

Functional MRI data were acquired using a Siemens Sonata 1.5 Tesla MRI system. The 342 echo-planar imaging (EPI) volumes in total were acquired over three runs of 114 volumes each. A T_1-weighted high-resolution anatomical image (176 sagittal slices, $TR=2.11$ s, $TE=4.38$ ms) was also collected for each subject. For the functional data, 21 axial EPI slices (in-plane resolution 3×3 mm; thickness 5 mm, echo time $TE=63$ ms to maximize BOLD contrast) were acquired covering most of the cortex, including the whole of the temporal lobes and frontal regions. Sparse imaging with clustered volume acquisition was used (Hall et al. 1999). The total volume acquisition time *TA* was 2.7 s. On each trial, there was a 6.3-s stimulus interval followed by the 2.7-s scanning interval, making a total repetition time of $TR=9$ s. Two items from the same condition were presented within one stimulus interval.

36.4.4 Data Analysis

All anatomical and functional data were analyzed using statistical parametric mapping (SPM5, http://www.fil.ion.ucl.ac.uk/spm). The preprocessing of the functional brain images included realignment of subject motion, normalization to a standard EPI template, and smoothing with a Gaussian filter of 5 mm full width at half maximum for all directions. Fixed-effects analysis was conducted across the whole group of 15 subjects (5,130 scans in total) using the general linear model.

36.4.5 Results

A summary of the activation maps in response to spectrally rotated speech is illustrated in Fig. 36.3. The top row (Fig. 36.3a) shows two contrasts based on a difference in acoustical features of the presented sounds. The areas marked in light grey are those regions that showed significantly more activation in response to natural speech than in response to the noise-only control condition. These two conditions clearly differ in temporal as well as spectral features, with the speech spectrum exhibiting a harmonic structure with prominent formant frequencies and distinct modulations of the envelope. Both superior and medial temporal gyrus, including parts of Heschl's gyrus, are showing significantly more activation for the speech signal than for noise. This is equally the case for both hemispheres.

The next contrast (darker grey in Fig. 36.3a) reflects the differences between the natural speech signal and the initial presentation of spectrally rotated speech. As demonstrated in the previous section, the latter stimulus is nearly unintelligible and sounds rather distorted, although overall temporal and spectral features are pretty similar for these conditions. Only regions below the surface of the temporal lobes, including superior temporal sulcus (STS) and medial temporal gyrus (MTG), exhibit a differential activation, and none of the primary auditory regions is involved. Again, activation patterns in the left and the right hemisphere look very similar, with no preference of the left hemisphere for the undistorted speech signals.

The bottom row of the figure (Fig. 36.3b) shows the isolated effect of the intelligibility of distorted speech, irrespective of acoustical features. Areas marked in light grey represent those regions that show significantly more activation in condition *RS2-0* (second presentation of rotated speech at 95% intelligibility) than in condition *RS1* (initial presentation of rotated speech at 10% intelligibility). Areas marked in darker grey show the similar contrast for the conditions *RS2-5* (50% intelligibility) and condition *RS1*. All conditions, *RS1*, *RS2-0,* and *RS2-5* included

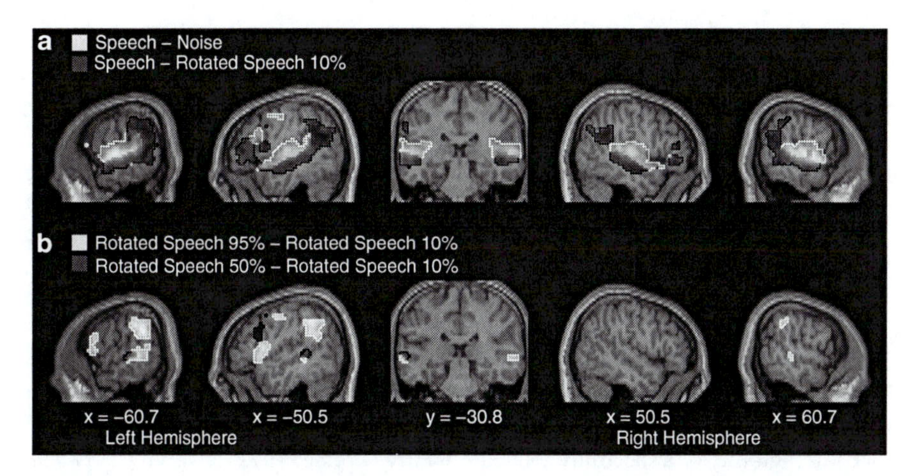

Fig. 36.3 Functional activation maps for natural and spectrally rotated speech material, group analysis for 15 normal-hearing listeners. (**a**) Contrast between natural speech and noise (*light grey*) and natural speech and 10%-intelligible rotated speech (*RS1*, *darker grey*). BOLD activation appears bilateral in superior temporal sulcus, especially posterior STS (Wernicke area), MTG, pSTG, SMG, IFG, precuneus, and in anterior STS and precentral gyrus (Broca area) in the left hemisphere. (**b**) Contrasts in activation for identical acoustic stimuli, i.e., spectrally rotated speech at different intelligibilities. *Light grey*: *RS2-0* (95% intelligible) vs. *RS1* (10% intelligible), activation in SMG, STS, *Darker grey*: *RS2-5* (50% intelligible) vs. *RS1* (10% intelligible). BOLD activation appears in SMG, STS. IFG, precentral gyrus and parts of MFG, mainly in the left hemisphere

the same distorted speech material, and only the intelligibility is varied due to the effect of presentation order relative to the respective undistorted sentences. The most important finding is that all auditory regions in the temporal lobes, which previously showed a significant difference in activation between the natural speech condition and unintelligible distorted speech, are now quiet. Activation is now largely restricted to four separate areas in the left hemisphere, including supramarginal gyrus (SMG), superior temporal sulcus (STS), inferior frontal gyrus (IFG), and precentral gyrus.

36.5 Discussion

The fMRI data indicate that speech stimuli in general cause activation in the temporal lobes of both hemispheres, although the contrast between natural and distorted speech only comes up in regions beyond primary auditory areas, mainly in superior temporal gyrus, bilaterally. This finding is in good agreement with previous work on speech-sound specific activation (Davis and Johnsrude 2003; Narain et al. 2003; Uppenkamp et al. 2006). The contrast between acoustically identical stimuli only differing in their intelligibility reveals several areas, largely in the left hemisphere and clearly separate from the temporal lobes.

These regions are interpreted to represent the processing of the meaning of intelligible speech, irrespective of its exact acoustic realization. Two of the activated regions in this study (STS and IFG) can be assigned to the classical language areas, Wernicke and Broca, which both appear to be heavily involved in the processing stream for the meaning of spoken language. This processing stream is lateralized in the left hemisphere. The greater the number of intelligible words, the more activation is found. For both hemispheres, there is no significant activation in the auditory cortex in the temporal lobes.

36.6 Conclusions

The intelligibility of a spectrally rotated sentence (Blesser 1972) is varying between "completely not intelligible" and "fully intelligible," depending on its exact position relative to the undistorted version of the same sentence. A general speech-specific activation was found bilaterally and mostly in the temporal lobes, based on the contrast between the clean speech and the unintelligible rotated-speech condition. This activation supports the idea of a dorsal and ventral processing stream for speech (Scott and Wise 2004; Davis and Johnsrude 2007; Hickok and Poeppel 2007). Using the effect of the interim presentation of an undistorted version of the same speech material, physically identical distorted speech stimuli could be employed to identify regions in human cortex that are specifically sensitive to the meaning of spoken language, irrespective of its acoustic realization. The results demonstrate that the strong lateralization of language comprehension to the left hemisphere does not occur before this stage.

Acknowledgment The authors are grateful for the support of Birger Kollmeier and the Medical Physics group in Oldenburg. This work was funded by Deutsche Forschungsgemeinschaft (Up 10/2-2).

References

Baumgart F, Kaulisch T, Tempelmann C, Gaschler-Markefski B, Tegeler C, Schindler F, Stiller D, Scheich H (1998) Electrodynamic headphones and woofers for application in magnetic resonance imaging scanners. Med Phys 25:2068–2070

Blesser B (1972) Speech perception under conditions of spectral transformation: I. phonetic characteristics. J Speech Hear Res 15:5–41

Byrne D, Dillon H, Tran K, Arlinger S, Wilbraham K et al (1994) An international comparison of long-term average speech spectra. J Acoust Soc Am 96:2108–2120

Davis MH, Johnsrude IS (2003) Hierarchical processing in spoken language comprehension. J Neurosci 23(8):3423–3431

Davis MH, Johnsrude IS (2007) Hearing speech sounds: top-down influences on the interface between audition and speech perception. Hear Res 229(1–2):132–147

Davis MH, Johnsrude IS, Hervais-Adelman A, Taylor K, McGettigan C (2005) Lexical information drives perceptual learning of distorted speech: evidence from the comprehension of noise-vocoded sentences. J Exp Psychol Gen 134(2):222–241

Hall DA, Haggard MP, Akeroyd MA, Palmer AR, Summerfield AQ, Elliott MR, Gurney EM, Bowtell RW (1999) “Sparse” temporal sampling in auditory fMRI. Hum Brain Mapp 7(3):213–223

Hervais-Adelman A, Davis MH, Johnsrude IS, Carlyon RP (2008) Perceptual learning of noise vocoded words: effects of feedback and lexicality. J Exp Psychol: Hum Percept Perform 34(2):460–474

Hickok G, Poeppel D (2007) The cortical organization of speech processing. Nat Rev Neurosci 8(5):393–402

Kollmeier B, Wesselkamp M (1997) Development and evaluation of a german sentence test for objective and subjective speech intelligibility assessment. J Acoust Soc Am 102(4): 2412–2421

Narain C, Scott SK, Wise RJS, Rosen S, Le A, Iversen SD, Matthews PM (2003) Dening a left-lateralized response specific to intelligible speech using fMRI. Cereb Cortex 13(12): 1362–1368

Oldeld RC (1971) The assessment and analysis of handedness: the Edinburgh inventory. Neuropsychologia 9(1):97–113

Remez RE, Rubin PE, Pisoni DB, Carrell TD (1981) Speech perception without traditional speech cues. Science 212(4497):947–949

Scott SK, Wise RJS (2004) The functional neuroanatomy of prelexical processing in speech perception. Cognition 92(1–2):13–45

Scott SK, Blank CC, Rosen S, Wise RJS (2000) Identification of a pathway for intelligible speech in the left temporal lobe. Brain 123(Pt 12):2400–2406

Uppenkamp S, Johnsrude IS, Norris D, Marslen-Wilson W, Patterson RD (2006) Locating the initial stages of speech-sound processing in human temporal cortex. Neuroimage 31(3): 1284–1296

Chapter 37
Intelligibility of Time-Compressed Speech with Periodic and Aperiodic Insertions of Silence: Evidence for Endogenous Brain Rhythms in Speech Perception?

Oded Ghitza and Steven Greenberg

Abstract This study was motivated by the hypothesis that low-frequency cortical oscillations help the brain decode the speech signal. The intelligibility (in terms of word error rate) of natural-sounding, synthetically-generated sentences was measured using a paradigm that alters speech-energy rhythm over a range of modulation frequencies. The material comprised 96 semantically unpredictable sentences (SUS), each approximately 2 s long, generated by a high-quality text-to-speech (TTS) synthesis engine. The TTS waveform was time-compressed by a factor of 3, creating a signal with a syllable rhythm three times faster than the original and whose intelligibility is poor (<50% words correct). A waveform with an *artificial* rhythm was produced by segmenting the time-compressed waveform into consecutive 40-ms fragments each followed by a *silent* interval. The parameters varied were the length of the silent interval (0–160 ms) and whether the intervals of silence were equal ("periodic") or not ("aperiodic"). The performance curve (word error rate as a function of mean duration of silence) was U-shaped. The lowest word error rate occurred when the silence was 80-ms long and inserted periodically. This was also the condition for which word error rate increased when the silence was inserted aperiodically. These data are consistent with a model ("TEMPO") in which low-frequency brain rhythms influence the ability to decode the speech signal. In TEMPO, optimum intelligibility is achieved when the syllable rhythm is within the range of the high-theta frequency brain rhythms (6–12 Hz), comparable to the rate at which segments and syllables are spoken in conversational speech.

Keywords Speech perception • Intelligibility • Time-compressed speech • Brain rhythms • Cortical oscillations

---

O. Ghitza (✉)
Sensimetrics Corporation and Boston University, Boston, MA, USA
e-mail: oded@sens.com

E.A. Lopez-Poveda et al. (eds.), *The Neurophysiological Bases of Auditory Perception*,
DOI 10.1007/978-1-4419-5686-6_37, © Springer Science+Business Media, LLC 2010

37.1 Introduction

Speech is inherently rhythmic and temporally structured. Most of its energy fluctuates between 3 and 20 Hz (e.g., Greenberg and Arai 2004). These fluctuations are not perfectly periodic; rather there are constraints on syllable duration and fluctuation patterns within and across prosodic boundaries. Long syllables are often followed (and preceded) by syllables shorter in duration. And conversely, short syllables are typically preceded and followed by longer ones. Syllable amplitude follows a similar pattern, as does fundamental frequency (intonation – e.g., Liberman 1975; Ladd 1996). This rhythmic variation is important for intelligibility and naturalness (e.g., van Santen et al. 2008; Schroeter 2008). Does it reflect some fundamental property, one internal to the brain?

In our view, many aspects of naturally spoken speech could reflect properties of higher-order cortical processing and auditory function, not just biomechanical and articulatory constraints. In particular, speech's temporal properties are likely constrained not only by how fast the articulators *can* move but also by how long certain phonetic and syllabic elements need to be for the signal to sound intelligible and natural. The range of time intervals (40–2,000 ms) associated with different levels of linguistic abstraction (phonetic feature, segment, syllable, word, metrical foot, and prosodic phrase) may reflect temporal constraints associated with neural circuits in the cerebral cortex, thalamus, hippocampus, and other regions of the brain. More specifically, certain endogenous neural rhythms could be the reflection of both local and longer-range, transcortical processing (e.g., von Stein and Sarnthein 2000; Buzsáki 2006). The modulation-frequency range over which such rhythms operate (0.5–80 Hz) may serve as the basis for hierarchical synchronization function through which the central nervous system processes and integrates sensory information (Lakatos et al. 2005; Freeman 2007). It may also reflect a hierarchy of topographic and neural scales, in which higher (and more abstract) levels of processing depend on information from more extensive cortical areas (von Stein and Sarnthein 2000).

Such neural rhythms could play an important role in spoken-language comprehension (e.g., Giraud et al. 2007). Of particular interest are the gamma and theta rhythms (e.g., Bastiaansen and Hagoort 2006; Giraud et al. 2007; Luo and Poeppel 2007). Theta activity (3–12 Hz) is most closely associated (linguistically) with the syllable (mean duration 200 ms, core range 100–300 ms; Greenberg 1999) and the segment (mean duration 80 ms, core range 60–150 ms; Greenberg et al. 1996), and is thought to involve some form of sensory-memory comparison process. Gamma oscillations (30–80 Hz) are most closely associated with neural processing of phonetic constituents and features. Finally, delta oscillations (0.5–3 Hz) may be involved in processing sequences of syllables and words embedded within the metrical foot and prosodic phrase, which could be important for certain aspects of linguistic processing (Roehm et al. 2004).

What is the relation between brain rhythms and naturally spoken speech? And why should we expect the perception of speech to be influenced by neural oscillations?

As discussed above, there is a close correspondence between the time scales of speech and the frequencies of endogenous brain oscillations. Is this temporal isomorphism a mere coincidence, or does it reflect something deeper? Although this question cannot be answered directly by this (or any other psychophysically-based) study, we can begin to investigate this possibility by perturbing the speech input in ways that potentially challenge the operation of brain rhythms and ascertain the impact on intelligibility. In our study, short sentences were interrupted with variable lengths of silence, both periodically and aperiodically. Interruption aperiodicity was used to gauge how underlying neural rhythms interact during the speech-decoding process.

37.2 Background

Miller and Licklider (1950) were among the first to systematically examine the temporal parameters associated with the perception of speech. In their study, monosyllabic words were presented using a variety of signal-processing conditions. Common to all conditions was the use of an analog gating device (a square-wave generator), which interrupted the speech signal over a range of periodic intervals (interruption frequencies ranging between 0.1 and 10,000 Hz). In the simplest set of conditions, the speech signal was gated on for a specific time (e.g., 50 ms) and then gated off for the same interval (i.e., an interruption frequency of 10 Hz in this example). Interruption frequencies above 100 Hz resulted in relatively little degradation (and which could be attributed to spectral distortion). For interruption rates between 1 and 10 Hz, intelligibility declined dramatically. Miller and Licklider speculated that under such conditions, intelligibility is governed by the number of "glimpses" associated with each phonetic segment or syllable in the word. When the interruption rate is extremely low, certain segments, syllables or words could not be glimpsed in their entirety, resulting in word-recognition difficulty.

Miller and Licklider also varied the "speech-time" ratio, the proportion of the gating cycle occupied by the acoustic signal. For example, for a speech-time ratio of 0.25 and an interruption rate of 10 Hz, 25 ms of speech would be followed by 75 ms of silence. As the speech ratio increased, so did intelligibility, which became increasingly sensitive to interruption frequency when the speech-time ratio was reduced from 50% to 25%. Such a result implies a complex interaction between the speech signal glimpsed and the interval of time over which the brain integrates the interleaved speech-silence signal.

Many issues regarding the temporal processing of speech were left unresolved by Miller and Licklider's study. For example, the signal was (in the listening conditions of interest) either *interrupted* or *masked* (depending on the specific condition), which means that some portion of the acoustic signal was either discarded or acoustically obscured, and hence withheld from the listener. Is this speech data loss important for intelligibility, or not?

In 1975, Huggins revisited the issue through an ingenious set of experiments. He sought to delineate the underlying temporal factors governing intelligibility. Instead of monosyllabic words, spoken passages (ca. 150 words long) were used. Listeners continuously "shadowed" this material (i.e., speaking the words at a comfortable, self-determined delay). Instead of varying the interruption rate, Huggins inserted silence ranging between 16 and 500 ms (the range over which intelligibility was most affected in Miller and Licklider's study). In contrast to Miller and Licklider's paradigm, no speech was discarded in Huggins' study. Word-error rate was measured as a function of speech-time and silence-time durations. With the duration of the speech interval fixed at (for example) 63 ms, shadowing performance was high for short silent gaps (<60 ms). However, when the silent gap was long (>150 ms), intelligibility was poor. Huggins suggested that intelligibility depends on "gap bridging" in these experiments. In his view, there is an ~180-ms-long echoic memory buffer. As long as adjacent speech fragments fall within this interval, the brain is able to extract sufficient detail from the acoustic signal to construct a coherent linguistic message. Huggins suggested that the main factor governing intelligibility was not phonetic glimpsing per se, but rather some internal time constraint on processing spoken material.

In the current study, we offer an alternative hypothesis – that the decline in intelligibility is the result of a disruption in the syllabic rhythm beyond the limits of what neural circuitry can handle. We test this hypothesis by conducting an experiment that extends Huggins' study in a number of important ways.

37.3 Experiment

A drawback of using semantically plausible material is the ability of listeners to guess words using contextual information. Semantically unpredictable sentences (SUS) make it more difficult for the listener to guess individual words on the basis of semantic criteria.

37.3.1 SUS Corpus

The corpus used comprised 96 SUS, each approximately 2 s in length (6–8 words per sentence). Each sentence conformed to standard rules of English grammar, but was composed of semantically anomalous word combinations (e.g., "Where does the cost feel the low night?"). Text word sequences were generated by Tim Bunnell's "SUSGEN" program, which produces sentences conforming to a specified grammar and uses a constrained vocabulary (Bunnell et al. 2005).

37.3.2 Stimulus Preparation

It is difficult for a human talker to speak semantically anomalous sentences in a natural way. For this reason, the AT&T Text-to-Speech (TTS) System (http://www.research.att.com/~ttsweb/tts/demo.php) was used to produce natural-sounding, highly intelligible spoken material. The AT&T system is based on a form of unit-selection, concatenative synthesis that uses high-quality, prerecorded voices (Schroeter 2008). Each TTS-generated sentence was evaluated carefully for intelligibility and naturalness.

Following synthesis, the waveforms were time-compressed by a factor of 3 using a pitch-synchronous, overlap and add (PSOLA) procedure (Moulines and Charpentier 1990) incorporated into PRAAT, a speech analysis and modification program (http://www.fon.hum.uva.nl/praat/). In the time-compressed signal, the formant patterns and other spectral properties are altered in duration; however, the fundamental frequency ("pitch") contour remains the same. Figure 37.1, panel (b) shows the time-compressed version of the original (shown in panel (a)).

The time-compressed signal served as the baseline waveform for the insertion of silence. First, the waveform was segmented into consecutive 40-ms-long intervals. This segmentation remained fixed throughout. The silences were then inserted; the main parameter was the duration of silent intervals. The stimulus conditions are summarized in Table 37.1. To reduce the effect of transients, the onset and offset of each speech fragment was multiplied with a 1-ms (rise/fall time) cosine-shaped window. In order to perceptually mask the discontinuities associated with the signal processing used to insert silence, a speech-spectrum-shaped noise was added to the entire signal after the insertions. The noise level was adjusted to an SNR of 30 dB relative to the power of the signal prior to silence insertions (i.e., condition x0). This background noise was kept intentionally low in order not to mask speech energy required to decode the signal. Panels (b), (c), and (d) in Fig. 37.1 show the waveforms for conditions x0, x40, and x80, respectively.

The conditions in Table 37.1 are the periodic class of conditions. An *aperiodic* set was also created, one in which for each condition the silence-time interval was of *variable* duration, chosen quasi-randomly from one of a set of four intervals (equal to 0.5, 0.83, 1.16, and 1.5 of a mean silence interval listed in Table 37.1).

37.3.3 Subjects

All five listening subjects were young adults (ages 22–27), educated in the U.S., with normal hearing and no history of auditory pathology.

37.3.4 Instructions to Subjects

Subjects performed the experiment in their home/office environment using headphones. There were two separate listening sessions, "Training" and "Testing," each lasting approximately 30 min. In the training session, the subject listened to the original, unprocessed (96) sentences. In the test session, the subject listened to the same 96 sentences, but this time in processed form. The 96 sentences were divided into 12 groups of eight sentences each, with six groups for periodic

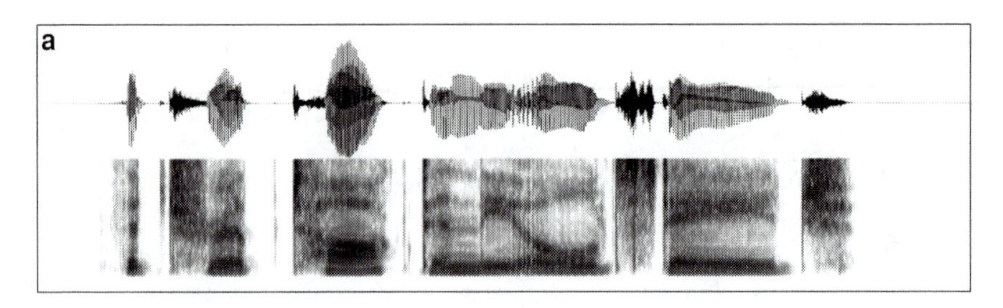

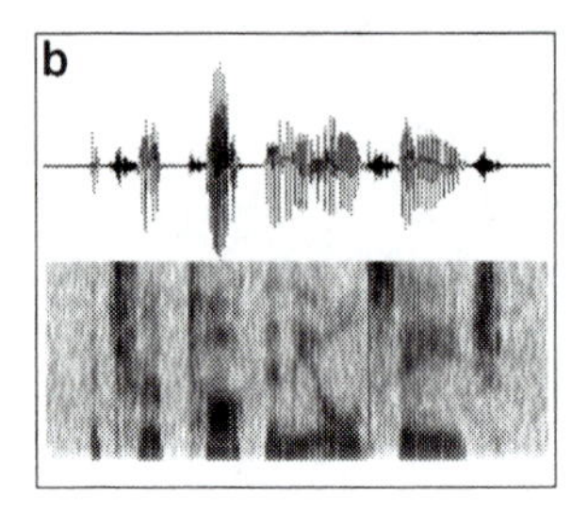

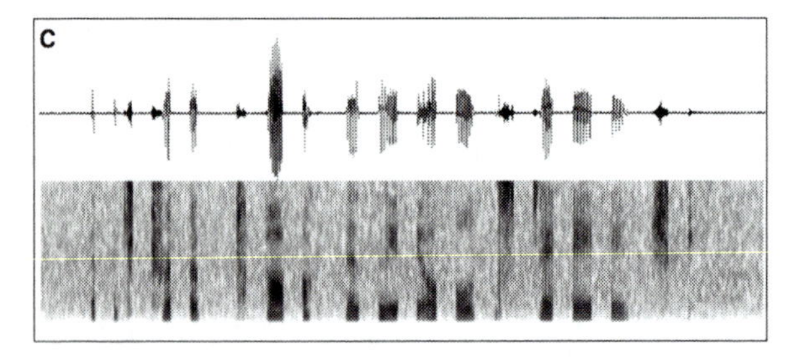

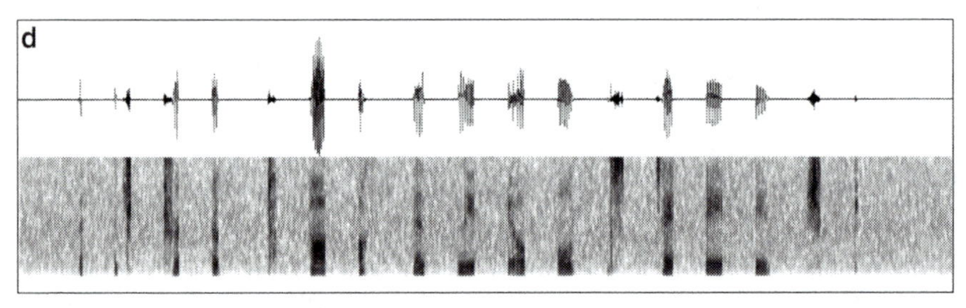

Fig. 37.1 (**a**) Waveform (*top*) and broadband spectrogram (*bottom*) of the sentence "the trip talked in the old stage". The waveform duration is 2.4 s, the upper frequency of the spectrogram 5 kHz; (**b**) Same as (**a**), time-compressed by a factor of 3; (**c**) Consecutive 40-ms speech intervals of (**b**), with 40-ms silence insertions. Note that the duration of the processed speech waveform is two-thirds the duration of the original signal (i.e., time-compressed by a factor of 1.5 relative to the original); (**d**) Same as (**c**) with 80-ms silence intervals. The waveform's duration is the same as the original (i.e., no time compression relative to the original). Note that the speech intervals are identical to those in (**c**). The background noise visible in the spectrogram was intended to mask discontinuities resulting from inserting silence into the acoustic signal

Table 37.1 Experimental conditions used in the study

Condition	Speech interval (ms)	Silence interval (ms)	Silence/speech	Speed
x0	40	0	0	3
x20	40	20	0.5	2
x40	40	40	1	1.5
x80	40	80	2	1
x120	40	120	3	0.75
x160	40	160	4	0.6

"Speed" refers to the duration of the signal relative to the original, uncompressed sentence

(covering the list of conditions in Table 37.1) and six for the aperiodic insertion conditions. Each sentence was listened to only once in the test phase. For each sentence, the subject was instructed to type the words heard (in the order presented) into an electronic file.

The experiment protocol was approved by the Institutional Review Board of Sensimetrics Corporation.

37.3.5 Results

37.3.5.1 Overall

In the training phase, the word error-rate was less than 2% for all subjects. Figure 37.2 shows the mean intelligibility in the test session (averaged over the five subjects) as a function of silence interval. Without silence insertions (condition x0), intelligibility is poor (<50% words correct). Intelligibility is equally poor (or worse) when the silence interval is 160 ms (condition x160). Interestingly, for silence intervals between 20 and 120 ms, intelligibility is far better. This is particularly true when the silence is 80 ms and inserted periodically. This is also the condition in which there is a significant difference in intelligibility between periodic and aperiodic insertion of silence (the error rate of the latter is nearly twice as high).

Two points are noteworthy. First, throughout all conditions, the spectro-temporal information of the speech fragments is time-compressed by a factor of 3. Thus, the U-shaped performance curve is an unexpected result difficult to explain in terms of conventional models of speech perception (see the discussion below). Second, the results indicate a performance advantage for a periodic syllabic rate (particularly for the silence interval condition of 80 ms). Such a result is also difficult to explain with conventional models.

37.3.5.2 Statistical Analysis

An analysis of variance (ANOVA) was used to compute the statistical significance of the data in Fig. 37.2. Two factors were used, insertion interval and type of insertion (i.e., periodic vs. aperiodic).

Mauchly's test for sphericity revealed that assumptions of sphericity were not violated. The omnibus repeated-measures 2-way (2-variable) ANOVA, with a significance (alpha) level of 0.05 (or 5%) showed that: (1) there is a significant main effect of insertion interval ($F(5,20)=24.163$, $p<0.0001$), (2) there is also a significant main effect of periodicity (i.e., aperiodic vs. periodic) ($F(1,4)=16.231$, $p<0.05$), and (3) there is no significant interaction between insertion interval and type of insertion (periodic/aperiodic) ($F(5,20)=2.371$, $p>0.05$). Post-hoc Tukey/Kramer tests revealed that there are significant differences across insertion interval conditions (collapsed across periodicity conditions): (a) the x0 condition differs from the x20, x40, x80, and x120 conditions, and that (b) the x160 condition differs from the x20, x40, x80, and x120 conditions.

37.4 Discussion

The data described in Sect. 37.2 show that intelligibility of the time-compressed speech is poor (ca. 50% word error) relative to the original (i.e., uncompressed) signal (<2% word error). Insertion of silent intervals markedly improves intelligibility – as long as the silence intervals are between 20 and 120 ms long (Fig. 37.2).

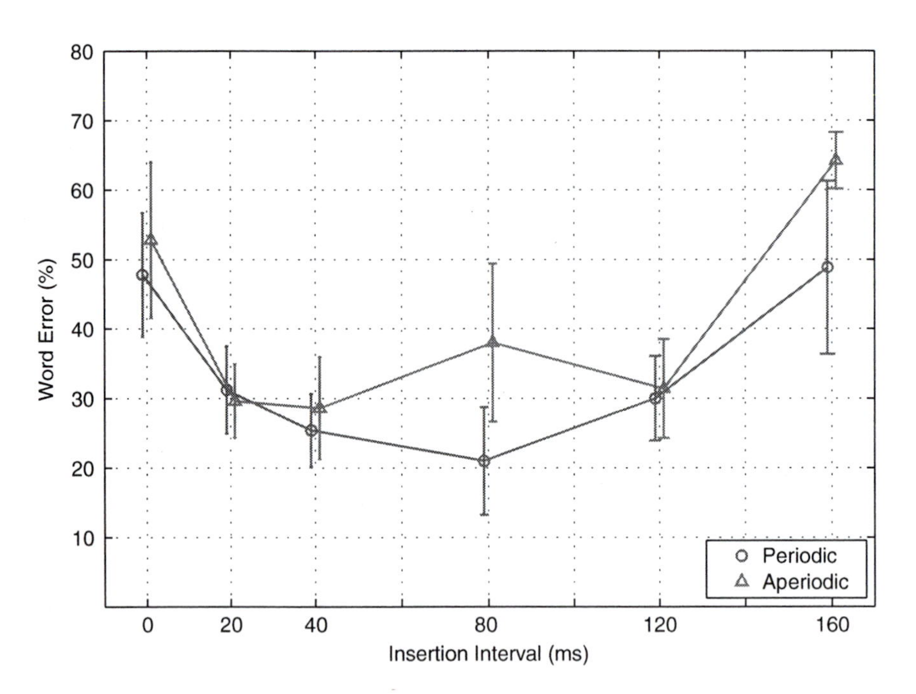

Fig. 37.2 Intelligibility of time-compressed speech as a function of the length of silence inserted. Word error rate is plotted as a function of silence interval (*error bars* represent the standard deviation of the mean). Speech was time-compressed by a factor of 3. Speech segments were consecutive 40-ms intervals and were the same for all conditions

Conventional models of speech perception have difficulty accounting for this result because they assume a strict decoding of the acoustic signal by the auditory system and higher neural centers.

Can the U-shaped intelligibility curve be explained simply by comparing the temporal properties of the distorted speech with the original, uncompressed waveform? In the 0-ms condition (i.e., no silence insertions), there was only a single distortion – linear time compression. This compression is sufficient to reduce intelligibility (from the original, unprocessed condition) by some 50%. Inserting fragments of silence 20–120 ms in length improves intelligibility dramatically. The best performance is observed when the silence is 80 ms long (and inserted periodically). In this condition, the packets of speech information (40-ms intervals of compressed speech) are aligned with the 120-ms intervals of the speech information in the original (uncompressed) speech.

Within the classical framework, intelligibility would not be expected to vary with the length of silence because insertions do not affect the acoustic signal directly, only the temporal distribution of speech information to the auditory system and the brain. However, it has been known since the studies of Miller and Licklider (1950) and Huggins (1975) that temporal packaging of the speech signal can exert an enormous impact on intelligibility. Miller and Licklider (1950) attributed most of the intelligibility decline to lost opportunities for glimpsing speech-relevant information in the signal. Specifically, they suggested that intelligibility depends on two parameters: (1) glimpsing rate, and (2) the amount of speech information transmitted per glimpse. In their experiment, the speech was uncompressed, and the amount of speech information deleted was reflected in the duty cycle. The shorter the duty cycle the smaller the ratio of speech to the overall interruption cycle. For example, for a fixed periodic interruption (i.e., glimpsing rate) of 5 Hz, the word error rate increases from ~30% for a 50% duty cycle to ~70% for 25% duty cycle. In the current experiment, the glimpsing rate was varied in a different way – by changing the duration of the silent interval. The amount of speech information per glimpse was held constant (it was determined by a speech-fragment duration of 40 ms and a time-compression ratio of 3).

In both our studies and in Miller and Licklider's, glimpsing appears to play an important role in intelligibility. However, there is a crucial difference between the two studies pertaining to the neural mechanisms involved in decoding speech. In Miller and Licklider's study, glimpsing rate and speech information per glimpse were confounded – the amount of speech information accessible to listeners was determined by the *interaction* of interruption rate and duty cycle. Hence, their results could be explained, at least in part, by the amount of speech information per glimpse available to the listener. In this sense, Miller and Licklider's study does not directly challenge conventional models of speech processing.

In contrast, the current study dissociated glimpsing rate from the amount of speech information per glimpse. Because no portion of the acoustic signal was discarded (only time-compressed), and because the speech intervals remained fixed throughout the experiment, the amount of speech information per glimpse was kept constant. Only the glimpsing rate varied, and its rate was directly tied to the length of inserted silence. Any change in intelligibility could therefore be attributed to glimpsing rate per se, rather than to the amount of information contained in the speech signal. This variation in intelligibility is much harder to explain in terms of the standard models of speech perception. It is not just the information in the acoustic signal that is important but also the timing of the information packets. This is a factor that the standard models do not address.

So, we pose the question: what are the neural mechanisms that distinguish glimpsing rates easy to linguistically decode from those that are not?

In our view, linguistically decodable glimpsing rates are likely controlled by endogenous brain rhythms. Neural circuitry involved in speech perception may "prefer" incoming sensory information in the form of temporally constrained packets that are compatible for pattern matching and which compensate for various forms of background interference (as well as variation in speaking rate). Normally, the operation of these cortical rhythms is "hidden" from view. Special methods are required to expose their influence. The aperiodic insertion conditions were designed to reveal the operation of such rhythms.

37.4.1 Does Intelligibility Reflect Short-Term Memory Limitations?

In Huggins' (1975) study, all of the speech information was preserved. Hence, the variation in intelligibility could only have been the result of the temporal distribution of speech information, analogous to what was observed in our own study. Huggins did not invoke brain rhythms as a potential explanation (his study was published in 1975). Instead, he suggested that the variation in intelligibility reflected limitations of short-term working memory. When the silence interval exceeded a certain duration, the time between adjacent speech fragments would be too long for the brain to integrate into a linguistically decodable stream.

If short-term memory was the primary factor affecting intelligibility, the word-error-rate pattern would have been very different. Performance would not necessarily have improved with the insertion of silence. Why should it, if the major parameter affecting intelligibility is the interval between successive speech fragments? In addition, performance would not have been different for the x80 periodic and the x80 aperiodic conditions, as the requirements pertaining to short-term memory are very similar. Although it is possible that short-term memory does play *some* role in decoding speech, it is unlikely to be the all-important factor suggested by Huggins.

37.4.2 Why Is the Intelligibility Curve U-Shaped?

What is particularly striking is that intelligibility *improves* so markedly when the signal is temporally distorted. This result demonstrates that intelligibility is not simply a matter of decoding the spectro–temporal pattern – something else is also going on.

One possible explanation is that the brain requires a time-window of a minimum duration in order to accurately decode the signal, and that this interval is within a period of time constrained by theta and alpha oscillations. Hence, our results may reflect the degree to which syllabic modulations are matched to those low-frequency rhythms internal to the brain. Theta rhythm, in particular, may reflect communication across distant brain regions (Buzsáki 2006). The hippocampus, important for short-term memory retrieval, also appears to be involved (Buzsáki 2006).

In order to gain insight into how brain rhythms affect speech decoding, we briefly describe a phenomenological model of speech processing, "TEMPO." In this model, the process of matching spectro–temporal patterns with phonetic and other types of linguistic elements is temporally controlled (and guided) by *nested* oscillators operating in the delta (<3 Hz), theta (3–12 Hz) and gamma (30–80 Hz) range. The delta frequency range has temporal properties commensurate with lexical- and phrasal-length units (400–2,000 ms), while theta oscillations are associated with individual segments (80–160 ms) and syllables (100–400 ms). Gamma oscillations (>25 Hz) possess temporal properties potentially relevant to rapid spectro–temporal (i.e., formant) transitions associated with diphone elements (i.e., consonant–vowel or vowel–consonant, 20–40 ms long). Although all three intervals are important for decoding speech, associated with information at the phonetic, segmental, syllabic, lexical, and phrasal levels, the current study focuses on rhythms in the theta range.

TEMPO encapsulates this multilevel property of speech, as shown in Fig. 37.3. Speech is processed using a model of the auditory periphery (e.g., Ghitza 2007), resulting in a spectro–temporal sensory representation. This multichannel information is then processed by a Template Matching Circuit (TMC) that matches "phonetic

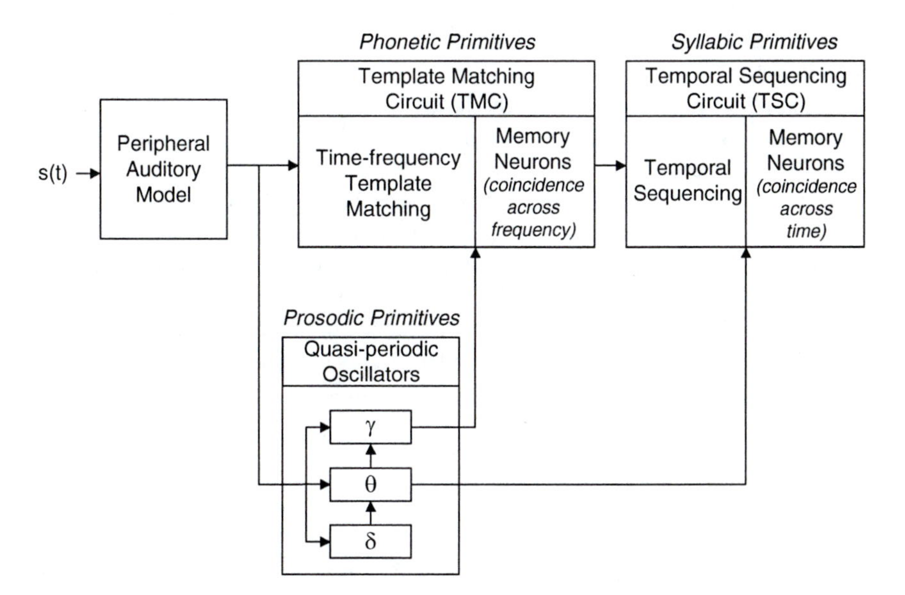

Fig. 37.3 A block diagram of the TEMPO model. See text for details

primitives" to "memory" neurons by computing coincidence in firing activity across frequency channels. At this level, time–frequency patterns are matched over relatively short time intervals (ca. 25–50 ms), which are usually associated with formant transitions and such phonetic features as place of articulation (important for distinguishing among consonants). In TEMPO, the pattern matching at this time scale is realized by a neuronal circuit regulated by gamma oscillations at its core (Ghitza 2007).

These memory neurons' firing patterns are temporally integrated by a Temporal Sequencing Circuit (TSC), with theta oscillations at their core. The theta oscillator operates on a time scale between 50 and 200 ms, and matches syllabic primitives to memory neurons by measuring coincidence in the firing activity of TMC memory neurons across time. The theta oscillations define the time windows in which this temporal integration occurs.

37.4.3 Condition x80: Why Does Intelligibility Deteriorate in the Aperiodic Condition?

Although the intelligibility of both the x80 periodic and x80 aperiodic conditions is higher than the baseline condition (i.e., x0, time compression with no silence insertions), the aperiodic condition is decoded more poorly than the periodic and is the only condition where the disparity in intelligibility is large. Moreover, the variability among listeners (as reflected in the standard deviation) is high. Recall that the silence intervals in the aperiodic condition x80 range between 40 and 120 ms (with a mean of 80 ms). Unlike condition x80, intelligibility in conditions x40 and x120 – periodic or aperiodic – is comparable. Moreover, these conditions are decoded far more accurately than the x80 aperiodic condition. Clearly, there is something important about randomly varying the length of the silence in this specific condition that presents problems for decoding speech (this is the case for all five subjects). What could this be?

We believe the differential intelligibility may be a reflection of brain rhythms involved in speech processing. If the speech-decoding process exploits an underlying synchronization mechanism in which theta oscillations act as a pacemaker, then a disruption in the input rhythm is likely to exert a negative impact on intelligibility. By varying the temporal distribution of acoustic information in a quasi-random fashion, it is possible that the correspondence between speech-input rhythm and the brain rhythms – at the heart of the linguistic-decoding process – has been disrupted.

Acknowledgments This study was funded by a research grant from the Air Force Office of Scientific Research. We thank Dr. Willard Larkin for his encouragement and valuable discussion as well as Professor Tim Bunnell for providing the SUSGEN sentence list. We also thank Dr. Ann Syrdal of AT&T who gave valuable advice about generating the stimuli and Dr. Udi Ghitza who provided valuable assistance with the statistical analyses.

References

Bastiaansen M, Hagoort P (2006) Oscillatory neuronal dynamics during language comprehension. Prog Brain Res 159:179–196

Bunnell HT, Pennington C, Yarrington D et al (2005) Automatic personal synthetic voice construction. In: Proceedings of Interspeech-2005, Lisbon, Portugal, pp 89–92

Buzsáki G (2006) Rhythms of the brain. Oxford University Press, New York

Freeman W (2007) My legacy: a launch pad for exploring neocortex. In: Brain network dynamics conference, Berkeley, CA, 26 January 2007

Ghitza O (2007) Using auditory feedback and rhythmicity for diphone discrimination of degraded speech. In: Proceedings of XVIth International Congress of Phonetic Sciences, Saarbrücken, Germany, pp 163–168

Giraud AL, Kleinschmidt A, Poeppel D et al (2007) Endogenous cortical rhythms determine cerebral specialisation for speech perception and production. Neuron 56:1127–1134

Greenberg S (1999) Speaking in shorthand – a syllable-centric perspective for understanding pronunciation variation. Speech Commun 29:159–176

Greenberg S, Arai T (2004) What are the essential cues for understanding spoken language? IEICE Trans Inf Syst E87:1059–1070

Greenberg S, Hollenback J, Ellis D (1996) Insights into spoken language gleaned from transcription of the Switchboard corpus. In: Proceedings of Fourth International Conference on Spoken Language Processing, Philadelphia, PA, pp S24–S27

Huggins AWF (1975) Temporally segmented speech. Percept Psychophys 18:149–157

Ladd R (1996) Intonational phonology. Cambridge University Press, Cambridge

Lakatos P, Shah AS, Knuth KH et al (2005) An oscillatory hierarchy controlling neuronal excitability and stimulus processing in the auditory cortex. J Neurophysiol 94:1904–1911

Liberman M (1975) The intonational system of English. Dissertation, MIT

Luo H, Poeppel D (2007) Phase patterns of neuronal responses reliably discriminate speech in human auditory cortex. Neuron 54:1001–1010

Miller GA, Licklider JCR (1950) The intelligibility of interrupted speech. J Acoust Soc Am 22:167–173

Moulines E, Charpentier F (1990) Pitch-synchronous waveform processing techniques for text-to-speech synthesis using diphones. Speech Commun 9:453–467

Roehm D, Schlesewsky M, Bornkessel I et al (2004) Fractionating language comprehension via frequency characteristics of the human EEG. Neuroreport 15:409–412

Schroeter J (2008) Basic principles of speech synthesis. In: Benesty J, Sondhi MM, Huang Y (eds) Handbook of speech processing. Springer-Verlag, Berlin, pp 413–428

van Santen JPH, Mishra T, Klabbers E (2008) Prosodic processing. In: Benesty J, Sondhi MM, Huang Y (eds) Handbook of speech processing. Springer-Verlag, Berlin, pp 471–487

von Stein A, Sarnthein J (2000) Different frequencies for different scales of cortical integration: from local gamma to long range alpha/theta synchronization. Int J Psychophysiol 38:301–313

Chapter 38
The Representation of the Pitch of Vowel Sounds in Ferret Auditory Cortex

Jan Schnupp, Andrew King, Kerry Walker, and Jennifer Bizley

Abstract According to ANSI, pitch is defined as that attribute according to which sounds can be ordered on a scale from low to high. However, most studies of pitch so far asked merely whether a subject can detect a pitch change, not whether he can also determine the direction of the pitch change. Using behavioural testing after operand conditioning we have shown that ferrets can discriminate rising from falling pitches in artificial vowels, although behavioural thresholds (with Weber fractions of ~0.4) may seem high compared to humans. We have further studied the cortical representation of these sounds in five distinct areas of ferret auditory cortex, A1, AAF, PPF, PSF, and ADF. Of over 600 units which responded to our artificial vowels, approximately half were sensitive to pitch changes. Thirty-eight percent of these exhibited high-pass, another 38% low-pass characteristics. The remaining 24% had non-monotonic pitch tuning curves. Our results did not reveal a clear 'pitch area'. Rather, pitch sensitive neurons could be found throughout all cortical areas studied. We used neurometric analysis to make the observed neural sensitivities more directly comparable to the animal's behavioural performance. Only a small number of units yielded neurometric curves which approached or matched the animal's psychometric functions. Furthermore a unit's neurometric could account for the animal's behavioural capabilities only around a small range of 'reference pitch' values around the steepest part of the unit's pitch tuning curves. We also developed a "population neurometric" analysis, with which we decoded the activity of up to 60 simultaneously recorded units. These population responses often performed substantially better than the individual constituent neurons. Populations also performed well over a wider range of reference pitches. Interestingly, spike count or spike latency based decoding of the population response yielded very similar neurometric curves.

Keywords Pitch • Complex sounds • Psychoacoustics • Auditory cortex • Physiology • Neural coding

---

J. Schnupp (✉)
Department of Physiology, Anatomy and Genetics, University of Oxford, Sherrington Building, Parks Road, Oxford, OX1 3PT, USA
e-mail: jan.schnupp@dpag.ox.ac.uk

E.A. Lopez-Poveda et al. (eds.), *The Neurophysiological Bases of Auditory Perception*,
DOI 10.1007/978-1-4419-5686-6_38, © Springer Science+Business Media, LLC 2010

38.1 Introduction

According to the American National Standards Institute (ANSI) 1994 definition, "pitch is that attribute of auditory sensation according to which sounds may be ordered on a scale extending from low to high." However, most studies of pitch discrimination so far, particularly those working with nonhuman animals (Dooling and Saunders 1975; Heffner et al. 1969; Nelson and Kiester 1978), have investigated merely whether a participant can detect a pitch change, not whether he can also determine the direction of the pitch change. On each trial in these go/no-go paradigms, a sound of constant pitch is presented repeatedly, and at a variable point in the sequence, the pitch of the sound changes by some small amount. The participant has to react to the change within a short reaction time window. However, the participants can perform well at this task as long as they are able to detect any change at all, and even if they were unable to classify this change as a "rise" or "fall" in pitch. Such tasks may therefore overestimate the subject's ability to perform "true" pitch discrimination in the sense of the ANSI definition. Indeed, a recent study of pitch discrimination in human children found significantly higher thresholds when the children had to judge the direction of a pitch change, rather than merely indicating that the pitch was no longer the same (Vongpaisal et al. 2006). We have therefore investigated the ability of ferrets, a commonly used animal model for mammalian hearing, to judge changes in pitch, not just as different, but as "higher" or "lower" than a given reference sound.

Most common everyday sounds are complex and contain numerous frequency components, so that at a neural level their pitch cannot simply be inferred tonotopically from the point of maximal excitation on the basilar membrane. Determination of the pitch of these sounds must instead rely either on an analysis of the sound's harmonic structure, or equivalently on periodic regularities ("periodicity") in its temporal waveform. The inverse of the fundamental period of such periodic sounds corresponds to their fundamental frequency (F_0), and the pitch of periodic sounds normally matches that of a pure tone with a frequency equal to F_0. In the work reported here, we used "artificial vowels," i.e., click trains that passed through a cascade of bandpass "formant" filters, as a representative class of easily parameterized yet spectrally complex stimuli, which evoke a strong pitch percept. There is still a great deal of uncertainty regarding the neural substrate of pitch judgments for such complex sounds. Some authors have suggested that this substrate might be provided by "periodotopic maps" (Schreiner and Langner 1988; Schulze et al. 2002) in the mammalian ascending auditory pathway, or that there may be cortical fields specialized for pitch processing (Bendor and Wang 2005). However, previous optical imaging studies of ferret auditory cortex have not yielded evidence for a systematic pitch map or cortical module for periodicity coding (Nelken et al. 2008). In an attempt to identify possible substrates for pitch discrimination in ferrets, we therefore performed an electrophysiological mapping of the sensitivity of neurons to changes in F_0 of five identified cortical fields (Bizley et al. 2005), including the two primary areas A1 and AAF, and three second-order, or belt areas, including

ADF, PPF, and PSF. We also asked whether neurons in ferret cortex, which were sensitive to changes in F_0 were specialized for pitch, in the sense that they were insensitive to changes in vowel timbre or spatial position. Finally, we recorded detailed F_0 tuning curves for hundreds of neurons throughout ferret auditory cortex, and subjected these to neurometric analyses in an attempt to identify whether, or how, these neurons might support the animal's psychoacoustic performance.

38.2 Ferret Behavioral Pitch Sensitivity

38.2.1 Psychoacoustic Methods

The behavioral training and testing methods are described in detail in Walker et al. (2009). Briefly, sound stimuli consisted of two artificial vowel sounds in rapid succession, a "reference" vowel followed by a "target sound." The artificial vowels consisted of click trains of varying interclick interval to vary F_0, sent through bandpass filters centered at 430, 2,132, 3,070, and 4,100 Hz, as described in Malcom Slaney's Auditory Toolbox (http://cobweb.ecn.purdue.edu/~malcolm/interval/1998-010/). The intensity of the sounds varied randomly over a 15 dB range. Ferrets were trained using operant conditioning with drinking water as a positive reinforcer. They initiated each trial by licking a central spout, which triggered the sound stimuli. They were then required to respond by moving to a water spout to the left if the pitch of the target was lower than that of the reference, and to move to the right if it was higher. Incorrect responses were negatively reinforced with a timeout. The F_0 of the reference was held constant throughout 1 week of training sessions, but was changed to a new reference frequency every week. The target sounds varied randomly and could take 1 of 30 possible values spanning just over an octave range around the reference frequency. Psychometric curves were derived by fitting cumulative Gaussians (probit fits) to the observed proportions of "higher" (right spout) responses as a function of stimulus F_0. Difference limens were calculated from the fitted curves as half the distance between the 69.15% point and the 30.85% point (equivalent to a d' of 1). Weber fractions were calculated as the ratio of difference limen over reference F_0.

38.2.2 Psychoacoustic Results

Figure 38.1a shows the psychometric pitch discrimination curves from one of four ferrets tested. Figure 38.1b plots discrimination thresholds, expressed as Weber fractions, as a function of reference F_0 for all four animals. Our results show that ferrets can discriminate rising from falling pitches in artificial vowels, although behavioral thresholds (with Weber fractions of ~0.4) may seem high compared to humans. Further experiments (not shown) indicated that the ferret's pitch discrimination performance with artificial vowels was not significantly different from that

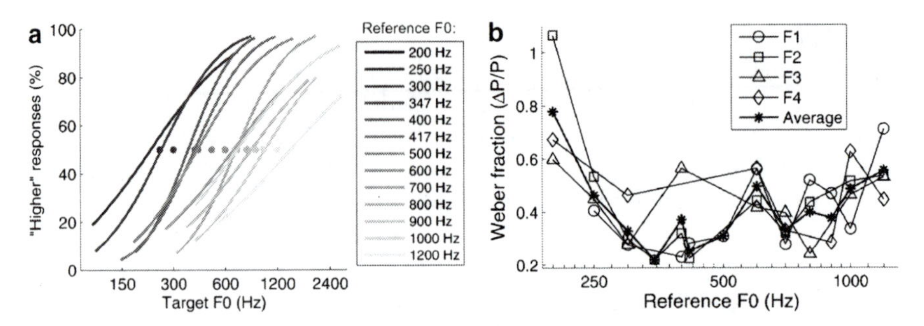

Fig. 38.1 (**a**) Psychometric pitch discrimination curves obtained from one of the four ferrets tested. The curves are probit fits to observed proportions of "higher" judgments for a probe sound with F_0 given on the *x*-axis following a reference sound as indicated by the grayscale in legend. (**b**) Pitch discrimination for four ferrets expressed as Weber fractions, as a function of reference F_0

of pure tones, and that their discrimination thresholds in a go/no-go pitch change discrimination task which does not require distinguishing "higher" from "lower" pitches are approximately tenfold lower.

38.3 Mapping of F_0 Sensitivity Across Cortex

38.3.1 Electrophysiological Methods

In an attempt to identify cortical areas, which may be particularly involved in processing the pitch of complex sounds, we performed electrophysiological mapping experiments in five ferrets under metatomidine/ketamine anesthesia. Using silicon multielectrode probes, we recorded responses to artificial vowels from 324 single units and 292 multiunits throughout areas A1, AAF, PPF, PSF, and ADF of ferret auditory cortex. The vowel sounds could take one of four F_0 values: 200, 336, 565, or 951 Hz. To test sensitivity to timbre or sound source direction, we presented these sounds in virtual acoustic space at virtual azimuths of 45°, −15°, 15°, or 45°, and varied the formants to correspond either to: /a/ with formant frequencies F1–F4 at 936, 1,551, 2,815, and 4,290 Hz; /e/ with formant frequencies at 730, 2,058, 2,979, and 4,294 Hz; /u/ with formant frequencies at 460, 1,105, 2,735, and 4,115 Hz; or /i/ with formant frequencies at 437, 2,761, 3,372, and 4,352 Hz. Responses to at least 20 presentations of each pitch/timbre/azimuth combination were recorded for each unit, and the poststimulus time histograms (PSTHs) were subjected to an ANOVA-style analysis to calculate the proportion of response variability that was attributable to changes in pitch, timbre, or azimuth. Our results did not reveal a clear "pitch area." Rather, F_0 sensitive neurons could be found throughout all cortical areas studied. For further details on the methodology, see Bizley et al. (2009).

38.3.2 Mapping Results (Sensitivity Maps)

Figure 38.2 shows the sensitivity maps obtained by embedding statistics describing the neural response variance explained onto a composite map of the surface of the ferret auditory cortex. Figure 38.2a serves for orientation, and provides a sketch of the position of the left auditory cortex of the ferret on the temporal cortex, as well as an inset showing the parcellation into physiologically defined areas, superimposed on a tonotopic map. Figure 38.2b–d show the proportions of response variances explained by azimuth, pitch and timbre, respectively, as Voronoi tessellations. Each "tile" is shown with a grayscale value corresponding to the mean variance explained value obtained from the neurons in a vertical electrode penetration made at the center of the tile. Comparing Figs. 38.2c and 38.2a, it is clear that highly F_0 sensitive neurons can be found in patches that are distributed widely across ferret auditory cortex, and are not confined to one or just a few of the cortical areas tested. The sensitivity for timbre and azimuth appeared similarly distributed. We also found that units were usually sensitive to more than one of the three stimulus attributes. Sensitivity to both F_0 and timbre was observed in 56% of all units, while 30% of all units were sensitive to both F_0 and azimuth, and 34% of units were sensitive to timbre and azimuth. These results suggest that at least in ferret cortex, the F_0 of complex sounds may be represented in a widely distributed fashion, rather than being confined to an anatomically circumscribed cortical module.

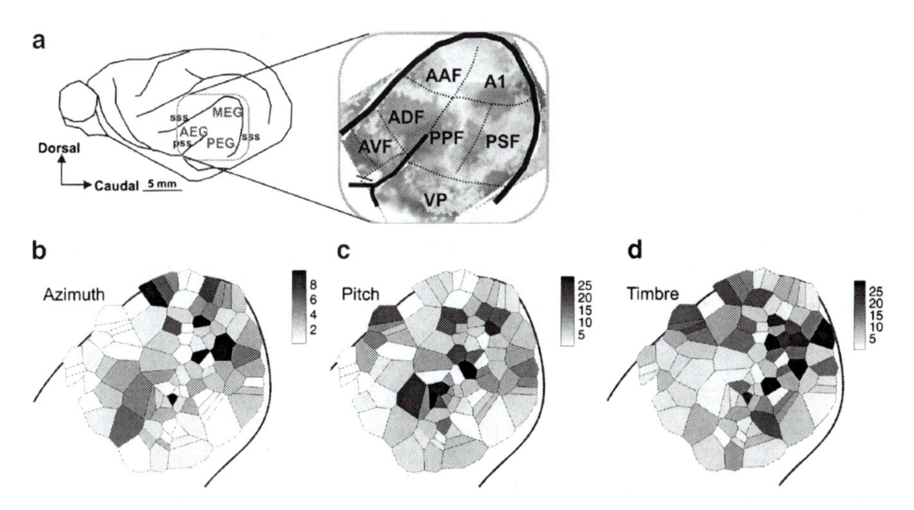

Fig. 38.2 (**a**) Schematic showing the position of ferret auditory cortex on the temporal lobe. The subdivision of auditory cortex into physiological subfields is also shown, superimposed on a tonotopic map derived from previous optical imaging studies (Nelken et al. 2004). *Darker gray* values indicate regions of cortex tuned to lower best frequencies. (**b–d**) Voronoi tessellations showing the distribution of sensitivity to sound source direction, pitch, and timbre, respectively. The *grayscale* of each tile shows the proportion of the neural response variance explained by variation in pitch, timbre, or azimuth, respectively

38.4 Neurometric Analysis: Putative Neural Codes for Pitch

38.4.1 Neurometric Methods

To investigate whether the apparent distributed representation of vowel pitch observed above can account for the animals' behavioral performance, we recorded more detailed F_0 tuning curves for 634 cortical units in 6 metatomidine/ketamine anesthetized and 1 awake, passively listening animal. Electrode penetrations were made throughout the same five cortical areas, using similar methods to the ones described in Sect. 38.3.1. Most recordings were carried out with 32-channel silicon probe multielectrodes (Neuronexus), allowing us to record from between 12 and 60 units simultaneously. The sound stimuli were artificial vowels identical to those used in the psychoacoustical experiments described in Sect. 38.2.1, with fundamental frequencies varying between 200 and 1,838 Hz.

38.4.2 Neurometric Results

Of the 634 units which responded to our artificial vowels, approximately half were sensitive to F_0 changes. The temporal discharge patterns observed in our recorded population were rich and varied, but temporal firing patterns were rarely substantially influenced by changes in stimulus pitch. Rather, 38% of F_0 sensitive units systematically increased their spike rate systematically as a function of stimulus F_0 ("high-pass" neurons), while in another 38%, firing rates decreased with increasing F_0 ("low-pass" neurons). The remaining 24% had nonmonotonic F_0 tuning curves.

We used neurometric analyses to make the observed neural sensitivities more directly comparable to the animal's behavioral performance. These analyses described performance on our pitch discrimination task by an observer of a neuron's activity. If the neuron was high-pass, the observer made "higher" and "lower" pitch judgments on each trial based on whether the neuron's firing rate in response to the target was higher or lower than in response to the reference, respectively (and vice versa for low-pass neurons). More sophisticated, for example ROC based, analyses did not improve the neurometrics over those obtained with this very simple decoder, and neurometrics calculated for individual units only rarely matched the animal's psychometric functions, and could account for the animal's behavioral capabilities only around a small range of reference F_0 values around the steepest part of the unit's F_0 tuning curves. To achieve a more robust decoding of the neural responses, we developed a population neurometric analysis, with which we decoded the activity of ensembles of simultaneously recorded units. This population decoding algorithm represented the neural response (typically the evoked spike rate of each neuron) as a vector, where each unit is a separate element, or dimension of the response vector. Figure 38.3a shows population response vectors from one simultaneously recorded population of 26

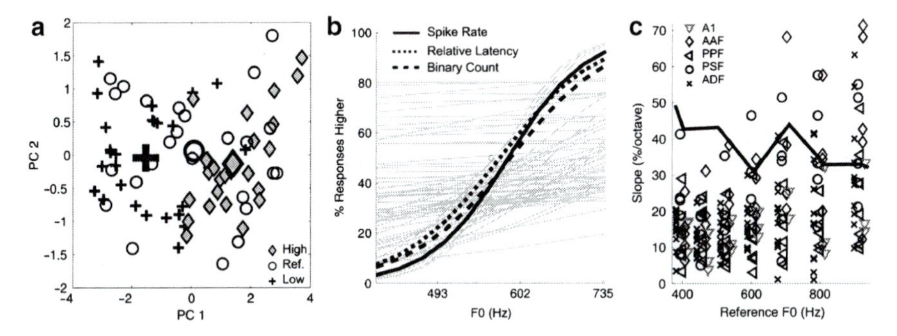

Fig. 38.3 (**a**) Population responses (spike count vectors) for a group of 26 simultaneously recorded units to artificial vowels with F_0 of 459.5, 919, and 1,838 Hz, plotted in principal component space. The *small symbols* show individual responses, the *large symbols* show the average over all 30 trials shown for each stimulus. (**b**) Neurometric curves for another population of 44 simultaneously recorded units. The *light gray curves* show individual neurometrics, while the *dark gray curves* give the population neurometrics based on spike counts (*continuous line*), relative spike latencies (*broken line*) or "binary coding" (*dotted line*). (**c**) Comparison of psychometric and neurometric sensitivity. The *black line* shows the mean psychometric slopes (compare Fig. 38.1b). The symbols plot the neurometric slopes for each of the 39 recorded neural populations. Different symbols are used for populations from different cortical areas, as shown on the legend

units for 30 reference sounds (F_0=919 Hz) as well as to target sounds one octave above and below the reference. It is readily apparent that the responses to the different stimuli form overlapping, but nevertheless distinguishable "clouds" in "response space." To aid visualization only, the spike rate vectors have been plotted in principle component space. For analysis purposes, distances between responses were measured in the raw, in this case 26 dimensional, response space. We decoded individual population responses in a two-stage process. Firstly, we asked whether the population activity could distinguish between the reference sound and the target sound. For each response to the target sound, we asked whether the response on that particular trial was "closer" to (i.e., had a smaller Euclidean distance from) the average response vector to the reference F_0, or closer to the average response for the corresponding target F_0 in response space. An individual response, which was found to be closer to the reference, was deemed indistinguishable from the reference sound and our algorithm randomly classified such a trial as either "higher" or "lower" with equal probability. More commonly, however, the individual trial was distinguishable from the reference, and the decoding algorithm decided whether the target was "higher" or "lower" based on whether the response was closer to the average response to sounds one octave above or one octave below the reference. Population neurometric curves were then derived by fitting cumulative Gaussian curves (probit fits) to the proportions of observed "higher" responses obtained from the classifier algorithm.

These population decoders often performed substantially better than the individual unit neurometrics, and their performance remained better across a range of reference pitches. Interestingly, spike count vectors were not the only response metric yielding good and robust neurometric thresholds. For almost all populations, it was possible to use relative response latency vectors (which encoded first spike latencies for each unit after the first spike fired in the population) or "binary response vectors" (which encoded simply whether a unit had fired within the first 150 ms after stimulus onset or not) to derive neurometrics, which were very similar to, and as sensitive as those based on spike rate vectors. This is illustrated in Fig. 38.3b, which shows rate-based, latency-based "binary" population neurometric curves (drawn in black) for another population of 44 units. For comparison, these are plotted against a background of spike rate neurometrics derived from each individual unit in the population (shown in gray).

The population neurometrics derived in this manner compared favorably with the psychometric curves described in Sect. 38.2.2, and were as sensitive to changes in F_0 or more so in over 2/3 of the cases. Highly sensitive, steep neurometric curves appeared to be particularly common in areas A1 and PPF, but the number of populations recorded is not sufficient to establish whether this trend is statistically significant, and neurometrics capable of matching psychoacoustic performance were encountered in all five cortical areas . Figure 38.3c summarizes these results. These neurometric results are therefore consistent with the mapping results shown in Fig. 38.2.

38.5 Discussion and Conclusions

Our psychoacoustic results, discussed in Sect. 38.2, show that ferrets are capable of judging the pitch of a complex sounds as either higher or lower than that of a given reference sound. As mentioned in the introduction, most previous work on pitch discrimination in nonhuman animals required them merely to detect changes in the fundamental frequencies, without the need of indicating the direction of the change. Some previous work in macaques (Brosch et al. 2004) and birds (Cynx 1995) has shown that these animals can be trained to discriminate rising from falling changes in pure tone frequency. Our experiments may be the first to show that nonhuman animals are also capable of low/high judgments for the pitch of complex sounds. The pitch discrimination thresholds obtained in ferrets with our 2AFC task (Weber fractions of 34% on average) are substantially higher than those typically seen in humans, and are also about tenfold higher than pure tone discrimination thresholds seen in go/no-go pure tone discrimination tasks in other animals. However, our thresholds are comparable with pure tone 2AFC thresholds of around 21% reported in chinchillas (Shofner 2000). We think it likely that the high thresholds we observed in this task reflect the fact that judging pitch along a scale of high to low and mapping it onto behavior is a genuinely difficult task for many animals, but not one that they are fundamentally incapable of.

In our search for possible neurophysiological substrates for these pitch judgments we constructed maps charting the distribution of sensitivity to pitch, timbre and sound source location across the surface of ferret auditory cortex. The results of this mapping shown in Fig. 38.2, yielded patchy, distributed representations. In the subsequent neurometric study, summarized in Fig. 38.3, we showed that such a distributed representation is in principle amply sufficient to support pitch discrimination at the behaviorally measured levels of sensitivity.

Overall, our results are therefore suggestive of a highly distributed and interdependent representation of derived, second-order perceptual features of sounds, such as pitch, timbre, and location. We do, however, acknowledge that our results so far are correlative in nature, i.e., they have identified likely neural codes for pitch in cortex, but are not sufficient to demonstrate conclusively that there is a causal link between the neural population codes for stimulus F_0 we have described and the animal's perceptual pitch judgments. Further work is therefore needed, but it may prove to be technically very difficult to subject the distributed representations we described here to highly targeted and specific experimental manipulations in order to demonstrate a causal link with perception. This relative inaccessibility to external manipulation would also make these distributed and highly redundant representations very resilient in the face of other potential sources of disruption, such as those that might stem from injury or disease.

References

Bendor D, Wang X (2005) The neuronal representation of pitch in primate auditory cortex. Nature 436:1161–1165

Bizley JK, Nodal FR, Nelken I, King AJ (2005) Functional organization of ferret auditory cortex. Cereb Cortex 15:1637–1653

Bizley J, Walker K, Silverman B, King A, Schnupp J (2009) Interdependent encoding of pitch, timbre and spatial location in auditory cortex. J Neurosci 29(7):2064–2075

Brosch M, Selezneva E, Bucks C, Scheich H (2004) Macaque monkeys discriminate pitch relationships. Cognition 91:259–272

Cynx J (1995) Similarities in absolute and relative pitch perception in songbirds (starling and zebra finch) and a nonsongbird (pigeon). J Comp Psychol 109:261–267

Dooling RJ, Saunders JC (1975) Hearing in the parakeet (Melopsittacus undulatus): absolute thresholds, critical ratios, frequency difference limens, and vocalizations. J Comp Physiol Psychol 88:1–20

Heffner HE, Ravizza RJ, Masterton B (1969) Hearing in primate mammals III: tree shrew (Tupaia glis). J Aud Res 9:12–18

Nelken I, Bizley JK, Nodal FR, Ahmed B, Schnupp JW, King AJ (2004) Large-scale organization of ferret auditory cortex revealed using continuous acquisition of intrinsic optical signals. J Neurophysiol 92:2574–2588

Nelken I, Bizley JK, Nodal FR, Ahmed B, King AJ, Schnupp JW (2008) Responses of auditory cortex to complex stimuli: functional organization revealed using intrinsic optical signals. J Neurophysiol 99(4):1928–1941

Nelson DA, Kiester TE (1978) Frequency discrimination in the chinchilla. J Acoust Soc Am 64:114–126

Schreiner CE, Langner G (1988) Periodicity coding in the inferior colliculus of the cat. II. Topographical organization. J Neurophysiol 60:1823–1840

Schulze H, Hess A, Ohl FW, Scheich H (2002) Superposition of horseshoe-like periodicity and linear tonotopic maps in auditory cortex of the Mongolian gerbil. Eur J NeuroSci 15:1077–1084

Shofner WP (2000) Comparison of frequency discrimination thresholds for complex and single tones in chinchillas. Hear Res 149:106–114

Vongpaisal T, Trehub SE, Schellenberg EG (2006) Song recognition by children and adolescents with cochlear implants. J Speech Lang Hear Res 49:1091–1103

Walker KMM, Schnupp JWH, Hart-Schnupp SMB, King AJ, Bizley JK (2009) Pitch discrimination by ferrets for simple and complex sounds J Acoust Soc Am 126(3):1321–1335

Chapter 39
Macroscopic and Microscopic Analysis of Speech Recognition in Noise: What Can Be Understood at Which Level?

Thomas Brand, Tim Jürgens, Rainer Beutelmann, Ralf M. Meyer, and Birger Kollmeier

Abstract A series of experiments and model developments were performed to quantitatively describe and predict speech recognition in listeners with normal and impaired hearing in quiet as well as in realistic, fluctuating, and spatially distributed noise environments. On a macroscopic level, classical speech-information-based models such as the Speech Intelligibility Index (SII) yield accurate predictions only for average intelligibility scores and for a limited set of acoustical situations. A binaural extension using a binaural preprocessing model provides surprisingly accurate predictions for a wide range of acoustically complex, spatial situations.

On a microscopic (i.e. phoneme-to-phoneme) scale, the combination of a psychoacoustically and physiologically motivated preprocessing model with a pattern recognition algorithm adopted from automatic speech recognition (ASR) technology allows for a detailed analysis of phoneme confusions and explains the "man-machine-gap" of approx. 12 dB in signal to noise ratio. This finding highlights the superiority of human world-knowledge-driven (top-down) speech pattern recognition in comparison to the training-data-driven (bottom-up) machine learning approaches.

Keywords Speech recognition • Fluctuating noise • Binaural • Auditory model • Normal hearing • Hearing imparied

39.1 Introduction

This is an overview on our work aiming at a quantitative, comprehensive, and model-based understanding of speech recognition in complex acoustical situations. The special difficulties that hearing-impaired listeners encounter in these situations provide a variety of data for developing and evaluating speech recognition models. This also highlights the need for improved models that can be exploited for hearing

T. Brand (✉)
Medical Physics, University of Oldenburg, Oldenburg, Germany
e-mail: thomas.brand@uni-oldenburg.de

E.A. Lopez-Poveda et al. (eds.), *The Neurophysiological Bases of Auditory Perception*,
DOI 10.1007/978-1-4419-5686-6_39, © Springer Science+Business Media, LLC 2010

instruments of the future. Model components operate on three levels. The *acoustical*, the *sensory*, and the *central* level are combined for predicting speech intelligibility in fluctuating, spatial noise situations. The three levels of modeling are combined to form a hierarchical approach with a minimum number of specific assumptions at the respective layer. The aim is to quantitatively model the influence of factors other than loss in audibility that specifically impair speech recognition in hearing-impaired subjects. This could eventually serve as a reference for various applications (such as speech processing and presentation, computer speech recognition, and hearing instruments).

The *acoustical level* (as defined here) incorporates the acoustical features of the stimulus entering the inner ear and is closely related to the *audibility* determined by the relation between the listener's absolute hearing threshold and the sound pressure level of the acoustical stimulus. There is an inevitable overlap between the acoustical level and the sensory level (see below) because some sensory concepts like auditory filters are also needed for the models on the acoustical level like, for instance, the Speech-Intelligibility-Index (SII). Nevertheless, the models described on the acoustical level are much more stimulus driven than the models described on the sensory level and the central level. Concerning the acoustical level, we concentrated our work on SNR-based speech perception measures (SII approaches) for speech recognition in silence and noise.

The *sensory level* incorporates all (linear and nonlinear) steps that the human auditory system performs by transforming the acoustical stimulus into the "internal representation". This level is considered as crucial in distinguishing between normal and sensorineural hearing-impaired listeners. Several features of the "internal representation" can be defined both experimentally and with models, such as loudness, binaural interaction, and representation of amplitude modulations. The hearing-loss-related distortions in these features provide a valid basis for modeling the altered processing in listeners with hearing loss. Here, we concentrate our work on the binaural interaction that increases the "effective" signal-to-noise ratio (Beutelmann and Brand 2006; Beutelmann et al. 2009, 2010).

Since the models mentioned so far work on the averaged frequency spectra of speech and noise and do not perform a detailed psychoacoustic modeling, we call them *macroscopic* models.

In order to model the interaction between (potentially) blurred internal representation and (imperfect, but yet working) object and speech recognition, an emphasis was put on experimental methods and comparatively simple models to quantitatively separate between performance deterioration due to sensorineural distortion or due to central processing and recognition deficiencies. The work on this *central level* concentrated on using a *microscopic model* of phoneme recognition (Jürgens and Brand 2009). We define "*microscopic*" modeling twofold: Firstly, the particular stages involved in the speech recognition of human listeners are modelled in a way typical of psychophysics based on a detailed "internal representation" of the speech signals. Secondly, the recognition rates and confusions of single phonemes are compared to those of human listeners. This twofold definition is in line with Barker and Cooke (2007), for instance.

39.2 Acoustical Level: SNR-Based Speech Perception Measures (SII Approaches)

For modeling the effect of audibility on speech intelligibility, the Speech-Intelligibility-Index (SII, ANSI S3.5-1997) was used. The SII predicts the speech recognition threshold (SRT) in silence of listeners with normal and with impaired hearing with high accuracy (standard deviation approx. 2 dB) (Brand and Kollmeier 2002). In the case of *steady-state background noise,* the correlation between observed and predicted SRT-values decreases significantly for two main reasons: Firstly, the SRTs of normal-hearing listeners in background noise are closer to the SRTs of listeners with impaired hearing. This reduced variance is the reason for a decrease in correlation between predicted and observed SRTs even though the absolute accuracy of the prediction may stay the same. Secondly, the interaction between hearing loss and background noise on the sensory level ("suprathreshold distortions") is not taken into account. If further factors like spatial separation between speech and noise or fluctuations of the background noise are included, the variance of the observed data increases. This allows higher correlations between predictions and observations as long as the combined effects of the auditory processing in these acoustical conditions and speech intelligibility could be modelled properly.

In the framework of the investigation of speech intelligibility in fluctuating noise using macroscopic models on an acoustical level, the SII served as a baseline, as it is a validated tool for predicting SRT values for listeners with normal hearing in stationary noise and in silence. The original SII and three short-term extensions of the SII were investigated: Each extension comprised an increasing deviation from the original SII concept. The simplest extension took only broadband fluctuations of the noise into account whereas the frequency spectra of noise and speech were assumed as constant (Brand and Kollmeier 2002). In this way, this approach stayed very close to the original SII. The next more elaborated extension took the frequency dependent fluctuations of the noise into account (Rhebergen and Versfeld 2005). The third extension took the frequency dependent fluctuations of the noise and the speech into account (Meyer et al. 2007a). The four approaches differ in their complexity with regard to the temporal information present in the speech and noise signals.

The four model versions were evaluated using 113 listeners with normal hearing and with different degrees and kinds of impaired hearing. Speech Reception Thresholds (SRTs, i.e., the SNR belonging to a recognition rate of 50%) were measured using sentences (Wagener et al. 1999a, b, c) in different kinds of noise with speech-like modulations (ICRA5-250 noise: Dreschler et al. 2001; Wagener et al. 2006). The correlation between measured and predicted SRTs was low for the standard SII ($r^2=0.23$) and increase to $r^2=0.40$ to $r^2\approx0.50$ for the three short-term approaches (see Fig. 39.1). This means that the predictions for fluctuating noise can be considerably improved using short-term SII procedures compared to the standard SII. However, the quality of the predictions did not improve significantly

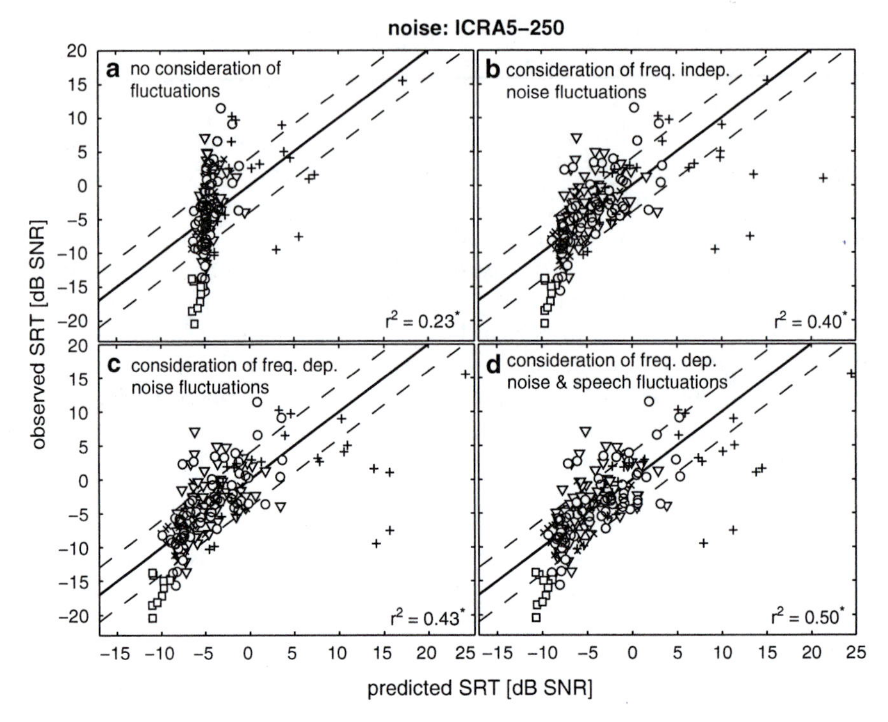

Fig. 39.1 Observed SRT values (ordinate) vs. predicted SRT values (abscissa) for ICRA5-250 noise: The results of all four model versions are shown ((**a**) Standard SII, (**b**) Frequency independent variation of the noise level, (**c**) Frequency dependent variations of the noise level, (**d**) Frequency dependent variations of the speech and noise level). The squared correlation coefficients between the predicted and observed SRT values are denoted in the corners of the scatter plots (*stars* denote highly significant correlations with $p<0.01$). The different groups of hearing loss are denoted in the following way: normal hearing: *squares*; flat hearing loss: *circles*; steep hearing loss: *downwards triangle*; hearing loss with at least one value over 85 dB HL: *plus*; uncategorized hearing loss: *x*

from the first to the third extension. Hence, the accuracy of this method appears to be limited if an adequate model of the subsequent sensory/peripheral processing level is not taken into account.

39.3 Sensory Level/Peripheral Processing: (Example: Binaural Interaction)

Humans frequently have to recognize speech in adverse listening conditions, including noise sources from different directions and adverse room acoustics. The binaural processing of the (intact) auditory system causes a substantial improvement of speech recognition in these adverse listening conditions. On the other hand, these are the most difficult conditions for listeners with hearing impairment.

A binaural speech intelligibility model (BSIM) was developed and evaluated using binaural speech intelligibility data of individual listeners (Beutelmann and Brand 2006). BSIM (see Fig. 39.2) consists of a gammatone filter bank (Hohmann 2002), an independent equalization-cancellation (EC) (Durlach 1963) process in each frequency band, a gammatone resynthesis, and the SII. The listener's hearing loss was simulated by adding two uncorrelated masking noise signals (according to the pure-tone audiogram) to the ear channels. SRTs were measured using eight listeners with normal hearing and 15 listeners with impaired hearing in three acoustic conditions (anechoic, office room, and cafeteria hall) and eight directions of a noise source (speech in front). Artificial EC processing errors derived from binaural masking level difference data using pure tones (vom Hövel 1984) were incorporated into the model. Except for an adjustment of the SII-to-intelligibility transformation, no further model parameter had to be fitted. The overall correlation between predicted and observed SRTs was very high ($r^2=0.91$) (see Fig. 39.3). The dependence of the SRT of an individual listener on the noise direction and on room acoustics was predicted with a median squared correlation coefficient of 0.82. The effect of individual hearing impairment was predicted with a median squared correlation coefficient of 0.88. However, for subjects with mild hearing loss, the release from masking was overestimated. All correlations given here were highly significant ($p<0.01$).

A complete revision of this model was based on an analytical expression of binaural unmasking for arbitrary input signals and is computationally very efficient (Beutelmann et al. 2010), while preserving the model's performance. An extension for nonstationary interferers was realized by applying the model to short-time frames of the input signals and averaging over the predicted SRT results similar to the approaches described above. The extended model predictions were compared to data from eight listeners with normal hearing and 12 listeners with impaired hearing, incorporating all combinations of four rooms, three source setups, and three noise types with different degrees of modulation. Depending on the noise type, the squared correlation coefficients between observed and predicted

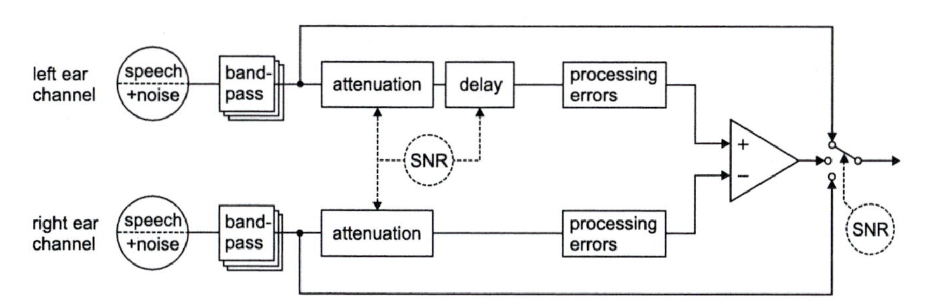

Fig. 39.2 Binaural processing using the modified, multifrequency channel EC-model according to vom Hövel (1984). The speech and noise signals are processed in the same way, but separately for exact SNR calculation. The noise signal part includes the internal masking noise. Attenuation is only applied to one of the channels, depending on which of them contains more noise energy compared to the other (from Beutelmann and Brand 2006, by permission of *J Acoust Soc Am*)

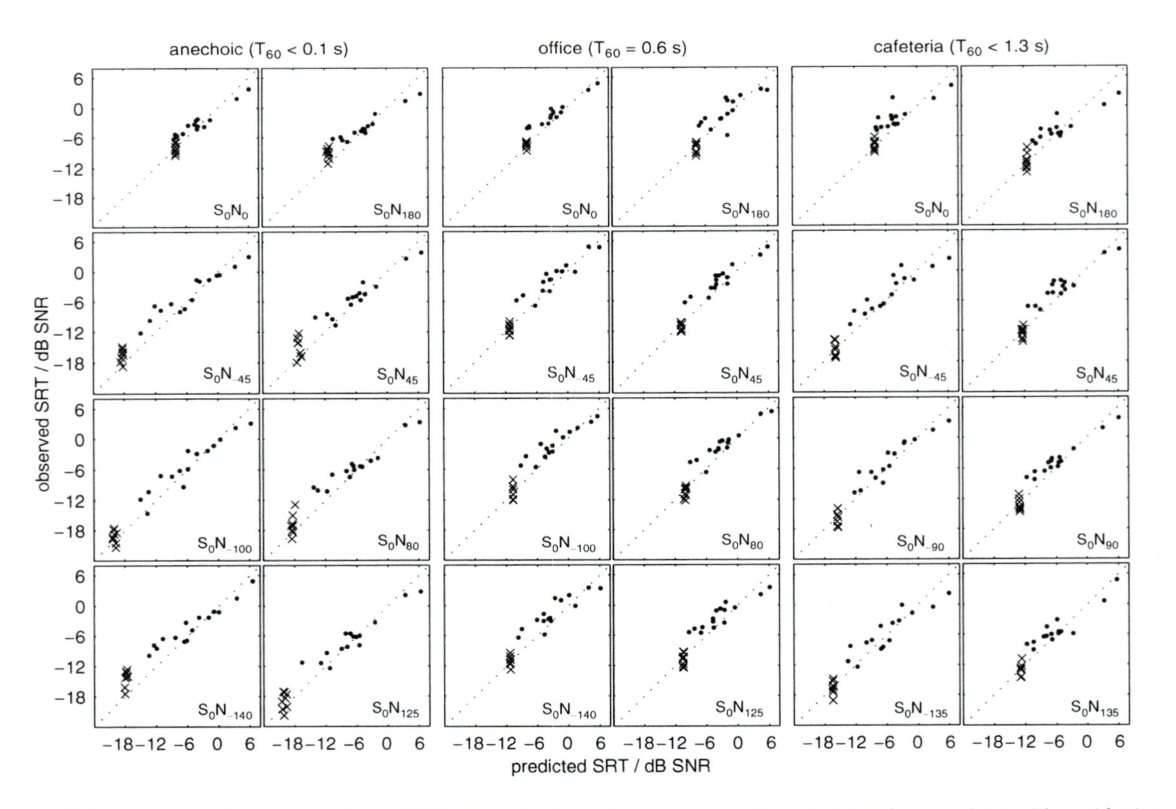

Fig. 39.3 Observed SRT values plotted against predicted SRT values. Each panel contains the SRTs of 15 hearing-impaired subjects (*dots*) measured at the noise source azimuth indicated in the lower right corner. There are two columns of panels for each room condition, marked by the respective room names at the top of the figure. The SRTs of 8 normal-hearing subjects (*crosses*) have been added for comparison (from Beutelmann and Brand 2006, by permission of *J Acoust Soc Am*)

SRTs were 0.65–0.83 ($p<0.01$) for subjects with normal hearing and 0.35–0.63 ($p<0.01$) for subjects with impaired hearing. The root mean square prediction error was between 3.6 and 5.7 dB. The model (which is based only on the pure-tone audiogram and so far neglects other factors) explains 70% of the variance of SRTs of listeners with impaired hearing.

In a further study (Beutelmann et al. 2009), it was investigated if it is justified that the BSIM model implicitly assumes completely independent EC processing in adjacent binaural frequency bands except the overlapping of the gammatone filters. We tested this hypothesis by determining the effective binaural auditory bandwidth for broadband signals like speech and noise. SRTs were measured for binaural conditions with frequency-dependent interaural phase differences (IPDs) of speech and noise. BSIM predictions were compared with the observed data. The SRT benefit decreased from 6 to 0 dB with decreasing frequency distance of IPD's sign changes. The observed results can be predicted very well by the model if it is assumed that the binaural analysis filter bandwidth is about the factor 2.3 wider than the monaural bandwidth (i.e. 2.3 ERB instead of 1 ERB).

39.4 Central Level: A Microscopic Model of Speech Recognition

For modeling phoneme recognition of listeners with normal hearing, a microscopic model of speech recognition was installed by combining a psychoacoustically and physiologically motivated auditory perception model (Dau et al. 1996) with a simple Dynamic-Time-Warp (DTW) speech recognizer (see Fig. 39.4). The model's recognition performance in speech-shaped noise was evaluated and compared to the performance of listeners with normal hearing. Contrary to an earlier version of this model approach by Holube and Kollmeier (1996), the current model was extended using a modulation filterbank to extract information about the modulation of the speech signal. The model was evaluated by presenting nonsense speech stimuli (CVCs or VCVs) in a closed-set paradigm. The model itself computes the perceptual distances between the "internal representations" of speech items and bases speech recognition on the similarity of these "internal representations". Overall recognition rates, phoneme recognition rates, and phoneme confusions of both, the model and normal-hearing listeners were compared using confusion matrices. The model itself computes the perceptual distances between the "internal representations" of speech stimuli and bases speech recognition on similarity of the "internal representation". The influence of different perceptual distance measures and that of the model's a-priori knowledge was investigated (Jürgens and Brand 2009). The results show that human performance can be predicted by this model using identical speech waveforms for both, training of the recognizer and testing. An exemplary result is shown in Fig. 39.5 in the form of consonant confusion matrices of normal-hearing listeners (panel a) and the model (panel b) at a SNR of −15 dB. The rate of correct phoneme recognitions (diagonal elements of the matrices) are very well predictable ($r^2=0.91$, $p<0.01$). The characteristic confusions of single phonemes, however, can only be

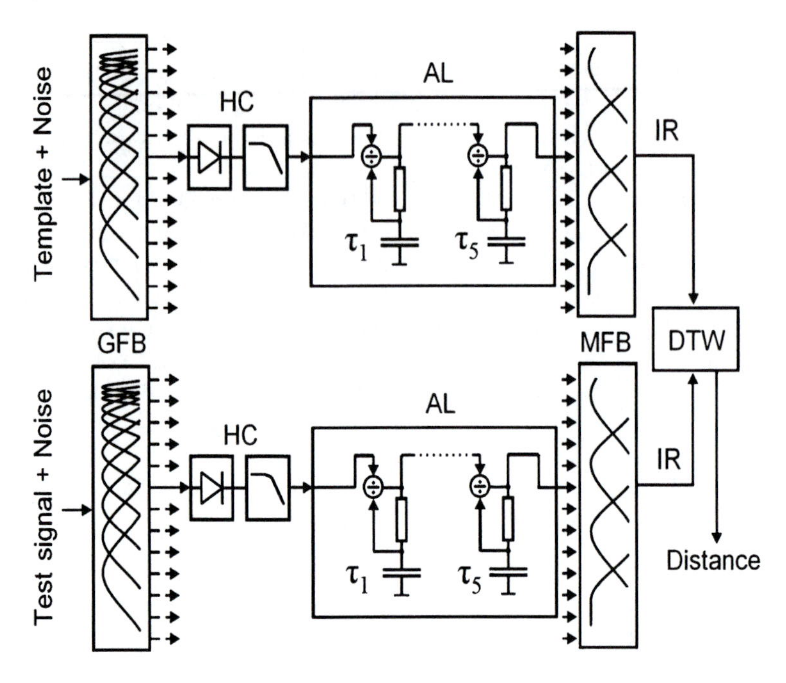

Fig. 39.4 Scheme of the perception model: The time signals of the template recording added with running noise and the time signal of the test signal added with running noise are preprocessed in the same effective "auditory-like" way. A gammatone filterbank (GFB), a haircell model (HC), adaptation loops (AL), and a modulation filterbank (MFB) are used. The outputs of the modulation filterbank are the internal representations (IR) of the signals. They serve as inputs to the Dynamic-Time-Warp (DTW) speech recognizer that computes the "perceptual" distance between the IRs of the test logatom and the templates (from Jürgens and Brand 2009, by permission of *J Acoust Soc Am*)

predicted partially. Different types of perceptual distance measures were investigated. The best model performance is yielded by distance measures, which focus mainly on small perceptual distances between the internal representations of the speech item to recognize and the trained speech item i.e. the best matching speech passages, and neglect poor matching passages. This goes in line with the hypothesis that the human auditory system focuses on the identifiable acoustic objects, which are associated with the desired speech objects and neglects acoustical objects which are probably related to the undesired background noise. Using this model, the SRTs in speech-shaped noise and in silence could be predicted with an accuracy of 1 to 2 dB. The recognition rates of consonant phonemes as well as vowel phonemes could be predicted in noise. However, the model still has difficulties predicting individual consonant or vowel recognition rates and confusions in silence. As a conjoined result with a research project on automatic speech recognition (ASR), it was shown that the human-machine gap in speech reception in noise amounts to 10 to

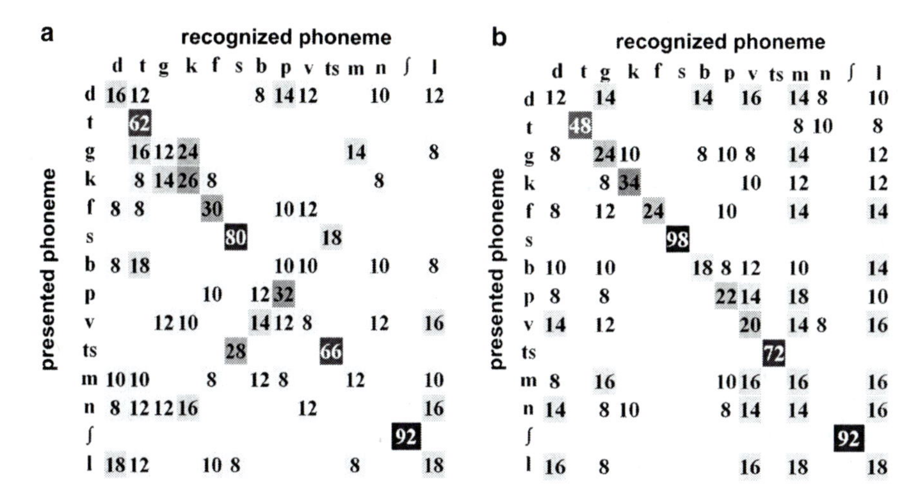

Fig. 39.5 Confusion matrices (response rates in %) for VCVs at –15 dB SNR of subjects with normal hearing (**a**) and of the model (**b**): *Row*: presented consonant phoneme, *column*: recognized consonant phoneme. *Grey scales* denote different grades of response rates. Response rates below 8% are not shown (from Jürgens and Brand 2009, by permission of *J Acoust Soc Am*)

12 dB (Jürgens et al. 2007; Meyer et al. 2007b). This means, if an ASR system should yield the same performance as a listener with normal hearing, it needs a 10 to 12 dB higher SNR. To model speech intelligibility appropriately, this gap can be bypassed (as done by the microscopic model) by using an optimal detector as speech recognizer (i.e. using internal representations of the same speech waveform for training as to recognize).

The perception model mentioned above included a "linear" gammatone filter bank. In a further approach we replaced this perception model by a new Computational Auditory Signal-processing and Perception (CASP) model (Jepsen et al. 2008) that includes a DRNL (dual resonance nonlinear) filter bank mimicking the complex nonlinear BM response behavior of the basilar membrane. The CASP model is capable of, e.g., predicting Growth-of-Masking (GOM) curves appropriately, which is a major advantage to the auditory model we used before. It can be shown (publication in preparation) that the SRTs of normal-hearing listeners in noise and in quiet can also be predictedusing the CASP model instead of the linear perception model. Additionally to a loss in audibility also a change in cochlear compression can be implemented using this model.

Taken together, a model was developed that allows to vary parametrically different aspects of sensorineural hearing loss and that allows to individually predict speech recognition. Thus, the model can be used in the future to systematically investigate how different aspects of hearing loss affect the individual's speech intelligibility in different auditory scenes.

39.5 Conclusions

Macroscopic modeling based on the SII allows modeling of speech recognition rates and speech reception thresholds in silence and steady state noises.

Extensions of this model for fluctuating noises and spatial noise situations were developed and evaluated, and allow a very high predictability of listeners' results, especially in case of spatial noise.

Using a microscopic model that mimics the auditory processing and uses a simple speech recognizer, it is also possible to predict recognition rates of single phonemes in silence and in steady state noise. Furthermore, extensions for listeners with impaired hearing seem to be feasible by adequate change of the auditory model.

Acknowledgment This work was supported by DFG SFB TRR 39, The active auditory system. CEC-Project Hearcom, and the Audiologie-Initiative Niedersachsen.

References

ANSI (1997) Methods for the calculation of the speech intelligibility index. American national standard S3.5 – 1997, Standards Secretariat, Acoustical Society of America, New York, USA

Barker J, Cooke M (2007) Modelling speaker intelligibility in noise. Speech Commun 49:402–417

Beutelmann R, Brand T (2006) Prediction of speech intelligibility in spatial noise and reverberation for normal-hearing and hearing-impaired listeners. J Acoust Soc Am 120:331–342

Beutelmann R, Brand T, Kollmeier B (2010) Revision, extension, and evaluation of a binaural speech intelligibility model (BSIM). J Acoust Soc Am (in press)

Beutelmann R, Brand T, Kollmeier B (2009) Prediction of binaural speech intelligibility with frequency-dependent interaural phase differences. J Acoust Soc Am 126(3):1359–1368

Brand T, Kollmeier B (2002) Vorhersage der Sprachverständlichkeit in Ruhe und im Störgeräusch aufgrund des Reintonaudiogramms (Jahrestagung der Deutsche Gesellschaft für Audiologie)

Dau T, Püschel D, Kohlrausch A (1996) A quantitative model of the 'effective' signal processing in the auditory system: I. model structure. J Acoust Soc Am 99:3615–3622

Dreschler WA, Verschuure H, Ludvigsen C, Westermann S (2001) ICRA noises: artificial noise signals with speech-like spectral and temporal properties for hearing instrument assessment. International collegium for rehabilitative audiology. Audiology 40:148–157

Durlach NI (1963) Equalization and cancellation theory of binaural masking-level differences. J Acoust Soc Am 35:1206–1218

Jepsen ML, Ewert SD, Dau T (2008) A computational model of human auditory signal processing and perception. J Acoust Soc Am 124:422–438

Jürgens T, Brand T, Kollmeier B (2007) Modelling the human-machine gap in speech reception: microscopic speech intelligibility prediction for normal-hearing subjects with an auditory model. In: Interspeech. Antwerp, Belgium, pp 410–413

Jürgens T, Brand T (2009) Microscopic prediction of speech recognition for listeners with normal hearing in noise using an auditory model. J Acoust Soc Am 126:2635–2648

Holube I, Kollmeier B (1996) Speech intelligibility prediction in hearing-impaired listeners based on a psychoacoustically motivated perception model. J Acoust Soc Am 100:1703–1716

Hohmann V (2002) Frequency analysis and synthesis using a gammatone filterbank. Acta acustica/ Acustica 88:433–442

Meyer RM, Brand T, Kollmeier B (2007a) Predicting speech intelligibility in fluctuating noise. In: 8th EFAS Congress/10th Congress of the German Society of Audiology, Deutsche Gesellschaft für Audiologie e.V., Heidelberg, CD-ROM

Meyer B, Brand T, Kollmeier B (2007b) Phoneme confusions in human and automatic speech recognition. In: Interspeech. Antwerp, Belgium, pp 1485–1488

Rhebergen K, Versfeld N (2005) A speech intelligibility index-based approach to predict the speech reception threshold for sentences in fluctuating noise for normal-hearing listeners. J Acoust Soc Am 117:2181–2192

vom Hövel H (1984) Zur Bedeutung der Übertragungseigenschaften des Außenohrs sowie des binauralen Hörsystems bei gestörter Sprachübertragung. Dissertation, Fakultät für Elektrotechnik, RTWH Aachen

Wagener K, Kühnel V, Kollmeier B (1999a) Entwicklung und Evaluation eines Satztests für die deutsche Sprache I. Zeitschrift für Audiologie 38(1):4–14

Wagener K, Brand T, Kollmeier B (1999b) Entwicklung und Evaluation eines Satztests für die deutsche Sprache II. Zeitschrift für Audiologie 38(2):44–56

Wagener K, Brand T, Kollmeier B (1999c) Entwicklung und Evaluation eines Satztests für die deutsche Sprache III. Zeitschrift für Audiologie 38(3):86–95

Wagener K, Brand T, Kollmeier B (2006) The role of silent intervals for sentence intelligibility in fluctuating noise in hearing-impaired listeners. Int J Audiol 45:26–33

Chapter 40
Effects of Peripheral Tuning on the Auditory Nerve's Representation of Speech Envelope and Temporal Fine Structure Cues

Rasha A. Ibrahim and Ian C. Bruce

Abstract A number of studies have explored how speech envelope and temporal fine structure (TFS) cues contribute to speech perception. Some recent investigations have attempted to process speech signals to remove envelope cues and leave only TFS cues, but the results are confounded by the fact that envelope cues may be partially reconstructed when TFS signals pass through the narrowband filters of the cochlea. To minimize this reconstruction, investigators have utilized large numbers of narrowband filters in their speech processing algorithms and introduced competing envelope cues. However, it has been argued that human peripheral tuning may be two or more times sharper than previously estimated, such that envelope restoration may be stronger than originally thought. In this study, we utilize a computational model of the auditory periphery to investigate how cochlear tuning affects the restoration of envelope cues in auditory nerve responses to "TFS speech." Both the envelope-normalization algorithm of Lorenzi et al. (Proc Natl Acad Sci USA 103:18866–18869, 2006) and the speech-noise chimaeras of Smith et al. (Nature 416:87–90, 2002) were evaluated. The results for the two processing algorithms indicate that the competing noise envelope of the chimaeras better reduces speech envelope restoration but does not totally eliminate it. Moreover, envelope restoration is greater if the cochlear tuning is adjusted to match Shera and colleagues' (Proc Natl Acad Sci USA 99:3318–3323, 2002) estimates of human tuning.

Keywords Speech Perception • Auditory Models • Cochlear Tuning • Envelope Cues • Temporal Fine Structure • Auditory Chimaeras • Spectro • Temporal Modulation Index

---

I.C. Bruce (✉)
Department of Electrical and Computer Engineering, McMaster University, Room ITB-A213, 1280 Main Street, West Hamilton, ONT L8S 4K1, Canada
e-mail: ibruce@ieee.org

E.A. Lopez-Poveda et al. (eds.), *The Neurophysiological Bases of Auditory Perception*,
DOI 10.1007/978-1-4419-5686-6_40, © Springer Science+Business Media, LLC 2010

40.1 Introduction

Speech signals can be characterized by two types of coding: temporal envelope (E) coding and temporal fine structure (TFS) coding (Smith et al. 2002; Lorenzi et al. 2006). Temporal envelope coding is the relatively slow variation in amplitude over time, while the fine structure coding is the rapid variations in the signal. Both types of temporal coding contain some information (cues) that can be used for speech perception. The envelope information is embedded in the variations of the average discharge rate of the auditory nerves (ANs), whereas the TFS cues are coded as the synchronization of the nerve spikes to a particular phase of the stimulus, which is known as the phase-locking phenomenon (Young 2008). It has been suggested that the E information is responsible for speech understanding, while the TFS information is linked to melody perception and sound localization. Recently, some studies pointed to the possibility that the TFS code has a role in speech perception, especially in complex background noise (Lorenzi et al. 2006). It has been observed that normal-hearing people perform much better than hearing-impaired people in fluctuating background speech intelligibility tests. This was linked to a lack of ability to use TFS cues in hearing-impaired people, which accordingly indicates that TFS has a significant role for speech intelligibility. Such possibilities may have some implications on the development of cochlear implants, which in the current systems do not provide TFS information and are concerned mainly with providing E information. However, it has been argued that these results are complicated by possible restoration of E cues by the passing of the TFS signal through the cochlear filters (e.g., Zeng et al. 2004). Indeed, the preliminary results of Heinz and Swaminathan (2008) from AN recordings in chinchillas indicate that envelope restoration is not fully eliminated even for 8 or 16 processing channels. Furthermore, envelope restoration may be more significant in humans, if human cochlear tuning is sharper than in other mammals, as has been recently suggested (Shera et al. 2002). In this study, we investigate the role of TFS coding in speech perception using a model of the auditory periphery and a cortical model of speech processing. To separate the TFS code from E information, speech signals are divided into frequency bands to extract the envelope and fine structure codes in each band. The input stimulus to our auditory model is either the TFS-only signal or auditory chimaeras. Auditory chimaeras (Smith et al. 2002) are created such that the E (TFS) is from the speech signal, while the TFS (E) is coming from a spectrally matched noise signal. The effects of human versus cat cochlear tuning on envelope restoration are explored.

40.2 Methods

40.2.1 The Auditory Periphery Model

Our model is based on Zilany and Bruce model for the cat auditory periphery (Zilany and Bruce 2006, 2007a). The model consists of several blocks representing different stages in the auditory periphery from the middle ear to the AN. The model can accurately represent the nonlinear level-dependent cochlear effects, and it provides an accurate description of the response properties of AN fibers to complex stimuli in both normal and impaired ears. However, the model is designed to simulate the auditory periphery system in cats, and there are some important differences between cat and human ears (Recio et al. 2002) that should be taken into account in the final model.

We have developed a version of the computational model for the ear that incorporates human cochlear tuning. The Q_{ERB} values for the human cochlear filter have been estimated in Shera et al. (2002) using stimulus frequency otoacoustic emissions and improved psychophysical measurements.

The cochlear frequency selectivity is represented by Q_{ERB}, defined as

$$Q_{\mathrm{ERB}}(\mathrm{CF}) = \frac{\mathrm{CF}}{\mathrm{ERB(CF)}} \tag{40.1}$$

The characteristic frequency (CF) is the center frequency of the filter, and ERB is the equivalent rectangular bandwidth, defined as the bandwidth of the rectangular filter that produces the same output power as the original filter when driven by white noise. In Fig. 40.1a, we show the human Q_{ERB} values as function of CF given in Shera et al. (2002). The Q_{ERB} values reported in Shera et al. (2002) are two or three times sharper than the previous behavioral measurements (Glasberg and Moore 1990) shown in the same figure. In our work, we mapped the Q_{ERB} values derived in Shera et al. (2002) to the corresponding Q_{10} values to set the tuning in the computational model. The mapping is illustrated in Fig. 40.1b, where Q_{10} and Q_{ERB} values are computed at each center frequency using the model's cochlear filter transfer function. A linear mapping between Q_{10} and Q_{ERB} is then estimated using least square curve fitting to obtain

$$\mathrm{Q}_{10} = 0.2085 + 0.505\mathrm{Q}_{\mathrm{ERB}}. \tag{40.2}$$

The cat Q_{ERB} versus CF curve is also shown in Fig. 40.1a, where we have calculated the cat Q_{ERB} values from the corresponding Q_{10} values in the Zilany and Bruce (2006, 2007a) model using the mapping above.

40.2.2 Speech Intelligibility Metric (STMI)

The output of the model is assessed to predict the speech intelligibility based on the neural representation of the speech. This is achieved through the spectro-temporal modulation index (STMI; Elhilali et al. 2003; Bruce and Zilany 2007; Zilany and Bruce 2007b). A simple model of the speech processing of the auditory cortex assumes an array of modulation selective filter banks, which are referred to as spectro-temporal response fields. The output of the AN model is represented by a time-frequency "neurogram." The neurogram is made up from the averaged discharge rates (over every 8 ms) from 128 AN fibers with CFs ranging from 0.18 to 7.04 kHz.

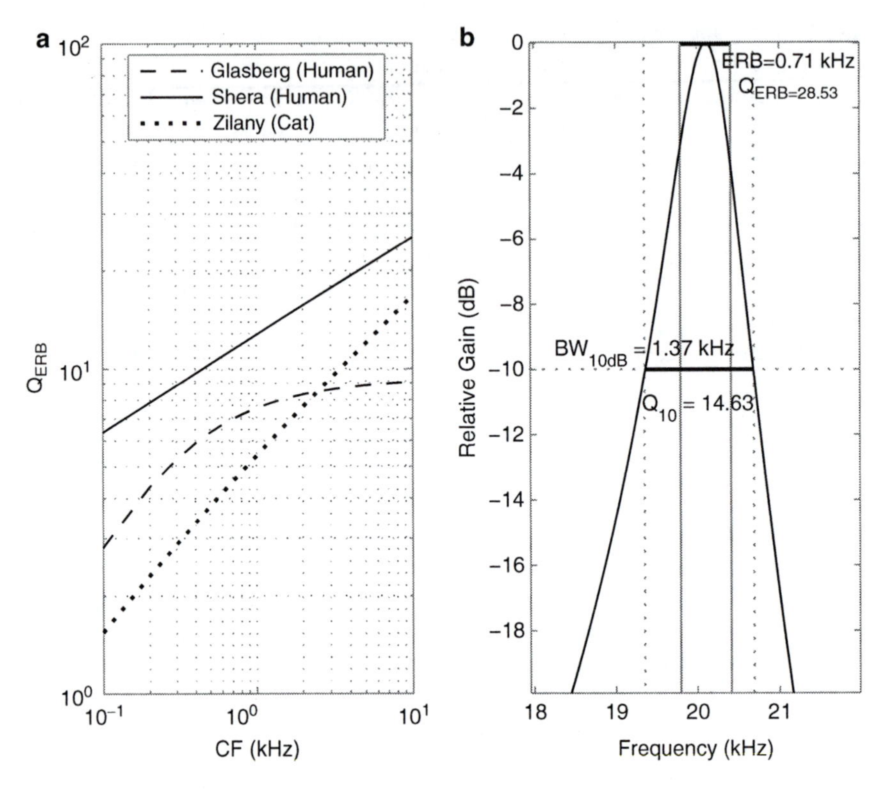

Fig. 40.1 (**a**) Comparison of the human Q_{ERB} values as a function of CF given in Shera et al. (2002), the earlier human Q_{ERB} data in Glasberg and Moore (1990), and the cat Q_{ERB} values in Zilany and Bruce (2006, 2007a). (**b**) Example illustrating Q_{10} to Q_{ERB} mapping for an AN filter at a CF of 20.107 kHz

This neurogram is processed by a bank of modulation selective filters to compute the STMI. The rates for temporal modulations of the filters range from 2 to 32 cyc/s (Hz), and the scales for spectral modulations are in the range from 0.25 to 8 cyc/oct. The STMI is computed using a template (the expected response) generated as the output (at the cortical stage) of the normal model to the stimulus at 65 dB SPL. The cortical output of the test stimulus is compared to the template, and the STMI is computed according to the formula,

$$\mathrm{STMI} = \sqrt{1 - \frac{\|T-N\|^2}{\|T\|^2}} \tag{40.3}$$

where, T is the cortical output of the template signal, and N is the cortical output of the test stimulus.

Because of the large time bins in the AN neurogram and the slow temporal modulation rates for the cortical filters, the STMI is only sensitive to spectral and temporal modulation in the neural response to speech – all phase-locking information about TFS cues is filtered out. Consequently, STMI values computed for TFS-only speech or speech TFS-noise E chimaeras are dependent only on the E cues present in the signal and any E cues restored by passing the TFS signal through the cochlear filters in the auditory periphery model.

40.2.3 Auditory Chimaeras

In Smith et al. (2002), a method to separate TFS from E cues was presented. Two acoustic waveforms are processed using a bank of band pass filters followed by Hilbert transforms to generate envelope-only and TFS-only versions of the signals. In each band, the envelope of one waveform is multiplied by the TFS of the other. The products are then summed across frequency bands to construct the auditory chimaeras. We may have speech–speech chimaeras, where both waveforms are sentences. We may also produce speech–noise chimaeras, where one waveform is the speech signal and the other is noise.

40.2.4 TFS-Only Signals

In Lorenzi et al. (2006), the role of TFS cues in speech perception is assessed by presenting TFS-only signals to a group of normal and hearing-impaired listeners and recording the intelligibility results over several sessions of training. TFS-only signals are generated in a similar method to that of Smith et al. (2002), as they both have the same technique for processing speech signals in each frequency band to separate TFS from E information. However, some distinctive differences exist between the two approaches. First, the number of frequency bands used in Lorenzi et al. (2006) is fixed (16 frequency bands), while in Smith et al. (2002) different choices are tested (from only 1 filter band up to 16 filter bands). Second, the speech TFS-only signal is used directly as the sound stimulus in Lorenzi et al. (2006), while in Smith et al. (2002) they use the speech TFS-only signal to first modulate a noise E-only signal creating auditory chimaeras, which are then used as the new stimulus.

40.3 Test Speech Material

In our work, we have used 11 sentences from the TIMIT database, randomly selected for different male and female speakers from eight major dialect regions of the United States. The sentences are used to create auditory chimaeras following the same procedure as in Smith et al. (2002). Each sentence signal is filtered using a number of band-pass filters. In our study, we have used different number of filters (1, 2, 3, 4, 6, 8, and 16) to divide the signal into frequency bands. These filters are designed as Butterworth filters of order 6, with cutoff frequencies determined such that the filters cover frequency range from 80 to 8,820 Hz with logarithmic frequency spacing (Smith et al. 2002). In each band, we compute the signal envelope using the Hilbert transform. Note that, when comparing our results to Lorenzi et al. (2006), we only use 16 frequency bands to separate the TFS and E signals.

To reproduce the stimulus signals created in Smith et al. (2002), we constructed a spectrally matched noise signal for each test sentence of the TIMIT database. The noise signal is processed in the same way as the sentence signals to produce the envelope and TFS for the noise waveform in each frequency band. The two waveforms, sentence signal and noise signal, are combined to form the speech–noise auditory chimaeras. For every sentence of the 11 TIMIT examples and for each choice of the number of frequency bands used, two sets of chimaeras are developed: speech E-noise TFS chimaeras, and speech TFS-noise E chimaeras. These chimaeras are provided to our auditory periphery model to compute the output neurogram, which is then assessed to evaluate the extent of speech intelligibility using the STMI metric. The experiment is repeated for each stimulus, and the results are averaged over all sentences in the same speech–noise chimaeras set. STMI values were computed both with the original cat cochlear tuning of Zilany and Bruce (2006, 2007a) and the human tuning of Shera et al. (2002).

40.4 Results

We have compared our STMI results to the intelligibility scores reported in Lorenzi et al. (2006). We computed the cat and human TFS-only STMI values for the case of 16 filter bands averaged over all test sentences. Our STMI result from the cat auditory model is 0.3806, while a value of 0.4849 is obtained from the human auditory model. In order to get a better understanding of these results, we computed the STMI of a white Gaussian noise (WGN) only stimulus. Our STMI results in this case are 0.2769 in cats and 0.298 for humans, indicating the lowest possible values for the STMI. It can be concluded therefore that even with 16 filter bands, there is still some restoration of E cues from the "TFS-only" speech of Lorenzi et al., and this restoration is enhanced with human cochlear tuning.

In order to reduce (or eliminate) any E cues that might be recovered by the TFS-only signal, we have generated speech TFS-WGN E auditory chimaeras. The average STMI results in this case are 0.3466 for cats and 0.4252 for humans. It can be seen that the average STMI values are reduced for these chimaeras from the TFS-only values, indicating that introducing the noise E cues does diminish somewhat the restoration of speech E cues from the speech TFS signal, but restoration is not completely eliminated.

We have computed STMI values for both cat and human tuning using the auditory chimaeras which we have generated as in Smith et al. (2002). In Fig. 40.2, we display our STMI results for cats and humans together with the intelligibility scores obtained in Smith et al. (2002). The STMI for speech E-noise TFS is monotonically increasing with the number of filter bands, while the speech TFS-noise E starts

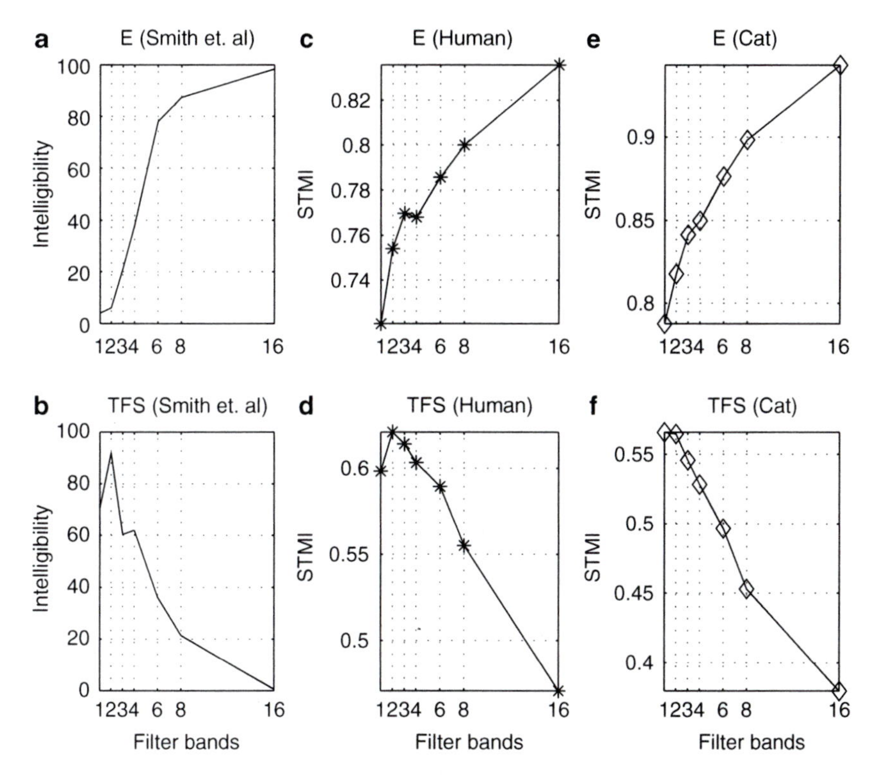

Fig. 40.2 Speech reception of sentences versus number of filter bands in (**a**) speech E-noise TFS chimaera and (**b**) speech TFS-noise E chimaera as in Smith et al. (2002). Average STMI values versus number of filter bands for (**c**) speech E-noise TFS chimaeras as the input to our human model and (**d**) speech TFS-noise E chimaeras. Average STMI values versus number of filter bands for (**e**) speech E-noise TFS chimaeras as the input to the cat model and (**f**) speech TFS-noise E chimaera

increasing with filter bands having a maximum value for two frequency bands, then it decreases with further increase in number of frequency bands. These results match well with the behavior of the intelligibility scores of Smith et al. (2002) as a function of number of filters, which are displayed in the same figure.

It is observed that the STMI values are higher for speech E-noise TFS than speech TFS-noise E over the entire range of numbers of filters. For the speech E-noise TFS signals, the STMI values for cat tuning are consistently higher than those for human tuning. This is due to the broader cat filters being less sensitive to degradation of the speech spectrum by the filter bank in the chimaeras algorithm. Comparing STMI values for cat and human tuning in the case of the speech TFS-noise E signals, scores are consistently higher with the human tuning than with the cat tuning. This observation is related to the narrower cochlear tuning incorporated in the human auditory periphery model. This narrow tuning implies better capability of the human auditory filters to restore E information from the TFS signal.

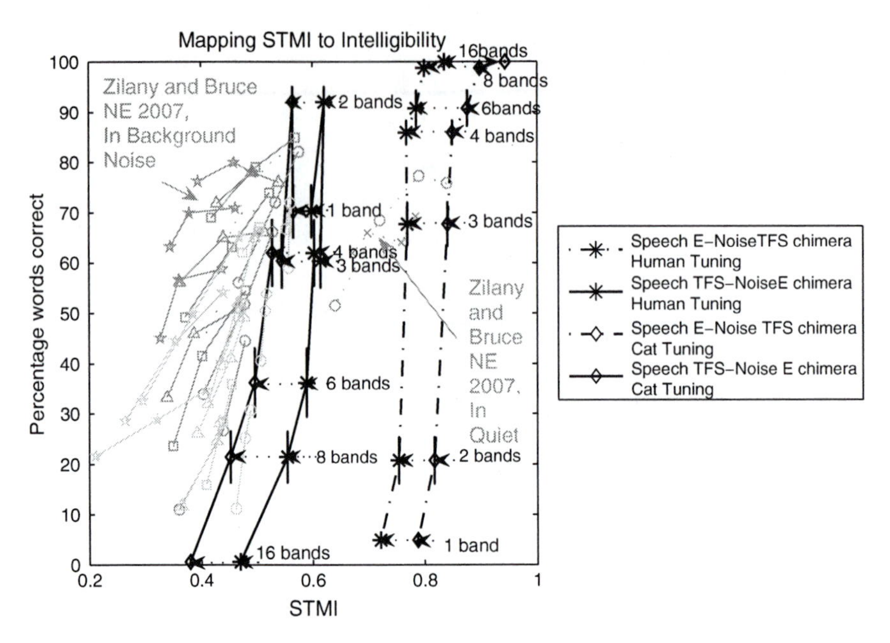

Fig. 40.3 Mapping curves between STMI and percent intelligibility explained in the legend (*thick black lines*), together with STMI-speech intelligibility mappings for cat tuning from Zilany and Bruce (2007b) for different signal-to-noise-ratio (SNR) values (*gray lines*) for comparison

Our STMI results for both cat and human tuning can be mapped to the corresponding intelligibility results obtained in Smith et al. (2002). Hence, for each (species) version of the model we have two mapping functions, one for the speech E-noise TFS chimaeras and the other for the speech TFS-noise E chimaeras. In Fig. 40.3, we display these STMI-intelligibility mapping curves (thick black lines), together with previous STMI-speech intelligibility mappings for cat tuning (Bruce and Zilany 2007; Zilany and Bruce 2007b) for different signal-to-noise-ratio (SNR) values (gray lines) for comparison. It can be observed that the mappings for the speech TFS-noise E signals with cat tuning are somewhere between the mappings obtained in Zilany and Bruce (2007b) for speech in noise and speech in quiet. The human mapping is shifted to the right, closer to the Zilany and Bruce (2007b) mappings for speech in quiet.

If the speech TFS-noise E intelligibility results of Smith et al. (2002) were due entirely to envelope restoration, then it might be expected that the mapping function for these signals would be identical to that for the speech E-noise TFS signals. This is clearly not the case for cat cochlear tuning. For human tuning, the mappings for 6–16 channels for both speech E-noise TFS and speech TFS-noise E signals do appear to be consistent with an extrapolation of the Zilany and Bruce (2007b) mappings for speech in quiet.

40.5 Conclusions

In this work, we show that STMI values for speech TFS-noise E chimaeras attain a maximum value for 1 and 2 frequency bands and then decline consistently with any further increase in bands. For 1 and 2 frequency bands, the narrowband human auditory filters can restore E cues from the TFS signal. Hence, the relatively high scores for 1 and 2 filter bands are obtained from the recovered E cues. Similar conclusions have been presented in Zeng et al. (2004), where it was argued that E recovery from TFS cues is the main reason for the relatively high intelligibility scores for small numbers of frequency bands. From our results, the STMI values obtained for the TFS-only case with 16 channels could almost completely explain the initial speech intelligibility scores for normal-hearing listeners in the Lorenzi et al. (2006) study, consistent with the observation of Heinz and Swaminathan (2008). The dependence of the envelope-restoration phenomenon on the number of filter bands in the processing algorithm and the bandwidth of the cochlear filters is illustrated by the STMI scores for the cat auditory model, where the cochlear filters are wider than the human model. In this case, the ability to recover E cues from TFS-only signals is reduced, and the STMI value is consequently less than the human tuning version. This observation is very important as it supports the theory that TFS information is used indirectly by the cochlea to recover E information, which is then used for speech understanding. This also explains the reduced ability of hearing-impaired people to benefit from TFS-only information as observed in Lorenzi et al. (2006). Since hearing-impaired people suffer from the broadening of the cochlear tuning, the recovery of E cues from TFS information is degraded and hence speech intelligibility is reduced.

However, a consistent mapping between STMI and speech intelligibility for the two types of chimaeras was not obtained for small numbers of channels. Preliminary results indicate that this may be due to the effects of the matched noise used in constructing the chimaeras on the model neural response. Future work should be concentrated on how the use of such matched noise, rather than an independent noise or a flat envelope, affects both STMI values and speech intelligibility in humans in the case of a small number of processing bands.

Acknowledgments The authors thank Dr. Zach Smith for providing speech intelligibility data. This work was funded by NSERC Discovery Grant 261736.

40.6 Comment by Michael Heinz

We recently quantified envelope recovery for chimaeric speech in recorded and modeled auditory-nerve responses and came to very similar conclusions as in your paper (Heinz and Swaminathan 2009). Your STMI model predictions provide an interesting complement ,in that they provide a prediction of recovered envelope cues at central levels, which could be present due to peripheral and/or central recovery of speech envelope cues from TFS cues. Have your STMI predictions provided any insight into the potential for envelope recovery to occur at central levels?

Heinz MG, Swaminathan J (2009) Quantifying envelope and fine-structure coding in auditory nerve responses to chimaeric speech. J Assoc Res Otolaryngol 10 (3).

40.7 Reply Rasha Ibrahim

Thanks for your comments. Our present results indicate Envelope recovery from TFS cues at the peripheral level. It is interesting to investigate if any Envelope recovery can occur at the central level also. This might be considered in our future work to complement our study.

References

Bruce IC, Zilany MSA (2007) Modeling the effects of cochlear impairment on the neural representation of speech in the auditory nerve and primary auditory cortex. Auditory signal processing in hearing-impaired listeners, International Symposium on Audiological and Auditory Research (ISAAR), Denmark,1–10, August 2007

Elhilali M, Chi T, Shamma SA (2003) A spectro-temporal modulation index (STMI) for assessment of speech intelligibility. Speech Commun 41:331–348

Glasberg BR, Moore BCJ (1990) Derivation of auditory filter shapes from notched-noise data. Hear Res 47:103–138

Heinz MG, Swaminathan J (2008) Neural cross-correlation metrics to quantify envelope and fine-structure coding in auditory-nerve responses. J Acoust Soc Am 123:3056

Lorenzi C, Gilbert G, Carn H, Garnier S, Moore BCJ (2006) Speech perception problems of the hearing impaired reflect inability to use temporal fine structure. Proc Natl Acad Sci USA 103:18866–18869

Recio A, Rhode WS, Kiefte M, Kluender KR (2002) Responses to cochlear normalized speech stimuli in the auditory nerve of cat. J Acoust Soc Am 111:2213–2218

Shera CA, Guinan JJ Jr, Oxenham AJ (2002) Revised estimates of human cochlear tuning from otoacoustic and behavioral measurements. Proc Natl Acad Sci USA 99:3318–3323

Smith ZM, Oxenham AJ, Delgutte B (2002) Chimaeric sounds reveal dichotomies in auditory perception. Nature 416:87–90

Young ED (2008) Neural representation of spectral and temporal information in speech. Philos Trans R Soc Lond B Biol Sci 363:923–945

Zeng FG, Nie KB, Liu S, Stickney G, Del Rio E, Kong YY, Chen HB (2004) On the dichotomy in auditory perception between temporal envelope and fine structure cues. J Acoust Soc Am 116:1351–1354

Zilany MSA, Bruce IC (2006) Modeling auditory-nerve responses for high sound pressure levels in the normal and impaired auditory periphery. J Acoust Soc Am 120:1446–1466

Zilany MSA, Bruce IC (2007a) Representation of the vowel /ε/ in normal and impaired auditory-nerve fibers: model predictions of responses in cats. J Acoust Soc Am 122:402–417

Zilany MSA and Bruce IC (2007b) Predictions of speech intelligibility with a model of the normal and impaired auditory-periphery. Proceedings of the 3rd International IEEE EMBS Conference on Neural Engineering, NJ, May 2007, pp 481–485

Chapter 41
Room Reflections and Constancy in Speech-Like Sounds: Within-Band Effects

Anthony J. Watkins, Andrew Raimond, and Simon J. Makin

Abstract The experiment asks whether constancy in hearing precedes or follows grouping. Listeners heard speech-like sounds comprising eight auditory-filter shaped noise-bands that had temporal envelopes corresponding to those arising in these filters when a speech message is played. The "context" words in the message were "next you'll get _to click on," into which a "sir" or "stir" test word was inserted. These test words were from an 11-step continuum that was formed by amplitude modulation. Listeners identified the test words appropriately and quite consistently, even though they had the "robotic" quality typical of this type of 8-band speech. The speech-like effect of these sounds appears to be a consequence of auditory grouping. Constancy was assessed by comparing the influence of room reflections on the test word across conditions in which the context had either the same level of reflections, or where it had a much lower level. Constancy effects were obtained with these 8-band sounds, but only in "matched" conditions, where the reflections' level was varied in the same bands in the context and test word. In "mismatched" conditions, no constancy effects were found. It would appear that this type of constancy in hearing precedes the across-channel grouping whose effects are so apparent in these sounds. This result is discussed in terms of the ubiquity of grouping across different levels of representation.

Keywords Hearing • Speech • Constancy • Room reflections • Grouping

41.1 Introduction

Although the visual systems of organisms respond to the physical properties of light, such as its wavelengths or its luminance level, the information that they provide is about meaningful "real-world" properties of things, such as their colour, or lightness.

A.J. Watkins (✉)
Department of Psychology, The University of Reading, Reading RG6 6AL, UK
e-mail: syswatkn@reading.ac.uk

E.A. Lopez-Poveda et al. (eds.), *The Neurophysiological Bases of Auditory Perception*, DOI 10.1007/978-1-4419-5686-6_41, © Springer Science+Business Media, LLC 2010

The physical properties of light from an object are affected by viewing conditions, such as the shade that affects objects' luminance in sunlight, but there is a compensation for effects of viewing conditions in visual processing, so that a light grey surface in dim shade can be distinguished from a black surface in full sunlight, even though the luminance of the surfaces might be the same. This ability to identify real-world properties in various viewing conditions is known as perceptual constancy, and vision uses information from the surrounding context in order to achieve this (Adelson 2000). Recently, visual researchers have investigated how perceptual grouping stands in relationship to constancy mechanisms, asking whether constancy precedes or follows grouping (Palmer et al. 2003). The present research asks the same question of hearing.

The physical property of a sound that is studied here is the array of temporal envelopes that arises in the frequency channels of devices such as spectrographs or vocoders, and at the earliest stages of auditory processing (Glasberg and Moore 1990). When speech signals are played, the information that hearing provides is predominantly about the real-world phonetic content of the message. Constancy arises when a speech message is played several metres away from the listener in a room, where it is usually heard to have much of the same phonetic content as it does when played nearby, even though the different amounts of reflected sound from the room's surfaces make the temporal envelopes of the two signals very different (Watkins and Makin 2007). This perceptual constancy would appear to result from the surrounding context being taken into account, as certain words recorded from the distant message are heard as other words when they are played in a context that is recorded nearby. This constancy seems to arise from a perceptual compensation for the effects of room reverberation, which is seen in experiments where room reflections in context-speech have compensatory perceptual effects on adjacent test-words (Watkins 2005).

The speech-like sounds used in the present experiment were obtained by signal processing, whereby the speech recording is passed through an 8-filter "bank" before obtaining the temporal envelope in each of these channels. Each envelope is then applied to a narrowband noise that has the channel's centre-frequency and bandwidth. When the filter-bank's centre-frequencies span the speech range, signals obtained by adding the processed bands together are heard to be distinctly speech-like, and the original message is quite intelligible (Shannon et al. 1995). This outcome seems to be a classic example of a grouping effect, as the speech-like quality of the summed-band "whole" is not at all apparent when any of the individual-band "parts" are played in isolation.

Compensatory perceptual effects of context-speech on a test-word could happen either before or after the bands are grouped, as described in Fig. 41.1 for an 8-band signal. A before-grouping mechanism might act within each band, so that the compensation is effected in a "band-by-band" manner. This idea is tested here by applying distant (10-m) room-reflection patterns to only half (4) of the context's bands, while holding the reflection-pattern in the other bands at a nearby distance (0.32 m). In matched conditions, the test word has 10-m reflection-patterns in bands that the context does, while in mismatched conditions it does not. The experiment tests the band-by-band idea by looking for reduced compensation in mismatched conditions.

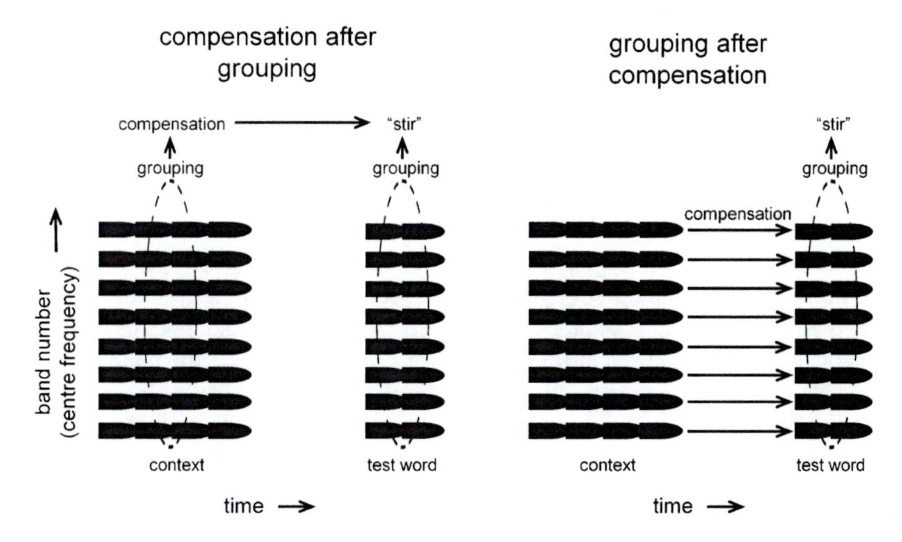

Fig. 41.1 Two ideas about the way that grouping stands in relation to constancy in the 8-band speech used in this experiment

41.2 Method

41.2.1 Speech Contexts and the Test-Word Continuum

The methods described by Watkins (2005) were used to obtain context phrases containing test-words from a continuum between "sir" and "stir". This method used the speech of one of the authors (AW) recorded with 16-bit resolution at a 48-kHz sampling rate using a Sennheiser MKH 40 P48 cardioid microphone in an IAC 1201 booth, giving "dry" speech. The context phrase was originally such a recording, of "next you'll get sir to click on"; with the "sir" test-word excised using a waveform editor. A recording of a "stir" test-word was also obtained in this context phrase. The durations of the context's first and second parts were both 685 ms, and the original recordings of the test words were both 577 ms long.

To form a test-word continuum, the wide-band temporal envelopes of "sir" and "stir" were obtained by full-wave rectification followed by a low-pass filter that had a corner frequency of 50 Hz. The envelope of "stir" was then divided (point wise) by the envelope of "sir" to give a modulation function, and clear "stir" sounds were obtained by amplitude modulating the waveform of "sir" with this function. The original "sir" along with the "stir" produced by the modulation were the 11-step continuum's end-points; nominally steps 0 and 10, respectively. The intermediate steps were produced from the recording of "sir" using appropriately attenuated versions of the modulation function.

Test words were re-embedded into the context parts of the original utterance. This re-embedding was performed by adding the context's waveform to the test word's waveform. Before the addition, silent sections were added to preserve temporal alignment, and to allow different reflection-patterns to be separately introduced into the test word and the context.

41.2.2 Category Boundaries

When room reflections obscure cues to the presence of a [t] in test words, they oppose the amplitude modulation that formed the continuum, so more of the continuum's steps will be identified as "sir" in conditions where this happens. To indicate differences between conditions in the number of steps that are identified as "sir," listeners' category boundaries were compared. The boundary is the step, or point between steps, where listeners switch from predominantly "sir" to predominantly "stir" responses.

Listeners were asked to identify four presentations of each of the continuum's steps played in the context, and category boundaries here were found from the total of number of "sir" responses across all 11 steps. This total was divided by four before subtracting 0.5, to give a boundary step-number between −0.5 and 10.5.

41.2.3 Room Reflections

The methods described by Watkins (2005) were also used to introduce room reflections into the dry contexts and test words by convolution with room impulse responses. This gives the effect of monaural real-room listening over headphones. The monaural impulse-responses were obtained in rooms using dummy-head transducers (a speaker in a Bruel and Kjaer 4128 head and torso simulator, and a Bruel and Kjaer 4134 microphone in the ear of a KEMAR mannequin), so that they incorporate the directional characteristics of a human talker and a human listener. To obtain signals at the listener's eardrum that match the signal at KEMAR's ear, the frequency-response characteristics of the dummy-head talker and of the listener's headphones were removed using appropriate inverse filters.

The room impulse responses were obtained in a disused office that was L-shaped with a volume of 183.6 m^3. To obtain different amounts of reflected sound in a "natural" way, different distances between the dummy-head transducers of the talker and listener were used, as the proportion of reflected sound, relative to direct sound energy, increases with source-to-receiver distance. The transducers faced each other, while the talker's position was varied to give distances from the listener of 0.32 m or 10 m. The amount of reflected sound at these distances is indicated by the time taken for the room's impulse-response energy to decay by 10 dB, "EDT" ISO 3382 (1997). At 10 m, the A-weighted EDT was 0.14 s, while at 0.32 m this EDT was less than 0.01 s.

41.2.4 8-Band Speech

The individual bands were narrow-band noises, each with the temporal-envelope fluctuations that arise in an auditory filter when speech is played. The impulse response of a filter was a "gammatone" function with the parameter $\eta=4$ and with the bandwidth appropriate for its centre frequency, as given by the "Cambridge ERB" (Glasberg and Moore 1990). The eight centre-frequencies were equally log-spaced across the speech range, starting at 250 Hz, and increasing by intervals of a musical fifth (7/12 octave). Bands were numbered from low to high centre-frequency, using a band number, $n=1,2, \ldots 8$.

To obtain one of these bands, the speech was played through an auditory filter, followed by a "signal correlated noise" operation, which involves reversing the polarity of a randomly selected half of the signal's samples. This operation gives a wideband signal, but it preserves the temporal envelope of the filter's output. The signal was then played through the auditory filter again, to eliminate frequencies outside the filter's band. The impulse response of this second filter was reversed in time to correct for delays introduced by the operation of the first filter.

The relative levels of the bands were adjusted with a "speech-shaping" filter, whose frequency response was the long-term average spectrum of the original speech-context. Room-reflection patterns were then added to each band, using distances appropriate for the experimental condition, and the eight bands were summed.

41.2.5 Design

The reflection pattern's distance was varied between 0.32-m and 10-m to give the different context-distance and test-word distance conditions. The 8-band speech was divided between interleaved sets of odd-numbered and even-numbered bands, and the distance manipulation was applied to some of these bands, while the others were held at 0.32-m. In a mismatched condition, the distance was varied in the odd-numbered bands of the context and in the even-numbered bands of the test word. In a matched condition for one of the listener groups, distance was varied in the even-numbered bands of both the context and the test-word. The other group had a different type of matched condition, where distance was varied in the odd-numbered bands of the context but in all eight bands of the test word. These conditions are illustrated in Fig. 41.2.

The 12 listeners were evenly divided between the two groups. Each group identified test-word continua in unprocessed speech conditions as well in their matched and mismatched 8-band conditions. All combinations of the context and test-word distance were presented in each of these conditions, and each listener received the trials in a different randomized order.

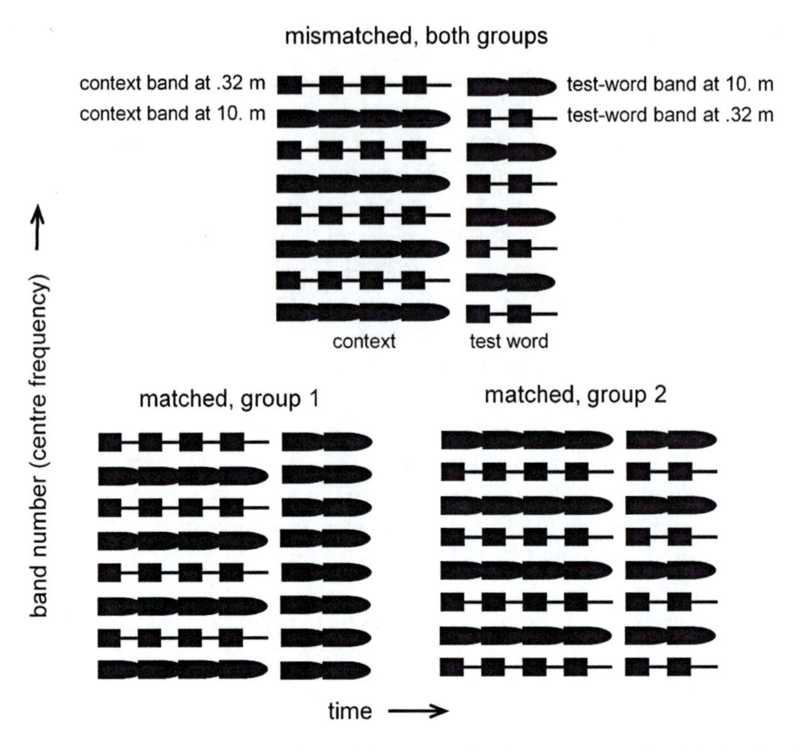

Fig. 41.2 Diagrammatic representations of the room-reflection content of the eight bands of the speech in conditions where the context's distance and test-word's distance were both 10 m

41.2.6 Procedure

Sounds were delivered to listeners at a peak level of 48 dB SPL through the left earpiece of Sennheiser HD480 headphones in the otherwise quiet conditions of the IAC booth. Before the experimental trials, listeners were informally given a few randomly-selected practice trials to familiarize them with the sounds and the set up, and to check that they could hear the 8-band sounds as speech. Trials were administered to listeners in individual sessions by an Athlon 3500 PC computer with Matlab 7.1 software and with an M-Audio Firewire 410 sound card. On each of these trials, a context with an embedded test-word was presented. Listeners then identified the test word with a click of the computer's mouse, which they positioned while looking through the booth's window at the "sir" and "stir" alternatives displayed on the computer's screen. The computer waited for the listener to respond before presenting the following trial.

41.3 Results

For each condition, category boundaries were pooled across the six listeners, and the resulting means are shown with their standard errors in Fig. 41.3.

Results with unprocessed speech replicate the compensation effects reported in earlier work (Watkins 2005). When the context is nearby, increasing the test word's distance causes more of the continuum's members to be heard as "sir," so there is a corresponding increase in the category boundary. However, when the context's distance is also increased to 10 m, there is a compensation effect, giving a reduction in the category boundary, which is indicated by the arrows in the leftmost panels of Fig. 41.3.

Results with 8-band speech depend on whether the test-word's bands are matched to those of the context in the crucial 10-m conditions. In mismatched conditions, there is no compensation effect, as indicated by the absence of any

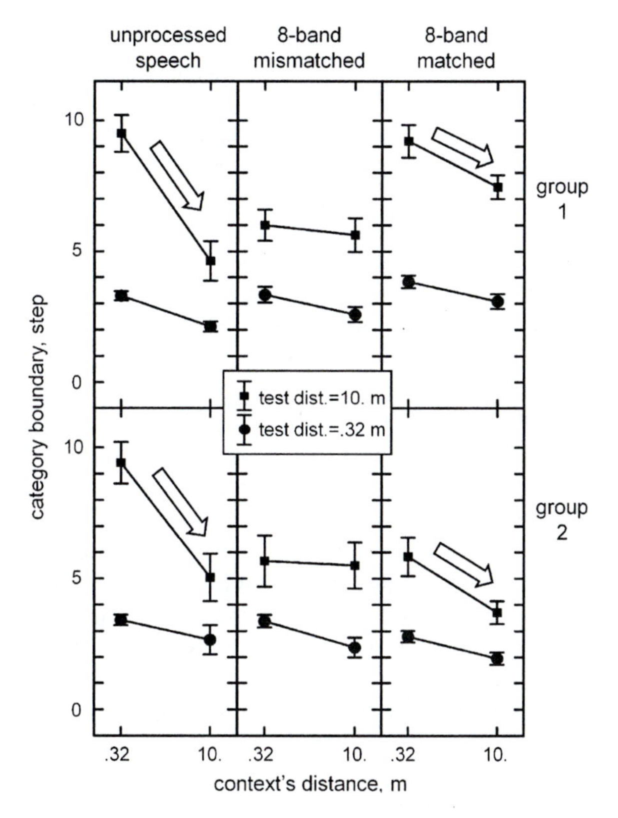

Fig. 41.3 Means and standard errors of category boundaries in the experiment's conditions. The arrows indicate compensation effects

arrows in the centre panels of Fig. 41.3. In matched conditions, however, there is a compensation effect, which is indicated by the arrows in the rightmost panels of Fig. 41.3.

This pattern of results supports the band-by-band hypothesis, which was tested statistically with a 4-way analysis of variance, using the combination of 2-level factors that describe the conditions in the centre and rightmost panels of Fig. 41.3. The 4-way interaction was not significant, but the crucial interaction was among the three factors; "test-word's distance," "context's distance," and "matched vs. mismatched," which was found to be significant with $F(1, 10)=20.7$, and $p<0.0012$.

41.4 Discussion

Room-reflections in both the even-numbered and the odd-numbered frequency-bands seem to have some effect on the test word. Increasing the reflections' distance in test-words' even bands increases the number of "sir" responses in both matched and mismatched conditions, while for group 1, increasing the distance of test words' odd bands in the matched conditions also increases the number of "sir" responses for the near contexts. This last effect from the odd bands appears as an increase over effects from the even bands alone. Therefore, information about the [s] vs. [st] distinction needed to classify the test word seems to be distributed across more than one of the frequency bands in these 8-band stimuli.

Both sets of bands also seem to bring about compensation, as for group 1, this effect is from the odd-numbered bands in the context, while it is from the even-numbered bands in group 2. However, this compensation effect is only seen in matched conditions, where the test-word and context share bands that have the 10-m room-reflection pattern.

Is perception governed by the less distorted, 0.32-m bands in these sounds? This might happen if hearing behaves like a "missing data" speech recogniser, and bases its decisions on the less distorted parts of the signal (Palomäki et al. 2002). Clearly, this could not be happening on a word by word basis with the listeners in this experiment, as there would be no effects of distance when only four of the test-word's bands are given the 10-m reflection patterns. A related idea is that the less distorted, 0.32-m bands are selected in the context and then "tracked" through the test word, perhaps by an attention-like process, while the other bands are effectively ignored. Such a tracking could account for results with group 2, as it would give the effect of distance in mismatched conditions, where the tracked bands would be the ones that are at 10-m in the test word. This sort of tracking could also give the compensation effect in group 2's matched conditions, where the tracked bands would become the four that are also at 0.32 m in the test word. However, the compensation found with group 1 is not consistent with this "band-tracking" idea, as compensation is obtained across conditions where all the test-word's bands are at 10 m.

A substantial part of the grouping that occurs when hearing these 8-band sounds is likely to be attentional in nature (Cooper and Roberts 2007). It is also possible that a more "primitive," pre-attentive grouping process operates as well; perhaps, one that is informed by amplitude modulation, and that exploits the correlations among the bands' temporal envelopes at the lower modulation frequencies in speech signals (Crouzet and Ainsworth 2001). Such a grouping may be "obligatory" (Roberts et al. 2002), so that an individual band is difficult to "track" independently of the others.

Whatever processes bring about the cross-channel grouping in these 8-band sounds, it can be said from the present results that the constancy mechanism precedes the grouping. This is not to say that grouping at other levels is not involved, in particular, a sequential within-band grouping might well operate prior to the constancy operation. This view is consistent with observations about diverse types of grouping in vision, where it has been concluded that perceptual grouping is ubiquitous, in that it occurs for each level of representation. Consequently, grouping in vision can occur before, after, and even during different types of constancy operation (Palmer et al. 2003).

Acknowledgements This work was supported by a grant to the first author from EPSRC. We are grateful to Amy Beeston and Guy Brown for discussion.

References

Adelson EH (2000) Lightness perception and lightness illusions. In: Gazzaniga M (ed) The new cognitive neurosciences, 2nd edn. MIT Press, Cambridge MA

Cooper HR, Roberts B (2007) Auditory stream segregation of tone sequences in cochlear implant listeners. Hear Res 225:11–24

Crouzet O, Ainsworth WA (2001) On the various influences of envelope information on the perception of speech in adverse conditions: an analysis of between-channel envelope correlation. In: CRAC workshop (consistent and robust acoustic cues for sound analysis). Aalborg, Scandinavia

Glasberg BR, Moore BCJ (1990) Derivation of auditory filter shapes from notched-noise data. Hear Res 47:103–138

ISO 3382 (1997) Acoustics – measurement of the reverberation time of rooms with reference to other acoustical parameters. International Organization for Standardization, Geneva

Palmer SE, Brooks JL, Nelson R (2003) When does grouping happen? Acta Psychol 114:311–330

Palomäki KJ, Brown GJ, Barker J (2002) Missing data speech recognition in reverberant conditions. In: Proceedings of 2002 International Conference on Acoustics, Speech, and Signal Processing (ICASSP2002), pp 65–68

Roberts B, Glasberg BR, Moore BCJ (2002) Primitive stream segregation of tone sequences without differences in fundamental frequency or passband. J Acoust Soc Am 112:2074–2085

Shannon RV, Zeng F, Kamath V, Wygonski J, Ekelid M (1995) Speech recognition with primarily temporal cues. Science 270:303–304

Watkins AJ (2005) Perceptual compensation for effects of reverberation in speech identification. J Acoust Soc Am 118:249–262

Watkins AJ, Makin SJ (2007) Perceptual compensation for reverberation in speech identification: effects of single-band, multiple-band and wideband contexts. Acta Acustica united Acustica 93:403–410

Chapter 42
Identification of Perceptual Cues for Consonant Sounds and the Influence of Sensorineural Hearing Loss on Speech Perception

Feipeng Li and Jont B. Allen

Abstract A common problem for people with hearing loss is that they can hear the noisy speech, with the assistance of hearing aids, but still they cannot understand it. To explain why, the following two questions need to be addressed: (1) What are the perceptual cues making up speech sounds? (2) What are the impacts of different types of hearing loss on speech perception? For the first question, a systematic psychoacoustic method is developed to explore the perceptual cues of consonant sounds. Without making any assumptions about the cues to be identified, it measures the contribution of each subcomponent to speech perception by time truncating, high/low-pass filtering, or masking the speech with white noise. In addition, AI-gram, a tool that simulates auditory peripheral processing, is developed to show the audible components of a speech sound on the basilar membrane. For the second question, speech perception experiments are used to determine the difficult sounds for the hearing-impaired listeners. In a case study, an elderly subject (AS) with moderate to severe sloping hearing loss, trained in linguistics, volunteered for the pilot study. Results show that AS cannot hear /ka/ and /ga/ with her left ear, because of a cochlear dead region from 2 to 3.5 kHz, where we show that the perceptual cues for /ka/ and /ga/ are located. In contrast, her right ear can hear these two sounds with low accuracy. NAR-L improves the average score by 10%, but it has no effect on the two inaudible consonants.

Keywords Hearing loss • Perceptual cue • Consonant

F. Li (✉)
ECE Department, University of Illinois at Urbana-Champaign, Urbana, IL, 61801, USA
e-mail: fli2@illinois.edu

E.A. Lopez-Poveda et al. (eds.), *The Neurophysiological Bases of Auditory Perception*, DOI 10.1007/978-1-4419-5686-6_42, © Springer Science+Business Media, LLC 2010

42.1 Introduction

People with hearing loss often complain about the difficulty of hearing speech in noisy environments. Depending on the type and degree of hearing loss, a hearing-impaired (HI) listener may require a more favorable signal-to-noise ratio (SNR) than the normal-hearing (NH) listeners, to achieve the same level of performance for speech perception. The gap between the HI and NH listeners usually is unevenly distributed across different speech sounds. In other words, an HI listener may have serious problems with certain consonants, yet show no problem at all with other consonants.

State-of-the-art hearing aids have little effect in diminishing the gap, because they amplify everything, including noise, instead of enhancing the speech sounds that the HI listeners have difficulty with. As a consequence, many HI listeners now can hear the speech, with the assistance of hearing aids, but still they cannot understand it. To explain this, the following two questions need to be addressed. First, what are the perceptual cues of speech sounds? Second, what is the impact of hearing loss on speech perception?

Past study on hearing-impaired speech perception has two major limitations. First, most researches have been focused on the use of pure tone audiometry. For example, Zurek and Delhorne (1987) showed that the difficulties in consonant reception can be fully accounted on average, by the shift in the pure tone threshold for a group of listeners with mild-to-moderate hearing loss. Second, the impact of hearing loss on speech perception can only be assessed for speech on average, not for individual sound, due to the lack of accurate information on speech cues. Speech banana, a tool of qualitative assessment used in audiological clinic, is based on the pure tone audiogram and the formant data of vowels and consonants, which accounts for only part of the perceptual cues for speech sounds.

It is well known that most sensorineural hearing loss can be attributed to the malfunctioning of outer hair cells (OHCs) and inner hair cells (IHCs) within the cochlea. Damage to the OHCs reduces the active vibration of the cell body that occurs at the frequency of the incoming signal, resulting in an elevated detection threshold. Damage to the IHCs reduces the efficiency of mechanical-to-electrical transduction, also results in an elevated detection threshold. The audiometry configuration is not a good indicator of the physiological nature of the hearing loss (Moore et al. 2000), specifically, subjects with OHC loss and IHC loss may show the same amount of shifting in hearing threshold, yet the impacts of the two types of hearing loss on speech perception are very different. In the past decade, Moore and his students developed a threshold equalized noise (TEN) test (Moore et al. 2000) and a psychoacoustic tuning curve (PTC) test (Moore and Alcántara 2001) for the diagnosis of cochlear dead regions, an extreme case of IHC loss, which provides a psychoacoustic way of partitioning the two types of hearing loss.

In this chapter, we investigate the impact of cochlear dead regions on consonant identification. Based on the analysis of a large amount of data, it is hypothesized that speech sounds are encoded by some time-frequency energy onsets called acoustic cues.

Perceptual cues (events), the representation of acoustic cues on the basilar membrane (BM), are the basic units for speech perception. The HI listeners have difficulty understanding noisy speech because they cannot hear the weak events, missing due to the hearing loss and the masking effect introduced by the noise. The research work can be divided into two parts: identification of events for consonant sounds and influence of hearing loss on event reception.

42.2 Identification of Perceptual Cues

Speech sounds are characterized by three properties: time, frequency, and amplitude (intensity). Event identification involves isolating the cues along the three dimensions.

42.2.1 Modeling Speech Reception

In very general terms, the role of cochlear is to decompose the sound wave received at the eardrum through an array of overlapping narrow-band critical filters along the BM, with the base and apex of BM being tuned to the high-frequency (20,000 Hz) and low-frequency (20 Hz) respectively. Once a speech sound reaches the cochlea, it is represented by some time-varying energy patterns across the BM. Some of the energy patterns contribute to speech recognition, others do not. The purpose of event identification is to isolate the specific parts of the psychoacoustic representation that are critical for speech perception.

To better understand how speech sounds are represented on the BM, AI-gram (Lobdell 2006), a what-you-see-is-what-you-hear, WISIWYH (/wisiwai/) tool that simulates the auditory peripheral processing, is applied for the visualization of the speech sound. Given a speech sound, AI-gram provides an approximate image of the effective components that are audible to the central auditory system. However, it does not implicate which component is critical for speech recognition. To find out which component entertains the most significance for speech recognition, it is necessary to correlate the results of the speech perception experiments to the AI-grams.

42.2.2 Principle of 3D Approach

In this research, we developed a 3D approach (Li et al. 2009) to explore the perceptual cues of consonants from natural speech. To isolate the events along time, frequency, and amplitude, speech sounds are truncated in time, high/low-pass filtered, or masked with white noise, before being presented to normal-hearing listeners. Suppose an acoustic cue that is critical for speech perception has been removed or masked, it would degrade the speech sound and reduce the recognition score significantly.

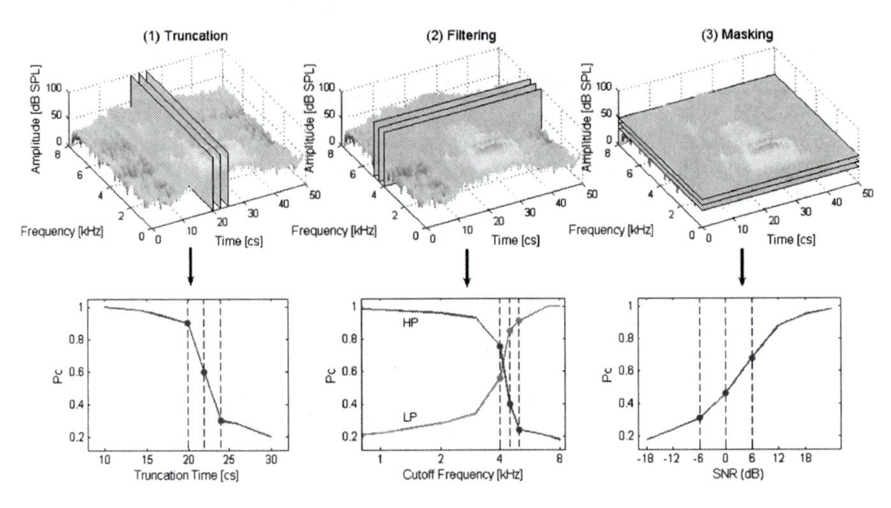

Fig. 42.1 3D approach for the identification of acoustic cues. The three plots on the *top row* illustrate how the speech sound is processed. The three plots on the *bottom row* depict the correspondent recognition scores for modified speech

For a particular consonant sound, the 3D approach (refer to Fig. 42.1) requires three experiments to measure the weight of subcomponent to speech perception. The first experiment determines the contribution of various time intervals by truncating the consonant into multiple segments of 5, 10, or 20 ms per frame depending on the duration of the sound. The second experiment divides the fullband into multiple bands of equal length on the BM and measures the importance of different frequency bands by using high-pass/low-pass filtered speech as the stimuli. Once the coordinates of the event are identified, the third experiment assesses the event strength by masking the speech at various signal-to-noise ratios.

42.2.3 Data Interpretation

The direct results of speech perception tests take the form of confusion patterns (CPs), which display the probabilities of all possible responses, including the target sound and the competing sounds, as a function of time, frequency, and amplitude. Since the probabilities of responses are not additive (Allen 1994; Fletcher and Galt 1950), in other words, the total contribution of multiple cues cannot be assessed by summing up the recognition scores of individual cues. Following the same idea of additivity of AI from multiple articulation bands, we defined a set of heuristic importance functions for the evaluation of contribution from different speech components.

Let $e_L(f_k)$ and $e_H(f_k)$ denote the low-pass and high-pass errors at the kth cutoff frequency f_k. The *frequency importance function* (FIF) is defined as

$$D_L(f_k) = \log_{e0} e_L(f_k) - \log_{e0} e_L(f_{k-1})$$

for the low-pass filtering case, or

$$D_H(f_k) = \log_{e0} e_H(f_{k+1}) - \log_{e0} e_H(f_k)$$

for the high-pass filtering case. The cumulative FIF is defined as the merge of the low-pass and high-pass values

$$D_F(f_k) = \max\,[D_L(f_k) \cup D_H(f_k)].$$

Generally, D_F should have its maximum value around the intersection point of the low-pass error e_L and high-pass error e_H that divide the full band into two parts of equal information.

The *event strength function* (ESF) is defined as

$$D_S(\text{snr}_k) = \log_{e0} e_S(\text{snr}_k) - \log_{e0} e_S(\text{snr}_{k-1})$$

where $e_S(\text{snr}_k)$ is the probability of error under the *k*th SNR condition. When a speech sound is masked by noise, the recognition score usually keeps unchanged until the noise hits a certain level snr_k (in dB) and has the event masked. Then, the recognition error e_S will increase dramatically and create a peak on the ESF at the corresponding position. Usually, the higher the peak, the more important the event is to the perception of the sound; the lower snr_k the more robust the sound is to noise.

The *time importance function* (TIF) is defined as the product of the probability of correctness, i.e., $1 - e_T(t_k)$ and the instantaneous AI $a(t_k)$ (Lobdell 2006)

$$D_T(t_k) = [1 - e_T(t_k)] \cdot a(t_k),$$

where $a(t_k)$ can be simply regarded as the energy of the speech signal at time t_k.

42.2.4 Perceptual Cues of Stop Consonants

In this section, we demonstrate how the events of consonants are identified by applying the 3D approach. To facilitate the integration of multiple information sources, the AI-gram and the importance functions are aligned in time or frequency, and depicted in a compact figure form, as shown by Fig. 42.2a.

Analysis reveals that the event of /ka/ is a mid-frequency burst around 1.6 kHz and articulated 50–70 ms before the vowel, as highlighted by the rectangular box in panel (a) of Fig. 42.2a. The event is strong enough to resist white noise at 0 dB SNR. Since the three importance functions all have a single sharp peak, which clearly shows where the event is, the process of event identification is straight forward. The TIF (panel (b) of Fig. 42.2a) has a distinct peak at $t = 165$ ms, and it drops to zero after that. As soon as the burst is truncated, the perceptual score for /ka/ drops drastically. It is perceived as /pa/ thereafter. The FIF (panel (d) of Fig. 42.2A) has a peak around 1.6 kHz, where the intersection point of the high-pass and low-pass recognition scores is located, indicating that the mid-frequency burst is critical for the perception of /ka/. The above analysis is verified by the

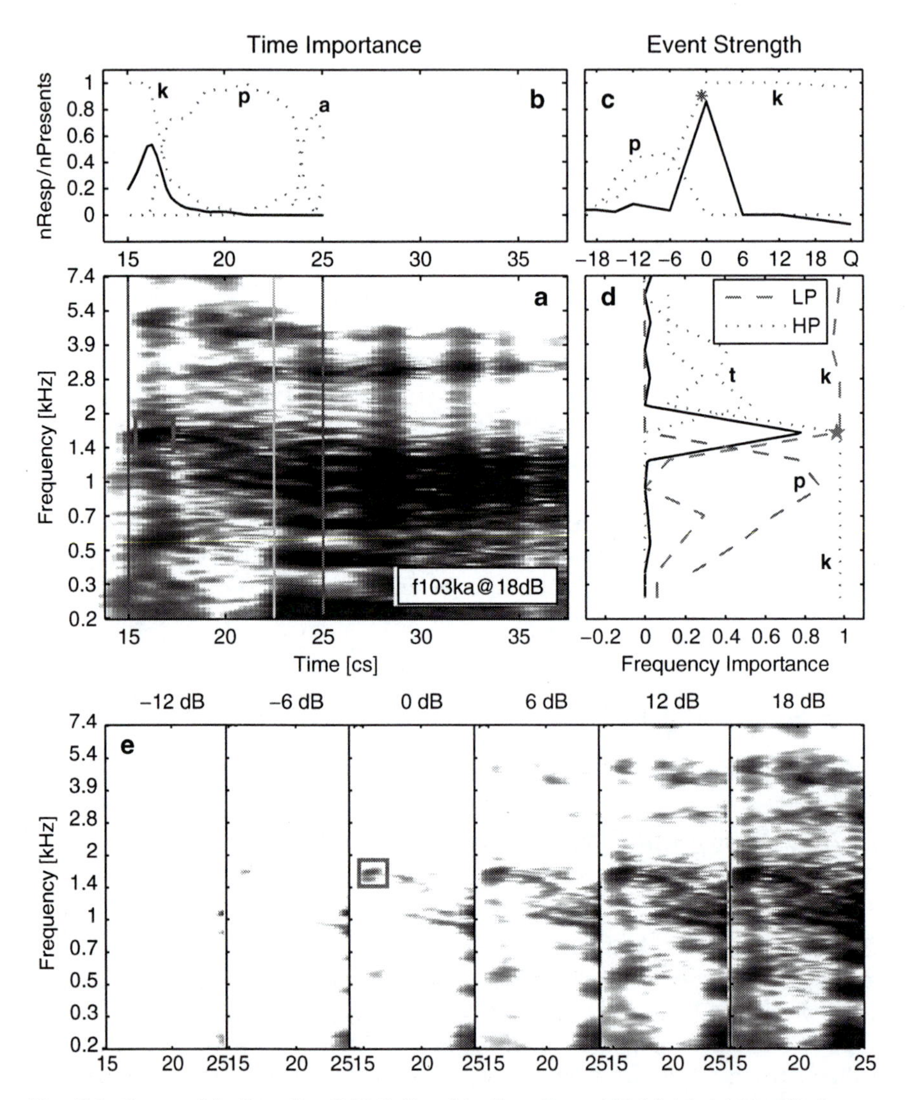

Fig. 42.2 Events of /ka/ by talker f103 (*left*) and /ga/ by talker m111 (*right*). (**a**) Identified events highlighted by *rectangular boxes* on the AI-gram. (**b**) TIF (*solid*) and CPs (*dotted*) as a function of truncation time. (**c**) ESF (*solid*) and CPs (*dotted*) as a function of signal-to-noise ratio. (**d**) FIF (*solid*) and CPs (*dotted*) as a function of cutoff frequency. (**e**) AI-grams of the consonant part at −12, −6, 0, 6, 12, 18 dB SNR

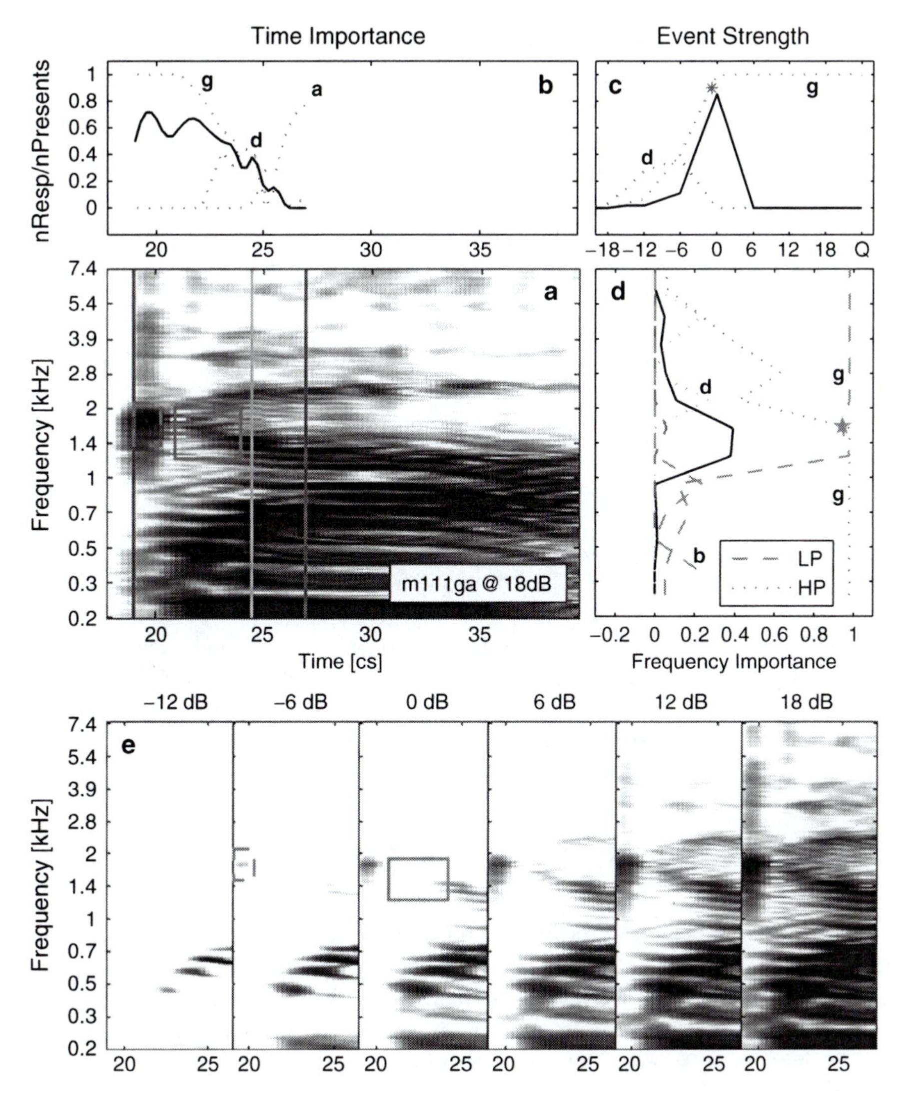

Fig. 42.2 (continued)

AI-grams under different SNRs (panel (e) of Fig. 42.2a), which shows that the /ka/ event becomes barely intelligible at 0 dB SNR. Correspondingly, ESF (panel (c) of Fig. 42.2a) has a sharp peak at the same SNR, where the recognition score of /ka/ begins to drop dramatically.

The events of /ga/ include a burst and a transition between 1 and 2 kHz as highlighted in panel (a) of Fig. 42.2b. The TIF in panel (b) of Fig. 42.2b shows two peaks at around $t = 190$ ms and $t = 230$ ms, where the burst and transition are located.

When the initial burst is truncated, the recognition score of /ga/ begins to drop, and the confusion with /da/ rapidly increases. The frequency importance function (panel (d) of Fig. 42.2B) has a peak at the intersection point of the high-pass and low-pass curves, which falls within the frequency range of the burst and the transition. Both the high-pass and low-pass scores show sharp decreases when the frequency range of 1–2 kHz is removed, indicating that the burst and the transition region are the perceptual cues for /ga/. This analysis is supported by the ESF and the AI-grams under various SNRs. When the F2 transition, the dominant cue, becomes masked, the recognition score of /ga/ from the masking experiment drops quickly and leaves a sharp peak on the ESF curve.

Through the same method, we also identified the events of other stop consonants preceding vowel /a/. Except for the two bilabial consonants, /pa/ and /ba/, which have a noise spike at the beginning, the stop consonants are characterized by a burst, caused by the sudden release of pressure in the oral cavity, and a transition in the second formant created by the jaw movement thereafter. Specifically, the events of /pa/ include a formant transition at around 1–1.4 kHz and a wide band noise spike at the beginning. /ba/ is characterized by a low frequency burst around 0.4 kHz and a transition around 1.2 kHz and a wide band noise spike in the range of 0.3–4.5 kHz. The event of /ta/ is a high frequency burst above 4 kHz. The /da/ event is a high frequency burst similar to the case of /ta/ in unvoiced sounds at around 4 kHz and an additional transition region at around 1.5 kHz. Generally, the noise spikes are much weaker than the bursts and transitions. The former usually disappear at 12 dB SNR, while the latter are still audible at 0 dB SNR.

42.3 Influence of Hearing Loss on Speech Perception

Depending on the nature of the hearing loss, an HI listener may have difficulty with a certain speech sounds. Finding the inaudible sounds that the HI listener cannot hear is the first step of solving the problem. With the identified events of consonants, the reason why an HI listener cannot hear certain sounds could be explained by integrating the speech events and the configuration of hearing loss.

42.3.1 Diagnosis of Hearing Loss

The degree of hearing loss is measured by the traditional pure tone audiometry. In addition to that, TEN test (Moore et al. 2000; Moore 2004) and PTC test (Moore and Alcántara 2001) are combined for the diagnosis of cochlear dead regions. Due to the time issue, the PTC test is used only to verify the existence of a cochlear dead region at certain frequencies suggested by the TEN tests. The procedures of the tests are controlled by a Matlab program.

42.3.2 *Quantification of Consonant Loss*

A speech perception experiment (SL07), using 16 nonsense CVs as the stimuli, is employed to collect the CPs across consonant sounds. There are two versions of the experiments: SL07a and SL07b. The only difference between the two versions is that the speech sounds are amplified by NAL-R in the latter in order to compensate for the hearing loss of the HI listener. NH Listeners only take the first test, while the HI listeners are instructed to take both of the tests. The results of each HI ear are compared to those of the average normal-hearing (ANH) listeners to determine the distribution of consonant loss. The detail of experiment SL07 is given below.

42.3.2.1 Speech Stimuli

Sixteen nonsense CVs: /p, t, k, f, T, s, S, b, d, g, v, D, z, Z, m, n/ + /a/ chosen from the LDC-2005 S22 corpus, were used as the common test material for both the HI and ANH listeners. The speech sounds were sampled at 16,000 Hz. Each CV has only six talkers, half male and half female. The speech sounds were presented through an ER-2 insert earphone to one ear of the subjects at the listener's most comfortable level. All experiments were conducted in a soundproof booth.

42.3.2.2 Conditions

Besides the quiet condition, speech sound were masked at five different signal-to-noise ratios [−12, −6, 0, 6, 12] using speech-shaped noise. In SL07b, the noisy speech was amplified with NAL-R before being presented to the HI listeners.

42.3.2.3 Procedure

A mandatory practice session was given to each subject at the beginning of the experiment. Speech tokens were randomized across the talkers, conditions, and consonants. Following each presentation, subjects responded to the stimuli by clicking on the button labeled with the CV that he/she heard. In case the speech was completely masked by the noise, or the processed token did not sound like any of the 16 consonants, the subject was instructed to click a "Noise Only" button. To prevent fatigue, the subjects were asked to take a break whenever they feel tired. Subjects were allowed to play each token for up-to three times. A PC-based Matlab program was created for the control of the procedure.

42.4 Results

An elderly female subject (AS), mentally agile, trained in linguistics, volunteered for the pilot study. The subject has been wearing a pair of ITE hearing aids for more than 2 years.

42.4.1 Hearing Loss

Pure tone audiometry (Fig. 42.3) shows that AS has a bilateral moderate sloping hearing loss. Her left ear and right ear have similar configurations of hearing threshold with the PTA values being equal to 40 and 42 dB HL respectively.

Results of TEN tests (Fig. 42.4) suggest that subject AS may have a cochlear dead region around 2–3.5 kHz in the left ear, as the absolute hearing threshold and TEN-masked hearing threshold have a gap of more than 10 dB SPL at the frequencies of 2 and 3 kHz. This result is confirmed by the PTC test, in that the tuning curve of 2 kHz barely has a tip, while the turning curve at 3 kHz has a tip displaced in frequency. In contrast, her right ear has no cochlear dead region in the mid-frequency range, as suggested by the results of the TEN and PTC tests. The right ear may have a cochlear dead region around 8 kHz, since the absolute hearing threshold is greater than 120 dB SPL. PTC tests at 1 and 2 kHz show tuning curves of normal shape.

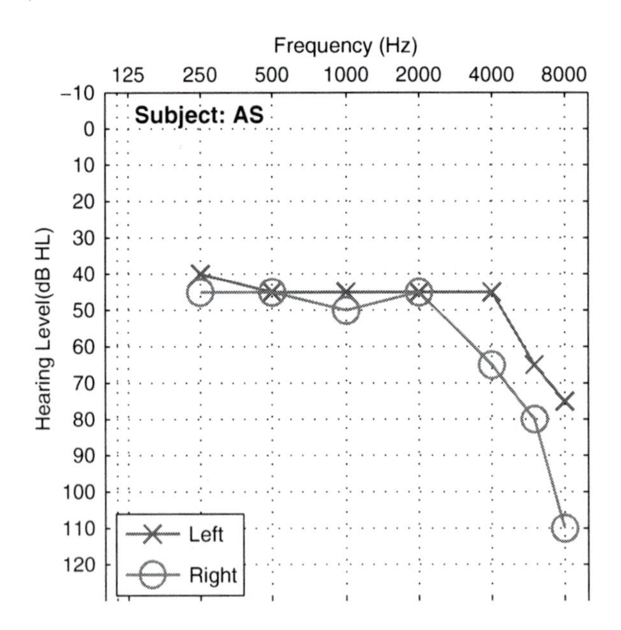

Fig. 42.3 Pure tone hearing threshold of AS

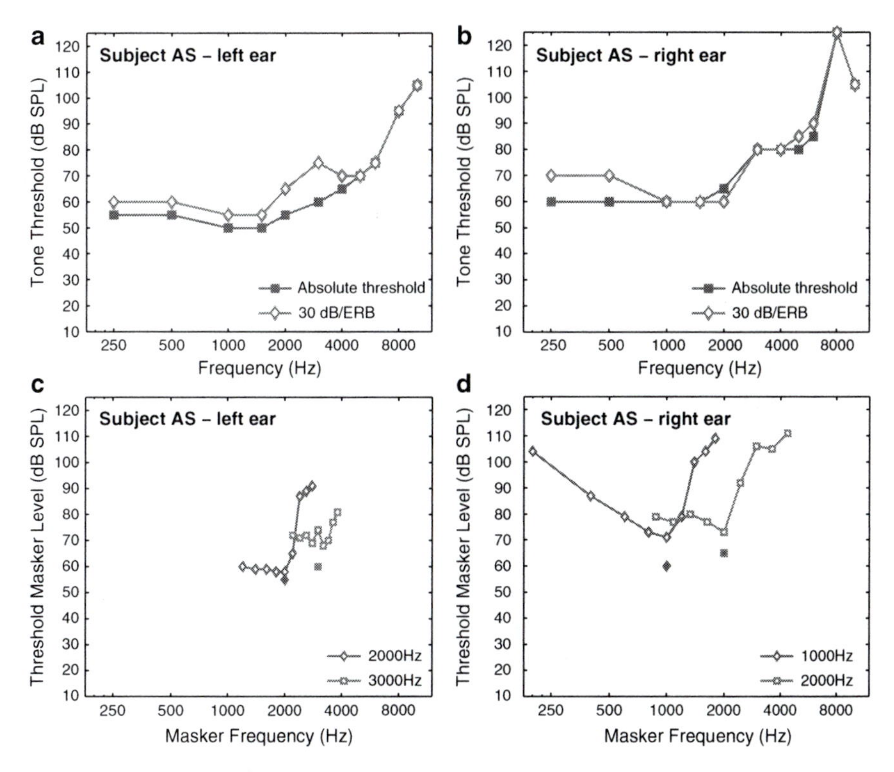

Fig. 42.4 Results of TEN and PTC tests for subject AS. *Upper panels* show the absolute and TEN-masked hearing thresholds in dB SPL. *Lower panels* show the PTCs for the frequencies of interest

42.4.2 Consonant Identification

The results of experiment SL07 indicate that cochlear dead regions may have a significant impact on the perception of speech sounds. Figure 42.5 depicts the recognition scores of subject AS for six stop consonants /pa, ta, ka, ba, da, ga/. Due to the cochlear dead region from 2 to 3.5 kHz, where the perceptual cues for /ka/ and /ga/ are located (Fig. 42.5), subject cannot hear these two sounds in the left ear. In contrast, her right ear can hear these two sounds with low accuracy, despite that the two ears have close configuration of hearing loss in terms of pure tone audiometry.

Confusion analysis shows that more than 80% of the /ka/s are misinterpreted as /ta/, while about 60% of the /ga/s are reported as /da/. A noteworthy fact revealed by the perceptual data is that NAL-R does not always help for the HI listeners. It increases the grand percent correctness (Pc) by about 10% for both ears in quiet, but it has no effect on the perception of the two inaudible consonants /ka/ and /ga/ for the left ear. Sometimes, it even degrades the speech sounds, for example, /ka/ for AS-R at 6, 12, and 18 dB SPL, due to unknown reasons.

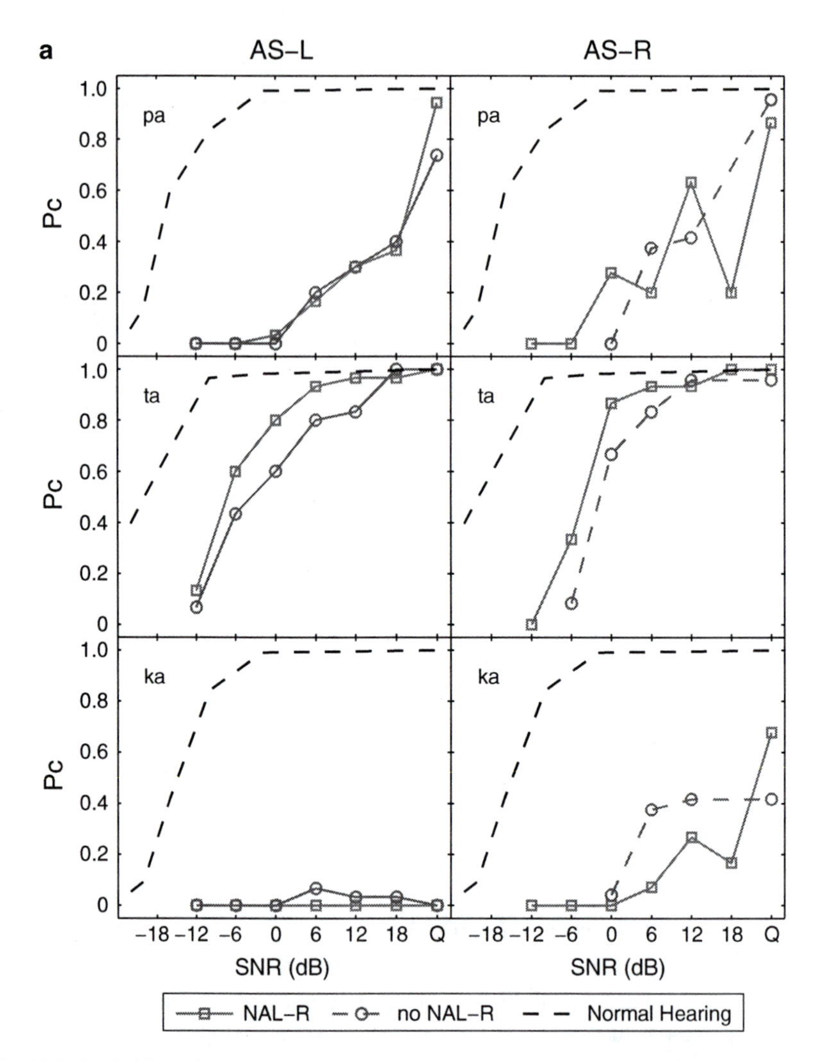

Fig. 42.5 Probability of correctness (Pc) for unvoiced stops /pa, ta, ka/ and voiced stops /ba, da, ga/ for subject AS. In both subfigures, panels on the *left column* show the data of the left ear (denoted by AS-L), and panels on the *right column* show the data of the right ear (denoted by AS-R). Perceptual data w/out NAL-R are depicted by the *square-masked* and *circle-masked curves* respectively. The recognition scores of ANH listeners are depicted by the *unmasked dashed curves* for comparison

42.5 Discussion

The pilot study presented above, in which the mentally healthy volunteer subject has similar configuration of audiometry for both ears with the left ear has a big cochlear dead region around 2–3.5 kHz, is an excellent case for the investigation of HI speech perception. The results support our hypothesis that the HI listeners have

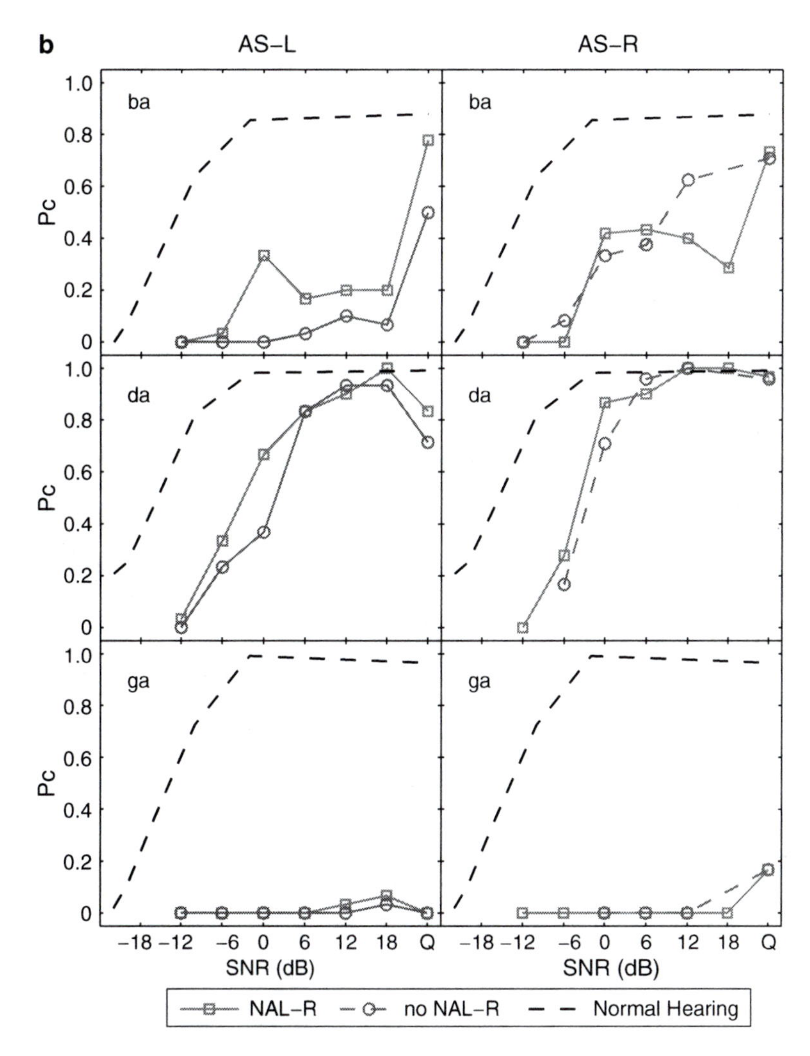

Fig. 42.5 (continued)

problem understanding speech because they cannot hear certain sounds whose events are missing because of their hearing loss or the masking effect introduced by the noise.

Other useful findings of the pilot study include: (1) Pure tone audiometry, and therefore the use of speech banana in audiological clinic, is a poor way of evaluating the problem of HI speech perception, especially for those who have cochlear dead regions; (2) cochlear dead regions have considerable impact on consonant identification and the effectiveness of NAL-R amplification; (3) speech enhancement can be regarded as a problem of optimization. As a static compensation scheme, NAL-R improves the recognition scores of certain sounds, but it hurts the perception of other sounds. An adaptive amplification scheme that accounts for the HI listener's hearing loss as well as the speech characteristics may produce a better performance.

References

Allen JB (1994) How do humans process and recognize speech? IEEE Trans Speech Audio Process 2(4):567–577

Fletcher H, Galt R (1950) The perception of speech and its relation to telephony. J Acoust Soc Am 22:89–151

Li F, Menon A, Allen JB (2009) Perceptual cues in natural speech for 6 stop consonants. J Acoust Soc Am 127

Lobdell BE (2006) Information theoretic tool for investigating speech perception. MS Thesis, University of Illinois at Urbana-Champaign, Urbana, IL

Moore BCJ (2004) Dead regions in the cochlea: conceptual foundations, diagnosis, and clinical applications. Ear Hear 25(2):98–116

Moore BCJ, Alcántara JI (2001) The use of psychophysical tuning curves to explore dead regions in the cochlea. Ear Hear 22(4):268–278

Moore BCJ, Huss M, Vickers DA, Glasberg BR, Alcántara JI (2000) A test for the diagnosis of dead regions in the cochlea. Br J Audiol 34:205–224

Zurek PM, Delhorne LA (1987) Consonant reception in noise by listeners with mild and moderate sensorineural hearing impairment. J Acoust Soc Am 82(5):1548–1559

Part VII
Auditory Scene Analysis

Chapter 43
A Comparative View on the Perception of Mistuning: Constraints of the Auditory Periphery

Astrid Klinge, Naoya Itatani, and Georg M. Klump

Abstract Harmonicity serves to group together frequencies from a single source to a perceived auditory object. The peripheral auditory system may exploit either spectral cues resulting in a specific spatial pattern of excitation or temporal cues that are due to the interaction of frequency components in the complex to evaluate whether a component does not belong to a harmonic complex, i.e., is mistuned. Which cues are useful for mistuning detection may depend on the anatomical and physiological constraints in the peripheral auditory system. Here we compare the perception of frequency shifts in harmonic complexes (i.e., mistuning) and in pure tones across species. Mongolian gerbils and birds are superior to humans in detecting small amounts of mistuning in sine phase harmonic complexes. This difference is reduced in the detection of mistuning in random phase harmonic complexes (but not in harmonic complexes with "frozen random phase"). Humans are superior to birds and gerbils in detecting pure-tone frequency shifts. The results suggest that species with a short cochlea and only few hair cells per critical band tend to rely more on temporal fine structure in the analysis of mistuning of components in harmonic complexes whereas excitation patterns may play a larger role in humans with a much longer cochlea. For the analysis of pure-tone frequency shifts, excitation patterns appear to play a more prominent role in all species. Exemplary neurophysiological data obtained in starlings support this view.

Keywords Mistuning detection • Harmonic complex • Mongoliangerbil • Starling human • Temporal processing

---

G.M. Klump (✉)
Animal Physiology and Behaviour Group, Institute for Biology and Environmental Sciences, 26111, Oldenburg, Germany
e-mail: georg.klump@uni-oldenburg.de

E.A. Lopez-Poveda et al. (eds.), *The Neurophysiological Bases of Auditory Perception*,
DOI 10.1007/978-1-4419-5686-6_43, © Springer Science+Business Media, LLC 2010

43.1 Introduction

Harmonicity is one of the most important grouping cues in the analysis of sound. To determine the effect of harmonicity on the grouping or the segregation of frequencies in a complex stimulus, the "mistuned harmonic" paradigm (Moore et al. 1984) can be used. Shifting the frequency of one component in a harmonic complex (i.e., "mistuning" the component) changes the percept of the whole complex for small frequency shifts and results in hearing the mistuned component as a separate object for large frequency shifts (Moore et al. 1985, 1986). Frequency difference limens (FDLs) for detecting the mistuning of a component in a harmonic complex vary considerably between humans and animal species such as gerbils, starlings, zebra finches, and budgerigars (Klinge and Klump 2009; Klinge and Klump submitted; Klump and Groß submitted; Lohr and Dooling 1998). This suggests that different mechanisms may be employed by the different species that will be discussed here.

There is an ongoing discussion whether temporal or spectral pattern recognition mechanisms play a greater role in detecting such mistuning. According to the spectral pattern recognition hypothesis (e.g., Goldstein 1973), the frequency of each component of a complex is determined and compared to a template pattern of harmonic frequencies to decide whether the frequency of the component belongs to the harmonic series or not. Models based on this concept (e.g., Duifhuis et al. 1982) imply that at least some frequency components of the harmonic complex are resolved and can be separately analyzed. Another class of models that was introduced by Licklider (1951) to account mainly for the processing of unresolved harmonics relies on exploiting temporal patterns created by the interacting components on the basilar membrane (e.g., Cheveigné 1998; Meddis and Hewitt 1991a, b). While in some studies separate mechanisms have been proposed for low and high frequencies, i.e., resolved and unresolved harmonics, respectively (e.g., Houtsma and Smurzynski 1990), others have suggested that one common mechanism based on temporal processing may be sufficient to explain the human psychophysical data (e.g., Meddis and O'Mard 1997; Gockel et al. 2004). Here we will discuss whether one mechanism may account for the variation in the animal data or whether there is evidence for different mechanisms. Furthermore, we will investigate the possible relation between differences in the frequency representation by the peripheral auditory system (e.g., as exemplified by the cochlear-map functions) of humans, gerbils and two bird species (starlings and budgerigars) and the perception of frequency shifts in components of a harmonic complex and of pure tones.

43.2 Detecting Frequency Shifts of Pure Tones

Before discussing the detection of mistuned harmonics, we will turn to the FDL of pure tones as a baseline measure of frequency discrimination. Two mechanisms have been suggested to represent pure-tone frequencies in the auditory periphery.

For low frequencies at which phase locking of auditory nerve (AN) fibres to the period of the pure-tone frequency is possible, frequency can be encoded in the temporal spiking pattern of the neurons. At frequencies higher than about 2–4 kHz, the ability of neurons in the AN to phase lock to the period of the pure-tone frequency degrades (Köppl 1997). Thus, a place mechanism on the cochlear frequency map is assumed to operate and frequency shifts are detected by the change in the distribution of neural activity in the peripheral auditory system across the AN fibres tuned to different frequencies (i.e., a rate-place code is used).

Let us first compare the pure-tone FDLs across species and then evaluate their relation to the species' cochlear frequency map. Gerbils show the poorest pure-tone frequency discrimination ability (Fig. 43.1a) if compared to starlings and humans. The gerbil's FDLs range from 21% (Weber fraction) at 200 Hz to 7% at 6,400 Hz (Klinge and Klump 2009) whereas the pure-tone FDLs of starlings range from 9% at 200 Hz to 1.5% at 4,800 Hz (Buus et al. 1995; Klump and Groß submitted). The budgerigar's performance in pure-tone frequency discrimination lies between that of starlings and humans decreasing from 2.8% at 500 Hz to about 0.7% for frequencies from 1,000 to 2,000 Hz and then increasing to 1.3% at 5,700 Hz (Dooling and Saunders 1975). In contrast to most animal data, humans show low FDLs ranging from 0.2% within the frequency range of 500–2,000 Hz and increasing to 1.2% at about 8,000 Hz (e.g., Moore 1973). How can these differences between pure-tone FDLs be related to the change in the peripheral excitation pattern and to the ability to represent frequencies by the temporal firing patterns of the AN fibres? For humans it has been suggested that in the frequency region in which phase locking of AN fibres is still possible, temporal mechanisms are exploited to reach the very low FDLs (e.g., Sek and Moore 1995). A similar performance would be expected for the starling and possibly also for the gerbil (assuming that it has a similar phase locking ability as the Guinea pig, see Köppl 1997) if they also would rely on temporal mechanisms. However, the much higher FDL in these species compared to the human FDL suggests that no such temporal mechanism is exploited at low pure-tone frequencies.

Based on the cochlear place-frequency functions derived by Greenwood (1990), Fay (1992) proposed that pure-tone FDLs, the critical bandwidth, the critical masking ratio bandwidth and the width of psychophysical tuning curves can all be related to the spatial frequency map on the basilar membrane. Thus, if the pure-tone FDLs increase with increasing frequency proportional to the cochlear-map functions derived on the basis of critical bandwidths, critical-ratio bandwidths or physiological frequency maps, then a place mechanism would explain the FDLs. Figure 43.1b shows the FDLs of humans, gerbils, starlings, and budgerigars as a function of frequency together with a function predicting the FDL (based on the Greenwood function) assuming that it represents a constant spatial shift on the cochlear-map function of these species. For gerbils and starlings, the functions of the observed FDLs are parallel and differ (i.e., are smaller) by a scaling factor of 1.5 for gerbils and a scaling factor between 5 and 7 for starlings from the critical-bandwidth functions calculated on the basis of the species' physiological data (Buus et al 1995; Klump and Groß submitted; Kittel et al. 2002). For budgerigars, only the FDLs for

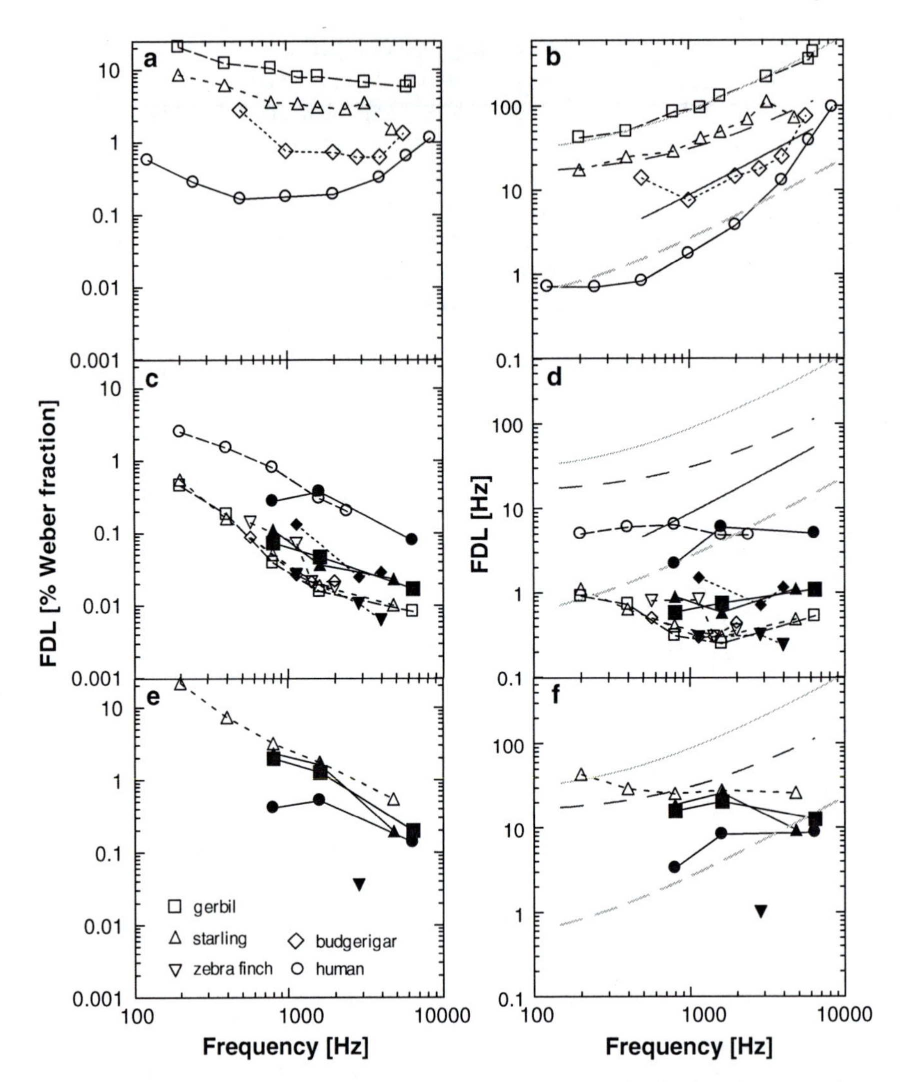

Fig. 43.1 Mean FDLs of pure tones (**a**, **b**), of mistuned harmonics in sine phase complexes (**c**, **d**), and of mistuned harmonics in random phase complexes (**e**, **f**) for different species (indicated by the *symbol shape*, see legend). In the panels on the left side the FDLs are displayed in % Weber fraction ($\Delta f/f$). On the right side the FDLs are shown in Hz in comparison to the prediction for pure-tone FDLs generated using cochlear-map functions of the different animal species (gerbil *light grey solid line*, budgerigar *black solid line*, starling *black dashed line*) and humans (*light grey dashed line*) based on the Greenwood (1990) functions. In (**c**)–(**f**) data obtained for complexes using low fundamentals (e.g., 200–285 Hz) are shown with *open symbols* whereas data for complexes with high fundamentals (e.g., 570–800 Hz) are shown with *filled symbols*. Gerbil data are from Klinge and Klump (2009), starling data from Klump and Groß (submitted), zebra finch and budgerigar data from Lohr and Dooling (1998), and human data from Moore (1973, pure-tone FDLs) and Klinge and Klump (2009, mistuned harmonic FDLs)

frequencies higher than 1 kHz fit the cochlear-map function estimated on the basis of the critical-ratio bandwidth and the average scaling factor is about 30 (however, Dooling and Saunders 1975, point out that the shapes of the FDL and CR functions deviate). The good fit of these functions to the pure-tone FDLs in the gerbil, the starling and for part of the frequency range in the budgerigar suggest that a pure place mechanism might explain the observed FDLs over most of the frequency range in these species. In humans, the FDLs do not follow the function predicted by the CBs over the total frequency range tested. An average scaling factor of 70 can be calculated on the basis of the data provided by Moore (1973) that is reduced to below 20 at 8 kHz. This high value may indicate that temporal mechanisms in addition to the proposed place mechanism are exploited in the human auditory system (e.g., Sek and Moore 1995).

It remains to be discussed why humans show so much better pure-tone FDLs at high frequencies at which they have to rely on place mechanisms similar to the other species described here. Fay (1992) proposed that these differences might arise from differences in the anatomy of the auditory periphery. The best frequency discrimination ability should be found in animals with a long cochlea and a narrow frequency range of hearing sensitivity. The lowest frequency discrimination acuity should be found in species with a short cochlea and a hearing sensitivity extending to high frequencies. The gerbil cochlea is only 11 mm long (Müller 1996) which is short in comparison to 34 mm reported for the human cochlea (e.g., Wright et al. 1987). The bird species considered here have a much shorter basilar papilla than the cochlea of the gerbil (2.9 mm for starlings, Buus et al. 1995; 2.5 mm for budgerigars, Manley et al. 1993). In mammals, the number of hair cells corresponds to the difference in the length of the cochlea, and thus is reduced from about 3,500 inner hair cells (IHCs) in the human cochlea (Wright et al. 1987) to about 1,400 IHCs in the gerbil cochlea (Plassmann et al. 1987). Birds have several rows of hair cells across the basilar papilla that function like mammalian IHCs and may provide a large afferent input in a critical band (total number of hair cells on the starling and budgerigar basilar papilla is 5,800 and 5,400, respectively, see Gleich and Manley 1988; Manley et al. 1993). If we now compare the hearing range between the different species, we find that gerbils have the largest hearing range extending from approximately 0.1 to 60 kHz (Ryan 1976). The human hearing range extends from about 0.01 to 16 kHz (ISO 389-7 1996) whereas the hearing range of budgerigars and starlings extend from about 0.1 to 6.1 and 6.3 kHz, respectively (Dooling and Saunders 1975; Buus et al. 1995). Comparing the bird species to the mammal species, we can estimate that one critical band covers the shortest distance in the cochlea of the birds (0.1 mm for the starling and 0.13 mm for the budgerigar, using parameters for the Greenwood function suggested by Buus et al. 1995; Fay 1992). The distance on the cochlear map covered by each critical band is about 0.2 mm in the gerbil (Kittel et al. 2002) and about 1.15 mm in humans (Fay 1992). Given these parameters of the species, we find that in the gerbil the number of hair cells in a critical band is the lowest from all four species, which suggests that due to the reduced afferent input only large changes in the excitation pattern may be detectable by the gerbil. Viewed on the basis of distance in the cochlea corresponding to the FDL, the gerbil needs the largest spatial shift to detect a frequency shift (0.13 mm), whereas in budgerigars, starlings, and humans the spatial shift corresponding to the pure-tone FDL is about an order of magnitude smaller than in the gerbil.

43.3 Detecting a Mistuned Component in Harmonic Complexes

The fundamental frequency, the number of frequency components in a harmonic complex, their resolvability, their phase relationship, and the shape of the resulting temporal waveform in different auditory filters may affect the FDL for mistuned harmonics in an otherwise harmonic complex. Some of these parameters are important for mechanisms of mistuning detection relying on the spatial pattern of excitation in the cochlea or on the temporal processing. By varying the fundamental frequency of a harmonic complex, the spacing between two adjacent harmonics and thus their resolvability and the period of the temporal waveform of the harmonic complex can be changed. Given the limited ability of the auditory system to follow fast temporal modulations of the stimulus (e.g., see review by Joris et al. 2004), temporal patterns corresponding to the 5-ms period in a 200 Hz complex should be much better represented by the temporal discharge of neurons in the auditory system than the patterns corresponding to the 1.25 ms period in an 800 Hz complex (see also Hartmann et al. 1990). The phase relationship between the frequency components of the complex will mainly be important for the temporal mechanisms. A starting phase of 0° (sine phase) for all harmonics in a complex results in a waveform with a non-varying and peaky temporal fine structure that will only change if mistuned harmonics are present. A randomized starting phase, however, degrades the distinctiveness of the temporal waveform and, if the phase is randomized for every presented stimulus, the constantly varying temporal fine structure makes it difficult to detect the change that is due to the mistuning. Thus, it is a hallmark of temporal mechanisms of mistuning detection that they are sensitive to the phase relation of the components of the harmonic complex.

43.3.1 Mistuning Detection in Sine Phase Harmonic Complexes

Contrary to the pattern observed for the pure-tone FDL, the ability to detect a frequency shift of a mistuned component in a sine phase harmonic complex was much better in gerbils and the three bird species than in humans. The animal species' FDLs for this stimulus condition were the lowest of all experiments described here (Fig. 43.1c in % Weber fraction and Fig. 43.1d in Hz). Humans showed about one order of magnitude higher thresholds for detecting a mistuned harmonic in a sine phase complex than the animal species. At a specific frequency, the FDLs for the harmonic complexes varied with the fundamental frequency in the animal species but not so much in the humans (e.g., Klinge and Klump submitted; Moore et al. 1985). In the gerbil and the three bird species, the FDLs for a mistuned component in a complex with a high fundamental (800 Hz for gerbil and starling, 570 Hz for budgerigar and zebra finch) were higher than the FDLs for a mistuned component in a complex with a low fundamental (200 Hz for gerbil and starling, 285 Hz for budgerigar and zebra finch). If we compare the prediction from the cochlear-map function (that fitted the pure-tone FDLs quite well) with the FDLs for mistuned harmonics in a sine phase complex, it becomes evident that mechanisms relying on the spatial pattern of excitation in the cochlea are unlikely to explain the data.

43.3.2 Mistuning Detection in Random Phase Harmonic Complexes

Also in the random phase condition, in which the spectral composition of the signal equals that of a sine phase complex but has an altered temporal waveform, the FDLs of the animal species are much smaller than their pure-tone FDLs (Fig. 43.1e, f). In starlings (Klump and Groß submitted) and gerbils (Klinge and Klump 2009), the FDLs for detecting a mistuned harmonic in an 800 Hz complex with all harmonics in random phase are similar and range from between approximately 2% for the first harmonic (800 Hz) to 0.2% for the sixth (4,800 Hz) or eighth (6,400 Hz) harmonic, respectively. FDLs for the 200 Hz complex obtained in starlings (Klump and Groß submitted) resulted in similar values as for the 800 Hz complex. As for pure-tone frequency discrimination, FDLs for mistuned harmonics in random phase complexes with a fundamental of 800 Hz are lower in humans (FDLs ranging from 0.5 to 0.1%) than in gerbils or starlings. However, in a study using the same stimuli and procedures these differences were only significant for the first harmonic (800 Hz, Klinge and Klump 2009). The human's FDLs in the random phase condition were not significantly different from those for detecting a mistuned harmonic in a sine phase complex. The zebra finch (diamond in Fig. 43.1e, f) has the lowest FDL of all species tested in the random phase condition (Lohr and Dooling 1998) which might be due to differences in the variability of the waveforms depending on how the stimulus is generated. To obtain FDLs for mistuned harmonics in a random phase complex with a fundamental frequency of 800 Hz, the humans, starlings, and gerbils in the study of Klinge and Klump (2009) were provided with a stimulus that had a randomized starting phase of every component for every presented stimulus. For the 200 Hz complex, the starlings were presented with a large set of at least 10 different random phase renditions of every test stimulus and of 30 random phase renditions of the reference stimulus that were newly produced before each session. Lohr and Dooling (1998), however, used only five different "frozen" random phase test stimuli and ten different "frozen" random phase reference stimuli and presented them to a subject throughout the experiment. The low number of different stimuli might have allowed the zebra finches to memorize the temporal waveforms of the stimuli and thus made detecting a change in the waveform due to a mistuned harmonic easier for the birds. In their recent study on starlings, Klump and Groß (submitted) found that the FDL decreased nearly to the FDL obtained for components in a sine phase harmonic complex if the birds were presented with a single "frozen" random phase stimulus as the reference for multiple sessions. Taken together these data suggest that also in the random phase condition the detection of mistuned harmonics may rely on temporal mechanisms.

43.3.3 Neural Basis of Mistuning Detection

The results of the different experiments described above showed that humans seem to process the information in a mistuned harmonic complex differently in their auditory system than the animal species. The gerbils and the birds showed much higher pure-tone FDLs than humans but were superior to humans when detecting a mistuned harmonic in a sine phase complex. In the animal species, the FDLs for the random phase complex were about one order of magnitude higher than the FDLs for the sine phase complex. In contrast, the FDLs in humans did not differ significantly in the random phase and the sine phase complex. This raises the question whether the processing mechanisms may differ between humans and the animal species that were tested. Observing substantially lower FDLs for zebra finches and budgerigars in comparison to humans, Lohr and Dooling (1998) suggested either temporal processing mechanisms in the birds and spectral mechanisms in the humans to detect a mistuned harmonic in sine phase or an enhanced ability of birds to discriminate the same cues as humans. Klinge and Klump (2009, submitted) and Klump and Groß (submitted) suggest that temporal mechanisms might also be employed by the gerbil and the starlings to detect a mistuned component in a harmonic complex for which the evidence was presented above.

We have argued above that the animal species having a short cochlea cannot rely on the spatial pattern of excitation on the basilar papilla but have to rely on the analysis of temporal discharge patterns of AN fibres to achieve the low FDLs for mistuned harmonics. Three temporal cues are available in the auditory periphery allowing the discrimination of a complex with a mistuned harmonic from a fully harmonic complex (e.g., Moore et al. 1985; Klinge and Klump submitted). The first cue that is available if more than one component of the complex drives the neural response is the slow amplitude modulation that is due to the beats created by the interaction of a mistuned component with a neighbouring harmonic frequency. We have observed a temporal neural response pattern in the starling auditory forebrain that indicates a change in the discharge rate with the period of the beating (Fig. 43.2). This periodic rate change could be used for mistuning detection. The lower limit for the FDL of mistuning detection will be set by the neurons' integration time window and the duration of the complex used for stimulation. For small amounts of mistuning, the change in the amplitude created by the beating of neighbouring frequencies throughout the stimulus duration may be too small to be usable for mistuning detection.

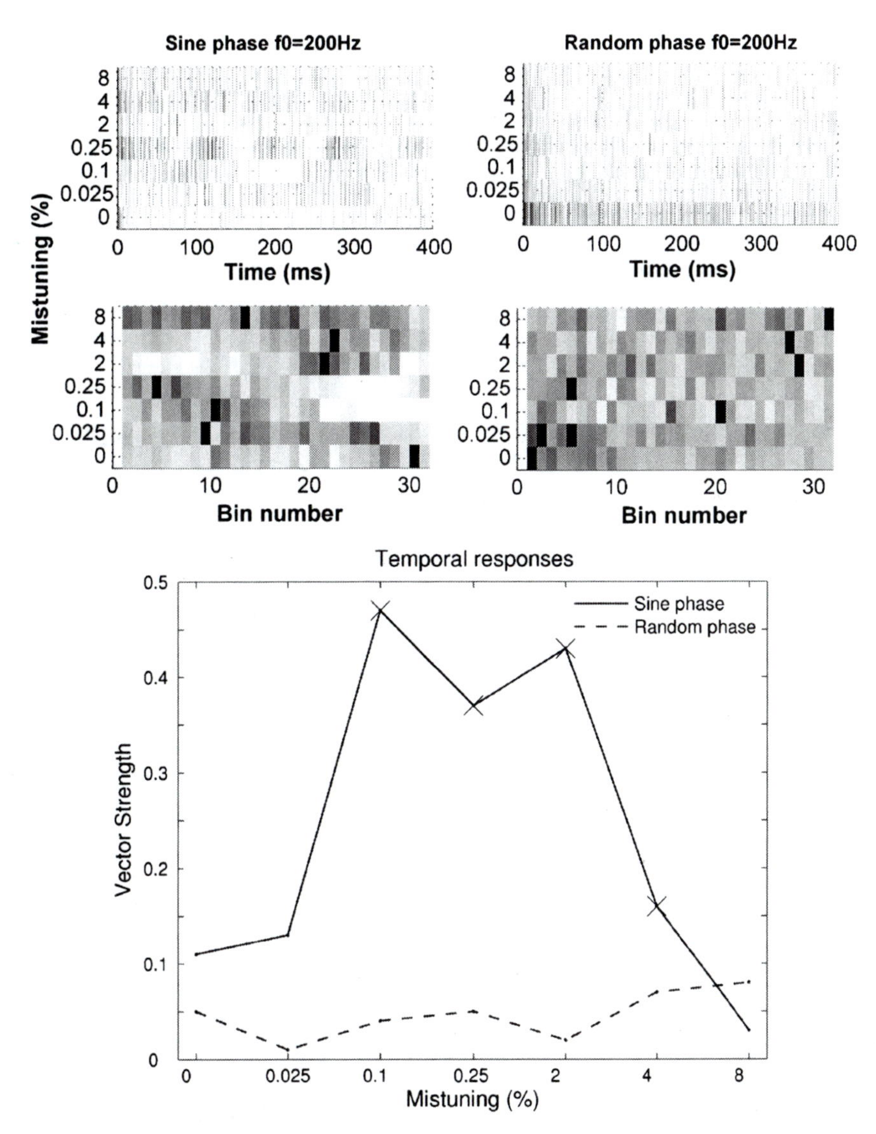

Fig. 43.2 Multiunit response observed in the starling forebrain when stimulated with a harmonic complex (fundamental 200 Hz, 48 harmonics, component at 4,800 Hz mistuned, level of components 60 dB SPL; excitatory frequency range of tuning curve at 60 dB SPL ranged from about 3,500 to 5,200 Hz). Normalized peri-stimulus time histograms (*top graphs*) and beat frequency period histograms (*middle graphs*, the first 30 ms of the response were excluded) of neuronal activity elicited by a mistuned harmonic are shown; the numbers on the ordinate of the graphs indicate the amount of mistuning (Weber fraction $\Delta f/f$ in %). The bottom graph shows the vector strength computed from the period histograms demonstrating a significant locking (marked with X) of the response to the beating period in the sine phase condition

The second available cue lies in the temporal fine structure of the waveform that results from the phase relationship of the components and which differs, for example, between the mistuned and the harmonic condition of sine phase complexes. As in the case of beating, more than one component of the complex must be able to drive the neural response to make this cue effective, i.e., at least two frequency components must lie in the frequency range to which the auditory neurons are tuned. In the sine phase condition, for example, the fine structure of the waveform changes throughout the mistuned stimulus while it is constant in the fully harmonic stimulus. Previous studies have demonstrated a sensitivity of auditory neurons in the brainstem to the shape of the periodic amplitude change of the waveform (i.e., ramped and damped sounds, see Pressnitzer et al. 2000). These neurons should also be sensitive to the change in the waveform occurring with mistuning in a sine phase complex. If the phase of the components of a complex is random and constantly changes from one stimulus to the next, this cue is not suited for mistuning detection.

The third cue that can be used for mistuning detection is provided by the phase of individual resolved components of the stimulus represented in different frequency channels. If the phase information is transmitted by the auditory neurons (e.g., in AN fibres this should be possible up to 2–4 kHz, see Köppl 1997), wide-band units functioning as coincidence detectors above the level of the auditory periphery would be able to detect a gradual phase shift occurring throughout the stimulus that is due to the mistuning. This gradual phase shift is found only in those auditory filters (and the respective neurons) that contain the mistuned harmonic and could be compared to the output driven by filters containing only in-tune harmonics of the complex. This cue and the second cue complement each other in the frequency range of hearing. In the low frequency range in which the harmonics are resolved in the response of the peripheral auditory system the direct evaluation of the phase is possible, whereas in the high frequency range with unresolved harmonics the direct interaction of components in the cochlea allows for an indirect evaluation of the phase relation between components.

References

Buus S, Klump GM, Gleich O, Langemann U (1995) An excitation-pattern model for the starling (*Sturnus vulgaris*). J Acoust Soc Am 98:112–124

de Cheveigné A (1998) Cancellation model of pitch perception. J Acoust Soc Am 103:1261–1271

Dooling RJ, Saunders JC (1975) Hearing in the parakeet (*Melopsittacus undulatus*): absolute thresholds, critical ratios, frequency difference limens, and vocalizations. J Comp Physiol Psychol 88:1–20

Duifhuis H, Willems LF, Sluyter RJ (1982) Measurement of pitch in speech: an implementation of Goldstein's theory of pitch perception. J Acoust Soc Am 71:1568–1580

Fay RR (1992) Structure and function in sound discrimination among vertebrates. In: Webster D, Fay RR, Popper A (eds) The evolutionary biology of hearing. Springer, New York, pp 229–263

Gleich O, Manley GA (1988) Quantitative morphological analysis of the sensory epithelium of the starling and pigeon basilar papilla. Hear Res 34:69–85

Gockel HE, Carlyon RP, Plack CJ (2004) Across-frequency interference effects in fundamental frequency discrimination: questioning evidence for two pitch mechanisms. J Acoust Soc Am 116:1092–1104

Goldstein JL (1973) An optimum processor theory for the central formation of the pitch of complex tones. J Acoust Soc Am 54:1496–1516

Greenwood DD (1990) A cochlear frequency-position function for several species – 29 years later. J Acoust Soc Am 87:2592–2605

Hartmann WM, McAdams S, Smith BK (1990) Hearing a mistuned harmonic in an otherwise complex tone. J Acoust Soc Am 88:1712–1724

Houtsma AJ, Smurzynski J (1990) Pitch identification and discrimination for complex tones with many harmonics. J Acoust Soc Am 87:304–310

ISO 389-7 (1996) Acoustics-Reference zero for the calibration of audiometric equipment - part 7: Reference threshold of hearing under free-field and diffuse-field listening conditions

Joris PX, Schreiner CE, Rees A (2004) Neural processing of amplitude-modulated sounds. Physiol Rev 84:541–577

Kittel M, Wagner E, Klump GM (2002) An estimate of the auditory-filter bandwidth in the Mongolian gerbil. Hear Res 164:69–76

Klinge A, Klump GM (2009) Frequency difference limens of pure tones and harmonics within complex stimuli in Mongolian gerbils and humans. J Acoust Soc Am 125:304–314

Klinge A, Klump GM (submitted) Mistuning detection and onset asynchrony in harmonic complexes in Mongolian gerbils

Klump GM, Groß S (submitted) Detection of frequency shifts and mistuning in complex tones in the European starling

Köppl C (1997) Phase locking to high frequencies in the auditory nerve and cochlear nucleus magnocellularis of the barn owl, *Tyto alba*. J Neurosci 17:3312–3321

Licklider JC (1951) A duplex theory of pitch perception. Experientia 7:128–134

Lohr B, Dooling RJ (1998) Detection of changes in timbre and harmonicity in complex sounds by zebra finches (*Taeniopygia guttata*) and budgerigars (*Melopsittacus undulatus*). J Comp Psychol 112:36–47

Manley GA, Schwabedissen G, Gleich O (1993) Morphology of the basilar papilla of the budgerigar *Melopsittacus undulatus*. J Morphol 218:153–165

Meddis R, Hewitt MJ (1991a) Virtual pitch and phase sensitivity of a computer model of the auditory periphery. I: Pitch identification. J Acoust Soc Am 89:2866–2882

Meddis R, Hewitt MJ (1991b) Virtual pitch and phase sensitivity of a computer model of the auditory periphery. II: phase sensitivity. J Acoust Soc Am 89:2883–2894

Meddis R, O'Mard L (1997) A unitary model of pitch perception. J Acoust Soc Am 102:1811–1820

Moore BC (1973) Frequency difference limens for short-duration tones. J Acoust Soc Am 54:610–619

Moore BC, Glasberg BR, Peters RW (1984) Frequency and intensity difference limens for harmonics within complex tones. J Acoust Soc Am 75:550–561

Moore BC, Peters RW, Glasberg BR (1985) Thresholds for the detection of inharmonicity in complex tones. J Acoust Soc Am 77:1861–1867

Moore BC, Glasberg BR, Peters RW (1986) Thresholds for hearing mistuned partials as separate tones in harmonic complexes. J Acoust Soc Am 80:479–483

Müller M (1996) The cochlear place-frequency map of the adult and developing Mongolian gerbil. Hear Res 94:148–156

Plassmann W, Peetz W, Schmidt M (1987) The cochlea in Gerbilline rodents. Brain Behav Evol 30:82–101

Pressnitzer D, Winter IM, Patterson RD (2000) The responses of single units in the ventral cochlear nucleus of the guinea pig to damped and ramped sinusoids. Hear Res 149:155–166

Ryan A (1976) Hearing sensitivity of the Mongolian gerbil, *Meriones unguiculatis*. J Acoust Soc Am 59:1222–1226

Sek A, Moore BC (1995) Frequency discrimination as a function of frequency, measured in several ways. J Acoust Soc Am 97:2479–2486

Wright A, Davis A, Bredberg G, Ulehlova L, Spencer H (1987) Hair cell distributions in the normal human cochlea. Acta Otolaryngol Suppl 444:1–48

Chapter 44
Stability of Perceptual Organisation in Auditory Streaming

Susan L. Denham, Kinga Gyimesi, Gábor Stefanics, and István Winkler

Abstract In everyday situations, we perceive sounds organised according to their source, and can follow someone's speech or a musical piece in the presence of other sounds without apparent effort. Thus, it is surprising that recent evidence obtained in the most widely used experimental test-bed of auditory scene analysis, the two-tone streaming paradigm, demonstrated extensive bistability even in regions of the parameter space previously thought to be strongly biased towards a particular organisation. This raises the question of what aspects of the rich natural input allow the auditory system to form stable representations of concurrently active sound sources. Here, we report the results of perceptual studies aimed at testing this issue. It is possible that the extreme repetitiveness of the alternating two-tone sequence, i.e. lack of change, causes perceptual instability. Our first experiment addressed this hypothesis by introducing random changes in the stimulation. It is also possible that under natural conditions, multiple redundant cues stabilise perception. The second experiment tested this hypothesis by adding a second cue which favoured one organisation. Much to our surprise, neither one of these manipulations stabilised the perception of two-tone streaming sequences. We discuss these experimental results in the light of our previous theoretical proposals and findings of significant differences between the first and later perceptual phases. We argue that multi-stability is inherent in perception. However, it is normally hidden by switches of attention, which allow the return of the dominant perceptual organisation resulting in the subjective experience of perceptual stability. In our third experiment, we explored this possibility by inserting short gaps into the sequences, since gaps have been shown to reset auditory streaming in a manner similar to switches in attention.

Keywords Auditory streaming • Bistability • Perceptual switching • Auditory scene analysis

---

S.L. Denham (✉)
Centre for Theoretical and Computational Neuroscience, University of Plymouth, Plymouth, Devon UK
e-mail: sdenham@plymouth.ac.uk

E.A. Lopez-Poveda et al. (eds.), *The Neurophysiological Bases of Auditory Perception*,
DOI 10.1007/978-1-4419-5686-6_44, © Springer Science+Business Media, LLC 2010

44.1 Introduction

The phenomenon of bistability in auditory perceptual organisation has been highlighted in a number of recent studies (Denham and Winkler 2006; Denham et al. 2008; Pressnitzer and Hupé 2005, 2006; Winkler et al. 2005). These investigations have shown, using the auditory two tone streaming paradigm, that bistability is found extensively even in regions of the parameter space previously thought to be strongly biased towards a particular organisation (Denham et al. 2008). Very similar characteristics of perceptual switching are observed in the visual and auditory modalities (Denham et al. 2008; Pressnitzer and Hupé 2006), suggesting that at some level of perceptual processing, generic mechanisms may be involved. Furthermore, a notable difference was identified between the first percept evoked by the auditory streaming sequence (first perceptual phase) and subsequent percepts (subsequent perceptual phases), with the duration of the first phase being stimulus-parameter dependent and an order of magnitude longer in duration than the parameter-independent subsequent phases (Denham et al. 2008).

These rather surprising results raised a number of further questions. Firstly, given that subjective impressions of the world tend to be rather stable, what factors are responsible for stabilising perceptual organisation? Is it the case that the extreme repetitiveness of alternating two tone sequences causes instability (e.g. by exhausting the neural networks specifically responding to the given stimuli), or is it that that these stimuli are inherently ambiguous, and that in such cases perception simply entertains the alternatives? To address these questions, in the first experiment, we introduced random changes in the frequency and timing of the tones, and in the second experiment a cue which favoured streaming was included. Finally, in a third experiment, we explored the possibility that switches in attention could trigger the return of the dominant perceptual organisation, and hence give rise to perceptual stability. In this experiment, we inserted short gaps into the two tone sequences, since gaps have been shown to reset auditory streaming in a manner similar to switches in attention (Cusack et al. 2004).

44.2 Experiment 1

The first experiment was designed to explore whether the extreme repetitiveness of the alternating two tone sequence was responsible for the observed bistability in perceptual organisation. In order to do this, we introduced random changes by jittering both the timing and the frequencies of the alternating tones.

44.2.1 Participants

Eleven young healthy volunteers (six female, 19–24 years of age, average: 21.0 years) participated in the first experiment. For all the experiments reported here, participants received modest financial compensation for their participation. The study was conducted in the sound-attenuated experimental chamber of the Institute for Psychology, Hungarian Academy of Sciences. It was approved by the Ethical Committee of the Institute for Psychology. After the aims and procedures of the study were explained to them, participants signed an informed consent form before starting the experiment. Participants were pre-selected on the basis of the results of clinical audiometry with the criteria that the hearing threshold between 250 and 6,000 Hz should not exceed 25 dB, and the difference between the two ears does not exceed 15 dB in the same frequency range.

44.2.2 Stimulus Paradigm

Participants were presented with 4-min long trains of the ABA-structure, where A and B were pure tones of 75 ms duration, including 5 ms linear onset and offset ramps. In separate trains, Δf was 1, 4, 7 or 10 semitones (ST) and stimulus onset asynchrony (SOA) was 75, 100, 150, or 200 ms. Each combination of the parameters was presented in one stimulus block (altogether, $4 \times 4 = 16$ stimulus conditions), delivered in an order randomised separately for each participant. The mean frequency of the lower-pitched, more frequent A tones was a nominal 400 Hz, and that of the B tones was higher by the Δf for the corresponding condition. The actual frequency of each tone in the sequence was chosen randomly from a uniform distribution centred on the nominal frequency ±10% of the Δf. The onset of each tone was jittered in the range ±20% of the SOA.

44.2.3 Procedure

The procedure for all experiments was as described here. While listening to the test sound sequences, participants sat in a comfortable chair in the experimental chamber. They were instructed to depress one response key so long as they experienced an integrated percept and the other key when they experienced a segregated percept. The role of the two keys was randomly assigned across participants. To eliminate possible confusion caused by the perception of rhythms other than the "galloping" rhythm, the notion of an integrated percept was generalised and defined for participants as hearing a repeating pattern, which contained both low and high tones, and the notion of a segregated percept was defined as hearing some repeating pattern(s) formed either exclusively of high or exclusively of low tones, with the possibility that multiple repeating segregated patterns (i.e. A---A---A… and B-B-B…) may be perceived concurrently. Participants were asked to mark their perception throughout the duration of the stimulus sequence and not to attempt hearing the sound according to one or another perceptual organisation. The experimenter made sure that participants understood the types of percepts they were required to report, using both auditory and visual illustrations. To avoid possible implications of exclusivity between the two potential organisations, subjects were explicitly told that it was possible that they may sometimes hear both types of patterns at the same time. However, they were also cautioned to be sure to release the button when they stopped hearing the corresponding pattern. When analysing the responses, we discarded all phases with duration shorter than 300 ms in order to avoid cases in which participants may simply have been slightly inaccurate in synchronising their button presses and releases.

44.2.4 Results

The principal finding was that the introduction of jitter in the frequencies and timing of the tones did not stabilise perceptual organisation. Once again perceptual switching was found in all conditions and for all subjects; the distribution of switching across conditions and subjects is shown in Fig. 44.1 together with the distribution of the phase durations.

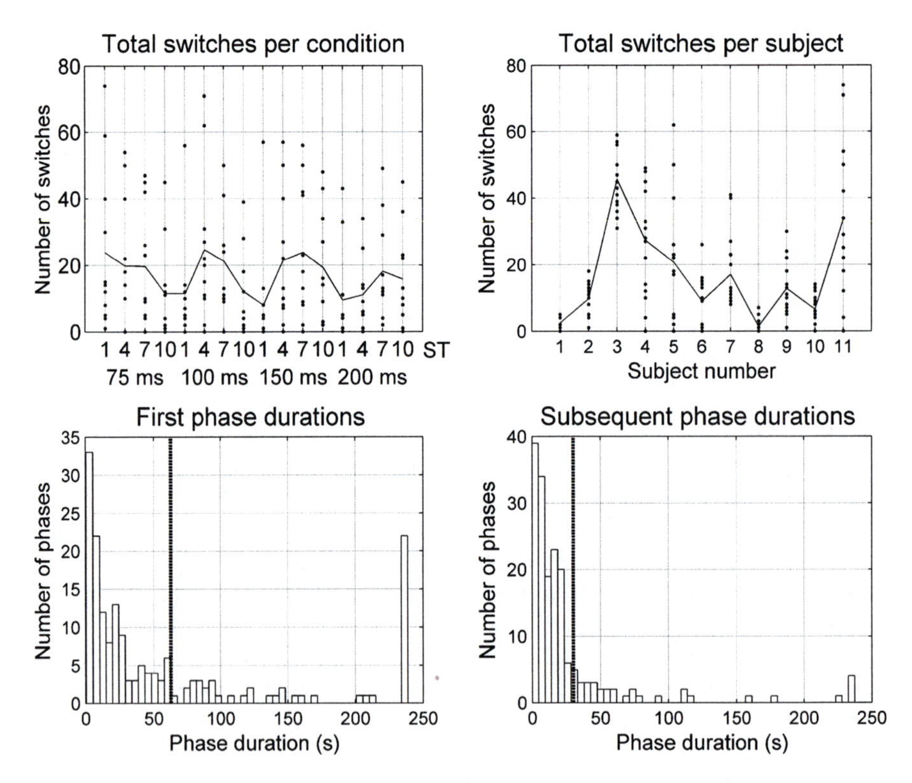

Fig. 44.1 Top: The number of perceptual switches; *left*: for each condition with the Δf and SOA indicated; *right*: for each subject. *Black dots* indicate individual subjects (*left*) or conditions (*right*); *solid lines* connect the mean for each condition/subject. Bottom: Distribution of phase durations for the first (*left*) and all subsequent phases (*right*). The means are marked by the thick dashed vertical lines

We have previously found a profound difference between the first phase duration, and that of subsequent phases, with first phases typically an order of magnitude longer than the mean of subsequent phases (Denham et al. 2008). An ANOVA test comparing the durations of the first and the mean of all subsequent phases (First vs. Subsequent Phases $\times \Delta f \times$ SOA) showed that the first phase was significantly longer than the mean of subsequent phases (63.40 vs. 35.69 s; $F(1,10) = 6.48$, $p < 0.05$ with the η^2 effect size being 0.73). This effect was stronger for small than large Δf's, as was shown by the significant interaction between the First vs. Subsequent Phases and the Δf factors ($F(3,30) = 4.25$, $p < 0.05$, $\eta^2 = 0.30$ with the ε Greenhouse–Geisser correction of the degrees of freedom being 0.61). The mean duration of subsequent phases was longer than those we found in our previous study. So, perhaps, random jittering of the stimulus parameters results in a less distinct representation of the regularities, which is reflected in increased duration of phases following the first percept.

The results of this experiment suggest that it is not change *per se* which stabilises perceptual organisation. There may be some tendency for less switching in the presence of varying stimulus parameters when compared to the case when stimulus parameters are kept constant throughout the stimulus, but it is not clear whether this is significant as the reduction in switching is well below the high inter-subject variability. Thus, we can conclude that the bistability of the auditory streaming sequence is not an artefact produced by the constancy of stimulus parameters.

44.3 Experiment 2

In the second experiment, we investigated whether perceptual bistability in auditory streaming occurred because of the inherent ambiguity of the tone sequences. In order to do so, we introduced a sound location cue supporting the segregated perceptual organisation. Our hypothesis was that this may stabilise perception since the integrated organisation would include tones coming from opposite directions.

44.3.1 Participants

Twelve young healthy volunteers participated in the experiment (five female, 19–24 years of age, average: 21.4 years).

44.3.2 Stimulus Paradigm

As in experiment 1, participants were presented with 4-min long trains of the ABA-structure, where the A and B were pure tones of 75 ms duration, including 5 ms linear onset and offset ramps. In separate trains, Δf was 0, 0.5, 2, 4 or 6 semitones (ST) and SOA was 125 or 150 ms. Each of the combinations except for the 0 ST condition was presented with and without the location cue. Location was simulated by increasing the level of the signal to one ear by 6 dB for all high tones and decreasing it by 5 dB for the other ear, while doing the opposite for all low tones. Altogether, $4 \times 4 + 2 = 18$ conditions were tested in separate blocks, the order of which was randomised separately for each participant. All other parameters and procedures were identical to Experiment 1.

44.3.3 Results

The principal finding was that the introduction of an additional unambiguous cue, which favoured segregation, did not stabilise perceptual organisation. Once again we found perceptual switching in all conditions and for all subjects; as illustrated in Fig. 44.2.

As was expected, adding the location cue significantly increased the probability of perceiving segregation (means: 0.32 and 0.53 without and with the location cue). The ANOVA (No vs. Location Cue $\times \Delta f \times$ SOA) showed significant effects of all three factors ($F(1,11)=52.20$, $p<0.0001$, $\eta^2=0.83$; $F(3,33)=29.92$, $p<0.00001$, $\varepsilon=0.68$, $\eta^2=0.73$; $F(1,11)=5.98$, $p<0.05$, $\eta^2=0.35$ for the three factors, respectively) with no significant interaction between them. Although a previous similar study using short tone sequences showed that Δf and location difference interacted in determining the probability of hearing the sound sequence as two segregated sound streams (Farkas et al. 2006), the current results did not show significant interaction between the two cues. The probability of segregation calculated by assuming independence of the two cues ($P_{\Delta f+\mathrm{loc}}(\Delta f, \mathrm{SOA}) = P_{\Delta f}(\Delta f, \mathrm{SOA}) + P_{\mathrm{loc}}(\mathrm{SOA}) - P_{\Delta f}$

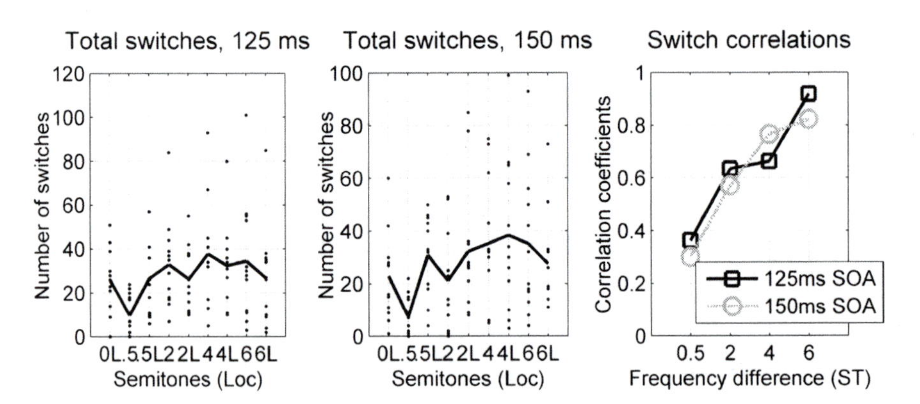

Fig. 44.2 The number of perceptual switches for each condition; (**a**) the 125 ms SOA conditions, and (**b**) the 150 ms SOA conditions. "L" indicates the presence of the location cue, no "L" indicates that the location cue was absent. *Black dots* are for individual subjects, and *solid lines* connect the mean of each condition. (**c**) Interactions between Δf and the location cue, showing the correlation coefficients between corresponding Δf conditions with and without the location cue

(Δf, SOA) × P_{loc}(SOA)) did not significantly differ from the probability of segregation measured when both cues were present (ANOVA: Independence-assuming model vs. Actual measurement × Δf × SOA; F(1,11) = 1.74 for the comparison between the model and the measurement). This result can be explained by our previous finding showing that stimulus parameters mainly affect the first perceptual phase of auditory stream segregation, but not so much the subsequent perceptual phases (Denham et al. 2008). However, independence of the frequency and locations cues can also be investigated by plotting the correlation of the number of switches between conditions with and without the location cue as a function of Δf and SOA. Figure 44.2c shows that the correlations between corresponding conditions with and without the location cue monotonically and dramatically increase at both SOA's with increasing Δf's; i.e. as Δf increases it begins to dominate perceptual organisation irrespective of the location cue. Thus, the two cues do not act independently of each other.

44.4 Experiment 3

In the final experiment, we tested the effects of short gaps on auditory streaming. The effects of gaps were assumed to model the effect of switching attention back to the sequence, as it has been shown that switching attention away and then back to the sound sequence, as well as the insertion of gaps as short as 500 ms in duration, restarted the build-up of auditory streaming (Cusack et al. 2004). In addition to testing the effects of gaps on the stability of auditory streams, we also tested (1) whether the effect was due to the introduction of a silent interval or whether any temporal violation would result in the reset of streaming, and (2) if the effect was indeed due to the introduction of silent gaps, what was the minimal duration of the gap needed to trigger a reset.

44.4.1 Participants

Eighteen young healthy volunteers participated in the experiment (six female, 18–25 years of age, average: 22.7 years).

44.4.2 Stimulus Paradigm

As in experiment 1, participants were presented with 4-min long trains of the ABA-structure, where the A and B were pure tones of 75 ms duration, including 5 ms linear onset and offset ramps. In separate trains, Δf was 4, 16 or 22 semitones (ST) and SOA was 150, 200 or 250 ms. In each stimulus block, two gaps were introduced. The first gap was inserted 90 s, and the second 180 s into the stimulus sequence. Each combination of Δf and SOA was delivered twice within the experimental session: once with the first gap being 500 ms and the second 150 ms long and once with the first gap being 200 ms and the second –200 ms long. The –200 ms value means that the onset of the ABA tone triplet following the "gap" was moved closer to the previous tone triplet by 200 ms. The –200 ms gap was used to test whether breaking the uniform rhythm of the sequence without introducing a silent interval would have an effect similar to that of the silent gaps. The other gap durations were selected on the basis of pilot studies suggesting that only gaps exceeding the circa 170 ms long temporal window of integration (for a review, see (Cowan 1984)) had a significant effect on auditory streaming.

44.4.3 Results

We found that (1) breaking the rhythm did not by itself affect auditory streaming, and (2) only silent gaps exceeding 150 ms had a significant effect on streaming. A comparison of the probability of integration in the 10 s long intervals preceding and following a gap, ANOVA (Before vs. After the gap × Gap duration × SOA × Δf), showed significant main effects of all four factors ($F(1,17)=20.72$, $p<0.001$, $\eta^2=0.55$, $F(3,51)=7.59$, $p<0.001$, $\varepsilon=0.84$, $\eta^2=0.90$, $F(2,34)=14.04$, $p<0.001$, $\varepsilon=0.77$, $\eta^2=0.45$, and $F(2,34)=48.28$, $p<0.001$, $\varepsilon=0.70$, $\eta^2=0.74$ for Before vs. After the gap, Gap duration, SOA, and Δf, separately). The probability of perceiving integration significantly increased after the gap compared to before the gap and with increasing gap duration. However, there was a significant interaction between Before vs. After the gap and Gap duration ($F(3,51)=8.53$, $p<0.0002$, $\varepsilon=0.80$, $\eta^2=0.33$). Figure 44.3 shows the probability of integration as a function of gap duration, separately before and after the gap. Whereas the duration of the gap obviously had no effect on the probability of integration during the 10-s interval preceding the gap, the probability of perceiving the integrated organisation after the gap increased with increasing gap duration. With the negative "gap" and the 150 ms gap, the difference between the pre- and post-gap probability of integration did not reach significance. In contrast, with the 200 ms long gap, there was a tendency for a difference ($p<0.08$, post-hoc Sheffé pair-wise comparisons with $df=51$ and $MSE=0.98$), which became significant by the 500-ms long gap ($p<0.0001$). Figure 44.3 also shows that the probability of perceiving integration after the gap increases rapidly between gap durations of 150 and 200 ms, suggesting that temporal integration plays an important role in this effect.

The influence of gaps on the stability of perception was studied by comparing the duration of the perceptual phase preceding and following the gaps. We found no significant increase in the duration of the perceptual phase immediately following the gap at any gap duration (ANOVA: Before vs. After the gap × Gap duration × SOA × Δf; $p<0.09$, $F(1,17)=3.36=0.09$ for the Before vs. After the gap

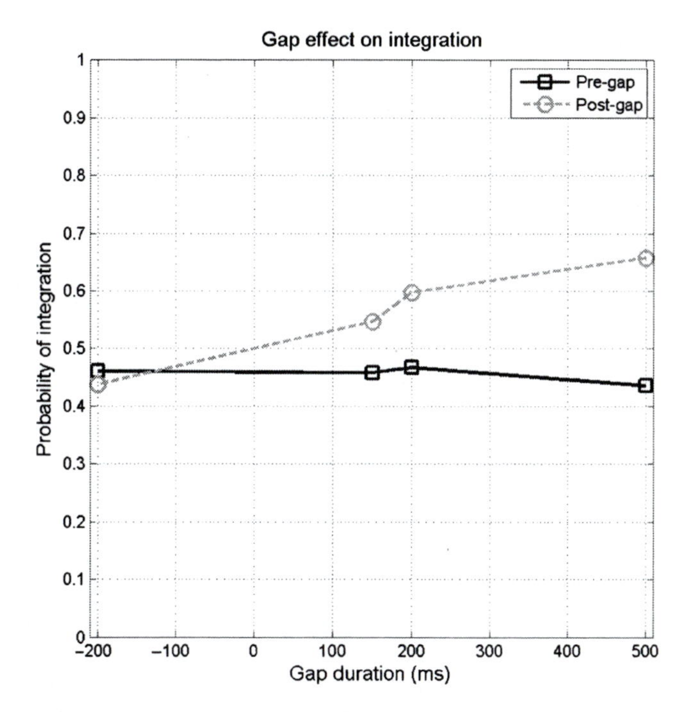

Fig. 44.3 Group-averaged ($n = 18$) probability of perceiving integration in the 10-s long periods before (*solid line*) and after (*dashed line*) the gap as a function of the gap duration

factor). This means that gaps do not have a stabilising effect on the perception of the auditory streaming sequence. Thus, either attentional switches do not help in stabilising auditory perception or, despite the similarities found between the effects of gaps and attention switching on the reset of streaming (Cusack et al. 2004), the two manipulations affect different processes.

44.5 Discussion

The experiments reported here were designed to investigate what aspects of natural sounds allow the auditory system to form stable representations of concurrently active sound sources, given the demonstration that auditory perceptual organisation can be bi- or even multi-stable. We tested (1) whether the extreme repetitiveness of the alternating two-tone sequence causes perceptual instability; (2) whether under natural conditions, multiple redundant cues stabilise perception; and (3) whether attentional switches lead to a reset of perceptual organisation thereby leading to apparently stable perception.

Much to our surprise, we found that randomly jittering the frequencies and timing of the tones did not stabilise perceptual organisation; i.e. neither constancy nor precise predictability of stimulus parameters are necessary for bistable perception. As long as a predictable *distribution* of stimulus features exists, perceptual switching can occur. There was a tendency for phase durations to increase with a decrease in the precision with which the regularity can be represented, but this cannot explain the perceptual stability experienced in real-life situations. Even so, this aspect of the results requires further investigation.

The introduction of an additional cue, which unambiguously favoured segregation rather than integration, similarly failed to stabilise perceptual organisation. Subjects continued to report hearing the integrated organisation even in the face of interaural level differences which gave the impression that the alternating tones came from different directions. However, the biasing cue did increase the probability of perceiving segregation, even though it did not prevent perceptual switching. The influence of this cue was strongly modulated by the strength of the primary cue to segregation, namely Δf. This is consistent with other auditory streaming experiments showing the dominance of the spectral composition of the alternating tones over other features such as pitch (Vliegen et al. 1999).

The final hypothesis we explored was whether switches in attention may lead to perceptual stability. We investigated this by introducing silent gaps into the tone sequence, as Cusack and colleagues have previously argued that gaps have a similar effect on the build up of streaming to switching attention (Cusack et al. 2004). The main conclusion that we can draw from this experiment is that the introduction of even very brief gaps can increase the probability of reporting integration. However, this does not amount to a complete return to the starting state, as evidenced by the lack of a long duration perceptual phase (a "first phase"), following the gap.

Violation of predictability is not sufficient for a reset to occur. The –200 ms "gap" condition violated the expected timing of the following triplet significantly, but did not increase the probability of a return to integration. A further requirement is that the gap should exceed the window of temporal integration. This finding suggests that temporal integration may play an important role in perceptual sound organisation. It is as if gaps of shorter durations are disregarded by the system, and possibly regarded as natural variation in timing. A more formal description of the effect could be based on extending Zwislocki's model of temporal integration (Zwislocki 1969) to expected sounds.

The results of the gap experiment can be compared to the effect of gaps in binocular rivalry experiments. It has been shown that when subjects view a bistable visual stimulus which is presented intermittently, then they often resume the perceptual organisation that they experienced before the interruption (Leopold et al. 2002). Frequently interrupting the stimulus can lead to prolonged periods of perceptual stability. Further experiments will be needed to investigate whether a similar effect holds in auditory streaming. Nevertheless, the increased tendency to report integration after a gap suggests that intermittent presentation may lead to increased perceptual stability.

Acknowledgements This work was supported by the European Research Area Specific Targeted Projects EmCAP (IST-FP6-013123), and SCANDLE (IST-FP7-231168).

References

Cowan N (1984) On short and long auditory stores. Psychol Bull 96:351–370
Cusack R, Deeks J, Aikman G, Carlyon RP (2004) Effects of location, frequency region, and time course of selective attention on auditory scene analysis. J Exp Psychol Hum Percept Perform 30:643–656
Denham SL, Winkler I (2006) The role of predictive models in the formation of auditory streams. J Physiol Paris 100:154–170
Denham SL, Gyimesi K, Stefanics G, Winkler I (2008) Perceptual bi-stability in auditory streaming: how much do stimulus features matter? Biological Psychology, under review
Farkas A, Dudás T, Horváth J, Sussman E, Winkler I (2006) Jelzmozzanatok integrációja hangláncok elválasztásában [Cue integration in separating auditory streams]. A Magyar Pszichológiai Társaság XVII., Országos Tudományos Nagygyılése, Budapest
Leopold DA, Wilke M, Maier A, Logothetis NK (2002) Stable perception of visually ambiguous patterns. Nat Neurosci 5:605–609
Pressnitzer D, Hupé JM (2005) Is auditory streaming a bistable percept? Forum Acusticum, Budapest, pp 1557–1561
Pressnitzer D, Hupé JM (2006) Temporal dynamics of auditory and visual bistability reveal common principles of perceptual organization. Curr Biol 16:1351–1357
Vliegen J, Moore BC, Oxenham AJ (1999) The role of spectral and periodicity cues in auditory stream segregation, measured using a temporal discrimination task. J Acoust Soc Am 106:938–945
Winkler I, Takegata R, Sussman E (2005) Event-related brain potentials reveal multiple stages in the perceptual organization of sound. Brain Res Cogn Brain Res 25:291–299
Zwislocki JJ (1969) Temporal summation of loudness: an analysis. J Acoust Soc Am 46:431–440

Chapter 45
Sequential and Simultaneous Auditory Grouping Measured with Synchrony Detection

Christophe Micheyl, Shihab Shamma, Mounya Elhilali, and Andrew J. Oxenham

Abstract Auditory scene analysis mechanisms are traditionally divided into "simultaneous" processes, which operate across frequency, and "sequential" processes, which bind sounds across time. In reality, simultaneous and sequential cues often coexist, and compete to determine perceived organization. Here, we study the respective influences of synchrony, a powerful grouping cue, and frequency proximity, a powerful sequential grouping cue, on the perceptual organization of sound sequences (Experiment 1). In addition, we demonstrate that listeners' sensitivity to synchrony is dramatically impaired by stream segregation (Experiment 2). Overall, the results are consistent with previous results showing that prior perceptual grouping can influence subsequent perceptual inferences, and show that such grouping can strongly influence sensitivity to basic sound features.

Keywords Auditory scene analysis • Stream segregation • Performance measures • Timing • Synchrony

45.1 Introduction

Research on auditory scene analysis has led to the identification of several "cues," which the auditory system can use to organize sounds perceptually (Bregman 1990; Darwin and Carlyon 1995). Synchrony, harmonicity, and frequency proximity are among the most important such cues. Synchronicity and harmonicity are used primarily to group simultaneous spectral components across frequency, whereas frequency proximity has an important role in binding sequential elements across time.

An important question currently facing psychophysicists is how these grouping cues interact with each other in order to determine the "correct" perceptual organization of acoustic scenes, which typically contain a multiplicity of such cues. Earlier studies have revealed that sequential grouping based on frequency proximity

C. Micheyl (✉)
Department of Psychology, University of Minnesota, Minneapolis, MN, USA
e-mail: cmicheyl@umn.edu

E.A. Lopez-Poveda et al. (eds.), *The Neurophysiological Bases of Auditory Perception*,
DOI 10.1007/978-1-4419-5686-6_45, © Springer Science+Business Media, LLC 2010

can counteract simultaneous grouping based on synchrony (e.g., Darwin et al. 1995; 1989; Shinn-Cunningham et al. 2007). For instance, a series of elegant experiments by Darwin and colleagues (e.g., Darwin et al. 1995; 1989) have demonstrated that "precursor" tones at the same frequency as a "target" component in a complex tone "captured" the target into a separate stream, thereby reducing its influence on the pitch or timbre of the complex.

The present study was inspired by these earlier findings and addressed two questions. The first question was whether sequential grouping affects listeners' ability to detect synchrony. The results of several previous studies suggest that listeners are unable to accurately perceive the temporal relationships between sounds across auditory streams. In particular, listeners cannot accurately discriminate the duration of temporal intervals between consecutive tones (Vliegen et al. 1999; Roberts et al. 2002), or correctly identify the temporal order of these tones (Bregman and Campbell 1971), under conditions where the tones are heard in separate streams. However, in all of these studies, the tones never overlapped in time. The situation might be quite different with synchronous tones, because synchrony detection appears to involve different mechanisms than temporal order identification, or temporal interval discrimination (Mossbridge et al. 2006). For instance, while synchrony detection could in principle be achieved using widely tuned neural coincidence detectors (Oertel et al. 2000), temporal interval discrimination require mechanisms for measuring the elapsed time between events. These various mechanisms, possibly taking place at different stages of processing in the auditory system, could be differently affected by sequential grouping. The finding that detrimental effects of sequential grouping generalize to synchrony detection would provide evidence that listeners' access to the output of coincidence detectors is strongly constrained by perceptual organization mechanisms.

The second question addressed in this study is whether across-frequency grouping based on synchrony predominates over sequential grouping based on frequency proximity. In order to answer this question, we measured listeners' thresholds for the detection of an asynchrony between two tones at different frequencies, A and B, preceded by a series of either synchronous or asynchronous "precursor" tones at the same two frequencies, A and B. We reasoned that, if across-frequency grouping due to synchrony predominates over segregation due to frequency separation, thresholds should be lower with synchronous precursors than with asynchronous precursors.

45.2 Experiment 1: Sequential Capture Overrides Synchrony Detection

45.3 Methods

Schematic spectrograms of the stimulus conditions tested in this experiment are shown in Fig. 45.1a. The basic stimulus elements were 100 ms pure tones at two frequencies, A, which was fixed at 1,000 Hz, and B, which was set 6 or 15 semitones above A. In the baseline, "No captor" condition (upper panel in Fig. 45.1a), only these two A and B tones were present. In one observation interval, the tones were synchronous; in the other, the B tone was delayed or advanced by Δt ms relative to the A tone. The task of the listener was to indicate in which observation interval the A and B tones were asynchronous.

Two other conditions were tested. In the "On-frequency captor" condition (middle panel in Fig. 45.1a), the A and B pair was surrounded by "captor" tones at the A frequency, with five captor tones before, and two captor tones after, the A–B pair. The captor tones were separated from each other, and from the target A tone, by a constant gap of 50 ms. Thus, in this condition, the target A tone formed part of a temporally regular sequence. In the final, "Off-frequency captor" condition (lower panel in Fig. 45.1a), the frequency of the captor tones was set to six semitones below that of the A tone. The listener's task was the same as in the baseline condition: to indicate in which of the two observation intervals presented on a trial the target A and B tones were asynchronous. A three-down one up adaptive procedure was used to measure thresholds, with Δt as the tracking variable. Each listener completed at least four threshold measurements in each condition. The data shown here are geometric mean thresholds across listeners.

In all experiments described here, the stimuli were generated digitally and played out via a soundcard (Lynx Studio L22) with 24-bit resolution and a sampling frequency of 32 kHz, and presented to the listener via the left earpiece of Sennheiser HD

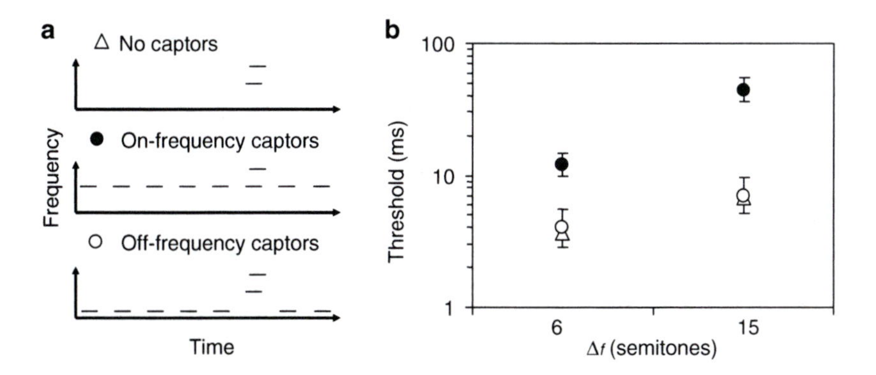

Fig. 45.1 (**a**) Schematic spectrograms of the stimuli in Experiment 1. The small horizontal bars represent 100-ms tones. In the baseline ("No captors") condition (*top panel*), the stimuli were a fixed 1,000-Hz tone, A, and a (6- or 15-semitone) higher-frequency tone, B. In one of the two observations intervals on a trial, the onset of the B tone was delayed (as shown here) or advanced (not shown) by Δt ms relative to that of the A tone; in the other observation interval, the two tones were synchronous (not shown). In the "On-frequency captors" condition (*middle panel*), the target A and B tones were preceded by five, and followed by two "captor" tones at the A frequency (1,000 Hz). Consecutive captors tones were separated from each other, or from the A tone, by a fixed, 50 ms silent interval. In the "Off-frequency captors" condition, the frequency of the captor tones was set to six semitones below that of the A tone, being equal to approximately 707 Hz. (**b**) Thresholds for the detection of an asynchrony between the target A and B tones in the different stimulus conditions shown on the left, for the two A–B frequency separations (6 and 15 semitones). Each data point was obtained by averaging thresholds across listeners. The error bars show geometric standard errors of the mean

580 headphones. Listeners were seated in a double-walled sound-attenuating chamber (Industrial Acoustics Company). The level of the tones was set to 60 dB SPL.

Eight listeners took part in this experiment. All had normal hearing (i.e., pure tone thresholds lower than 15 dB HL at octave frequencies between 500 and 6 kHz).

45.4 Results and Discussion

The results of this experiment are shown in Fig. 45.1b. Significantly larger thresholds were observed in the "On-frequency captors" condition than in both the "No-captor" [$F(1, 7) = 19.96$, $p = 0.003$], and "Off-frequency captors" [$F(1, 7) = 13.47$, $p = 0.008$] conditions. In fact, at the largest (15-semitone) A–B frequency separation, thresholds in the presence of the on-frequency captors were occasionally at ceiling (100 ms). Thresholds in the "Off-frequency captors" and "No-captors" conditions were not statistically different, and generally low (3–5 ms), consistent with earlier findings (e.g., Zera and Green 1993). Finally, thresholds were generally larger at the largest (15 semitones) A–B frequency separation (Δf) than at the smaller one (6 semitones) [$F(1, 7) = 81.11$, $p < 0.0005$], an effect that was larger in the "On-frequency captors" condition than in the other two conditions [$F(1, 7) = 22.11$, $p = 0.002$].

The results of this experiment are consistent with earlier findings, which used "capture" effects to demonstrate an influence of sequential grouping (based on frequency proximity) on the perception of temporal relationships between sounds (Bregman and Campbell 1971; O'Connor and Sutter 2000). The current results reveal that the detection of synchrony (or asynchrony) is no more immune to sequential grouping influences than other temporal discrimination abilities (e.g., Broadbent and Ladefoged 1959; Roberts et al. 2002, 2008; Vliegen et al. 1999). Thus, while there is physiological evidence for the existence of "coincidence detectors" or "synchrony detectors" in the auditory system (Oertel et al. 2000), the present findings suggest that listeners' conscious access to the outputs of these detectors is constrained by perceptual organization mechanisms.

45.5 Experiment 2: Synchrony Overrides Sequential Grouping

45.6 Methods

The stimuli used in this experiment are illustrated schematically in Fig. 45.2a. They were sequences of A and B tones, where A and B represent different frequencies. The frequency of the A tone was kept constant at 1,000 Hz. The frequency of the B tone was set 6, 9, or 15 semitones above that of the A tone. Each sequence consisted of five "precursor" tones at each frequency (i.e., five A tones and five B

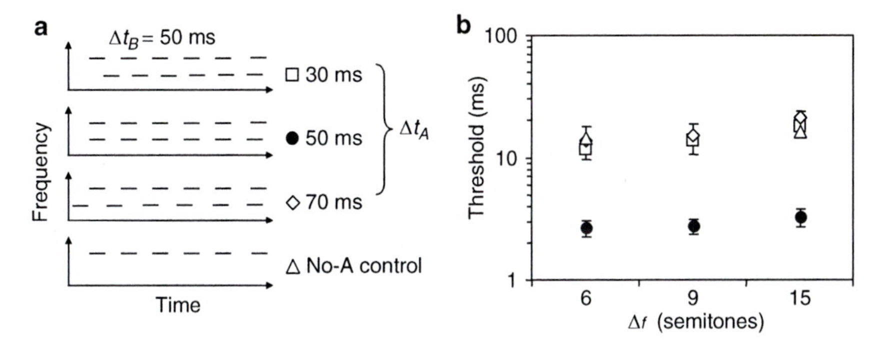

Fig. 45.2 (**a**) Schematic spectrograms of the stimuli in Experiment 2. The stimuli were sequences of 100-ms pure tones (shown here as small horizontal bars) at two different frequencies, A and B, separated by 6, 9, or 15 semitones. Except for the last two, tones at the higher (B) frequency were separated by a constant ITI, Δt_B, of 50 ms. Depending on the condition, tones at the lower (A) frequency were either present and separated by an ITI, Δt_A, of 30 ms (*top panel*), 50 ms (*second panel from top*), or 70 ms (*second panel from bottom*), or they were absent (No-A control condition, *lower panel*). In the condition where Δt_A and Δt_B were both equal to 50 ms, all A and B tones except the last two were synchronous. Depending on the observation interval, the last ("target") A and B tones were either asynchronous (as shown here) or synchronous (not shown here). In the former case, they were separated by a variable delay, Δt_{AB}, which was controlled by the adaptive threshold-tracking procedure. The task of the listener was to indicate the observation interval containing the delay Δt_{AB}. (**b**) Thresholds in Experiment 2. Thresholds measured in the different conditions illustrated in Fig. 45.2a are shown using different symbols (as indicated in Fig. 45.2a). Each data point was obtained by averaging thresholds across listeners. The error bars show geometric standard errors of the mean

tones), followed by two "target" tones (i.e., one A tone and one B tone). Each tone was 100 ms in duration, including 10-ms raised-cosine onset and offset ramps. The duration of the silent interval between two consecutive B precursors, Δt_B, was fixed at 50 ms. The inter-tone interval (ITI) between A precursors, Δt_A, varied across conditions; it was equal to 50 ms (in which case, the A and B precursors were synchronous, as shown in the middle panel of Fig. 45.2a), 30, or 70 ms (in which cases, the A and B precursors were asynchronous; see upper and lower panels in Fig. 45.2a).

The precursors were followed by a pair of "target" A and B tones, which were either synchronous, or asynchronous. In the latter case, the B tone randomly led or lagged the A tone by an amount, Δt, which was varied adaptively by the tracking procedure used to measure the threshold. Fig. 45.2a illustrates the case of a lagging B tone. In all cases, the interval between the A target and the preceding A precursor was the same as that between two consecutive A precursors. Importantly, the A and B sequences were always positioned in time relative to each other in such a way that the target A and B tones were synchronous in one of the two observation intervals presented during a trial, and shifted by plus or minus Δt in the other observation interval. In addition, a control condition was run, in which the A tones were turned off, and the B tones were generated in exactly the same way as described above.

Thresholds for the detection of an asynchrony between the target A and B tones were measured using a two-interval, two-alternative forced-choice (2I-2AFC) procedure with an adaptive three-down one-up rule. Listeners had to indicate, on each trial, which of the two presented tone sequences (separated by a silent gap of 500 ms) contained asynchronous target A and B tones at the end. The order of presentation of the two sequences was randomized. At the beginning of each adaptive run, the tracking variable, Δt, was set to 20 ms. It was divided by a factor c after three consecutive correct responses, and multiplied by that same factor after each incorrect response. The value of c was set to four at the beginning of the adaptive run; it was reduced to two after the first reversal in the direction of tracking (from decreasing to increasing), and to $\sqrt{2}$ after a further two reversals. The procedure stopped after the sixth reversal with the $\sqrt{2}$ step size. Threshold was computed as the geometric mean of Δt at the last six reversal points. Each listener completed at least four threshold measurements in each condition. The data shown here are geometric mean thresholds across listeners.

Nine listeners with normal hearing (i.e., pure-tone hearing thresholds of 15 dB HL or less at octave frequencies between 500 and 8,000 Hz) took part in this experiment.

45.7 Results and Discussion

The results of this experiment are shown in Fig. 45.2b. In the condition in which the A and B precursor tones were synchronous (50 ms ITI at both frequencies), thresholds (indicated by filled circles) were generally small (around 3 ms), even at the largest A–B frequency separation tested (15 semitones). In contrast, in conditions in which the nominal duration of the ITI in the A-tone stream was shorter (30 ms) or longer (70 ms) than that the ITI in the B-tone stream, so that the precursor A and B tones were presented asynchronously and at different tempi, thresholds were considerably larger (10–20 ms). The difference was highly statistically significant [$F(1, 10)=7.394$, $p<0.001$].

The finding of relatively large thresholds in the conditions in which the precursor A and B tones were asynchronous is consistent with other results in the literature (e.g., Bregman and Campbell 1971; Vliegen et al. 1999; Roberts et al. 2002), which indicate that listeners cannot accurately judge the relative timing of sounds across streams. In fact, the thresholds measured in those two conditions were not significantly different from those measured in the control condition, in which the A tones were turned off (upward pointing triangles), and the only cue available for task performance was a temporal irregularity in the B stream (i.e., a longer or shorter ITI between the last two B tones than between previous tones). This suggests that in conditions in which the A and B tones formed two separate streams, performance was based on a within-stream cue, rather than on across-stream timing comparisons.

The finding of consistently low thresholds in conditions involving synchronous precursor tones indicates that in this condition, listeners were able to make accurate timing judgments between the A and B tones. This suggests that in these conditions, the A and B tones formed a single stream. Thresholds increased somewhat with the A–B separation [$F(1, 10)=5.73, p<0.001$], indicating that synchrony-based grouping did not completely override the effect of frequency separation. However, even at the largest frequency separation tested (15 semitones), they were still quite small, and considerably smaller than in the other (asynchronous precursors or no-A-tones control) conditions. This result is noteworthy, because it demonstrates that spectral components separated by more than an octave can still be grouped in perception if they are synchronous or quasi-synchronous; this provides further evidence that the auditory system can accurately detect synchrony, or lack thereof, across widely separated frequencies (Mossbridge et al. 2006).

45.8 Conclusions

The results of this study indicate that while sequential grouping can prevent synchrony-based grouping, and dramatically impair listeners' ability to detect asynchrony (Experiment 1), on the other hand, synchrony can prevent stream segregation based on frequency separation, and lead to perceptual grouping of spectral components at remote frequencies (Experiment 2). How can these apparently contradictory results be reconciled?

A possible explanation is based on Bregman's (1990) "old plus new" heuristic. According to this explanation, whichever grouping cue is introduced first dominates the subsequent organization of the auditory scene. Thus, in situations where sequential grouping cues precede simultaneous grouping cues, sequential grouping overrides simultaneous grouping (Experiment 1); conversely, in situations where simultaneous grouping cues (such as synchrony) are introduced before sequential grouping cues, simultaneous grouping predominates (Experiment 2).

Some findings in the psychoacoustic literature suggest that the "old plus new" heuristic cannot be the whole story, however. For instance, Dau et al. (2009) found that comodulation masking release (CMR) was eliminated when the flankers were followed by spectrally similar "post-cursors," with which they formed a sequential stream, separate from the signal that the listener had to detect. This suggests that if we had presented only postcursor tones (no precursors) in Experiment 1, sequential capture may have also occurred, and qualitatively similar (albeit perhaps weaker) results would have been obtained.

Because grouping effects induced by postcursors cannot be explained simply in terms of the old-plus-new heuristic, a more general account of perceptual grouping is needed. Elhilali et al. (this volume) describe a computational model, which can account for these and other psychophysical results on simultaneous and sequential grouping in auditory scene analysis.

Acknowledgments This work was supported by the National Institutes of Health (R01 DC 07657). Cynthia Hunter is acknowledged for assistance with data collection. More details concerning the methods and results of Experiment 2 may be found in Elhilali et al. (2009). Experiment 1 formed part of a broader study, the results of which are described in Micheyl et al (2010).

References

Bregman AS (1990) Auditory scene analysis: the perceptual organization of sound. MIT Press, Cambridge

Bregman AS, Campbell J (1971) Primary auditory stream segregation and perception of order in rapid sequences of tones. J Exp Psychol 89:244–249

Broadbent DE, Ladefoged P (1959) Auditory perception of temporal order. J Acoust Soc Am 31:151–159

Darwin CJ, Carlyon RP (1995) Auditory grouping. In: Moore BCJ (ed) Hearing. London, Academic Press, pp 387–424

Darwin CJ, Hukin RW, al-Khatib BY (1995) Grouping in pitch perception: evidence for sequential constraints. J Acoust Soc Am 98:880–885

Darwin CJ, Pattison H, Gardner RB (1989) Vowel quality changes produced by surrounding tone sequences. Percept Psychophys 45:333–342

Dau T, Ewert S, Oxenham AJ (2009) Auditory stream formation affects comodulation masking release retroactively. J Acoust Soc Am 125:2182–2188

Elhilali M, Ma L, Micheyl C, Oxenham AJ, Shamma AS (2009) Temporal coherence in the perceptual organization and cortical representation of auditory scenes. Neuron 61:317–329

Micheyl C, Hunter CH, Oxenham AJ (2010) Auditory stream segregation and the perception of across-frequency synchrony. J Exp Psychol Hum Percept Peform. In press

Mossbridge JA, Fitzgerald MB, O'Connor ES, Wright BA (2006) Perceptual-learning evidence for separate processing of asynchrony and order tasks. J Neurosci 26:12708–12716

O'Connor KN, Sutter ML (2000) Global spectral and location effects in auditory perceptual grouping. J Cogn Neurosci 12:342–354

Roberts B, Glasberg BR, Moore BC (2002) Primitive stream segregation of tone sequences without differences in fundamental frequency or passband. J Acoust Soc Am 112:2074–2085

Roberts B, Glasberg BR, Moore BC (2008) Effects of the build-up and resetting of auditory stream segregation on temporal discrimination. J Exp Psychol Hum Percept Perform 34:992–1006

Oertel D, Bal R, Gardner SM, Smith PH, Joris PX (2000) Detection of synchrony in the activity of auditory nerve fibers by octopus cells of the mammalian cochlear nucleus. Proc Natl Acad Sci U S A 97:11773–11779

Shinn-Cunningham BG, Lee AK, Oxenham AJ (2007) A sound element gets lost in perceptual competition. Proc Natl Acad Sci U S A 104:12223–12227

Vliegen J, Moore BC, Oxenham AJ (1999) The role of spectral and periodicity cues in auditory stream segregation, measured using a temporal discrimination task. J Acoust Soc Am 106:938–945

Zera J, Green DM (1993) Detecting temporal onset and offset asynchrony in multicomponent complexes. J Acoust Soc Am 93:1038–1052

Chapter 46
Rate Versus Temporal Code? A Spatio-Temporal Coherence Model of the Cortical Basis of Streaming

Mounya Elhilali, Ling Ma, Christophe Micheyl, Andrew Oxenham, and Shihab Shamma

Abstract A better understanding of auditory scene analysis requires uncovering the brain processes that govern the segregation of sound patterns into perceptual streams. Existing models of auditory streaming emphasize tonotopic or "spatial" separation of neural responses as the primary determinant of stream segregation. While partially true, this theory is far from complete. It overlooks the involvement of and interaction between both "sequential" and "simultaneous" grouping mechanisms in the process of scene analysis.

Here, we describe a new neuro-computational model of auditory streaming. Inspired by recent psychophysical (cf. abstract by Micheyl et al.) and physiological findings, this model is based on the premise that perceived segregation results from spatio-temporal incoherence, rather than just tonotopic separation. While tonotopic separation still plays an important role in this model, it is an indirect one: tonotopic overlap tends to reduce temporal incoherence, which in turn impedes segregation. The model simulates responses at the level of the primary auditory cortex and performs a correlative analysis of cortical responses in order to assess how different sound elements evolve in time in relation to each other. An eigenvector decomposition of this coherence analysis is used to predict how the input stimulus is organized into streams. The model is evaluated by comparing its neural and perceptual predictions under various stimulus conditions to physiological and psychophysical results.

Keywords Auditory streaming • Temporal coherence • Auditory cortex • Multirate model

---

M. Elhilali (✉)
Department of Electrical and Computer Engineering, Johns Hopkins University, Baltimore, MD, USA
e-mail: mounya@jhu.edu

E.A. Lopez-Poveda et al. (eds.), *The Neurophysiological Bases of Auditory Perception*,
DOI 10.1007/978-1-4419-5686-6_46,

46.1 Introduction

A well established Gestalt principle that has been often evoked in visual perception is that of *common fate*; i.e., the tendency to group together objects that move together with the same motion pattern and speed (Blake and Lee 2005). In the auditory domain, this principle simply translates into the observation that features which "move" together in time will likely group together perceptually (Bregman 1990). While simple enough in its postulate, this idea has not been explored in studies of neural correlates of streaming. Until now, the prevalent view, based on data recorded mostly in the primary auditory cortex, has focused on a "spatial" (i.e., tonotopic) explanation of how the brain solves the segregation problem (Fishman et al. 2004, 2001; Micheyl et al. 2005). This view postulates that neuronal populations with spatially segregated average responses will likely give rise to perceptually segregated streams. While this principle holds for simple sequential organization conditions such as alternating tone sequences, it does not generalize to other stimuli. In particular, this principle does not address the interaction between synchrony and sequential grouping cues. Mounting perceptual evidence, most recently from the accompanying paper by Micheyl et al. (this volume; see also: Micheyl et al. 2010), indicates that these principles do indeed interact in guiding how our brain segregates sound.

Based on these results, we explore the idea of *temporal coherence* as a new framework for understanding the neural correlates of streaming. We present physiological experiments from recordings in single units, which support a spatio-temporal basis of stream segregation. In addition, we propose a model that successfully validates the perceptual data using tone sequences, based on response properties in cortical neurons.

46.2 Neurophysiological Basis of Stream Organization in AI

We set out to explore the neurophysiological basis of the organization of streams as guided by perceptual grouping principles. In this study, we focused on the organization of synchronous and sequential tones at the level of primary auditory cortex (AI), and explored the nature of the neural code to both stimulus types in order to account for their very different percepts.

In this experiment, we examined the distribution of responses to two pure tones, the frequencies of which were adjusted relative to the best frequency (BF) of an isolated single unit in AI of awake ferrets in five steps (labeled 1–5), where positions 1 and 5 correspond to one of the tones being at BF. The frequency separation (ΔF) between the tones was fixed at 1, 0.5, or 0.25 octaves, corresponding to 12, 6, and 3 semitones, respectively. Tones A and B were shifted coherently relative to BF, with tone B starting at the BF and tone A ending at the BF. ΔF between the tones was 0.25, or 0.5, or 1 octave, which was fixed within a trial and varied among different trials. The total number of conditions was: 5 positions $\times 3$ $\Delta F \times 2$ modes.

The results from a population of 122 units in the AI of four ferrets are shown in Fig. 46.1. We analyzed the average rate profiles of each unit to each tone sequence under all frequency separations and frequency position. When the tones are far apart ($\Delta F = 1$ octave; right panel of Fig. 46.1), and when either tone is near BF, responses are strong (positions 1 and 5); they diminish considerably when the BF is midway between the tones (position 3), suggesting relatively good spatial separation between the representations of each tone. When the tones are closely spaced ($\Delta F = 0.25$ octave; left panel of Fig. 46.1), the responses remain relatively strong at all positions, suggesting that the representations of the two tones are not well separated. More importantly, the average rate profiles are similar for both presentation modes; in all cases, the responses are well-segregated with significant dips when the tones are far apart ($\Delta F = 1$ octave), and poorly separated (*no* dips) when the tones are closely-spaced ($\Delta F = 0.25$ octaves). Thus, based on average rate responses, the neural data mimic the perception of the asynchronous but not the synchronous tone sequences. Therefore, the distribution of average rate responses does not appear to represent a general neural correlate of auditory streaming.

Overall, the results from the physiological experiments in awake ferrets reveal that a simple rate profile of responses in primary auditory cortical neurons is not sufficient to explain the perceptual difference between synchronous and alternating tone sequences. Clearly, a model that is successfully able to predict perception from these neural data will need to incorporate the time dimension.

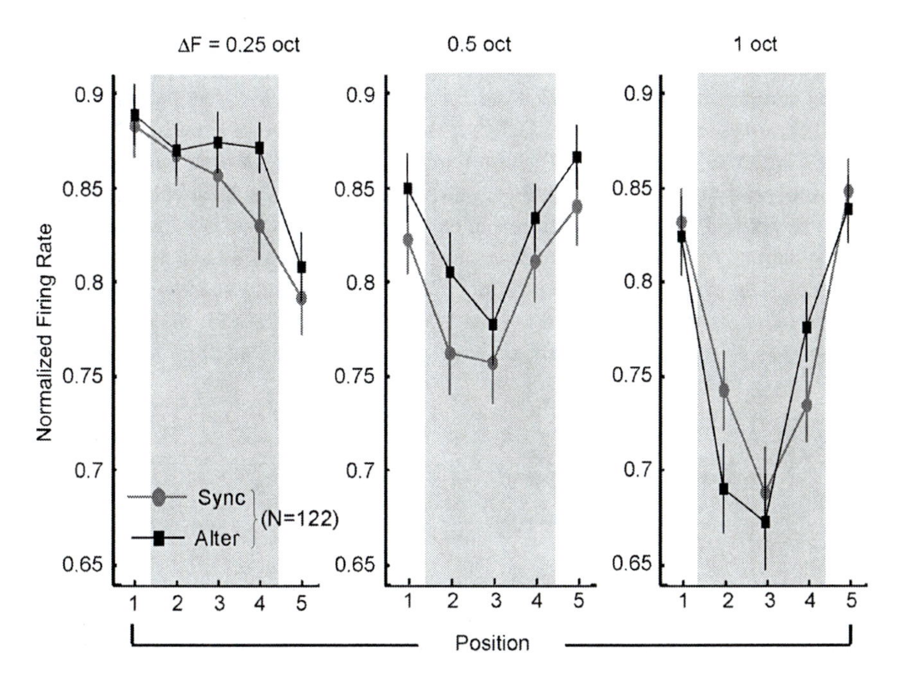

Fig. 46.1 Single unit recordings in AI using synchronous and alternating tone sequences

46.3 Spatio-Temporal Coherence Model

In this work, we emphasize the need for incorporating the temporal dimension in any model of stream organization. The basic premise of the model is that temporal coherence of sound features is an important principle for organizing sound mixtures into streams.

46.3.1 Auditory Processing from Periphery to Cortex

It is well established that acoustic signals undergo a series of transformations as they journey up the auditory system starting at the periphery all the way to cortex, mapping signals into a higher-dimensional representation. In order to capture this mapping in a mathematical formulation, we employ a model of auditory processing which abstracts from existing physiological data in animals and psychoacoustical data in human subjects as explained in details by Chi et al. (1999, 2005) and Elhilali et al. (2003).

The *early auditory stages* process the incoming acoustic signal through a sequence of stages representing cochlear filtering, hair cell transduction, and lateral inhibition to yield a final auditory spectrogram output (Fig. 46.2). This sequence of operations effectively computes a spectrogram of the signal using a bank of constant-Q filters, with a bandwidth tuning Q of about 12 (or just under 10% of the center frequency of each filter).

The *central cortical stages* further analyze the auditory spectrum into more elaborate representations and separate the different cues and features associated with different sound percepts. Electrophysiological evidence shows that cortical neurons are tuned to a variety of sound features, including BF, spectral bandwidth, and temporal dynamics. In the present study, we focus on the temporal integration tuning of cortical neurons. The time-scales of cortical dynamics are commensurate with dynamics of stimuli used in streaming experiments, as well as the dynamics of speech (Chi et al. 1999; Elhilali et al. 2003), musical melodies, and many other sensory percepts (Carlyon and Shamma 2003; Viemeister 1979). Mathematically, this analysis is achieved via an affine wavelet analysis of the auditory spectrogram.

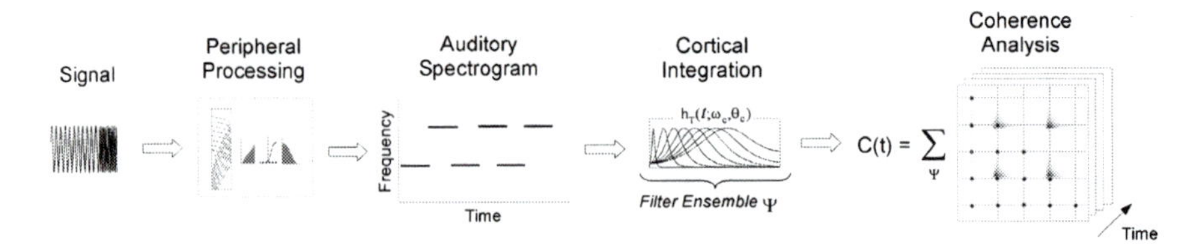

Fig. 46.2 Schematic of the spatio-temporal coherence model

The cortical temporal model estimates the temporal modulation content of the auditory spectrogram via a bank of modulation-selective filters (the wavelets) centered at each frequency along the tonotopic axis. Each filter is tuned ($Q=1$) to a range of temporal modulations (also referred to as rates or velocities (in Hz)), and is constructed by a temporal gamma function. This mother wavelet is scaled and shifted at different rates (Wang and Shamma 1995). Effectively, the model analyzes the time-sequence from each frequency-scale channel by convolving it with a temporal receptive field, effectively integrating the signal energy over a multiple scales of time ranging from 4 Hz to 64 Hz in logarithmic steps. It is worth noting that 64 Hz is a relatively high upper limit to known cortical dynamics. It is however used in the present study to ensure that short sounds (of the order of tens of milliseconds, such as one short tone) do indeed induce a response through the cortical integration stage.

46.3.2 Coherence Analysis

The focal proposition of this model is that features are grouped based on their coherence over time. The premise is extensible to a range of dimensions, and should be valid when applied to acoustic features, including frequency, spectral shape (or timbre), pitch, spatial location, etc. For details of correlation analysis based on spectral shape, (Elhilali and Shamma 2007) describes an analysis of informational masking stimuli.

In the present paper, we focus on applying the model along the frequency dimension. Specifically, a signal is processed through the peripheral processing stage (described in sect. 46.3.1) yielding a time-frequency representation. The outcome is then passed through the cortical temporal analysis described in sect. 46.3.1, where each spectral channel is integrated through multirate analysis windows. A correlation analysis is then performed on the channel against each other, as described in the steps below:

1. Map the signal $x(t)$ into a time-frequency spectrogram $y(t,x)$
2. Perform the cortical multirate analysis: $r(t,x;\omega_c,\theta_c) = y(t,x) *_t h_T(t;\omega_c,\theta_c)$
3. For each

$$R(t_0,\omega_c,\theta_c) = \left[r(x_1;t_0,\omega_c,\theta_c), y(x_2;t_0,\omega_c,\theta_c),\ldots, y(x_N;t_0,\omega_c,\theta_c)\right]^T,$$

   perform a correlation analysis:RR* (where * denotes complex-conjugate).
4. Integrate all matrices over the range of rate filters Ψ:

$$C(t_0) = \sum_{\Psi} R(t_0,\omega_0,\theta_0)$$

46.3.3 Decomposing the Coherence Matrix

The matrix C captures the degree of coherence in the neural responses at different frequency locations along the tonotopic axis. A high correlation value between two channels indicates a strong coherent activity at these two locations, while a low correlation value indicates lack of coherent activity. In order to determine the optimal factorization of the matrix in terms of maximally correlated channels, we perform an Eigen Value Decomposition (EVD). The structure of the significant eigenvectors is informative about which channels should be grouped together as one stream, and which should belong to a different stream. If a sound mixture contains only mutually coherent activity, its EVD will yield one strong eigenvalue corresponding to the channels that are maximally coherent. If, instead, a sound mixture contains two streams, their uncorrelated activity over time will emerge in the coherence matrix as two main directions, which will give rise to two strong eigenvalues. In the simulations presented in this work, we will use the ratio of second to the first eigenvalue as a correlate of this segregation: The smaller the ratio, the more likely the original sound contains *one* stream. In order to explore the actual structure of the streams associated with these eigenvector, we use the corresponding eigenvectors as weights on the different frequency channels.

46.3.4 Model Validation

In order to test the model's performance, we simulate a range of stimuli consisting of tone sequence with various spectro-temporal organizations. In the first simulation, we vary the degree of synchrony between 2 tone sequences, spanning the continuum from fully synchronous to fully asynchronous. In the second and third simulations, we explore the interaction between synchrony and sequential grouping cues, following the perceptual studies presented in the accompanying study by Micheyl et al. (2009).

46.3.4.1 Varying Degrees of Synchrony

In the first simulation, we use a sequence of a low A tone fixed at 300 Hz, and a high B tone at 952 Hz. Both tones were 75 ms long, with 10 ms onset and offset raised cosine ramp. We vary the onset to onset delay between the A and B tones from $\Delta T=0\%$ (for fully synchronous) to $\Delta T=100\%$ (fully asynchronous) with graduate steps in between. Figure 46.3a shows the ratio of eigenvalues as a function of ΔT. At the lowest end ($\Delta T=0\%$), the coherence matrix maps to almost one main eigenvalue, hence the eigen-ratio is very small correlating with a percept of one stream (Elhilali et al. 2009). In this case, the coherence matrix can be mapped onto one main dimension, which yields an almost zero second eigenvalue. At the other

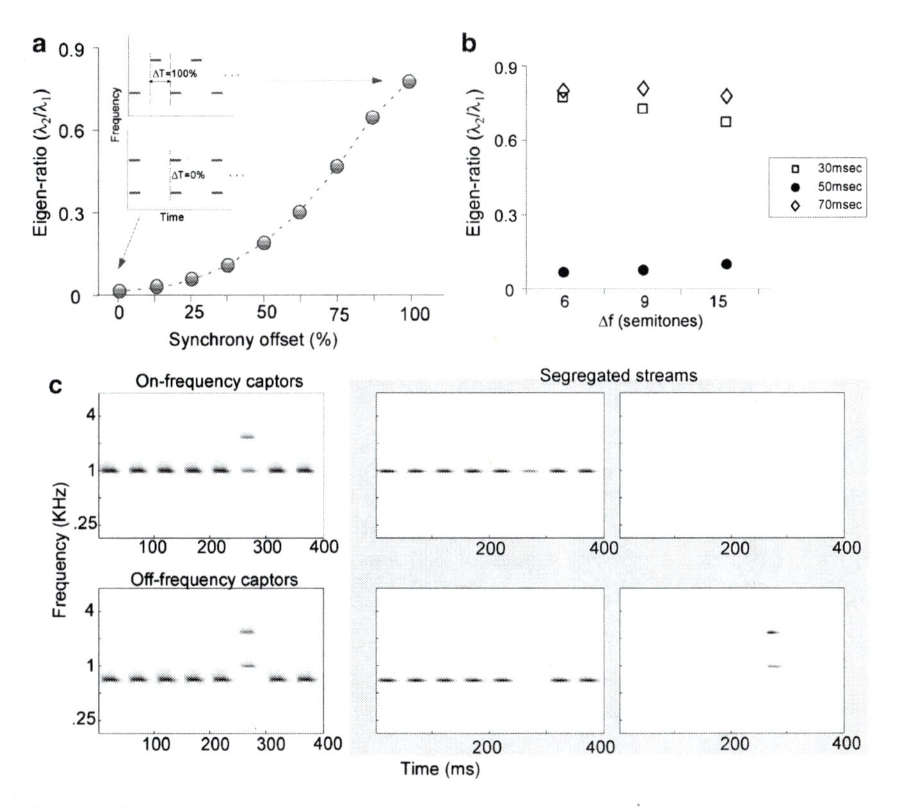

Fig. 46.3 Model simulation results

end of the continuum, the relative ratio of λ_2 to λ_1 reaches a high value indicating that both λ_1 and λ_2 are almost of equal value. In this case, the coherence matrix is in fact almost a rank 2 matrix, which can be mapped onto two main dimensions. In between these two extreme cases, we gradually vary the degree of synchrony between the two sequences. In this case, the relative ratio of λ_2 to λ_1 increases gradually; hence, allowing us to parametrically follow the influence of degree of asynchrony on grouping of two frequency streams, thereby allowing us to predict the transition between the percepts of one and two streams.

46.3.4.2 Experiment I: Synchrony Overrides Sequential Grouping

Next, we test the model using the same Experiment I paradigm used by Micheyl et al. in the accompanying paper (Micheyl et al. 2009). The stimuli consist of sequences of A and B, where A was fixed at 1,000 Hz and B was set at $\Delta F = 6$, 9, or 15 semitones above the A tone. Each tone was 100 ms in duration, including 10 ms raised-cosine onset and offset ramps. The silence interval between consecutive B tones was fixed at $\Delta T_B = 50$ ms. The silence between consecutive A tones

(ΔT_A) was varied across conditions. It was equal to 50 ms (in which case, the A and B precursors were synchronous), 30, or 70 ms (in which case, the A and B precursors were asynchronous).

In this simulation, we test how segregated are the two sequences under all variations of ΔF and ΔT_A. The psychoacoustic experiments show that the synchronous condition yields small thresholds even at the largest frequency separation (15 semitones). This low threshold is explained by the subjects' ability to make timing judgments within-stream, which is consistent with other results that synchronous tone sequences do indeed form a single perceptual stream even at large frequency separations exceeding 1 octave. In contrast, the asynchronous conditions where $\Delta T_A = 30$ or 70 ms yield larger thresholds, which in turn is consistent with an across-stream judgment (Bregman and Campbell 1971). The model simulations for these conditions are shown in Fig. 46.3b. The plot reveals that the synchronous condition (dark filled circles) does indeed yield a low eigen-ratio. This result is consistent with the perceptual finding that the 50 ms condition does indeed result from a percept of a single stream. It is worth noting that the eigen-ratio is low for all three frequency separation values of 6, 9, and 15 semitones. In contrast, the asynchronous conditions at $\Delta T_A = 30$ or 70 ms yield considerably higher eigen ratios, consistent with the perceptual findings.

46.3.4.3 Experiment II: Sequential Capture Overrides Synchrony Detection

In the next experiment, we explore the effect of sequential capture on synchrony judgments. This paradigm follows the design of Experiment II by Micheyl et al. (2009). The stimuli consisted of 3 tones, a single A tone at 1,000 Hz, a B tone at 6 or 15 semitones above A, and a C tone ("captor") at the same frequency as A ("On-frequency captor") or 6 semitones below A ("Off-frequency capture"). All tones were again 100 ms long with 10 ms raised-cosine onset and offset ramps. In the "On-frequency captor" condition, the A and B pair was surrounded by "captor" tones at the A frequency, with five captor tones before, and two captor tones after, the A–B pair. The captor tones were separated from each other, and from the target A tone, by a constant delay of 50 ms (Fig. 46.3c). In this condition, the target A tone formed part of a temporally regular sequence. In the "Off-frequency captor" condition, the frequency of the captor tones was set 6 semitones below the A tone, hence affecting the sequential grouping cues. The simulation results for this experiment are shown in Fig. 46.3c. Here, we show the results for frequency separation between A and B set at 15 semitones. In the first row, we show the simulation of the "On-frequency captors" condition. The leftmost panel shows the actual input analyzed by the model. After the coherence matrix is generated, we use the eigenvector structure to weigh the different frequency channels and group all coherent activity into one stream, and anticorrelated activity into a second stream (middle and rightmost panels). As shown in the figure, the sequential grouping cues override the presence of the synchronous A–B token, and groups the A tone with the preceding captor C tones. It is important to note that this result is only possible because of the cortical integration stage in the model, which gives the segregation of the streams inertia to look over relatively longer time scales; hence, allowing sequential cues to supersede the rules of synchrony. By this same inertia, a portion of the B tone is also "grabbed" along, though its energy is very weak because its presence was not long enough to drive responses from the cortical filters. The remaining energy of tone B is left in stream 2, though it does not show clearly in the spectrogram of the second stream. In contrast, the "off-frequency captors" condition results in a different organization (Fig. 46.3c, second row). In this case, the coherent activity from the C channel has no reason to group the A–B tones, hence segregating them into a separate stream.

46.4 Conclusions

Overall, the physiological data supports the proposal that our current thinking of streaming in the auditory system needs to incorporate the temporal axis as a key principle in organizing acoustic scenes. It is however important to emphasize that this principle does negate the rule of spatial segregation. If two alternating tones are too close together in frequency, their activation pattern will not be distinct enough to be able to see their anti-correlated temporal coherence. Hence, the overall principle is truly a *spatio-temporal* model of stream segregation, as tested directly by our computational model. An outstanding question remains as to the exact biological mechanisms involved in the process of "matrix decomposition" (i.e., *detection* of temporal coherence), and whether it is indeed a process that occurs at the level of the primary auditory cortex or beyond.

Acknowledgments We thank Pingbo Yin and Stephen David for their assistance with physiological recordings. More details concerning the methods and results of the physiological experiment may be found in Elhilali et al. (2009). This work is supported by grants from the National Institute on Deafness and Other Communication Disorders (R01 DC 07657) and the National Institute on Aging, through the Collaborative Research in Computational Neuroscience program (R01 AG 02757301).

References

Blake R, Lee SH (2005) The role of temporal structure in human vision. Behav Cogn Neurosci Rev 4:21–42

Bregman AS (1990) Auditory scene analysis. (Cambridge MIT Press), MA

Bregman AS, Campbell J (1971) Primary auditory stream segregation and perception of order in rapid sequences of tones. J Exp Psychol 89:244–249

Carlyon RP, Shamma S (2003) An account of monaural phase sensitivity. J Acoust Soc Am 114:333–348

Chi T, Gao Y, Guyton MC, Ru P, Shamma S (1999) Spectro-temporal modulation transfer functions and speech intelligibility. J Acoust Soc Am 106:2719–2732

Chi T, Ru P, Shamma SA (2005) Multiresolution spectrotemporal analysis of complex sounds. J Acoust Soc Am 118:887–906

Elhilali M, Chi T, Shamma SA (2003) A spectro-temporal modulation index (STMI) for assessment of speech intelligibility. Speech Commun 41:331–348

Elhilali M, Ma L, Micheyl C, Oxenham AJ, Shamma S (2009) Temporal coherence in the perceptual organization and cortical representation of auditory scenes. Neuron 61:317–329

Elhilali M, Shamma SA (2007) The correlative brain: a stream segregation model. In: Kollmeier B, Klump G, Hohmann V, Langemann U, Mauermann M, Uppenkamp S, Verhey J (eds) Hearing: from Sensory processing to perception. Springer, New York

Fishman YI, Arezzo JC, Steinschneider M (2004) Auditory stream segregation in monkey auditory cortex: effects of frequency separation, presentation rate, and tone duration. J Acoust Soc Am 116:1656–1670

Fishman YI, Reser DH, Arezzo JC, Steinschneider M (2001) Neural correlates of auditory stream segregation in primary auditory cortex of the awake monkey. Hear Res 151:167–187

Micheyl C, Shamma S, Elhilali M, Oxenham A (2010) Sequential and simultaneous auditory grouping measured with synchromy detection. In: E.A. Lopez-Poveda, A-R. Palmer, R. Meddis (eds) The neurophysiological baser of auditory perfection. Springer, New York

Micheyl C, Tian B, Carlyon RP, Rauschecker JP (2005) Perceptual organization of tone sequences in the auditory cortex of awake macaques. Neuron 48:139–148

Viemeister NF (1979) Temporal modulation transfer functions based upon modulation thresholds. J Acoust Soc Am 66:1364–1380

Wang K, Shamma SA (1995) Spectral shape analysis in the central auditory system. IEEE Trans Speech Audio Process 3:382–395

Chapter 47
Objective Measures of Auditory Scene Analysis

Robert P. Carlyon, Sarah K. Thompson, Antje Heinrich, Friedemann Pulvermuller, Matthew H. Davis, Yury Shtyrov, Rhodri Cusack, and Ingrid S. Johnsrude

Abstract We describe objective measures of two aspects of auditory scene analysis (ASA). We show that the build-up of auditory streaming can be measured using a behavioural task – detection of a temporal shift of one tone relative to another – that is easiest when the two tones are part of the same stream. This paradigm is used to show that performance can be improved by requiring participants to briefly divert their attention away from the tones, a manipulation that has previously been shown to increase the number of "one stream" judgements in a subjective task. A physiological measure of streaming build-up is also obtained using the mismatch negativity paradigm. In contrast to the strong effects of attention on streaming, we show, using fMRI measures of the brain response to veridical and illusory vowels, that the continuity illusion does not depend strongly on attention.

Keywords Auditory scene analysis • Auditory streaming • fMRI • Mismatch negativity • Continuity illusion • Speech

47.1 Introduction

In everyday life our ears are bombarded with a mixture of sounds arising from multiple sources. In order to extract the signal of interest – such as the voice of a given talker – from this mixture, our auditory systems must perform a number of tasks which, collectively have been termed "Auditory Scene Analysis (ASA)". These tasks include the segregation of temporally overlapping frequency components into two or more sources ("auditory grouping"), the sequential organisation of sounds into one or more streams ("auditory streaming"), and the perceptual completion of sounds that are briefly obliterated by other stimuli. Our ability to

R.P. Carlyon (✉)
MRC Cognition and Brain Sciences Unit,
15 Chaucer Rd, Cambridge, UK
e-mail: bob.carlyon@mrc-cbu.cam.ac.uk

E.A. Lopez-Poveda et al. (eds.), *The Neurophysiological Bases of Auditory Perception*,
DOI 10.1007/978-1-4419-5686-6_47, © Springer Science+Business Media, LLC 2010

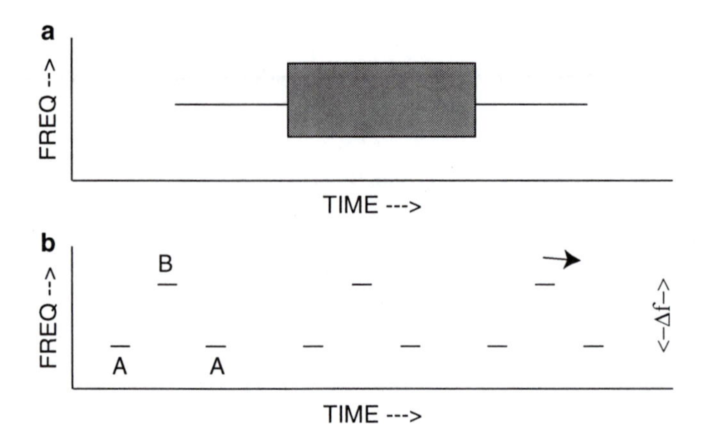

Fig. 47.1 Schematic spectrograms of (**a**) the continuity illusion, and (**b**) auditory streaming. The *arrow* in part (**b**) illustrates the temporal delay in the RHYTHM condition of experiment 1

perform this last task is convincingly demonstrated by the "continuity illusion", in which a sound is heard as continuous, even when a portion of it is deleted and replaced by a brief burst of noise (Vicario 1960; Houtgast 1972 – see Fig. 47.1a).

Although many decades of research have succeeded in elucidating the stimulus characteristics that are important for each of these tasks, both the neural bases of ASA and the involvement of higher-order cognitive processes, such as attention, have only recently become the focus of sustained experimental study. At this juncture, it is worth emphasising that the answers to both questions are likely to depend on the particular aspect of ASA under question. For example, as Carlyon et al. (2001) pointed out, attention may well have a smaller effect on the fusion of brief sounds based on common onset than on the ability to track one melody, over several seconds, in the presence of a competing tune. Here, we focus on two different aspects of ASA: auditory streaming and the continuity illusion.

47.2 Auditory Streaming

The sequential organisation of sounds ("auditory streaming") is commonly studied using stimuli similar to the one shown in Fig. 47.1b, in which "ABA" triplets of tones are repeated for several seconds. Listeners tend to hear a galloping rhythm, corresponding to a single-stream percept, when the repetition rate is slow and when the frequency difference (Δf) between the A and B tones is small. At faster rates and/or larger Δf, listeners hear two separate streams, each consisting of A or B tones. Intriguingly, listeners are also more likely to report hearing two streams at the end than at the beginning of a given sequence. Carlyon et al. (2001) argued that this build-up is modified by attention: if attention is initially diverted to a competing auditory task, then, when listeners start attending to the tones again, they report hearing a single stream, as if the sequence had just started. However, those measurements were obtained using purely subjective reports, and it has been argued that the effects of attention could have resulted from a response bias (Macken et al. 2003). Although we have presented multiple arguments against such an interpretation (Carlyon et al. 2001, 2003; Cusack et al. 2004), the more general issue remains as to whether attention can have an *obligatory* effect on streaming. To test this, we first developed a performance-based measure of the build-up of streaming, which we then used to test the effects of attention. We also obtained a physiological measure of the build-up of streaming using the mismatch negativity (MMN) paradigm.

47.2.1 Experiment 1: Objective Measures of the Build-Up of Streaming

47.2.1.1 Methods

Stimuli were sequences of repeating tones in an ABA-ABA-pattern (Fig. 47.1b) where A and B were 50-ms sine tones, separated by a 75-ms silent gap; a gap of 125 ms completed each triplet, giving a total ABA – triplet length of 500 ms. Each sequence contained 25 triplets and therefore had a total duration of 12.5 s. The frequency of the "B" tone was roved on a trial-by-trial basis within ±0.5 octaves of 800 Hz, while the "A" frequency was co-varied with the B frequency and was 4 or 8 semitones lower, depending on the condition. All stimuli were presented diotically via Sennheiser HD250 headphones at a level of 55-dB SPL. Eight subjects listened individually in a sound-insulated booth.

To obtain an objective measure of streaming build-up, it was necessary to use a task where performance differed over time, but where this variation could not be attributed to other factors such as listener fatigue. To do this, we developed two tasks, one which we predicted to be easiest when the sequence was heard as a single stream, and one that should be easiest when heard as two streams. In the RHYTHM task, listeners had to detect a delay on the "B" tone, leading to a change described as a "skipping", irregular rhythm compared to the standard (arrow in Fig. 47.1b). There were four levels of delay: 15, 20, 30 and 40 ms. Based on previous reports of poor between-stream timing judgements, we expected this task to be easiest when the sequence was heard as a single stream (Vliegen et al. 1999). The opposite prediction held for the PITCH task, where the signal was a 1, 2, 3 or 4% upward shift in the frequency of the "B" tone. In both tasks, the change to be detected lasted for four triplets. The change was introduced on *either* the 5th (EARLY) or 20th (LATE) triplet. The two tasks and the two frequency separations (4 and 8 semitones) were performed in separate blocks of trials.

Seven data points per subject were collected in each of the above experimental conditions, leading to a total of 56 trials in which a deviant was present. In addition, 14 trials in each block did not contain either an early or a late deviant. The trials were self-paced. There were four blocks in total, according to task and frequency separation, with the order counterbalanced across subjects. The instruction in all cases was to click on a "Yes" button with a mouse when they heard a specific change in the sequence, or to click "No" at the end of the sequence if they had not detected a change. For signal detection analysis, therefore, a hit was a positive response to a sequence that contained a deviant, either early or late. False alarms were "Yes" responses to sequences that did not contain a deviant: no distinction was made between Early and Late false alarms, and the false alarm rate was averaged for the purposes of calculating *d*-prime values within the block.

47.2.1.2 Results

The results for the RHYTHM task are shown by the bold solid lines in Fig. 47.2a. Scores were higher when the delayed tones occurred early rather than late in the sequence, consistent with the build-up of streaming impairing performance. This effect was confirmed by a two-way (position × Δf) ANOVA ($F = 12.232$; $p < 0.01$). There was also an apparent increment in performance for the 4 semitone condition compared to the 8 semitone condition, although this effect did not quite reach significance (solid vs. dashed lines: $F = 4.338$, $p = 0.07$).

In contrast to the effect of position in the RHYTHM task, performance in the PITCH task was better for targets presented *late* in the sequence ($F = 6.575$; $p = 0.037$; faint dashed lines in Fig. 47.2a). Although the absence of an effect of Δf in this task means that we cannot conclude that it indexes streaming, the opposite effect of position in the two tasks strongly suggests that the position effects observed for the RHYTHM task were not due to non-sensory factors such as the subject getting bored or tired towards the end of each sequence. The difference between the two tasks was underlined by a three-way (task × position × Δf) ANOVA, which revealed a significant interaction between task and position ($F = 17.957$; $p < 0.005$).

47.2.2 *Experiment 2: Effect of Attention on Streaming*

47.2.2.1 Rationale and Method

Experiment 2 was similar to the RHYTHM task of experiment 1, with the modification that the attentional state of the listener was manipulated. Each tone sequence had a duration of 13.5 s and was presented only to the left ear. For the first 10 s of each sequence, a series of 400-ms noise bursts was presented to the right ear, with the interval between the start of successive bursts drawn from a rectangular distribution spanning 0.9–1.1 s. Each noise burst could either increase or decrease linearly in amplitude throughout its duration, and was similar to either the "approach" or "depart" bursts described by Carlyon et al. (2001). The targets consisted of a single triplet in which the B tone was delayed by 50 ms, and could

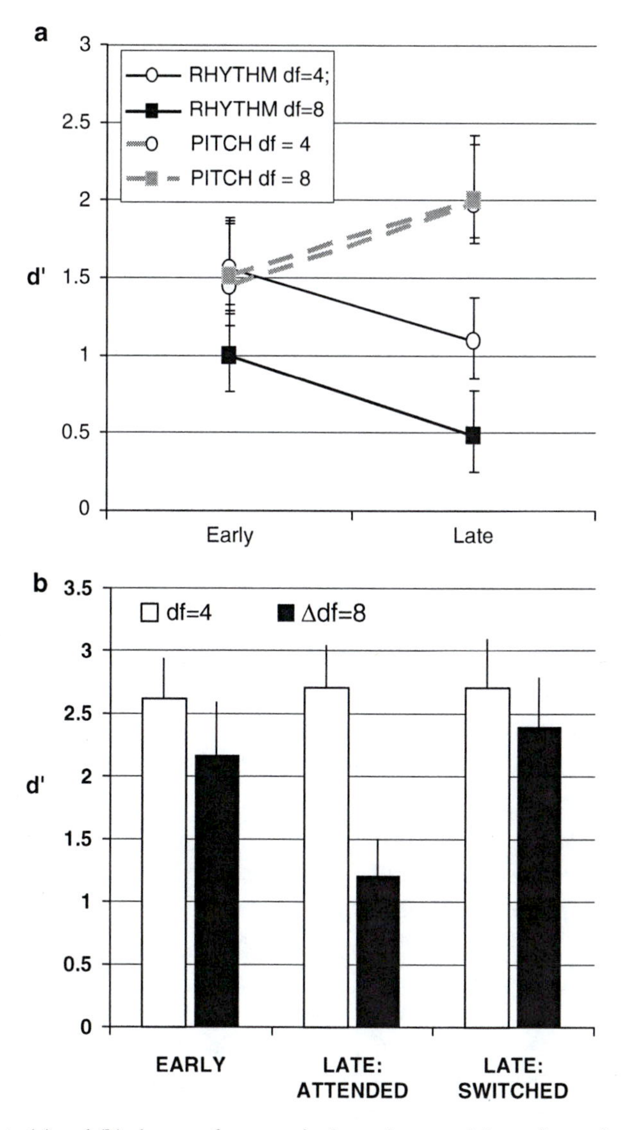

Fig. 47.2 Parts (**a**) and (**b**) show performance in the various conditions of experiments 1 and 2, respectively

start either 2.5 s ("early") or 12.5 s ("late") into the sequence. Delays could appear early or late in the sequence, with a 50% probability of appearing at either position, regardless of the task that the subject was doing. There were two blocks with 48 trials in each block A hit was defined as a "yes" response following presentation of a delayed tone; any response that occurred after 2.5 s and before 12.5 s was defined as an "early" response, while any later response was presumed to be a late response. If a response occurred in a window in which no deviant had been presented, it would be counted as a false alarm. Where sequences contained no deviant, any response that occurred before the midpoint of the sequence, 6.75 s, was counted as an early false alarm, and any that appeared later was a late false alarm.

The most important manipulation concerned the task instructions. In the "attend" condition, listeners ignored the noise bursts and paid attention to the tones throughout. In the "switch" condition, listeners initially categorised each noise burst as either "approach" or "depart" by using a mouse to press one of two virtual buttons on a computer screen. After 10 s the screen displayed a visual instruction to start attending to the tones. The prediction was that if attention affects streaming, that performance should be *better* in this switch condition than for the late targets in the attend condition, and that this effect should be greater at wider Δf, where the build up of streaming is greatest

47.2.2.2 Results

The results of experiment 2 are shown in Fig. 47.2b. It can be seen that, at $\Delta f=8$ semitones, detection of targets late in the sequence is significantly better in the SWITCH than in the ATTEND condition. The fact that this difference occurs only at $\Delta f=8$ semitones is consistent with the faster build-up of streaming at the wider frequency separation. A two-way ANOVA (attention $\times \Delta f$) performed on the detection of late targets confirmed this finding, showing a significant interaction ($F=18.93$, $p<0.005$). A t-test performed subsequently only at $\Delta f=8$ semitones revealed that the effect of attention was indeed significant ($t=-4.870$; $p=0.002$)

47.2.3 Experiment 3: Electrophysiological Measure of Streaming Build-Up

47.2.3.1 Rationale and Method

Experiment 3 used the MMN paradigm to determine whether an electrophysiological measure of streaming build-up could be obtained, without requiring subjects to make explicit judgements of streaming. The MMN is a negative wave measured in response to a rare deviant stimulus in a sequence of more common standards; in this study the negativity was measured relative to the response to stimuli identical to the deviants, but presented in a separate block without standards – the so-called "identity MMN". Each 10-s sequence consisted of 20 ABA triplets similar to those used in the RHYTHM task of experiment 1, with a deviant triplet (delay = 50 ms) presented at random both in an early position (5th, 6th, 7th, 8th, or 9th triplet) and in a late position (15th–19th triplet). The listener was not required to respond to these deviants but was instead instructed to attend to a rare "target", consisting of a single triplet attenuated by 12 dB, which occurred in an extra 10% of sequences. Sequences containing the target sounds were excluded from the analysis.

It is worth noting that, based on the results of experiment 1, we would expect the MMN to a delayed B tone to be largest when the sequence is heard as a single stream. This differs from the approach adopted in several previous electrophysiological studies of streaming (e.g. Sussman et al. 2007) and may be important for two reasons. First, it is known that tones separated by more than a few semitones are segregated at many levels of the auditory system, from cochlea to cortex. Hence, it may be more revealing to observe a cortical response that requires those tones to be "re-integrated" than to observe a measure that may simply reflect this early segregation. Second, because the MMN requires some minimum number of preceding standards, and because we predict that the MMN should be greatest *earlier* in the sequence, such a finding cannot be attributed to the number of preceding standards. As a further test for position effects unrelated to streaming, two values of Δf – 0 and eight semitones – were used. The prediction was that the MMN to the deviants should decrease over time only for the 8-semitone separation. There were three blocks of 112 sequences each, which contained sequences at both frequency separations. The first two sequences in each block were dummy trials included to orient the listener to the task and were not included in the analysis. The experiment ended with a block of 90 "identity" sequences (in which every triplet had a delayed tone), at each of the two values of Δf. Fifteen normal-hearing subjects took part in the experiment, but three were discarded from the final analysis, as the recordings from these subjects were very noisy ($SNR<2$).

The recording and analysis methods were broadly similar to those described by Carlyon et al. (2009). Briefly, subjects sat in a sound-isolated and electrically shielded room and listened diotically through Sennheiser HD 250 headphones. The EEG was recorded with Ag/AgCl electrodes mounted in an extended 10–20 system cap (EasyCaps, Falk Minow, Germany) with a 64-channel EEG set-up (Compumedics Neuroscan). Cz was used as the reference electrode for recordings, with electrode AFz as the ground. Horizontal and vertical eye movements were monitored using two bipolar electro-oculogram electrodes. Signals were amplified, sampled at a rate of 500 Hz, and bandpass filtered between 0.1 and 100 Hz. They were then processed off-line, starting with bandpass filtering (2–20 Hz, 24 dB/octave slopes). Event-related potentials were obtained by averaging epochs, which started at the beginning of each triplet and ended 500 ms later, at the time when the subsequent triplet began. Epochs containing voltage variations in excess of 100 μV at any EEG or EOG channel were discarded, as were epochs corresponding to a standard stimulus that just followed a deviant. Subsequently, the average across channels was subtracted from all traces (average-referencing). Finally, the data were re-referenced to a baseline of 50 ms at the beginning of each triplet.

47.2.3.2 Results

The results of experiment 3 are shown by the identity MMNs in Fig. 47.3 for electrode FCz; traces were similar over several fronto-central electrodes and the topography of the responses were typical of the MMN, including polarity inversion

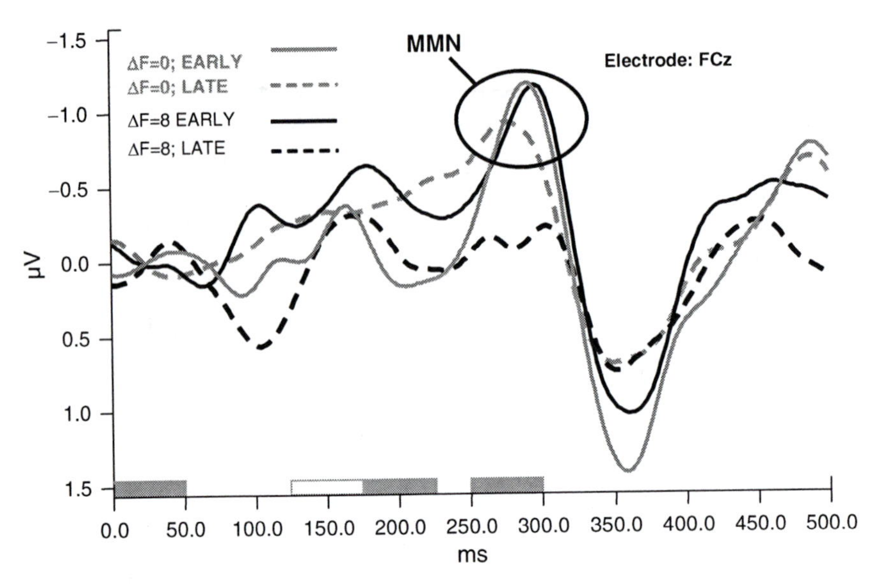

Fig. 47.3 Mismatch negativity (*circled*) in the four conditions of experiment 3. The *filled rectangles* near the abscissa show the timing of the tones in a deviant triplet

below the mastoids. It can be seen that, as predicted, there is an MMN both for the early and late deviants at $\Delta f = 0$, but only for the early deviants at $\Delta f = 8$ semitones. Statistical analysis of the mean amplitude of this negativity at this electrode (FCz), in a time window of 50 ms around the mean maximal negativity across all conditions (284 ms, or 109 ms following the onset of the deviant B tone) showed a significant interaction between frequency separation and position in sequence ($F = 9.716$; $p = 0.01$).

47.3 The Continuity Illusion

47.3.1 A Correlate of the Continuity Illusion Obtained Using fMRI

In a recent study (Heinrich et al. 2008), we obtained a neural correlate of the continuity illusion in a region of the brain – bilateral middle temporal gyrus (MTG) – that is preferentially activated by speech stimuli. To do so, we exploited the behavioural finding that, although listeners find it hard to identify two-formant vowels in which the formants alternate over time (Fig. 47.4c), performance can be improved by filling the silent gaps in each frequency region with bursts of noise (Fig. 47.4d: Carlyon et al. 2002). We have argued that the improvement arises from the fact that the noise bursts cause each formant to be heard as continuous and that, therefore, the two formants are heard as simultaneous. When the formant level is

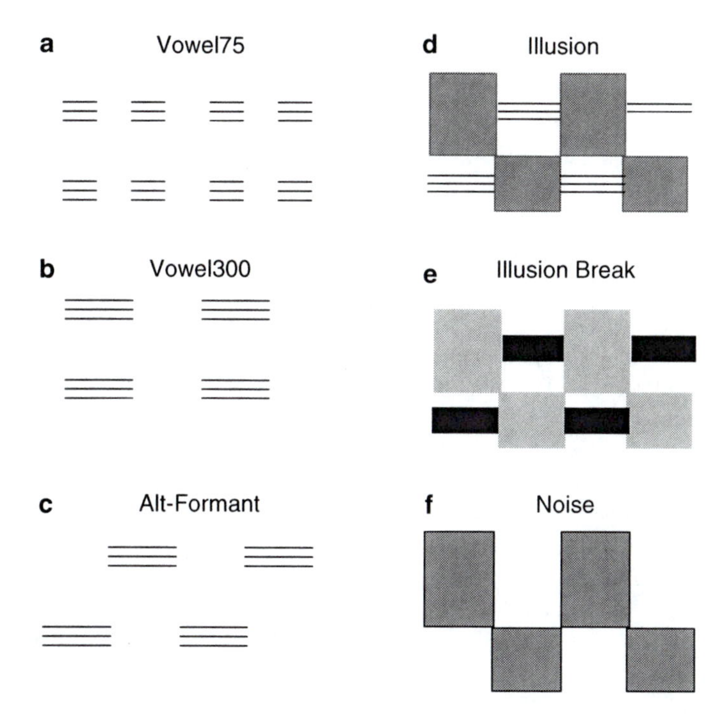

Fig. 47.4 Schematic spectrograms of stimuli used by Heinrich et al. and in experiment 4

increased, and the noise level decreased, the continuity illusion for single formants breaks down, and, in this "illusion break" condition, the alternating-formant stimuli no longer sound speech-like (Fig. 47.4e: Heinrich et al. 2008). To obtain a neural correlate of the illusion, we first identified regions of the brain that showed greater activation for intact vowels than for non-speech sounds (noise bursts and alternating formants; Fig. 47.4c and f, respectively). Two such regions, in left and right posterior MTG respectively, showed a significant difference when corrected at the whole-brain level; these regions are close to "speech sensitive" regions identified by other researchers (e.g. Uppenkamp et al. 2006). Importantly, activation in these regions was also significantly greater for the "illusion" (Fig. 47.4d) than for the "illusion break" (Fig. 47.4e) stimuli. This is striking since speech formants are presented at higher amplitude in the illusion break than in the illusion condition, and yet activation is reduced. Furthermore, this difference was not due to some stimuli being more effective overall at eliciting neural activity: the opposite pattern of results was obtained in a region of primary auditory cortex (slightly lateral and posterior to Heschl's gyrus), which we had identified as being preferentially activated by sounds with more onsets (i.e by the "Vowel 75" compared to the "Vowel 300" stimuli shown in Fig. 47.4a and b). We therefore concluded that activation in MTG can reflect the operation of the continuity illusion. In contrast, activity in primary auditory regions may reflect detection of the onset of sounds; these onsets are absent when formants are perceived as continuing behind bursts of noise.

47.3.2 *Experiment 4: Effect of Attention on the Continuity Illusion*

47.3.2.1 Rationale and Method

The identification of neural responses to illusory vowels paved the way for an investigation of whether the continuity illusion depends on attention. To do so, we presented similar illusory vowel and non-vowel stimuli in conditions that differed only in the instructions given to the participants. The methods of stimulus generation, presentation and image acquisition/analysis were similar to those described by Heinrich et al. (2008). The main hypothesis under test was whether activation in a vowel-sensitive region was more dependent on attention for responses to illusory than to veridical vowel stimuli.

Three sets of stimuli were presented concurrently in each scanning session. One set consisted of vowel stimuli similar to those used by Heinrich et al. in the "Vowel 300", "Vowel 75", "Illusion" and "Break" conditions. Each of these sounds consisted of two synthetic formants, which together formed one of the four vowels /ɑ: /, / ɛ: /, / ɔ: / and / ɜ: /. Each 8.4-s sequence consisted of a mixture of these vowels, which could also differ in F0, and the sounds were presented with a variable ISI. They were low-pass filtered at 4,000 Hz, and mixed with a sequence of twelve 400-ms noises (ISI mean = 300 ms, SD 49 ms), band-pass filtered between 4,000 and 5,000 Hz, and with "approach" envelopes (Sect. 47.2.2). At the same time, participants watched a visual sequence of ellipses, each filled by diagonal stripes, and whose orientation alternated between 0° and 90° every 150 ms (cf. Carlyon et al. 2003).

Before each trial, participants were instructed to attend to one of the three stimulus types, which could contain, on 12.5% of sequences, a target that the participant had to identify. In the vowel, noise and visual attention conditions, these targets were, respectively: two consecutive tokens attenuated by 9 dB; one or two noises with "departing" envelopes; one or two ellipses in which the diagonal lines were broken up (cf. Carlyon et al. 2003). It is important to note that sequences containing targets were not included in the analyses.

Auditory stimuli were presented via Nordic Neuro Labs electrostatic headphones and visual stimuli were projected onto a screen that could be viewed by participants using an angled mirror placed inside the head coil. Imaging data were acquired by a Magnetom Trio 3-Tesla MRI system with head gradient coil, using a sparse imaging technique. 324 echo-planar-imaging volumes were acquired in six 10-min scanning runs for each of the 22 participants. Acquisition was transverse oblique, angled to avoid the eyeballs and to cover the whole brain; where this was not possible the very top of the parietal lobe was omitted.

47.3.2.2 Results

Although we observed activation in many brain areas that differed significantly as a function of either sound stimulus and/or attention, here we discuss only that activation corresponding to the "vowel-sensitive" and "onset" areas described by

Heinrich et al. (2008). A whole-brain-corrected contrast between the mean of the two vowel conditions and the break condition revealed peak activations in voxels in the left MTG (co-ordinates: −60 −32 2; p(FDR) <0.05) and right superior temporal sulcus ("STS": 54 −22 −6; p(FDR) <0.05) which were within 8 and 17 mm of those observed in our previous study, respectively. Furthermore, a comparison of the illusion minus the break conditions revealed peaks in the left posterior superior temporal gyrus ("STG": −64 −36 14; p(FDR) <0.05) and right STG (52 −24 8; p(FDR) <0.05) that were within 11 and 13 mm of the peak "vowel sensitive" voxels described by Heinrich et al. (2008). Finally, a comparison of activation in the vowel 75 minus vowel 300 conditions revealed a peak in each of the left (−58 −18 6; p(FDR) <0.001) and right (48 −16 8; p(FDR) <0.001) hemispheres that were within 8 and 12 mm, respectively, of the peak onset responsive region in the previous study.

The issue of greatest interest is whether the activation in the vowel-sensitive areas is modified by attention, and crucially whether any such effect is greater for the illusory than for the intact vowels. The top and bottom parts of Fig. 47.5 show activation for each of the four stimulus conditions as a function of attention, in the peak "vowel-sensitive" voxels in the left and right hemispheres, respectively. It is clear that although activation is strongest when attention is paid to the vowels, and weakest when participants attend to the visual stimuli, these effects are similar for all four stimuli. A repeated-measures ANOVA with three attention and four sound conditions on the two peak voxels revealed main effects of attention (LH: $F(2, 42)=13.96$, $p<0.001$, RH: $F(2, 42)=17.47$, $p<0.001$) and sound (LH: $F(3, 63)=11.35$, $p<0.001$, RH: $F(3, 63)=8.20$, $p<0.001$), but no interaction ($F(6, 126)=1$) in either case.

47.4 Summary

The experiments described here have used objective measures to probe the effects of attention on two aspects of ASA, with very different results. First, we showed that attention can have obligatory effects on streaming build-up. As we have pointed out previously (Cusack et al. 2004), a crucial remaining question is whether streaming fails to build up in the absence of attention, or, alternatively, builds up but is "reset" as soon as attention is re-directed to a set of sounds. This question is under active consideration in our laboratory. Second, we showed that, in contrast, the response of vowel-sensitive brain regions to illusory vowels is not more dependent on attention than that to physically intact vowels. Hence, unlike streaming build-up, the continuity illusion does not seem to be strongly dependent on attention. The present study shows that, although attention can affect activation in speech-sensitive areas, the increased activation for illusion over break stimuli in these areas does not depend on attention. This extends previous electrophysiological work (Micheyl et al. 2003), showing that the illusion is at least partly complete even when participants are instructed to attend to a silent movie, to a more rigorous experimental paradigm in which attentional state is explicitly controlled.

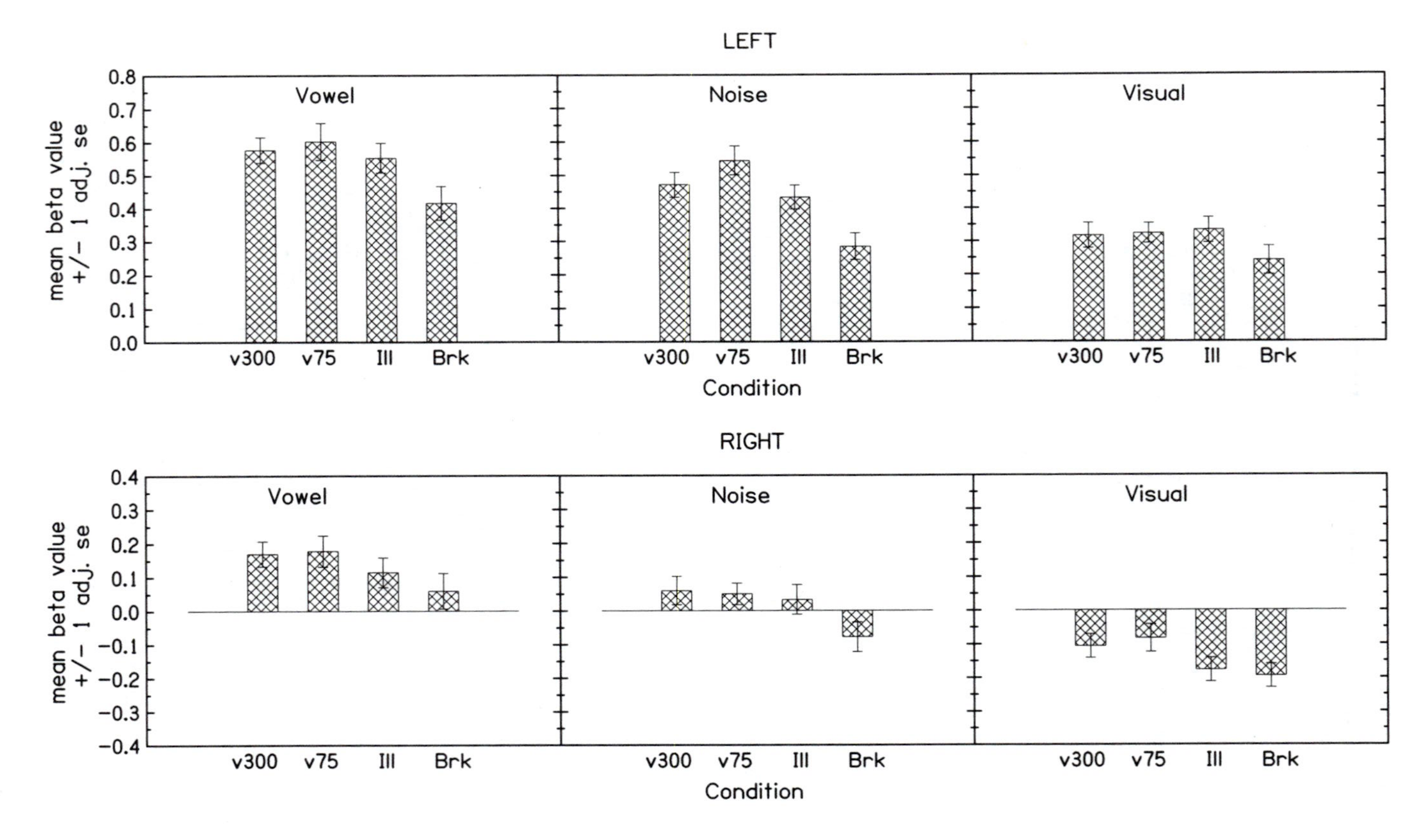

Fig. 47.5 *Top* and *bottom* parts show mean BOLD signals, *re* a silent baseline, in the peak vowel-sensitive voxels in the left and right hemispheres, respectively

47.5 Comment by Daniel Oberfeld-Twistel

A brief sort of historical note: The first use of a time discrimination task as a diagnostic for perceptual organisation appears to be due to Jones et al. (1995).

References

Carlyon RP, Cusack R, Foxton JM, Robertson IH (2001) Effects of attention and unilateral neglect on auditory stream segregation. J Exp Psychol Hum Percept Perform 27:115–127

Carlyon RP, Deeks JM, Norris D, Butterfield S (2002) The Continuity illusion and vowel identification. Acta Acustica united with Acustica 88:408–415

Carlyon RP, Deeks JM, Shtyrov Y, Grahn J, Gockel HE, Hauk O, et al. (2009) Changes in the perceived duration of a narrowband sound induced by a preceding stimulus. J. Exp. Psychol Hum Percept Perform 35:1898–1912

Carlyon RP, Plack CJ, Fantini DA, Cusack R (2003) Cross-modal and non-sensory influences on auditory streaming. Perception 32:1393–1402

Cusack R, Deeks J, Aikman G, Carlyon RP (2004) Effects of location, frequency region, and time course of selective attention on auditory scene analysis. J Exp Psychol Hum Percept Perform 30:643–656

Heinrich A, Carlyon RP, Davis MH, Johnsrude IS (2008) Illusory vowels resulting from perceptual continuity. A functional magnetic resonance imaging study. J Cogn Neurosci 20:1–16

Houtgast T (1972) Psychophysical evidence for lateral inhibition in hearing. J Acoust Soc Am 51:1885–1894

Jones MR, Jagacinski RJ, Yee W, Floyd RL, Klapp ST (1995) Tests of attentional flexibility in listening to polyrhythmic patterns. J Exp Psychol Hum Percept Perform 21(2):293–307

Macken WJ, Tremblay S, Houghton RJ, Nicholls AP, Jones DM (2003) Does auditory streaming require attention? evidence from attentional selectivity in short-term memory. J Exp Psychol Hum Percept Perform 29:43–51

Micheyl C, Tian B, Carlyon RP, Rauschecker J (2003) The neural basis of stream segregation in the primary auditory cortex. Assoc Res Otolaryngol Abs 26: 195***

Sussman ES, Horvath J, Winkler I, Orr M (2007) The role of attention in the formation of auditory streams. Percept Psychophys 69:136–152

Uppenkamp S, Johnsrude IS, Norris D, Marslen-Wilson W, Patterson RD (2006) Locating the initial stages of speech-sound processing in human temporal cortex. Neuroimage 31:1284–1296

Vicario G (1960) L'effetto tunnel acustico. Riv Psicol 54:41–52

Vliegen J, Moore BCJ, Oxenham AJ (1999) The role of spectral and periodicity cues in auditory stream segregation, measured using a temporal discrimination task. J Acoust Soc Am 106:938–945

Chapter 48
Perception of Concurrent Sentences with Harmonic or Frequency-Shifted Voiced Excitation: Performance of Human Listeners and of Computational Models Based on Autocorrelation

Brian Roberts, Stephen D. Holmes, Christopher J. Darwin, and Guy J. Brown

Abstract Keyword identification in one of two simultaneous sentences is improved when the sentences differ in F0, particularly when they are almost continuously voiced. Sentences of this kind were recorded, monotonised using PSOLA, and re-synthesised to give a range of harmonic ΔF0s (0, 1, 3, and 10 semitones). They were additionally re-synthesised by LPC with the LPC residual frequency shifted by 25% of F0, to give excitation with inharmonic but regularly spaced components. Perceptual identification of frequency-shifted sentences showed a similar large improvement with nominal ΔF0 as seen for harmonic sentences, although overall performance was about 10% poorer. We compared performance with that of two autocorrelation-based computational models comprising four stages: (i) peripheral frequency selectivity and half-wave rectification; (ii) within-channel periodicity extraction; (iii) identification of the two major peaks in the summary autocorrelation function (SACF); (iv) a template-based approach to speech recognition using dynamic time warping. One model sampled the correlogram at the target-F0 period and performed spectral matching; the other deselected channels dominated by the interferer and performed matching on the short-lag portion of the residual SACF. Both models reproduced the monotonic increase observed in human performance with increasing ΔF0 for the harmonic stimuli, but not for the frequency-shifted stimuli. A revised version of the spectral-matching model, which groups patterns of periodicity that lie on a curve in the frequency-delay plane, showed a closer match to the perceptual data for frequency-shifted sentences. The results extend the range of phenomena originally attributed to harmonic processing to grouping by common spectral pattern.

B. Roberts (✉)
Psychology, School of Life and Health Sciences,
Aston University, Birmingham B4 7ET, UK
e-mail: b.roberts@aston.ac.uk

E.A. Lopez-Poveda et al. (eds.), *The Neurophysiological Bases of Auditory Perception*,
DOI 10.1007/978-1-4419-5686-6_48,

Keywords Simultaneous grouping • Speech perception • Harmonic relations • Autocorrelation models

48.1 Introduction

A mixture of two harmonic stimuli on different fundamental frequencies (F_0s) is typically heard as two separate entities, each with its own pitch and timbre. The perceptual segregation of such a mixture has long been modelled by the concept of a harmonic sieve or template (e.g. Duifhuis et al. 1982). However, Roberts and his colleagues have proposed that the perceptual segregation of sources based on deviations from a common F_0 is a special case of a more general sensitivity to deviations from a common pattern of spectral spacing (see Roberts 2005, for a review). Roberts and Bregman (1991) provided the first demonstration of grouping by spectral regularity; they added a single even harmonic to a complex tone comprising only odd harmonics and found that the even harmonic was typically more salient than its odd neighbours. Subsequent studies have shown that a single mistuned component tends to segregate from a complex tone irrespective of whether the complex is harmonic or has been frequency shifted from harmonic relations by adding a constant increment to the frequency of each component (Roberts and Brunstrom 1998, 2001).

A limitation of the studies of grouping by spectral regularity reported so far is that they have focussed exclusively on the perceptual segregation of individual components from complex tones. Studies of grouping by harmonic relations, however, have explored the ability of the auditory system to use deviations from a common F_0 under a wide variety of conditions. These conditions include concurrent complex tones with distinct but overlapping spectra and dynamic stimuli such as mixtures of speech sounds. Indeed, the idea of harmonic selection as a means of separating target speech from interfering speech has a long history (e.g. Parsons 1976). Carlyon and Gockel (2008) recently posed the question of whether performance in tasks requiring the segregation of concurrent complex tones is specific to grouping by harmonic relations, or can also be explained in terms of sensitivity to deviations from a more general form of spectral regularity. Roberts (2005) suggested an approach to answering this question, which involved testing whether mixtures of frequency-shifted stimuli can be heard with their own, separate, timbres and pitches. Synthetic speech sounds offer a convenient stimulus for measuring the extent to which the timbres of separate sources can be retrieved perceptually from a mixture.

It is well established for harmonic stimuli that perceptual identification of keywords in one of two temporally overlapping sentences is substantially improved when the sentences differ in F_0 (Brokx and Nooteboom 1982). The effect of small ΔF_0s results primarily from cues arising from low-frequency beating (Culling and Darwin 1994), but these cues cannot account for the additional improvement observed for larger ΔF_0s. Indeed, when pairs of sentences are almost continuously voiced, target identification can improve from 20 to 80% when ΔF_0 increases from 0 to 8 semitones (Bird and

Darwin 1998). The enhanced effect of ΔF_0 for these stimuli arises from the absence of unvoiced segments and from the almost complete absence of abrupt onsets and offsets associated with closures in stops and affricates, which may themselves act as grouping cues. The experiment reported here extends the study of Bird and Darwin (1998) by measuring the effect of ΔF_0 on the perceptual identification of pairs of frequency-shifted sentences as well as pairs of harmonic sentences. The perceptual data for both the harmonic and shifted conditions has undergone computational modelling to assess the extent to which traditional approaches to the separation of target and interfering speech, based on autocorrelation, can account for human performance.

48.2 Experiment

The same 80 sentences as used by Bird and Darwin (1998) were recorded, spoken by Darwin on a flat pitch contour, and processed using Praat (Boersma and Weenink 1996). After low-pass filtering at 5 kHz, the harmonic stimuli were created by re-synthesising all sentences in a monotonised form on six different F_0s using the Pitch Synchronous Overlap and Add method (Moulines and Charpentier 1990). The frequency-shifted stimuli were created by additional re-synthesis using LPC with the LPC residual frequency shifted by 25% of F_0. This preserved the filter characteristics of the vocal tract but changed the excitation source from harmonic to one that was inharmonic but with regularly spaced components. Frequency shifting a harmonic source by 25% of F_0 effectively produces the greatest deviation from harmonic relations (a shift of 50% of F_0 would transform an all-harmonic stimulus into an odd-harmonic stimulus). When heard alone, the shifted sentences had an unusual timbre, but evoked a clear pitch and were similarly intelligible to their harmonic counterparts.

The 80 sentences were ranked in duration (range = 1.25–2.93 s) and then divided in half, giving 40 targets (shorter) and 40 interferers (longer). Two of the six sets of targets were used, corresponding to F_0s of 90 and 160 Hz (ΔF_0 = 10 semitones); all six interferer F_0s were used to give a range of ΔF_0s when targets and interferers were combined. For a target F_0 of 90 Hz, interferer F_0s were set to 0, 1, 3, or 10 semitones above (corresponding to 90, 95, 107, and 160 Hz, respectively). For a target F_0 of 160 Hz, interferer F_0s were set to 0, 1, 3, or 10 semitones below (corresponding to 160, 151, 135, and 90 Hz, respectively). Both directions of F_0 separation were tested to differentiate between the effects of ΔF_0 and those of the interferer's absolute F_0. For both targets and interferers, the nominal F_0s for the frequency-shifted stimuli were the same as for their harmonic counterparts. A fixed onset asynchrony of 400 ms was used when a target and interferer were combined, always with the target delayed. Pairings were pre-selected to ensure that the target always ended before the interferer.

The experiment used a mixed design, with one between-subjects factor (target F_0, 2 levels) and two within-subjects factors (ΔF_0, 4 levels; harmonicity, 2 levels). 16 listeners took part; 8 per target F_0. Harmonic targets were always paired with harmonic

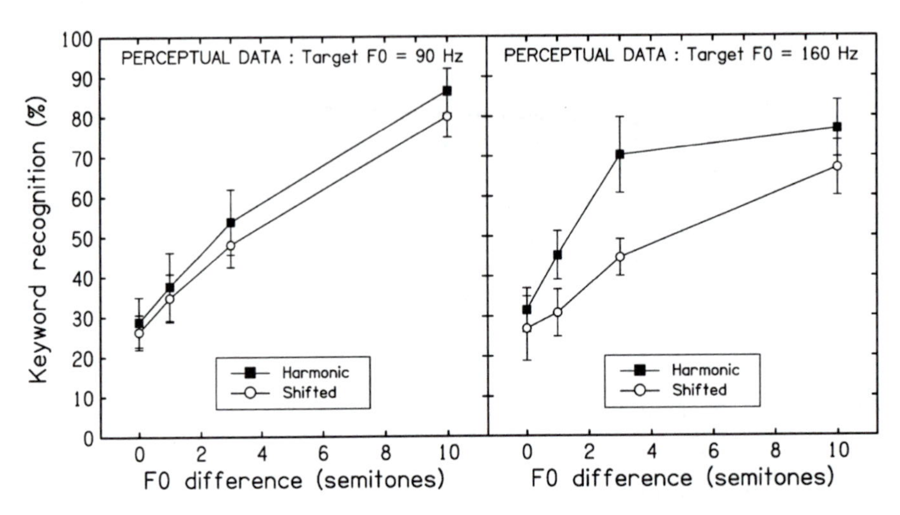

Fig. 48.1 Keyword recognition performance for human listeners. Adapted with permission from "Roberts B, Holmes SD, Darwin CJ, Perception of concurrent sentences: within and across-formant grouping in harmonic and frequency-shifted speech. J Acoust Soc Am (submitted)"

interferers and shifted targets were always paired with shifted interferers. Every listener heard all combinations of ΔF_0 and harmonicity (4×2=8) for the target F_0 to which they were allocated. For each listener, eight sets of five sentence pairs were rotated across all combinations of harmonicity and ΔF_0 in a balanced design, giving 40 trials in total. The order of each repetition of the eight combinations was randomised anew. Each listener only heard any particular sentence on one trial during the experiment. All individual sentences were normalised beforehand to give a presentation level of 70 dB SPL, prior to mixing to form sentence pairs. Listeners were seated in a sound-attenuating booth and the stimuli were presented diotically (sampling frequency = 11 kHz, 16-bit resolution) over Sennheiser HD 480-13-II earphones.

On each trial, listeners heard the selected sentence pair twice. During the first presentation, they were instructed to focus on the longer sentence (the interferer) and to type in any words that they could hear. During the second presentation, they were instructed to focus on and transcribe the shorter sentence (the target). Listeners did not receive any further repetition of the stimuli or any feedback on their responses. They were not informed that the constituent sentences of a pair were designated as target and interferer and that it was their responses to the target that were of primary interest. The complete set of trials in the main experiment took about an hour. Beforehand, listeners received training using a small number of representative pairs of sentences that were not used in the main experiment.

The intelligibility of the target sentences in the presence of interferers was quantified in terms of percentage correct keywords in the response to the second presentation of a sentence pair; tight scoring was used. The results are displayed in Fig. 48.1. Analysis of variance showed significant main effects of ΔF_0 [$F(3,42)=49.855$, $p<0.001$] and of harmonicity [$F(1,14)=20.385$, $p<0.001$]; there was also a significant interaction between target F_0 and harmonicity [$F(1,14)=5.378$, $p=0.036$]. The effect of ΔF_0 is consistent with that observed before for sentence pairs (Brokx and Nooteboom 1982), especially when mainly voiced (Bird and Darwin 1998). Perceptual identification improved progressively as ΔF_0 increased up to 10 semitones. Of particular note is the finding that performance for frequency-shifted sentences showed a similar large improvement with increasing nominal ΔF_0 as observed for harmonic sentences, although overall performance was about 9% worse. As far as we are aware, this is the first demonstration of concurrent segregation of multi-component complexes based on spectral pattern rather than on harmonic relations. It is also worth noting that, just as for harmonic stimuli, introducing a ΔF_0 for a pair of frequency-shifted sentences creates a clear impression of one sentence spoken on a lower pitch and one on a higher pitch.

The effect of target F_0 emerges only in its interaction with harmonicity. Performance for harmonic and shifted stimuli becomes more divergent at intermediate ΔF_0s when target F_0 is increased from 90 to 160 Hz; this difference largely disappears when $\Delta F_0 = 10$ semitones, where the mean improvement in performance for the shifted stimuli relative to the $\Delta F_0=0$ case was about 40%. One might have predicted that increasing target F_0 from 90 to 160 Hz would improve performance in the harmonic condition for smaller ΔF_0s, because a higher F_0 lowers the harmonic number of components in the region of the first and second formants, increasing their resolvability. However, this might actually impair performance in the shifted condition, because the lower-numbered components show a greater deviation from harmonic relations. Whatever the cause of this interaction, the most striking finding overall is the broad similarity in perceptual identification for harmonic stimuli and their shifted counterparts.

48.3 Computational Models

The performance of human listeners was compared with that of two autocorrelation-based computational models, using a template-based approach to speech recognition. A simple model of the auditory periphery was used, consisting of a bank of 30 fourth-order linear gammatone filters with centre frequencies equally spaced in ERB-rate between 50 and 4,500 Hz. The output of each filter was half-wave rectified to give a crude simulation of auditory nerve firing rate. The correlogram was computed from the simulated response according to:

$$\mathrm{ACF}(t, f, \tau) = \left[\sum_{k=1}^{M} x[t-k, f]\, x[t-k-\tau, f]\, h[k] \right]^{1/2} \tag{48.1}$$

where t represents time, f is frequency channel and M is the duration (in samples) of a Hann integration window $h[k]$. M was set to 880 samples, corresponding to a time frame of 20 ms duration for the up-sampled stimuli (44 kHz). The autocorrelation lag τ was varied between 0 and 20 ms, so that the lowest F_0 detectable was 50 Hz. A square root was applied in (48.1) so that the autocorrelation function had the same dynamic range as the original signal. Pitch analysis was based on the summary autocorrelation function (SACF), given by:

$$\mathrm{SACF}(t,\tau) = \sum_{f=1}^{30} \mathrm{ACF}(t,f,\tau) \tag{48.2}$$

For a single source, the largest peak in the SACF is a reliable indicator of the pitch period. For a mixture of two sources, one or more major peaks will occur in the SACF depending on the difference between the F_0s.

A simple template-based approach was used for speech recognition, which uses dynamic time warping (DTW) to align an observed time series of acoustic features with a stored template (overlapping 20-ms frames, sampled every 10 ms). A template was computed for each entire sentence, and so matching was performed on sentences rather than on individual keywords. DTW computes the cost of time-aligning each template with the observed acoustic features; the template that has the minimal cost is then chosen. We used a Matlab implementation of DTW (Ellis 2003), in which the distance between two feature vectors was computed as their normalised dot product. Two versions of the recogniser were used, for which the templates were based on spectral or temporal features. For the recogniser based on spectral features, the templates were based on an energy measure corresponding to the value of the correlogram according to (48.1) at zero delay (τ=0). The template for each target sentence was obtained by computing for each time frame the energy in each channel for the harmonic and shifted versions on each of the six available F_0s, and taking the average over all 12 stimuli.

The second recogniser was based on temporal information; it was a multi-frame version of the "timbre region" approach described by Meddis and Hewitt (1992). For each time frame, the short-lag part of the SACF was used as the acoustic feature vector. Meddis and Hewitt refer to this as the timbre region of the SACF, because it encodes information about spectral shape rather than pitch. More specifically, the templates were given by:

$$\mathrm{timbre}(t,\tau) = \mathrm{SACF}(t,\tau) \text{ for } 0 \le \tau \le 200 \tag{48.3}$$

The longest lag used (200 samples) corresponds approximately to 4.5 ms and a lower limit in frequency of 220 Hz, which is above the highest F_0 used here (160 Hz). The template for each target sentence was formed by averaging the timbre regions over all 12 versions. To avoid any influence of overall level on the pattern matching, the templates were also normalised, so that the timbre region for each time frame had a zero mean and unity variance.

48.4 Modelling Studies: Results

The model based on spectral pattern matching is an extension of the scheme proposed by Assmann and Summerfield (1990) for double-vowel separation, but differs in that it exploits the 400-ms delay of the onset of the target sentence by comparing the correlograms in the interferer-only and target-plus-interferer regions of the signal. It is reasonable to suppose that listeners also exploit this difference in onset time between the target and interferer.

First, the SACFs for time frames 10–30 of the signal (i.e., 100–300 ms) were computed and averaged to obtain a robust estimate of the interferer F_0. The same was then done for frames 50–70 (500–700 ms), which were occupied by the target and interferer. The two average SACFs were then normalised by the value of the largest peak that occurs after a lag of 100 samples (the short-lag part was excluded because the peak at $\tau=0$ will always be largest). The normalised average SACF for the interferer region was then subtracted from the normalised average SACF for the target-plus-interferer region, giving a difference function $d(\tau)$. If the target and interferer sentences differ in F_0, then the largest peak in $d(\tau)$ will occur at the period of the target. If the target and interferer have the same F_0, then the two SACFs will largely cancel out, and $d(\tau)$ will have a small value. Accordingly, an amplitude threshold ϕ was used to determine whether a second F_0 was present. The value of ϕ was set to 0.3 by inspection. If a target F_0 was not detected, then pattern matching proceeded using the autocorrelation values at zero lag, which correspond to the energy measure in (48.1). If a target F_0 was detected, then for each frame a pseudo-spectrum was formed for the target sentence by sampling the correlogram at the lag corresponding to the F_0 period of the target. The pseudo-spectrum is given by (48.1) when τ equals the F_0 period of the target.

Results for the spectral model are shown in the upper panels of Fig. 48.2. Performance for the harmonic condition is qualitatively similar to the listener data, in that: (a) the model shows monotonically increasing recognition with increasing F_0 difference; (b) performance in the 160-Hz condition is below that for the 90-Hz condition; (c) performance in the 160-Hz condition flattens off between ΔF_0s of 3 and 10 semitones; (d) performance for the harmonic stimuli is, in most cases, above that for the shifted stimuli. However, for shifted stimuli, the model failed to reproduce at either target F_0 the observed monotonic increase in listener performance with increasing ΔF_0. Furthermore, in contrast with the perceptual data, the overall performance of the model was much worse for shifted than for harmonic stimuli. Indeed, model performance for shifted stimuli in the 160-Hz condition fell almost back to baseline ($\Delta F_0=0$) when ΔF_0 was increased to 10 semitones.

The temporal approach was based on channel selection, rather than sampling the correlogram at a particular lag, because the former has proven to be more successful in modelling the increase in double-vowel identification performance with increasing ΔF_0 (Meddis and Hewitt 1992). We used an extension of their double-vowel model that performed pattern matching based on timbre region templates. Again, an average SACF was computed for the interferer and target-plus-interferer regions of the signal, and a similar threshold-based approach was used to determine whether a second (target) F_0 was present. If the target F_0 was not detected, then an SACF was formed from all channels of the correlogram, and the short-lag (up to 200 samples) portion was taken as the timbre region for each frame. The sequence of timbre regions was then matched against the template timbre regions using DTW.

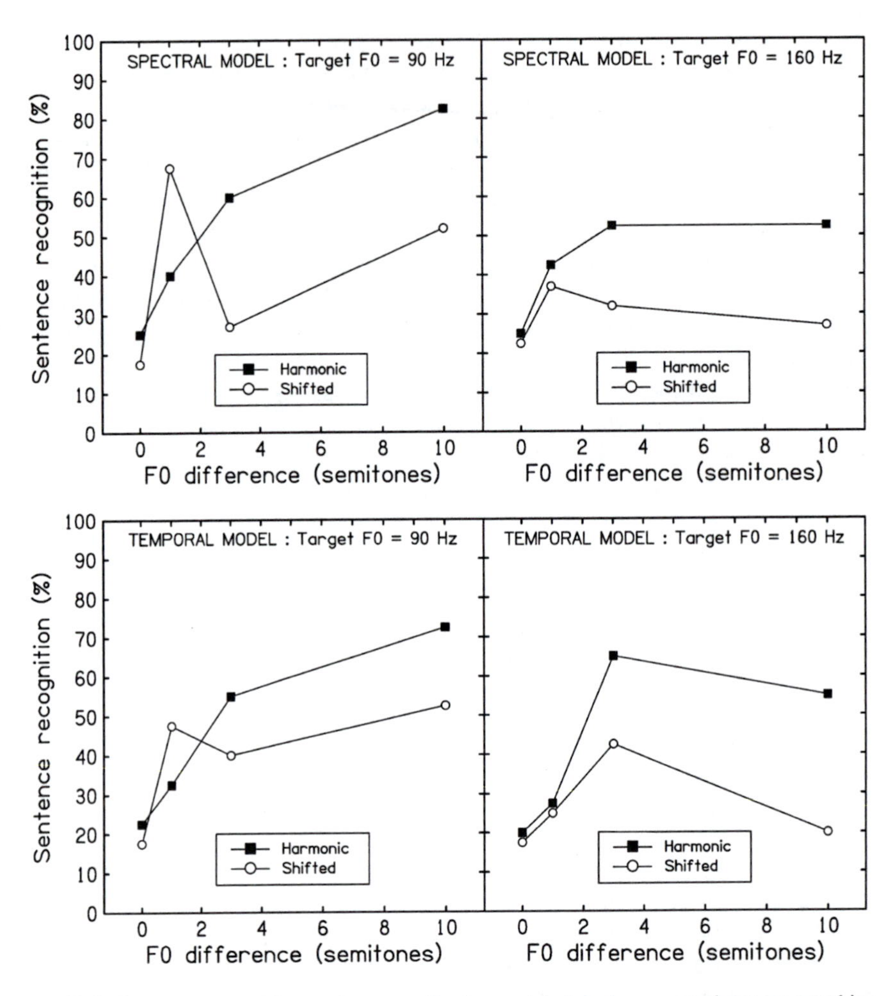

Fig. 48.2 Sentence recognition performance for the models based on spectral pattern matching (*upper panels*) and based on temporal ("timbre region") pattern matching (*lower panels*)

If a target F_0 was detected, then the channels of the correlogram dominated by the interferer F_0 were removed to reveal the target. More specifically, if the F_0 period of the interferer is τ_i, then channel f is removed from the correlogram if:

$$\frac{\mathrm{ACF}(t,f,\tau_i)}{\mathrm{ACF}(t,f,0)} > \theta \qquad (48.4)$$

where θ is a threshold, set by inspection to 0.8. Brown and Wang (1997) found this threshold-based approach to be more reliable than channel selection based on peak identification, as used by Meddis and Hewitt (1992). An SACF was formed from the remaining channels (which should mostly contain the target, since they were not dominated by the interferer), and the timbre regions for the set of time frames were matched against the timbre templates. Sentence recognition performance for the temporal model is shown in the lower panels of Fig. 48.2. These show a broadly similar pattern to the performance of the spectral model. Again, for the 90-Hz harmonic target, model performance increases monotonically with increasing ΔF_0 between the target and interferer. In the 160-Hz case, the model also shows a flattening-off (actually a small decrease) in performance for a ΔF_0 of 10 semitones. Once again, the model does not correctly predict listeners' responses to the shifted stimuli. This discrepancy with the perceptual data is especially marked for the 160-Hz case, where model performance for a ΔF_0 of 10 semitones returned almost completely to baseline. It is also worth noting for the shifted stimuli in the 90-Hz condition that the spectral and temporal models both showed an anomalous peak in recognition performance when $\Delta F_0 = 1$ semitone, for which there is no counterpart in the perceptual data. This may relate to the greater number of overlapping inharmonic components in the first-formant region for these stimuli.

48.5 Modelling Studies: Limitations and Future Directions

Our spectral and temporal matching models can both provide a reasonable qualitative account of listeners' performance with pairs of harmonic sentences, but both models considerably underestimate the perceptual benefit of an increasing difference in nominal F_0 observed for pairs of frequency-shifted sentences. This is perhaps unsurprising, given that both models are based on detecting peaks in the SACF. This approach assumes a consistent "spine" in the correlogram, which only occurs for harmonic stimuli. If a harmonic complex is frequency shifted, then the peaks of the spine do not line up across frequency channels, especially at low frequencies. Consequently, a major peak in the SACF associated with the frequency-shifted target speech may not occur; even if it does, an approach based on sampling the correlogram at a particular lag will fail to select all the components of the target speech. Figure 48.3 suggests a possible mechanism for the perceptual segregation of frequency-shifted stimuli; perhaps auditory grouping mechanisms can exploit any coherent pattern along the frequency-delay axis, such as a curve, as proposed by Roberts and Brunstrom (2001).

We tested a revised version of our spectral model that simulated grouping based on spectral pattern by sampling the channels of the correlogram through a curve in the frequency-delay plane. The revised version produced a much closer match in model performance between the harmonic and shifted conditions, and hence a better match to human performance. However, the challenge is to produce a model that can operate autonomously and give similar results, without being given a priori information about the target F_0 or degree of frequency shift. In terms of automation, a cancellation-based model (de Cheveigné 1993) might be successful if it operated within independent frequency bands, rather than assuming a common periodicity across all channels. The general approach would be to estimate the harmonic frequency of the interferer in each separate band and then to cancel that frequency from the target region.

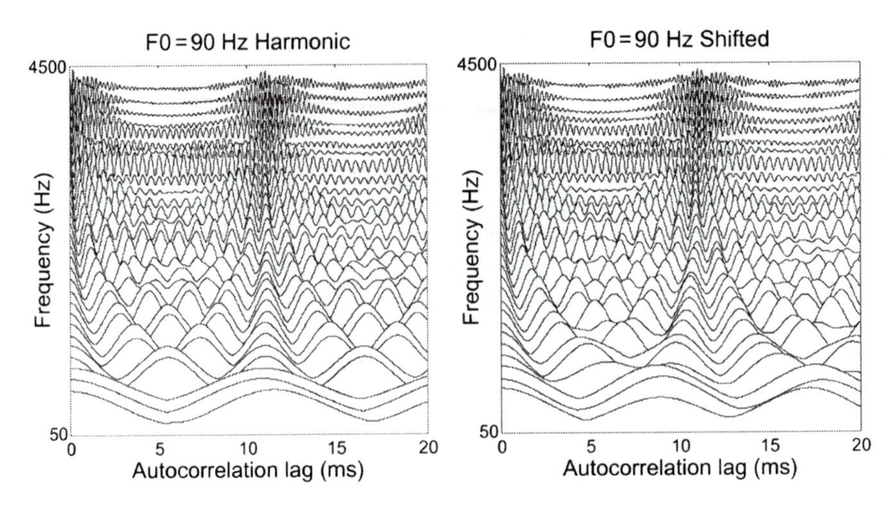

Fig. 48.3 Correlograms for one frame of a sentence with a nominal F_0 of 90 Hz, which is harmonic (*left*) and frequency shifted (*right*)

There are a number of other ways in which the simulations reported here could be improved. First, they could be run using a nonlinear filterbank (Lopez-Poveda and Meddis 2001) rather than a gammatone filterbank. Second, performance could be improved if only the portion of the signal containing the target was used for recognition; currently, the whole length of the signal is passed to the recogniser, after the subtraction of the interferer's contribution based on the model's estimate of its F_0. It should be possible to pre-select automatically the portion of the signal containing the target based on a comparison of SACFs in the earlier and later parts of the signal. Third, the recogniser could be modified to perform keyword recognition, which would allow a more direct comparison between the performance of listeners and of our models. We used recognition based on whole sentences for ease of implementation; searching for every keyword in every sentence would require a more sophisticated template matching approach and would be computationally intensive. Notwithstanding the possible benefits of these modifications, none of them is likely to impact greatly on the discrepancy in the performance of both models for harmonic and for shifted stimuli.

48.6 Summary and Conclusions

The ability of listeners to identify keywords in a target sentence mixed with interfering speech improves with increasing ΔF_0 not only when the voiced excitation source is harmonic, but also when it is frequency-shifted. We compared the ability of two computational models to account for the perceptual data. One model sampled the correlogram at the F_0 period of the target and performed spectral matching; the temporal model deselected channels dominated by the interferer and performed matching on the short-lag portion of the residual SACF ("timbre region"). Both models gave a good qualitative match to listener performance for the harmonic stimuli, in that speech recognition typically increased with increasing ΔF_0. Neither of the standard models predicted listener performance well for the shifted stimuli. A revised spectral model that groups patterns of periodicity that lie on a curve in the frequency-delay plane showed a closer match to the perceptual data. However, this model is not autonomous; currently, it is provided with a priori information about the target F_0 and degree of frequency shift. The results extend the range of phenomena originally attributed to harmonic processing to grouping by a more general form of spectral regularity. Current models based on harmonic separation are not adequate to account for the human data, but can in principle be modified to do so.

References

Assmann PF, Summerfield Q (1990) J Acoust Soc Am 88:680–697
Bird J, Darwin CJ (1998) In: Palmer AR et al (eds) Psychophysical and physiological advances in hearing. Whurr, London, pp 263–269
Boersma P, Weenink D (1996) Praat, a system for doing phonetics by computer. Institute of Phonetic Sciences, University of Amsterdam
Brokx JPL, Nooteboom SG (1982) J Phonet 10:23–36
Brown GJ, Wang DL (1997) Neural Netw 10:1547–1558
Carlyon RP, Gockel HE (2008) In: Yost WA et al (eds) Auditory perception of sound sources. Springer, New York, pp 191–213
Culling JF, Darwin CJ (1994) J Acoust Soc Am 95:1559–1569
de Cheveigné A (1993) J Acoust Soc Am 93:3271–3290
Duifhuis H, Willems LF, Sluyter RJ (1982) J Acoust Soc Am 71:1568–1580
Ellis D (2003) http://www.ee.columbia.edu/~dpwe/resources/matlab/dtw/
Lopez-Poveda EA, Meddis RM (2001) J Acoust Soc Am 110:3107–3118
Meddis R, Hewitt MJ (1992) J Acoust Soc Am 91:233–245
Moulines E, Charpentier F (1990) Speech Commun 9:453–467
Parsons TW (1976) J Acoust Soc Am 60:911–918
Roberts B (2005) Acta Acust Acust 91:945–957
Roberts B, Bregman AS (1991) J Acoust Soc Am 90:3050–3060
Roberts B, Brunstrom JM (1998) J Acoust Soc Am 104:2326–2338
Roberts B, Brunstrom JM (2001) J Acoust Soc Am 110:2479–2490

Part VIII
Novelty Detection, Attention and Learning

Chapter 49
Is There Stimulus-Specific Adaptation in the Medial Geniculate Body of the Rat?

Flora M. Antunes, Ellen Covey, and Manuel S. Malmierca

Abstract Neurons in the auditory cortex (AC) show a reduced response to a repeated stimulus, but briefly resume firing if a novel stimulus is presented (Nat Neurosci 6:391–398, 2003). This phenomenon is called stimulus-specific adaptation (SSA). This same study concluded that SSA is absent in the medial geniculate body (MGB), and is therefore a unique feature of the AC. More recently, however, SSA has been observed in the inferior colliculus (IC) of the rat (Eur J Neurosci 22:2879–2885, 2005; J Neuroci 29:5483–5493, 2009). Since the MGB receives its main inputs from the AC and IC, both of which show SSA, we reexamined the issue of whether neurons in the MGB also show SSA. We used a protocol similar to that of Ulanovsky et al. (Nat Neurosci 6:391–398, 2003) to record extracellular single unit responses in the MGB of the urethane-anesthetized rat. Our data demonstrate that SSA is indeed present throughout the MGB, being more prominent in the dorsal and medial subdivisions. Our results taken together with those from the IC indicate that SSA and enhanced responses to novel stimuli are present at every level from the IC on, and we hypothesize that SSA may be shaped, at least in part, through a bottom-up process.

Keywords Thalamus • Frequency change detection • Odd ball stimulus paradigm • Stimulus history • Novelty detection

49.1 Introduction

A novel or unexpected sound may represent potential danger for an animal, compromising its survival. In order to respond adequately, it is necessary to rapidly distinguish what is novel from what is familiar. However, despite the obvious

---

M.S. Malmierca (✉)
Auditory Neurophysiology Unit, Institute for Neuroscience of Castilla y Leon, University of Salamanca, No 1, 37007 Salamanca, Spain
e-mail: floraa@usal.es

E.A. Lopez-Poveda et al. (eds.), *The Neurophysiological Bases of Auditory Perception*,
DOI 10.1007/978-1-4419-5686-6_49,

importance of detecting and responding to novel sounds, the underlying neuronal mechanisms remain unclear. Novelty detection has been extensively studied using the non-invasive technique of recording mismatch negativity (MMN) (Näätänen 1992). However, the identity and location of the neuronal generator of the MMN in the cortex are still unknown, and there is a general lack of knowledge about novelty processing at subcortical levels (Deouell 2007).

In a dynamic environment where changes in the auditory scene occur continuously, it would be adaptive for animals to be able to continuously modulate the state of their nervous system (Mountcastle 1995; Ulanovsky et al. 2004; Wark et al. 2007). These changes would likely be reflected in changes in neuronal firing patterns, particularly any population of neurons involved in the detection of novel sounds occurring in a background of familiar and uninteresting sounds.

Neural adaptation is a ubiquitous process in biological systems (Ranganath and Rainer 2003; Wallach et al. 2008), and it is well known to occur in various forms in the olfactory (Best et al. 2005), visual (Kohn 2007) and auditory systems (Dean et al. 2008; Eytan et al. 2003; Fritz et al. 2007; Yu et al. 2008). However, Ulanovsky and colleagues (2003, 2004) were the first to show that auditory cortex (AC) neurons show a reduced response to a repeated stimulus (*standard*), and briefly resume firing if a novel stimulus (*oddball*) is presented. This phenomenon is referred to as stimulus-specific adaptation (SSA) and has been proposed as a single neuron correlate of the MMN in humans (Nelken and Ulanovsky 2007; Ulanovsky et al. 2003, 2004).

The approach used by Ulanovsky et al. (2003) employed a probabilistic stimulus approach, and demonstrated that AC neurons in the cat experience a very rapid form of adaptation that is highly sensitive to stimulus statistics and occurs with a long latency. When they tested neurons in the medial geniculate body (MGB) of the cat using the same stimulus approach, they did not observe any sign of SSA. On this basis, they concluded that SSA is generated by cortical processing. Subsequently, however, SSA has been demonstrated in the inferior colliculus (IC) of both the rat (Malmierca et al. 2009; Perez-Gonzalez et al. 2005) and the barn owl (Reches and Gutfreund 2008).

The IC is the target for most of the auditory pathways that originate in the lower brainstem, and it is also the principal source of ascending input to the MGB (for review see (Malmierca 2003)). The MGB in turn, represents the main auditory center of the thalamus, and is therefore the last center for auditory processing before inputs reach the AC (Malmierca 2003; Winer et al. 2005; Lee and Winer 2008). All MGB subdivisions receive massive *descending* projections from the AC that are about ten times more abundant than the *ascending* projections from the IC. As in the case of other corticothalamic projections, the descending pathways have been implicated in gain control, signal filtering, and other dynamic functions (Lee and Winer 2005; Winer et al. 2001; Malmierca and Ryugo 2009).

Since both the IC (Malmierca et al. 2009; Perez-Gonzalez et al. 2005) and the AC (Ulanovsky et al. 2003, 2004) contain neurons that exhibit SSA, one would expect MGB neurons to inherit this property from the IC, the AC, or both. The purpose of the current study was to reexamine this issue, using a probabilistic stimulus paradigm similar to that of Ulanovsky et al. (2003), recording from a larger population of neurons throughout all subdivisions of the MGB. Our results demonstrate that SSA is indeed present throughout the MGB in the rat, being more prominent in the dorsal and medial subdivisions. We also discuss how SSA in the MGB differs from that of the AC and IC in an attempt to understand the mechanisms that govern SSA and whether it is simply a property inherited through excitatory descending projections from the AC or whether a bottom-up process could be involved.

49.2 Materials and Methods

49.2.1 Surgical Procedures, Acoustic Stimuli, and Electrophysiological Recording

Experiments were performed on adult rats weighting between 160 and 280 g. All experiments were carried out at the University of Salamanca with the approval, and using methods conforming to the standards, of the University of Salamanca Animal Care Committee. Full details of surgical preparation were as described elsewhere (Malmierca et al. 2005, 2008, 2009; Perez-Gonzalez et al. 2005). Here, it suffices to mention that surgical anesthesia was induced and maintained with urethane (1.5 g/kg, i.p.), with supplementary doses (0.5 g/kg, i.p.) given as needed. The trachea was cannulated, and atropine sulfate (0.05 mg/kg, s.c.) was administered to reduce bronchial secretions. Body temperature was maintained at 38 ± 1°C. The animal was placed in a stereotaxic frame in which the ear bars were replaced by hollow specula that accommodated a sound delivery system.

A craniotomy was performed to expose the cerebral cortex overlying the MGB. A tungsten electrode (1–2 MΩ; Merryll and Ainswoth 1972) was lowered through the cortex and used to record extracellular single unit responses in the MGB. Neuron location in the MGB was based on stereotaxic coordinates, response properties, and histological verification using electrolytic lesions (5–10 μA for 5–10 s) to mark recording sites and identify subdivisions (Winer et al. 1999).

Stimuli were delivered through a sealed acoustic system (Malmierca et al. 2008; Perez-Gonzalez et al. 2005; Rees 1990) using two electrostatic loudspeakers (TDT-EC1) driven by two TDT-ED1 modules. Pure tone bursts were delivered to one or both ears under computer control using TDT System 2 (Tucker-Davis Technologies) hardware and custom software. Typically, tones were 75 ms duration with a 5 ms rise/fall time. Selected parameters were varied one at a time during testing. Action potentials were recorded with a BIOAMP amplifier (TDT), before passing through a spike discriminator (TDT SD1). Spike times were displayed as dot rasters ordered by the acoustic parameter varied during testing. Search stimuli were pure tones or noise bursts. The frequency response area (FRA), i.e., the combination of frequencies and intensities capable of evoking a response was obtained automatically using a randomized stimulus presentation paradigm.

49.2.2 Stimulus Presentation Paradigms

For all neurons, stimuli were presented in an oddball paradigm similar to that used to record MMN responses in human studies (Näätänen 1992) and more recently in the cat auditory cortex (Ulanovsky et al. 2003, 2004) and the rat IC (Malmierca et al. 2009). Briefly, we presented two stimuli consisting of pure tones at two different frequencies (f_1 and f_2) that elicited a similar firing rate at a level of 10–40 dB above threshold. Both frequencies were within the excitatory FRA previously determined for the neuron. We presented a train of 400 stimuli containing both frequencies in a probabilistic manner at a specific repetition rate. One frequency (f_1) was presented as the *standard* (i.e., 90% probability within the sequence); interspersed randomly among the standards were the *oddball* stimuli (i.e., 10% probability) at the second frequency (f_2). After obtaining one data set, the relative probabilities of the two stimuli were reversed, with f_2 as the *standard* and f_1 as the *oddball*. The responses to the *standard* stimulus and *oddball* stimulus were normalized to account for the different number of presentations in each condition, due to the different probabilities.

The same paradigm was repeated varying the inter-stimulus interval (ISI=250 and 500 ms), or the frequency contrast between the standard and oddball. The frequency contrasts were chosen to be as close as possible to values that have been used in other studies, i.e., $\Delta f=0.37$, 0.10 and 0.04; where $\Delta f=(f_2-f_1)/(f_2\times f_1)^{1/2}$ is the normalized frequency difference (Ulanovsky et al. 2003, 2004). To quantify the amount of SSA that occurred, we calculated two different forms of the stimulus-specific adaptation index (SSA index) using the method described by Ulanovsky et al. (2003, 2004). One was the frequency-specific index $\mathrm{SI}(f_i)$ where $i=1$ or 2, defined for each frequency f_i as $\mathrm{SI}(f_i)=[d(f_i)-s(f_i)]/[d(f_i)+s(f_i)]$, where $d(f_i)$ and $s(f_i)$ are responses (i.e. normalized spike counts) to frequency f_i when it was *oddball* or *standard*, respectively. The other SSA index was the neuron-specific index defined as $\mathrm{NSSI}=[d(f_1)+d(f_2)-s(f_1)-s(f_2)]/[d(f_1)+d(f_2)+s(f_1)+s(f_2)]$, where $d(f)$ and $s(f)$ are responses to each frequency f_1 or f_2 when they were the *oddball* or *standard* stimuli, respectively. These indices reflect the extent to which a neuron's response to the *standard* was suppressed and/or response to the *oddball* was enhanced. The possible range of NSSI values is from −1 to +1, being positive if the response to the *oddball* stimulus was greater and negative if the response to the *standard* stimulus was greater. To determine the conditions that elicited SSA and novelty responses in a given neuron, the indices were analyzed as a function of Δf and ISI.

49.3 Results

We recorded from 63 single neurons throughout the three main subdivisions of the MGB, while presenting the *oddball* stimulus paradigm. Our data indicate that about half of the MGB neurons responded more strongly to the *oddball* stimulus than to the *standard*, regardless of whether it was f_1 or f_2. Histological localization of

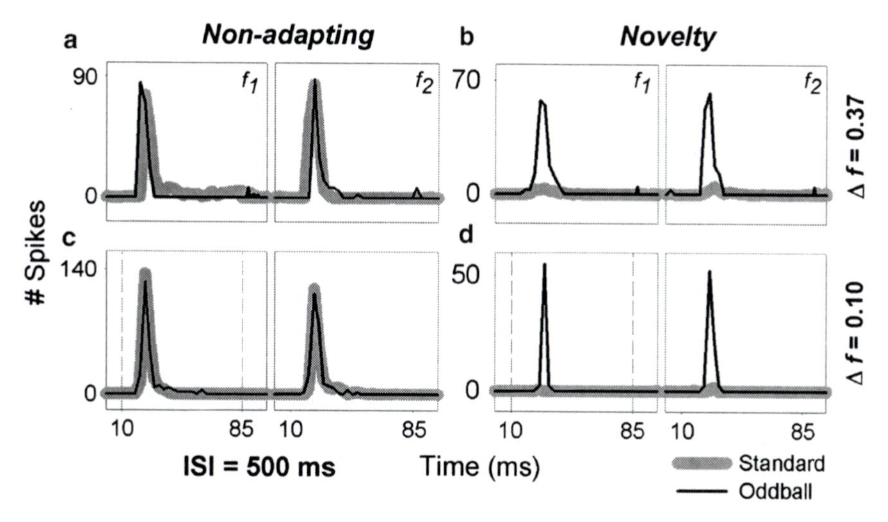

Fig. 49.1 Examples of the responses of four neurons to the oddball stimulus paradigm at 500 ms ISI (425 ms inter-tone pause), for $\Delta f=0.37$ and $\Delta f=0.10$. *Dotted vertical lines* indicate the duration of the stimulus (75 ms). *Black and grey lines* represent the neuronal activity, i.e., number of spikes normalized to 100 trials as a function of time, in response to the *oddball* and *standard* stimulus, respectively. (**a**, **c**) Nonadapting neurons. Activity of these neurons is similar for the *oddball* and *standard* frequency. NSSI = –0.09 and NSSI = 0.02 for (**a**) and (**c**), respectively. (**b**, **d**) Novelty neurons. Activity of these neurons is stronger to the *oddball* than to the *standard* frequency. NSSI = 0.97 and NSSI = 0.96 for (**b**) and (**d**), respectively

recording sites within the three main subdivisions of the MGB showed that SSA is indeed present throughout the MGB in the rat, although it is more prominent in the dorsal and medial subdivisions.

Figure 49.1 shows examples of the responses of four MGB neurons to the oddball stimulus paradigm, at two different frequency contrasts ($\Delta f=0.37$ and 0.10), and a 425 ms inter-tone duration. Two of these neurons did not show any degree of SSA (nonadapting, Fig. 49.1a, c) while two others exhibited robust SSA (novelty, Fig. 49.1b, d). Black and grey lines represent the number of spikes normalized to 100 trials as a function of time, in response to the *oddball* and *standard* stimulus, respectively. For each neuron, we show the neuronal responses to f_1 and f_2. By definition, the responses of the nonadapting neurons to each frequency were similar regardless of whether the frequency was presented as the *standard* or the *oddball*, whereas for the novelty neurons, the responses were significantly stronger to either frequency when it was the *oddball* than when it was the *standard*.

The examples shown in Fig. 49.1 represent neurons with extreme SSA values. However, across the MGB population as a whole, the distribution of NSSI values forms a continuum covering the entire range between the extremes of 0 and +1.

Figure 49.2 illustrates population data across all neurons, and illustrates the continuous nature of NSSI distribution. Because each neuron was tested under multiple conditions, we calculated an average NSSI value across conditions. Since not all neurons were tested with all conditions, we considered only those neurons

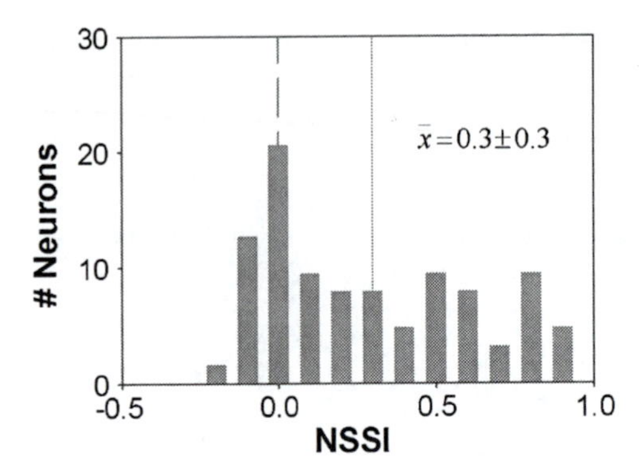

Fig. 49.2 Population number of neurons as a function of the NSSI, averaged for all NSSI values obtained across conditions for each neuron. The majority of neurons (86%) had positive NSSI values ranging between 0 and 1. Mean NSSI = 0.3 ± 0.3, significantly larger than zero (*t*-test for mean > 0: $p < 0.0001$)

with two or more conditions for this analysis ($n = 59$ neurons). These results show that 86% of the neurons had positive average NSSI values, and that there was a continuous distribution all the way to 0.99. The mean NSSI of the whole population is positive (mean = 0.34 ± 0.33) and significantly larger than zero (*t*-test for mean > 0: $p < 0.0001$). Only a few neurons had negative values ($n = 9$; 14%, ranging between NSSI = −0.17 and −0.006), meaning that their overall response to the *standard* was stronger than that to the *oddball*. However, all but one (NSSI = −0.17) of these negative values were not significantly different from zero (*t*-test: $p < 0.05$). For descriptive purposes, we have classified MGB neurons into three different types: nonadapting neurons (if all NSSI < 0.25 under all conditions), novelty neurons (if all NSSI > 0.5 under all conditions), and partially adapting neurons (if NSSI > 0.5 in some but not all conditions). Nonadapting neurons made up nearly half of the population (49%; $n = 29$, including the negative values; mean NSSI = 0.05 ± 0.1). The next largest class was novelty neurons (36%; $n = 21$; mean NSSI = 0.74 ± 0.14) followed by partially adapting neurons (15%; $n = 9$; mean NSSI = 0.36 ± 0.08).

Figure 49.3 shows scatter plots of the frequency-specific SSA index, SI(f_1) vs. SI(f_2) for the neurons, tested at three different frequency contrast values ($\Delta f = 0.37$, 0.10 and 0.04) and two different inter-stimulus intervals (ISI = 500 and 250 ms). The majority of points lie above the diagonal, in the upper right quadrant of the scatter plot, for almost all conditions, confirming that SSA is present at the population level. However, at the minimum frequency contrast tested ($\Delta f = 0.04$) and the highest ISI (500 ms), points are distributed around zero both above and below the diagonal. It is noteworthy that, at $\Delta f = 0.37$ and 0.10, there is a cluster of points in the right upper corner of the scatter plot at both ISI. These neurons have an SI > 0.6 for both f_1 and f_2, and represent the highest degree of novelty selectivity. Even at $\Delta f = 0.04$, three neurons maintained a high degree of novelty selectivity for both ISI, suggesting that a small percentage of neurons in the MGB exhibit SSA at this very small frequency contrast, corresponding to 0.57 octaves.

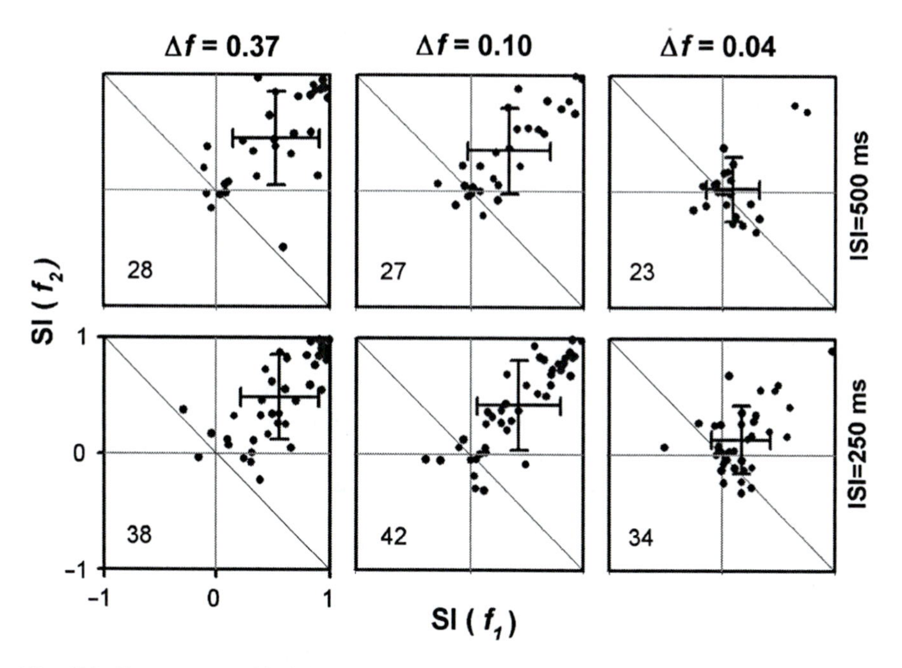

Fig. 49.3 Frequency-specific indices SI(f_1) vs. SI(f_2) values obtained separately for the different frequency contrast values (Δf=0.37, 0.10 and 0.04) and ISI (250 and 500 ms). Neurons that were tested for more than one condition are represented in more than one panel. Each point represents data from one neuron. Number in the *lower left quadrant* is the number of neurons for each condition. *Crosses* indicate the mean and standard deviation for each axis. Neurons above the diagonal indicate the presence of SSA. There are neurons with high SI(f_i) values in all conditions tested, being more pronounced for Δf=0.37 and 0.10

Therefore, SSA is present at the population level in the MGB, with a high percentage of neurons showing strong SSA. The degree of SSA varies with the stimulus repetition rate and the frequency difference between f_1 and f_2.

49.4 Discussion

The present study demonstrates that a large percentage of neurons in the MGB of the anesthetized rat show some degree of SSA. Although a large number of neurons had high NSSI values for the largest Δf tested, it is also clear that a few neurons also exhibit high SSA for Δf as small as 0.04, i.e., 0.57 octaves. Our results are in agreement with previous findings from the AC and IC in that the amount of SSA is positively correlated with both the ISI and the magnitude of the frequency contrast (Ulanovsky et al. 2003; Malmierca et al. 2009).

Before discussing our results in a functional context and comparing them with previous studies, a few methodological issues need to be considered.

Special care was taken to choose two frequencies (f_1 and f_2) that were within the excitatory portion of the FRA and elicited similar spike counts when presented randomly during measurement of the FRA. By doing so, we could be certain that the recorded SSA response was related to the probability of occurrence and not to frequency interactions. Additionally, we only used files with a similar SI for both frequencies, excluding those in which [$SI(f_1)–SI(f_2)>0.5$].

In the present study, we recorded neurons with very high values of $SI(f_i)$ ([$SI(f_1)\approx SI(f_2)\approx 0.9$], see examples in Fig. 49.1b, d), which genuinely reflects stronger neuronal adaptation to the *standard* than to the *oddball* stimuli. MGB neurons seem to be similar to those in the AC of the cat (Ulanovsky et al. 2003) and the IC of the rat (Malmierca et al. 2009) with regard to the parameter ranges over that which we tested SSA in the MGB neurons. Since the MGB receives inputs from both the IC (Malmierca 2003; Oliver et al. 1999) and the AC (Lee and Winer 2005; Malmierca and Ryugo 2009; Winer et al. 2001) it is not surprising that MGB neurons show SSA. A previous study using the cat and based on a rather small sample ($n=27$, Ulanovsky et al. 2003) failed to demonstrate SSA in the MGB using $\Delta f=0.10$ and an ISI of 736 ms (i.e., an inter-tone pause of 506 ms, for a tone duration of 230 ms). But using similar parameters ($\Delta f=0.10$ and inter-tone pause of 425 ms) we found neurons with extreme NSSI values in the MGB of the rat. These differences may be due to interspecies differences in SSA processing and/or other factors such as anesthesia, recording procedures, or merely sample size.

An important finding is that MGB neurons do not seem to be organized into discrete groups according to their SSA sensitivity. Rather, MGB neurons show a gradient in the amount of SSA throughout all the MGB subdivisions. Nevertheless, there are some regional differences. Neurons in the ventral division show some degree of SSA (NSSI values can be as large as 0.48), but the highest NSSI values (up to 0.99) were found in neurons from the medial and dorsal divisions. Similarly, SSA observed in the IC was more robust in the dorsal, lateral and rostral cortices (Perez-Gonzalez et al. 2005; Malmierca et al. 2009). These findings taken together suggest that SSA may be more important in the paralemniscal auditory pathway, which is known to be implicated in polysensory interactions and the processing of other complex features of sound such as fear conditioning or auditory learning (Malmierca et al. 2002; Malmierca 2003).

Another important similarity between SSA in MGB and IC neurons is related to neuronal response type and the latency for SSA to reduce spike counts within a response. As in the case of the IC (Perez-Gonzalez et al. 2005; Malmierca et al. 2009), we found that high SSA neurons in the MGB were mainly onset responders, with short latencies for both the *standard* and the *oddball* stimuli. This argues against the idea of an additional processing time required for a cortical modulation effect to occur. In this sense, SSA in the thalamus and in midbrain seems to differ from SSA in the AC, where SSA is expressed mostly in the late portion of sustained responses, suggesting the contribution of intracortical processing (Ulanovsky et al. 2003). It remains unclear to what extent SSA observed in the MGB is due to thalamic processing or to what extent it reflects processing that has already occurred in the IC or AC.

In conclusion, our data demonstrate for the first time that SSA is present throughout the MGB of the rat. The comparison of our MGB results with those from the IC and AC studies reveal many similarities between the MGB and the IC and AC in terms of SSA. This leads us to hypothesize that SSA, at least in part, is created at the midbrain level and transmitted in a bottom-up process through the ascending pathway. What is unclear so far is to what extent the AC modulates SSA subcortically through the descending pathway. This issue needs to be explored in future experiments.

Acknowledgments Financial support was provided by the Spanish MEC (BFU2009-07286), EU (EUI2009-04083) and JCYL-UE (GR221) to MSM, the NIH (NIDCD R01 DC-00287) and the NSF (IOS-0719295) to EC. Support for software development was provided by NIH (NIDCD P30DC004661); FAM held a fellowship from the Spanish MEC (BES-2007-15642)).

References

Best AR, Thompson JV, Fletcher ML, Wilson DA (2005) Cortical metabotropic glutamate receptors contribute to habituation of a simple odor-evoked behavior. J Neurosci 25: 2513–2517

Dean I, Robinson BL, Harper NS, McAlpine D (2008) Rapid neural adaptation to sound level statistics. J Neurosci 28:6430–6438

Deouell LY (2007) The frontal generator of mismatch negativity revisited. J Psychophysiol 21:188–203

Eytan D, Brenner N, Marom S (2003) Selective adaptation in networks of cortical neurons. J Neurosci 23:9349–9356

Fritz JB, Elhilali M, David SV, Shamma SA (2007) Auditory attention – focusing the searchlight on sound. Curr Opin Neurobiol 17:437–455

Kohn A (2007) Visual adaptation: physiology, mechanisms, and functional benefits. J Neurophysiol 97:3155–3164

Lee CC, Winer JA (2005) Principles governing auditory cortex connections. Cereb Cortex 15:1804–1814

Lee CC, Winer JA (2008) Connections of cat auditory cortex: I. Thalamocortical system. J Comp Neurol 507:1879–1900

Malmierca MS (2003) The structure and physiology of the rat auditory system: an overview. Int Rev Neurobiol 56:147–211

Malmierca MS, Ryugo DK (2009) Cortical descending projections to auditory midbrain and brainstem. In: Winer JA, Schreiner CE (eds) The auditory cortex. Springer, New York

Malmierca MS, Merchan MA, Henkel CK, Oliver DL (2002) Direct projections from cochlear nuclear complex to auditory thalamus in the rat. J Neurosci 22:10891–10897

Malmierca MS, Hernandez O, Rees A (2005) Intercollicular commissural projections modulate neuronal responses in the inferior colliculus. Eur J Neurosci 21:2701–2710

Malmierca MS, Izquierdo MA, Cristaudo S, Hernandez O, Perez-Gonzalez D, Covey E, Oliver DL (2008) A discontinuous tonotopic organization in the inferior colliculus of the rat. J Neurosci 28:4767–4776

Malmierca MS, Cristaudo S, Perez-Gonzalez D, Covey E (2009) Stimulus-specific adaptation in the inferior colliculus of the anesthetized rat. J Neurosci 29(17):5483–5493

Merryll EG, Ainswoth A (1972) Glass-coated platinum coated tungsten microelectrodes. Med Biol Eng 10:662–672

Mountcastle V (1995) The evolution of ideas concerning the function of the neocortex. Cereb Cortex 5:289–295

Näätänen R (1992) Attention and brain function. Lawrence Erlbaum, Hillsdale, New Jersey

Nelken I, Ulanovsky N (2007) Mismatch negativity and stimulus-specific adaptation in animal models. J Psychophysiol 21:214–223

Oliver DL, Ostapoff EM, Beckius GE (1999) Direct innervation of identified tectothalamic neurons in the inferior colliculus by axons from the cochlear nucleus. Neuroscience 93:643–658

Perez-Gonzalez D, Malmierca MS, Covey E (2005) Novelty detector neurons in the mammalian auditory midbrain. Eur J Neurosci 22:2879–2885

Ranganath C, Rainer G (2003) Neural mechanisms for detecting and remembering novel events. Nat Rev Neurosci 4:193–202

Reches A, Gutfreund Y (2008) Stimulus-specific adaptations in the gaze control system of the barn owl. J Neurosci 28:1523–1533

Rees A (1990) A close-field sound system for auditory neurophysiology. J Physiol 430:2

Ulanovsky N, Las L, Nelken I (2003) Processing of low-probability sounds by cortical neurons. Nat Neurosci 6:391–398

Ulanovsky N, Las L, Farkas D, Nelken I (2004) Multiple time scales of adaptation in auditory cortex neurons. J Neurosci 24:10440–10453

Wallach A, Eytan D, Marom S, Meir R (2008) Selective adaptation in networks of heterogeneous populations: model, simulation, and experiment. PLoS Comput Biol 4(2):e29

Wark B, Lundstrom BN, Fairhall A (2007) Sensory adaptation. Curr Opin Neurobiol 17:423–429

Winer JA, Kelly JB, Larue DT (1999) Neural architecture of the rat medial geniculate body. Hear Res 130:19–41

Winer JA, Diehl JJ, Larue DT (2001) Projections of auditory cortex to the medial geniculate body of the cat. J Comp Neurol 430:27–55

Winer JA, Miller LM, Lee CC, Schreiner CE (2005) Auditory thalamocortical transformation: structure and function. Trends Neurosci 28:255–263

Yu X, Xu X, Chen X, He S, He J (2008) Slow recovery from excitation of thalamic reticular nucleus neurons. J Neurophysiol 101(2):980–987

Chapter 50
Auditory Streaming at the Cocktail Party: Simultaneous Neural and Behavioral Studies of Auditory Attention

Mounya Elhilali, Juanjuan Xiang, Shihab A. Shamma, and Jonathan Z. Simon

Abstract We present a pair of simultaneous behavioral-neurophysiological studies in human subjects, in which we manipulate subjects' attention to different features of an auditory scene. In the first study, we embed a regular acoustic target in an irregular background; in the second study, we pair competing simultaneous regular acoustic streams. Our experimental results reveal that attention to the target, rather than to the background or unattended stream, correlates with a sustained increase in the neural target representation, as measured by magnetoencephalography (MEG), beyond auditory attention's well-known transient effects on onset responses. The enhancement originates in core auditory cortex and covaries with both behavioral states. Furthermore, for the slower streams, where the rhythmic rate is commensurate with that of speech prosody, the target's perceptual detectability improves over time, correlating strongly, within subjects, with the target representation's neural buildup.

Keywords Attention • Magnetoencephalography • MEG • Buildup • Auditory scene analysis

50.1 Introduction

Due to limited processing capacity of the auditory system, only a subset of the information available at the ear can be attended and processed in more detail at the high level of the auditory system. The cognitive process involved in this selection process is called as selective attention. Recent psychoacoustic studies on selective attention have demonstrated that human listeners can allocate attention not only to a particular location, but also to a particular feature, such as modulation rate

---

J.Z. Simon (✉)
Department of Electrical and Computer Engineering, University of Maryland, College Park, MD, USA
e-mail: jzsimon@umd.edu

E.A. Lopez-Poveda et al. (eds.), *The Neurophysiological Bases of Auditory Perception*, DOI 10.1007/978-1-4419-5686-6_50,

(Grimault et al. 2002), pitch (Vliegen and Oxenham 1999) or timbre (Cusack and Roberts 2000; for review, see Moore and Gockel 2002).

The neural correlates of the auditory spatial attention have been extensively investigated using dichotic paradigms, where subjects attend to a series of tone pips in one ear and ignored concurrent tone pips in the other ear (Giraud et al. 2000; Hillyard et al. 1973; Woldorff and Hillyard 1991). On the other hand, there have been a limited number of studies (e.g., Bidet-Caulet et al. 2007; Gutschalk et al. 2008) that attempted to examine the neural correlates of feature-based auditory attention. Here, we take advantage of the auditory Steady-State Response (aSSR), an electrophysiological signature of modulated sounds. In this pair of studies, either one stream in a background of maskers, or two concurrent streams at different rhythmic rates, were diotically presented to human listeners. Each stream elicits an aSSR at the corresponding modulation rate. By requiring subjects to selectively attend to one stream, the modulatory effects of rate-based selective attention can be examined.

The first study employs stimuli consisting of a tone sequence repeated at 4 Hz in the midst of random maskers (Fig. 50.1a). The second employs stimuli consisting of a two parallel tone sequences with different rhythmic rates, 4 and 7 Hz (Fig. 50.1b). For each study, in separate tasks with identical stimulus ensembles, subjects are asked to detect deviants, either in a rhythmic stream or the background maskers while magnetoencephalography (MEG) responses are recorded. The evolution of neural representation of an attended stream is correlated with the behavioral performance. In addition, the neural source locations of the aSSR are assessed.

The stimuli, and hence the tasks, between studies are very different. In the first study, the rhythmic component's percept ranges from imperceptible to quiet-but-clear. In the second study, both streams are of similar loudness: each easily perceptible at all times.

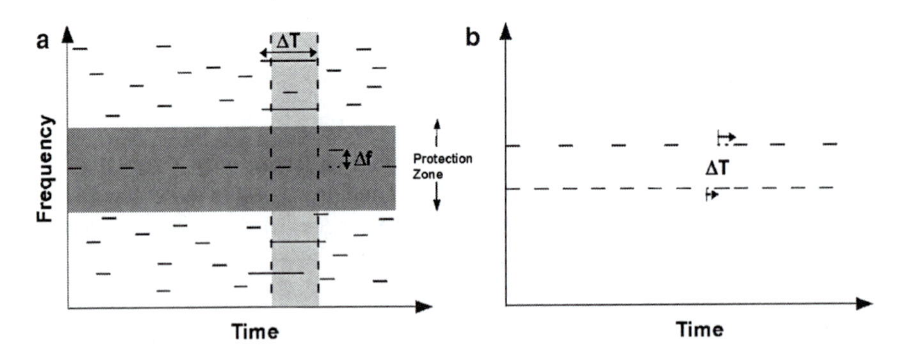

Fig. 50.1 Stimulus description. Cartoon spectrograms of typical stimuli. (**a**) The stimulus consists of a repeating target note embedded in random interferers. A 16 semitone spectral protection zone surrounds the target frequency (*gray band*). In the target task, subjects are instructed to detect a frequency shifted (Δf) deviant in the repeating target notes. In the masker task, subjects are instructed to detect a sudden temporal elongation (ΔT) of the masker notes. (**b**) The stimulus consists of two repeating target notes separated by 8 semitones. In each task, subjects are instructed to detect a temporally jittered (ΔT) deviant present in that stream

50.2 Methods

Subjects: 18 subjects participated in the first study; 28 in the second. Four subjects (respectively 2) were excluded from further analysis due to nonneural electrical artifacts or an inability to perform the tasks.

Stimuli: In the first study, sounds were 5.5 s in duration. Each stimulus contained a 75 ms target note, repeating at 4 Hz, with frequency randomly chosen in the range 250–500 Hz. The background consisted of random 75 ms tones uniformly distributed over time and log-frequency. The frequencies were randomly chosen from five octaves centered at 353 Hz, except for a 16-semitone protection zone. Fifteen exemplar stimuli were generated for each of the four condition types: no deviants; one target deviant per stimulus; one masker deviant per stimulus; and one target deviant and one masker deviant, at independent times, per stimulus. Each target deviant was the displacement (up or down) by 2 semitones. Each masker deviant was a single 500 ms time-window, in which all masker tones were elongated to 400 ms. The deviants were randomly distributed in time.

In the second study, the duration of sounds were randomly chosen from 5.25, 6.25, or 7.25 s. The spectral distance between the two streams was ±8 semitones, where the specific frequencies of each stream were randomly chosen in the range of 250–500 Hz. The loudness of each stream was approximately equal. Duration was 75 ms. Twelve exemplar stimuli were generated (and presented twice each) for each of the three condition types: no deviants; one deviant per 4 Hz stream; and one deviant per 7 Hz stream. Each deviant was the temporal displacement of a target tone by 70 ms (4 Hz stream) or 40 ms (7 Hz stream) from the regular interval. All subjects performed both tasks: T4 (4 Hz deviant detection) and T7 (7 Hz deviant detection).

In both studies, each subject performed both tasks (with order counterbalanced across subjects), and subjects were instructed to press a button held as soon as they heard the appropriate deviant.

MEG recording: Subjects were placed horizontally in a dimly lit magnetically shielded room. The signals were delivered with sound tubing attached to foam plugs inserted into the ear-canal, and presented at a comfortable loudness of approximately 70 dB SPL.

MEG recordings were made with a 160-channel whole-head system (Kanazawa Institute of Technology, Kanazawa, Japan). Neural channels were denoised twice with an adaptive filter: using external reference channels (Ahmar and Simon 2005), and using the two channels with the strongest cardiac artifacts (Xiang et al. 2005).

Behavioral Analysis: The task performance was assessed by calculating a d-prime measure of performance (Kay 1993).

To investigate the buildup of the target object during a task, we divided the deviant trials according to temporal locations of the deviants. Because of the temporal uncertainty in the false alarm trials, we calculated an average false alarm rate, and combined it with the time-specific hit rate to derive a d-prime measure for each segment.

Neural Data Analysis: The concatenated neural responses were characterized by the magnitude and phase of the frequency component at the tone presentation rate and were used for localization. The remainder of the analysis was based on the normalized neural responses, defined to be the squared magnitude of the frequency component at the tone presentation rate divided by the average squared magnitude of the frequency components between 1 Hz above and below that rate, averaged over the 20 channels with the strongest normalized neural responses. The same analysis is also done at the two adjacent frequency bins, ±1/4.25 Hz or ±1/4 Hz.

To investigate the buildup of the target object in target task, the analysis epochs were divided into temporal segments, 750 ms duration for the first study, 1,000 ms for the second, extracted and concatenated. The first segment began at 1,250 ms post stimulus.

The behavioral curves for each subject were interpolated to match the sampling rate of the neural data. Subsequently, these two curves were then fitted by a line to derive the slope relating them. The slope was transformed into an angle, and combined across subjects using circular statistics to yield an angular mean (Fisher 1993). Confidence measures were then derived from the bootstrap statistics (Efron and Tibshirani 1993).

Neural Source Localization: Source localization for the neural response to the target was obtained by calculating the complex current-equivalent dipole best fitting the complex magnetic field configuration at 4 Hz peak, in each hemisphere (Simon and Wang 2005). Significance of the relative displacement between the (previously obtained) M100 and target dipole sources were determined by a two-tailed paired *t*-test in each of three dimensions.

50.3 Results

Depending on listeners' attentional focus, the neural representations of the streams mirror the percept of the scenes.

In the first study, during the performance of the target task (mean d-prime: 3.0), the rhythm of the stream emerges as a strong 4 Hz component in the neural signal of an individual subject (Fig. 50.2a, left panel). In contrast, during the performance of the masker task (mean d-prime: 3.0), the cortical response entrained at 4 Hz is relatively suppressed (Fig. 50.2a, right panel).

In the second study, during the T4 task (mean d-prime: 2.9), the rhythm of the slow stream emerges as a strong 4 Hz component for an individual subject (Fig. 50.2b, top left panel). In contrast, during the T7 task (mean d-prime: 1.8), the cortical response entrained at 4 Hz is relatively suppressed (Fig. 50.2a, top right panel). This effect is correspondingly reversed for the cortical representations of the fast stream: the neural response at 7 Hz is stronger in the T7 task than the T4 task (Fig. 50.2a, lower panels).

The neural response change between tasks was averaged across all subjects (Fig. 50.2c, d). In the first study, a significant positive (bootstrap across subjects, $p < 10^{-4}$) change at the 4 Hz aSSR is observed, reflecting enhanced phase-locked,

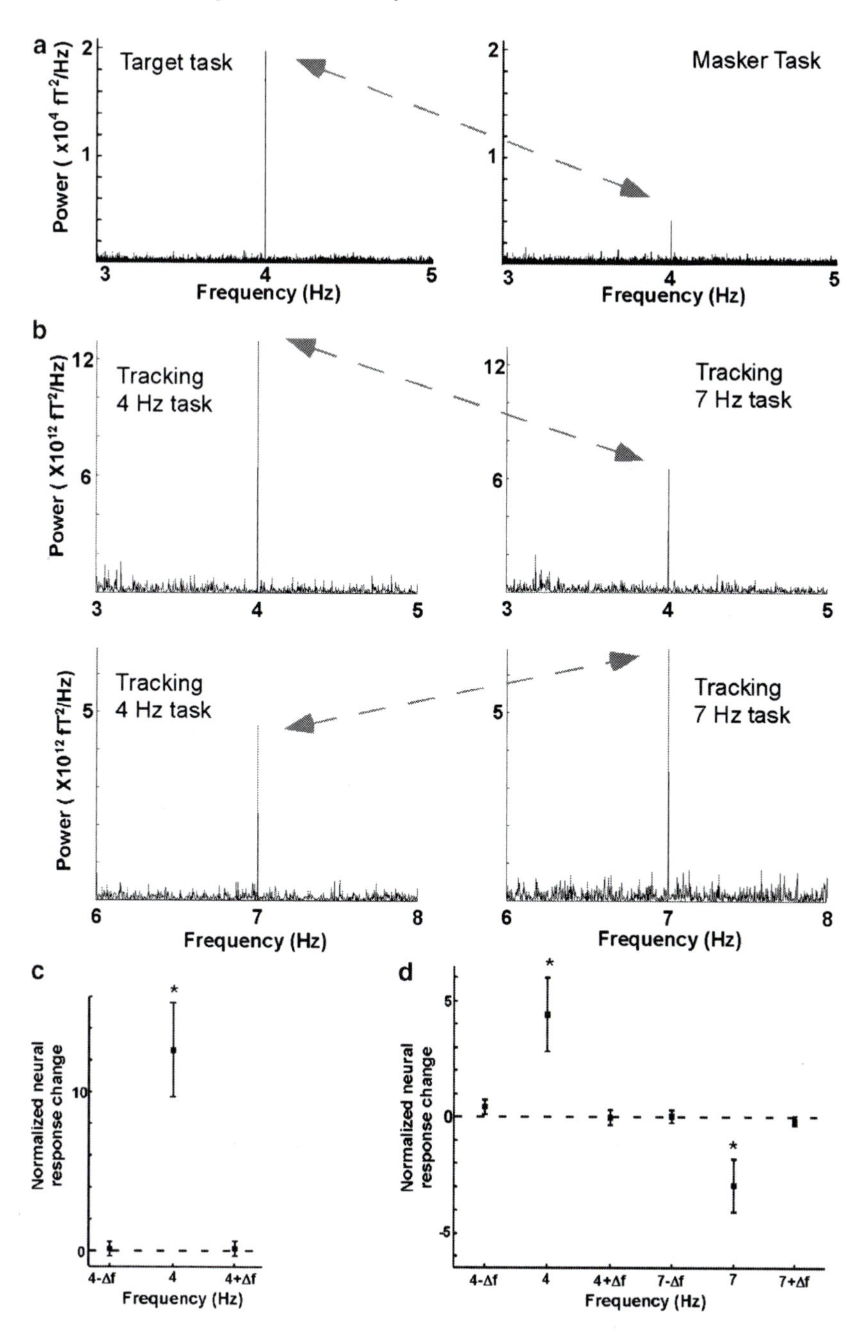

Fig. 50.2 Neural responses. (**a**) Power spectral density of MEG responses for a single subject in the first study in target (*left*) and masker (*right*) tasks, averaged over 20 channels. (**b**) Power spectral densities of MEG responses for a single subject in the second study in T4 (*left column*) and T7 (*right column*) tasks, averaged over 20 channels. (**c**, **d**) Normalized neural responses of one task relative to the other task ((**c**) target task minus masker task; (**d**) T4 task minus T7 task) shows enhancement exclusively at the frequency of the target rhythm. Each data points represents the average difference between normalized neural responses; *error bars* represent standard error

sustained activity when subjects' attention is directed toward the target stream. In the second study, a significant positive ($p < 10^{-3}$) change at the 4 Hz aSSR and a significant negative ($p < 0.002$) change at the 7 Hz aSSR are observed, reflecting an enhanced phase-locked, sustained activity when subjects' attention is directed toward the target stream. In contrast, there is no significant change in normalized neural response at adjacent frequencies, confirming that this feature-based selective attention precisely modulates the cortical representation of the specific feature, but not overall neural activities.

The neural sources of all the target rhythm response components originate in auditory cortex. In the first study, the neural source's mean displacement from the source of the auditory M100 response (Naatanen and Picton 1987) was significantly different (for the left auditory cortex only) (two-tailed *t*-test; $p = 0.017$) by 14 ± 5 mm in the anterior direction. In the second study, the total significant displacement (two-tailed *t*-test; $p = 0.016$) was 19 ± 6 mm in the anterior direction (for both hemispheres combined). Assuming an M100 origin of *planum temporale*, this is consistent with an origin for the neural response to the target rhythm in Heschl's gyrus, the site of core auditory cortex, and *not* consistent with the aSSR arising from a concatenation of periodic M100 responses.

For the 4 Hz targets in both studies, subjects' performance during the target task improves over several seconds as shown in Fig. 50.3a, b (solid lines). Moreover, the neural response to the target rhythm also displays a statistically significant buildup (Fig. 50.3a, b, dashed line) closely aligned with the behavioral curve, and, consequently, decoupled from the actual acoustics. The remarkable correspondence between these two measures strongly suggests that the enhanced perception of the target over time is mediated by an enhancement of the neural signal representation. In the second study, no such buildups are present in the 7 Hz task, which remains a puzzle (particularly so since the 4 Hz tasks across studies are so different).

We also note that the sub-segments over which the neural buildup is measured are required to span several rhythmic periods (at least three). There is no buildup using intervals with shorter durations, despite sufficient statistical power, implicating temporal phase coherence (in contrast to spatial phase coherence) as critical to the buildup of the neural target representation. The need for a longer time window cannot be attributed to increased power at 4 Hz; since this explanation would imply that using one rhythmic period would also show buildup. We have constructed a computational model (not shown) that further supports this interpretation.

50.4 Discussion

These studies build on previous work in stream segregation (Fishman et al. 2001; Gutschalk et al. 2005; Micheyl et al. 2005; Snyder et al. 2006) but keeping the physical parameters of the stimulus fixed while manipulating only the attentional state of the listeners. One major finding is that auditory attention strongly modulates the sustained neural representation of the target. This neural representation is

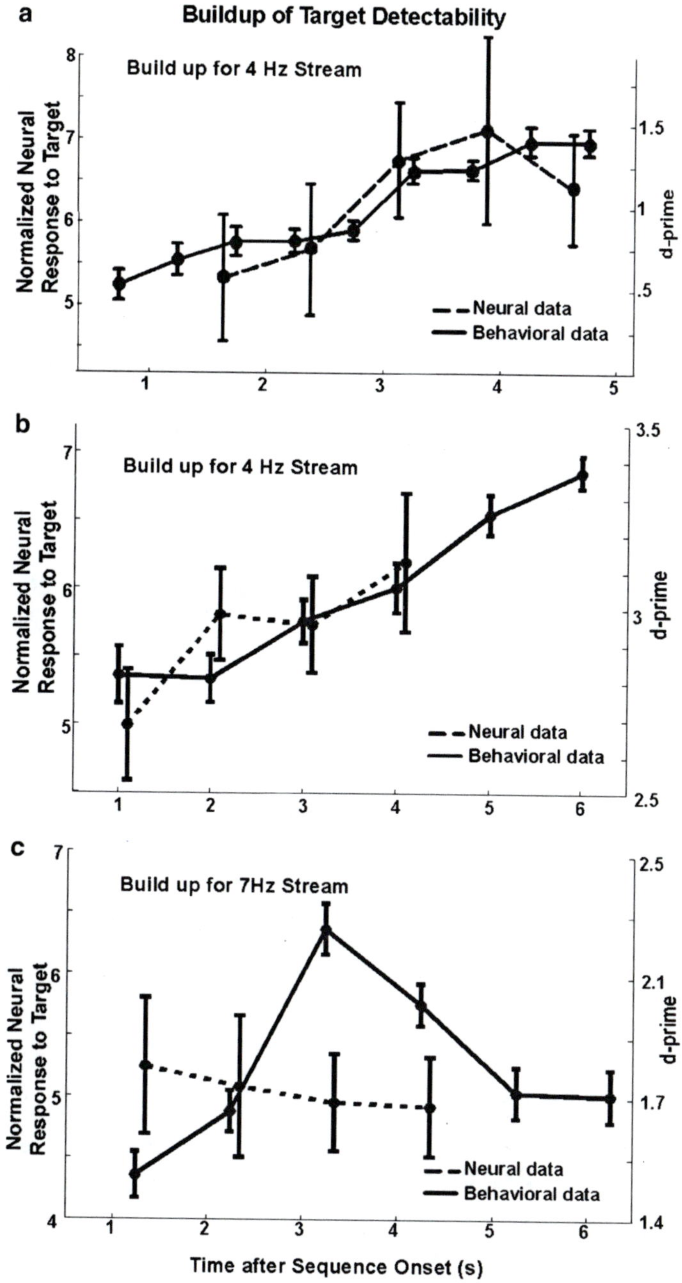

Fig. 50.3 Buildup over time of behavioral and neural responses in target task averaged over subjects (*error bars* represent standard error). (**a**) Normalized neural response to target rhythm, and behavioral performance, in the first study. (**b**, **c**) Normalized neural response to target rhythms, and behavioral performance, for the two tasks and their respective rates, in the second study

not merely a repeated M100 response, since its location is inconsistent with that interpretation, rather its location is consistent with core auditory cortex.

Finally, this study offers the first demonstration of the top-down mediated buildup over time of the neural representation of a target signal that also follows the same temporal profile of the buildup based on listeners' detectability performance in the same subject, *as long as the target is slow rather than fast* (4 Hz rather than 7 Hz). Using the current experimental paradigm, we are able to monitor the evolution in time of attentional processes as they interact with the sensory input. Many studies overlook the temporal dynamics of the neural correlates of attention, either by using cues that prime subjects to the object of attention (thereby stabilizing attention before the onset of the stimulus), or by explicitly averaging out the buildup of the neural signal in their data analysis (focusing instead on the overall contribution of attention in different situations, and not monitoring the dynamics by which the process builds up). Our findings reveal that even though the sensory target signal is unchanged, attention allows its neural representation to grow over time, closely following the time-course of the perceptual representation of the signal.

Acknowledgments Support has been provided by NIH grants R01DC008342, 1R01DC007657 and (via the CRCNS NSF/NIH joint mechanism) 1R01AG027573. We thank Jonathan Fritz and David Poeppel for comments and discussion. We are grateful to Jeff Walker for excellent technical support.

References

Ahmar N, Simon JZ (2005) MEG adaptive noise suppression using fast LMS. In: International IEEE EMBS conference on neural engineering 2005

Bidet-Caulet A, Fischer C, Besle J, Aguera PE, Giard MH, Bertrand O (2007) Effects of selective attention on the electrophysiological representation of concurrent sounds in the human auditory cortex. J Neurosci 27(35):9252

Cusack R, Roberts B (2000) Effects of differences in timbre on sequential grouping. Percept Psychophys 62:1112–1120

Efron B, Tibshirani R (1993) An introduction to the bootstrap. Chapman & Hall/CRC, New York

Fisher NI (1993) Statistical analysis of circular data. Cambridge University Press, New York

Fishman YI, Reser DH, Arezzo JC, Steinschneider M (2001) Neural correlates of auditory stream segregation in primary auditory cortex of the awake monkey. Hear Res 151:167–187

Giraud AL, Lorenzi C, Ashburner J, Wable J, Johnsrude I, Frackowiak R et al (2000) Representation of the temporal envelope of sounds in the human brain. J Neurophysiol 84(3):1588–1598

Grimault N, Bacon SP, Micheyl C (2002) Auditory stream segregation on the basis of amplitude modulation rate. J Acoust Soc Am 111:1340–1348

Gutschalk A, Micheyl C, Melcher JR, Rupp A, Scherg M, Oxenham AJ (2005) Neuromagnetic correlates of streaming in human auditory cortex. J Neurosci 25:5382–5388

Gutschalk A, Micheyl C, Oxenham AJ (2008) Neural correlates of auditory perceptual awareness under informational masking. PLoS Biol 6(6):e138

Hillyard SA, Hink RF, Schwent VL, Picton TW (1973) Electrical signs of selective attention in the human brain. Science 182(4108):177–180

Kay SM (1993) Fundamentals of statistical signal processing: estimation theory, Prentice-Hall, Inc. Upper Saddle River, NJ, USA

Micheyl C, Carlyon RP et al (2005) Performance measures of auditory organization. Auditory Signal Processing. Physiology, Psychoacoustics, and Models. D. Pressnitzer, A. de Cheveigne, S. McAdams and L. Collet. New York, NY, Springer: 203–11

Moore BCJ, Gockel H (2002) Factors influencing sequential stream segregation. Acta Acustica-Acustica 88:320–333

Naatanen R, Picton T (1987) The N1 wave of the human electric and magnetic response to sound – a review and an analysis of the component structure. Psychophysiology 24:375–425

Simon JZ, Wang Y (2005) Fully complex magnetoencephalography. J Neurosci Methods 149(1):64–73

Snyder JS, Alain C, Picton TW (2006) Effects of attention on neuroelectric correlates of auditory stream segregation. J Cogn Neurosci 18:1–13

Vliegen J, Oxenham AJ (1999) Sequential stream segregation in the absence of spectral cues. J Acoust Soc Am 105:339–346

Woldorff MG, Hillyard SA (1991) Modulation of early auditory processing during selective listening to rapidly presented tones. Electroencephalogr Clin Neurophysiol 79(3):170–191

Xiang J, Wang Y, Simon JZ (2005) MEG responses to speech and stimuli with speechlike modulations. In: International IEEE EMBS conference on neural engineering 2005

Chapter 51
Correlates of Auditory Attention and Task Performance in Primary Auditory and Prefrontal Cortex

Shihab Shamma, Jonathan Fritz, Stephen David, Mounya Elhilali, Daniel Winkowski, and Pingbo Yin

Abstract Auditory experience can reshape cortical maps and transform receptive field properties of neurons in the auditory cortex of the adult animal in a manner that depends on the behavioral context and the acoustic features of the stimuli. This has been shown in physiological and behavioral experiments, in which auditory cortical cells underwent rapid, context-dependent changes of their receptive field properties so as to sculpt the most effective shape for accomplishing the current auditory task. Here, we extend these findings to new behavioral paradigms (utilizing either positive or negative reinforcement) and explore the possible role of top-down signals from prefrontal cortex (PFC) in modulating plasticity in the primary auditory cortex. We also combine physiological experiments with microstimulation in PFC to test if it modulates cortical responses and receptive fields.

Keywords Auditory cortex • Prefrontal Cortex • Attention • Rapid plasticity

51.1 Introduction

Auditory experience can have profound global effects by reshaping cortical maps and significant local effects by transforming receptive field properties of neurons in the primary auditory cortex (A1) (King 2007; Weinberger 2007; Edeline 1999). The exact form of this remarkable plasticity is determined by the salience or task-relevance of the spectral and temporal characteristics of the acoustic stimuli, and may also reflect the behavioral state of the animal in relation to the dimensions of expectation, attention, motivation, motor response, and reward (Recanzone 2000; Kilgard et al. 2001a, b; Kilgard and Merzenich 2002; Knudsen 2007;

S. Shamma (✉)
Institute for Systems Research, Electrical and Computer Engineering Department, University of Maryland, College Park, MD, USA
e-mail: sas@umd.edu

E.A. Lopez-Poveda et al. (eds.), *The Neurophysiological Bases of Auditory Perception*, DOI 10.1007/978-1-4419-5686-6_51, © Springer Science+Business Media, LLC 2010

Kacelnik et al. 2007; Rutkowski and Weinberger 2005). Consistent with findings in other neural systems (Nicolelis and Fanselow 2002; "Gilbert and Sigman 2007; Womelsdorf et al. 2008), auditory cortical cells undergo rapid, short-term, and context-dependent changes of their receptive fields and responses whenever an animal engages in a new auditory behavioral task that has different requirements and stimulus feature salience. In this kind of adaptive plasticity, top-down signals from higher cortical areas associated with engagement in behavioral repertoires may lead to changes in receptive fields that may enhance performance on the relevant sensory tasks (Fritz et al. 2001, 2004, 2007a; Li et al. 2004; see Sect. 51.1).

Cortical receptive fields are situated at the focal juncture of this process, depicted by the simplified and highly schematized model in Fig. 51.1a. During behavior in a trained animal, receptive fields adapt so as to enhance behavioral *performance*, monitored through external (reward or aversive) feedback signals. The auditory cortex receives behaviorally relevant acoustic stimuli (e.g., warning sounds associated with positive or negative reward, and safe sounds associated with positive reward), and generates corresponding *sensory representations* that are ultimately categorized and associated with meaning in the prefrontal cortex (PFC), resulting subsequently in the appropriate motor behavior. This sensory-motor mapping defines a specific learnt *task or behavioral context* (Blake et al. 2006). We hypothesize that this process conceptually involves a series of steps as follows: (a) When an animal engages in auditory behavior, a "behavioral gate" opens that allows specific A1 responses to pass on to higher cortical levels and the PFC (see Sect. 51.2); (b) PFC

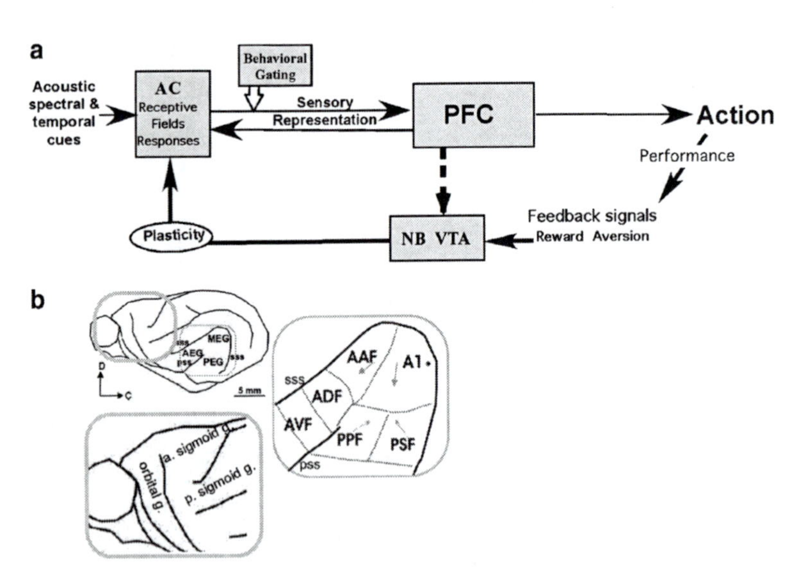

Fig. 51.1 Rapid plasticity during auditory behavior. (**a**) Schematic of cortical and subcortical interactions during auditory behavioral tasks. See text for details. (**b**) Layout of ferret auditory and prefrontal cortex. Recordings are focused in A1, anterior sigmoid gyrus, and dorsal-medial regions of orbital gyrus

responses categorize and encode the meaning of the sounds; (c) these responses [directly top-down or indirectly via nucleus basalis (NB)] induce plasticity in A1, but *only* when paired with the responses to the acoustic stimulus that induced them (see Sect. 51.3.2).

In this report, we shall explore this hypothesis by illustrating some results of combined physiological/behavioral experiments, in which we rapidly and comprehensively characterize cortical response properties in a given cell, e.g., tuning curves and spectrotemporal receptive fields (STRFs), while the animal engages in a series of auditory tasks, and compare these response measures across tasks or during passive listening. We shall illustrate responses during behavior in two cortical regions (Fig. 51.1b): A1, and PFC. We then explore the relationship between responses in these two areas. Our overarching scientific challenge is to explain how the rules and goals of behavioral tasks in combination with the salient acoustic cues dictate the form and extent of cortical plasticity.

51.2 Rapid Plasticity in A1 Receptive Fields

We hypothesize that receptive fields in A1 adapt during behavior so as to enhance processing of salient acoustic information that in turn optimizes behavioral outcomes (i.e. maximize positive reward and/or minimize negative consequences for the animal). This idea can be restated in terms of the various components of a task as follows (Fig. 51.1a): Adaptive changes reflect the nature of stimulus features relevant for performance (spectral, temporal, or spectrotemporal), task objectives (whether the sounds are *targets* or *references, i.e., foreground or background*), and rules of the task (aversive or appetitive rewards, and other task design constraints). Furthermore, while selective attention enhances plasticity to selective acoustic features, globally attending to a complex acoustic stimulus is sufficient to render all its features effective in inducing plasticity. Figure 51.2a illustrates these concepts in the context of several simple tasks that we have tested, highlighting in particular tone detection and tone discrimination. In all tasks, a trial consists of a random number of similar *reference* sounds (blue) followed by a *target* (red). In tone detection, the reference signals are TORC noise (especially designed spectro-temporally modulated broadband noise used to measure the STRFs; see right panel of Fig. 51.2a; Klein et al. 2006), followed by a target tone. In tone discrimination, the reference and target signals are tones of different frequencies; however, both targets and references have TORCs attached to them in order to measure the STRFs (Fig. 51.2a; Fritz et al. 2005a, b). All other tasks have similar structures. In the *aversive* (conditioned avoidance) version of these tasks, animals were trained to lick through the reference epoch, and to refrain from licking for a short period immediately following the target tone in order to avoid a mild shock (Fritz et al. 2003; Heffner and Heffner 1995). In the *appetitive* version of the task, the animals must avoid licking the spout during the reference stimuli, and only lick after they hear the target sound in order to receive a reward.

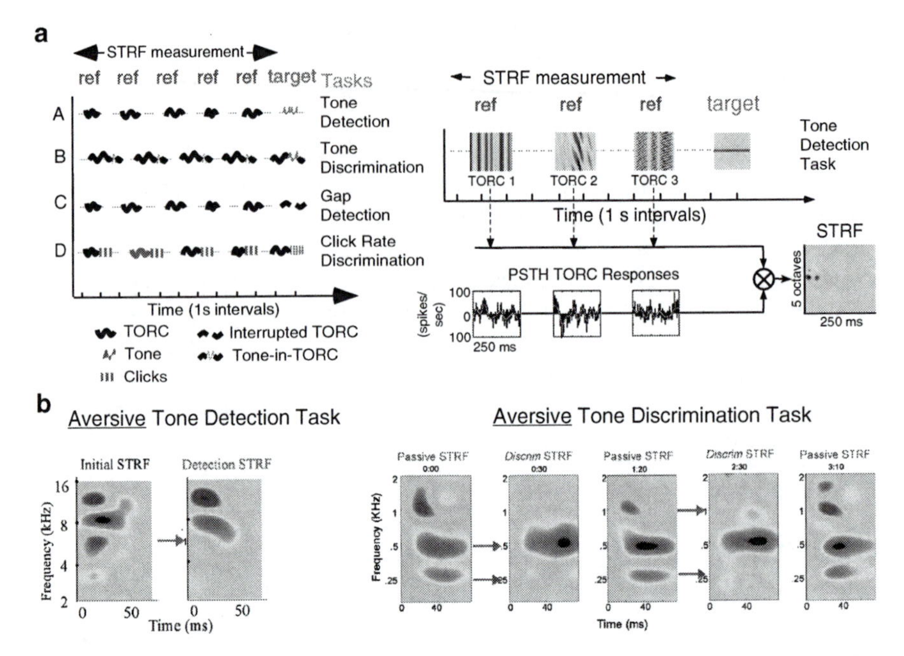

Fig. 51.2 Behavioral physiology and STRF adaptations. (**a**, *left*) Structure of behavioral tasks, illustrating a *reference* epoch (*blue*), followed by a *target* sound (*red*); (*right*) STRF measurements are performed *only* during the reference epoch using the responses to the TORCs, which are specially designed modulated noises. Therefore, target and reference tones and other cues are used to define the behavioral objectives of the task variants, while responses to the *same* TORCs are used to measure the STRFs under different behavioral states. (**b**) STRFs change when the animal engages in a task. Examples of single units in tone-detection task (*left*) and tone-discrimination (*right*) tasks. In these aversive tasks, target sounds induce facilitatory effects while reference tones induce suppression

51.2.1 STRF Plasticity in A1 During Aversive Tone Detection and Discrimination Tasks

STRFs measured before, during, and after performance of aversive tone detection and discrimination tasks exhibit characteristic changes that can be summarized as follows (Fig. 51.2b): Target tones (indicated by *red* arrows) induced enhanced sensitivity in the STRF at their frequency, whereas reference sounds (indicated by blue arrows) induced suppression. These changes are illustrated by the single-unit examples in Fig. 51.2b for the two-tone tasks. We have also found that two spectral factors increase the strength of these adaptations: high performance levels (Fritz et al. 2003; Atiani et al. 2009) and proximity of the target and reference tones to the center of the STRF being observed (Fritz et al. 2007a).

51.2.2 Contrasting Effects of Aversive and Appetitive Tasks

A separate group of ferrets were also trained on the behavioral "inverse" of the aversive (conditioned avoidance) tasks described above, namely on a positive reinforcement (or an appetitive go/no-go) paradigm in which they withheld licking during a (random) number of reference sounds (i.e., the TORCs), and licked only after onset of the target tone, as illustrated in Fig. 51.3a. These two behavioral paradigms form an excellent counterpoint to one another since the *rules and actions* are reversed while the stimuli and the sensory categories remain identical. The comparison of the neural responses in these two "inverse" behavioral paradigms

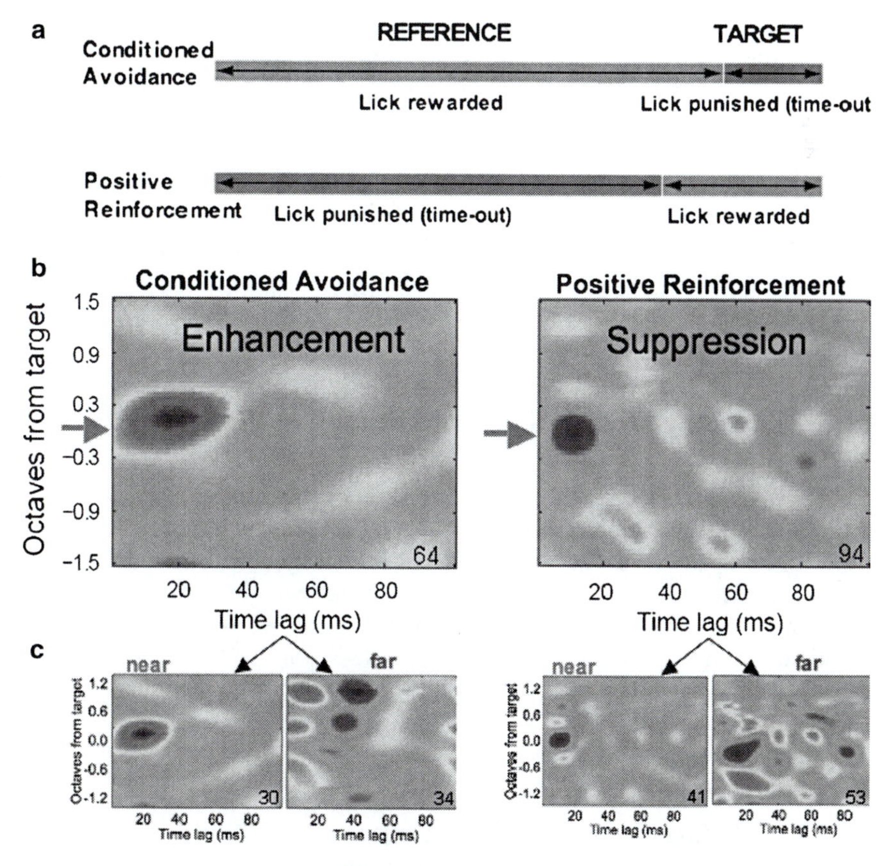

Fig. 51.3 Contrasting average STRF changes during aversive and appetitive tone-detection tasks. (**a**) Structure of the two behavioral tasks. (**b**) Changes from a population of cells reveal enhancement in *aversive* (*left*) and suppression in *appetitive* (*right*) conditions. The averaged changes are computed by aligning the STRF differences from all cells at the target frequency (*red arrow*), and then summing the results. (**c**) In both conditions, targets located *near* the STRF have the greatest impact on plasticity

allowed us to explore the effects of task rules on the responses in the PFC and potentially on the extent of adaptive changes in the primary auditory cortex.

Figure 51.3b contrasts the average plasticity in two populations of neurons in animals engaged in the two types of tone-detection experiments. All STRF changes during the appetitive tasks were measured in exactly the same way as in the aversive task (Fig. 51.2; Fritz et al. 2003). The neural results differed in an important respect. As illustrated above, STRFs in the conditioned avoidance task developed facilitation at the frequency of the target tone (red arrow), while exactly the opposite pattern emerged in the positive enhancement task. These results suggest that, in addition to salient target and reference features, motor and reward contingencies associated with the discrimination task play a central role in determining plasticity effects in A1. Intuitively, the results of Fig. 51.3b may be understandable if one interprets target enhancement as increased sensitivity. Such increased sensitivity in the aversive task would help the animal avoid the shock, but may also increase the "false alarm" rate (i.e. more frequent, but low cost, withdrawals during the reference epoch in aversive tasks). The opposite effect would occur during the appetitive task. Here, incorrectly detecting the target tone during the reference epoch would lead to a costly time-out, and hence could be avoided by *suppressing* target sensitivity. Another intuitively plausible "explanation" is that the sign of the target plasticity arises in relation to an inhibition of on-going behavior – in positive reinforcement, we are inhibiting restraint from licking, whereas in conditioned avoidance, we inhibit on-going licking, so one would expect the sensory-motor linkages to be opposite to one another, and also the plasticity to be opposite in sign.

51.3 Encoding of Task Rules and Stimuli in Prefrontal Cortex

If PFC responses are the source of top-down attentional influences that induce rapid A1 plasticity, then we predict that they should reflect some of its characteristic and distinctive properties such as being strongly contingent on the behavioral context, sensitive to the stimulus categories, dependent on task rules (aversive or appetitive), and would be modulated by task performance. Here, we summarize the basic properties of PFC responses during the same detection and discrimination tasks described above, and then extend the results to behaviors with multimodal (visual and auditory) stimuli.

51.3.1 PFC Responses During Aversive (or Conditioned Avoidance) Tasks

PFC responses in the ferret during task performance exhibit a wide range of properties that encode target onset and offset and other task events and contingencies. The most striking aspect of these responses is their near total dependence on the behavioral

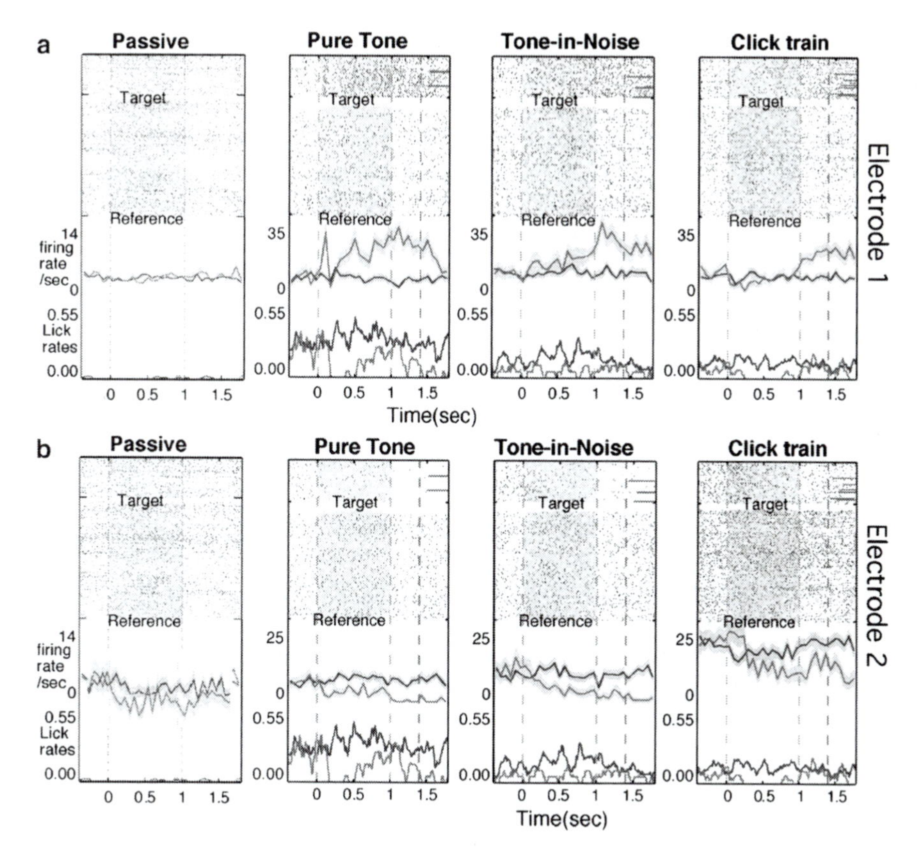

Fig. 51.4 PFC responses in two units recorded simultaneously during three aversive detection tasks. Only target response modulations (*red*) are strong, but *not* during passive state (first panel). (**a**) Response modulations in are all excitatory and roughly similar regardless of the nature of target sound, thus reflecting the meaning of the sound as a target. (**b**) Response modulations are all similar and suppressive in this cell, reflecting again only the meaning of the sound as a target

context. This is illustrated by typical responses shown in Fig. 51.4a, where initially the animal was in the passive state (leftmost panel), and then performed a series of tasks (each separated by a passive pretask stimulus presentation for each task condition): tone discrimination, tone pure tone detection, a tone-in-noise detection, and a click train detection. During the passive state, no significant responses were observed to the target (red) or reference (blue) sounds (first panel from the left). However, when the animal entered a behavioral context, performing the tone-detection task using exactly the same stimuli as in passive state, the identical target tone now induced vigorous excitatory responses. Reference TORCs, by contrast, did not show any change in responses. During the tone-in-noise task (third panel), responses were again strong only to the target (red). Finally, when the target was a click train (fourth panel), the target response buildup remained similar (red), demonstrating again that, in the PFC, the physical, acoustic parameters of the target are not as important as its behavioral *meaning*. Figure 51.7b displays responses from an adjacent electrode in the same experiment. Although the context of the responses was similar, in that they occurred only during behavior, they showed an opposite polarity (targets *suppressed* the firing rate – red PST).

Finally, to highlight the dependence of PFC responses on stimulus meaning and not its specific physical properties, we performed a sequence of two behaviors. The first was a tone-detection task, in which the target tone was one of two randomly alternating tones (2.2 kHz and 550 Hz). Both tones elicited strong responses (red PST) that built up rapidly until the onset of the shock (1.4 s; Fig. 51.5a; second panel), The second task was a two-tone discrimination task using exactly the same tones (2.2 kHz and 550 Hz tones), but in which only one tone acted as target (2.2 kHz), while the other one acted as reference (550 Hz). The results (Fig. 51.5b) were quite compelling, showing that previously vigorous responses to the 550 Hz tone vanished when it played the role of *reference* (blue), while the responses to the other (2.2 kHz) tone remained significant since it continued to play the role of target (red). This type of result demonstrates clearly the context dependency of the PFC responses and the rapidity of adaptive changes in PFC during task switching.

Figure 51.6 displays responses from a population of 200 behaviorally modulated cells to gain a broader view of the different types of PFC responses observed. Panels in Fig. 51.6a depict the responses *before* (left), *during* (center), and *after* (right) a tone-detection task. In the figure, PSTH responses for each cell to the target tone are ordered according to the polarity (excited or inhibited) and latency of the *peak response* during the target tone (center panel). The data reveal several

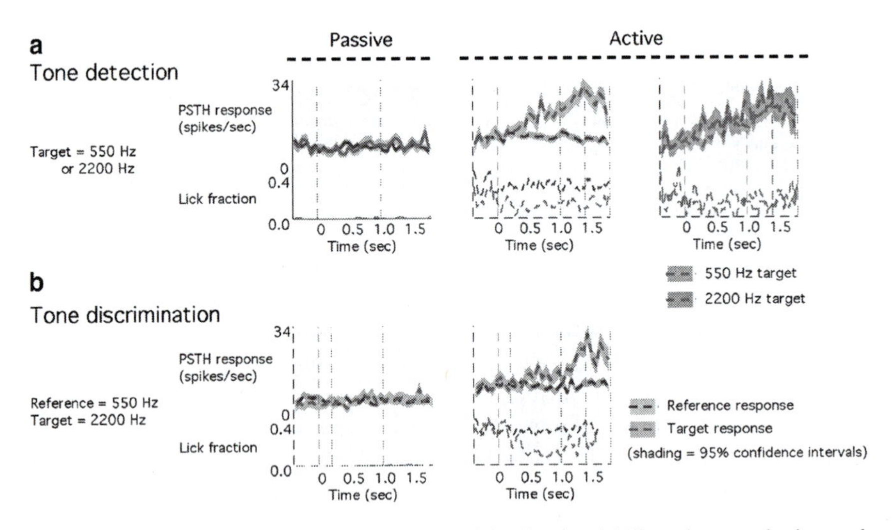

Fig. 51.5 Responses reflect the behavioral context of the stimulus. (**a**) The unit responds vigorously to both tones (550 and 2,200 Hz) when they are targets in a tone-detection task. (**b**) When the 550 Hz tone now serves as a reference tone in a two-tone discrimination task, responses to it cease completely while those to the 2,200 Hz target tone remain strong

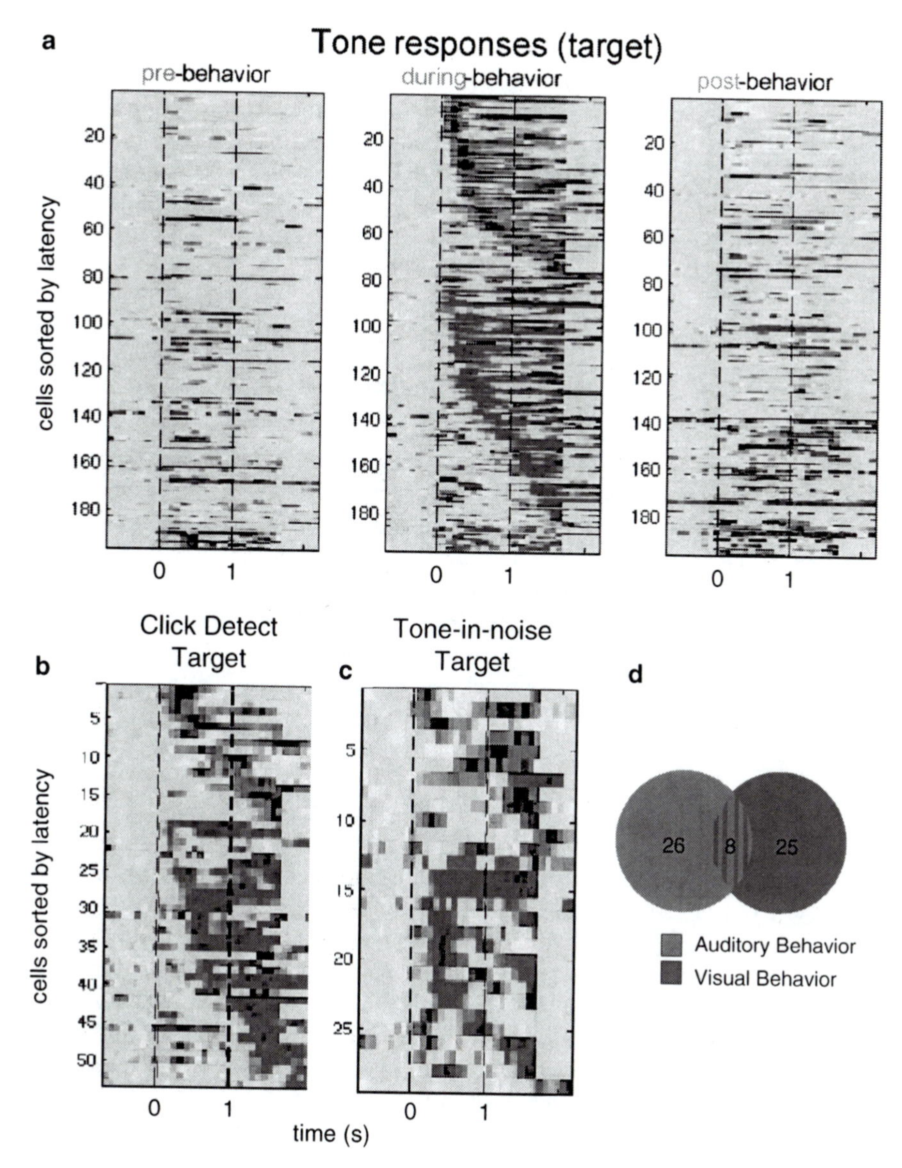

Fig. 51.6 PFC population and multimodal responses during aversive detection tasks. (**a**) PST responses from 200 cells organized according to their latency. Prebehavior passive responses are weak, compared to during the behavior. Persistent weaker responses remain postbehavior. Both excitatory and suppressive modulations of firing rates are found. (**b**, **c**) Similar responses are found with click train and tone-in-noise targets. (**d**) Cells are mostly segregated by modality (auditory and visual), although some overlap exists, thus demonstrating a class of PFC neurons that encode a task target independent of stimulus modality

characteristics common to most cells in PFC: (1) *Responses are context-dependent*, being the strongest *during* task performance, and barely measurable prior to that; (2) cells come in two flavors of *response polarity* to the target, in approximately equal proportions; (3) onset and latency of the peak response relative to onset of target tone are widely and uniformly distributed from about 30 ms to 2 s; (4) responses can be phasic or sustained in roughly equal proportions. Similar trends are seen in data from detection tasks with click trains (Fig. 51.6b) and tone-in-noise (Fig. 51.6c) targets.

51.3.2 PFC Responses During Appetitive Tasks

Basic response properties during appetitive tone-detection tasks were similar in the behavioral contingency, polarity, and latencies. One key difference was the abundance of strong responses to the reference (TORC) sounds. Target responses remained strong, and often of the same polarity in a given cell (enhanced or suppressed). A possible explanation for the stronger responses to reference sounds is their enhanced behavioral significance in the appetitive tasks. This is especially true of the first reference TORC since it was the cue to the animal that the trial has commenced and that they must cease licking (much like the target tone in the aversive tasks).

51.3.3 PFC Responses in Tasks with Visual Stimuli

Ferrets were also trained on a simple visual discrimination task, that paralleled the auditory detect task, using the conditioned avoidance paradigm. They learned to lick freely to a sustained (1 s) dim light – the safe visual stimulus, and refrain from licking after presentation of a bright flashing light (4–8 Hz for 1 s) – the warning, target visual stimulus. As was the case with auditory tasks, PFC responses were selective for the target stimuli, and showed a similar time course. Some PFC neurons responded only to visual targets, others only to auditory targets, and some responded to both visual and auditory targets (see Fig. 51.6d).

51.4 Relationship Between A1 and PFC Responses

We also explored aspects of the relationship between responses in A1 and prefrontal cortices by *simultaneously recording* and measuring the correlations in the activity between these two regions during behavior and also by *microstimulating PFC* neurons so as to induce plasticity in A1. Given the direct and substantial indirect connections between PFC and A1, it is plausible that their responses and local field potentials (LFPs) are dynamically coherent and functionally related, and that microstimulation in PFC (in conjunction with paired acoustic stimuli) could induce forms of plasticity in A1 receptive fields that would be consistent with observed plasticity changes elicited during behavior. Examples of preliminary findings along those directions are shown below.

51.4.1 Analysis and Coherence of Local Field Potentials

Synchrony of neural activity between the PFC and A1 could be indicative of functional connections between them, and also of the distributed representation of attended stimuli (Barcelo et al. 2000; Fuster et al. 1985; Gazzaley et al. 2004; Womelsdof et al. 2007). In order to understand cortical network activity on a large scale, we recorded LFPs simultaneously with single-unit recordings in A1 and PFC. LFPs were acquired from the single-unit electrodes by low-pass filtering the recording below 1 kHz; spikes were measured by band-pass filtering from 1 to 6 kHz.

51.4.1.1 Within PFC and A1 Correlations

Thus far, preliminary analysis of LFP spectrograms *within* PFC during task performance has revealed that (Fig. 51.7a, b): (a) *During* tone detection in the aversive task, the spectral power in two commonly studied LFP bands – β (10–30 Hz) and γ (50–80 Hz) – change systematically relative to the *passive* state. Namely, the β-band always *decreases,* whereas the γ-band *increases* during behavior (Fig. 51.7a). (b) Significant changes also occur in the *relative* power of

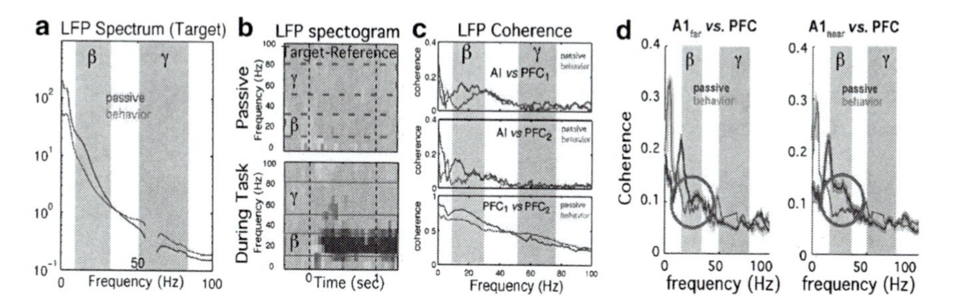

Fig. 51.7 LFP spectrograms and coherence. (**a**) Average LFP spectrum during target stimuli during passive and active states. The β- and γ-bands are indicated by the *shaded ranges*. Break in *curve* is the 60 Hz line filter. (**b**) Spectrogram of LFP's induced by target tone (relative to reference TORCs) highlighting dynamics of different bands of PFC activity during passive and aversive tone-detection task performance. β- and γ-bands are indicated by the *dashed* and *solid red lines*. (**c**) Coherence of LFPs between AI and two locations in PFC (*top two panels*), compared to locations within PFC (*bottom panel*). (**d**) LPF coherence before and during behavior between PFC and two BF locations in AI – *near* target tone (*right*) and *far* from it (*left*)

the target and reference responses in these two bands during different phases of the tasks. For example, the (power) spectrogram of the "target-reference" (Fig. 51.7b) shows modest changes during the *passive* state (top panel), but substantial changes during the *aversive* tone-detection task (lower panel), where the relative power in the β-band was severely depressed following onset of the target. The γ-band exhibited only a weaker decline (Fig. 51.7b).

51.4.1.2 Coherence Between PFC and A1

Another key objective of the experiments in this aim is to measure the coherence of activity *during* the task between simultaneously recorded signals in AC and PFC. Our preliminary data suggest changes in coherence during behavior between these two regions.

For example, in one of several simultaneous recordings, we contrasted the LFP coherence between A1 and PFC (top two panels of Fig. 51.7c) during target duration of the appetitive tone-contour discrimination task (green curve) and passive state (blue curve). This was further contrasted with coherence within the PFC (between the two PFC channels) in the same two states (bottom panel of Fig. 51.7c). Coherence in the β-band (gray shaded) between A1 and PFC decreased dramatically during the task, compared to the (relatively) smaller changes within the PFC. *Within* the PFC, and *within* A1 (not shown), behavior brought about a small decrease in the power of the β-band (gray shade Fig. 51.7c), and a small increase in γ-band (pink shade Fig. 51.7c), a pattern that is seen in all experiments. These coherence patterns, however, were location and task dependent. For instance, in another experiment using the *aversive tone-detection* task (Fig. 51.7d), we found the same substantial β-band coherence decrease *only* between PFC and BF locations in A1 with frequency tuning *near* to the target frequency (Fig. 51.7d: within red circle in right panel), and not in those *far* from the target (Fig. 51.7d: left panel). Full interpretation of these coherence patterns must await more data and further study, but they nevertheless suggest the *hypothesis* that interareal communication and synchrony patterns may become dramatically modulated in the β-band and γ-band during behavior.

51.4.2 Microstimulation in PFC Modulates Receptive Fields in A1

If the PFC is a significant source of the top-down signals responsible for adapting auditory cortical receptive fields and responses, then it might be possible to simulate its natural action during behavior with microstimulation that is paired with the appropriate stimuli. Figure 51.8 illustrates the results of one such preliminary experiment in a *naïve* animal, in which the *reference* stimuli consisted of random tones that covered a range of frequencies surrounding the BF and ending with a

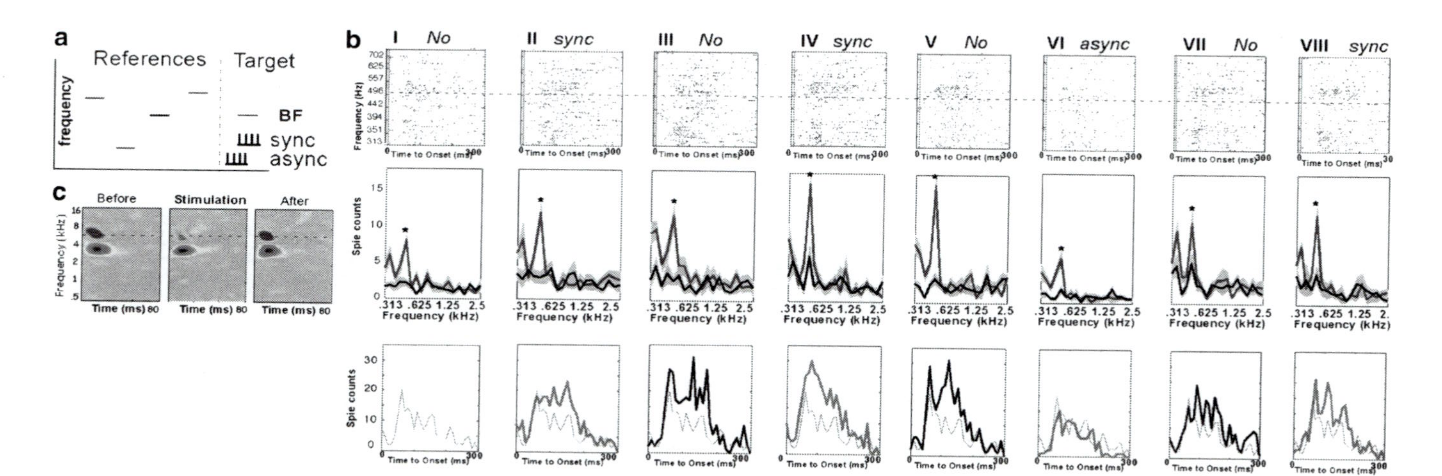

Fig. 51.8 PFC microstimulation alters AI receptive fields. (**a**) Stimuli and stimulation mode. (**b**) Changes in a cell's responses through a sequence of tests (*panel columns I–VIII*) alternating between no stimulation, and PFC *synchronous* or *asynchronous* stimulation. (*Top*) Response rasters; *dashed line* is at BF; (*Mid*) tuning curves (in *blue*) and spontaneous activity (in *black*) *during* the tests; (*Bottom*) PSTs at BF. Initial measurement in *panel I* (gray), *during* stimulation (*red*) and no stimulation (*black*). (**c**) STRF enhancement induced by PFC stimulation paired with the target tone (at *dashed line*) in a passive tone-detection paradigm

150 ms *target* tone set at BF (495 Hz). An electrode inserted in the PFC was used to deliver a series of current pulses that were paired (synchronously or asynchronously; Fig. 51.8a) with the *target* BF tone in alternating blocks of trials. Initially, the unit was tuned near 500 Hz as shown in panel I raster responses (top), tuning curve (middle), and PSTH histogram of the BF tone (bottom). *During* synchronous stimulation with the *target* BF tone (panels II), the unit became more responsive to the BF tone (*asterisk*), and the changes persisted during the following passive epoch (panels III). A second block of stimulation considerably enhanced the responses and tuning around the BF (panels IV), changes which persisted afterward (panels V). In the following block (panels VI), the stimulation pulses *asynchronously* preceded the BF tone by 75 ms. The result was a striking resetting of the cell's responses to its initial state (as in panels I). Subsequent passive measurements showed only a partial rebound (panels VII), but another synchronous stimulation session led to a reemergence of the enhanced tuning and BF responses (panels VIII). In Fig. 51.8c, we illustrate the results from a different animal under *passive* conditions, where we presented the stimuli of the tone-detection paradigm (P1) of TORCs and target tone to measure A1 STRFs simultaneous with microstimulation in PFC, paired with the target tone at 6 kHz. In this case, substantial enhancement occurred at the tone frequency *during* stimulation (dashed line) that altered its shape, only to return partially to its original shape afterward. In summary, PFC stimulation paired with tone stimuli can induce *rapid, reversible,* or *persistent frequency STRF* changes, mimicking those seen with behavior.

51.5 Summary and Discussion

In a series of experiments over the last few years, we have sought to elaborate the role of A1 in expressing rapid plasticity during behavior in various auditory tasks, and to delineate the limitations and requirements of the overall process illustrated in Fig. 51.1. For instance, we found that plasticity does not occur in a naïve *or* nonbehaving trained animal (Elhilali et al. 2007) as that effectively breaks the upper and lower loops of Fig. 51.1a, and with it, the feedback into the A1. Furthermore, we confirmed that enhancement of the acoustic spectral or temporal difference between warning and safe sounds was the key to shaping plasticity (Fritz et al. 2005a, 2005b, 2007a), that STRF changes were significant when salient features of the stimuli were *within* its receptive field boundaries, and were the largest when these features were near the center of the receptive field (Fritz et al. 2007a). We also explored in detail the contribution of task difficulty to the extent of plasticity, and the breakdown of STRF changes into pure gain and (orthogonal) shape changes (Atiani et al. 2009). Finally, we confirmed the hypothesis that "spectral" targets and references induce plasticity that mirrors their spectral structure in a manner consistent with a contrast filter (Fritz et al. 2007a). These findings have suggested new hypotheses that we are currently investigating especially concerning the *role of attention* on STRF plasticity in A1 and the exact nature and *function of the top-down signals* from the PFC (Fritz et al. 2007b). For example, our current hypothesis is that attention plays a global role in these auditory tasks, that target (warning) and reference (safe) stimuli essentially "imprint" their appropriate spectrotemporal signature upon the STRFs during an active task, and that the PFC provides the necessary top-down feedback *during* behavior to initiate and maintain the adaptive processes that change the A1 STRFs. Finally, it should be stressed that in the highly schematized and simplified account of Fig. 51.1a, we have left out many other important factors in initiating, directing, and modulating rapid plasticity such as the neuromodulatory influences on the auditory cortex arising from subcortical structures including the amygdala, NB, ventral tegmental area, and locus coeruleus, the concurrent plasticity of the task itself (or of the sensory-motor map) in the auditory, motor, and premotor cortices (Li et al. 2000; Singh and Scott 2003; Stefan et al. 2004; Selezneva et al. 2006; Frost et al. 2003).

The scientific challenge we face now is to understand how this broadly distributed orchestration of multiple changes in the attentional network, dynamically modulates sensory processing in concert with achieving behavioral goals.

References

Atiani S, Elhilali M, David S, Fritz J, and Shamma S (2009) Task difficulty and performance induce diverse adaptive patterns in gain and shape of primary auditory receptive fields, Neuron 61:467–480

Barcelo F, Suwazono S, Knight RT (2000) Prefrontal modulation of visual processing in humans. Nat Neurosci 3:399–403

Blake DT, Heiser M, Caywood M, Merzenich MM (2006) Expereince-dependent adult cortical plasticity requires cognitive association between sensation and reward. Neuron 52:371–381

Edeline J (1999) Learning-induced physiological plasticity in thalamo-cortical sensory systems: a critical evaluation of receptive field plasticity, map changes and their potential mechanisms. Prog Neurobiol 57:165–224

Elhilali M, Fritz J, Chi T-S, Shamma S (2007) Auditory cortical receptive fields: stable entities with plastic abilities. J Neurosci 27(39):10372–10382

Fritz J, S. Shamma, M. Elhiliali, and D. Klein, (2003) Rapid task-dependent plasticity of spectrotemporal receptive fields in primary auditory cortex, Nature Neuroscience, 6(11): 1216–1223

Fritz J, Shamma S, Elhiliali M, Klein D (2003) Rapid task-dependent plasticity of spectrotemporal receptive fields in primary auditory cortex. Nat Neurosci 6(11):1216–1223

Fritz J, Elhiliali M, Shamma S (2005a) Differential dynamic plasticity of A1 receptive fields during multiple spectral tasks, J Neurosci 25:7623–7635

Fritz J, Elhiliali M, Shamma S (2005b) Active listening: task-dependent plasticity of receptive fields in primary auditory cortex. Hear Res 206:159–176

Fritz J, Elhilali M, Shamma S (2007a) Adaptive changes in cortical receptive fields induced by attention to complex sounds. J Neurophysiol 98:2337–2346

Fritz J, Elhiliali M, Shamma S (2007b) Does attention play a role in dynamic receptive field adaptation to changing acoustic salience in A1? Hear Res 229(1–2):186–203

Frost S, Barbay S, Friel K, Plautz E, Nudo R (2003) Reorganization of remote cortical regions after ischemic brain injury: a potential substrate for stroke recovery. J Neurophys 89:3205–3214
Fuster JM, Bauer RH, Jervey JP (1985) Functional interactions between inferotemporal and prefrontal cortex in a cognitive task. Brain Res 330:299–307
Gazzaley A, Rissman J, D'esposito M (2004) Functional connectivity during working memory maintenance. Cogn Affect Behav Neurosci 4:580–599
Gilbert C, Sigman M (2007) Brain states: top-down influences in sensory processing. Neuron 54:1
Heffner HE, Heffner RS (1995) Conditioned avoidance. In: Klump GM et al (eds) Methods in comparative psychoacoustics, vol 6, BioMethods. Birkhauser Verlag, Switzerland, pp 79–94
Kacelnik O, Nodal F, Parsons C, King A (2007) Training-induced plasticity of auditory localization in adult mammals. PLoS Biol 4:e71
Kilgard M, Merzenich M (2002) Order-sensitive plasticity in adult primary auditory cortex. Proc Nat Acad Sci USA 99:3205–3209
Kilgard M, Pandya P, Vazquez J, Rathbun DL, Engineer N, Moucha R (2001a) Spectral features control temporal plasticity in auditory cortex. Audiol Neurootol 6:196–202
Kilgard M, Pandya P, Vazquez J, Gehi A, Schreiner C, Merzenich M (2001b) Sensory input directs spatial and temporal plasticity in primary auditory cortex. J Neurosci 86:326–337
King AJ (2007) Auditory neuroscience: filling in the gaps. Curr Biol 17(18):R799–R801
Klein DJ, Depireux DA, Simon JZ, Shamma SA (2006) Linear stimulus-invarient processing and spectrotemporal reverse-correlation in primary auditory cortex. J Comput Neurosci Springer Netherlands ISSN 0929-5313 (Print) 1573-6873 (Online) 20(2):111–136 DOI 10.1007/s10827-005-3589-4
Knudsen E (2007) Fundamental components of attention. Annu Rev Neurosci 30:57–78
Li W, Poech V, Gilbert C (2004) Perceptual learning and top-down influences in primary visual cortex. Nat Neurosci 7(6):651
Nicolelis MA, Fanselow EE (2002) Dynamic shifting in thalamocortical processing during different behavioural states. Philos Trans R Soc Lond B Biol Sci 357(1428):1753–1758
Recanzone G (2000) Response profiles of auditory cortical neurons to tones and noise in behaving macaque monkeys. Hear Res 150:104–118
Rutkowski RG, Weinberger NM (2005) Encoding of learned importance of sound by magnitude of representational area in primary auditory cortex. Proc Natl Acad Sci USA 102(38):13664–13669
Selezneva E, Scheich H, Brosch M (2006) Dual time scales for categorical decision making in auditory cortex. Curr Biol 16:2428–2433
Singh S, Scott S (2003) A motor learning strategy reflects neural circuitry for limb control. Nat Neurosci 6:399–403
Stefan K, Wycislo M, Classen J (2004) Modulation of associative human motor cortical plasticity by attention. J Neurophysiol 92:66–72
Weinberger NM (2007) Auditory associative memory and representational plasticity in the primary auditory cortex. Hear Res 229(1):54–68
Womelsdof T, Mathijs J, Oostenveld R, Singer W Desimone R, Engel A, Fries P (2007) Modulation of Neuronal Interactions Through Neuronal Synchronization Science 316:1156–1160
Womelsdorf T, Anton-Erxleben K, Treue S (2008) Receptive field shift and shrinkage in macaque middle temporal area through attentional gain modulation. J Neurosci 28(36):8934–8944

Chapter 52
The Implicit Learning of Noise: Behavioral Data and Computational Models

Trevor R. Agus, Marion Beauvais, Simon J. Thorpe, and Daniel Pressnitzer

Abstract A prerequisite for the recognition of natural sounds is that an auditory signature of the sources should be learnt through experience. Here, we used random waveforms to investigate whether implicit learning occurs with repeated exposure. Listeners were asked to discriminate between 1-s samples of running noise and two seamlessly repeated 0.5-s samples of noise (repeated noise; RN). Unbeknownst to them, one particular exemplar of repeated noise was presented in several trials interspersed in a block (reference repeated noise; RefRN). A first experiment showed that RefRN was discriminated more efficiently than RN, even though listeners did not know that memorizing noise samples across trials could be beneficial, nor which trials should be memorized. Most of the learning occurred rapidly within ten presentations. A second experiment confirmed that the effect was a genuine increase in sensitivity. Preliminary investigations of computational models of cochlear and cortical activity failed to distinguish noises that produced large learning effects, so it is unclear which noise aspects were learnt. The findings show neural learning mechanisms of acoustic information that are unsupervised, resilient to interference, and fast-acting.

Keywords Memory • Auditory psychophysics • Perceptual learning • Plasticity

52.1 Introduction

One basic goal of auditory perception is to recognize the plausible physical cause of incoming sensory information. In order to do so, listeners must learn recurring features or templates of complex sounds and associate them to specific sound sources. How such templates are formed and stored in memory is still largely unknown

D. Pressnitzer (✉)
CNRS & Université Paris Descartes & École normale supérieure,
29 rue d'Ulm, 75005 Paris, France
e-mail: Daniel.Pressnitzer@ens.fr

E.A. Lopez-Poveda et al. (eds.), *The Neurophysiological Bases of Auditory Perception*,
DOI 10.1007/978-1-4419-5686-6_52, © Springer Science+Business Media, LLC 2010

(Demany and Semal 2008). Here, we investigated whether repeated exposure to a long-duration (0.5 s) random noise sample would be sufficient to cause some learning of the sample. It was reasoned that random noises may be especially suited stimuli with which to probe the formation of new templates, as the listener would not have been exposed to a given sample before the experiment. In addition, before learning occurs, random noises are acoustically diverse, yet perceptually homogenous.

Listeners can distinguish Gaussian noise samples in a same-different task (Hanna 1984). Moreover, if the same noise samples are used throughout a block of trials, listeners' ability to distinguish them improves (Goossens et al. 2008). This indicates that listeners can learn some aspects of random noise samples when the experimental task requires them to do so. It is unclear, however, whether the mechanisms recruited in such tasks apply to more realistic situations in which learning has to occur without explicit repetition of the sound to be learnt and with intermingled distracting sounds.

We used an indirect measure to probe the effect of repeated exposure. Listeners were either presented with a 1-s sample of running noise (noise condition, N), or two seamlessly repeated 0.5-s samples of noise (repeated noise, RN). They were asked to discriminate between the two conditions. Several studies show that listeners are able to perform such a repetition-detection task (Guttman and Julesz 1963; Warren et al. 2001; Kaernbach 2004), although with just one repetition and a noise duration of 0.5 s, performance is expected to be only slightly above chance (Kaernbach 2004). In the present experiment, unbeknownst to the listeners, one particular exemplar of the RN condition was repeated identically in several trials interspersed throughout an experimental block (reference repeated noise, RefRN). Note that listeners were not told that memorizing the RefRN trials might be beneficial, and in any case, they could not have predicted when the RefRN trials would occur. Thus, any difference in performance between RefRN and RN on the repetition-detection task is likely to reflect implicit learning caused by repeated exposure.

52.2 Experiment 1

52.2.1 Method

Stimuli. The stimuli were formed from Gaussian noises, generated as sequences of normally-distributed random numbers at 44.1 kHz. The N condition was a 1-s sample of noise. The RN and RefRN were both formed from a 0.5-s sample of noise, concatenated to an identical copy of itself without any intervening silence. Both N and RN were generated anew for each presentation. RefRN was identical within an experimental block.

Procedure. Each block consisted of 100 N trials, 50 RN trials, and 50 RefRN trials. Thus, half of the trials featured repeated noise, and half were unrepeated noise. The trials were pseudorandomly ordered, with the restriction that the RefRN was never presented on two consecutive trials. After each stimulus presentation, listeners had to report whether or not they heard a repetition in a "yes-no" task. Each listener completed four blocks of 200 trials, with each block based on a different RefRN. No feedback was given.

Apparatus. Stimuli were played through an RME Fireface digital-to-analog converter at a 16-bit resolution and a 44.1-kHz sample-rate. They were presented to both ears simultaneously through Sennheiser HD 250 Linear II headphones. Presentation level was at 70 dB(A). Listeners were tested individually in a double-walled IAC sound booth.

Participants. There were 12 listeners with self-reported normal hearing, aged between 19 and 55. They had not previously participated in experiments involving repeated-noise stimuli. Six of these "naïve listeners" returned as "trained listeners" and repeated the experiment with different RefRNs.

52.2.2 Results

Figure 52.1a shows the mean sensitivity to repetitions for RN and RefRN conditions for the naïve listeners. Performance was better for RefRN (mean $d'=1.0$) than RN (mean $d'=0.4$; $t_{11}=4.12$, $p=0.002$). Since RN and RefRN were created by the same process, this difference cannot be attributed to acoustical differences. The patterns persisted when six of the naïve listeners repeated the experiment as trained listeners (Fig. 52.1b). Trained listeners improved their ability to detect the RN (mean $d'=0.5$ versus 0.1 for this subgroup of listeners in the first part of the experiment; $t_5=3.82$, $p=0.01$), but the improvement was even larger for the new RefRN (mean $d'=1.4$ versus 0.6; $t_5=3.9$, $p=0.01$). Figure 52.1c summarizes the benefit of repeated exposure for RefRN by displaying the difference in d's between RefRN and RN. All listeners but one showed a benefit. The benefit increased for all listeners that repeated the experiment. Finally, across both sets of results, the benefit was positively correlated with RN performance ($r_{16}=0.54$, $p=0.02$). In other words, the better performers on the repetition-detection tasks displayed the greater benefit to repeated exposure.

Figure 52.1d shows the mean hit-rates plotted in the order of the individual trials within each block, averaged across the four blocks for each of the 12 listeners. Initially the hit-rates are the same for RN and RefRN. This was expected because RN and RefRN are drawn from the same random process. After that, an increase in the hit-rate for the RefRN is observed. Three-parameter exponential curves were fitted to the RefRN data by the least-squares method. The exponential model provided a significantly better description of the data than a model that assumed a constant hit-rate at the mean, even taking into account the additional two parameters ($F_{2,47}=8.29$, $p=0.001$) (see Motulsky and Christopoulos 2004, p. 142). This demonstrates that the hit-rate for RefRN improved significantly during the block. Based on the time-constant of the exponential (mean half-life = 3.9 trials), most of the learning occurred within the first ten presentations of RefRN. A similar analysis was also performed on the RN data and it showed that the hit-rate for RN decreased during the block, with a similar time course (mean half-life = 6.6 trials).

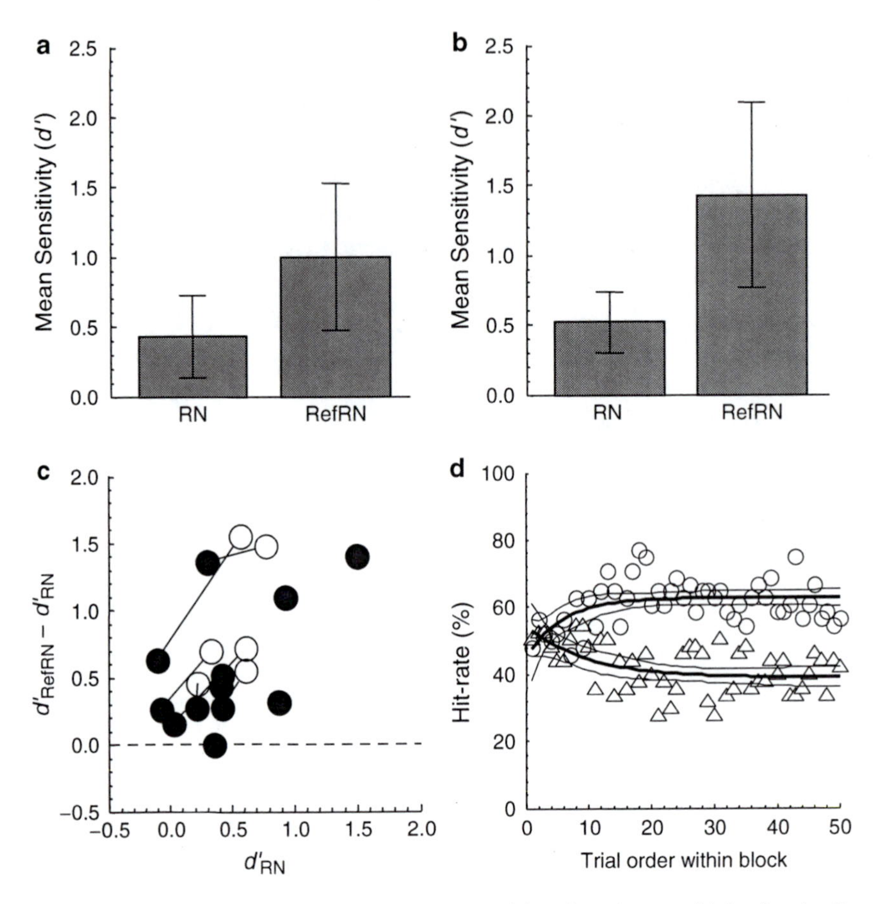

Fig. 52.1 Results for Experiment 1. (**a**) Mean repetition-detection sensitivity for the first visit of all 12 naïve listeners for the RN and RefRN. Error bars are 95% confidence intervals. (**b**) The same, but for the second session of the six listeners who repeated the experiment as trained listeners. (**c**) The benefit for RefRN plotted against RN sensitivity for naïve listeners (*filled circles*) and trained listeners (*open circles*) with the lines linking the results of the same listeners. Results above the *dotted line* indicate a benefit for RefRN. (**d**) The variation of RN and RefRN hit-rates throughout the block averaged over all blocks for naïve and trained listeners. The *thick lines* show the best-fit exponential lines to each of the data, and the *thin lines* show the 95% confidence intervals of these fitted exponentials

52.3 Experiment 2

The advantage of RefRN over RN in Experiment 1 could have been due to a genuine increase in sensitivity to RefRN or to a reduced sensitivity to RN during the block. In Experiment 2, RN and RefRN performance were measured in separate blocks ("RN-only" and "RefRN-only", respectively) as well as in combination ("Mixed", as in Experiment 1). Following Experiment 1, a benefit for RefRN is expected in the

Mixed condition. Comparing performance between the Mixed and RN-only shows if RN performance is affected by the inclusion of the RefRN.

52.3.1 Method

Stimuli. The N, RN, and RefRN stimuli were generated by the same process as in Experiment 1. Different RefRNs were generated for each block.

Procedure. There were three conditions: Mixed, RN-only, and RefRN-only. The blocks of the Mixed condition were identical to those of Experiment 1 (100 N, 50 RN, and 50 RefRN trials). The RN-only condition was equivalent, except for the RefRN trials replaced by RN trials (100 N and 100 RN); conversely, in the RefRN-only condition, the RN trials were replaced by RefRN trials (100 N and 100 RefRN). There were four Mixed blocks, two RN-only blocks, and two RefRN-only blocks. Thus, there were an equal number of RN trials in the RN-only and Mixed conditions, and the same number of RefRN trials in the RefRN-only and Mixed conditions. The blocks were grouped by condition, and the ordering of the three conditions formed a Latin square over all the listeners. Listeners were not made aware of the differences between the conditions. There were six listeners, four of whom had previously taken part in Experiment 1.

52.3.2 Results

Figure 52.2a shows the sensitivities and criteria for Experiment 2. In the Mixed condition, a benefit for RefRN was observed over RN, replicating the result from the first experiment ($t_5=7.94$, $p=0.001$). Importantly, performance for RN was the same in both the Mixed and RN-only conditions ($t_5=0.18$, $p=0.99$).

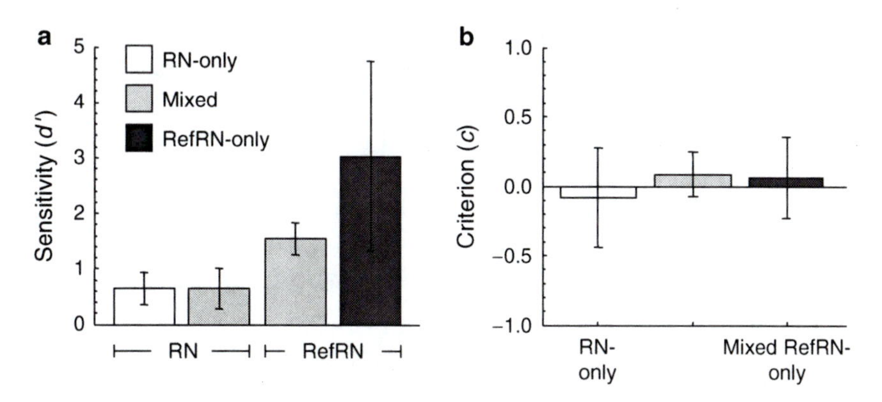

Fig. 52.2 Results for Experiment 2. (**a**) Mean sensitivity for RN and RefRN in each of the conditions. (**b**) Mean criteria for each of the conditions. Error bars are 95% confidence intervals

This shows that the inclusion of RefRN trials does not impair sensitivity for RN trials. We also observed a trend for a better sensitivity for RefRN in the RefRN-only condition compared to the Mixed condition ($t_5=2.24$, $p=0.08$). It may have been easier to learn the RefRN trials when they occurred at a greater rate, and sometimes on subsequent trials. Finally, as illustrated in Fig. 52.2b, there was no effect on criterion across conditions ($F_{2,10}=1.06$, $p=0.39$). A single criterion was computed for the Mixed condition, as subjects are unable to maintain separate criteria when performing interleaved detection tasks (Gorea and Sagi 2000).

52.4 Computational Models

Listeners often report detecting the repetition in RN stimuli by means of particular features that are revealed by the repetition (Guttman and Julesz 1963; Kaernbach 2004). It is therefore possible that noises which, by chance, exhibit more-salient features would be easier to memorize. We performed a preliminary investigation of this hypothesis by trying to correlate features extracted by computational modeling with performance in the RefRN detection task. All of the RefRN stimuli used in the experiments were processed by the model of Chi et al. (2005) and various statistics were extracted. The simplest kind of salient feature might be a large peak in the spectrotemporal pattern of excitation at the output of the cochlea. We computed the average and standard deviation for cochleagrams of white noise by running 1,000 samples through the model and then computed *z*-scores for all RefRNs. The maximum *z*-score, however, was not correlated with the sensitivity (d') to each RefRN ($r_{70}=0.01$, $p=0.91$). Features might also be expressed as more complex patterns in the spectrotemporal plane, such as brief frequency glides or sudden amplitude variations. Such patterns would be captured by the "cortical" representation of the model (Chi et al. 2005). No correlation was found between the maximum *z*-score (along the dimensions of time, frequency, spectral modulation, and temporal modulation) and behavioral performance ($r_{70}=-0.01$, $p=0.95$). It could be that different feature-extraction methods would produce better correlations, or as suggested by behavioral studies, that listeners do not all use the same feature to detect repetitions (Limbert 1984; Kaernbach 2004).

52.5 Discussion

52.5.1 *Repeated Exposure Produced Learning of Noise Samples*

The results showed that listeners were better able to detect RefRN stimuli, which re-occurred throughout a block, even though they were intermingled with RN. The effect was truly one of increased sensitivity to RefRN, shown in Experiment 2, and not a decreased sensitivity to RN. This is consistent with the observation in

Experiment 1 that the increase in hit-rate with successive presentations of RefRN was mirrored by a decrease in hit-rate for RN: as the sensitivity to RefRN increases while listeners maintain their overall criterion (Fig. 52.2b; Gorea and Sagi 2000), hit-rates to RN must decrease during the block. Repeated exposure thus improved the sensitivity to RefRN.

It is not a novel result per se that the use of the same noise token over multiple trials improves performance. For example, pure-tone detection in a noise masker improves if the same noise token is used throughout a block (Pfafflin 1968; Wright and Saberi 1999). Pfafflin attributed this benefit to the learning of the noise stimuli. However, using tonal maskers, Wright and Saberi (1999) showed that the reduced external variability across trials when a small set of maskers are used could explain part of the benefit. In Experiment 1, the variability of stimuli was little reduced by the inclusion of RefRN: the variability in any pair of trials was as great as if no RefRN had been used because RefRN trials were never presented consecutively. Across all trials, the variability was reduced by at most 25%, since only one in four of the trials were RefRN. This does not seem sufficient to account for the doubling in sensitivity observed in RefRN. It is also unclear how an overall reduction in variance would explain the progressive increase observed for only the RefRN hit-rate (Fig. 52.1d). The current results seem better described as an effect of learning than one of reduced variability.

The learning that was achieved was relatively robust to interference. Figure 52.3 shows the hit-rate for RefRN trials re-analyzed according to the number of intervening trials (N or RN) between two successive presentations. Performance did not decay with increasing number of intervening trials. The average duration of a trial was 2.3 s, so no decay in performance was observed for more than 15 s between RefRN presentations, even though listeners could not tell a priori which trials were to be learnt and which were potentially interfering with the learning process.

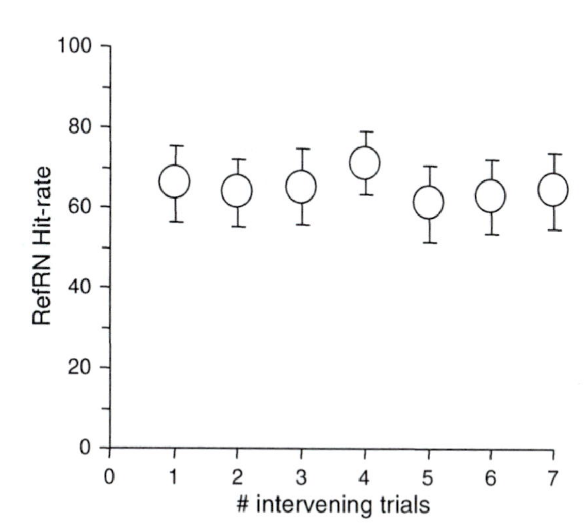

Fig. 52.3 Hit-rate for RefRN in Experiment 1, re-analyzed as a function of the number of intervening trials since the previous RefRN trial within the block

52.5.2 Forming Noise Templates or Memorizing Single Features

There are at least two conceivable forms of learning that could explain the results. It might be that a template of the neural activity produced by the RefRN sample was somehow imprinted by the repeated exposure, facilitating further processing of the sample when it re-occurred. It is also possible that listeners discovered a particular feature in the RefRN trials at the beginning of the block, and attended to the corresponding frequency region and time in the stimulus for subsequent trials. It is difficult to discard either of those possibilities on the basis of the present results alone, but a few elements make the single-feature hypothesis less likely. First, in line with previous studies (Limbert 1984; Kaernbach 2004), our computational model failed to exhibit a simple salient feature that would explain the amount of learning. If listeners indeed memorized a single feature, it is likely to be a different one for different listeners. Second, if listeners focused their attention on one particular feature of the RefRN noise, this should have had an effect on the sensitivity to RN stimuli. An increase or a decrease in sensitivity could be anticipated: focused attention on a subpart of the stimulus might increase sensitivity by reducing the amount of information to process (Goossens et al. 2008), or it could impair sensitivity by missing a useful feature of the new RN noises. However, we observed no effect on RN sensitivity due to the presence of RefRN trials.

52.5.3 Constraints for Neural Mechanisms

The RN and RefRN trials were acoustically identical on average, with no obvious features in terms of the distribution of energy across frequency channels. Thus, any features that were learnt must have been more subtle than those that would be required to distinguish many everyday sounds. Furthermore, the learning occurred implicitly, by means of repeated exposure at random times within a block, and with interspersed interferers. This is therefore a computationally difficult problem. A recent modeling study has shown that spike-time dependent plasticity (STDP) can show such learning behavior, with the absence of both explicit exemplars and salient features. Masquelier et al. (2008) presented random patterns of activity embedded in noise to neurons with simulated STDP mechanisms. They invariably observed the emergence of selectivity to the repeated pattern, if enough repetitions were presented to the STDP neuron. A further constraint suggested by our results is that the learning is fast. Such fast-acting plasticity, with adaptation to the auditory task, has been reported in auditory cortex (Fritz et al. 2007).

Finally, whatever the underlying neural mechanism, the features of the learning that we observed for noise samples (unsupervised, resistant to interference, and fast-acting) would be highly desirable for learning about the structure of the acoustic environment in realistic situations.

Acknowledgments This work was supported by grant ANR-06-NEUR-O22-01. We thank Timothée Masquelier for sharing pilot data using a similar paradigm (Masquelier & Thorpe, unpublished observations).

References

Chi T, Ru P, Shamma SA (2005) Multiresolution spectrotemporal analysis of complex sounds. J Acoust Soc Am 118(2):887–906

Demany L, Semal C (2008) The role of memory in auditory perception. In: Yost WA, Popper AN, Fay RR (eds) Auditory perception of sound sources. Springer, New York

Fritz JB, Elhilali M, David SV et al (2007) Does attention play a role in dynamic receptive field adaptation to changing acoustic salience in A1? Hear Res 229(1–2):186–203

Goossens T, van de Par S, Kohlrausch A (2008) On the ability to discriminate Gaussian-noise tokens or random tone-burst complexes. J Acoust Soc Am 124(4):2251–2262

Gorea A, Sagi D (2000) Failure to handle more than one internal representation in visual detection tasks. Proc Natl Acad Sci USA 97(22):12380–12384

Guttman N, Julesz B (1963) Lower limits of auditory analysis. J Acous Soc Am 35(4):610

Hanna TE (1984) Discrimination of reproducible noise as a function of bandwidth and duration. Percept Psychophys 36(5):409–416

Kaernbach C (2004) The memory of noise. Exp Psychol 51(4):240–248

Limbert C (1984) The perception of repeated noise. Dissertation, Cambridge University

Masquelier T, Guyonneau R, Thorpe SJ (2008) Spike timing dependent plasticity finds the start of repeating patterns in continuous spike trains. PLoS Comput Biol 1:e1377

Motulsky H, Christopoulos A (2004) Fitting models to biological data using linear and nonlinear regression. Oxford University Press, New York

Pfafflin SM (1968) Detection of auditory signal in restricted sets of reproducible noise. J Acoust Soc Am 43(3):487–490

Warren RM, Bashford JA Jr, Cooley JM et al (2001) Detection of acoustic repetition for very long stochastic patterns. Percept Psychophys 63(1):175–182

Wright BA, Saberi K (1999) Strategies used to detect auditory signals in small sets of random maskers. J Acoust Soc Am 105(3):1765–1775

Chapter 53
Role of Primary Auditory Cortex in Acoustic Orientation and Approach-to-Target Responses

Fernando R. Nodal, Victoria M. Bajo, and Andrew J. King

Abstract The role of the primary auditory cortex in acoustic orientation and sound localization was explored by measuring the accuracy with which ferrets turned toward and approached the source of broadband sounds in the horizontal plane. In two animal groups trained by positive conditioning to perform this task, we either made small, bilateral aspiration lesions in the middle ectosylvian gyrus, where the primary auditory cortex is located, or implanted over it a polymer that released over a period of several weeks the $GABA_A$ agonist muscimol to silence the underlying cortex. Neither aspiration nor inactivation of the primary auditory cortex prevented the animals from orienting toward or approaching the source of the sound, but did produce a modest deficit in performance, as reflected by a reduction in accuracy in the localization of brief sounds (40-200 ms), especially for lateral locations. The accuracy of orienting responses was essentially unaltered, and only a small increase in variability for lateral sound locations was observed for short-duration stimuli. The close similarity in the behavioral responses observed after each manipulation suggests that these results reflect the true contribution of the primary auditory cortex to auditory localization. Given the modest nature of the observed deficits, our findings further suggest that additional areas within the auditory cortex must be also involved in spatial hearing.

Keywords Sound localization • Head movement • Orienting response • Auditory space • Azimuth • Mutual information

F.R. Nodal (✉)
Department of Physiology, Anatomy and Genetics, University of Oxford, Sherrington Building, Parks Road, Oxford OX1 3PT, UK
e-mail: fernando.nodal@dpag.ox.ac.uk

E.A. Lopez-Poveda et al. (eds.), *The Neurophysiological Bases of Auditory Perception*,
DOI 10.1007/978-1-4419-5686-6_53,

53.1 Introduction

The role of auditory cortex in sound localization has been examined in different mammalian species (rats, Kelly and Kavanagh 1986; ferrets, Kavanagh and Kelly 1987; Smith et al. 2004; cats, Jenkins and Merzenich 1984; Beitel and Kaas 1993; Malhotra et al. 2004; monkeys, Heffner and Masterton 1975; humans, Adriani et al. 2003). In general, it is accepted that, at least in primates and carnivores, damage to or inactivation of the auditory cortex has a disruptive effect on sound localization accuracy in the contralateral spatial hemifield. However, the magnitude of the deficits that have been reported depends on multiple factors, including the exact region and extent of the cortex affected and the method employed to measure localization performance. The behavioral approaches used to assess sound localization abilities either involve training animals to approach the location of the sound source (e.g., Kavanagh and Kelly 1987; Jenkins and Merzenich 1984) or measuring acoustic orienting responses (Thompson and Masterton 1978; Beitel and Kaas 1993). Some studies have required the animals to select between two loudspeakers (Heffner and Masterton 1975; Kavanagh and Kelly 1987), whereas others have been based on multiple choice paradigms, with the loudspeakers spanning different regions of space (Jenkins and Merzenich 1984; Smith et al. 2004; Malhotra et al. 2004). These methodological differences have complicated conclusions about the precise role played by the auditory cortex in sound localization.

We have previously characterized sound localization behavior in ferrets by measuring both the head orienting and approach-to-target responses made by the animals following the presentation of broadband noise bursts from 1 to 12 loudspeakers covering the full 360° of sound azimuth (Parsons et al. 1999; Kacelnik et al. 2006; Nodal et al. 2008). Both measures of performance varied in accuracy with sound direction, but were affected in different ways when the duration of the stimulus was varied (Nodal et al. 2008). Nevertheless, we observed a close correlation between the head orienting and approach-to-target responses, raising the possibility that the same neural circuitry for translating auditory signals into motor commands may contribute to these two behaviors. In the present study, we explored this possibility further by investigating the effects of either aspirating the primary auditory cortex (A1) or reversibly inactivating it bilaterally in trained ferrets on the accuracy of both the initial head orienting movement and the subsequent approach-to-target response.

53.2 Methods

Seven adult pigmented ferrets (*Mustela putorius*) were used in this study. The apparatus and methods used to train the ferrets have been described in detail in previous reports and are outlined briefly below (Parsons et al. 1999; Kacelnik et al. 2006; Nodal et al. 2008). All procedures involving animals were performed following local ethical review committee approval and under license from the UK Home Office in accordance with the Animal (Scientific Procedures) Act (1986).

53.2.1 Sound Localization Testing

The localization task was carried out in a circular arena of 70 cm radius enclosed by a hemispheric mesh dome, located inside a double-walled testing chamber. Twelve loudspeakers, positioned at 30° intervals around the perimeter of the arena and hidden from the animal by a muslin curtain, were used to deliver the sound stimuli once the animal was positioned on a platform at the center of the arena. Training the animals to lick a waterspout located in front of this platform ensured that the head was consistently oriented toward the loudspeaker at 0° when the stimuli were delivered.

All stimuli were broadband noise bursts (with a low-pass cut off frequency of 30 kHz) generated afresh each time using Tucker-Davis Technologies (Alachua Fl) System 2 hardware. The stimuli were filtered using the inverse transfer function for each speaker in order to obtain a flat spectrum, and matched for overall level across the different speakers. The animals were initially trained to approach the speakers using continuous noise. Once the animals had learned the task, data were collected using sound durations of 2,000, 1,000, 500, 200, 100, and 40 ms. In each testing session, the sound duration was kept constant, while the level was roved pseudo-randomly from trial to trial in 7 dB steps from 56 to 84 dB SPL.

The animals were trained to approach the speaker that delivered the stimulus in order to receive a water reward. Each of the 12 loudspeakers had a separate reward spout located beneath it and only correct responses were rewarded. To avoid bias toward particular speaker locations, every incorrect response was followed by a correction trial (same stimulus and location) up to two times. If the animal continued to mislocalize the sound, an easy trial (in which continuous noise was delivered from the same loudspeaker) was presented. Neither the responses from the correction nor the easy trials were included in the analysis.

The horizontal angle of the head was measured during the first second after the onset of the stimulus by tracking at a 50 Hz frame rate the position of a self-adhesive reflective strip attached along the midline using an overhead infrared-sensitive camera and video contrast detection device. From the coordinates of the head strip, we calculated the angular extent of the acoustic orienting response relative to the initial head position (see Nodal et al. 2008 for details). Each testing period lasted for a maximum of 14 days, during which at least 300 trials were performed at each of the stimulus durations.

53.2.2 Inactivation of the Auditory Cortex

All animals were extensively trained in the localization task before any cortical manipulation was carried out, both to ensure that any later observed deficit could not be attributed to a learning problem and to provide pre-lesion control data. Bilateral aspiration lesions of A1 were made in three ferrets under ketamine/medetomidine anesthesia. Briefly, a craniotomy was made over the ectosylvian gyrus, the dura was retracted and electrophysiological recordings made to help determine the ventral limit of A1. Together with visible cortical landmarks, these recordings were used to judge how much of the cortex to remove. In each case, we attempted to preserve the underlying white matter. The aspirated neural tissue was replaced with sterile fibrin sponge, both to reduce the risk of hemorrhage and to provide stability for the replaced piece of cranium during the healing process. To inactivate A1, sheets of 200-μm thick Elvax slow-release polymer containing 75 mM of the $GABA_A$ agonist muscimol were prepared as described previously (Smith et al. 1995; Smith et al 2004) and placed subdurally over the posterior part of the middle ectosylvian gyrus (MEG) in a further four ferrets. In all animals, the temporal muscles were repositioned and the scalp sutured. Following further behavioral testing, the Elvax implants were removed and the sound localization performance of the animals assessed again. One of these ferrets subsequently received bilateral A1 lesions, and the resulting data were pooled with the data from the other lesioned animals.

In contrast to the animals with cortical lesions, no recordings were made prior to Elvax placement in order to avoid the risk of damaging A1. We have previously shown, however, that the cortex beneath the muscimol-Elvax implants is silenced for several weeks following implantation and that spread of inactivation to adjacent regions of the cortex is quite limited, particularly in the deeper cortical layers (Smith et al. 2004; Bizley et al. 2007).

53.2.3 Histology

All animals were perfused transcardially under terminal anesthesia, the brain was extracted from the skull, cryoprotected with 30% sucrose and cut in 50-μm coronal sections. In the ferrets with A1 lesions, serial sections at the level of the ectosylvian gyrus were Nissl stained with 0.2% cresyl violet. In the Elvax-muscimol cases, three sets of serial sections were obtained and stained to visualize Nissl substance, SMI_{32} neurofilament, and the neuron-specific nuclear protein NeuN. The sections were analyzed with a Leica DMR microscope (Leica Microsystems) and a digital camera (Microfire™, Olympus America Inc.) and histological reconstructions were performed using Neurolucida 7 software (MBF Bioscience, MicroBrightField Inc.).

53.2.4 Data Analysis

Data from the approach-to-target and head orienting response measurements were exported to Excel (Microsoft Corporation, Redmond, WA) and Matlab (Mathworks, Natick, MA) for further analysis and presentation. The algorithms used to measure head movement latency and accuracy were implemented in advance and carried out automatically, therefore avoiding any possibility of subjective variations in the data analysis. The statistical analysis was done with SPSS software (SPSS Inc., Chicago, IL).

53.3 Results

53.3.1 Anatomy

In all four animals with bilateral aspiration lesions, the region of cortex that had been removed was restricted to the posterior half of the MEG (Fig. 53.1a, b) without encroaching on the pseudosylvian sulcus. The lesions were typically quite superficial near their edges and gradually extended down toward the underlying white matter in their centers. In two cases, the lesions were symmetric on the left and right sides and, based on the surrounding cortical landmarks and recording data, were deemed to occupy most of A1. In the other two cases, the lesion was larger on the right side than on the left.

Visual examination of the cortex showed that the muscimol-Elvax sheets were positioned over the MEG and centered over A1 (Fig. 53.1c, d). Usually, the implants reached the dorsal part of the pseudosylvian sulcus and partially covered the dorso-posterior part of the suprasylvian sulcus, but did not extend over the anterior part of the middle EG. These placements are consistent with the position of A1, as judged in previous electrophysiological (Bizley et al. 2005) and intrinsic optical imaging (Nelken et al. 2004) studies from our laboratory. Neurons stained for Nissl substance, NeuN and/or SMI_{32} in the MEG were normal in size and morphology. Also, the thickness and appearance of the layers in the region of cortex previously covered by the Elvax implants did not show any differences from sections taken at the same level in other normal cases taken from our archive. We therefore observed no indication of any damage to the cortex in these animals.

53.3.2 Approach-to-Target Sound Localization

Prior to aspiration or inactivation of A1, the sound localization behavior of all the ferrets used in this study was consistent with our previous reports in this species (Parsons et al. 1999; Kacelnik et al. 2006; Nodal et al. 2008). The accuracy of the approach-to-target responses, measured as the percentage of correct trials, declined as the duration of the stimulus was reduced (Fig. 53.2a, b). For the long sound durations (≥500 ms), localization performance was independent of target location, with comparably high scores obtained at all 12 loudspeaker locations (see Fig. 53.3a, first panel for control data obtained with a stimulus duration of 1,000 ms). However, for the shorter sound durations (≤200 ms), sound localization accuracy varied with target location. Performance in the anterior region of space was only slightly different from that obtained with longer stimuli, whereas localization accuracy was much worse at lateral and posterior locations (compare the first and second rows in Fig. 53.3).

Although fewer correct responses were made when brief noise bursts were presented from peripheral sound directions, the magnitude of the errors remained

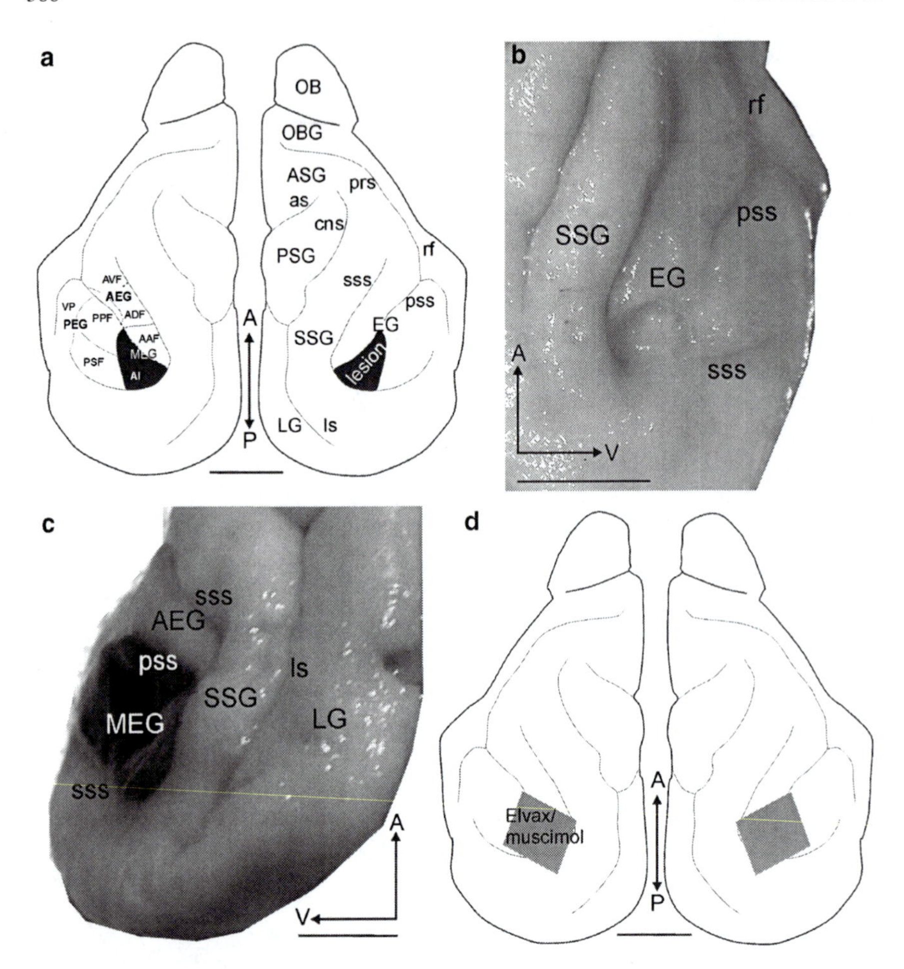

Fig. 53.1 Location of the lesions and implanted Elvax polymers. (**a**) Dorsal view of the ferret cortex showing the location of the different auditory areas that have been described physiologically and, in black, the location of the primary auditory cortex, A1. (**b**) Photograph showing the aspiration lesion of the right A1 in animal 0206. (**c**) Photograph showing a sheet of muscimol-Elvax in place over the MEG, where the primary auditory areas are found. (**d**) Diagram showing the location of the Elvax polymer over the MEG on both sides of the brain. Abbreviations: *OB* olfactory bulb, *OBG* orbital gyrus; *ASG*, anterior sigmoid gyrus, *PSG* posterior sigmoid gyrus, *LG* lateral gyrus, *SSG* suprasylvian gyrus, *EG* ectosyslvian gyrus, *MEG* middle ectosyslvian gyrus, *PEG* posterior ectosyslvian gyrus, *AEG* anterior ectosyslvian gyrus, *prs* presylvian sulcus, *cns* coronal sulcus, *as* anterior sigmoid, *ls* lateral sulcus, *rf* rhinal fissure, *sss* suprasylvian sulcus, *pss* pseudosylvian sulcus, *A1* primary auditory cortex, *AAF* anterior auditory field, *PPF* posterior pseudosylvian field, *PSF* posterior suprasylvian field, *VP* ventral posterior, *ADF* anterior dorsal field, *AVF* anterior ventral field

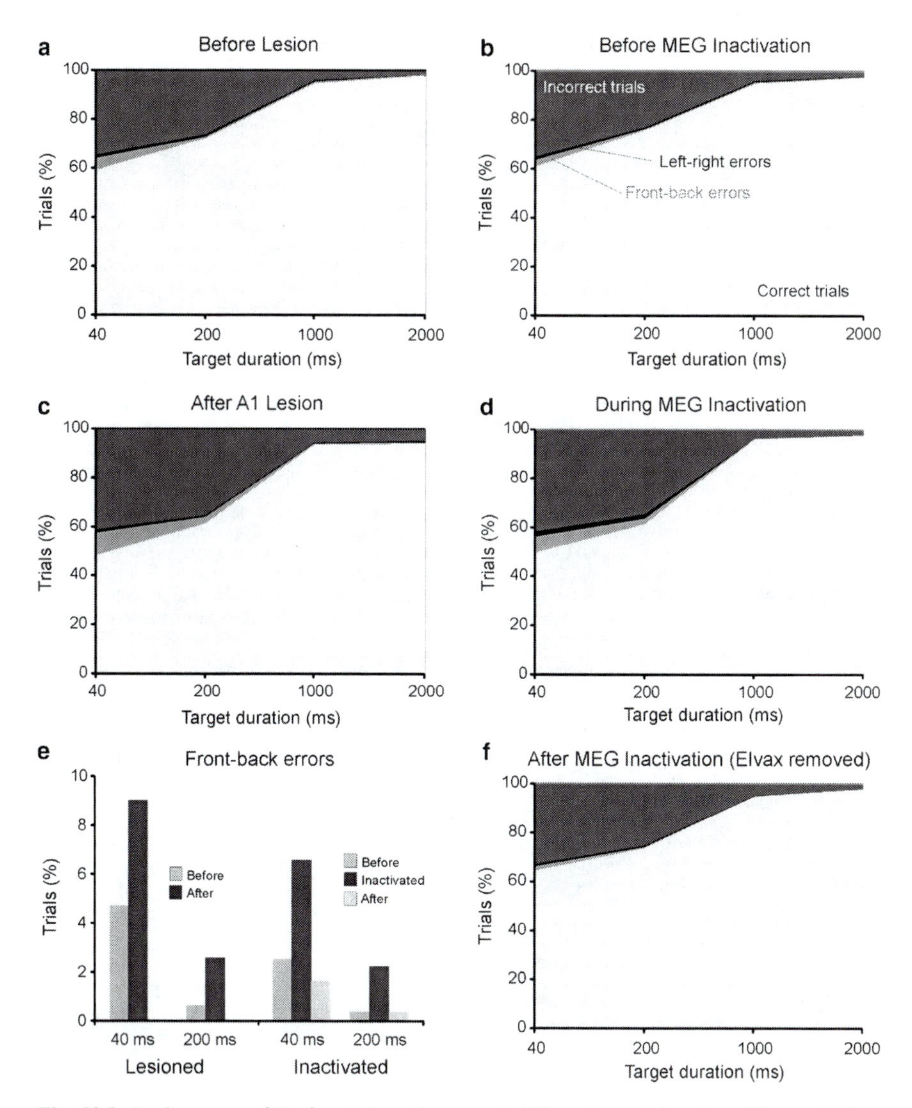

Fig. 53.2 Performance of the ferrets in each group at different sound durations. The stacked area plots show the percentages of each type of response made by the animals: correct responses (*light gray*), front-back errors (*medium gray*), left-right errors (*black*) and unclassified incorrect responses (*dark gray*). Performance declines as the duration of the stimulus is reduced. (**a, c**) Data obtained before (**a**) and after (**c**) bilateral aspiration of A1. (**b, d, f**) Data obtained before (**b**), during inactivation with muscimol-Elvax (**d**), and after the muscimol Elvax had been removed (**f**). Both groups performed less well at short sound durations after silencing neurons in A1. (**e**) Bar chart showing the increase in the proportion of front-back errors made in the approach-to-target task following aspiration lesion or inactivation of A1 at two different stimulus durations

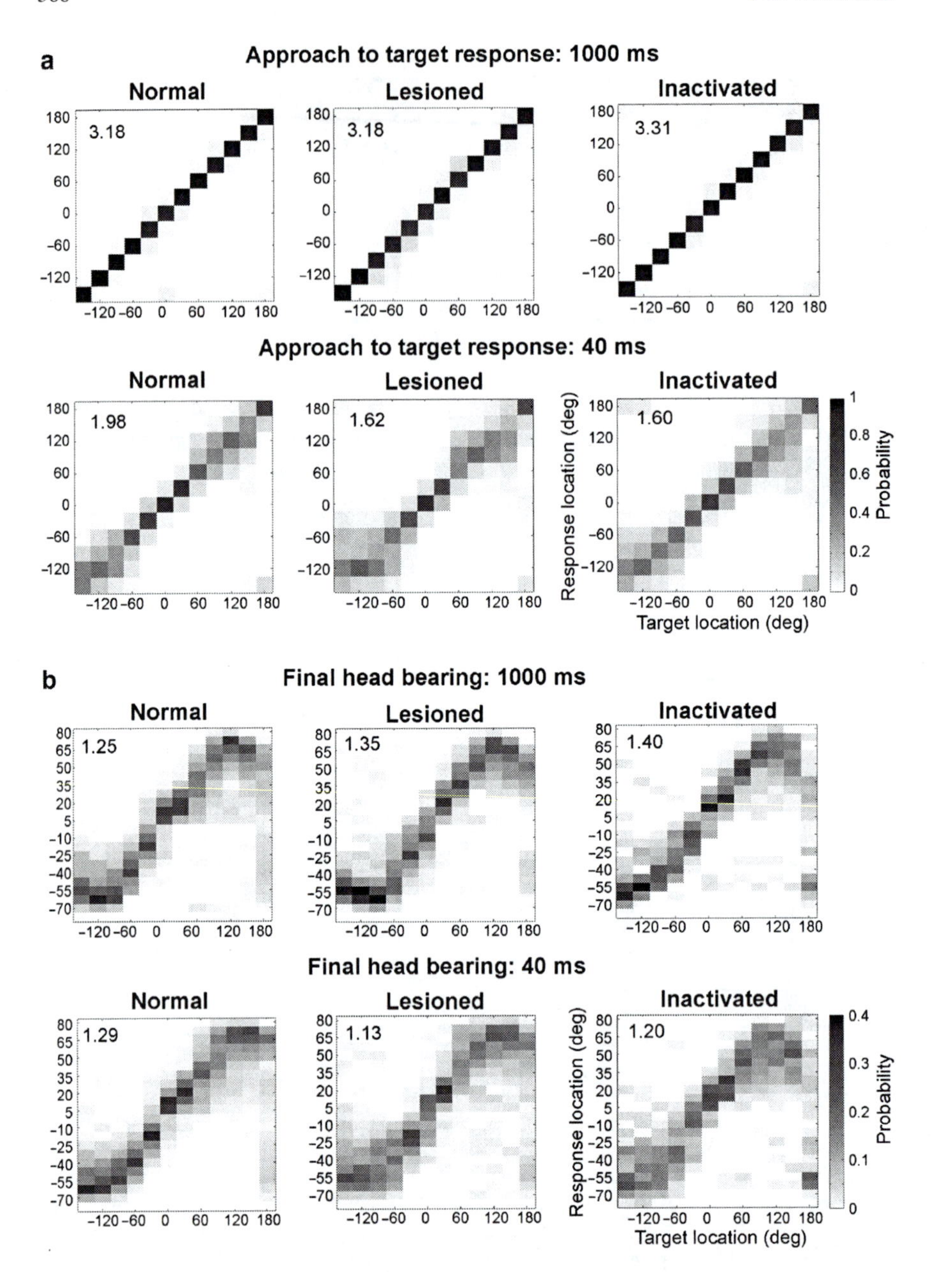

Fig. 53.3 Conditional probability plots for approach-to-target responses (**a**) and final head bearing in 7.5° bins (**b**) at the target locations shown on the *x*-axis for sound durations of 40 ms (*second* and *fourth rows*) and 1,000 ms (*first* and *third rows*). Presurgery (normal) performance is shown in the left column. Sound localization behavior after bilateral A1 lesions is shown in the middle column and during muscimol inactivation in the right column. Approach-to-target responses and final head bearing frequencies were normalized for each target location, so that the resulting values could be read as estimates of the conditional probability of a particular response or final head

fairly constant across different sound durations, with around 80% of all errors being 30° in size, the smallest value that could be recorded with our 12-loudspeaker setup. For those trials in which larger errors were made, the magnitude of the errors did vary with target location. Errors of 60° were more frequent in lateral positions, whereas when they occurred, larger errors were typically made in response to sounds presented from more frontal locations. Left-right errors (responses made to the hemifield contralateral to the target) were extremely rare, and so the larger errors were usually front-back errors (i.e., where sounds presented in the anterior hemifield were mislocalized to the ipsilateral posterior hemifield or vice versa). About 5% of all trials in which 40 ms noise bursts were presented were classified as front-back errors in ferrets with an intact auditory cortex, and this proportion declined with increasing stimulus duration (Fig. 53.2e).

Aspiration or pharmacological inactivation of A1 had a very similar effect on acoustic approach-to-target behavior. At longer sound durations, the performance of the animals was almost unchanged, particularly in the animals with muscimol-Elvax implants, as shown by the percentage of correct responses (Fig. 53.2c, d) and the amount of transmitted information between response and target location (Fig. 53.3a, first row). However, at shorter sound durations, the animals clearly localized less accurately following either lesion or inactivation of A1 than they had prior to surgery (Fig. 53.2c, d). Nevertheless, these deficits were fairly modest and varied with the direction of the target. Thus, performance in the anterior region of space (±30°) was barely affected, whereas the ability of the animals to localize brief noise bursts presented from more lateral angles was more disrupted (Fig. 53.3a, second row). This was associated with a reduction in the mutual information between response and target location (Fig. 53.3a, second row) and an increase in the incidence of front-back errors (Fig. 53.2c–e). Following removal of the muscimol-Elvax implants, the localization deficits measured for short duration noise bursts disappeared (Fig. 53.2f) and the performance of the animals closely resembled that observed prior to cortical inactivation (compare Fig. 53.2 b and f).

The time taken by the animals to reach the reward spout depended on whether a correct response was made or not. Correct responses were faster than incorrect ones. This relationship was observed both before and after either of the cortical manipulations, suggesting that the decision making process was not affected by loss of activity in A1.

Fig. 53.3 (continued) bearing given a particular target location. We used those distributions to calculate the mutual information between the response location or final head bearing and the target location using the formula: $\mathrm{MI}(r;s) = \sum_{r,s} p(r,s)\log_2\left(\frac{p(r,s)}{p(r)p(s)}\right)$ where r is the response or final head bearing, s is the stimulus location, $\mathrm{MI}(r;s)$ is the mutual information between r and s, $p(r, s)$ is the joint probability of r and s (which was obtained from the conditional probability values and is equivalent to $p(r|s)\,p(s)$, where $p(s)$ equals 1/12, as there are 12 equiprobable speakers, and $p(r)$ is obtained from the overall distribution of head bearings (or equivalently from summing $p(r,s)$ across all s)

53.3.3 Head Orienting Responses

We also recorded the head movements made during the first second after the onset of the stimulus as the animals turned toward the target. Consistent with the lack of left-right errors in the approach-to-target responses, the animals usually turned toward the appropriate side of the midline (Fig. 53.3b). The head movements followed a sigmoidal curve in which both the slope of the exponential phases and the asymptote (final bearing) varied systematically with the horizontal angle of the target. A linear relationship was found between target location and final head bearing for stimuli presented within the frontal hemifield, irrespective of their duration. Nevertheless, a consistent undershoot was observed at all speaker positions (Fig. 53.3b), and the head movement amplitude recorded was usually restricted to a range of ±60°. Stimuli presented behind the animal gave rise to much more variable head movements than those presented in front. In contrast to the approach-to-target responses, which became less accurate for lateral and posterior targets as the duration of the stimulus was reduced (Fig. 53.3a), the accuracy of acoustic orienting behavior was less dependent on stimulus duration and similar mutual information values between head bearing and target location were obtained for both long and short duration sounds (Fig. 53.3b).

Neither ablation nor inactivation of A1 altered the overall pattern of head orienting responses observed prior to surgery, with a clear correlation still present between final head bearing and target location (Fig. 53.3b). Indeed, the head movements recorded following the presentation of long duration noise bursts appeared to be completely unchanged, with no reduction in the mutual information between head bearing and target location (Fig. 53.3b, top row). However, at short sound durations the amplitude of the head turns was more variable after A1 had been aspirated or inactivated, resulting in a reduction in the mutual information values (Fig. 53.3b, bottom row). As with the change in the accuracy of the approach-to-target behavior, the increase in the variability of the acoustic orienting responses following loss of activity in A1 was particularly pronounced for lateral and posterior target locations.

53.4 Discussion and Conclusions

These data confirm the results of previous studies (e.g., Heffner and Masterton 1975; Kavanagh and Kelly 1987; Jenkins and Merzenich 1984; Malhotra et al. 2004) in showing that aspiration lesions or reversible inactivation of A1 impairs the ability of ferrets to localize brief sounds in the horizontal plane. In each case, however, the ferrets could still perform the localization task and the deficits observed were smaller than those typically reported in other studies. Because we obtained nearly identical results in the animals in which aspiration lesions were made and in those in which the cortex was reversibly inactivated with muscimol-Elvax, it seems unlikely that this difference is due to the method used for silencing the cortex. We cannot, of course, rule out the possibility that activity in part of A1 was preserved, or conversely that regions within the adjacent cortical fields were affected as well. The behavioral changes were, however, very consistent across different animals and therefore suggest that while A1 is necessary for normal sound localization, ferrets retain some ability to judge the location of a sound source in the horizontal plane in its absence. This is consistent with the results of an investigation into the role of A1 in vertical localization (Bizley et al. 2007).

Kavanagh and Kelly (1987) reported that bilateral lesions restricted to A1 in ferrets produced smaller localization deficits than those which included neighboring cortical fields.

The magnitude of the localization deficit has also been found to be correlated with the size of the cortical lesion in monkeys (Heffner and Masterton 1975). These findings are consistent with behavioral (Malhotra et al. 2004) and electrophysiological (Harrington et al. 2008; Bizley et al. 2009) data indicating that multiple areas within the auditory cortex contribute to sound localization.

One of the novel aspects of our study is that we measured both the initial head turn made by the animals in response to the sound, as well as their locomotor response toward the perceived location of its source in order to receive a reward. Although lesion and inactivation studies have consistently shown that the auditory cortex plays a critical role in the accuracy of acoustic approach-to-target responses in carnivores and primates, the effects of cortical lesions on reflexive head orienting responses are less clear cut (Thompson and Masterton 1978; Beitel and Kaas 1993). This implies that different neural circuits might be responsible for reflexive orientation of the head toward a sound source and the ability to perceive a sound as coming from a particular location in space. However, the approach-to-target behavior that we measured was almost invariably preceded by a head orienting response, and so it seems reasonable to assume that both are guided by the same neural processing of auditory localization cues. Indeed, we have previously shown that when ferrets with an intact auditory cortex make an incorrect approach-to-target response, the preceding head turns are more closely correlated with the direction approached by the animal than with the actual direction of the target. By showing that head orienting and approach-to-target responses are disrupted in a very similar fashion following aspiration or inactivation of A1, the present results support this possibility and suggest either that the neural circuits involved in each aspect of sound localization behavior are the same or that there is a close interaction between them.

53.5 Comment by Catherine Carr

Your lesion study reminded me of studies from both the Knudsen and Wagner groups that showed that barn owls can localize sounds without their midbrain space map. Complete or partial lesions of the map render the owl incapable of localizing sound in the selectively damaged areas. The owl's ability to localize returns, however, and is only irretrievably compromised when a second pathway, from IC to forebrain by way of the thalamus, is lesioned (Wagner 1993; Knudsen et al. 1993).

53.6 Reply Fernando Nodal

There are similarities between our lesion study and those in the barn owl as each study has tried to unravel the different participation or contribution to sound localization of pathways intrinsic to the midbrain (inferior colliculus to superior colliculus) and those ascending to the thalamus and cortex (in mammals) or to the forebrain (in owls). It is often assumed that midbrain pathways are sufficient to allow reflexive orienting movements toward a sound source, whereas processing within the cortex gives rise to the percept of where that source is. This is why we measured the contribution of the auditory cortex to both the initial head orienting response and the selection made by the ferrets of which source location to approach – the first time that both measures of performance have been obtained in the same animal. We also decided to lesion the cortex bilaterally because unilateral lesions in the ICx in the barn owl result in a temporary deficit only, which could potentially be explained by the function being recovered either by the intact forebrain pathway or by the contralateral midbrain pathway.

Our results are broadly consistent with those in the barn owl as lesions of the ferret auditory cortex produce a permanent deficit in sound localization that does not recover over time even with training. They also show that an intact auditory cortex is required for normal sound-evoked orienting behavior as well as the accuracy of the response when the animal has to approach the target. While the midbrain auditory space map undoubtedly plays a role in sound localization (as shown by the barn owl data and by the results of similar experiments in mammals), processing of information in the auditory cortex appears to be critical for both measures of spatial hearing. It seems likely that descending projections from the cortex to the superior colliculus (see Bajo et al. 2007, Fig. 11, for data in the ferret) are involved, at least in guiding orientation behavior. It is therefore probably more appropriate to think about spatial information being processed serially, involving both midbrain and forebrain circuits, rather than each level being exclusively responsible for a different aspect of sound localization, insofar as this can be assessed by measuring behavior in animals.

Acknowledgments We are grateful to Susan Spires, Jenny Bizley, Peter Keating, and Dan Kumpik for assistance with the data collection and to Pat Cordery for making the muscimol-Elvax. This study was supported by the Wellcome Trust through a Wellcome Principal Research Fellowship to A. J. King.

References

Adriani M, Maeder P, Meuli R, Thiran AB, Frischknecht R, Villemure JG, Mayer J, Annoni JM, Bogousslavsky J, Fornari E, Thiran JP, Clarke S (2003) Sound recognition and localization in man: specialized cortical networks and effects of acute circumscribed lesions. Exp Brain Res 153:591–604

Bajo VM, Nodal FR, Bizley JK, Moore DR, King AJ (2007) The ferret auditory cortex: descending projections to the inferior colliculus. Cereb Cortex 17:475–491

Beitel RE, Kaas JH (1993) Effects of bilateral and unilateral ablation of auditory cortex in cats on the unconditioned head orienting response to acoustic stimuli. J Neurophysiol 70:351–369

Bizley JK, Nodal FR, Nelken I, King AJ (2005) Funcional organization of ferret auditory cortex. Cereb Cortex 15:1637–1653

Bizley JK, Nodal FR, Parsons CH, King AJ (2007) Role of auditory cortex in sound localization in the midsagittal plane. J Neurophysiol 98:1763–1774

Bizley JK, Walker KMM, Silverman B, King AJ, Schnupp JWH (2009) Interdependent encoding of pitch, timbre and spatial location in auditory cortex. J Neurosci 29(7):2064–2075

Harrington IA, Stecker GC, Macpherson EA, Middlebrooks JC (2008) Spatial sensitivity of neurons in the anterior, posterior, and primary fields of cat auditory cortex. Hear Res 240:22–41

Heffner H, Masterton B (1975) Contribution of auditory cortex to sound localization in the monkey (*Macaca mulatta*). J Neurophysiol 38:1340–1358

Jenkins WM, Merzenich MM (1984) Role of cat primary auditory cortex for sound-localization behavior. J Neurophysiol 52:819–847

Kacelnik O, Nodal FR, Parsons CH, King AJ (2006) Training-induced plasticity of auditory localization in adult mammals. PLoS Biology 4:e71

Kavanagh GL, Kelly JB (1987) Contribution of auditory cortex to sound localization by the ferret (*Mustela putorius*). J Neurophysiol 57:1746–1766

Kelly JB, Kavanagh GL (1986) Effects of auditory cortical lesions on pure-tone sound localization by the albino rat. Behav Neurosci 100:569–575

Knudsen EI, Knudsen PF, Masino T (1993) Parallel pathways mediating both sound localization and gaze control in the forebrain and midbrain of the barn owl. J Neurosci. 13:2837–2852

Malhotra S, Hall AJ, Lomber SG (2004) Cortical control of sound localization in the cat: unilateral cooling deactivation of 19 cerebral areas. J Neurophysiol 92:1625–1643

Nelken I, Bizley JK, Nodal FR, Ahmed B, Schnupp JWH, King AJ (2004) Large-scale organization of ferret auditory cortex revealed using continuous acquisition of intrinsic optical signals. J Neurophysiol 92:2574–2588

Nodal FR, Bajo VM, Parsons CH, Schnupp JWH, King AJ (2008) Sound localization behavior in ferrets: comparison of acoustic orientation and approach-to-target responses. Neuroscience 154:397–408

Parsons CH, Lanyon RG, Schnupp JWH, King AJ (1999) Effects of altering spectral cues in infancy on horizontal and vertical sound localization by adult ferrets. J Neurophysiol 82:2294–2309

Smith AL, Cordery PM, Thompson ID (1995) Manufacture and release characteristics of Elvax polymers containing glutamate receptor antagonists. J Neurosci Meth 60:211–217

Smith AL, Parsons CH, Lanyon RG, Bizley JK, Akerman CJ, Baker GE, Dempster AD, Thompson ID, King AJ (2004) An investigation of the role of auditory cortex in sound localization using muscimol-releasing Elvax. Eur J Neurosci 19:3059–3072

Thompson GC, Masterton RB (1978) Brain stem auditory pathways involved in reflexive head orientation to sound. J Neurophysiol 41:1183–1202

Wagner H (1993) Sound-localization deficits induced by lesions in the barn owl's auditory space map. J Neurosci 13:371–386

Part IX
Hearing Impairment

Chapter 54
Objective and Behavioral Estimates of Cochlear Response Times in Normal-Hearing and Hearing-Impaired Human Listeners

Olaf Strelcyk and Torsten Olaf Dau

Abstract Derived-band auditory brainstem responses (ABRs) were obtained in 5 normal-hearing and 12 sensorineurally hearing-impaired listeners. The latencies extracted from these responses as a function of the derived-band center frequency served as objective estimates of cochlear response times. In addition, two behavioral measurements were carried out. In the first experiment, differences between frequency-specific cochlear response times were estimated, using the lateralization of pulsed tones, interaurally mismatched in frequency. In the second experiment, auditory-filter bandwidths were estimated using a notched-noise masking paradigm. The correspondence between objective and behavioral estimates of cochlear response times was examined. An inverse relationship between the ABR latencies and the filter bandwidths could be demonstrated as the result of the larger across-listener variability among the hearing-impaired listeners, as compared to the normal-hearing listeners. The results might be useful for a better understanding of how hearing impairment affects the spatiotemporal cochlear response pattern in human listeners.

Keywords Brainstem responses • Cochlear response times • Lateralization • Auditory filters • Impaired listeners

54.1 Introduction

The cochlea separates a sound into its constituent tonal components and distributes their responses spatially along its length by the distinctive spatial and temporal vibration patterns of its basilar membrane (BM). For example, the vibration pattern evoked by a single tone appears as a travelling wave (TW). This propagates down the cochlea and reaches maximum amplitude at a particular point, before slowing

O. Strelcyk (✉)
Centre for Applied Hearing Research, Department of Electrical Engineering, Ørsteds Plads, Building 352, 2800 Kgs Lyngby Denmark
e-mail: os@elektro.dtu.dk

E.A. Lopez-Poveda et al. (eds.), *The Neurophysiological Bases of Auditory Perception*,
DOI 10.1007/978-1-4419-5686-6_54, © Springer Science+Business Media, LLC 2010

down and decaying rapidly. The lower the frequency of the tone, the further its waves propagate down the cochlea. Thus, each point along the cochlea has a characteristic frequency (CF) to which it is most responsive. This tonotopic map is an important organizational principle of the primary auditory pathway and is preserved all the way to the auditory cortex. At the level of the auditory nerve, the frequency of a tone is encoded both spatially, by its CF location, and temporally, by the periodicity of the responses in the nerve fibers that innervate this CF. Several studies have suggested that the extraction of spatiotemporal information, i.e., the combination of phase-locked responses and systematic frequency-dependent delays along the cochlea (associated with the TW), is important in the context of pitch perception (e.g., Loeb et al. 1983), localization (e.g., Shamma 2001), speech formant detection (e.g., Deng and Geisler 1987), and tone-in-noise detection (Carney et al. 2002).

It has been proposed that a distorted spatiotemporal response might be, at least partly, responsible for the problems of hearing-impaired listeners to process temporal fine-structure information across frequencies (e.g., Moore 1996), which might be one of the reasons for their difficulties to understand speech in noise. Hence, it would be important to gain a better understanding of how (different types of) hearing impairment affects the spatiotemporal response pattern. However, so far, empirical evidence for spatiotemporal information processing in humans is lacking since BM response patterns are difficult to monitor.

In this study, as a starting point, it was attempted to measure one important component of the spatiotemporal response pattern: the cochlear response time (CRT) for various signal frequencies. The CRT can be considered as the sum of the cochlear transport time, reflecting the delay of a broadly tuned linear BM filter, and a filter build-up time reflecting the delay introduced by the cochlear amplifier sharpening the BM tuning (Don et al. 1998). Here, behavioral and objective estimates of the CRT in normal-hearing (NH) and hearing-impaired (HI) listeners were considered.

In the first experiment, the measurement paradigm was closely related to the concept of (across-ear) spatiotemporal information processing. Differences between frequency-specific response times were estimated via perceptual lateralization of pulsed tones that were interaurally mismatched in frequency (Zerlin 1969). The results were compared with the objective estimates of CRT (experiment 2) in the same listeners, using the paradigm of derived-band auditory brainstem responses (ABRs). The particular focus in this experiment was to investigate the effect of cochlear hearing impairment on the obtained latency values. Only very few studies previously addressed CRTs in HI listeners and their results were equivocal. Donaldson and Ruth (1996) measured derived-band ABRs in HI listeners and found no alterations as a consequence of hearing loss (in the group without Meniere's disease). In contrast, Don et al. (1998), using a similar method, reported a tendency toward shorter response latencies with increasing hearing loss, consistent with the notion of a decreased auditory-filter tuning. In the third experiment conducted in this study, auditory-filter bandwidths were estimated behaviorally in the same listeners, using a notched-noise masking paradigm. This was done in order to explicitly study the relation between (objectively estimated) CRTs and (behaviorally estimated) auditory-filter bandwidths across the individual listeners. It was also expected that the variability of estimates across the HI listeners would make it easier to investigate the relation between CRT and frequency tuning.

54.2 Lateralization of Mismatched Tones

54.2.1 Methods

Three female NH listeners participated in this experiment. Short trains of tone bursts with interaurally mismatched frequencies f_1 and f_2 were presented to the two ears, at equal levels of 50 dB SPL (similar results were obtained when the tones were balanced in loudness). The interaural time difference (ITD) that produced a unified percept centered at the midline was measured. In the following, the notation $f_1|f_2$ is used, where f_1 represents the frequency of the tone presented in one ear and f_2 the frequency of the tone presented in the other ear. The considered tone frequencies were: 400|480 Hz, 800|900 Hz, 1,000|900 Hz, and 1,400|1,550 Hz. Each tone burst had a total duration of 40 ms, including an exponential onset ramp with a rise time of 10 ms and a 10-ms raised-cosine shaped offset ramp. Each train consisted of six tone bursts, separated by 40-ms silent gaps. Its lateralization was varied by introducing a waveform delay to one of the ears, giving rise to an ITD.

A two-interval, two-alternative, forced-choice (2I-2AFC) task was used. The first interval always contained the diotic reference tone-burst train, consisting of both tones in both ears, while the second interval contained the $f_1|f_2$ target train. Listeners were instructed to indicate if the latter was lateralized to the left or right side relative to the reference train. If the target train was lateralized to the right, the ITD was adjusted such that the percept would move further to the left in the next presentation, and vice versa. Following the adaptive procedure for subjective judgments introduced by Jesteadt (1980), two sequences of trials were interleaved, tracking 71 and 29% lateralization to the right. The mean of the two resulting ITDs was taken as an estimate of the ITD yielding a centered percept. The final centering ITD was estimated as the arithmetic mean over at least four interleaved runs.

In order to compare the behavioral ITDs with objective CRT estimates, derived-band ABR latencies were obtained for the same listeners at comparable sensation levels (see Sect. 54.3.1 for methodological details).

54.2.2 Results and Discussion

The results of the lateralization measurements for the three NH listeners are presented in Table 54.1. It shows the ITDs that led to centered percepts of the tones with interaurally mismatched frequencies f_1 and f_2. The frequency-mismatched tones with zero ITD were always lateralized toward the ear receiving the higher-frequency

Table 54.1 The ITDs yielding centered percepts of the tones with interaurally mismatched frequencies f_1 and f_2, for three NH listeners and different stimulus conditions (the numbers in brackets represent the standard error)

$f_1 \vert f_2$ (kHz)	ITD (μs) – NH_1	ITD (μs) – NH_2	ITD (μs) – NH_3	$\Delta\tau_{ABR}$ (μs)
0.4|0.48	340 (40)	396 (8)	335 (5)	[609 (36)]
0.8|0.9	264 (13)	232 (10)	207 (3)	259 (4)
1.0|0.9	185 (13)	262 (3)	180 (9)	216 (4)
1.4|1.55	138 (9)	129 (22)	–	158 (8)
0.8–1.0	449 (18)	494 (11)	387 (9)	474 (7)

Also, the corresponding wave-V latency differences $\Delta\tau_{ABR}$ are given (in brackets: standard deviation across listeners). The value in square brackets is based on an extrapolation beyond the range of measured frequencies

tone. The obtained ITDs were generally consistent and well reproducible. Therefore, the standard errors of the ITD estimates were relatively small. In addition to the behavioral ITD data, the objective ABR wave-V latency differences $\Delta\tau_{ABR}$ between the frequencies f_1 and f_2 are given in Table 54.1. They were calculated on the basis of individual fits to derived-band ABR data (described in detail in Sect. 54.3). Since these latency differences were very similar for the three listeners, only the average $\Delta\tau_{ABR}$ values are given (the lowest derived-band frequency was 700 Hz, and therefore the extrapolation to lower frequencies should be regarded with caution). The behavioral ITD-based measure and the objective ABR-based measure yielded very similar results. This suggests that the behavioral ITDs indeed reflected differences in CRT between remote positions on the BM, supporting the principal feasibility of Zerlin's (1969) paradigm.

The behavioral ITDs reflected *interaural* time differences, whereas the ABR latency differences $\Delta\tau_{ABR}$ reflected *monaural* time differences. Hence, part of the remaining deviation between these two could have been due to differences in CRT between the left and right cochleae. Therefore, the ITDs for the 800|900-Hz and 1,000|900-Hz tone pairs were added (see last row in Table 54.1). Since these tone pairs shared the common reference frequency of 900 Hz, the sum estimated the time difference between 800 and 1,000 Hz in the ABR test-ear. However, similar deviations from the ABR latencies as for the single-tone-pair ITDs were observed for these "monaural" time differences. Hence, the remaining deviations did not seem to be attributable to asymmetries between the left and right cochleae.

All three listeners had more difficulties with the lateralization task for the high-frequency tones (for 1,400|1,550 Hz, no consistent ITDs were obtained for listener NH_3) than for the lower-frequency tones. None of the listeners could perform the task reliably at frequencies higher than 1.5 kHz. Here, the sound image could not be lateralized with reasonable precision. It was perceived as rather diffuse and often did not cross the midline.

Assuming that the observed ITDs and latency differences $\Delta\tau_{ABR}$ reflected travel times on the BM, the corresponding TW velocities were estimated, based on the cochlear frequency-position map supplied by Greenwood (1961). Figure 54.1 shows the TW velocity estimates for the three NH listeners, based on

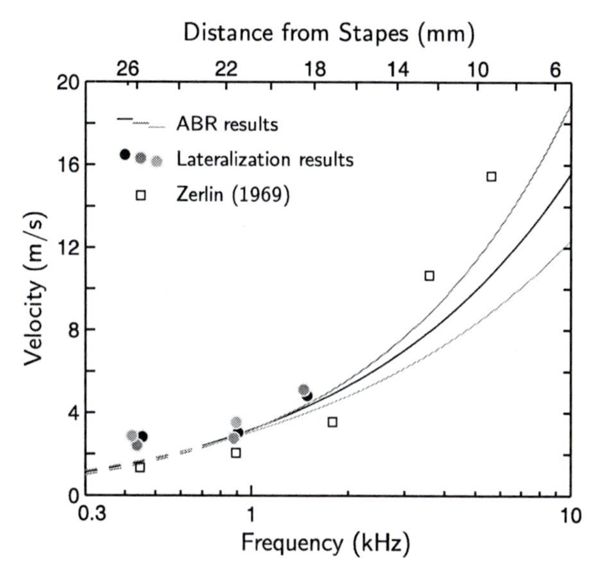

Fig. 54.1 Traveling-wave velocity as a function of frequency/distance from stapes for three NH listeners. The bullets denote the estimates based on the mismatched-tone ITDs (experiment 1). They are slightly horizontally displaced for better visibility. The *squares* are corresponding estimates based on the ITDs reported by Zerlin (1969). The *solid curves* represent the individual velocity estimates derived from the derived-band ABR latencies (experiment 2)

the ABR latencies (curves) and the behavioral ITDs (circles). For direct comparison, the open squares indicate velocities that were derived from Zerlin's (1969) ITDs. The ITD-based velocity estimates were consistent with the ABR-based velocity estimates. In both measures, velocities increased with increasing frequency. Only little inter-individual differences were observed up to 2 kHz. At these frequencies, the estimates from the present study and those of Zerlin (1969) are roughly comparable. At high frequencies, however, it was impossible to reproduce Zerlin's findings.

There are certain fundamental limitations in the lateralization paradigm with interaurally mismatched tones. In principle, a large frequency mismatch $|f_2-f_1|$ between the tones would be desirable to increase the accuracy of the ITD estimate. However, with increasing frequency mismatch, it becomes increasingly difficult to attribute a fused position although the two tones are still perceived as a single auditory object (in time). More importantly, the lateralization threshold, i.e., the ITD for which the position of a noncentered sound object can just be distinguished from that of a centered object, increases strongly as soon as the interaural frequency mismatch exceeds a value that corresponds to the critical bandwidth at that frequency (e.g., Scharf et al. 1976). In order to be measurable, however, the centering ITD reflecting CRT disparity needs to be larger than the corresponding lateralization threshold. In the present study, the tones were chosen such that their frequency mismatch did not exceed the critical bandwidth at the corresponding frequencies. In contrast, the frequency mismatches for all tone pairs used by Zerlin (1969) exceeded the critical bandwidths, and the reported ITDs for the 3,200|4,000-Hz and 5,000|6,300-Hz tone pairs in that study clearly fall below the corresponding lateralization thresholds given by Scharf et al. (1976). Furthermore, Scharf, Florentine, and Meiselman emphasized the importance of controlled tone-onset phases for tone frequencies below about 2 kHz: without controlling the onset phase, their ITD data were inconsistent and the observed ITD detection thresholds became substantially larger. Zerlin (1969) did not control onset phases. Hence, the validity of his results appears questionable both at low and high tone frequencies.

Since the main goal was to study CRT over a wide range of frequencies, the lateralization of mismatched tones was not pursued further here, due to its limitation to frequencies below about 1.5 kHz. In the following, objective derived-band ABR measurements in normal-hearing (NH) and hearing-impaired (HI) listeners are presented and related to behavioral measures of frequency selectivity.

54.3 Auditory Brainstem Responses

54.3.1 Methods

ABRs were measured in response to 93-dB pe SPL high-pass masked clicks, for a total of 17 female listeners. Five of the listeners had normal hearing and 12 had a sensorineural hearing impairment. The HI listeners showed sloping audiograms, with mild-to-moderate hearing losses at high frequencies. Narrow-band contributions to the ABR were derived by means of the derived-band technique (Don and Eggermont 1978), for center frequencies of 0.7, 1.4, 2.8, 5.7, and 9.5 kHz. Subsequently, wave-V latencies were extracted. For the further analysis of the wave-V latencies, the following latency model was adopted from Neely et al. (1988):

$$\tau(f) = a + b f^{-d} \tag{54.1}$$

where f represents derived-band frequency, normalized to 1 kHz, and a, b, and d are fitting parameters.

54.3.2 Results and Discussion

Figure 54.2 shows the measured (symbols) and fitted (curves) wave-V latencies for the NH (black) and HI (gray) listeners as a function of derived-band frequency. Generally, the latencies decreased with increasing frequency, in agreement with results reported in the literature (e.g., Eggermont and Don 1980). For 8 of the 12 HI listeners, shorter latencies were observed than for the NH listeners (at 1.4 and 2.8 kHz), consistent with Don et al. (1998). Furthermore, the across-listener variance was larger for the HI group than for the NH group. For the lowest

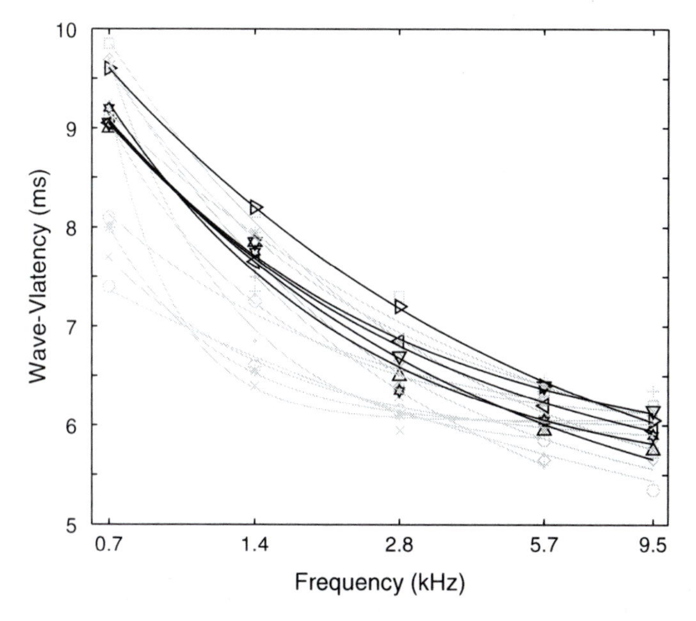

Fig. 54.2 Measured ABR wave-V latencies for the NH (*black symbols*) and HI listeners (*gray symbols*) in response to 93-dB clicks presented in high-pass masking noise, as a function of the derived-band center frequency (experiment 2). The *solid curves* show individual model fits according to (54.1)

frequency of 700 Hz, a wave-V latency could not be identified for one of the HI listeners. For the highest frequency of 9,500 Hz, latencies could not be reliably identified for four of the HI listeners.

Based on a nonlinear mixed-effects model (NMEM), which followed the latency model given in (54.1), an ANOVA was performed. The effect of derived-band frequency on the measured wave-V latencies was significant [$p<0.0001$], while the effect of listener group was marginally significant [$p=0.05$].

In the next experiment, auditory-filter shapes were estimated in the same NH and HI listeners in order to examine if individual filter tuning could account for part of the observed across-listener variability in the ABR latencies.

54.4 Auditory-Filter Bandwidth

54.4.1 Method

Auditory-filter shapes were determined at 2 kHz for the ABR test-ear using a notched-noise paradigm (Patterson and Nimmo-Smith 1980). The 2-kHz target tones of 440-ms duration were temporally centered in the 550-ms noise maskers. Target tones and maskers were presented with 50-ms raised-cosine ramps. Five symmetric ($\Delta f/f_0$: 0.0; 0.1; 0.2; 0.3; 0.4) and two asymmetric notch conditions ($\Delta f/f_0$: 0.2|0.4; 0.4|0.2) were used, where Δf denotes the spacing between the inner noise edges and the signal frequency f_0. For the NH listeners, the tones were presented at a fixed level of 40 dB SPL. For some of the HI listeners, a level of 40 dB SPL would have resulted in a sensation level of less than 15 dB. In order to obtain reliable filter estimates, the tone level was increased in these cases to ensure a minimum sensation level of 15 dB. The average tone level for the HI listeners was 46.5 (SD 8.4) dB. A 3I-3AFC up-down method was applied and each threshold was estimated as the average of at least three runs. A rounded-exponential filter was fitted (in the least-squares sense) to the threshold data, and the equivalent rectangular bandwidth (ERB) was calculated.

54.4.2 Results and Discussion

The rounded-exponential filter provided a good description of the notched-noise threshold data, with a residual root-mean-square fitting error of 0.5 (SD 0.3) dB, averaged across all listeners. Although the HI listeners showed, on average, larger bandwidths than the NH listeners by a factor of 1.2 (0.9, 1.8) [15th and 85th percentiles], the difference in bandwidths between the two groups was not significant [$p=0.19$]. This was due to the large spread of results within the group of HI listeners and the fact that six of the HI listeners showed bandwidths within the range of the NH listeners.

In order to test if the bandwidth estimates at 2 kHz could account for the across-listener variability in the ABR latencies, the NMEM (from Sect. 54.3.2) was extended. In addition to the significant effect of derived-band frequency, the filter bandwidth in terms of the ERB was included, following a power-law dependence with exponent e (as new model parameter):

$$\tau(f) = a + b\, f^{-d}\,\mathrm{ERB}^{-e} \tag{54.2}$$

The ERB was found to be significant [$p=0.005$], with an estimated value of 0.5 for the exponent e. This implied that listeners with broader auditory filters at 2 kHz showed generally shorter derived-band latencies. Due to the inclusion of the ERB, the effect of listener group was no longer significant.

Figure 54.3 illustrates the inverse relation between the objective ABR-based CRT estimates and the behaviorally estimated filter bandwidths. The black squares indicate the results for the NH listeners, while the individual results for the HI listeners are shown with the gray bullets. In order to obtain the CRT, a synaptic delay, assumed to be 0.8 ms, and the individual derived-band wave I-V delay (cf. Don et al. 1998) were subtracted from the wave-V latency observed in the 2.8-kHz derived band. The trend across listeners is apparent: the estimated CRT decreased with increasing filter bandwidth. As can be seen, the inclusion of the HI listeners was crucial in order to study the relation between CRT and auditory filter bandwidth. While the HI listeners provided a relatively large span of bandwidths, the variability among the NH listeners alone would have been too small.

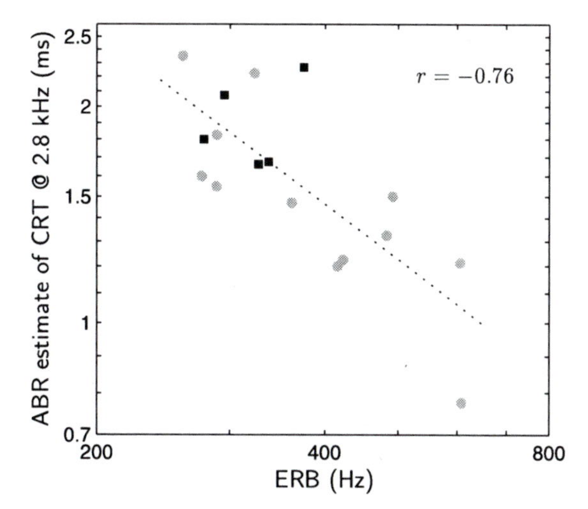

Fig. 54.3 Individual CRT (estimated by subtracting a synaptic delay of 0.8 ms and the derived-band wave I-V delay from the 2.8-kHz derived-band wave-V latency) for the NH (*black squares*) and HI listeners (*gray bullets*), as a function of the ERB

For the group of HI listeners, a significant correlation was found between the ERB at 2 kHz and the individual hearing threshold (estimated using a 3I-3AFC task) at this frequency [r=0.65, p=0.02]. Hence, the predictive power of the ERB (for estimating CRT) could have been due to its correlation with the hearing threshold rather than being an effect of filter bandwidth per se. Therefore, in a further step, the significance of the individual pure-tone audiogram (PTA, defined as the average pure-tone threshold at 0.5, 1, 2, and 4 kHz) was examined in a type-III ANOVA, by allowing the parameter b in the NMEM to be an exponential function of the PTA. However, the PTA did not reach significance, while the effect of bandwidth remained significant.

54.5 Summary and Conclusion

The behavioral paradigm (experiment 1) introduced by Zerlin (1969) yielded estimates of CRT differences (across remote BM positions) as well as TW velocity for frequencies up to 1.5 kHz. The estimates were reasonably accurate, in terms of variability across measurements, and consistent with objective estimates based on derived-band ABRs. For frequencies above 1.5 kHz, no consistent results could be obtained because of limitations of the measurement paradigm. However, the behavioral method might prove particularly useful for the estimation of CRT differences at very low frequencies (≤500 Hz), where the accuracy of objective methods is rather limited.

A trend toward shorter derived-band ABR latencies was observed in the HI listeners (experiment 2), as previously reported by Don et al. (1998). A larger across-listener variability was observed in the HI listeners over the NH listeners. Individual auditory-filter bandwidth, which was estimated via the notched-noise paradigm, accounted for part of this across-listeners variability (experiment 3).

Listeners with broader auditory filters showed shorter ABR latencies. To the authors' knowledge, this is the first time that direct empirical evidence was found for a relation between auditory filter tuning and CRT in human listeners.

As a consequence, the present study demonstrated that frequency-specific ABR measurements could in principle provide an objective tool for estimating frequency selectivity. However, as can be seen in Fig. 54.3, a substantial amount of across-listener variability remained unexplained. It is unclear to what extent this was due to (random) measurement errors, or due to methodological limitations in the estimation of CRTs and auditory filter shapes.

Besides the possibility to study other aspects of the spatiotemporal BM response pattern than CRT (such as response *amplitude*), a next step could be to investigate possible relations between the individual estimates of CRT and the performance in other behavioral tasks that have been discussed in terms of spatiotemporal processing (e.g., pitch perception and tone-in-noise detection). Here, the inclusion of HI listeners should be crucial in order to obtain sufficient across-listener variability (as observed in experiments 2 and 3). Alternatively, one might model the effect of CRT alterations due to hearing impairment within the framework of existing spatiotemporal models.

Acknowledgments The authors thank Dimitrios Christoforidis and Dr. James Harte for technical support and valuable scientific discussions about various aspects of this project. Part of this work was supported by the Oticon foundation.

References

Carney LH, Heinz MG, Evilsizer ME, Gilkey RH, Colburn HS (2002) Auditory phase opponency: a temporal model for masked detection at low frequencies. Acta Acustica 88:334–347

Deng L, Geisler CD (1987) A composite auditory model for processing speech sounds. J Acoust Soc Am 82:2001–2012

Don M, Eggermont JJ (1978) Analysis of the click-evoked brainstem potentials in man using high-pass noise masking. J Acoust Soc Am 63:1084–1092

Don M, Ponton CW, Eggermont JJ et al (1998) The effects of sensory hearing loss on cochlear filter times estimated from auditory brainstem response latencies. J Acoust Soc Am 104: 2280–2289

Donaldson GS, Ruth RA (1996) Derived-band auditory brain-stem response estimates of traveling wave velocity in humans: II. Subjects with noise-induced hearing loss and Meniere's disease. J Speech Hear Res 39:534–545

Eggermont JJ, Don M (1980) Analysis of the click-evoked brainstem potentials in humans using high-pass noise masking. II. Effect of click intensity. J Acoust Soc Am 68:1671–1675

Greenwood DD (1961) Critical bandwidth and the frequency coordinates of the basilar membrane. J Acoust Soc Am 33:1344–1356

Jesteadt W (1980) An adaptive procedure for subjective judgments. Percept Psychophys 28:85–88

Loeb GE, White MW, Merzenich MM (1983) Spatial cross-correlation. A proposed mechanism for acoustic pitch perception. Biol Cybern 47:149–163

Moore BCJ (1996) Perceptual consequences of cochlear hearing loss and their implications for the design of hearing aids. Ear Hear 17:133–161

Neely ST, Norton SJ, Gorga MP et al (1988) Latency of auditory brain-stem responses and otoacoustic emissions using tone-burst stimuli. J Acoust Soc Am 83:652–656

Patterson RD, Nimmo-Smith I (1980) Off-frequency listening and auditory-filter symmetry. J Acoust Soc Am 67:229–245

Scharf B, Florentine M, Meiselman CH (1976) Critical band in auditory lateralization. Sens Processes 1:109–126

Shamma S (2001) On the role of space and time in auditory processing. Trends Cogn Sci 5:340–348

Zerlin S (1969) Traveling-wave velocity in the human cochlea. J Acoust Soc Am 46:1011–1015

Chapter 55
Why Do Hearing-Impaired Listeners Fail to Benefit from Masker Fluctuations?

Joshua G.W. Bernstein

Abstract Hearing-impaired (HI) listeners do not receive as much benefit to speech intelligibility from fluctuating maskers, relative to stationary noise, as normal-hearing (NH) listeners. Investigators have focused on reduced audibility, deficits in spectral or temporal resolution, or limited cues for target-source separation as possible underlying causes of the reduced FMB. An alternative possibility is that the FMB differences may arise as a consequence of differences in the signal-to-noise ratio (SNR) at which HI and NH listeners are tested. The Extended Speech Intelligibility Index (ESII) was fit to NH data, and then used to make FMB predictions for a variety of results in the literature. Using this approach, reduced FMB for HI listeners and NH listeners presented with distorted speech was accounted for by SNR differences in many cases. HI listeners may retain more of an ability to listen in the gaps of a fluctuating masker than previously thought.

Keywords Speech intelligibility • Intensity importance function • Masking release • Modulated noise • Interfering talker

55.1 Introduction

While normal-hearing (NH) listeners demonstrate better speech intelligibility in the presence of a fluctuating masker than a stationary noise, hearing-impaired (HI) listeners generally receive little or no fluctuating-masker benefit (FMB). Most of the research investigating this problem has focused on the possible roles of limited

J.G.W. Bernstein (✉)
Army Audiology and Speech Center, Walter Reed Army Medical Center, Washington, DC 20307, USA
e-mail: joshua.g.bernstein@us.army.mil

E.A. Lopez-Poveda et al. (eds.), *The Neurophysiological Bases of Auditory Perception*,
DOI 10.1007/978-1-4419-5686-6_55,

audibility (e.g., Eisenberg et al. 1995; Dubno et al. 2003; Summers and Molis 2004; George et al. 2006), reduced spectral or temporal resolution (e.g., Dubno et al. 2003; George et al. 2006; Jin and Nelson 2006), or an inability to use auditory source-segregation cues (e.g., Lorenzi et al. 2006b; Hopkins et al. 2008) in limiting the ability to "listen in the gaps" of a fluctuating masker.

An alternative explanation for the limited FMB in HI listeners relates to the signal-to-noise ratio (SNR) used to measure speech intelligibility. NH listeners typically show a larger FMB at unfavorable SNRs and a smaller FMB at more favorable SNRs (Oxenham and Simonson 2009). The FMB is generally estimated by comparing the speech-reception threshold (SRT; the SNR required for 50% correct sentence or word identification) for stationary and fluctuating maskers. Because HI listeners require a higher stationary-noise SNR to yield intelligibility comparable to NH listeners, effects of hearing loss on the FMB may be confounded by differences in the stationary-noise SRT.

Bernstein and Grant (2009) investigated the role of SNR effects in reducing the FMB for HI listeners. NH and HI listeners were presented with sentence materials in a stationary-noise and two fluctuating-masker backgrounds (an interfering-talker and a noise modulated by the envelope of a single talker). When the FMB was calculated by comparing the stationary and fluctuating-masker SRTs, NH listeners showed a substantial FMB for both fluctuating maskers, while HI listeners did not, consistent with previous studies. When the FMB was instead calculated at the same baseline stationary-noise SNR, differences between listener groups were greatly reduced. Therefore, HI and NH listeners may be more similar in their ability to listen in the gaps of a modulated masker than previously thought.

SNR effects may also account for the reduced FMB for HI listeners observed in other studies (e.g., Festen and Plomp 1990; Hygge et al. 1992; Versfeld and Dreschler 2002; George et al. 2006; Lorenzi et al. 2006a, b; Wilson et al. 2007) and for NH listeners presented with speech processed to simulate aspects of hearing loss. This includes processing to reduce spectral contrasts (ter Keurs et al. 1993; Baer and Moore 1994), remove temporal fine-structure (TFS) information (Qin and Oxenham 2003; Hopkins et al. 2008), or limit audibility for portions of the speech spectrum or dynamic range via filtering (Oxenham and Simonson 2009) or noise masking (Bacon et al. 1998). Each of these studies has concluded that a particular aspect of hearing loss limits the ability to listen in the dips of a fluctuating masker. However, listener groups or processing conditions with the least FMB generally had baseline stationary-noise SRTs at the highest (least negative) SNRs. The appearance of a reduced FMB could result from the higher stationary-noise SNR needed to overcome a general speech intelligibility deficit, rather than an impaired ability to listen in the gaps of a fluctuating masker.

As discussed by Freyman et al. (2008), the instantaneous level of a fluctuating masker will vary about the average level, masking and unmasking speech information relative to the stationary-noise case. The net FMB will depend on the relationship between the masker level and the intensity importance function (IIF) describing the distribution of speech information across the dynamic range. Rhebergen et al. (2006)

proposed a modification to the Speech Intelligibility Index (ANSI 1997) to better predict intelligibility for speech presented in fluctuating backgrounds. Bernstein and Grant (2009) found that the Extended Speech Intelligibility Index (ESII) could account for the SNR-dependent variation in FMB across NH and HI listeners, if it was assumed that the IIF for speech presented in a fluctuating masker differs from the IIF for a stationary noise.

In the study presented here, a variety of results from the literature were examined to determine the extent to which reductions in the FMB could be ascribed to the confounding effects of SNR differences. The approach was to model the SNR dependence of the FMB for NH data using the ESII, then to apply the model predictions to other results in the literature. To the extent that the model predicts the reduced FMB reported for HI listeners and for NH listeners presented with distorted speech, confounding SNR effects may account for the limited FMB. Any reduction in FMB exceeding that predicted by the model would indicate a real inability to listen in the gaps due to hearing loss or stimulus manipulation.

55.2 Procedure

55.2.1 Experimental Data

FMB estimates across a range of stationary-noise SNRs for NH listeners from Bernstein and Grant (2009) were used to parameterize the assumed IIFs underlying the ESII calculation. To test the model, data was taken from a variety of sources in the literature. There were three requirements for inclusion of data from a particular study in the analysis. First, only studies that estimated intelligibility in backgrounds of stationary noise and either an opposite-gender interfering talker or speech-modulated noise were considered. It would be of interest to apply this approach to other type of fluctuating maskers, such as interrupted noise (e.g., Dubno et al. 2003; Jin and Nelson 2006), sinusoidally amplitude-modulated noise (e.g., Eisenberg et al. 1995; Lorenzi et al. 2006a), or multitalker babble (Wilson et al. 2007). However, estimates of the FMB across stationary-noise SNR for NH listeners were not available to parameterize the model for other masker types. Second, only data allowing the characterization of the FMB in terms of dB differences were considered. A model prediction of the FMB in terms of a percentage-correct improvement would require an estimate of the effects of suprathreshold deficits associated with hearing loss on the transformation from intelligibility index to a percentage-correct score. Third, it was essential that the SRT for the stationary-noise condition was reported or could be estimated from the data.

Previous studies that met these criteria are summarized in the legend of Fig. 55.3. Most of the studies examined cases where the FMB was reduced relative to NH listeners presented with unprocessed speech. This includes situations involving HI listeners (Festen and Plomp 1990; Hygge et al. 1992; Bacon et al. 1998; Versfeld and Dreschler 2002; Bernstein and Grant 2009), elderly listeners (Versfeld and Dreschler 2002) or NH listeners presented with processed speech (ter Keurs et al. 1993; Bacon et al. 1998; Qin and Oxenham 2003; Oxenham and Simonson 2009). Spectral smearing (ter Keurs et al. 1993) and noise masking (Bacon et al. 1998) were implemented specifically to simulate aspects of hearing loss. Although the removal of TFS information (i.e., fast fluctuations in the stimulus waveform) was intended as a simulation of cochlear-implant processing (Qin and Oxenham 2003), HI listeners might also have a reduced ability to use TFS information (Buss et al. 2004; Lorenzi et al. 2006b; Moore et al. 2006; Hopkins et al. 2008). Filtering (Oxenham and Simonson 2009) was implemented as a way to reduce the strength of pitch cues, although it could be thought of as a simulation of a complete loss of audibility in certain frequency regions. Two of the studies (Hygge et al. 1992; Bernstein and Grant 2009) also examined conditions involving audiovisual (AV) speech, where the availability of visual cues increased the FMB relative to the audio-alone case.

55.2.2 Model Description

The ESII methodology was as described by Rhebergen et al. (2006). The SNR was estimated in 4-ms time frames over 21 critical bands for the speech materials used by Bernstein and Grant (2009). An intelligibility score was computed for each frame and band by estimating the proportion of the assumed speech dynamic range audible above the masker. The ESII was calculated by time-averaging the intelligibility scores in each band and computing a frequency-weighted average of the scores across bands. The FMB estimate was obtained by comparing the SNR in a fluctuating-masker condition needed to achieve the same ESII as the stationary-noise condition at a given SNR.

Bernstein and Grant (2009) found that the ESII overestimated the magnitude of the FMB, even for NH listeners, when the same IIF was assumed for stationary and fluctuating maskers. They proposed that the IIF might be different for fluctuating-masker conditions. Based on the results of Lorenzi et al. (2006a) indicating variability across consonant features in the effects of masker modulation rate on the FMB, they reasoned that speech information made audible near the minimum of a fluctuating-masker valley might have insufficient duration to contribute to intelligibility. The portion of the IIF influenced by the putative duration limitation would vary with SNR, leading to an SNR-dependent fluctuating-masker IIF.

The IIF for stationary noise was assumed to be distributed according to Studebaker and Sherbecoe (2002). Hypothetical fluctuating-masker IIFs were generated by reducing the assumed contribution to intelligibility of low-level portions of the dynamic range (Fig. 55.2). The lower tail of the stationary-noise IIF was shifted horizontally (in dB), while not allowing the IIF value to exceed the stationary IIF at any level within the dynamic range. Two parameters described the SNR-dependent fluctuating-masker IIF. SNR_0 (dB) described the upper SNR limit at which the fluctuating-masker and stationary-noise IIFs were identical. *R* (dB/dB) described the ratio between the magnitude of the IIF shift and the change in SNR. Using this approach, Bernstein and Grant (2009) showed that the ESII could account for the FMB SNR dependence across NH and HI listeners and audio-alone and AV conditions.

55.2.3 Model Parameterization

The mean NH FMB was derived for a range of SNRs from the data of Bernstein and Grant (2009). For each stationary-noise SNR tested (in 3-dB steps), an estimate was made of the fluctuating-masker SNR required to yield the stationary-noise performance level. Only SNRs where listeners identified between 5 and 95% of keywords were considered, due to the uncertainty in the FMB estimate associated with the saturation of the performance functions outside of this range. Figure 55.1 shows the resulting FMB functions for an interfering-talker (large squares) and speech-modulated noise (large circles).

To account for NH FMB data across SNR, it was necessary to assume a different IIF for the fluctuating-masker conditions. Here, the two free parameters describing the SNR-dependent fluctuating-masker IIF were adjusted to fit the data. The IIFs associated with the best-fitting parameter values ($SNR_0 = 28$ dB, $R = 0.46$ dB/dB) are shown in Fig. 55.2. The resulting FMB functions are plotted as solid (interfering talker) and dashed curves (speech-modulated noise) in Fig. 55.1. The model captured the dependence of the FMB on SNR. The model also accounted for the larger FMB for an interfering talker than for speech-modulated noise (e.g., Festen and Plomp 1990; Qin and Oxenham 2003; Oxenham and Simonson 2009; Bernstein and Grant 2009). This was likely the result of the spectral and temporal gaps available for the interfering-talker case, yielding more time–frequency combinations with relatively high SNRs.

55.2.4 Application to Earlier Results

ESII FMB predictions were compared to the FMB estimates from each study. The FMB was taken to be the difference (in dB) between the SRTs for the stationary-noise condition and a particular fluctuating masker, averaged across listeners. In most cases, SRTs were estimated for stationary and fluctuating maskers using an adaptive tracking technique (e.g., Plomp and Mimpen 1979). For studies that instead measured speech intelligibility across a range of SNRs (Oxenham and Simonson 2009; Bernstein and Grant 2009), SRTs were derived by fitting sigmoid functions to the reported performance functions.

An ESII FMB prediction was generated for each combination of listener group and stimulus condition. The prediction was determined from the curves in Fig. 55.1 based on the SNR associated with the stationary-noise SRT. Ideally, ESII predictions would be generated from the exact signals used in each study. As these were not generally available, a simpler approach was taken as a first approximation. Due to stimulus and procedural differences, there was considerable variation across studies (Fig. 55.1, small symbols) in the FMB for the baseline condition (broadband, unprocessed, audio-alone stimuli presented to NH listeners) relative to the model predictions. The ESII-generated FMB predictions were normalized for each study and masker type by subtracting the experimentally measured FMB for this baseline condition. This normalized value represents the reduction in the FMB (at the estimated 50% correct SRT) relative to the FMB for a NH listener presented with unprocessed audio-alone speech.

55.3 Results

The experimentally measured reduction in FMB (as estimated at the 50% correct SRT) is plotted as a function of the model-predicted FMB reduction (interfering talker: Fig. 55.3a; speech-modulated noise: Fig. 55.3b). Symbols represent combinations of processing condition, listener group or modality (audio-alone or AV) and study.

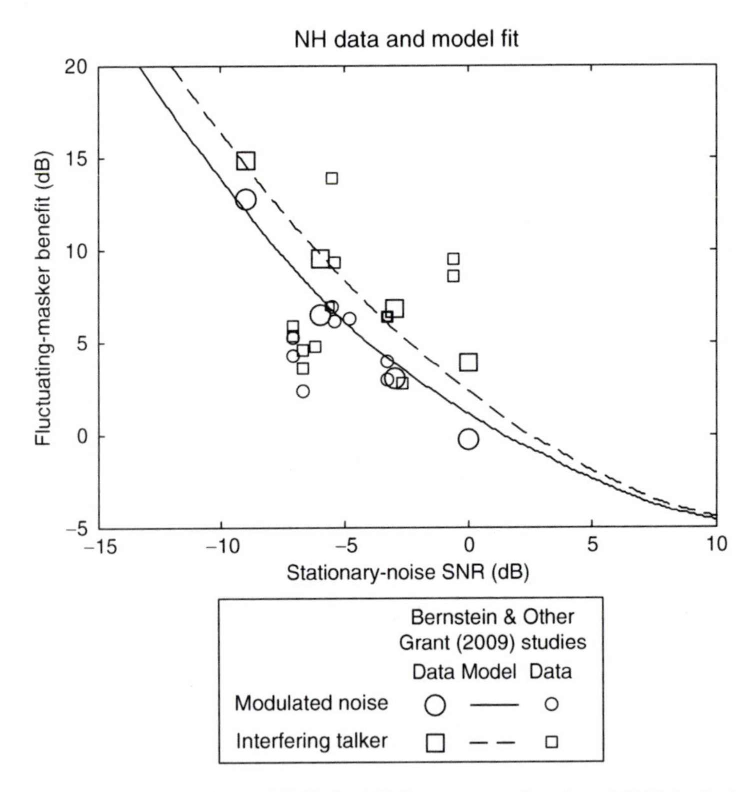

Fig. 55.1 Experimentally estimated FMB for NH listeners as a function of SNR (*points*) and the associated best-fitting model predictions (*curves*)

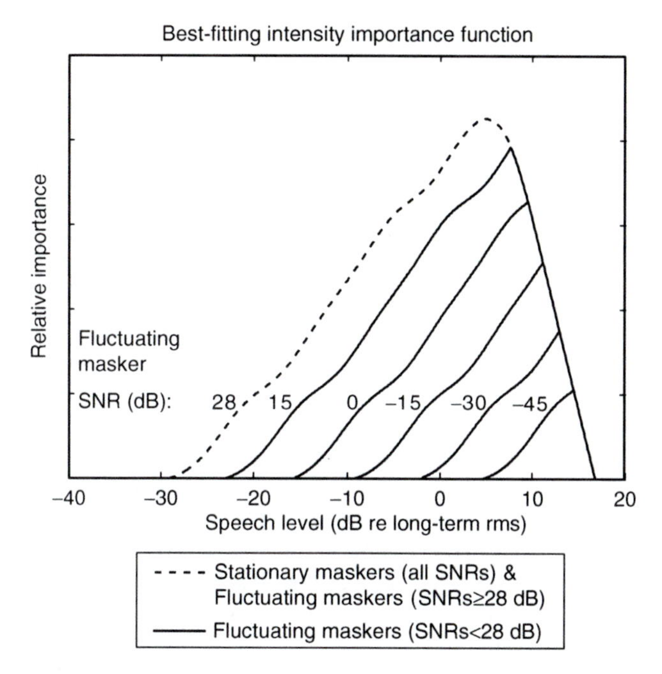

Fig. 55.2 The best-fitting family of stationary-noise and fluctuating-masker IIFs yielding the predictions shown in Fig. 55.1

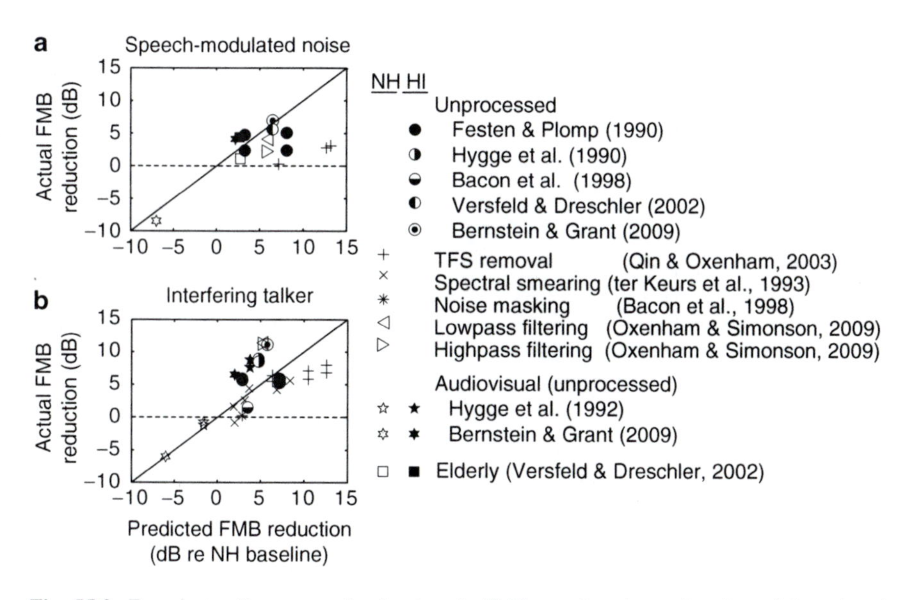

Fig. 55.3 Experimentally-measured reductions in FMB are plotted as a function of the reduced FMB predicted due to differences in the baseline stationary-noise SRT for (**a**) speech-modulated noise and (**b**) interfering-talker conditions

Multiple copies of the same symbol represent different stimulus conditions in the same general category. For example, ter Keurs et al. (1993) tested multiple degrees of spectral smearing. Festen and Plomp (1990) tested different combinations of talker and masker gender and forward and backward masking speech.

Points falling above the horizontal dotted line indicate a reduced FMB relative to the NH baseline. Almost all of the data falls into this category, confirming the claims that the FMB (estimated at the SRT) is reduced by hearing loss (filled symbols), age (squares), and various stimulus manipulations (e.g., spectral smearing, TFS removal, filtering). The main exception is the AV condition for NH listeners (open five- and six-pointed stars; Hygge et al. 1992; Bernstein and Grant 2009), where listeners received a larger FMB than for audio alone. Other exceptions near this horizontal line (i.e., no reduction in FMB) were a few conditions with NH listeners presented with processed speech (ter Keurs et al. 1993; Qin and Oxenham 2003) or an additional masking noise (Bacon et al. 1998).

The main question posed here is whether the apparent reduction in FMB in each of these cases is predicted by differences in the stationary-noise SRT. The diagonal line indicates the predicted reduction in the FMB based on the stationary-noise SRT. Above this line, the FMB was still reduced after effects of SNR were controlled. Below this line, listeners performed *better* in the fluctuating-masker conditions relative to stationary noise than would be predicted based on the stationary-noise SRT. The vast majority of the points fall very close to, or below the diagonal line. Thus in most instances, the reduction in FMB may be attributable to SNR effects, with no evidence of a reduced ability to listen in the gaps of a fluctuating masker. The main exceptions to this finding were interfering-talker conditions (Fig. 55.3a) involving HI listeners (filled symbols; Festen and Plomp 1990; Hygge et al. 1992; Bernstein and Grant 2009) or filtered speech presented to NH listeners (left- and right-pointing triangles; Oxenham and Simonson 2009), where the FMB was reduced by about 5 dB more than predicted. In these cases, listeners may have indeed been impaired in their ability to benefit from masker fluctuations.

55.4 Discussion

Reduced FMB for HI listeners and for NH listeners presented with distorted speech has been repeatedly reported in the literature. These findings have been interpreted as an indication of (a) a reduced benefit from momentary dips in the level of a fluctuating masker, and (b) a role for impaired suprathreshold hearing acuities (such as fine spectral resolution and the ability to use TFS information) in causing the reduced FMB. By controlling for SNR effects, the analysis presented here questions these assertions. First, HI listeners show a reduced FMB only for interfering-talker conditions. For speech-modulated noise, the appearance of a reduced FMB was accounted for by the higher stationary-noise SRT. Second, the reduced FMB associated with most of the various signal processing manipulations was also accounted for by stationary-noise SRT differences. There was no evidence that spectral smearing, TFS removal or noise masking caused any reduction in the ability of NH individuals to listen in the gaps of a fluctuating masker. The appearance of an increased FMB for AV conditions was also accounted for by stationary-noise SRT differences.

The only cases that still showed a reduced FMB after controlling for SNR effects were interfering-talker conditions for HI listeners (Festen and Plomp 1990; Hygge et al. 1992; Bernstein and Grant 2009) and for NH listeners presented with filtered speech (Oxenham and Simonson 2009). Broadband speech might be required to receive the maximum possible FMB for an interfering-talker condition, with the full speech spectrum not audible for HI listeners. Audibility analyses (Festen and Plomp 1990; Bernstein and Grant 2009) indicated that, on average, the target speech was inaudible to HI listeners for frequencies above about 2 kHz. One possible reason for a full-bandwidth requirement is that listeners could take maximum advantage of the redundancy of speech information across the spectrum (Oxenham and Simonson 2009), with dips in the masker level occurring at different times in different frequency regions for the interfering talker but not the modulated noise. Alternatively, across-frequency comodulation in the broadband signal might help facilitate source separation.

It remains an open question whether SNR effects could account for the reduced FMB observed for NH listeners and processed stimuli for fluctuating maskers not presented here, such as interrupted (e.g., Dubno et al. 2003; Jin and Nelson 2006) or amplitude-modulated noise (e.g., Eisenberg et al. 1995; Lorenzi et al. 2006a). In particular, Jin and Nelson (2006) found substantial intelligibility differences between NH and HI listeners for speech presented in interrupted noise, even though there were no measurable differences between NH and HI listeners in the stationary SRT. Audibility differences between the two groups might be more likely to impact the FMB for interrupted noise, where signals are effectively presented in quiet during the off cycle. FMB estimates across a range of SNRs for NH listeners would be required to apply the same analysis to other forms of fluctuating masker.

The modified ESII was used as a tool to control for possible confounding effects of SNR differences on the FMB. It should be noted that the use of the model was not essential to this analysis. The SNR dependence of the FMB could have been characterized by simply fitting a curve to the empirical NH data, possibly yielding slightly different results. Nevertheless, this result highlights the crucial role of the assumed IIF. To capture the SNR dependence of the FMB in NH listeners, it was necessary to assume that not all of the speech information made audible by dips in the fluctuating-masker level contributed to intelligibility. The particular fluctuating-masker IIF configuration in Fig. 55.2 is just one possibility that can account for the SNR dependence of the FMB in NH listeners. Direct empirical measurements of the IIF in fluctuating noise are needed to validate or modify this assumption.

55.5 Conclusions

The benefit to speech intelligibility yielded by a fluctuating masker depends largely on SNR. This relationship is accounted for by an existing model of speech intelligibility in fluctuating maskers (ESII), but only when the distribution of usable speech information across the audible dynamic range is assumed to differ for stationary and fluctuating maskers. In many cases, reduced FMB reported in published literature may be attributable to confounding SNR effects. The only exceptions were interfering-talker conditions for NH listeners presented with filtered stimuli and for HI listeners, with 5 dB of the FMB reduction unaccounted for by SNR effects. The reduced FMB for HI listeners in this instance may result from a reduced effective signal bandwidth. In summary, HI listeners and NH listeners presented with distorted speech may retain more of an ability to listen in the gaps of a fluctuating masker than previously thought.

Acknowledgments This work was supported by an Oticon Foundation grant. The opinions or assertions contained herein are the private views of the author and are not to be construed as official or as reflecting the views of the Department of the Army or Department of Defense.

References

ANSI (1997) Methods for calculation of the speech intelligibility index, S3.5. American National Standards Institute, New York

Bacon SP, Opie JM, Montoya DY (1998) The effects of hearing loss and noise masking on the masking release for speech in temporally complex backgrounds. J Speech Lang Hear Res 41: 549–563

Baer T, Moore BCJ (1994) Effects of spectral smearing on the intelligibility of sentences in the presence of interfering speech. J Acoust Soc Am 95:2277–2280

Bernstein JGW, Grant KW (2009) Auditory and auditory-visual speech intelligibility in fluctuating maskers for normal-hearing and hearing-impaired listeners. J Acoust Soc Am 125:3358–3372

Buss E, Hall JW, Grose JH (2004) Temporal fine-structure cues to speech and pure tone modulation in observers with sensorineural hearing loss. Ear Hear 25:242–250

Dubno JR, Horwitz AR, Ahlstrom JB (2003) Recovery from prior stimulation: masking of speech by interrupted noise for younger and older adults with impaired hearing. J Acoust Soc Am 113:2084–2094

Eisenberg LS, Dirks DD, Bell TS (1995) Speech recognition in amplitude-modulated noise of listeners with normal and listeners with impaired hearing. J Speech Hear Res 38:222–233

Festen JM, Plomp R (1990) Effects of fluctuating noise and interfering speech on the speech-reception threshold for impaired and normal hearing. J Acoust Soc Am 88:1725–1736

Freyman RL, Balakrishnan U, Helfer KS (2008) Spatial release from masking with noise-vocoded speech. J Acoust Soc Am 124:1627–1637

George ELJ, Festen JM, Houtgast T (2006) Factors affecting masking release for speech in modulated noise for normal-hearing and hearing-impaired listeners. J Acoust Soc Am 120:2295–2311

Hopkins K, Moore BCJ, Stone MA (2008) Effects of moderate cochlear hearing loss on the ability to benefit from temporal fine structure in speech. J Acoust Soc Am 123:1140–1153

Hygge S, Rönnberg J, Larsby B, Arlinger S (1992) Normal-hearing and hearing-impaired subjects' ability to just follow conversation in competing speech, reversed speech, and noise backgrounds. J Speech Hear Res 35:208–215

Jin SH, Nelson PB (2006) Speech perception in gated noise: the effects of temporal resolution. J Acoust Soc Am 119:3097–3108

Lorenzi C, Husson M, Ardoint M, Debruille X (2006a) Speech masking release in listeners with flat hearing loss: effects of masker fluctuation rate on identification scores and phonetic feature reception. Int J Audiol 45:487–495

Lorenzi C, Gilbert G, Carn H et al (2006b) Speech perception problems of the hearing impaired reflect inability to use temporal fine structure. Proc Natl Acad Sci U S A 103:18866–18869

Moore BCJ, Glasberg BR, Hopkins K (2006) Frequency discrimination of complex tones by hearing-imparied subjects: evidence for loss of ability to use temporal fine structure. Hear Res 222:16–27

Oxenham AJ, Simonson AM (2009) Masking release for low- and high-pass-filtered speech in the presence of noise and single-talker interference. J Acoust Soc Am 125:457–468

Plomp R, Mimpen AM (1979) Improving the reliability of testing the speech reception threshold for sentences. Audiology 18:43–53

Qin MK, Oxenham AJ (2003) Effects of simulated cochlear-implant processing on speech reception in fluctuating maskers. J Acoust Soc Am 114:446–454

Rhebergen KS, Versfeld NJ, Dreschler WA (2006) Extended speech intelligibility index for the prediction of the speech reception threshhold in fluctuating noise. J Acoust Soc Am 120: 3988–3997

Studebaker GA, Sherbecoe RL (2002) Intensity-importance functions for bandlimited monosyllabic words. J Acoust Soc Am 111:1422–1436

Summers V, Molis MR (2004) Speech recognition in fluctuating and continuous maskers: effects of hearing loss and presentation level. J Speech Lang Hear Res 47:245–256

ter Keurs M, Festen JM, Plomp R (1993) Effect of spectral envelope smearing on speech reception. II. J Acoust Soc Am 93:1547–1552

Versfeld NJ, Dreschler WA (2002) The relationship between the intelligibility of time-compressed speech and speech in noise in young and elderly listeners. J Acoust Soc Am 111:401–408

Wilson RH, Carnell CS, Cleghorn AL (2007) The words-in-noise (WIN) test with multitalker babble and speech-spectrum noise spectra. J Am Acad Audiol 18:522–529

Chapter 56
Across-Fiber Coding of Temporal Fine-Structure: Effects of Noise-Induced Hearing Loss on Auditory-Nerve Responses

Michael G. Heinz, Jayaganesh Swaminathan, Jonathan D. Boley, and Sushrut Kale

Abstract Recent psychophysical evidence suggests that listeners with sensorineural hearing loss (SNHL) have a reduced ability to use temporal fine-structure cues. These results have renewed an interest in the effects of SNHL on the neural coding of fine structure. The lack of convincing evidence that SNHL affects within-fiber phase locking has led to the hypothesis that degraded across-fiber temporal coding may underlie this perceptual effect. Spike trains were recorded from auditory-nerve (AN) fibers in chinchillas with normal hearing and with noise-induced hearing loss. A spectro-temporal manipulation procedure was used to predict spatiotemporal patterns for characteristic frequencies (CFs) spanning up to an octave range from the responses of individual AN fibers to a stimulus presented with sampling rates spanning an octave range. Shuffled cross-correlogram analyses were used to quantify across-CF fine-structure coding in terms of both a neural cross-correlation coefficient and a characteristic delay. Neural cross-correlation for fine-structure decreased and the estimated traveling-wave delay increased with increases in CF separation for both normal and impaired fibers. However, the range of CF separations over which significant correlated activity existed was wider, and the estimated traveling-wave delay was less for impaired AN fibers. Both of these effects of SNHL on across-CF coding have important implications for spatiotemporal theories of speech coding.

Keywords Auditory nerve • Sensorineural hearing loss • Across-fiber coding • Temporal fine structure • Traveling wave delay

---

M.G. Heinz (✉)
Department of Speech, Language, and Hearing Sciences, Purdue University, West Lafayette, IN 47907, USA
e-mail: mheinz@purdue.edu

E.A. Lopez-Poveda et al. (eds.), *The Neurophysiological Bases of Auditory Perception*,
DOI 10.1007/978-1-4419-5686-6_56,

56.1 Introduction

Recent psychophysical studies suggest that listeners with sensorineural hearing loss (SNHL) have a reduced ability to use temporal fine-structure cues, which is correlated with their reduced understanding of speech in complex backgrounds (Lorenzi et al. 2006; Hopkins and Moore 2007). These perceptual results have renewed an interest in the effects of SNHL on neural coding of temporal fine structure, both within single auditory-nerve (AN) fibers and across fibers with different characteristic frequencies (CFs). There is conflicting evidence as to whether within-fiber encoding of fine-structure (i.e., phase locking) is degraded following SNHL (Harrison and Evans 1979; Woolf et al. 1981; Miller et al. 1997). Thus, it has been hypothesized that degraded across-fiber temporal coding due to broader tuning and associated shallower phase responses could underlie these perceptual deficits, e.g., as implicated in spatiotemporal theories of speech coding (e.g., Shamma 1985). However, effects of SNHL on across-CF coding have been difficult to examine because of experimental limitations associated with sparse CF sampling in AN population studies and variability in CF estimates (Chintanpalli and Heinz 2007).

The present study compared the effects of noise-induced hearing loss on within- and across-fiber coding of temporal fine structure. Across-fiber variability was minimized by using responses of individual AN fibers to frequency-shifted stimuli to predict responses of a population of AN fibers with differing CFs to a single stimulus. Shuffled auto- and cross-correlograms were used to quantify across-CF temporal coding in terms of both a neural cross-correlation coefficient and a characteristic delay (CD) that estimates the traveling-wave delay between two CFs.

56.2 Methods

56.2.1 Experimental Procedures

All procedures were approved by the Purdue Animal Care and Use Committee. Neural recordings were made from AN fibers in two anesthetized chinchillas using standard procedures (e.g., Heinz and Young 2004; Chintanpalli and Heinz 2007). Spike times were measured with 10-μs resolution. Isolated fibers were characterized by an automated tuning-curve algorithm to determine fiber CF, threshold, and Q_{10}. Impaired-fiber CFs were chosen by hand near the steep high-frequency slope, which better estimates the original CF prior to SNHL (Liberman 1984). Spontaneous rate was determined over 20 s and PST histograms were measured to verify AN responses based on latency. Noise-induced hearing loss was produced in one chinchilla by presenting a 50-Hz-wide noise band centered at 2 kHz continuously for 4 h at 115 dB SPL, after which the animal recovered for 6 weeks. Consistent with previous studies in which noise exposure produced mixed outer- and inner-hair cell damage (Liberman 1984; Heinz and Young 2004), a moderate hearing loss was produced with thresholds elevated by ~30–50 dB and broadened tuning in all fibers (Q_{10}s below the normal range for chinchillas).

Neural responses were recorded to both a broadband noise and a speech sentence. Both stimuli were 1.7-s in duration at the baseline sampling frequency of 33 kHz. Positive and negative polarity versions of both stimuli were presented at 7 or 9 different sampling frequencies spanning a range of up to 1 octave. All stimuli were presented in an interleaved manner, with a new stimulus presented every 2.9 s. Both stimuli were presented to each AN fiber at 10 or 20 dB above stimulus threshold for that fiber, as determined by measured rate-level functions. Stimuli were repeated until ~2,000 spikes were recorded for all stimuli, or until the fiber was lost. Data are presented from 17 normal-hearing fibers and 19 hearing-impaired fibers.

56.2.2 Predicting Spatiotemporal Patterns from Individual AN Fibers

The ability to quantify across-CF temporal coding is significantly limited by sparse sampling and across-fiber variability inherent in AN population studies, as well as by variability in CF estimates from automated tuning-curve algorithms (Chintanpalli and Heinz 2007). These limitations are particularly true with SNHL. To overcome these limitations, a spectro-temporal manipulation procedure (STMP) was used to predict the spatiotemporal response of a population of AN fibers with a range of CFs responding to a single stimulus from responses of a single AN fiber to frequency-shifted stimuli (Heinz 2007). The STMP relies on scaling invariance in cochlear mechanics and is similar to procedures that have been used to study spatiotemporal coding of pitch (Larsen et al. 2008). Although some cochlear properties are not scaling invariant (e.g., roll-off in phase locking, refractoriness, adaptation), these effects are likely to be negligible in comparisons between normal and impaired responses over ±0.5 octaves (Larsen et al. 2008).

56.2.3 Within-CF and Across-CF Temporal Analyses

Shuffled correlogram analyses (Louage et al. 2004; Joris et al. 2006; Heinz and Swaminathan 2009) were used to quantify within- and across-CF fine-structure coding from single AN-fiber responses to broadband noise and speech. Within-fiber temporal coding was evaluated based on normalized shuffled auto correlograms (SACs, thick lines, Fig. 56.1a, b), which were computed by comparing spike times between all possible pairs of stimulus presentations for a given effective CF from the STMP. For each pair, intervals between every spike in the first spike train and every spike in the second spike train were tallied with a 50-μs binwidth to create a shuffled all-order interval histogram. For each AN fiber, SACs were computed for each effective CF from the STMP. Figure 56.1 shows correlogram analyses for two effective CFs separated by 0.5 octaves based on spike trains recorded in response to broadband noise. Responses to positive and negative polarity versions of each stimulus were recorded because polarity inversion inverts stimulus fine-structure while not affecting stimulus envelope. Cross-polarity auto correlograms (XpACs,

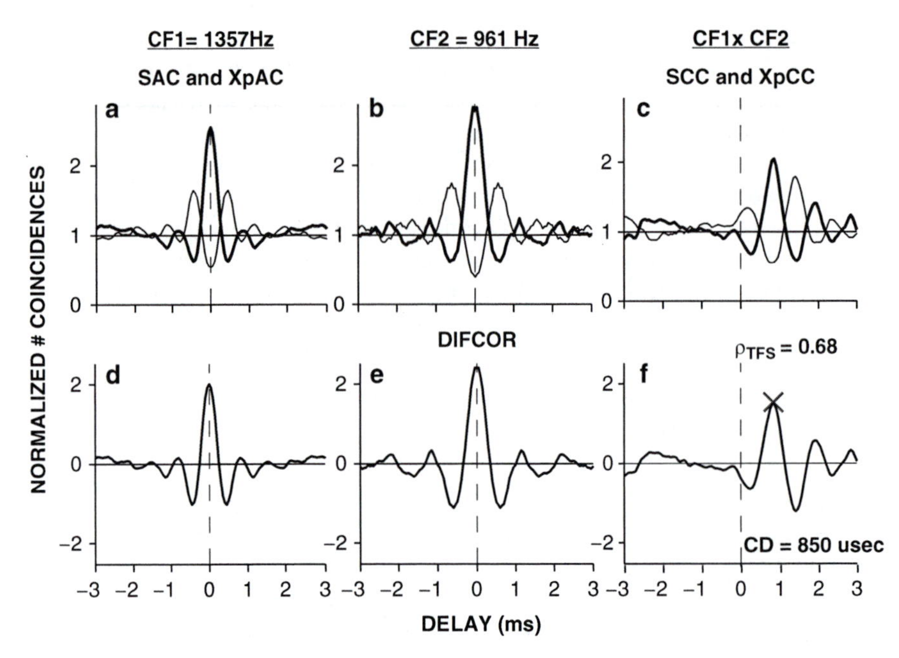

Fig. 56.1 Within- (cols. 1–2) and across-CF (col. 3) temporal coding based on shuffled correlograms. (**a**, **b**) Auto correlograms: SACs (*thick line*), XpACs (*thin line*). (**c**) Cross correlograms: SCC (*thick line*), XpCC (*thin line*). (**d–f**) DIFCORs emphasize fine structure by subtracting XpAC from SAC (or XpCC from SCC). Auto-correlogram DIFCOR peak heights quantify within-fiber fine structure. Across-CF coding was quantified with neural cross-correlation coefficients (ρ_{TFS}), computed as the ratio of cross-correlogram DIFCOR peak height (**f**) to the geometric mean of auto-correlogram DIFCOR peak heights (**d**, **e**). Characteristic delay (CD) (× in panel **f**) estimates traveling-wave delay between the two effective CFs 0.5 octaves apart. Spike trains recorded from one impaired AN fiber responding to a broadband noise with two sampling rates that differed by 0.5 octaves. STMP was used to predict responses of two effective CFs 0.5-octaves apart. CF = 1.36 kHz, thresh. = 49 dB SPL, Q_{10} = 0.9, spont. rate = 64 spikes/s

thin lines, Fig. 56.1a, b) were computed by tallying intervals between all spikes in response to positive and negative polarity versions of the stimulus. DIFCORs computed by subtracting the XpAC from the SAC thus emphasize fine structure coding, which was significant for both effective CFs in Fig. 56.1d, e.

Across-CF fine-structure coding was evaluated based on shuffled cross correlograms (SCCs, thick line, Fig. 56.1c) and cross-polarity, cross correlograms (XpCCs, thin line, Fig. 56.1c), which were computed by comparing spike trains across a pair of effective CFs from the STMP. For each effective CF pair for one AN fiber, the cross-correlogram DIFCOR was used to evaluate across-CF fine-structure coding with two metrics. A neural cross-correlation coefficient (ρ_{TFS}) was used to represent the degree of similarity between two spike-train responses (Heinz and Swaminathan 2009), and was computed as the ratio of the peak height of the cross-correlogram DIFCOR (Fig. 56.1f) to the geometric mean of the auto-correlogram DIFCOR peak heights (Fig. 56.1d, e). A significant benefit of this self-normalized similarity metric is that the degree of cross correlation is evaluated relative to the strength of within-fiber fine-structure coding for each fiber individually, which varies with differences in CF, spontaneous rate, and stimulus level (Louage et al. 2004). The computed value of $\rho_{TFS}=0.68$ indicates significant common temporal fine-structure in these hearing-impaired responses for effective CFs separated by 0.5 octaves. CD (× in Fig. 56.1f) of the cross-correlogram DIFCOR provides an estimate of the traveling wave delay between the two cochlear locations represented by these CFs (Joris et al. 2006). A CD of 850 μs was estimated for the two effective CFs separated by 0.5 octaves (Fig. 56.1f).

56.3 Results

Figure 56.2 illustrates the single-fiber analyses performed on each AN fiber. A normal-hearing fiber is compared to an impaired fiber with a similar CF (1.3 kHz) in terms of their tuning curves (Fig. 56.2a), the predicted effect of CF separation on cross-CF correlation (Fig. 56.2b) and CD (Fig. 56.2c). The normal-hearing tuning curve represents a low-threshold, high-spontaneous rate fiber with sharp tuning. The impaired tuning curve shows broad tuning without a defined tip. This tuning curve is representative in shape of all impaired fibers in this study, which had thresholds ranging from 40 to 60 dB SPL and CFs ranging from 0.7 to 5 kHz.

The effect of CF separation on cross-CF correlation for broadband noise is shown in Fig. 56.2b for both AN fibers. Neural cross-correlation coefficients (ρ_{TFS}) were computed for all effective CF pairs derived from the STMP for each AN fiber. Seven effective CFs predicted for the normal-hearing fibers produced 21 pairs with CF separations ranging from 0.05 to 0.55 octaves. For impaired fibers, 36 pairs with CF separations ranging from 0.05 to 1.0 octaves were obtained from nine effective CFs. The variation in ρ_{TFS} with CF separation was fit with fourth-order polynomials constrained to equal 1.0 at a CF separation of 0 octaves. Neural cross correlation decreased monotonically with increasing CF separation for all normal-hearing fibers. The example shown in Fig. 56.2b decreased to ~0.3 for the maximum CF separation of 0.55 octaves. Impaired AN fibers also showed a decrease in ρ_{TFS} as CF separation increased; however, the decrease was often less steep and sometimes did not drop below 0.6 for the largest CF separation of 1.0 octaves (especially for the speech stimulus). Based on the fitted lines, the width of the correlated region was estimated by the smallest CF separation at which ρ_{TFS} fell below 0.6. This normal-hearing fiber demonstrated correlated activity above $\rho_{TFS}=0.6$ out to a 0.34-octave CF separation, whereas the impaired fiber demonstrated a much wider CF-separation range (0.81 octaves) of correlated activity.

The increase in CD with increased CF separation is shown in Fig. 56.2c for the same two AN fibers. CD derived from the cross-correlogram DIFCORs is represented in units of CF cycles and is plotted as a function of CF separation for all effective-CF pairs. For all normal-hearing and hearing-impaired fibers, CD increased very systematically across the entire range of CF separations and was

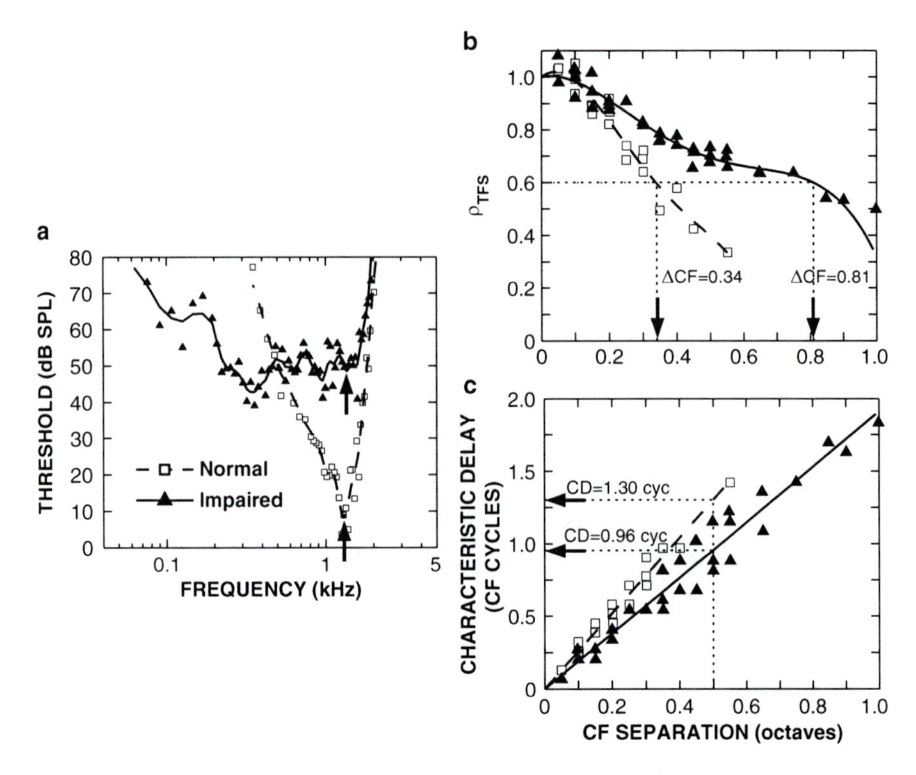

Fig. 56.2 Effect of CF separation on across-CF fine-structure coding for a normal-hearing and a hearing-impaired AN fiber with similar CFs. (**a**) Tuning curves. (**b**) Neural cross-correlation coefficients (ρ_{TFS}) as a function of CF separation. The smallest CF separation (ΔCF) at which ρ_{TFS} dropped to 0.6 was computed based on a fourth-order polynomial fit. (**c**) Characteristic delay (CD) increased linearly as a function of CF separation. CD was measured (in ms) from cross-correlogram DIFCORs and converted to CF cycles by multiplying by CF in kHz. The CD at a CF separation of 0.5 octaves was computed based on linear fits. AN fibers: normal (*open squares, dashed lines*): CF = 1.29 kHz, thresh. = 8 dB SPL, Q_{10} = 4.1, spont. rate = 91 spikes/s; impaired (*filled triangles, solid lines*): CF = 1.36 kHz, thresh. = 49 dB SPL, Q_{10} = 0.9, spont. rate = 64 spikes/s. Noise level: 10 dB above threshold for each fiber

well fit by a linear function constrained to equal 0 for no CF separation. The rate of increase in CD as CF separation increased was less for the impaired AN fiber than for the normal-hearing fiber (Fig. 56.2c). Thus, for all effective-CF separations the traveling-wave delay was predicted to be reduced following SNHL, consistent with broader tuning and the associated shallower phase transition. To quantify this effect, the CD at a 0.5-octave separation was computed for each AN fiber based on the fitted lines. For the examples shown, the CD at a 0.5-octave separation was 1.3 cycles for the normal-hearing fiber and 0.96 cycles for the impaired fiber, i.e., more than a quarter-cycle difference.

The normal-hearing and hearing-impaired populations of AN fibers are compared in Fig. 56.3 in terms of both within- and across-CF coding of temporal fine structure for broadband noise and speech responses. Auto-correlogram DIFCOR

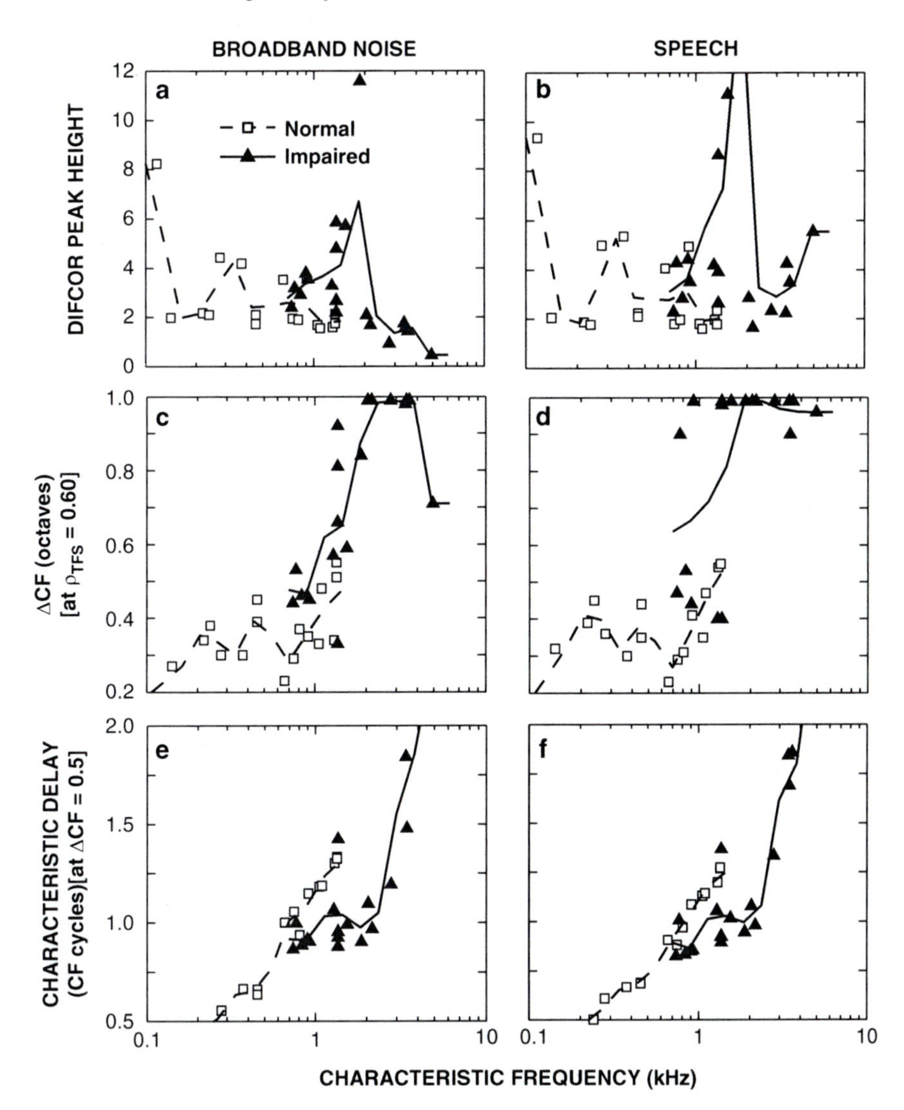

Fig. 56.3 Comparison of fine-structure coding between normal-hearing and hearing-impaired AN-fiber populations for broadband noise (*left*) and speech (*right*). (**a**, **b**) Within-fiber fine-structure coding represented by DIFCOR peak heights. (**c**, **d**) Smallest CF separation (ΔCF) at which ρ_{TFS} dropped to 0.6 represents the width of correlated activity. (**e**, **f**) Characteristic delay (CD) at a CF separation of 0.5 octaves estimates phase delay (in CF cycles) across two cochlear locations 0.5 octaves apart. Lines are weighted moving averages from a 0.7-octave-wide triangular window in steps of 0.35 octaves. All stimuli: 10 or 20 dB above stimulus threshold for each fiber

peak heights represent the strength of within-fiber fine-structure coding (Fig. 56.3a, b) and were not reduced in the hearing-impaired population for either broadband noise or speech. In fact, DIFCOR peak heights were slightly higher on average in the impaired population within the CF region from 0.7 to 1.5 kHz, where both normal and impaired data existed in these limited populations. A few fibers in the hearing-impaired population showed much larger DIFCOR peak heights than most of the normal-hearing data. All fibers with DIFCOR peak heights above 6 were low-spontaneous rate fibers, which typically have larger DIFCOR peak heights (Louage et al. 2004) and were more prevalent in the impaired population, consistent with previous studies (e.g., Heinz and Young 2004). Thus, there was no observed degradation in the strength of within-fiber coding of temporal fine structure, consistent with several previous studies (Harrison and Evans 1979; Miller et al. 1997).

Degradations were observed in across-CF coding of temporal fine-structure. The range of effective-CF separations over which correlated activity existed above $\rho_{TFS}=0.6$ is compared between the normal-hearing and hearing-impaired populations in Fig. 56.3c, d. For most normal-hearing AN fibers, correlated activity existed over a CF separation range between 0.2 and 0.5 octaves for both broadband noise and speech. For impaired fibers with CFs between 0.7 and 1.5 kHz, the width of correlated activity for broadband noise was 0.1–0.2 octaves wider than for the normal-hearing fibers, as indicated by the trend lines. For speech responses, this degradation was more significant, with many impaired fibers showing correlated activity that did not drop to $\rho_{TFS}=0.6$ over the entire 1.0-octave range of effective-CF separations.

A decrease in CD between effective CFs was observed in the impaired population for both broadband noise and speech responses (Fig. 56.3e, f). CD at a CF separation of 0.5 octaves is plotted against fiber CF for both populations. For the normal-hearing population, the delay in CF cycles increased systematically from 0.25 cycles for fiber CFs ~ 150 Hz to more than 1.25 cycles for CFs just above 1 kHz. This trend is consistent with sharper tuning and increased cochlear delays (in cycles) with increased CF as inferred from otoacoustic emissions (Shera et al. 2002). CDs were reduced by ~0.25 cycles in impaired AN fibers with CFs between 0.7 and 1.5 kHz. Impaired CD was roughly constant around 1 cycle for CFs from 0.7 to 2 kHz, and increased at higher CFs. Note that unlike the cross-correlation effects (Figs. 56.3c, d), CD effects were remarkably similar between broadband noise and speech responses. The reduction in CD with SNHL was smaller for CFs below 1 kHz than for those above 1 kHz; however, this CF-dependence may simply result from the specific noise-induction procedure used (i.e., 2-kHz exposure frequency), which produces the most significant hearing loss above 1 kHz (Heinz and Young 2004). Further study is necessary to evaluate SNHL effects at low CFs, given that listeners with high-frequency hearing loss and near-normal thresholds at low CFs have been shown to have a perceptual TFS deficit for lowpass filtered speech (Lorenzi et al. 2009).

It should be noted that the effect of SNHL on cochlear phase delays was not to eliminate the traveling-wave delay, but simply to reduce the across-CF delay by roughly 0.25-cycles in the present data for a moderate hearing loss. The size of this effect is consistent with level-dependent changes in the relative phase above and below CF in AN fiber responses (Palmer and Shackleton 2009). The relative phase for a 0.5-octave frequency difference around CF can vary by a quarter to a half cycle over a 40–50 dB range of tone level, which is presumably related to normal outer-hair-cell function associated with nonlinear cochlear tuning. Although the size of this effect is small relative to the overall phase delay of ~1 cycle for the 0.5-octave CF separation considered here, a quarter cycle phase shift (e.g., sine to cosine) represents the difference between in-phase and uncorrelated activity. Thus, characteristic-delay changes of this size would be significant in terms of any mechanism that relied on across-CF correlation at a fixed delay (e.g., cross-CF coincidence detection). Note that ρ_{TFS} represents a different effect, in that it quantifies the maximum correlation across all delays (i.e., computed at CD).

56.4 Discussion

The most significant effects of SNHL on fine-structure coding in AN fibers were in terms of across-CF coding rather than within-fiber coding, for which no degradation was observed. Across-CF coding was degraded in terms of both an increase in the cross-CF correlation and a decrease in CD between effective CFs. Broadening of the CF region over which correlated activity exists with SNHL could be perceptually significant for complex sounds because it would reduce the number of available independent neural information channels. A reduction in traveling-wave delay across CF with SNHL would result in a more coincident representation of temporal features across fibers that could degrade normal spatiotemporal response patterns. These patterns have been hypothesized to provide robust neural cues for a range of perceptual phenomena, including the coding of speech, pitch, and intensity, as well as tone detection in noise (Shamma 1985; Heinz et al. 2001; Carney et al. 2002; Heinz 2007; Larsen et al. 2008). Changes in across-CF delays would also have implications for binaural theories that rely on cochlear disparities as a source for interaural delays (Shamma et al. 1989; Joris et al. 2006).

Thus, these preliminary data suggest that the effects of SNHL on across-CF coding are significant and need to be considered when interpreting the reduced perceptual ability of listeners with SNHL to use fine-structure cues (e.g., Lorenzi et al. 2006; Hopkins and Moore 2007). If these physiological effects were perceptually relevant, they would suggest the need for new avenues into improving strategies for auditory prostheses, which currently do not attempt to restore normal spatiotemporal response patterns.

Acknowledgments This work was supported by grants from NIH (R03-DC07348) and the National Organization for Hearing Research Foundation.

References

Carney LH, Heinz MG, Evilsizer ME, Gilkey RH, Colburn HS (2002) Auditory phase opponency: a temporal model for masked detection at low frequencies. Acustica-Acta Acustica 88:334–347

Chintanpalli A, Heinz MG (2007) The effect of auditory-nerve response variability on estimates of tuning curves. J Acoust Soc Am 122:EL203–EL209

Harrison RV, Evans EF (1979) Some aspects of temporal coding by single cochlear fibres from regions of cochlear hair cell degeneration in the guinea pig. Arch Otorhinolaryngol 224:71–78

Heinz MG (2007) Spatiotemporal encoding of vowels in noise studied with the responses of individual auditory nerve fibers. In: Kollmeier B, Klump G, Hohmann V, Langemann U, Mauermann M, Uppenkamp S, Verhey J (eds) Hearing – from sensory processing to perception. Springer-Verlag, Berlin, pp 107–115

Heinz MG, Swaminathan J (2009) Quantifying envelope and fine-structure coding in auditory-nerve responses to chimaeric speech. J Assoc Res Otolaryngol 10:407–423

Heinz MG, Young ED (2004) Response growth with sound level in auditory-nerve fibers after noise-induced hearing loss. J Neurophysiol 91:784–795

Heinz MG, Colburn HS, Carney LH (2001) Rate and timing cues associated with the cochlear amplifier: level discrimination based on monaural cross-frequency coincidence detection. J Acoust Soc Am 110:2065–2084

Hopkins K, Moore BC (2007) Moderate cochlear hearing loss leads to a reduced ability to use temporal fine structure information. J Acoust Soc Am 122:1055–1068

Joris PX, Van de Sande B, Louage DH, van der Heijden M (2006) Binaural and cochlear disparities. Proc Natl Acad Sci USA 103:12917–12922

Larsen E, Cedolin L, Delgutte B (2008) Pitch representations in the auditory nerve: two concurrent complex tones. J Neurophysiol 100:1301–1319

Liberman MC (1984) Single-neuron labeling and chronic cochlear pathology. I. Threshold shift and characteristic-frequency shift. Hear Res 16:33–41

Lorenzi C, Gilbert G, Carn H, Garnier S, Moore BC (2006) Speech perception problems of the hearing impaired reflect inability to use temporal fine structure. Proc Natl Acad Sci USA 103:18866–18869

Lorenzi C, Debruille L, Garnier S, Fleuriot P, Moore BC (2009) Abnormal processing of temporal fine structure in speech for frequencies where absolute thresholds are normal. J Acoust Soc Am 125:27–30

Louage DH, Van Der Heijden M, Joris PX (2004) Temporal properties of responses to broadband noise in the auditory nerve. J Neurophysiol 91:2051–2065

Miller RL, Schilling JR, Franck KR, Young ED (1997) Effects of acoustic trauma on the representation of the vowel /ε/ in cat auditory nerve fibers. J Acoust Soc Am 101:3602–3616

Palmer AR, Shackleton TM (2009) Variation in the phase of response to low-frequency pure tones in the guinea pig auditory nerve as functions of stimulus level and frequency. J Assoc Res Otolaryngol 10:233–250

Shamma SA (1985) Speech processing in the auditory system. I: The representation of speech sounds in the responses of the auditory nerve. J Acoust Soc Am 78:1612–1621

Shamma SA, Shen NM, Gopalaswamy P (1989) Stereausis: binaural processing without neural delays. J Acoust Soc Am 86:989–1006

Shera CA, Guinan JJ Jr, Oxenham AJ (2002) Revised estimates of human cochlear tuning from otoacoustic and behavioral measurements. Proc Natl Acad Sci USA 99:3318–3323

Woolf NK, Ryan AF, Bone RC (1981) Neural phase-locking properties in the absence of cochlear outer hair cells. Hear Res 4:335–346

Chapter 57
Beyond the Audiogram: Identifying and Modeling Patterns of Hearing Deficits

Ray Meddis, Wendy Lecluyse, Christine M. Tan, Manasa R. Panda, and Robert Ferry

Abstract The choice of a hearing aid and its tuning parameters should benefit from a more detailed assessment of a patient's hearing than that occurs at present in clinical practice. The process of fitting and tuning might be facilitated even more if these data were used to generate a computer model of the patient's hearing. A number of hearing impaired patients have been tested using a range of tests that were devised to be easy to administer under automatic computer control and easy for the, often elderly, listeners to use. The principal finding is that there is an unexpected variation among patients in the patterns of impairment that are revealed by the tests. This is true even when the patients have similar audiograms. The data have been used to develop individual computer models of the patients' hearing which we call "hearing dummies." These dummies have been successfully evaluated using the same automated tests that were used to collect the original data. The creation of these dummies raises interesting questions about the basis of different forms of hearing pathology and point to a future where all forms of hearing impairment and their treatment might be diagnosed in terms of the underlying pathology rather than the symptomatology as at present.

Keywords Hearing impairment • Computer models

57.1 Introduction

The audiogram is the main basis for the fitting of hearing aids. However, experience shows that patients with the same audiogram may experience different degrees of benefit from the same hearing aid. It is likely that supplementary information, such as frequency selectivity and measures of residual compression, could improve the prescription process or drive the development of more appropriate hearing aid algorithms. Unfortunately, this additional information will not immediately help

---

R. Meddis (✉),
University of Essex, Colchester, Essex, UK
e-mail: rmeddis@essex.ac.uk

E.A. Lopez-Poveda et al. (eds.), *The Neurophysiological Bases of Auditory Perception*,
DOI 10.1007/978-1-4419-5686-6_57,

audiologists to prescribe existing aids because the manufactures fitting guidance does not take this information into account. Nor will it necessarily be of direct help to manufacturers unless some method is found for predicting the outcome for a particular patient of different aids, algorithms, or settings.

To address the "data poverty" problem, we are exploring methods for extending the range of measures that might usefully be deployed in a clinical context with an emphasis on the speed of data collection and the ease of use from the patient's point of view. Using a single-interval, up-down automated procedure, we have been able to measure frequency-selectivity in impaired, often elderly, patients over time periods that are clinically viable. The same approach has been used to measure residual compression. When this additional information is combined with other routine assessments, we obtain a rich audiological profile that reveals surprisingly diverse patterns across patients.

We have also begun to explore the problem of predicting hearing aid outcomes by developing computer models ("hearing dummies") of the hearing of individual patients. The models simulate the physiological processes underlying successive signal processing stages of the auditory periphery. Individual deficiencies are modeled by introducing pathological features such as reduced endocochlear potential, impaired outer hair cell (OHC) functioning, dead regions, etc. The model is, in effect, a hypothesis concerning the underlying pathology of the auditory periphery responsible for the deficit. The validity of the computer model is evaluated by testing it using the *same* software and procedures that are used with the patient. The ultimate aim of the project is to use the dummies to predict outcomes in an individual patient for a range of different hearing aids or different settings for a given aid.

The results reported below illustrate data collection and modeling results using data from one normal and three impaired listeners and our attempts to model their thresholds and frequency selectivity. The illustrations are restricted for reasons of space to absolute thresholds and frequency selectivity measured using iso-forward masking contours (IFMCs). The aim is to illustrate the general thrust of our ongoing work and to stress that the driving motivation of the research is to characterize the hearing of an individual patient rather than making scientific statements about the types of hearing impairment.

57.2 Methods

57.2.1 Psychoacoustic Profiles of Normal and Impaired Listeners

57.2.1.1 Psychoacoustic Measures

Absolute thresholds: Detection thresholds were measured where possible for 500-ms pure tones (raised cosine onset and offset times of 4 ms) at frequencies 250, 500, 1,000, 2,000, 4,000, and 8,000 Hz.

Frequency selectivity was assessed using IFMCs: In this measurement, the patient's task is to report the presence or absence of a probe tone that follows a masking tone after a gap of 10 ms. The probe had a duration of 8 ms (including 4-ms ramps) and was always presented at 10 dB SL. 100-ms masking tones were varied to find the masker level that resulted in 50% detection of the probe. Masker frequencies were 0.5, 0.7, 0.9, 1, 1.1, 1.3, and 1.6 times the probe frequency. Masker thresholds were measured for probe frequencies 250, 500, 1,000, 2,000, 4,000, and 6,000 Hz. No efforts were made to prevent off-frequency listening because this is a useful indicator of nonfunctioning regions of the cochlea. For the impaired listeners, not all frequencies could be tested because of very high absolute thresholds . In some instances, additional intermediate frequencies were tested. Each IFMC threshold is generally the mean of three measurements.

57.2.1.2 Threshold Estimation Procedure

The basic procedure for measuring absolute thresholds is based on an adaptive yes–no paradigm previously evaluated by Lecluyse and Meddis (2009). A stimulus is presented to the participant who simply responds by pressing a "yes" or "no" button to show whether or not he hears the target. The level is changed from trial to trial using a one-up, one-down adaptive procedure. After an initial series of stimuli using a large step size to find the threshold region, the step size is reduced to 2 dB and the run then continues for ten more trials not including an additional 20% of catch trials presented at random among the regular trials. These are trials where no stimulus is presented and the participant is expected to say "no." On the few occasions when the participant is "caught out," the run is stopped and restarted; possibly after resting the patient and giving further instructions. Patients are encouraged to be conservative in their judgments and false alarms are, in practice, very rare. Our experience is that this method is fast, reliable, and acceptable to often elderly patients who need little (or sometimes no) training before producing useful results. Similar methods have recently been explored by Leek et al. (2000) in a variety of audiological research contexts.

The threshold is estimated at the end of the run by fitting a psychometric function of the form $p=(1+\exp(-k(L-Th)))^{-1}$ to the responses, where p is the proportion of "yes" responses, L is the level of the stimulus (dB SPL), k is a slope parameter, and Th is the threshold, the level of the stimulus at which the proportion of yes-responses is 0.5. The function is fitted using the least-squares, the best-fit procedure, with Th and k as free parameters.

Stimuli were generated using a sampling rate of 96,000 Hz, with 24-bit resolution, and presented monaurally via circumaural Sennheiser HD600 headphones.

57.2.1.3 Listeners

The behavioral data presented in this report was obtained from one normal listener and three impaired listeners. The normal listener was 21 years old with no history of hearing problems. The three impaired listeners were aged between 68 and 76.

57.2.2 Computer Modeling

The models were developed using an existing published computer model of the auditory periphery (Meddis 2006). Its parameters were adjusted iteratively to produce an approximate match to the data for an individual patient. The model consists of a cascade of signal processing operations representing successive processing stages in the auditory periphery. The input is an acoustic signal and the output is a multichannel, multi-fiber representation of high spontaneous rate (HSR) spiking activity in the auditory nerve (see Fig. 57.1). This activity forms the input (on a tonotopic basis) to more models of sustained-chopping cells in the cochlear nucleus with low (~10 spikes/s) spontaneous firing rates. The output from these units is used as the input to a further layer of modeled units representing inferior colliculus (IC) cell coincidence-detectors (bottom panel, Fig. 57.1b). These were parameterized to have no spontaneous activity. Any activity in one or more of these cells was used to indicate that an acoustic stimulus had been detected.

The parameters of the model were fixed at values consistent with the physiological literature, and each stage of the model has been separately evaluated to be consistent with published data obtained using small mammals. The parameters were then adjusted as little as possible to give a good representation of a single listener with good hearing (see below). The parameters for this "normal" model were used as the starting point of the explorations used to model individual patterns of abnormal hearing again by making as few changes as possible. The illustrations below are restricted to a single change with no further adjustments to produce more flattering fits to the patient data.

The model was evaluated using *exactly* the same procedures and software used to collect the participant's data. The model was harnessed to the psychometric software, so that it "pressed the yes-button" if one or more spikes were registered in the IC unit during the presentation of the stimulus probe. If no spikes occurred, it "pressed the no-button."

57.3 Results

57.3.1 Normal Data and Model

Figure 57.2a shows a hearing profile of a 21-year-old male listener (CMa) with excellent hearing. His hearing is the best of all normal subjects tested in the laboratory. All thresholds are very low but within normal limits. Normal IFMCs were found at all frequencies. The computer model was carefully tuned to fit these data. Figure 57.2b shows the results of submitting the model to the same test procedures as the human listener. This model will be used as the baseline model for simulating impaired hearing. The impaired models will be described in terms of how they differ from the normal model.

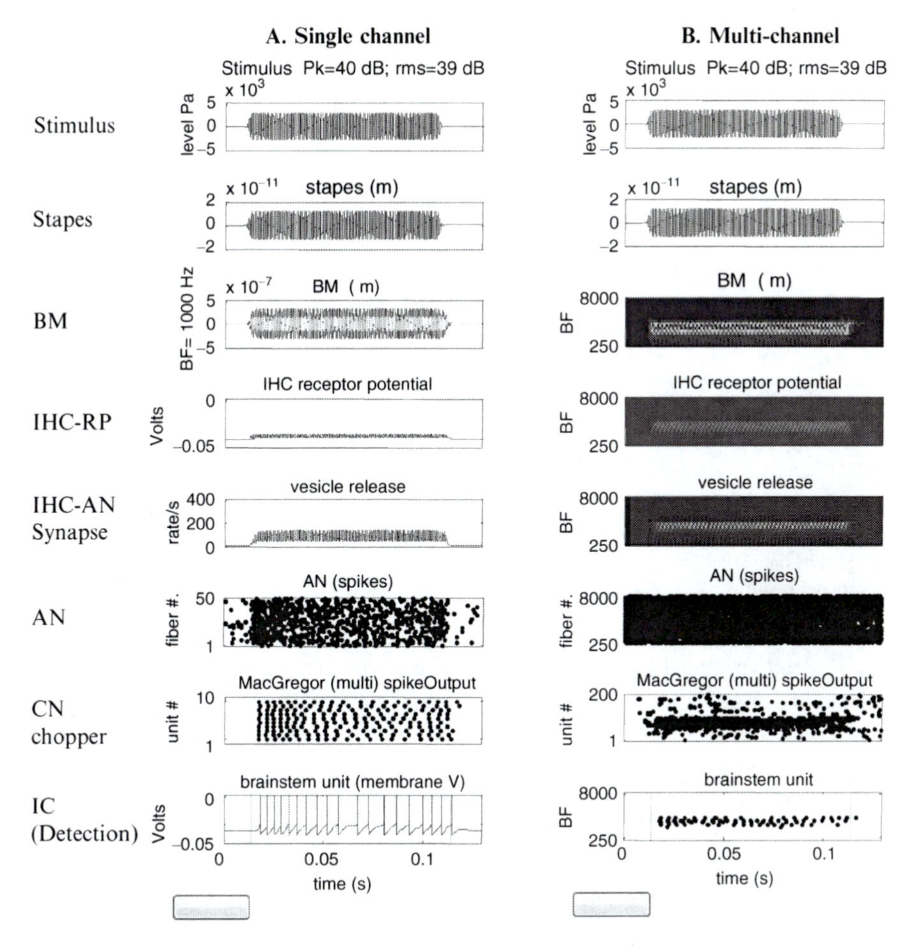

Fig. 57.1 Successive processing stages of the auditory model. Detection decisions are based on the presence or absence of at least one action potential in the final (IC) stage during the presentation of the stimulus. The stimulus is a 100-ms, 1,000-Hz pure tone presented at a level of 40 dB SPL (29 dB SL). (**a**) single channel model (*BF* = 1,000 Hz). (**b**) 40-channel model whose *BFs* range from 250 to 8,000 Hz. In this illustration, the auditory nerve stage of the model has 50 AN fibers per channel. The CN stage of the model has 10 units per channel. The IC stage has one unit per channel. The IC units have no spontaneous activity and any activity in any of these units is taken to be indicative of the detection of an acoustic stimulus

57.3.2 Impaired Data and Models

57.3.2.1 Profile 1 (Participant ECr)

ECr is a 76-year-old male with bilateral moderate-severe sensory-neural hearing loss with normal middle ear function. His absolute thresholds are raised at low frequencies and are difficult to measure at frequencies above 1 kHz. The IFMCs

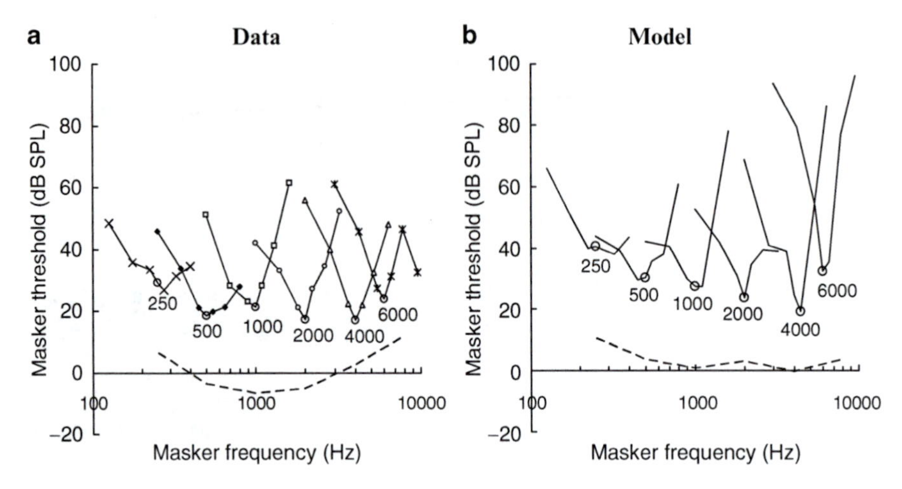

Fig. 57.2 Absolute thresholds (dashed line) and IFMCs (full lines). For the IFMCs, the masker threshold when the masker and probe frequency are the same is represented by the large open circle. The data labels represent the probe frequency. (**a**) Data for a normal listener (CMa, *left ear*). (**b**) model results

show greatly reduced frequency selectivity and follow the contour of the absolute thresholds (Fig. 57.3a).

To develop a suitable model, we hypothesized that ECr retains only one functioning location along the basilar membrane (BM) that has a best frequency (*BF*) of 250 Hz and that the response is linear. The results in Fig. 57.3b were produced using a reduced version of the normal model illustrated earlier in Fig. 57.2b. All BM locations in the normal model other than the 250 Hz location were disabled and the nonlinear path of the dual resonance nonlinear (DRNL) simulation of BM activity in the remaining 250 Hz channel was also disabled leaving *only a single linear filter*. While the only remaining location has a *BF* of 250 Hz, the lowest threshold occurs at 500 Hz because of the high-pass contribution of the outer-middle ear, which attenuates all frequencies below 1,000 Hz. The IFMCs for probes greater than 500 Hz form simple diagonals along the contour of absolute thresholds as expected for probes in a nonfunctioning region.

57.3.2.2 Profile 2 (Participant JEV)

JEv is a 69-year-old male with a bilateral, moderate, sensory-neural, sloping hearing loss with normal middle ear function. The absolute thresholds show a moderate loss up to 1,000 Hz and an increasing loss at higher frequencies. The loss at 8,000 Hz is approximately 80 dB. IFMCs are shallower than normal at all frequencies especially from 2,000 Hz onwards, but still clearly V-shaped (Fig. 57.4a). Raised thresholds particularly at high frequencies combined with some preservation of tuning are consistent with a hypothesis that the deficit results from a low

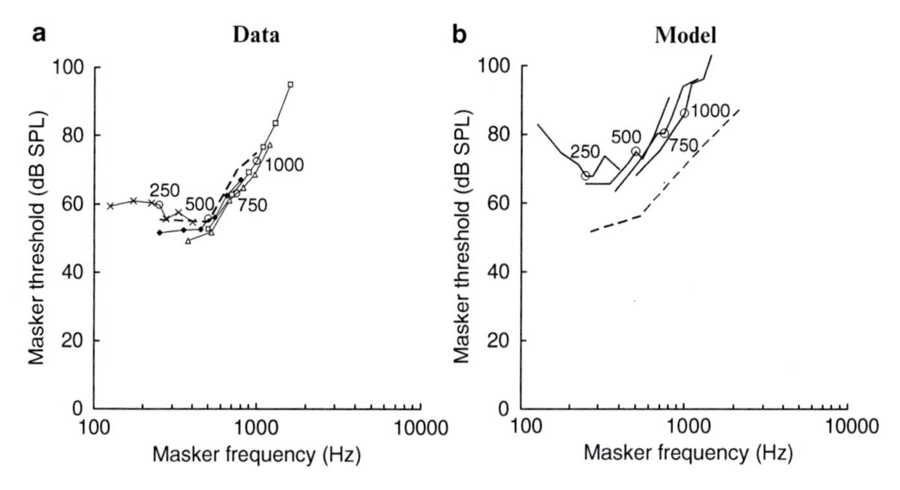

Fig. 57.3 (**a**) Data for an impaired listener (ECr, *left ear*). See Fig. 57.2 for more details. (**b**) The "impaired" model based on the normal model (Fig. 57.2b) with all channels disabled except $BF = 250$ Hz (see text)

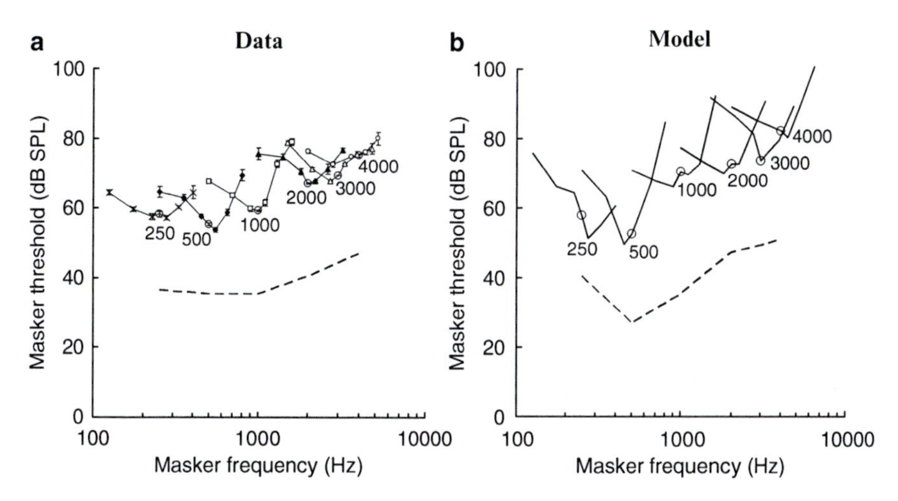

Fig. 57.4 (**a**) Data for impaired listener (JEv, *left ear*). See Fig. 57.2 for more details (**b**) the "impaired" model based on the normal model (Fig. 57.2b) with Ep reduced from 0.1 to 0.09 V (see text)

endocochlear potential (Ep). Schmiedt et al. (2002) has shown that thresholds for high frequency tones are more affected than low tones by loss of Ep.

Figure 57.4b shows the assessment of a model that is the same as the normal model in all respects except that Ep has been reduced from −0.1 V to −0.09 V.

The raised thresholds are caused by a reduced response in the inner hair cells (IHC) resulting from the reduction in Ep. While the tuning curves are broader than the normal subject, the IFMCs are what we would expect from a normal listener if tested using a more intense probe tone (i.e., at the same masker levels as used for JEv) (Nelson et al. 1990). The model that reproduces Schmiedt's observations of greater threshold increases at "higher frequencies" as an "emergent property."

Note that the model does not reproduce the flatter, asymmetric IFMCs at 3,000 and 4,000 Hz. This suggests that an improved model might be realized by disabling all BM locations above 2,700 Hz. The simpler model has been retained here because it illustrates the effect of a single parameter change to the normal model. The emerging pattern of results is typical of what might be expected from an uncomplicated case of presbyacusis.

57.3.2.3 Profile 3 (Participant JJo)

JJo is a 68-year-old man with a moderate bilateral, sensory-neural, ski-slope loss with normal middle ear function but no detectable acoustic reflex. Below 2,000 Hz, thresholds and IFMCs are unexceptional (Fig. 57.5a). Above this, thresholds rise steeply and the IFMCs roughly follow the absolute threshold contour. This is consistent with the hypothesis that the BM is severely compromised above 1,800 Hz.

This hypothesis is illustrated in Fig. 57.5b where the model is the same as the normal model except that all channels with *BFs* above 1,800 Hz have been disabled. Naturally, this results in raised thresholds for high frequency probes because they are heard only through lower-frequency channels. More interestingly,

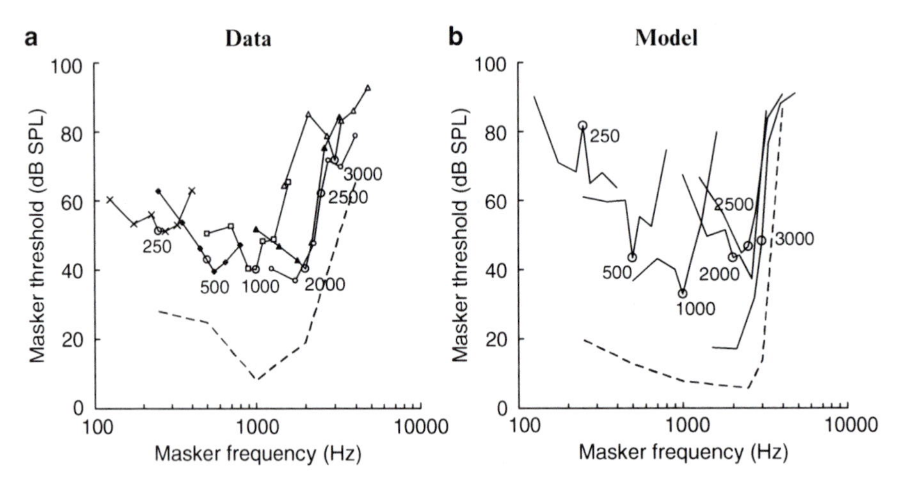

Fig. 57.5 (**a**) Data for an impaired listener (JJo, *left ear*). See Fig. 57.2 for more details. (**b**) The "impaired" model based on the normal model (Fig. 57.2b) with all channels with *BFs* > 1,800 Hz disabled (see text)

it gives rise to a pattern of IFMCs characteristic of non-functioning regions where the IFMC follows the contour of the absolute thresholds.

57.4 Discussion

While these results remain preliminary, they are encouraging in a number of respects. It is clear that the additional data obtained from the IFMCs allow us to identify different patterns of response in hearing impaired individuals. The addition of IFMC data helps to narrow the range of hypotheses that might account for the impairment. We also collect temporal masking curves (TMCs) and DPOAE data (not shown), and these help to clarify the picture further. We have been surprised by the variety of patterns of impairment in our group of volunteer (self-selected) participants in the study.

The results indicate that hypotheses concerning the underlying pathology can be tested for consistency with the data using computer models. As we gain confidence in the usefulness of the models and gain further experience in the admittedly black art of finding appropriate hypotheses and appropriate model parameters, we hope to be able to evaluate the effectiveness of different kinds of hearing aid algorithms and to make predictions as to their effectiveness for individual patients. To this end, we are linking the model to automatic speech recognition devices. We acknowledge that the main complaint of patients involve difficulty in following conversations in noisy environments. We hope that our modeling efforts will eventually contribute to improved prostheses to minimize this problem.

57.5 Comment by Michael Heinz

Can you elaborate on the source of the high-frequency hearing loss in your model with EP reduction? Schmiedt et al (2002) hypothesized that reduced EP produces a sloping high-frequency hearing loss due to the effects of reduced EP on OHC function and higher cochlear-amplifier gain at high frequencies. This hypothesis is consistent with numerous OHC-like effects in BM and AN responses with reduced EP from administration of furosemide (Sewell 1984; Ruggero and Rich 1991). However, your model appears to only implement EP reduction in terms of IHCs, but not OHCs.

57.6 Reply by Ray Meddis

This is a useful comment. Any change in EP is likely to affect both IHC and OHC activity. In the paper we show only the effect on IHC. The reason for this is that we decided to study one effect at a time and began with the IHC effect. We were surprised that this was sufficient to show threshold increases at all frequencies but particularly so at high frequencies. This produces the sloping loss characteristic of presbyacusis. Clearly, we also need to include the effect of EP changes to OHC function and show reductions in the BM response as well. At the time of writing, we are not sure exactly what the quantitative relationship should be but we are currently exploring the issue.

References

Lecluyse W, Meddis R (2009) A simple single-interval adaptive procedure for estimating thresholds in normal and impaired listeners. J Acoust Soc Am 126:2570–2579

Leek MR, Dubno JR, He N et al (2000) Experience with a yes–no single-interval maximum-likelihood procedure. J Acoust Soc Am 107:2674–2684

Meddis R (2006) Auditory-nerve first-spike latency and auditory absolute threshold: a computer model. J Acoust Soc Am 119:406–417

Nelson DA, Chargo SJ, Kopun JG et al (1990) Effects of stimulus level on forward-masked psychophysical tuning curves in quiet and in noise. J Acoust Soc Am 88:2143–2151

Ruggero MA, Rich NC (1991) Furosemide alters organ of Corti mechanics: Evidence for feedback of outer hair cells upon the basilar membrane. J Neurosci 11:1057–1067.

Schmiedt RA, Lang H, Okamura H et al (2002) Effects of furosemide applied chronically to the round window: a model of metabolic presbyacusis. J Neurosci 22:9643–9650

Schmiedt RA, Lang H, Okamura HO, Schulte BA (2002) Effects of furosemide applied chronically to the round window: A model of metabolic presbyacusis. J Neurosci 22:9643–9650.

Sewell WF (1984) The effects of furosemide on the endocochlear potential and auditory-nerve fiber tuning curves in cats. Hear Res 14:305–314.

Index

CPSIA information can be obtained at www.ICGtesting.com
233789LV00001B/27/P

9 781441 956859